Molecular and Cellular Biology of Viruses

This fully revised second edition of *Molecular and Cellular Biology of Viruses* leads students on an exploration of viruses by supporting engaging and interactive learning. All the major classes of viruses are covered, with separate chapters for their replication and expression strategies, and chapters for mechanisms such as attachment that are independent of the virus genome type. Specific cases drawn from primary literature foster student engagement. End-of-chapter questions focus on analysis and interpretation with answers being given at the back of the book. Examples come from the most-studied and medically important viruses such as SARS-CoV-2, HIV, and influenza. Plant viruses and bacteriophages are also included. There are chapters on the overall effect of viral infection on the host cell. Coverage of the immune system is focused on the interplay between host defenses and viruses, with a separate chapter on medical applications such as antiviral drugs and vaccine development. The final chapter is on virus diversity and evolution, incorporating contemporary insights from metagenomic research. The second edition has updated suggestions for primary literature to discuss along with each chapter. New to this second edition, a supplementary chapter, freely available for download, looks at how virology intersects with public health, and uses the COVID-19 pandemic as a notable example.

Key Features

- Readable but rigorous coverage of the molecular and cellular biology of viruses
- Molecular mechanisms of all major groups, including plant viruses and bacterio-phages, illustrated by example
- Host–pathogen interactions at the cellular and molecular level emphasized throughout
- Medical implications and consequences included
- Quality illustrations available to instructors
- New to this second edition, interactive quiz questions hosted online

Molecular and Cellular Biology of Viruses

Second Edition

Phoebe Lostroh

CRC Press
Taylor & Francis Group

A GARLAND SCIENCE BOOK

Designed cover image: Illustration by David S. Goodsell, RCSB Protein Data Bank; doi: 10.2210/rcsb_pdb/goodsell-gallery-026. Published under CC BY 4.0.

Second edition published 2024
by CRC Press
2385 NW Executive Center Drive, Suite 320, Boca Raton, FL 33431

and by CRC Press
4 Park Square, Milton Park, Abingdon, Oxon, OX14 4RN

CRC Press is an imprint of Taylor & Francis Group, LLC

© 2024 Phoebe Lostroh

First edition published by Garland Science 2019

Library of Congress Control Number: 2023952040

ISBN: 9781032732107 (hbk)
ISBN: 9781032732121 (pbk)
ISBN: 9781003463115 (ebk)

DOI: 10.1201/9781003463115

Typeset in Minion Pro
by Evolution Design and Digital Ltd (Kent)

Access the Instructor and Student Resources: www.routledge.com/cw/lostroh

. . . my greetings to all of you, Yeasts,
Bacteria, Viruses,
Aerobics and Anaerobics:
A Very Happy New Year
to all for whom my ectoderm
is as Middle-Earth to me.

(From "A New Year Greeting" by W. H. Auden)

I dedicate the second edition of this book to my sister, Ginger Howell.

Contents

9 GENE EXPRESSION AND GENOME REPLICATION IN THE SINGLE-STRANDED DNA VIRUSES   249

12   VIRUS–HOST INTERACTIONS DURING LYTIC GROWTH   **311**

13   PERSISTENT VIRAL INFECTIONS   **329**

16   MEDICAL APPLICATIONS OF MOLECULAR AND CELLULAR VIROLOGY   405

17   VIRAL DIVERSITY, ORIGINS, AND EVOLUTION   437

18   VIRUSES AND PUBLIC HEALTH (available online at www.routledge.com/cw/lostroh)

Preface

I teach undergraduate students, most of whom are biology, molecular biology, biochemistry, or neuroscience majors. Most undergraduate virology textbooks do not have enough molecular and cellular content for my virology elective. In contrast, most virology textbooks with substantial molecular and cellular content are written for more advanced audiences. There is a singular book from the 1990s that fits this niche, namely *The Biology of Viruses* by Dr. Bruce A. Voyles. But by 2017, it was time for a book of similar scope and focus but incorporating contemporary research.

The book is organized according to the stages of a virus replication cycle, with a few additional chapters to take into account themes in molecular and cellular virology that are especially interesting to undergraduates. These include immunology as it pertains to viral infections and medical applications of molecular and cellular virology. In our age of rising antibiotic resistance and a renaissance in phage research spurred by CRISPR-Cas and other biotechnological innovations, I decided that it would be best to keep the phages front and center even though other authors may have relegated them to an appendix or online materials.

The book is intended to give faculty choices about customizing the assigned reading. For example, faculty whose course goals emphasize molecular biology may want to use all the content about the gene expression and genome replications strategies. In contrast, faculty whose course goals emphasize cell biology may want to use only the simplest model viruses in those chapters so that they can instead assign more content about autophagy, signal transduction, apoptosis, and the cell cycle.

I have not made any attempt to be comprehensive; such a book would have defeated the purpose of its utility for a typical undergraduate course. Instead, I have focused on a few models to give undergraduate readers a taste of the diversity of virology and to lay down examples for comparison. Throughout this book I have favored HIV as an example because its molecular biology has been worked out in intensive detail, using experimental procedures that are understandable by current undergraduates. I also chose to write so much about HIV in order to encourage LGBTQ students and students from the Global South to identify with molecular biology. When HIV was not a good choice, I selected models that most interest my students. Throughout, I wrote about ideas derived from contemporary research findings in order to inspire students to love the creation of new knowledge.

While the first edition argued that more research on the molecular biology of coronaviruses would be prudent in case a new virulent coronavirus were to emerge, the second edition was produced during the COVID-19 pandemic. Because of the pandemic, biology students are more informed about the molecular and cellular aspects of viruses than ever before. This second edition draws on SARS-CoV-2 as a frequent example. I have also added an online chapter addressing viruses and public health.

For this second edition in particular, I want to thank the readers who enjoyed the book and reached out to tell me so.

I hope that faculty and students will enjoy using this book, and I look forward to hearing from readers.

Phoebe Lostroh

Acknowledgments

I would like to thank the students of Colorado College for teaching me about the roles of listening and compassion in good teaching. Thank you especially to my laboratory research students, who helped me keep the faith that students would be interested in such a book. I thank the anonymous reviewers who made this book so much better by their diligence, and I apologize for any remaining errors, which are mine alone. Thank you to Dr. C. Lee, without whom I would never have become a professor, let alone an author.

Thank you to the entire Department of Molecular Biology at Colorado College for being my intellectual home. I thank Dr. Olivia Hatton for help on the sections of this book pertaining to latent infections, cancer, and immunology. Special thanks to Dr. Anne Hyde, who began the tradition of Scholarly Writing and Research Mornings at Colorado College. Thank you to Dr. Tip Ragan for suggesting that I use Scrivener software and for being a great mentor. Thanks to Dr. Kristine Lang, a wonderful collaborator who encouraged me to keep working on the book while we also worked together on biophysics research projects, and whose collaboration enabled me to have a full school year without teaching responsibilities. Dr. Rebecca Tucker's support from the Crown Faculty Center allowed me to meet my editors in person at a crucial point, while Anna Naden was a thorough and kind research assistant. Thank you to Dr. Leslie Gregg-Jolly, Dr. Tomi-Ann Roberts, Dr. Barbara Whitten, and the other feminist scientists who inspired me to write with women and LGBTQ undergraduates in mind. Thanks to my sister Ginger Lostroh Howell, who continues to teach me about veterinary vaccines, and who believed in my dream of becoming an author. Thanks to Dr. Fran Pilch and Phyllis Dunn for the many hours playing bridge so that my mind could rest in between bouts of virology.

I thank the most incredible editors, Liz Owen and Jordan Wearing, who are endowed with an unending supply of superhero attributes, such as the patience of saints and the ability to tolerate American spelling. If every author were blessed with such editors, there would be many more books.

Thank you to Chuck Crumly, a wonderful editor with a sense of humor that saw us through the transition from one publisher to the next, and to the team of hard-working people at CRC Press/Taylor & Francis including Linda Leggio and Evolution Design and Digital Ltd, who smoothed the way to a printed book. I am also grateful to Patrick Lang at ScEYEnce Studios, who turned my drawings into informative, engaging illustrations, and to David Goodsell for permission to use some of his drawings, including the Ebola virus on the cover.

And finally, thanks beyond measure to my wife Dr. Amanda Udis-Kessler, preacher, writer, songwriter, scholar, and cat-whisperer. She improved the book tremendously through countless invigorating conversations, through editing every word, and through never losing faith that this book would appear in the flesh.

Author

Professor Phoebe Lostroh earned a BA from Grinnell College and a PhD in Microbiology and Molecular Genetics from Harvard University. She is a molecular microbiologist whose research has focused on bacterial "sex." She has recruited a diverse team of undergraduate researchers at Colorado College where she has won multiple awards, such as the Theodore Roosevelt Collins Outstanding Faculty award for teaching, mentoring, and advising students of color and first-generation students. Her research has been supported by the NSF (2009–2018), NIH (1994–2003), and the Keck Foundation (2001–2003). She served as a Program Officer at the NSF (2019–2021). She is a volunteer comedienne with Science Riot, which brings science to the public through stand-up routines.

Instructor and Student Resources

The following resources are available for students and instructors, and can be accessed at www.routledge.com/cw/lostroh.

Interactive quiz questions – These are interactive self-assessment questions for students to test their understanding of chapter content. The quiz makes use of various question formats: single-answer and multi-answer MCQs, true/false, fill-in-the-blanks, and matching-type questions. At the end of the quiz the student can see their overall score.

Animations – A small number of animations have been generated for this edition with the aim of bringing particular figures, concepts, and mechanisms to life. These animations are signposted in the book and can be viewed on the website. The animations are as follows:
- Protease activity on nascent proteins in polio
- Discontinuous minus strand synthesis in coronaviruses
- RNA editing in Ebola virus
- Influenza cap snatching
- Discontinuous reverse transcription in HIV

Figure slides – All the figures in this book have been made available for instructors for use in lecture slides and teaching materials. The figures have been provided in two convenient formats: PowerPoint and PDF.

The Fundamentals of Molecular and Cellular Virology

Virus	Characteristics
Human immunodeficiency virus (HIV)	Retrovirus responsible for the AIDS pandemic.
Variola virus	Cause of historical smallpox epidemic; eliminated through the use of vaccinia virus in prophylactic immunization.
Rabies virus	Pasteur intentionally attenuated this virus and subsequently employed the attenuated version for therapeutic immunization.
Bacteriophages	Viruses that infect bacteria and archaea; used as models for discovering the fundamental processes of molecular biology.
Influenza A virus	Cause of annual epidemics; occasional cause of global pandemics with high lethality.
Severe acute respiratory syndrome coronavirus 2 (SARS-CoV-2)	Cause of the coronavirus disease 2019 (COVID-19) pandemic.

The viruses you will meet in this chapter and the concepts they illustrate

In this book, you will learn how miniscule viruses enter, take over, and kill our cells. A molecular understanding of these processes provides insights into how both viruses and their host cells function. It is an exciting time to study viruses because of the breadth and depth of techniques available to investigate them, and the increasing number of known viruses and viral genome sequences (including those for viruses that have never been cultivated in the laboratory). In addition, there are many practical applications of virology. For example, basic research on the molecular biology of the human immunodeficiency virus (HIV) ultimately led to the first treatments that moved HIV infection away from being a death sentence to a chronic, controllable illness (**Figure 1.1**). Virology research allows us to track the evolution of severe acute respiratory syndrome coronavirus 2 (SARS-CoV-2), the etiological agent of coronavirus disease 2019 (COVID-19), and to predict the effects of that evolution on clinical interventions such as monoclonal antibody treatments. Many applications such as these attract people to the field of virology.

In this chapter, we examine the origins of virology in order to explain how molecular and cellular biology fit into the broader discipline of virology. Molecular biology is fundamentally concerned with how macromolecules, especially proteins and nucleic acids, function to control the structure and behavior of cells. By extension, molecular and cellular virology studies focus

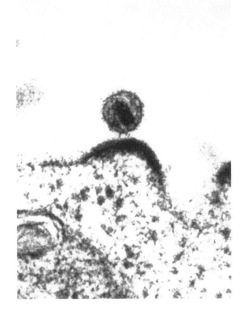

Figure 1.1 HIV. This micrograph shows a single HIV virion exiting its host cell. (Courtesy of the National Institute of Allergy and Infectious Diseases (NIAID). Published under CC BY 2.0.)

DOI: 10.1201/9781003463115-1

particularly on the interactions among viral proteins, viral nucleic acids, cellular proteins, cellular nucleic acids, and cellular organelles. In **Technique Box 1.1**, we will see that some of the consequences of viral infection can be observed with a light microscope. After, we consider the characteristics shared by all viruses (**Section 1.5**). We will then discuss viral diversity (**Section 1.6**), especially with respect to their genomes and the mechanisms by which they synthesize mRNA (**Section 1.7**). We will also explain, in **Section 1.7**, how diverse viruses have been named and classified. We will encounter the general method of propagating viruses in a laboratory setting in **Section 1.8**. The chapter continues with a consideration of the abundance of viral sequences in the human genome (**Section 1.9**); indeed, DNA of viral origin is found in almost every known cellular genome, where it contributes to the evolution of organisms, including humans. The chapter concludes with a consideration of how sequences of viral nucleic acids and proteins can be used to generate hypotheses about the evolution, structure, and function of viruses and their component parts.

Our goal in this chapter is to prepare you for the rest of the book. **Chapter 2** explains how we will divide the virus replication cycle into several parts. **Chapters 3–11** will address each of these parts of the cycle in turn, with **Chapters 5–10** focused on the different Baltimore classes of viruses and how they express and replicate their genomes. In **Chapter 12**, we will learn how viruses generally interact with host processes such as translation and apoptosis, and in **Chapter 13**, we will see how viruses can cause integrated and persistent infections that can last for the entire life span of their hosts. In **Chapters 14** and **15** we will examine how hosts fight back against viral infections. **Chapter 16** is about clinical applications of virology, such as vaccines, gene therapy, and antiviral drugs. **Chapter 17** concludes with a discussion of the diversity and evolution of viruses.

1.1   Molecular and cellular virology focuses on the molecular interactions that occur when a virus infects a host cell

Molecular and cellular virology investigates the molecular and cellular aspects of virus infection. For example, a typical research project in molecular and cellular virology would be to determine the function of each protein encoded by a viral genome or identify the cellular receptor that enables a virus such as SARS-CoV-2 to enter a target cell. Another typical project would be to determine the subcellular localization of virus assembly in infected cells and the cellular factors that are required for that subcellular localization. Molecular and cellular virology studies have provided a molecular explanation for why HIV infects cells of the immune system but not other types of cells. Other studies conducted through the lenses of molecular and cellular biology have helped us begin to understand why the related severe acute respiratory syndrome (SARS) and Middle East respiratory syndrome (MERS) coronaviruses, including SARS-CoV-2, cause similar yet distinct diseases (**Figure 1.2**).

Molecular and cellular virology also has practical applications such as the design of **antiviral medicines** and **vaccines** (**Figure 1.3**). Antiviral medications typically interfere with viral enzymes not normally found in human cells, whereas vaccines cause a protective immune response to develop without making the treated person sick. Furthermore, viruses can be genetically engineered to serve as agents of **gene therapy**, which includes the introduction of a functional gene to reverse the effects of nonfunctional or dysfunctional inherited genes. Although we will not address in detail how

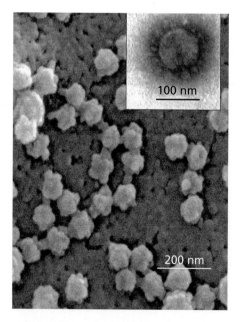

100 nm

200 nm

Figure 1.2 The virus that causes severe acute respiratory syndrome. This scanning electron micrograph shows many of the coronaviruses with a close-up on one of the particles.

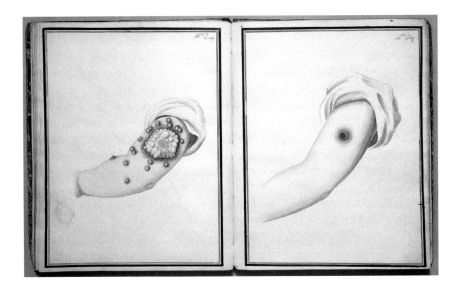

Figure 1.3 Early vaccination. In this watercolor image from 1802, the person on the left was inoculated with a small amount of smallpox virus; the person on the right was inoculated with cowpox virus instead. The person on the right will become immune to smallpox because of vaccination with the cowpox virus. (Courtesy of Wellcome Library, London. Published under CC BY 4.0.)

translational scientists take findings from basic research and turn them into treatments or vaccines, the book will point out many instances where basic research has been used to better human health.

1.2 The discipline of virology can be traced historically to agricultural and medical science

The meaning of the Latin word *virus* is poison. The name indicates that viruses were first discovered because of their role in producing disease in agriculturally important plants or animals, such as tobacco or livestock. In the late nineteenth century, European microbiologists invented a filter with pores smaller than the diameter of a bacterium that could remove all bacteria from a liquid suspension (**Figure 1.4**). Using this device to study tobacco mosaic disease revealed that the infectious agent could not be removed by filtration, and so the infectious substance must be smaller than a bacterium. Soon thereafter, other microbiologists showed that viruses were particles, not liquids, and that a virus also causes the agriculturally important foot-and-mouth disease, which can infect and even kill not only cattle but also sheep, goats, and pigs. Through this work, the word virus came to mean any infectious agent so small that it could be neither observed by light microscopy nor removed from a solution by filtration through the smallest pores that humans could manufacture. Although it is true that most viruses are too small to be observed directly with light microscopy, viral infections typically cause pathogenic alterations to host cells that can be observed using light microscopy (**Technique Box 1.1**).

Figure 1.4 A Chamberland filter. Charles Chamberland (1851–1908) discovered that warm porcelain can retain fine particles in solution and, along with Louis Pasteur, showed that bacteria can be filtered out of a solution using a porcelain filter in a glass tube such as the one depicted here. Viruses were first understood as microbes so tiny that they could pass through a Chamberland filter. (Courtesy of Wellcome Library, London. Published under CC BY 4.0.)

TECHNIQUE BOX 1.1 LIGHT MICROSCOPY

Viruses are too small to be seen by the human eye, and most are even too small to be seen by **light microscopy** (**Figure 1.5**). Nevertheless, light microscopy is an important tool in virology because it can be used to visualize the effects of virus infection on host cells. Light microscopy uses the physics of lenses and electromagnetic waves to magnify and resolve features too small to see unassisted. **Resolution** is the ability to distinguish two features that are very close together, and it is directly related to the wavelength used to illuminate the sample; contemporary light microscopy can resolve features that are separated by 0.22 μm (or 220 nm).

However, many viruses are 20–100 nm in diameter, smaller than the resolution achievable through light microscopy. Nevertheless, light microscopy is a useful tool, especially for studying viruses that infect eukaryotic cells. Animal cells in culture typically form **confluent** carpets in which cells introduced into a tissue culture flask attach to and spread out on the specially prepared plastic. As the cells undergo mitosis, the cell population increases in number until all the cells are touching one another without overlapping (**Figure 1.6A**); at this point, contact

inhibition (touching other cells on all sides) prevents further population increase. When viruses infect these cells, the cells typically display a variety of **cytopathic effects** that depend on the specific host cell and virus. For example, many viruses cause infected cells to **round up** and detach from the tissue flask as they die (**Figure 1.6B**).

Some viruses cause the cells to form **syncytia**, which are large, multinucleate cells (**Figure 1.7**). Other viruses cause abnormal internal structures visible when the host cells are stained; such **inclusion bodies** can form in the cytoplasm or nucleus, depending on the virus (**Figure 1.8**). In most cases that have been investigated more closely, these inclusions turn out to be the sites of viral gene expression, genome replication, assembly, or some combination of these. Still other viruses can **transform** the cells, setting them on the path toward becoming cancer cells. Transformed cells exhibit morphological changes detectable by light microscopy, such as a loss of contact inhibition (**Figure 1.9**).

Fluorescence microscopy is a variation of light microscopy in which specific subcellular, cellular, or viral features have been manipulated to make them fluorescent,

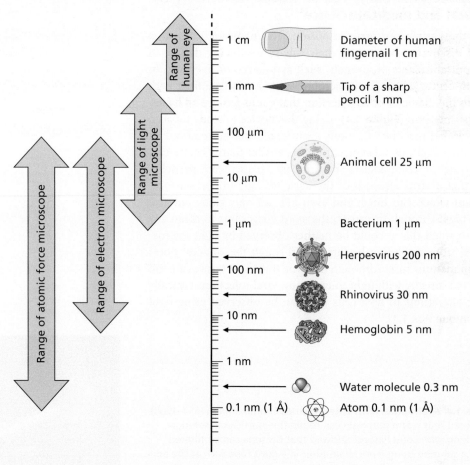

Figure 1.5 The sizes of cells, viruses, and molecules. This logarithmic scale depicts the relative sizes of eukaryotic cells, bacterial cells, viruses, proteins, and atoms, with techniques typically used to image these entities noted to the left of the scale.

meaning that they can absorb short wavelengths and emit them as longer wavelengths. Fluorescence optics, in combination with computer-assisted image analysis, can make internal structures of living cells visible, enabling the detection of the effects of virus infection on specific organelles or supramolecular structures (such as the cytoskeleton). Because of its importance, fluorescence microscopy is explained in more detail in **Chapter 2**.

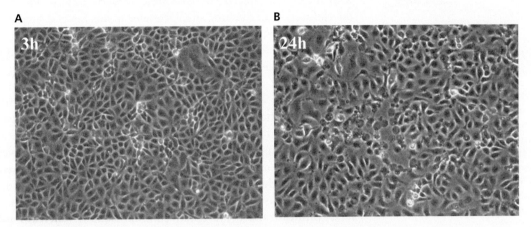

Figure 1.6 Cytopathic effects of viral infection on animal cells. Cells that have been infected by pancreatic necrosis virus exhibit rounding up as they die. Panel (A) shows 3 h postinfection and panel (B) shows 24 h postinfection. (Courtesy of Øystein Evensen. Published under CC BY 4.0.)

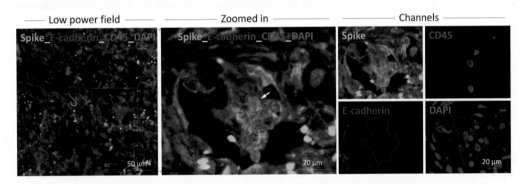

Figure 1.7 A syncytium in lung biopsy tissue from a patient with COVID-19. In this immunofluorescence micrograph, nuclei are blue while two different host proteins are labeled purple or red. SARS-CoV-2 spike protein is green. An example of a syncytium containing many nuclei has an arrow pointing to it in the middle panel. (From Zhang Z et al. 2021. *Cell Death Differ* 28:2765–2777. doi: 10.1038/s41418-021-00782-3. With permission from Springer Nature.)

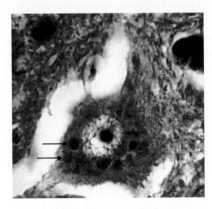

Figure 1.8 Inclusion bodies. Cells infected with rabies virus have been stained in such a way that inclusion bodies are visible in the cytoplasm (two are pointed out by arrows). (CDC/Dr. Daniel P. Perl.)

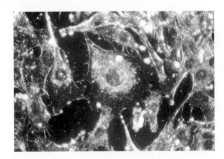

Figure 1.9 Transformed cells. Cancerous cells such as these grow on top of each other instead of responding to contact inhibition as normal cells would. In this image, the cells around the central one are piled on top of one another. (National Cancer Institute/Dr. Cecil Fox.)

Figure 1.10 Mosaic of the first person treated with antirabies therapeutic immunization. Detail from a mosaic in the tomb of Louis Pasteur showing Jean-Baptist Jupille fighting with a rabid dog and thereby contracting rabies. Pasteur saved Jupille's life with a therapeutic vaccine. (© Institut Pasteur, Musée Pasteur. With permission.)

Even before viruses were recognized as tiny infectious particles, they were already part of medical history. In the late eighteenth century, for example, Edward Jenner developed medical immunization against smallpox without ever observing the smallpox virus (variola) itself. Jenner inoculated children with infectious material derived from cowpox lesions, thereby protecting the children from catching smallpox later in life. Several generations later, in the late nineteenth century, Louis Pasteur became convinced that rabies was caused by an infectious agent smaller than a bacterium. He had also learned that infecting certain laboratory animals such as rabbits with rabies, then taking infected nerves from the animal and infecting another animal with the nerves over and over again, could lead to a weakening of the virus. This weakened or **attenuated virus** could subsequently be used to treat someone who had been bitten by a rabid animal and would otherwise die of rabies (**Figure 1.10**). Injecting such a person with the attenuated rabies virus in a rabbit's nerves dramatically improved their chances of surviving provided that the individual received the injection before symptoms of rabies developed. Jenner's use of the cowpox virus is an example of preexposure **prophylactic immunization** (preventative immunization), whereas Pasteur's use of attenuated rabies is an example of postexposure **therapeutic immunization**.

The use of viruses to prevent or cure infectious diseases is thus much older than molecular and cellular biology, which did not become firmly established until the mid- to late twentieth century. It is still common for twenty-first century virologists to be cross-trained not only in virology and general molecular and cellular biology, but also in immunology. Immunological techniques, such as those using antibodies, also remain commonplace in virology labs.

1.3 Basic research in virology is critical for molecular biology, both historically and today

Early molecular biologists were often former physicists or were inspired by the physics of the twentieth century. As such, they were not as interested in clinical problems as they were in understanding life at its most molecular level, all the while seeking general principles in molecular biology that would explain, for instance, how nucleic acids are passed from one generation to the next, how DNA encodes proteins, or how proteins fold into their own marvelous shapes. They were seeking the general principles that underlie the similarity among all living things, using an approach called **reductionism**. The reductionist physicists-cum-biologists sought the simplest systems available in order to focus on biological fundamentals with as few variables as possible. Thus, many scientists converged upon the **bacteriophages**, viruses that infect bacteria (**Figure 1.11**). Bacteriophages are much simpler than cells but nevertheless turn cells from relatively chaotic entities into factories synthesizing only a handful of viral nucleic acids and proteins, and they do so in less than an hour. Bacteriophages thus became the first models for examining the processes of DNA replication, transcription, gene regulation, and translation. As these reductionists had hoped, the study of viruses has been essential to the discovery of the fundamental general principles of molecular biology, such as the finding that DNA encodes proteins or the discoveries of tRNA and mRNA.

In molecular and cellular biology, a **model** is an organism, cell, virus, or macromolecule singled out for special attention because it is particularly amenable to experimental investigation and because results about its structure and function can be generalized. For example, studies of mice have taught us much about the vertebrate immune system and studies of

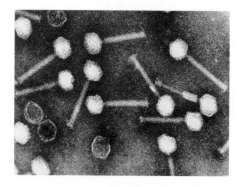

Figure 1.11 Bacteriophages. An electron micrograph depicting many T2 bacteriophages, which have prominent heads and tails. (© Institut Pasteur, Musée Pasteur. With permission.)

hemoglobin have revealed the principles of protein folding. Here, mice serve as a model for how vertebrate immune systems work and hemoglobin serves as a model for how proteins fold. This book focuses on the model viruses about which we have the clearest molecular understanding, most of which are important human pathogens or are closely related to important human pathogens. We will study, for example, the molecular and cellular biology of viruses that cause serious human disease such as COVID-19, polio, influenza, acquired immunodeficiency syndrome (AIDS), hepatitis, rabies, cervical cancer, and herpes. We will also learn about additional human pathogens and a few model viruses that infect bacteria or plants, not only because they are important in their own right but also because doing so is a reminder that this book covers only a small sample of viruses and there is much remaining to be learned.

Even though typical model viruses are associated with human disease, most virologists are engaged primarily in basic research. Basic research is focused on understanding the biological world for its own sake rather than on discovering practical applications that might improve human welfare. For basic researchers, viruses are endless sources of fascination because they are a study in contrasts. For example, viruses cause ugly human infections, yet are themselves quite beautiful (**Figure 1.12**). Although any individual virus may possess fewer than 10 genes, collectively viral genes are the most abundant genes in the biosphere. Another ironic juxtaposition is that although viruses can't reproduce without host cells, the outcome of viral infection is almost always death of the very host cell needed for the virus to replicate. A final example is that viruses can remain inert and infectious on the surface of a doorknob for days, yet inside a host cell can cause the production of hundreds of progeny in a matter of hours.

Influenza A is a model virus that provides examples of some of these intriguing contrasts (**Figure 1.13**). Although a single influenza virus encodes fewer than 20 proteins, it is able to kill a person—indeed, to kill half a million people every year—even though human beings have tens of thousands of genes. This ability raises the question, what does influenza do with its tiny number of proteins? Some activities associated with its proteins include attaching to and invading a host cell, taking over transcription and translation, interfering with the immune response, causing the host cell to produce hundreds of influenza progeny per infected cell, enabling the release of these progeny into the respiratory tract of its human host to allow spread to a new host, and ultimately killing the host cell. As molecular and cellular virologists, we want to understand exactly what each viral gene and protein does in order to understand the totality of how this virus and, by extension, all viruses function on a molecular and cellular level. Viruses are so much simpler than cells in terms of the number of genes that this holy grail seems entirely possible, or at least more possible than understanding any single cell in such detail.

Nevertheless, the more closely we examine viruses, the clearer it becomes how much more remains to be learned. HIV is an example. HIV's importance as the cause of the AIDS pandemic has led to billions of research dollars being invested in understanding its biology. Although functions for its 15 proteins have been known for a long time, we are only now coming to understand that the HIV replication cycle necessarily involves not only proteins encoded by the virus, but also proteins encoded by the host cell. For example, viruses frequently take over existing cellular proteins, as when HIV commandeers host proteins normally used in translation for copying the virus genome. Although the 15 HIV proteins are understood in tremendous detail, the ways that they interact with hundreds of host proteins are still not

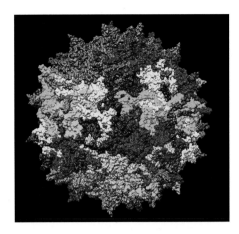

Figure 1.12 Crystal structure of a virus. The crystal structure of adeno-associated virus has been colored to reveal the patterns in the proteins displayed on the surface of the virus. (Courtesy of Jazzlw. Published under CC BY-SA 4.0.)

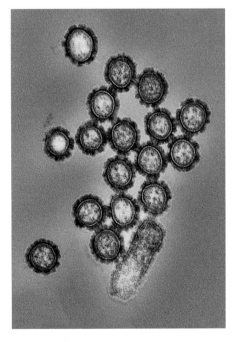

Figure 1.13 Influenza virus. This is a false-color electron micrograph of avian influenza, which has the potential to evolve and cause epidemics in human beings.

completely clear. Nevertheless, virologists remain committed to understanding every consequential molecular interaction that occurs between viral nucleic acids and proteins and host cells in order to create a holistic molecular picture of how any given virus replicates. Although every new discovery brings new questions, the allure of understanding every aspect of a simple virus's molecular biology is too tempting, and arguably too valuable, to give up.

1.4   Viruses, whether understood as living or not, are the most abundant evolving entities known

Viruses are infectious entities that are much simpler than the cells they infect. They are not, in and of themselves, alive. Microbiologists have to be very specific in defining what makes a cell alive, because it is not as obvious as it is in the case of macroscopic animals and plants. A simple test to determine if a person is alive is whether the brain and heart are still functioning. Cells don't have brains or hearts, so microbiologists need a definition of life that applies to all cells. For example, a microbe is undoubtedly alive if it can reproduce, but of course some cells might be alive yet never give rise to any offspring. One trait shared by all living microbes, whether or not they reproduce, is that they have an energized membrane across which the concentration of ions, especially protons, is different in the two compartments separated by that membrane. For a microbial cell, dissipation of the ion gradients is the definitive sign that the cell is no longer alive. By this definition there are no living viruses, because even though some viruses have a lipid bilayer, called a **viral envelope** (Figure 1.14), the viral envelope is not energized. SARS-CoV-2 and Epstein–Barr virus are two examples of enveloped viruses.

Whether or not they are classified as alive, viruses are the most abundant evolving entities; estimates suggest that there are at least 10 times more viruses than there are cells on Earth. Given that there are at least 10^{30} bacterial and archaeal cells, the number of viruses and viral genes is mind-boggling. Even if each virus encoded only 10 genes, that comes to 10^{32} viral

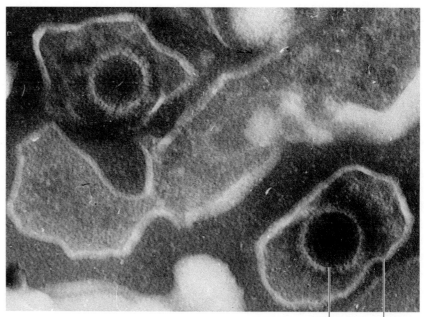

Figure 1.14 Enveloped virus. The Epstein–Barr viruses depicted here have spherical capsids and prominent envelopes.

Capsid   Envelope

genes. In contrast, astronomers estimate there are only about 10^{23} stars in the universe. There are, at a minimum, 10 million more viral genes than that. Take a moment to appreciate these numbers: awe is a great motivation for studying virology.

1.5   Viruses can be defined unambiguously by four traits

Despite this overwhelming abundance, viruses share four fundamental traits. In combination, these four traits define viruses unambiguously by excluding cells and some other noncellular nucleic acids, such as plasmids and transposons (see **Chapter 17**). First, viruses are obligate intracellular parasites that have an infectious extracellular stage. Most virologists refer to this extracellular stage as a **virion**. Until the virion makes contact with a susceptible host cell, the virion's macromolecules are inert, unable to catalyze any chemical reactions. Some forms of cellular life can give rise to metabolically inert spores, however, so the virion's state of metabolic inactivity cannot alone define viruses as different from cells. Neither does being obligate intracellular parasites alone unambiguously define viruses, because there are many cellular obligate intracellular parasites, such as *Chlamydia* and *Wolbachia*. Thus, additional traits are necessary to specify which microbes are viruses.

Second, all viruses encode at least one **capsomere** protein. Capsomeres cover and protect the nucleic acid in a virion, and all viruses carry genetic instructions for synthesizing at least one capsomere. The existence of an extracellular stage and the encoding of capsomeres that are component parts of this extracellular stage separates viruses from the many selfish DNA molecules that provide no known benefit to their cellular hosts yet replicate inside the cell. Furthermore, cells do not encode capsomeres, except in cases where cellular DNA of viral origin encodes one. There are subviral particles such as satellite viruses, conjugative plasmids, and prions (see **Chapter 17**) that also do not encode a capsomere, but these entities are not, strictly speaking, viruses.

All viruses replicate, not by growing larger and dividing as do cells, but by assembly. That is, cells infected by viruses synthesize the component parts of the virion, and then the parts, once synthesized, spontaneously assemble into new virions without further input of energy. Reproduction by assembly is therefore the third defining characteristic of viruses.

The fourth defining characteristic of viruses is the capacity to evolve. Because the minimum composition of viruses is nucleic acids and protein units called capsomeres, viral populations can change through typical evolutionary processes that alter those nucleic acids in a heritable way. HIV and influenza are examples of rapidly evolving viruses. HIV is probably less than 100 years old, and although influenza infections in humans are ancient, different versions of influenza virus emerge annually (see Figures 1.1 and 1.13). In a mere 100 years, HIV has become tremendously abundant; for example, each year more than 1.5 million people are killed by HIV infection, which is a significant decrease from the 2.7 million AIDS-related deaths in 2007. Meanwhile, between 3 and 5 million people become infected with influenza virus every year, causing up to 500,000 deaths. HIV only recently became a human pathogen, so AIDS is an example of an **emergent infectious disease**. Increasing human populations, the encroachment of humans into ever more environmental habitats, and global climate disruption continue to accelerate the pace of the emergence of viruses that have newly acquired the ability to infect humans or to spread through human populations. A more recent

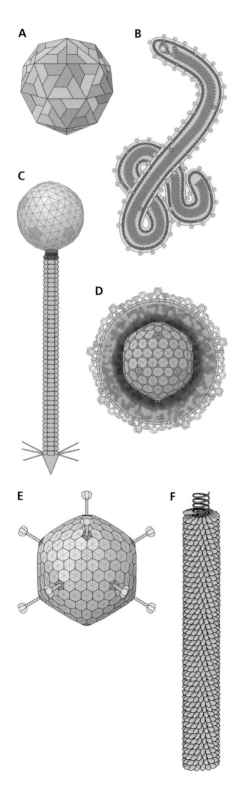

example of this phenomenon is the emergence of SARS-CoV-2 and the disease it causes, COVID-19. Thus, viruses evolve and then emerge and cause new diseases, and established viruses such as influenza also evolve and continue infecting humans over long spans of time.

Much has been made of the issue of whether viruses are alive. As we have already seen, virions are metabolically inert, lacking an energized membrane. In this strict sense, virions are not alive. Yet, when virions or the nucleic acids they contain enter a host cell, could the entire host cell not be considered an extension of the virus as it reproduces? Does the ability of viruses to evolve place them into the category of living things? The answers to these questions are not straightforward and there are legitimate arguments for or against viruses being alive. These questions are important to biologists because a universally accepted definition for life does not exist. Counting viruses might also rely upon defining which viruses are alive. As an analogy, we would not count acorns in order to determine the number of living oak trees. In this vein, however, some evolutionary virologists have questioned whether we ought to treat virions as seedlike and instead enumerate viruses by counting the number of active host-cell infections at any given time. Many other questions arise when considering how to define life; for the most part, we leave these to the philosophers.

1.6 Virions are infectious particles minimally made up of nucleic acids and proteins

At a minimum, virions are made up of nucleic acids protected by capsomeres. The protein coat of capsomeres that surrounds the nucleic acids is called a **capsid**. This minimal definition, however, does not adequately describe the structural diversity of virions (**Figure 1.15**). Because of their diversity, several schemes for dividing them into groups have been devised. For example, virions can be separated into two classes based on whether they have an external layer consisting of a proteinaceous lipid bilayer. Enveloped virions have a lipid bilayer, whereas **naked virions** do not. Another way to separate them into two groups is to consider whether the capsid is spherical or helical. Spherical capsids are actually icosahedrons, which approximate the volume of a sphere yet are constructed from repeated subunits. Spherical capsids are somewhat rigid, whereas helical capsids can be rigid or flexible, depending on the virus. Some bacteriophages combine icosahedral and helical elements, so that their heads, full of nucleic acids, are icosahedral but their tails are helical assemblages of specialized tail proteins. In other cases, the nucleic acid genome and the capsomeres are so intimately associated that their structure is termed a **nucleocapsid**. Many viruses are less than 200 nm (10^{-9} m) in diameter (for spherical viruses) or length (for helical ones), but some are much larger. As in most cases in biology, there are exceptions to this rule.

Figure 1.15 Diversity of virions. (A) Rhinovirus; approximately 30 nm in diameter. (B) Ebola virus; 970 nm long and 80 nm in cross-section. (C) Bacteriophage λ; the head is approximately 60 nm in diameter. (D) Herpesvirus; approximately 200 nm in diameter. (E) Adenovirus; capsid is 90 nm in diameter excluding spikes. (F) Tobacco mosaic virus; approximately 18 nm in cross-section but much longer than depicted. (A, From [2005] *PloS Biol* 3:e430. doi: 10.1371/journal. pbio.0030430. Published under CC BY 2.5. B–F, Courtesy of Philippe Le Mercier, ViralZone, © SIB Swiss Institute of Bioinformatics.)

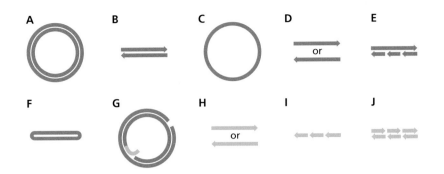

Figure 1.16 Virus and cellular genome architecture. Host cells have genomes composed of circular (A) or linear (B) double-stranded DNA. Although some viruses also have linear or circular DNA genomes, there are a range of other types of genomes that viruses can possess (C–J).

The newly discovered mimiviruses, for example, are 500 nm in diameter. Mimiviruses can even be infected by their own viruses, such as the Sputnik virophage. Two of the larger medically important viruses are smallpox and vaccinia, its relative, which are 200 nm in diameter and are enveloped with a complex internal anatomy. HIV is also an exception to the more typical capsid forms—its nucleocapsid is trapezoidal. Influenza nucleocapsids have a particular name, viral ribonucleoprotein, or vRNP, which reflects their singular structure and the intimacy of the capsid–RNA interactions.

One last intriguing feature of virions is that they contain a **genome** made of RNA or DNA but not both, and the arrangement of these nucleic acids is much more diverse than the usual double-stranded DNA found in host cells. For example, the viral RNA or DNA genomes can be single-stranded or double-stranded (**Figure 1.16**). They can also be linear, circular, or segmented. Although these generalities are useful for introducing the field of virology, keep in mind that virology is an evolving field; there are often discoveries that contradict rules that applied to all previously known viruses. An example is the hepatitis B virus, which has a small amount of RNA as a component of its otherwise DNA genome (see **Figure 1.16G**).

1.7 Viruses can be classified according to the ways they synthesize and use mRNA

Biologists have been naming viruses much longer than they have been attempting to classify them in a logical way that might reflect their evolutionary relatedness. Virus naming schemes are not systematic and can seem whimsical, if not actually haphazard. Some of the first viruses discovered were associated with terrible infectious disease, so those viruses are named for the diseases they cause. Examples include poliovirus and influenza virus; a more contemporary example is severe acute respiratory syndrome coronavirus 2 (SARS-CoV-2). Other viruses are named for the symptoms of the diseases they cause such as Crimean–Congo hemorrhagic fever virus. Some viruses are named for the parts of the human body they infect, such as the common cold virus rhinovirus (*rhino* means nose in Greek) or hepatitis (from the Greek word for liver). Still others are named for the geographical location where they first emerged, such as Marburg virus. Finally, viruses have been named for the properties of their virions. For example, the picornaviruses are tiny (pico) and have RNA genomes, whereas the geminiviruses have twin capsids (*gemini* is Latin for twins).

Classifying viruses by their evolutionary relatedness has proved much more challenging than naming them. Although viruses evolve, they may not share a single last universal common ancestor (see **Chapter 15**), which substantially complicates attempts to organize them by evolutionary history.

Table 1.1 An example of classifying the HIV-1 virus using taxonomic categories.

Order	Retrovirales
Family	*Retroviridae*
Subfamily	Orhovirinae
Genus	Lentivirus
Species	Human immunodeficiency virus 1
Subspecies	Group M

Within certain virus groups, however, obvious evolutionary relatedness exists, and so some viruses have been classified by order (-virales), family (-viridae), subfamily (-virinae), genus (-virus) and species. Some viruses also have subspecies (**Table 1.1**). Remember that taxonomic decisions are always in flux; the International Committee on Taxonomy of Viruses ultimately is responsible for keeping up with innovations in virus naming and taxonomy (see www.ictvonline.org).

We will sometimes use the **Baltimore Classification System** for viruses (**Figure 1.17**), named for the virologist David Baltimore who proposed it. It organizes viruses according to the way they encode and use mRNA. The system is thus particularly appropriate for molecular and cellular virology because it focuses on genes and the proteins they encode.

- **Class I viruses** have double-stranded DNA genomes, which are transcribed into mRNA.
- **Class II viruses** have single-stranded DNA genomes, which must first be turned into double-stranded DNA within a host cell before the transcription of mRNA can begin.
- **Class III viruses** have double-stranded RNA genomes, which are used as the templates for mRNA synthesis.
- **Class IV viruses** have single-stranded RNA genomes that are called (+) **strand** genomes because the genomes have the same sequence as mRNA. The genomes themselves are translated upon entry into a host cell, but the production of many viral mRNAs requires first copying the (+) strand genome into (−) **strand antigenomes**, which are then used as templates for synthesis of mRNA.
- **Class V viruses** have (−) strand RNA genomes, which are complementary to viral mRNAs. A (−) strand genome serves as a template for mRNA synthesis.
- **Class VI viruses**, also known as **retroviruses**, have single-stranded (+) orientation RNA genomes, but those genomes must be reverse transcribed into DNA before viral mRNA transcription.
- **Class VII viruses**, sometimes known as gapped dsDNA viruses or reverse-transcribing dsDNA viruses, or sometimes casually called reversiviruses, have double-stranded DNA genomes, which at first led to their classification as Class I viruses. Although these genomes can be used as a template to synthesize mRNA, their mechanism of genome replication involves the use of mRNA as a template and reverse transcriptase to synthesize new DNA genomes, making them very different from the Class I and Class II DNA viruses.

The Baltimore Classification System does not necessarily reflect the evolutionary history of viruses, nor does it correlate with the structural features of virions. For example, there are Class I viruses that are helical and enveloped,

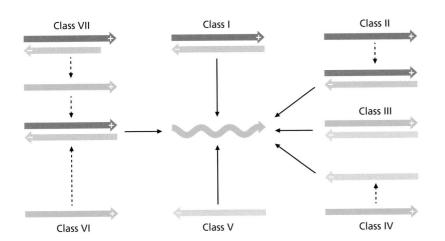

Figure 1.17 Baltimore Classification System. The Baltimore Classification System is derived from the mechanism by which the viral genome is related to mRNA. Dashed arrows indicate intermediate nucleic acids required for the virus to synthesize mRNA, and solid arrows show the synthesis of mRNA from a viral nucleic acid template.

and others that are naked and icosahedral. Class V viruses also include helical enveloped virions and both Classes III and IV include examples of naked, icosahedral virions, further illustrating that the shape of the nucleocapsid and the presence of an envelope does not correlate with any specific Baltimore class. Viruses with similar genomes often use similar strategies for gene expression and genome replication, however, so in this book we have used the Baltimore Classification System to organize **Chapters 4–10**, which are concerned with viral gene expression and genome replication.

1.8 Viruses are propagated in the laboratory by mixing them with host cells

Because viruses are **obligate intracellular parasites**, the only way to propagate them in the lab is to provide them with an appropriate host; because our focus is on molecular and cellular virology, we will focus on propagation in host cells specifically (as opposed to propagation in a laboratory animal such as a ferret). Virologists studying bacteriophages pioneered the methods for propagating viruses using **axenic** virus stocks (**Technique Box 1.2**). The result of mixing bacteriophages and host cells together is a hole or **plaque** in the **lawn** of host cells. The amount of virus in an axenic stock is typically expressed as plaque-forming units (PFUs) per milliliter.

The formation of plaques can also be used to detect animal and plant viruses, as long as the host cells provided are susceptible to viral infection. Because culturing animal and plant cells is not as simple as culturing bacteria, we have developed ways to propagate viruses using whole plants or animals. For example, a plant virus can be propagated by inoculating the leaves of a plant, whereas many animal viruses can be grown by inoculating fertilized eggs that contain a living chicken embryo. It is much more typical today, however, to work with cultured eukaryotic cells, such as **HeLa (human cervical cancer) cells** or Vero (African green monkey kidney) cells, which can be propagated indefinitely, or with **primary tissue culture cells** separated from a human tissue sample obtained through biopsy, autopsy, or following sacrifice of a lab animal.

Propagation of viruses in the lab using host cells allows us to recognize distinct though overlapping stages in the process of viral replication. The viral replication cycle can be divided into six stages: attachment to host cells; penetration and uncoating, where viral nucleic acids enter the host cell; synthesis of early proteins; synthesis of late proteins and new genomes; assembly of new virions; and release of new virions. The replication cycle is discussed more fully in **Chapter 2**.

TECHNIQUE BOX 1.2 THE PLAQUE ASSAY

One of the most fundamental techniques for studying the molecular and cellular biology of viruses is the **plaque assay** (**Figure 1.18**). A bacterial lawn is formed by plating so many bacteria on nutrient agar in a petri dish that the bacteria cannot form separate colonies and instead form a contiguous carpet or lawn. If a small number of viruses is mixed with the bacteria before plating, and then the mixture of viruses and bacteria is applied to the surface of nutritional agar in the petri dish, the bacteria will grow and form a lawn, with one exception. Wherever a virus attached to a bacterial host and reproduced, resulting in death and lysis of the host cell, a clear patch in the lawn will develop. This clearing in the bacterial forest is full of viruses; the term **plaque**, proposed by the famous phage biologist Felix d'Herelle, refers to this clearing. During the time it takes for the bacteria to grow to form a continuous lawn, typically 18–24 h, the viruses reproduce too, each one at first infecting a single cell. Then the 50–200 progeny virions are released from the dead, lysed cell, and they infect the surrounding bacteria. This cycle will repeat itself approximately once or twice per hour, so that after overnight incubation of the agar plate, there will be a macroscopic plaque wherever a microscopic bacteriophage had initially infected a cell.

The number of these plaques is proportional to the number of viruses added to the culture of bacteria in the first place, at the time of **inoculation**. A plaque assay can therefore be used to determine the number of **plaque-forming units** or PFUs per mL of a suspension of viruses. It also can be used to compare mutant viruses with their wild-type parental counterparts to find out whether the mutant bacteriophages have a phenotype such as altered plaque morphology, which usually indicates an altered interaction with the host cell.

Although the plaque assay was developed to study bacteriophages, it can be applied to the study of animal and plant viruses as well. Most bacteriophages form one PFU per particle in the suspension, but many suspensions of animal viruses contain 10 or more particles per PFU in order to form a single plaque. Strictly speaking, the term virion refers to an infectious particle and not to any other nonfunctional particles in a suspension. Because the PFU/mL is always proportional to the concentration of infectious viruses in a suspension, it is a useful measure of the infectious particles in that suspension.

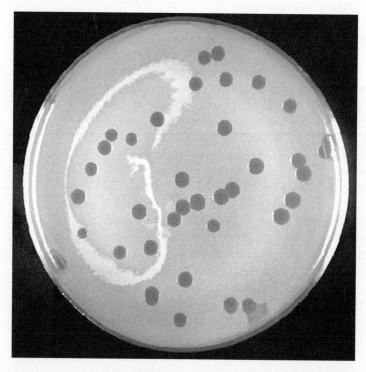

Figure 1.18 The plaque assay. Viruses and cells are mixed and then plated on a solid substrate under conditions in which the host cells are sufficiently abundant to form a lawn. Over the course of incubation, a lawn of bacteria becomes visible. Infected cells die, releasing progeny virions. These progeny go on to initiate further rounds of infection in neighboring cells, which ultimately results in a circular plaque (hole) in the lawn. The figure shows many areas of plaque and each plaque is full of viruses.

GenBank is a vast database of nucleic acid and protein sequences maintained by a collaboration of the European Molecular Biology Laboratory (EMBL) Data Library from the European Bioinformatics Institute (EBI), the DNA Data Bank of Japan (DDBJ), and the National Center for Biotechnology Information (NCBI) in the United States. As of 2016, the GenBank release 212.0 contained 207 billion nucleotide bases in more than 190 million different sequences, and the size of the database continued to double every 10 months, a metric that gets shorter and shorter as time goes by. The database is annotated not only to provide the raw sequence information, but also to transcribe and translate those sequences *in silico* (by computer), and then provide a predicted function for the encoded proteins.

The database includes many homologous viral proteins that typically serve similar functions, even if they are found in viruses separated by millions of years and billions of generations of evolution. Thus, comparing the **primary sequence** (the order of amino acids) of a novel protein with that of all known proteins is one of the first steps in determining the novel protein's function. Faster and more accurate algorithms for determining whether a protein **query** is similar to a statistically significant **hit** in a database are constantly being developed. An investigator will try to determine whether the function of any of the hundreds or even thousands of **homologous** proteins has been studied in order to associate a particular protein with a function. If so, it is possible that the newly discovered protein probably has a function similar to that of the protein or proteins in the databank. Evolution, however, makes these predictions somewhat tricky. Homologous proteins evolve separately in different viruses, and so they often have similar but not identical biological functions. A prominent example is that of the spike proteins that some viruses use to attach to host cells. Using protein–protein comparisons, it was

straightforward to identify the spike protein of the newly discovered and sequenced MERS coronavirus, a relative of SARS coronavirus (**Figure 1.19**). It was not, however, possible to use protein comparisons alone to discover the host protein to which this spike bound—instead, virologists had to use wet lab experiments to find the host receptor.

The Global Initiative on Sharing All Influenza Data (GISAID) maintains a database designed to deposit millions of influenza virus genome sequences to monitor the risk of an influenza epidemic. Virologists have also used GISAID to share sequences from the 2014–2015 Ebola virus outbreak and the SARS-CoV-2 pandemic. As of March 2023, there were more than 15 million SARS-CoV-2 genome sequences in this database.

In addition to uncovering shared ancestry, similarities in protein alignments can also reveal a great deal about the proper folding and function of a protein. Amino acids that are identical across an alignment covering billions of generations of viral replication are typically essential for that protein to fold properly, or they are essential for the protein's interactions with other molecules (such as host proteins or viral nucleic acids). If the protein is an enzyme, a subset of the conserved amino acids is critical for catalysis. Protein alignments can be used to hypothesize about the functional significance of conserved amino acids; these hypotheses can be tested using a variety of approaches. For example, a hypothesis might be that certain amino acids in the spike are essential for hydrogen bonding with a host receptor molecule. It is possible to alter those amino acids using genetic engineering and subsequently compare the properties of the altered and normal proteins. If the altered proteins have no attachment defects, then the hypothesis must be rejected.

The dissimilar amino acids in an alignment can also be informative and suggest hypotheses. For example, virus-spike proteins are often targeted by the host **immune system**.

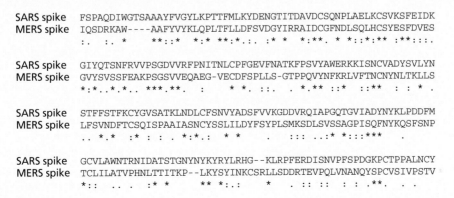

Figure 1.19 Alignment of coronavirus spike proteins. Alignment of an excerpted sequence from the coronavirus spike proteins from SARS and MERS. Asterisks indicate identical amino acids, whereas the other symbols indicate amino acids with very similar (:) or somewhat similar (.) side chains (R groups).

Host **antibodies** that bind to a virus-spike protein can block infection but it takes 1 to 2 weeks for a vertebrate to manufacture antibodies that can bind to a spike protein. The vertebrate immune system therefore exerts selective pressure on virus-spike genes, selecting against viruses that encode spikes that are so invariable that the antibodies always bind. Variations in surface-exposed amino acids on a virus spike, then, can indicate how fast the virus evolves in response to the immune system and can reveal regions of the protein that can vary substantially without affecting its function. Vaccine candidates that cause an immune response against these variable regions are likely destined to fail because the vaccine will result in antibodies that bind well to only a small subset of the virus proteins in circulation in human populations. Instead, **rational vaccine design** targets invariable, yet exposed, parts of surface proteins so that an antibody response can be effective against all naturally occurring variations of a given virus. Finding such constant surface-exposed virion proteins requires a combination of approaches including not only protein alignments but also X-ray crystallography and related techniques that can discover the structure of individual proteins or virions.

1.9 Viral sequences are ubiquitous in animal genomes, including the human genome

Advances in DNA sequencing technology and the sequencing of hundreds of animal and plant genomes have revealed that the genomes of multicellular organisms contain viral DNA and DNA that is complementary to viral RNA. **Technique Box 1.3** explains how sequences of nucleic acids or proteins can be used to discover evolutionary relatedness, and therefore to know that certain DNA in the human genome originated from a virus. For example, the human genome has over 100,000 partial segments of human endogenous retroviral (HERV) DNA in it. Endogenous retroviral sequences are DNA complementary to the genome of a retrovirus. That DNA is now inherited through the germ-line cells (sperm and egg) and most sequences are no longer able to encode infectious virions. These DNA sequences may have a function. The developmental program to build a mammalian **placenta** depends on a protein encoded by an **endogenous retrovirus**. The placenta is a temporary organ made of both fetal and maternal tissue and is essential for mammalian development in the uterus. Infection by a retrovirus made the emergence of placental mammals possible by providing a viral protein that enables formation of the placenta. The placenta is a single example; other functional consequences of endogenous retroviral DNA may be discovered. Often, novel endogenous retroviruses are found in animals that arose through rapid evolutionary radiation, an evolutionary process in which many species seem to appear from a common ancestor in a short period of time on an evolutionary time scale. An example of such rapid diversification is hominid radiation, when the genera *Australopithecus*, *Paranthropus*, and *Homo* emerged (**Figure 1.20**). The acquisition of endogenous retroviruses may have contributed to this diversification through an as yet unknown mechanism.

This evolutionary importance of retroviruses can even be detected in real time by looking at evolution in other animals. For example, there is evidence for the emergence of an endogenous retrovirus in Australian koalas. This particular retrovirus is still active as a virus that creates infectious virions, and it has been circulating among koalas for generations. The virus has somehow infected the germ line of a particular population of koalas in the northern states and territories of Australia. Koala retrovirus (KoRV) is the only known example of a retrovirus apparently in the process of becoming an endogenous retrovirus through infecting germ-line cells, where it may someday persist as DNA that does not direct synthesis of virions. Whether

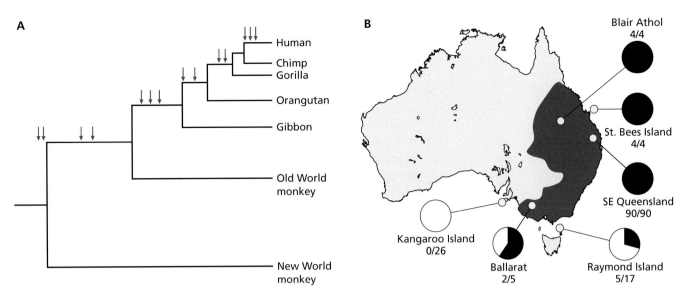

Figure 1.20 Endogenous retrovirus sequences and human evolution. (A) Family tree using arrows to show the entry of endogenous retroviruses into specific conserved sites in the depicted primate genomes. (B) Koala retrovirus (KoRV) is in the process of invading the germ line of koalas, and it can be found in both endogenous and infectious forms in different individual koalas. Areas of Australia with koalas that have KoRV infections are shown in red. The pie charts show the proportion of the population that carries the virus (black) or does not carry the virus (white). The numbers near the pie charts indicate the number of positive samples and the number of samples tested; for example, 0/26 means 0 koalas carried KoRV out of 26 koalas tested. (A, From Lebedev YB et al. 2000. *Gene* 247:265–277. doi: 10.1016/s0378-1119(00)00062-7. With permission from Elsevier. B, From Tarlinton RE et al. 2006. *Nature* 442:79–81. doi: 10.1038/nature04841. With permission from Springer Nature.)

this integration will lead to speciation of the koalas or to the infected koalas ultimately causing the extinction of the uninfected ones remains to be seen. It would be particularly interesting if KoRV became an endogenous retrovirus that causes an evolutionary radiation of koala species. These infections in koalas serve as a model to explore how retroviruses may have been necessary for the evolutionary emergence of *Homo sapiens*.

Retroviruses are not the only viruses that affect organismal evolution. Viral nucleic acids contribute to the evolution of bacterial pathogens, such as the bacteria that cause cholera, whooping cough, and dysentery. In these cases, genes that manipulate the human immune system or other tissues are part of the genomes of bacteriophages that themselves can chronically infect the bacteria, which in turn together infect human beings. Without their phages, the bacteria would not be able to cause such severe disease.

Essential concepts

- Viruses are abundant, diverse biological entities that infect all types of cells.
- Virology has historically been intertwined with the disciplines of molecular biology and immunology. This close interconnectedness continues today.
- Human viruses, their close relatives, and bacteriophages are the most commonly studied model viruses that can provide insights about viruses in general.
- Viruses all share the following four traits: they are obligate intracellular pathogens, they encode at least one capsomere, they reproduce by assembly, and they can evolve. No other biological entities share these four characteristics.

- Virions are the extracellular infectious particles that carry viral nucleic acids from one host cell to another. All virions are made up of, at a minimum, protein capsomeres enclosing nucleic acids.
- Virions can be icosahedral or helical, and they can be naked or enveloped.
- Virus genomes are diverse, can be composed of DNA or RNA, and can be arranged as linear or circular molecules that are either double- or single-stranded and either segmented or nonsegmented.
- The Baltimore Classification System organizes viruses into seven classes based on the way they produce and use mRNA; it is a useful system even though it does not reflect the evolutionary relatedness of different groups of viruses.
- Viruses are grown in the laboratory by mixing them with suitable host cells. A plaque assay can be used to determine the number of infectious plaque-forming units in a liquid suspension.
- Endogenous retroviruses are associated with major evolutionary events, including the origin of *Homo sapiens*, although the consequences of this association, if any, are hotly contested. Retroviral infections are likely causing evolutionary change today.

Questions

1. Compare and contrast a Baltimore Class I naked spherical virus with a linear genome to a human epithelial cell.
2. Compare and contrast an enveloped Baltimore Class V helical virus to an enveloped Baltimore Class VII spherical virus.
3. Viruses are not alive, yet they can evolve. Explain.
4. Why might it be important to decide if viruses are alive or not?
5. You dilute a stock of virus in a tenfold series and ultimately mix 100 μL of diluted virus with host cells. The 10^7 dilution results in a plate that is almost entirely lysed, whereas the 10^8 dilution has 243 plaques. The 10^9 dilution has 31 plaques. How many plaque-forming units/mL are there in the stock? Explain your reasoning.
6. About how big is a virus relative to the resolution of a light microscope?
7. Can you think of a circumstance in which invariable amino acids in an alignment might not necessarily be informative in that they do not clearly reveal the amino acids that must be conserved for both folding and function to occur?

 Interactive quiz questions

In addition to the questions provided above, this edition has a range of free interactive quiz questions for students to further test their understanding of the chapter material. To access these online questions, please visit the book's website: www.routledge.com/cw/lostroh.

Further reading

General virology

Baltimore D 1971. Expression of animal virus genomes. *Bacteriol Rev* 35:235–241.

Zimmer C 2015. *A Planet of Viruses*, 2nd ed. University of Chicago Press.

https://gisaid.org: Online archive of virus genome sequences used to share those sequences rapidly and equitably.

www.nature.com/scitable/blog/viruses101: A blog introducing undergraduates to virology.

https://pdb101.rcsb.org/: Structural biology of viruses and their component proteins.

https://viralzone.expasy.org/: Online archive of most viruses, with extensive information about their structure, proteins, and nucleic acids.

www.who.int/topics/infectious_diseases/en/: World Health Organization's information on infectious disease, including viral diseases.

Molecular biology

Cairns J, Stent G & Watson J 2007. *Phage and the Origins of Molecular Biology*, the centennial ed. Cold Spring Harbor Press.

Echols HG & Gross CA 2001. *Operators and Promoters: The Story of Molecular Biology and Its Creators*, 1st ed. University of California Press.

www.nobelprize.org: Online archive of all Nobel Prizes, including talks delivered by awardees. Virus-related Nobel Prizes include those for the crystallization of tobacco mosaic virus, the discovery and utility of restriction enzymes, the first sequenced genome, the discovery of intron splicing, the discovery that human papillomavirus causes cervical cancer, the discovery that HIV causes AIDS, and gene silencing by double-stranded RNA (an antiviral response).

Applied virology

Gilbert S & Green Hodder C 2021. *Vaxxers: The Inside Story of the Oxford AstraZeneca Vaccine and the Race Against the Virus*. Stoughton.

Lewis R 2013. *The Forever Fix: Gene Therapy and the Boy Who Saved It*. St Martin's Griffin.

Influenza

Spinney L 2017. *Pale Rider: The Spanish Flu of 1918 and How It Changed the World*. Public Affairs.

HIV

France D 2016. *How to Survive a Plague: The Inside Story of How Citizens and Science Tamed AIDS*. Knopf.

Shilts R 2011. *And the Band Played On: Politics, People, and the AIDS Epidemic*. Souvenir Press.

Stillwaggon E 2005. *AIDS and the Ecology of Poverty*. OUP, USA.

SARS-CoV-2

Zhu N, Zhang D, Wang W, Li X, Yang B et al. 2020. A novel coronavirus from patients with pneumonia in China, 2019. *N Engl J Med* 382:727–733.

https://ourworldindata.org/coronavirus: Database of statistics about COVID-19 such as cases and case fatality rates.

The Virus Replication Cycle

Virus	Characteristics
Bacteriophages	Models used to discover the six general stages of the virus replication cycle.
Influenza A virus	Model for six stages of viral replication in an animal cell.
Human immunodeficiency virus (HIV)	Example of persistent virus that causes simultaneous lytic and latent infections.
Human herpesvirus 1	Model of latent viral infection in humans.

The viruses you will meet in this chapter and the concepts they illustrate

A typical virus is composed of nothing more than a few proteins and some genetic material; as such, they are very small, but they can be visualized with techniques such as scanning electron microscopy (SEM), transmission electron microscopy (TEM), and atomic force microscopy (AFM) (**Technique Box 2.1**). This structural simplicity contrasts with the myriad molecular means by which the viruses subvert host-cell processes and turn those host cells into virus manufacturing centers. Viral replication can be divided into a few universal stages that apply to most viruses regardless of host. This chapter introduces universal features of the viral replication cycle, which were first elucidated through studies of bacteriophages and are illustrated in this chapter using bacteriophages and the influenza virus. Although most virus replication cycles are rapid and kill the host cell within days (or minutes, for bacteriophages), some viruses such as herpes simplex virus (now formally called human herpesvirus 1) have an alternative strategy that includes long-term infection of host cells without necessarily forcing the production of infectious virions. Although the stages of viral replication are nearly universal, each individual virus exhibits its own particularities at each step, much like individual musicians might play the same tune, always recognizable, but with individual flair. This chapter will help you recognize the themes in the midst of tremendous variation, setting the scene for detailed discussions of each stage of the replication cycle in **Chapters 3–11**.

DOI: 10.1201/9781003463115-2

TECHNIQUE BOX 2.1   ELECTRON MICROSCOPY AND ATOMIC FORCE MICROSCOPY

Although most are too small to be detected using light microscopy, viruses can be magnified and resolved using **electron microscopy** because electrons have much smaller wavelengths than visible light. For electron microscopy, biological samples must first be **fixed** and then **stained**, meaning that they are coated with heavy metals that adsorb or reflect the electron beam. Electron microscopy is useful for resolving tiny objects (0.1–1 μm) such as viruses (20–400 nm in diameter), but it can only detect features of viruses and cells after they have been stained with heavy metals. Samples for electron microscopy can also be stained with antibodies that are attached to metal nanoparticles (typically gold) so that the specific proteins recognized by the antibodies can be localized in a host cell by detecting the gold nanoparticles.

There are many forms of electron microscopy. **Scanning electron microscopy (SEM)** is used to see the exposed surfaces of microbes including virions (**Figure 2.1A**). **Transmission electron microscopy (TEM)** can be used with thin (<100 nm) sections that reveal the internal features of viruses (**Figure 2.1B**). There are continuous improvements to these techniques, pushing the limits of what we can see.

One of the most important innovations for studying virus structure is **cryo-electron microscopy (cryo-EM)**. Cryo-EM uses very cold temperatures (less than –161°C) to preserve the structure of biological samples without fixing. A variation, termed **cryo-electron tomography**, takes two-dimensional images of samples tilted at set angles within the microscope, after which a computer constructs a composite three-dimensional rendering of the virus using hundreds of two-dimensional images taken from different perspectives (**Figure 2.1C, D**). In some cases, the resolution afforded by cryo-electron tomography

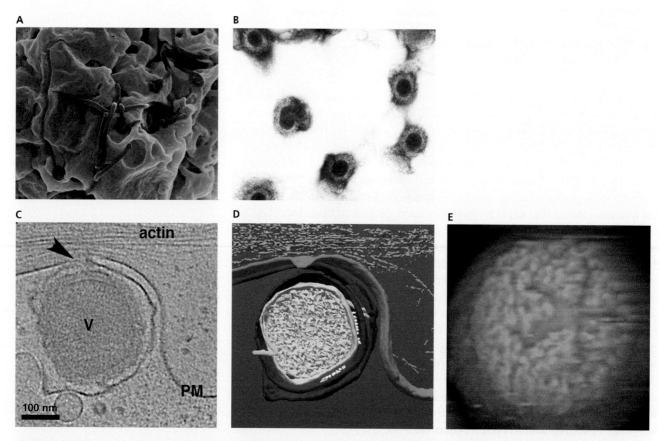

Figure 2.1 Viruses imaged using electron microscopy or atomic force microscopy (AFM). (A) SEM of Ebola virus budding from the surface of a Vero cell (African green monkey kidney epithelial cell line). (B) TEM of cytomegalovirus (CMV), a herpesvirus. (C) Cryo-EM showing an extracellular vaccinia virion (V) attached to the plasma membrane (PM). (D) Surface rendered representation of the virion bound to the plasma membrane depicted in (C). (E) AFM of mimivirus, a large virus covered by surface fibers, with a distinctive star-shaped depression. (A, Courtesy of NIAID. B, Courtesy of CDC/Sylvia Whitfield. C, D, From Cyrklaff M et al. 2007. *PLOS ONE* 2: e420. doi: 10.1371/journal.pone.0000420. Published under CC BY 4.0. E, From Kuznetsov YG et al. 2010. *Virology* 404:127–137. doi: 10.1016/j.virol.2010.05.007. With permission from Elsevier.)

approaches that of **X-ray crystallography** for visualizing virion structures, and often the two techniques are used in tandem to examine virus structure. X-ray crystallography is addressed in **Chapter 3**.

Images made by electron microscopy are two-dimensional. They can have very good resolution in the X- and Y-planes, but they cannot provide truly three-dimensional information. To get a three-dimensional view of virus particles with quantitative information about the vertical axis, a form of **scanned probe microscopy** termed **atomic force microscopy** (**AFM**) can be used. In AFM, a tiny probe (which can have a diameter less than 1 nm)

is literally dragged back and forth over the surface of a sample, and a computer constructs a three-dimensional image of the object, given the positional information available from the probe. AFM can make images of unstained virions and has nanometer resolution; in some specialized applications, it can even be used to image individual atoms (**Figure 2.1E**). Although most AFM examines dried specimens, it is increasingly common to image biological samples in aqueous solution, including enveloped viruses. AFM is typically used by physicists, however, and virological use of AFM is not as common as use of SEM or TEM.

2.1 Viruses reproduce through a lytic virus replication cycle

Most viruses cause **lytic infections** in which the host cell releases hundreds of new viruses and then dies as a consequence of viral infection. Usually, the host cell bursts open, or lyses, and thus the name for these infections. Careful study of bacteriophages provided the first clues to the events during a virus replication cycle. In the famous **one-step growth experiment**, Ellis and Delbrück mixed bacteriophages with host cells under conditions in which each cell was infected by a single virus and the initiation of infection was synchronous. The ratio of plaque-forming units (PFUs) to host cells is called the **multiplicity of infection (MOI)**. Because the infection was synchronous, they could keep careful track of the PFU per milliliter in the broth surrounding the host cells and the events inside the host cell, such as synthesis of RNA, proteins, and DNA.

Figure 2.2 shows the results from an idealized one-step growth experiment. The curve summarizes such experiments performed on a variety of model viruses, including animal viruses. At the time of inoculation (0 on the X-axis), virions were mixed with host cells at an MOI of 1. For some period of time after inoculation, it is impossible to detect extracellular virions. The time between the initial infection and the appearance of new virions is called the **eclipse period**, which is analogous to the way the sun can seemingly vanish from the sky when the moon's shadow is cast upon the Earth in just the right way. As mentioned in **Chapter 1**, the eclipse period is unique to viruses;

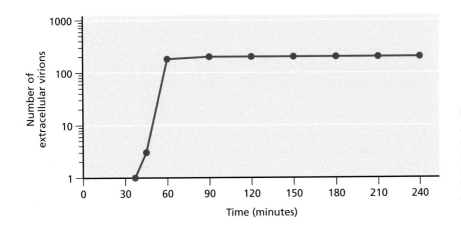

Figure 2.2 One-step growth curve. After a single virus is added to its host cell, extracellular viruses become undetectable during the eclipse period. Then, in an abrupt change, the number of extracellular viruses increases exponentially. Eventually, the number of viruses levels off, and there is no further increase in the population size.

Figure 2.3 Synthesis of viral components during an infection. Accumulation of viral mRNA, viral proteins, viral genomes, intracellular viruses, and extracellular viruses.

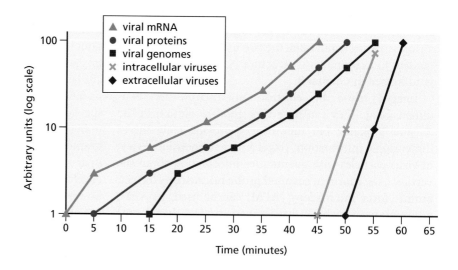

no cells reproduce in this manner. The virus appears to vanish because it is not possible to detect intact infectious PFU in the culture during the eclipse period. Then, all at once, the number of detectable PFU increases exponentially. The short period of time when the detectable virions increase by two or more orders of magnitude is called the **burst**. After that explosive increase, the number of viruses levels off and does not change. Unless more host cells are added to the culture, the PFU/mL then remains constant. The Delbrück research group named the one-step growth experiment because the offspring phages appear suddenly and abundantly, seemingly in a single step, compared with the many steps of any cell division cycle.

The development of techniques for synchronizing viral infections allowed not only for the one-step growth experiment, but also for closer examination of the virus and host cells during the specific inoculation, eclipse, and burst events (**Figure 2.3**). After the virus attaches to its host cell, the viral genome enters the cell and the eclipse phase begins. The eclipse phase may seem calm, but within the cells there is a frenzy of activity. First, the cell fills with viral mRNA; in the case of typical bacteriophage infections, cellular mRNA synthesis is completely shut off within just a few minutes. As viral mRNA increases, so do the levels of viral proteins needed for gene expression and genome replication. Next, viral genomes begin to accumulate and the capsomeres are synthesized. Finally, just before lysis, the cell fills with intracellular virions, which have assembled from their component parts.

2.2 Molecular events during each stage of the virus replication cycle

The one-step growth experiment revealed that a typical virus replication cycle can be divided into six steps:

1. **Attachment**: The virion attaches to the host cell.
2. **Penetration** and **uncoating**: The virion's genome enters the host cell; in doing so, the virus as a discrete virion ceases to exist, beginning the eclipse period. In some cases, the genome may enter along with the capsid or components of the capsid.

3. Synthesis of **early proteins**: The proteins expressed early in an infection often have one of three functions: to shut down the synthesis of host proteins, to regulate expression of viral genes, or to synthesize viral nucleic acids.

4. Synthesis of new viral genomes and **late proteins**: There are usually regulatory events that cause a shift from synthesis of early mRNA to the making of new genomes. Late proteins are those expressed after genome replication has begun. The proteins made at this phase are usually **structural proteins**, meaning that they will become components of the progeny virions. In some viruses, especially those with particularly small genomes, stages 3 and 4 may overlap.

5. **Assembly**: This is the stage when the component parts of the virion assemble into completed virions.

6. **Release**: This stage, the release of the newly assembled viruses, is the last step. The release of bacteriophages is sudden because lysis of their host cells must occur in order for the viruses to escape. Shedding of viruses from animal cells can be gradual and may not lyse the host cells; nevertheless, these viral infections also ultimately kill the host cells.

It is increasingly common to visualize events that occur during each stage using fluorescence microscopy (**Technique Box 2.2**). This form of microscopy is particularly suitable for observing interactions between host-cell macromolecules and membranes and viral proteins and nucleic acids. It can also be used to observe the movement of host and virus components over time, providing insights into the dynamic nature of viral infections.

2.3   The influenza virus is a model for replication of an animal virus

The influenza virus serves as a model for observing the virus replication cycle in an animal virus. Influenza virus is enveloped, and viral proteins protruding from its surface mediate attachment to host cells. The penetration and uncoating steps are somewhat complicated. First, the entire virus enters the cell in an **endosome**. Subsequently, the envelope of the influenza virus fuses with the endosomal membrane, releasing the viral genome into the cytoplasm. In this case, the influenza virus has eight genome segments consisting of single-stranded RNA closely associated with nucleocapsid proteins and enzymes (PB1, PA, and PB2) in a structure known as **viral ribonucleoprotein**, or **vRNP** (**Figure 2.4**). The vRNPs travel to the nucleus using the cell's cytoskeleton and motor proteins. The viral genome segments must enter the nucleus through nuclear pore complexes before viral gene expression can occur. Viral gene expression and genome replication ultimately result in the formation of new vRNPs in the nucleus. The vRNPs are exported through the nuclear pores to the cytoplasm so that they can associate with other virus proteins and acquire an envelope, after which they exit the cell through a process called **budding**. All of these stages, and some of the experiments that discovered them, are covered more fully later in this book (see **Chapters 3, 7, and 11**). The host cell will eventually die as a consequence of the viral infection, in part because the cell's adenosine triphosphate (ATP), amino acids, and nucleotides are depleted, leaving the cell unable to maintain homeostasis. In other cases, the host cell will die as part of an immune reaction against the virus and the cells it has infected.

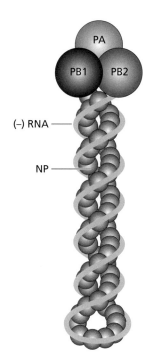

Figure 2.4 A vRNP from influenza. The vRNP is composed of several different proteins (PB1, PA, PB2, and NP) and a single strand of negative-sense viral RNA.

TECHNIQUE BOX 2.2   FLUORESCENCE MICROSCOPY

The use of **fluorescence** has revolutionized imaging of live cells by improving the contrast between various structures and the rest of the cell and by increasing the limits of resolution for light microscopy using computerized enhancement techniques in combination with lasers to illuminate the sample (**Figure 2.5**). Fluorescent molecules such as green fluorescence proteins absorb light at one wavelength and emit light at another, so samples are illuminated with one wavelength of light, after which the emitted light is collected through a filter that removes most **background signal**, providing a clear image. Confocal scanning laser microscopy illuminates a thin slice of a living sample using a tight laser beam for illumination and then takes many images of the same structure, moving the laser in small increments through the sample in a process known as **optical sectioning**. A computer can then take all of the scans to construct a three-dimensional image of the sample, or it can collapse all of the images into a two-dimensional view of the subject.

Fluorescent stains that have a high affinity for nucleic acids or membranes can stain the nucleus or internal membranes, respectively. More importantly, genetic engineering can be used to construct chimeric proteins (also known as tagged proteins) that have a **green fluorescent protein** (**GFP**) as one of the components of the **chimera**. The other half of the chimera is any other protein of interest, such as an enzyme encoded by a virus or a host protein that always localizes to a specific organelle, such as the endoplasmic reticulum (ER). The purpose of the chimeric construct is to enable viewing the abundance and localization of the chimera, which usually reproduces the abundance and localization of the protein of interest. There are variations of GFP that have different fluorescence properties, such as fluorescing yellow, blue, or red, and we are continually searching for other fluorescent proteins that might serve as fluorescent tags in living cells.

With chimeric fluorescent proteins, we can see the internal structures of living cells that are made visible. For example, a chimera between an endoplasmic reticulum protein such as calreticulum and GFP will cause the lumen of the ER to fluoresce green, marking the position of the ER in the living cell. Fluorescence microscopy using GFP chimeras can detect whether a virus alters a cell's normal cytoskeleton or organelles, or whether infected cells contain compartments where viral proteins necessary for genome synthesis are located, forming a virus factory. Because the cells are alive during imaging, it is also possible to assemble movies of dynamic processes, such as the movement of viruses inside a host cell. An example is the movement of vaccinia virus using actin comet tails to propel it through the cytoplasm and even into neighboring uninfected cells. Finally, because viruses are exquisitely

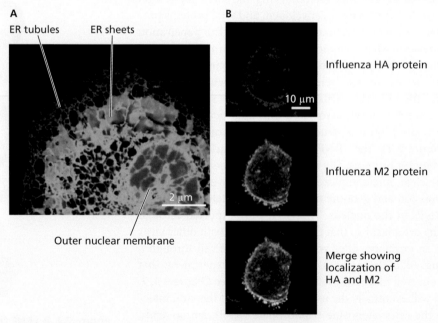

Figure 2.5 Fluorescence microscopy. (A) An animal cell in tissue culture that was genetically engineered to express an ER-membrane protein fused to a fluorescent protein. (B) Host cells infected with influenza were immunostained to detect the viral spike protein HA (red) and viral matrix protein M2 (green). The bottom panel shows a merged image, in which the combination of red and green results in the color yellow. Yellow areas show that the two viral proteins often co-localize; these are the sites where influenza virions will bud from the host cell. (A, Courtesy of Patrick Chitwood & Gia Voeltz.)

selective in terms of which cells they infect within an animal, viruses expressing GFP can also be used to mark the location and abundance of their specific host cells. For example, causing a neurotropic virus to express GFP during infection can enable a researcher to follow a single infected neuron throughout the body of an experimental animal.

Here we have focused on techniques that have become common in molecular and cellular virology laboratories. There are always new variations on fluorescence

microscopy, such as super-resolved fluorescence microscopy, in development. Super-resolution fluorescence microscopy allows images to be taken with greater resolution than ought to be possible defined by the wavelength of light used to illuminate the sample. The 2014 Nobel Prize in Chemistry was awarded for the development of super-resolved fluorescence microscopy. Similarly, there are always innovations in the development of fluorescent proteins and other fluorophores, expanding the utility of fluorescence-based microscopy even further.

2.4   The host surface is especially important for attachment, penetration, and uncoating

At each stage of the replication cycle, components of the virus interact with specific components of the host cell (Table 2.1; Figure 2.6). For example, the surface-exposed molecules of both the virion and the host cell are of critical importance for the attachment and penetration stages. For the virus, the molecules that mediate host attachment and penetration are typically proteins or glycoproteins, and the host molecules used for attachment might be proteins, glycoproteins, lipids, or glycolipids. The prefix glyco- on a macromolecule name indicates that carbohydrate groups have been covalently attached to it. The interactions between the virus and host for penetration and uncoating depend on whether the host cell has a cell wall and whether the host is eukaryotic. Both plant cells and bacterial cells have cell walls, so the viruses that infect them must have some means of overcoming this substantial barrier. Most bacterial viruses degrade the cell wall or take advantage of a preexisting proteinaceous organelle such as a pilus to pass through the cell wall, whereas most plant viruses rely on mechanical damage to cell walls to reach the plasma membrane. For animal and plant viruses, internalized virions often interact with endosomes and lysosomes during uncoating. Furthermore, because eukaryotes compartmentalize DNA replication and transcription in the nucleus, uncoating for viruses that infect eukaryotes might involve multiple steps that are necessary to introduce their genomes into the nucleus, as in the influenza example. It is more common, however, for eukaryotic RNA viruses to replicate entirely in the cytoplasm. DNA viruses are most often those in which uncoating culminates in viral nucleic acids entering the nucleus.

Table 2.1 Virus–host cell interactions for a cytoplasmic virus.

Stage of Virus Replication Cycle	Typical Virus Component	Typical Host-Cell Component
Attachment	Spike or capsomer	Glycoprotein or protein
Penetration and uncoating	Spike or capsomer	Plasma membrane or endomembrane
Expression of regulatory proteins	Viral mRNA	Ribosomes
Genome replication	Viral replicase	Varies
Expression of structural proteins	Viral mRNA	Ribosomes
Release	New virions	Plasma membrane

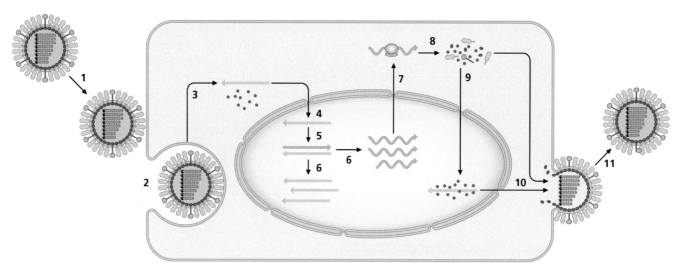

Figure 2.6 The influenza replication cycle. The virus attaches to particular sugars on the surface of the host cell (1). The virus is internalized by endocytosis during the penetration step (2). During uncoating, the genome segments and associated proteins (vRNPs) are released into the cytoplasm (3), where they are transported to the nucleus (4). The influenza genome is copied to make a double-stranded template (5) that can be used for production of mRNA and new genomes (6). The mRNAs are exported to the cytoplasm (7) and translated (8). Some influenza proteins re-enter the nucleus to assemble with new genome segments (9), whereas others assemble at sites of future budding. The vRNPs assemble with other component parts of the virus (10). The virus exits the host cell through budding (11), assisted by a viral enzyme that degrades the host surface sugars that would otherwise tether the virus to the cell surface.

2.5 Viral gene expression and genome replication take advantage of host transcription, translation, and replication features

The expression of viral regulatory proteins and enzymes results from an interaction between viral nucleic acids, proteins, and the host transcription and translation systems. Bacterial and eukaryotic mRNA molecules have quite different structural features (**Figure 2.7**), which are required for proper translocation, translation, and stability. These features have a profound effect on virus gene expression strategies. For example, transcription from a DNA template occurs in the cytoplasm in the Bacteria, but in the nucleus in the Eukarya. Furthermore, the mRNA of bacteria has internal **ribosome-binding sites**, also known as **Shine–Dalgarno sequences**, and is commonly **polycistronic**, meaning that it encodes more than one protein (see Figure 2.7). In contrast, the mRNA of eukaryotes has a 5′ cap, necessary for ribosome binding and for transport out of the nucleus; has a poly(A) tail, also necessary for export from the nucleus and for stability; and usually encodes just one protein.

Figure 2.7 Structural features of mRNA from Bacteria, compared with that from the Eukarya. (A) Bacterial mRNA, such as this one encoding the three proteins of the *lac* operon, can be polycistronic, with ribosome-binding sites adjacent to each start codon. Stop codons are also indicated. (B) Eukaryotic mRNA, such as this one encoding interferon α (IFNA) encodes just one protein with a single start codon and stop codon, has a 5′ methylated cap, and has a poly(A) tail.

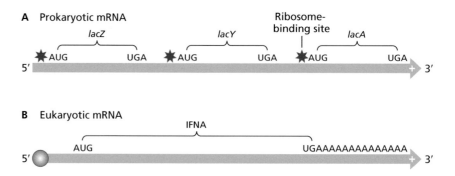

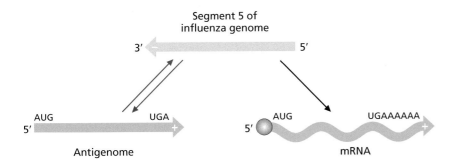

Figure 2.8 A viral genome, antigenome, and mRNA. The influenza virus is used as an example. The segmented viral genome is made up of (–) RNA, as illustrated by segment 5 in the figure. Antigenomes are perfect copies of the (–) RNA genome, lacking the 5′ cap and polyadenylation found in mRNA. Synthesis of antigenomes and genomes is indicated by the red arrows. The mRNA is a copy of the information in the genome, but it is also capped and polyadenylated; its synthesis is indicated by the black arrow.

Viruses must transcribe mRNA that has the genuine features of normal cellular mRNA specific for its particular host cell type, or must mimic or functionally replace these features so that the viral mRNA can be translated. In animal viruses, soluble proteins are translated by cytosolic ribosomes, whereas transmembrane proteins must be translated by ribosomes attached to the rough endoplasmic reticulum, using the same system the host cells use to translate their own transmembrane proteins. Another critical aspect of the host transcription machinery in all cells is that it exclusively uses a double-stranded DNA template to polymerize RNA. Viruses with RNA genomes must have a viral enzyme that can use that RNA as a template, whether to produce **antigenomes** (complementary copies of the genome, used as a template to synthesize new genomes), mRNA, or new genomes (**Figure 2.8**). In the cases of retroviruses and reversiviruses (gapped dsDNA reverse-transcribing viruses), a viral enzyme is needed to make a DNA copy of an RNA molecule.

2.6   The host cytoskeleton and membranes are typically crucial during virus assembly

Virus assembly is mainly a matter of interactions between the viral structural proteins themselves and the viral genome, but in many cases the virus proteins must still interact with specific host membranes in order to complete assembly. In eukaryotic viruses, assembly factories known as **virus replication complexes** (**VRC**) are associated with various membranes, such as the endoplasmic reticulum or Golgi body, in addition to the plasma membrane. Some animal viruses begin assembly in an intact nucleus, so that proteins translated in the cytoplasm have to be transported through nuclear pore complexes and subsequently immature virions must be transported out through the nuclear pore complexes, which exclude diffusion of particles larger than 60 kDa. There are also viruses such as herpesviruses that exit the nucleus by budding. Diffusion of particles larger than 500,000 Da, such as viruses (which are at least 3 million Da), is quite slow in the crowded cytoplasm, which has a viscosity that can be likened to that of wet sand. In eukaryotic cells, virus proteins or immature virions interact with the cytoskeleton and motor proteins in order to make directed progress to the surface of the cell and subsequently to mature and exit the host cell. Motor proteins specialize in moving cargo along microfilaments that are constructed from polymerized actin or along microtubules that are composed of polymerized tubulin. The cytoskeletal MreB (actin homolog) and FtsZ (tubulin homolog) proteins of bacteria have been investigated for a much shorter period of time than the cytoskeleton of eukaryotes, but it is increasingly apparent that the bacterial cytoskeleton and plasma membrane can also be critical for organizing bacteriophage assembly.

2.7   Host-cell surfaces influence the mechanism of virus release

Release also depends upon the interaction between viral proteins and host-cell components, such as the cell wall (if there is one) and the cell membrane, or, in some eukaryotes, other membranes, such as the nuclear membrane, endoplasmic reticulum, Golgi body, and secretory vesicles. Bacteriophages usually encode cell-wall-degrading enzymes so that the phages can escape through **osmotic lysis** of the cells, in which the weakened cell walls cannot squeeze back against the pressure of water rushing in to equalize the concentration of solutes on either side of the plasma membrane. In contrast, plant viruses are carried from one plant to another by biting arthropods, whose mouth parts break through the plant cell wall.

2.8   Viruses can also cause long-term infections

Although lytic infections are common, viruses can also cause longer-term **persistent infections**. Persistent infections were once considered unusual, but it is now apparent that most cells contain nucleic acids of viral origin, indicating that viral persistence as nucleic acids is extremely common. Among the bacteriophages, viruses that can cause long-term infections are called **temperate phages**. Temperate phages can persist in their bacterial hosts as **prophage** DNA (**Figure 2.9**) for generations, without producing virions; they are temperate in the sense that they don't always kill their host cells immediately following the initial infection. Among animal viruses, persistent long-term infections are very common; for example, most adult people have at least one long-term infection caused by a member of the *Herpesviridae* family. During some **chronic infections**, viral replication is continual, though the immune system holds clinical disease in check. **Latent infections** are those in which there is little or no production of virions, yet the virus persists, typically as viral DNA in the nucleus of a differentiated cell, analogous to a prophage. Latent infections are typical of the *Herpesviridae*.

Human immunodeficiency virus (HIV) disease is an example of a chronic viral infection. Over the course of a typical HIV infection, the person experiences a period of time when HIV virions are found in very low levels in their blood (**Figure 2.10**). This feature of the infection gave rise to the hope that HIV would turn out to be a latent infection in which viral replication was not ongoing. Unfortunately, it turned out that HIV instead persists as a chronic infection during which viral replication is ongoing. Within weeks

Figure 2.9 A prophage genome in a bacterial chromosome. When this bacterium containing a prophage reproduces, both of its clonal offspring will contain the prophage. In this way, the prophage can be passed from one bacterium to the next, infecting all of the offspring of the original parent. The prophage can be triggered to switch to a lytic reproductive cycle, which is not illustrated.

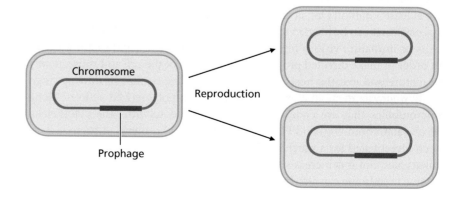

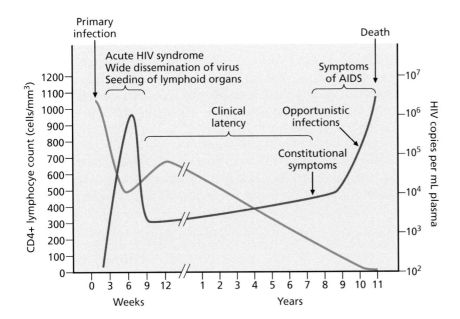

Figure 2.10 Clinical progression from HIV infection to acquired immunodeficiency syndrome (AIDS). This graph depicts the concentration of HIV virions (red line) in blood and the concentration of CD4+ lymphocytes (blue line) in blood. The CD4+ lymphocytes are hosts for HIV.

of infection, the population of host cells (CD4+ lymphocytes) in the blood declines because of high levels of virus replication. However, the immune system rebounds to control viral replication to some extent; it takes years for the host cells to become extremely depleted and, therefore, levels of virions in the bloodstream remain low. Because the host cells are required for normal immunity, the loss of host cells beyond a certain threshold, generally recognized as 200 mL^{-1}, and the rise of viral RNA in the bloodstream are associated with acquired immunodeficiency syndrome (AIDS), in which the patient becomes susceptible to many infections and eventually succumbs.

Although the immune response keeps levels of HIV in the bloodstream low throughout most years of an untreated HIV infection, the virus still replicates continuously in sites such as the gastrointestinal tract and bloodstream. Because of this continuous replication, in combination with a high rate of mutation, the population of HIV viruses inside the infected person can evolve, ultimately escaping the immune response, as well as the effects of antiviral drugs. In order to slow the evolution of resistance, HIV-positive patients typically take three different antiviral medications at any given time (see **Chapter 16**). If HIV becomes resistant to all treatment, the patient is at a high risk of developing AIDS and dying.

From data such as those plotted in Figure 2.10, it was clear that HIV causes a chronic infection. In a somewhat surprising turn of events, however, it turns out that HIV also infects some immune cells latently. The cells with latent infections cannot be eliminated by treatment with antiviral medications because the medications target the viral proteins needed to synthesize new virions. These proteins are not synthesized by latently infected cells, so the medications have no effect. Latently infected cells serve as a reservoir for new viruses even when a patient is taking long-term **combination antiretroviral therapy** (**cART**), which is a collection of drugs that a patient takes at the same time in order to keep lytic replication very low. Latently infected cells present a serious barrier when attempting to develop a cure for HIV, which would require eliminating all the HIV nucleic acids from a person's body.

2.9   Herpesvirus is a model for latent infections

Herpesviruses are the classic example of a latent viral infection in human beings. The herpes simplex virus (formally known as human herpesvirus 1) infects both epithelial tissues and terminally differentiated neurons. The epithelial infections typically cause genital or oral sores characteristic of a herpes infection, but the virus can also develop a quiescent infection in neurons without producing virions for the lifetime of an infected person. Latently infected neurons appear normal in tissue culture. Sporadically or in response to stress, however, the latent virus reactivates and the infected neuron begins to produce infectious particles. These particles are released and subsequently infect epithelial cells enervated by that particular nerve, which ultimately causes more herpes lesions with the potential to spread the virions to other people.

2.10   Research in molecular and cellular virology often focuses on the molecular details of each stage of the replication cycle

This chapter has provided an overview of how viruses interact with host cells. But of course, the particular molecular interactions between any given virus and its specific host cell are exquisitely specific. Typical research questions in the field of molecular and cellular virology often focus on these specific interactions. Examples of typical research questions in the field include the following. Exactly how does HIV attach to its host cells, and does that attachment explain the types of cells that can be infected by HIV? How do influenza vRNPs find their way to the nucleus and then pass through the nuclear pores? Why are they shaped in such a striking way? Does their structure indicate something about influenza gene expression or genome replication? Where does HIV assemble in the cell and is the location related to the components of its membranous envelope? Why is there no vaccine to prevent HIV? Why is it necessary to get a different influenza vaccine every year? How do antivirals work, and is there any way to prevent viruses from evolving resistance to them? Answers to these questions, and many more, require a thorough understanding of the molecular interactions between viruses and their host cells and are addressed in the following chapters. The next chapters cover the six stages of lytic infection in more detail, as well as the ways that we use a combination of experimental genetics, microscopy, biochemistry, molecular genetics, and bioinformatics to investigate virus–host cell interactions.

Essential concepts

- The one-step growth experiment was essential for discovering the phases of the virus replication cycle, which include aspects unique to viruses such as an eclipse period.
- The virus replication cycle has six distinct stages: attachment, penetration and uncoating, expression of viral early proteins, replication of the viral genome and expression of viral late proteins, assembly, and release of virions. In some cases, expression of certain viral proteins overlaps with replication of the genome, but it is easier to learn about the two processes when they are considered separately.
- The influenza virus provides an example of a lytic viral infection in humans.

- Particular interactions between certain subcellular components of host cells and the virus are responsible for advancing the virus through the replication cycle.
- Some eukaryotic viruses synthesize mRNA and new genomes in the nucleus, whereas others complete these processes in the cytoplasm.
- Virus replication compartments are typically found in close association with cellular membranes, which are typically modified to contain viral proteins.
- Viruses can also cause longer-term persistent infections, which differ in whether virions are constantly produced (lytic infections) or not produced at all (latent infections).
- HIV causes both chronic lytic and latent infections, whereas herpesvirus infection of differentiated neurons is an example of a latent infection.

Questions

1. Looking at **Figure 2.2**, how long is the eclipse period for this particular virus?
2. Looking at **Figure 2.2**, how large is the burst size? The burst size is defined as the number of progeny virions per PFU in the inoculum.
3. In order to perform a one-step growth experiment, we cannot use microscopy to measure the PFU/mL. Why not? What technique should be used instead?
4. List the processes that occur during the eclipse period.
5. List all of the events that occur during penetration and uncoating of the influenza virus.
6. Compare and contrast chronic infections with latent infections.
7. Consider an animal virus infection and list the organelles that are important for each of the six specific stages of the viral replication cycle.

Interactive quiz questions

In addition to the questions provided above, this edition has a range of free interactive quiz questions for students to further test their understanding of the chapter material. To access these online questions, please visit the book's website: www.routledge.com/cw/lostroh.

Further reading

Bacteriophages

Ellis EL & Delbrück M 1939. The growth of bacteriophage. *J Gen Physiol* 22(3):365–384.

Nobelprize.org. Physiology or Medicine 1969—Press Release. Nobel Media AB 2014. http://www.nobelprize.org/nobel_prizes/medicine/laureates/1969/press.html

Influenza

Krulwich R 2009. Flu Attack! How A Virus Invades Your Body. http://www.npr.org/sections/krulwich/2011/06/01/114075029/flu-attack-how-a-virus-invades-your-body

Sexually transmitted viruses (HIV and herpes)

Ebel C & Wald A 2007. *Managing Herpes: Living and Loving with HSV*, 1st ed. American Social Health Association.

http://HIV.gov

Reulas DS & Greene WC 2013. An integrated overview of HIV-1 latency. *Cell 155(3)*:519–529.

Scarleteen sex ed: http://www.scarleteen.com. Sex ed for the real world: Inclusive, comprehensive, supportive sexuality and relationships info for teens and emerging adults.

United States of America Centers for Disease Control & Prevention Sexually Transmitted Diseases (STDs). http://www.cdc.gov/std/

World Health Organization Health topics: Sexually transmitted infections. https://www.who.int/health-topics/sexually-transmitted-infections

Fluorescence microscopy

Nobelprize.org. The Nobel Prize in Chemistry 2008 Illustrated Presentation. http://www.nobelprize.org/nobel_prizes/chemistry/laureates/2008/illpres.html

Nobelprize.org. The Nobel Prize in Chemistry 2014 Illustrated Summary. https://www.nobelprize.org/prizes/chemistry/2014/summary/

Attachment, Penetration, and Uncoating

Virus	Characteristics
Rabies virus	Example of a virus with a broad host range.
Human immunodeficiency virus (HIV)	Example of a virus with a narrow host range; requires a cellular receptor and co-receptor; model for viral fusion peptide activated by protein–protein interactions.
Rhinovirus	Model of attachment for naked viruses; virion has canyons that bind to host intercellular adhesion molecule 1 (ICAM-1) receptor; model for pore formation during uncoating.
Influenza A virus	Model of attachment for enveloped viruses; virion has hemagglutinin (HA) spikes that attach to host sialic acids and mediate envelope fusion; model for acidification that activates the viral fusion peptide.
Adenovirus	Model of virus with many individual steps during uncoating.

The viruses you will meet in this chapter and the concepts they illustrate

This chapter is about first encounters: what happens when a virus collides with a susceptible host cell? The first stage of the replication cycle is **attachment**. Attachment is important not only because of its molecular significance during replication but also because it influences the susceptibility of an animal or cell within that animal to infection. To attach to host cells, naked viruses use their capsomers, structures that project from the main body of the capsid, or structures formed by interacting capsomers. Both naked and enveloped viruses use their **spike proteins** (if present on naked capsids) for attachment. In this chapter, we will learn several different ways to identify the cellular **receptor** that the virus uses during attachment. The **penetration** and **uncoating** stage follows attachment. For enveloped viruses, this stage necessarily includes membrane fusion catalyzed by a protein in the virion. Fusion proteins are interesting because they undergo significant conformational changes during penetration and uncoating. Once the penetration and uncoating stage is complete, the viral genome is ready to be expressed, meaning that new viral mRNA or proteins begin to be synthesized.

3.1 Viruses enter the human body through one of six routes

In order for a virus to infect an animal such as a human, the virion must enter the body through one of six major **routes of infection**. First, viruses can enter

DOI: 10.1201/9781003463115-3

through the respiratory tract, where they might infect target host cells in the ears, nose, mouth, throat, or lungs. Second, viruses can enter through the gastrointestinal tract, where they might infect target host cells in the mouth, throat, stomach, small intestine, or large intestine. Third, viruses can enter through the genitourinary tract, where they might infect the urethra, ureters, bladder, kidneys, vagina, cervix, uterus, or penis, including the immune cell-rich tissue of the foreskin. Fourth, viruses can enter the bloodstream directly when aided by some mechanical event such as a cut or scrape; the bite of an insect, arthropod, or other animal; vaginal or rectal sexual intercourse; or use of biomedical devices such as needles or scalpels. The fifth site of entry is the eyes, which can be infected when a virion on the hands gains entry when the eyes are rubbed. A neonate's eyes can become infected when passing through the vagina during birth. Sixth, herpesviruses and papillomaviruses can enter through the skin, especially where tiny lesions provide access to the lower (living) levels of the epithelium.

If a virus adapted for one route of infection enters the body through a different pathway it will usually not be able to cause an infection. For example, a respiratory virus that gets ingested will typically be degraded in the gastrointestinal tract. A notorious exception is that of blood-borne pathogens, which are usually also able to spread through sexual activity (and vice versa). Even Ebola virus, well known as an infection of the blood, can be spread sexually through semen for several months after a man has survived the infection and is symptom-free. After a virus enters the human body, the chance that it will encounter a susceptible **permissive host cell** is directly related to the abundance of those host cells at the site of entry and to the abundance of receptors on those host cells. The person's immune status also has a profound influence on the outcome of a virion's entry into the human body. If the immune system has encountered the virus, or its proteins in the past, the immune response will likely destroy the virus before it can cause a noticeable infection. Immunization mimics a natural infection by providing a safe first encounter so that subsequent exposure to the real pathogen does not result in disease.

If the human is **naive**, meaning that she has never before encountered the virus or its components, the virus may cause an active infection. The minimum number of viruses necessary to cause an infection is called the **infectious dose**. The infectious dose depends on the particular virus. For example, the infectious dose for influenza is about 800 inhaled viruses, while the infectious dose for smallpox is fewer than 100. The infectious dose for severe acute respiratory syndrome coronavirus 2 (SARS-CoV-2) is between 100 and 1,000.

3.2 The likelihood of becoming HIV+ depends on the route of transmission and the amount of virus in the infected tissue

In the case of HIV, which has been studied in detail, transmission rates are often defined in terms of the number of events that expose someone to the virus, rather than being defined by the infectious dose alone. These rates are derived by studying discordant couples in which one person was known to be HIV-positive (HIV+) while the other was known to be HIV-negative (HIV–) prior to transmission. For example, in discordant heterosexual couples, transmission occurs 8–9 times per 1,000 coital acts. In contrast, in discordant heterosexual couples where the HIV+ person has a long-term infection that results in low levels of virus in the blood, transmission to the uninfected sexual partners drops to 1–2 per 1,000 coital acts. Transmission through rectal intercourse is 15–20 times higher than these values, whereas transmission through a contaminated needle used for medical purposes is as high as 65 per 1,000 injections. Healthcare providers exposed to HIV through an accidental needle stick have up to a 2.4% chance of becoming HIV+ in the absence of preventative antiviral treatment. It is standard practice for healthcare providers accidentally exposed through a needle stick to take a short course of antiviral medications, referred to as **postexposure prophylaxis**, to prevent the virus from establishing itself in their bodies. Making this treatment available to survivors of sexual assault is also the standard of care wherever hospitals have access to enough medication. Hemophiliacs and others exposed through a contaminated blood transfusion acquire HIV 85%–100% of the time per contaminated transfusion. Fortunately, the blood supply in most countries is screened for viruses, including HIV, so that transfusions no longer transmit HIV as they did during the 1980s. Understanding the molecular and cellular interactions that result in these different chances of transmission that depend on the details of exposure may lead to better prevention strategies. Note that employing correct medical procedures, avoiding IV drug use, using condoms during vaginal and rectal intercourse, and using antiretrovirals as **preexposure prophylaxis** remain effective prevention strategies. Each strategy is associated with particular reductions in risk, and individuals should consult with their physicians regarding the most up-to-date information on these practices.

3.3   Viruses are selective in their host range and tissue tropism

The **host range** of a virus is the variety of different species that the virus can infect. The attachment stage can be largely responsible for host range. For example, rabies virus infects not only humans, but many other vertebrates, including dogs and bats (**Figure 3.1**). The broad host range occurs because the rabies virus can attach to human, dog, and bat host cells, and subsequently infect them. In contrast, poliovirus is restricted to apes (including humans) and monkeys. This restriction results from amino acid differences in the poliovirus receptor that prevent poliovirus from attaching to other mammals' cells.

The **tissue tropism** of a virus is the different tissue or cell types that a virus infects once it is inside a susceptible animal. Here, too, the attachment stage of infection is very important in explaining tissue tropism. The rabies virus has narrow tissue tropism, infecting only muscle and nervous tissue, even though it can infect these tissues in dogs, humans, bats, and other mammals. In contrast, HIV has a narrower host range (infecting only humans) but much wider tissue tropism within a person's body, infecting not only CD4+ helper T cells, but also many other CD4+ immune cells such as macrophages.

Although tissue tropism and host range are often determined by the ability of a virus to attach to, penetrate, and uncoat in a certain cell, there are examples in which host cells prevent virus replication at a later stage. An example is HIV, which can attach to and enter chimp immune cells but cannot replicate inside them. The chimp cells produce **host restriction factors**

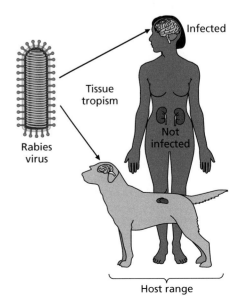

Figure 3.1 Comparing host range with tissue tropism. The host range of a virus is the range of species it can infect; in this case, the host range of rabies includes dogs and people. The tissue tropism of a virus is the specific tissues that support virus replication inside an infected host animal. The tissue tropism for rabies is the neurons of the central nervous system. In contrast, rabies does not infect most other organs, for example the kidneys.

that block HIV replication after the penetration and uncoating stage. Thus, although attachment is important for determining the host range and tissue tropism of a virus, it is not always the only important factor.

3.4   The virion is a genome delivery device

A virion is a genome delivery device. In many cases, if a virus genome can be delivered artificially into the right compartment of almost any cultured animal cell, the cell will produce offspring virions. This artificial method of propagation indicates that attachment to host cells is a significant determinant of host range and tissue tropism, because experimentally bypassing the attachment stage broadens the host range and tissue tropism of the virus. There are various experimental techniques that can bypass the attachment step, such as **transfecting** a host cell with DNA that encodes the viral proteins. Viruses are commonly studied using host cells engineered to express virus proteins and to produce the virion genome, which bypasses the attachment step and results in a convenient population of host cells uniformly producing viruses. Most approaches useful for studying the molecular biology of attachment and penetration require techniques to separate macromolecules such as nucleic acids and proteins (**Technique Box 3.1**).

Alteration of the host range of a virus or tissue tropism can sometimes be accomplished in the lab by **pseudotyping**, which changes the host range or tissue tropism of an enveloped virus by including spike proteins from more than one virus in the virion (**Figure 3.2**). The presence of these spike proteins allows the pseudotyped virus to attach to host cells that it would normally never infect in nature. This technique can be useful when the host cells normally infected by a virus are unsafe, when they are difficult or impossible to culture in the lab, or when we want to combine the properties of two viruses for some pragmatic purpose, such as gene therapy. Here, too, rabies virus and HIV provide an important example. HIV, like all lentiviruses, is very efficient at inserting its genome into a host chromosome. This property could be very useful for inserting a therapeutic gene into a patient's cells. However, the HIV spike proteins mediate attachment and entry exclusively into immune cells, which means that normal HIV virions can only be used for genetic alteration of these immune cells. In contrast, rabies is very efficient at infecting neuronal tissue, but it has a (−) RNA genome and cannot alter the nuclear genes in its host cells. It could be quite useful clinically to be able to use altered lentivirus virions to genetically alter neurons, for example to treat brain cancer or Parkinson's disease. Therefore, gene therapy labs are working on pseudotyping altered HIV viruses with rabies spike proteins in order to deliver the altered genome of the engineered HIV virus into nervous tissue. This genetic engineering procedure culminates in insertion of engineered therapeutic cDNA into one of a neuron's chromosomes.

Figure 3.2 Pseudotyping an enveloped virus. (A) The enveloped lentivirus uses its spike protein to infect white blood cells. (B) This enveloped virus uses its spike protein to infect neurons. (C) This pseudotyped lentivirus can now attach to and infect neurons because it has spikes from B. Oftentimes, a pseudotyped virus will also include a reporter gene such as one encoding green fluorescent protein (GFP). In that case, successful attachment and penetration results in fluorescent host cells that are easy to quantify.

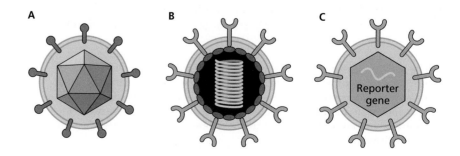

3.5 The genomic contents of a virion are irrelevant for attachment, penetration, and uncoating

Attachment, penetration, and uncoating do not depend upon the sequence of the genome inside any given virion. Artificial virions that contain a genome never found in nature can be constructed and readily infect the same cells that normal virions can infect. This property of virions makes pseudotyping a common method to study viral spike proteins. Pseudotyped viruses often encode a reporter gene that is readily quantified, such as *GFP* encoding green fluorescent protein. The amount of reporter gene expressed reflects attachment and penetration catalyzed by the viral spike. Pseudotyping is a common method for studying variations in the SARS-CoV-2 spike protein and their impact on attachment, penetration, and transmission.

In nature, virions can also deliver nonviral genes or only a subset of virus genes. Among bacteria and archaea, for example, such defective bacteriophages can carry fragments of a host cell's chromosome into another host cell in a form of **horizontal gene transfer** known as **transduction** (**Figure 3.3A**). Horizontal gene transfer (the acquisition of genes independently of inheritance from parents) also occurs in animals; an example is the many endogenous retroviral genomes found in the human genome (**Figure 3.3B**). If a horizontal gene transfer event occurs into the germ line of an animal, those horizontally acquired genes will become inherited in a vertical manner, passed from parents to offspring. Viruses are very abundant; even if transduction is rare, as 1 in 10^6 infections, transduction is an important source of genetic diversity, which in turn is important for evolution (see **Chapter 15**).

Defective viruses that deliver only a subset of viral gene are probably also common, a phenomenon that has been studied among animal viruses. Among animal viruses, the ratio of infectious virions to total virions is lower than 1:1 and can even be as low as 1:10,000. Explanations for this phenomenon depend on the exact virus. In some cases, the virion particles are released without a complete set of genes, which may be especially common among viruses with segmented genomes (such as influenza), or particles lacking any genome at all are present (for example, cytomegalovirus). Another explanation is that some virions carry genomes that acquired a null mutation relative to the parent genome and can no longer function to direct an entire virus replication cycle. Co-infection with another virion carrying a genome with different mutations may be required for production of offspring virions; dengue virus is an example of this situation. When two virions with different null mutations co-infect the same cell and together cause a productive infection that each virion alone could not do so, this is termed **complementation**.

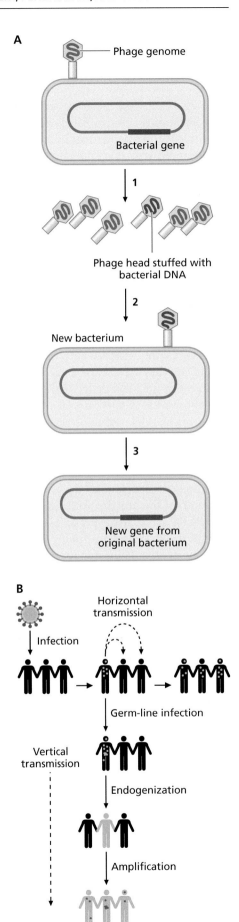

Figure 3.3 Horizontal gene transfer by viruses. (A) When a bacteriophage infects a host cell, sometimes one of the offspring phages will be accidentally filled with bacterial DNA (such as the DNA shown in red), instead of with phage genomic DNA (1). When that phage attaches to a new host cell (2), it introduces the bacterial DNA into the cytoplasm, where it can recombine with the host (3), thus transferring genes from one bacterium to another. This process is called transduction. (B) Humans can acquire endogenous retroviral DNA. Retrovirus infection requires that viral cDNA be inserted into a host chromosome. If viral cDNA inserts into the chromosome of a germ-line cell, the endogenous viral cDNA is passed vertically through the generations so that it is found in all of the descendants of the person who originally acquired the germ-line infection. (B, From Dewannieux M & Heidmann T 2013. *Curr Opin Virol* 3:646–656. With permission from Elsevier.)

TECHNIQUE BOX 3.1   SEPARATION AND ANALYSIS OF MACROMOLECULES

The simplest viruses are composed of nucleic acids and proteins, whereas more complex viruses include lipids and carbohydrates. Techniques to separate, identify, and analyze macromolecules are therefore central to the discipline of virology. Cell lysates can be fractionated by organelle, for example, separating the nuclear contents from the rest of the cell. Nuclear contents can then be further fractionated, for example into chromatin and nucleoplasm, or even into separate fractions containing predominantly DNA, RNA, or proteins. Such fractions can be analyzed for the presence of specific macromolecules using electrophoresis. During electrophoresis, a sample of different nucleic acids or proteins is loaded onto a gel of polymerized agarose or polyacrylamide and then exposed to an electrical field. Nucleic acids in such an electrical field migrate to the positive pole according to size because they have the same charge per unit length (conferred by the phosphoryl groups associated with each nucleotide). Proteins are typically mixed with a solution of sodium dodecyl sulfate (SDS) and heated prior to applying them to a gel, in order both to denature (unfold) the proteins and to coat the proteins with approximately the same charge per unit length (conferred by the sulfate groups in the SDS molecules). Following separation, there are many ways to visualize the nucleic acids or proteins in the gel. There are also techniques for visualizing only specific nucleic acids or proteins. Specific nucleic acids can be visualized because they have a unique sequence, whereas specific proteins can be stained because they contain unique epitopes.

In order to stain a gel to detect a single type of macromolecule, the contents of the gel are first transferred or **blotted** onto a sturdier substrate,

typically a nylon membrane coated with a positively charged polymer (nitrocellulose), in such a way that the macromolecules retain their relative positions. That is, the membrane is a faithful replica of the gel. Nitrocellulose membranes can be subjected to various procedures in order to stain an individual nucleic acid or protein. To detect nucleic acids, we would prepare a **probe**, which is complementary to the desired target sequence and has an indicator molecule, such as a fluorophore, attached. Incubating the membrane blot with the probe ultimately allows detection of specific DNA or RNA molecules

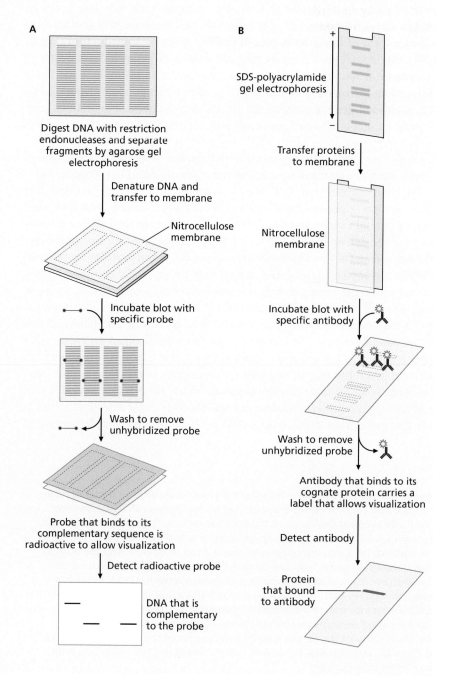

Figure 3.4 Blotting to detect specific macromolecules. (A) Southern blotting detects DNA that hybridizes to a probe. (B) Western blotting (or immunoblotting) detects proteins that bind to specific antibodies.

complementary to the probe (**Figure 3.4A**). The detection of DNA using this method is called a **Southern blot** named for its inventor, Edwin Southern. Detection of RNA is called a **northern blot** in tribute to Dr. Southern. For proteins, a two-stage process is used in which the blot is first submerged in a solution of **primary antibodies** that bind to an epitope on the protein of interest. After washing away excess primary antibodies, **secondary antibodies** that bind to the primary antibodies and are covalently attached to an indicator molecular are added to the blot. In this way, the signal from a small amount of target protein is amplified by use of the secondary antibodies, and the relative size and abundance of the protein of interest can be determined. The procedure for proteins is either called an **immunoblot**, reflecting the use of antibodies, or a **western blot** (**Figure 3.4B**), again in tribute to the eponymous Southern blot.

It is also typical to purify viral proteins in order to study them in more detail, whether in isolation or in combination with only a few other molecules. For example,

how would one discover the molecular mechanism by which the HIV integrase enzyme binds to DNA? To collect a large amount of protein for further study, it is more typical to use chromatography (**Figure 3.5A**) rather than electrophoresis, in part because chromatography does not require subjecting the proteins to denaturing conditions (SDS and heating). There are several types of chromatography (**Figure 3.5B**), all of which separate proteins according to the proteins' distinctive physical and chemical properties (conferred upon them by their unique amino acid sequence relative to the sequence of any other specific protein). Using genetic engineering, a protein can even be tagged with a small amino acid sequence that confers selective binding properties upon the protein chimera. For example, the addition of 6–12 histidine amino acids to a protein increases the binding affinity of the modified protein to a chromatography column loaded with inert beads that have nickel on their surface. The histidine tag has affinity for the nickel and enables use of affinity chromatography in which the protein of interest is purified using

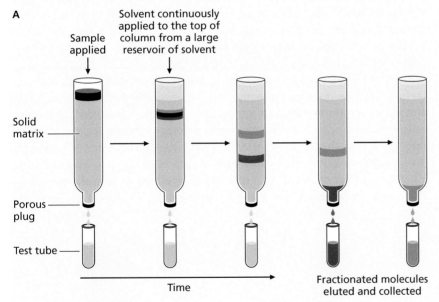

A

Sample applied

Solvent continuously applied to the top of column from a large reservoir of solvent

Solid matrix

Porous plug

Test tube

Time

Fractionated molecules eluted and collected

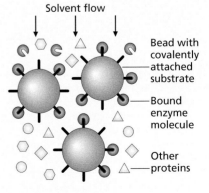

Figure 3.5 Column chromatography. (A) General principle of column chromatography. (B) Three different types of matrices, including one used for affinity chromatography. (From Alberts B et al. 2014. *Molecular Biology of the Cell*, 6th ed. With permission from W.W. Norton.)

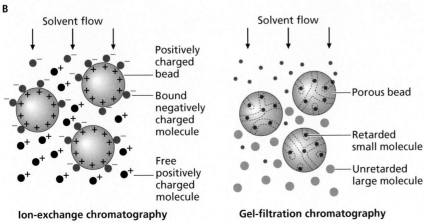

B

Solvent flow

Positively charged bead

Bound negatively charged molecule

Free positively charged molecule

Ion-exchange chromatography

Solvent flow

Porous bead

Retarded small molecule

Unretarded large molecule

Gel-filtration chromatography

Solvent flow

Bead with covalently attached substrate

Bound enzyme molecule

Other proteins

Affinity chromatography

its affinity for the nickel. As another example, epitopes recognized by specific antibodies can be added to a protein through genetic engineering, after which a chromatography column is loaded with a solid substrate that is attached to the antibody used to fish out the chimeric recombinant protein. It is also common to clone the cDNA encoding a viral protein into a bacterial plasmid that also encodes a convenient tag so that translation of

the protein in the bacteria attaches the tag to the protein. The bacterial lysates are then collected to obtain large quantities of the viral protein using affinity chromatography. Two other common forms of chromatography used in virology research are ion-exchange and gel-filtration chromatography. Ion-exchange chromatography separates proteins based on their differential surface charges and gel-filtration chromatography separates them by size.

3.6 Animal viruses attach to specific cells and can spread to multiple tissues

Although there are only six routes of infection, there are about 200 different cell types in an adult human. Some viruses cause simple infections in which they attach to and replicate in host cells encountered directly at the site of infection and do not spread to other cells in the body but remain confined to the site of infection. Influenza is an example of such a virus, which is acquired through the respiratory route and infects respiratory epithelial cells. Other viruses infect one cell type at the site of primary infection and progeny viruses spread to secondary sites, infecting other host types at those sites. An example is provided by human herpesvirus 3 (HHV-3), more commonly known by its older name, varicella–zoster virus (VZV). VZV causes varicella (chicken pox) and zoster (shingles) (**Figure 3.6**). VZV enters the body through the respiratory or ocular routes, where infection of the mucosal epithelium spreads the virus to lymph nodes associated with the head, neck, and upper respiratory tract. In the lymph nodes, the virus can infect other cells, such as T cells. After about a week, **primary viremia** occurs, in which infected T cells shedding many viruses enter the bloodstream. The infected T cells subsequently invade many tissues, such as the liver, spleen, and skin, allowing another round of viral replication, resulting in **secondary viremia** (the viruses released in the second round enter the bloodstream). When

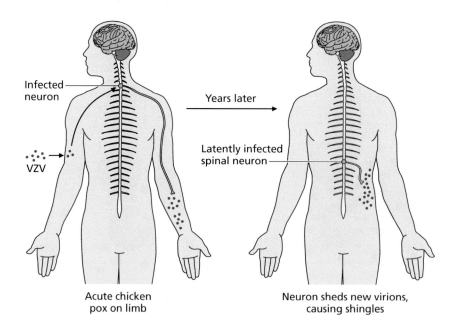

Figure 3.6 An acute, latent, and reactivated herpesvirus infection. Varicella–zoster virus (VZV) causes chicken pox the first time it infects someone, as shown on the left. Some of the offspring viruses establish a latent infection in spinal nerves, as indicated by the green cell nucleus. Years later, the virus can reactivate and cause an infection of the epithelium innervated by the latently infected neuron, causing shingles, shown on the right.

infected T cells invade the skin and infect there, they cause the rash characteristic of chicken pox. In the skin, some of the offspring viruses will enter the termini of sensory neurons, and the virus will then establish a long-term latent infection in the dorsal root ganglia located in the spinal column. Infected T cells might also deliver the virus to neurons. As might be expected from its more complex interactions with the human body, VZV attachment is more complex than that of influenza and may involve using different receptors on different cell types. Currently, the receptors for VZV are unknown.

3.7   Noncovalent intermolecular forces are responsible for attaching to host cells

Proteins generally interact with other molecules through noncovalent intermolecular forces such as ionic bonding, hydrogen bonding, and temporary dipole–dipole interactions, sometimes referred to as London dispersion forces (**Figure 3.7**). Because these types of interactions occur exclusively over short distances, the atoms participating in the bonds must be in very close proximity. Shape complementarity between the virus protein and its target host receptor is thus very important. Intermolecular forces can be disrupted experimentally by increasing the temperature, altering the pH, or increasing the ionic strength of the solution surrounding the virion and its target host cell.

The first stage of attachment is **reversible binding** to host cell surfaces through nonspecific electrostatic interactions between the virion surface and the host cells. The plasma membranes of animal cells are heavily glycosylated on their exterior surfaces, providing a net negative charge. It is likely that numerous nonspecific electrostatic interactions between virions and these sugars help restrict the diffusion of the virions to two dimensions in the plane of the membrane, making it more probable that the virion will encounter its specific receptor, which in turn enables irreversible attachment. Some viruses have even been observed surfing on waves of actin-filled filopodia on the surface of host cells prior to penetration (**Figure 3.8**).

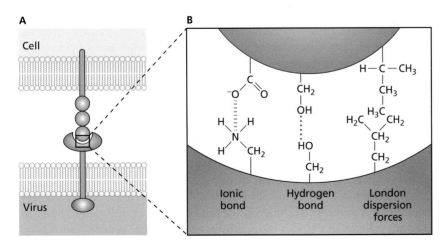

Figure 3.7 Intermolecular forces that enable a virus to attach to a host cell. (A) A virus spike protein (bottom) binds to its cellular receptor (top). (B) Close-up on regions of spike and receptor that interact, showing typical noncovalent intermolecular forces that hold the spike and receptor together such as ionic bonds, hydrogen bonds, and London dispersion forces (sometimes known as van der Waals interactions).

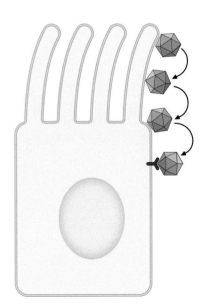

Figure 3.8 Some viruses browse on the surface of host cells before the irreversible binding step of attachment. The virus browses along the surface of an epithelial cell until it encounters its receptor, where it binds with high avidity.

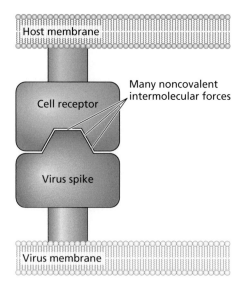

Figure 3.9 Shape complementarity is essential for attachment. The virus protein is bound to its cellular receptor through noncovalent intermolecular forces that are additively strong and can occur because of shape complementarity. Tight binding such as this is irreversible because it soon triggers the host cell to internalize the virus.

The second stage of attachment is **irreversible binding** to a specific receptor molecule on the host cells; this step is termed irreversible because the consequences of irreversible attachment, which are penetration and uncoating, do not ever proceed backwards. The protein components of virions must interact closely and specifically with specific host receptor molecules in order for irreversible attachment to occur. The strength of an interaction between two molecules (A and B) can be described by the **dissociation constant** (K_D), which is the ratio of [A][B] to [AB] at equilibrium. Strongly interacting molecules in a cell typically have a K_D of 1×10^{-9} M or less. But, when there are multiple interactions among two or more macromolecules, the receptor and its ligand are said to have **avidity**, which can be defined as the collective strength of multiple individual noncovalent intermolecular forces during an interaction among macromolecules. Because any one single noncovalent bond between a virus attachment site and its receptor increases the likelihood that other noncovalent bonds will form, avidity can be an order of magnitude or more than the sum of the affinities of individual interactions.

For naked viruses, the capsid itself has the surface that serves as a ligand for the host receptor; this surface can be an indentation or canyon in the capsid, or it can be a protrusion from the capsid, including spike proteins. For enveloped viruses, spike proteins that protrude from the lipid surface are used for attachment. Although reversible binding is easy to disrupt with small changes in pH or ionic strength, once irreversible attachment has occurred, the virion is much harder to remove from the host cell surface. Although noncovalent intermolecular forces are used for both reversible and irreversible binding, high-avidity binding relies on close shape complementarity between the virion and its specific receptor (**Figure 3.9**). The shape complementarity allows for greater total numbers of ionic bonds, hydrogen bonds, and London dispersion forces between the virion and the host cell, compared with the smaller numbers of such attractive forces during reversible binding.

Molecular modeling has become an important strategy for examining viral spike–host receptor interactions. During molecular modeling, software to model the three-dimensional structure of the proteins can be used to examine whether a virus spike protein is likely to attach to a host receptor. This technique was very important early during the COVID-19 pandemic. Virologists used the solved structure of the human SARS-CoV-2 receptor, known as ACE2, and the genome sequence of many other animals to model the ACE2 protein from many animals. Next, they used these structures to look at potential interactions between the solved SARS-CoV-2 spike protein and the various ACE2 proteins, and thereby predicted which mammals could likely be infected by the virus.

3.8   Most animal virus receptors are glycoproteins

Although there are viruses that can use glycolipids as their specific receptors (for example, SV40), most viruses attach to specific host proteins. These proteins serve a purpose in the cell's normal physiology; through binding to a certain cellular receptor, a virus subverts that normal protein for its own purpose. Proteins that protrude from the surface of animal cells are typically glycosylated, so it is more strictly true to refer to most animal virus receptors as glycoproteins. The indigestible oligosaccharides found in breast milk represent an interesting evolutionary response to the many virus receptors that are glycoproteins. The role of **human milk oligosaccharides** (HMOs) in protecting nursing babies from infection through breast milk is best understood using HIV as a model, but they probably protect infants from other infections as well.

Although human breast milk from an untreated HIV+ mother contains HIV (ranging from <200 viruses mL^{-1} to more than 40,000 viruses mL^{-1}), transmission through breastfeeding is considered inefficient compared with sexual transmission. Most transmission of HIV from mother to child occurs perinatally, when the fetus is still in the womb and before the baby is 6 weeks of age, and is typically not attributable to breastfeeding. If an infant is HIV– at 6 weeks of age, 80%–95% who are breastfed up to 18 months will remain HIV– despite continuous exposure to HIV in the breast milk over a long period of time.

During those 18 months, HMOs are important for preventing mother-to-child transmission of HIV from breast milk to the baby through oral ingestion of the virus. In general, HIV virions can be phagocytosed by specialized dendritic cells and from there passed on to CD4+ T cells in a process analogous to regurgitation. This process is very efficient and contributes substantially to the spread of HIV throughout the body once someone has become infected. Dendritic cell regurgitation of HIV is also the route by which a baby can become infected through breastfeeding. Dendritic cell phagocytosis of HIV virions relies upon a receptor protein called dendritic cell-specific intercellular adhesion molecule-3 grabbing non-integrin (DC-SIGN), which interacts with sugars on the HIV gp120 envelope spike glycoprotein.

As in adults, a baby's digestive tract contains dendritic cells. HIV virions ingested by an infant can stick to the DC-SIGN on the baby's dendritic cells and from there can be passed from the digestive tract to the baby's T cells, infecting them with HIV. HMOs block this process, probably because they coat the DC-SIGN receptors, physically occluding them so that they are not available to bind to virions (**Figure 3.10**). The virions then pass through the infant's gut without causing an infection.

The protection afforded by HMOs, though complex in mechanism and not perfect, is beneficial because long-term daily breastfeeding transmits HIV from untreated mothers to their babies 5%–20% of the time (rather than 100% of the time, according to the World Health Organization). Nevertheless, there are much better ways of reducing the risk of transmission from mother to child than relying on HMOs. For example, the use of antiviral drugs administered to the mother before and during labor and to both mother and baby for the duration of nursing are effective. Ideally, all HIV+ people would be diagnosed and treated soon after they first became HIV+, which would dramatically reduce mother-to-child transmission, but a major barrier to this medical goal is that many people who are HIV+ are unaware of their status, making breast milk HMOs all the more important.

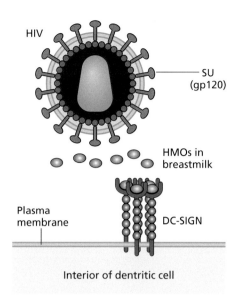

Figure 3.10 Human milk oligosaccharides interfere with HIV transmission through ingestion of breast milk. HMOs physically occlude DC-SIGN receptors on the surface of dendritic cells so that viruses such as HIV-1 in the breast milk cannot bind to and infect the DC-SIGN cells. The HIV-1 gp120 spike protein cannot bind to the receptor on the DC-SIGN cells in the presence of the HMOs.

3.9 Animal virus receptors can be identified through genetic, biochemical, and immunological approaches

Many techniques are used to investigate the attachment phase of the virus replication cycle. A common research goal is to identify the cellular receptor on a virus that is used for irreversible attachment. In addition to being of intrinsic interest because of its importance for the virus life cycle, discovery of a cellular receptor can in principle lead to development of antiviral medications and also enhances our ability to use viruses as gene therapy agents to target cells that have those receptors. Such discovery can also suggest novel antiviral therapies, for example by using a drug that reduces the amount of receptor available to the virus. In order to have high confidence that a virus receptor has been identified through one experimental technique, other techniques are then used to test the hypothesis that the putative receptor is the functional receptor for virus entry. The following three techniques (use of cDNA libraries, use of affinity chromatography, and use of antibodies to block infection) are described in **Sections 3.10–3.12** in an idealized way; the specific details of actual experiments can be found in the primary literature (see **Further reading**). Use of techniques derived from different disciplines (genetics, biochemistry, and immunology) to identify viral receptors also provides an excellent example of how we employ a diversity of methodological approaches.

3.10   Animal virus receptors can be identified through molecular cloning

A common genetic approach for discovering a candidate virus receptor is to use molecular genetics. First, two types of cells are identified: one that the virus can infect and a second that the virus cannot infect because of its inability to attach to the cells. Host cells that can support lytic virus replication are often called permissive cells to distinguish them from **nonpermissive cells**, which cannot support virus replication. A subset of nonpermissive host cells typically cannot bind irreversibly to the virion, though they can allow viral replication if the viral genome is provided artificially, for example, through transduction. These are the particular nonpermissive cells used in a cDNA library procedure. A **cDNA library** is constructed using mRNA collected from permissive cells as a template for reverse transcription (**Figure 3.11**); selective collection of mRNA takes advantage of techniques to separate macromolecules (see Technique Box 3.1). One of these cDNA molecules may encode the normal host protein that the virus uses as a receptor. The procedure is then to express the cDNA in the population of nonpermissive cells to which the virus cannot normally attach and then to **screen** the genetically altered cells to identify ones to which the virus can now attach. Various molecular techniques can subsequently be used to identify which cDNA was expressed in the cells that gained the ability to attach to the virus; once this cDNA has been sequenced, the candidate host receptor can be characterized.

Use of cDNA libraries to identify candidate receptors is not always successful; one reason the technique can fail is that a virus receptor may be

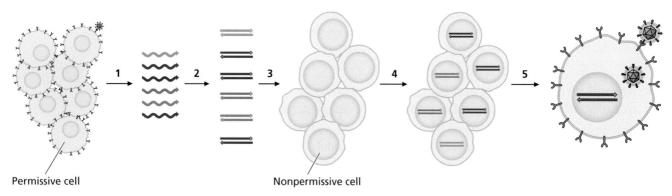

Permissive cell   Nonpermissive cell

Figure 3.11 Use of a cDNA library to identify a virus receptor. First, mRNA is collected from the permissive cells (1) and converted into a collection (library) of cDNAs (2), one of which encodes the receptor protein. The library is introduced into the nonpermissive cells (3) and some of the recipient cells now contain cDNA encoding the receptor protein (4). One of the previously resistant cells has become susceptible to the virus because it expresses the receptor protein from the cDNA (5), which can be recovered and sequenced in order to identify the receptor protein.

made up of different polypeptides. In that case, transfection of a single cDNA encoding a single polypeptide component of the receptor would be inadequate to allow viral attachment. Even in the case in which cDNA transfection is successful in allowing virus attachment, it is important to use additional strategies such as affinity chromatography to increase the likelihood that the candidate receptor is the actual receptor used by the virus during a natural infection.

3.11 Animal virus receptors can be identified through affinity chromatography

Biochemical methods can be used to discover a virus receptor when a suitable population of nonpermissive host cells is not available or as an independent strategy. A typical biochemical approach to identifying a receptor is **affinity chromatography** (**Figure 3.12; Technique Box 3.2**). In this case, the virions themselves (for naked viruses) or the virion spike proteins are purified and attached to an inert substance, such as agarose beads, which is subsequently used to load a chromatography column. A protein extract from the plasma membrane of permissive host cells is applied to the column under conditions that allow both nonspecific and specific binding between the virion (or spike) and the host cell membrane proteins. Next, the column is washed, typically using buffers of increasing ionic strength. Proteins that elute from the column under conditions of lower ionic strength were likely not associated specifically with the virion or spike. Lower ionic conditions that do not cause polypeptides to unfold nevertheless specifically disrupt weak intermolecular forces. The proteins that elute using buffers with higher

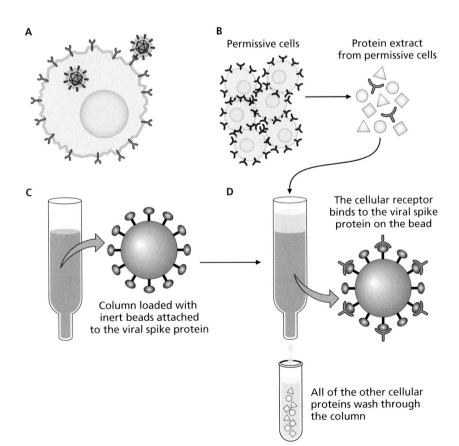

A

B
Permissive cells

Protein extract
from permissive cells

C

Column loaded with
inert beads attached
to the viral spike protein

D

The cellular receptor
binds to the viral spike
protein on the bead

All of the other cellular
proteins wash through
the column

Figure 3.12 Affinity chromatography to identify a virus receptor. (A) A permissive cell makes receptors that render it susceptible to infection by the enveloped virus with spike proteins. (B) A protein extract from the permissive cells contains many proteins, including the receptor. (C) A chromatography column is loaded with inert beads that have the viral spike protein covalently attached to them. (D) After applying the cellular extract to the column, the receptor binds to the spike protein on the beads, while all of the other proteins wash through. (Not shown: the beads with the bound receptor can then be washed and the receptor can be eluted by varying the salt concentrations or ionic conditions to favor dissociation. The purified receptor protein can be identified using a variety of biochemical techniques.)

ionic strength are then selected for further study, to determine whether they function as the virus receptor in intact host cells. Buffers with higher ionic strength are needed to dislodge the receptor from the spike because the normal intermolecular forces holding the two together are substantial given the high avidity of a normal virus–receptor interaction. Employing additional strategies such as cDNA transfection of nonpermissive host cells and others increases the likelihood that the candidate receptor is the actual receptor used by the virus during a natural infection.

TECHNIQUE BOX 3.2   USE OF ANTIBODIES TO INTERFERE WITH VIRUS–HOST CELL INTERACTIONS

Antibodies are large (~15 nm) multisubunit proteins that are produced during a vertebrate humoral immune response (**Figure 3.13**). The simplest human antibody, IgG, consists of a Y-shaped molecule with a constant region and two arms, ending in an epitope-binding site. Epitopes are small peptides (8–17 amino acids) that have a distinctive shape and noncovalent bonding properties, such that they can form extremely specific chemical interactions with the epitope-binding domain at the tip of an antibody's arm. There are also nonlinear conformational epitopes comprising amino acids that are adjacent to each other in the tertiary structure of an antigen, but not in its primary sequence. Epitopes that differ by a single amino acid typically cannot bind with high affinity to the same antibody. The human body has the capacity to produce more than 1 billion antibodies with different binding sites, allowing recognition of more than 1 billion different epitopes. All mammals can generate such diverse antibody responses, but in practice mice and rabbits are used to produce antibodies for research use.

When an antibody binds to its cognate epitope, the rest of the antibody protrudes from the surface of the protein antigen. Imagine that an antibody has bound to an epitope on a virion capsid protein. Because the rest of the antibody is large and bulky, the antibody may prevent the capsid protein from interacting with its target cell surface receptor (**Figure 3.14A**). Similarly, an antibody bound selectively to the surface of a cellular receptor that binds to a virus capsid protein may prevent viral attachment (**Figure 3.14B**). This use of antibodies is common for probing the attachment step of the viral life cycle. For example, antibodies raised against the receptor for the severe acute respiratory syndrome (SARS) coronavirus can be mixed with target cells and virus, and they will block SARS attachment. The same antibodies do not, however, block infection by the closely related Middle East respiratory syndrome (MERS) coronavirus. This experiment was part of a series of manipulations that determined that the MERS and SARS coronavirus receptors are distinct.

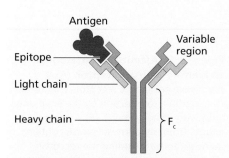

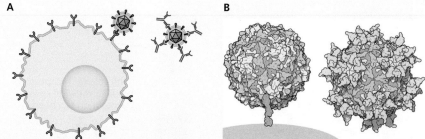

Figure 3.13 Structure of an IgG antibody. An IgG antibody is structurally the simplest antibody made by the human body. It consists of a constant region (F_c) and a variable region that includes the epitope-binding site. An IgG antibody has four separate polypeptide chains: two identical heavy chains and two identical light chains. The heavy and light chains are so named to reflect their relative masses.

Figure 3.14 An antibody physically occludes a receptor so that the virus cannot attach to the host cell. (A) Monoclonal antibodies that bind to an epitope on the viral spike protein prevent the virus from getting close to and binding to its receptor. (B) Antibodies can neutralize a virus by blocking its ability to bind to a receptor. On the left, the virus has bound to its receptor using a site adjacent to one of the yellow stars. On the right, the same virus has been covered with antibodies, in light blue, which bind to epitopes on and near the yellow stars. Only the epitope-binding region of the much larger antibody molecules has been shown in order to emphasize where the antibodies bind to the virion. The binding of antibodies blocks the virion from binding to its receptor. (B, From PDB-101 [http://pdb101.rcsb.org]. Courtesy of David Goodsell. doi: 10.2210/rcsb_pdb/mom_2001_8.)

Such experiments require a choice between using polyclonal antisera or monoclonal antibodies. Polyclonal serum contains antibodies that bind to a variety of epitopes on the protein used to immunize the research animal (such as a mouse or rabbit). Advantages to using polyclonal antibodies include their greater affordability and their ability, in theory, to cover the entire surface of a target protein, which is useful when the specific parts of a protein that interact with a virus are unknown. Monoclonal antibodies, in contrast, recognize one specific epitope. They are derived using a technique that begins with plasma cells collected from an immunized research mammal (**Figure 3.15**). Plasma cells are antibody secretion factories, each producing antibodies that bind to just one epitope. Plasma cells can be separated into individual culture dishes, **immortalized**, and cultured *in vitro* to produce monoclonal antibodies for research purposes. The advantage of monoclonal antibodies is their specificity. Both polyclonal and monoclonal antibodies can be used in other applications, such as western blotting (or immunoblotting; see Technique Box 3.1).

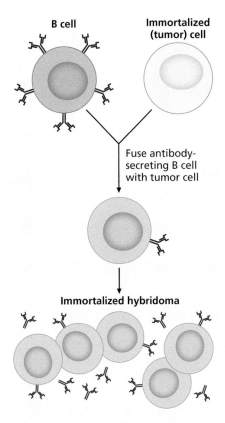

Figure 3.15 Production of monoclonal antibodies. A plasma cell (B cell) that makes just one type of antibody can be fused to an immortalized cell, creating a hybridoma that manufactures the monoclonal antibody indefinitely.

3.12 Antibodies can be used to identify animal virus receptors

Because the history of virology is inextricably intertwined with that of immunology, many immunological techniques are common in virology. An immunological approach that can be used to identify virus receptors is to start with a protein extract from permissive host cells and use this protein extract to immunize vertebrate laboratory animals such as mice. Such an immunization can ultimately result in production of monoclonal antibodies that recognize (bind to) the proteins in the extract (see Technique Box 3.2).

Next, the collection of monoclonal antibodies is screened for an ability to block virus infection of the permissive cells. The experiment entails incubating the permissive host cells with each monoclonal antibody and then adding the virions after the antibodies have had the opportunity to bind to their targets on the host cell surface. Monoclonal antibodies that reduce the number of plaque-forming units per milliliter are then selected for further study, to determine whether they bind to surface proteins that might function as the host cell receptor. Using additional strategies, such as cDNA transfection of nonpermissive cells and affinity chromatography, increases the likelihood that the candidate receptor is the actual receptor used by the virus.

All three of these experimental approaches have limitations. For example, the cDNA experiments can be done only if there are nonpermissive cells in which the only barrier to replication is attachment. Another barrier to success in using cDNA libraries is that the mRNA encoding the receptor may be rare, making it difficult to find in a library of thousands of cDNAs. Experiments with affinity chromatography can fail if a virus receptor is present only at low copy number on the permissive cell line or if the procedure used to solubilize the cellular membrane glycoproteins also disrupts the structure of the virus receptor. Sometimes a virus receptor is a multiprotein complex, which can prevent its identification using cDNA or affinity chromatography. Antibody-blocking experiments can also lead us astray when an antibody binds to an abundant surface glycoprotein that is not the virus's actual receptor, but the protein is so abundant that antibodies binding to it physically occlude the other surface proteins, including the actual cellular receptor. Finally, we have described experiments using cultured cell lines; it is therefore always a possibility that receptors identified on cultured cells are not the receptors used during an actual infection of a whole animal. It is essential to use a combination of techniques and to investigate whether a receptor identified on cultured cells is present on tissues at the site of viral replication in the animal, and if possible to find a method that proves the use of that receptor during a real infection. In the most straightforward case, an animal model can be genetically altered to knock out or knock down the levels of the candidate receptor in the tissue infected by the virus, but in many cases cellular receptors serve an essential purpose in the animal's physiology, and so reducing or completely preventing the presence of the protein in the animal's body is not possible.

3.13   Rhinovirus serves as a model for attachment by animal viruses lacking spikes

Rhinovirus (**Figure 3.16**), the **etiological agent** (cause) of the common cold, is a member of the picornaviruses. It has a small, naked icosahedral capsid lacking in spike proteins. Its attachment to host cells has been studied intensively, and it serves as a model for understanding how naked virions attach to their host cells. The virion has 60 copies of each of four different capsomers: VP1, VP2, VP3, and VP4. VP1, VP2, and VP3 form the outside surface of the capsid. At the fivefold axis of symmetry, the interaction of five VP1 molecules forms a five-pointed star that is raised somewhat from the surface of the rest of the virion and is surrounded by a groove, which is commonly called the canyon when discussing picornaviruses (see Figure 3.16). The VP4 protein is arranged in a layer below the surface, with no portion of its sequence exposed on the surface.

Rhinovirus can serve as a model for attachment by naked virions because of its use to pioneer techniques for identifying candidate host cell receptors. It also is a model for demonstrating how solving the structure of a virus and its candidate receptor through X-ray crystallography (**Technique Box 3.3**) can be informative. Furthermore, because the structure is known and there are many different isolates of the virus, it provides a model for using bioinformatics to compare the sequences of many different isolates in order to investigate natural selection. Selection restricts the range of amino acids that can be tolerated at positions in the viral proteins that interact with the host cell receptor.

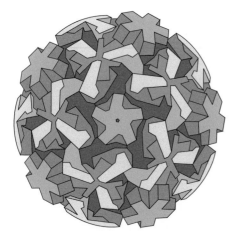

Figure 3.16 Rhinovirus. A three-dimensional drawing of rhinovirus, highlighting the five-pointed star surrounded by a canyon.

TECHNIQUE BOX 3.3   STRUCTURAL BIOLOGY

The interrelated nature of structure and function is a unifying concept in biology, applying as well to elephant trunks as to individual proteins. The structure of a leaf, the human thumb, and the protein hemoglobin all confer a function upon those structures; when something is amiss with the structures, the function is typically impaired. The field of structural biology, however, typically concerns itself not with structures that are visible to the human eye but rather with biological structures on the order of 100 nm or smaller. Examples are the surface-exposed portion of the HIV envelope (Env) protein (**Figure 3.17**) and the HIV protease enzyme bound to a drug that inhibits it (**Figure 3.18**).

Most techniques in structural biology are optimized to study individual molecules such as proteins. One of the most common techniques is X-ray crystallography, in which a purified protein is subjected to physical and chemical conditions that promote crystallization, analogous to the crystallization of salt molecules in a dehydrating tidal pool. Because proteins are much more complicated in structure than salt, discovering the exact conditions that will induce crystallization continues to be a great challenge that has even been extended to attempting crystallization on the International Space Station. Once crystals have been obtained, the next step is to subject the crystals to a beam of X-rays. X-rays are electromagnetic waves with a wavelength of 0.01–10 nm, encompassing the size of most proteins (average of 1–2 nm in diameter). Therefore, when the X-rays collide with the atoms in a protein crystal, they undergo **diffraction**, or scattering. After detecting the diffraction pattern formed by the X-rays that interacted with the crystal, computerized analysis of the diffraction pattern uses a mathematical process called **Fourier transformation** to reconstruct the three-dimensional structure of the protein from the diffraction pattern. Structures obtained by X-ray crystallography must be treated as hypothetical and subjected to verification to determine if the purified, dried, crystallized protein accurately reflects the protein's configuration *in vivo*.

Nuclear magnetic resonance (NMR) spectroscopy is a chemical technique that can solve the structures of purified proteins in aqueous solution. However, NMR spectroscopy functions best for small molecules and cannot be used to resolve the structure of larger proteins. But NMR can provide information on the variety of conformations assumed by flexible proteins in solution, a state that more closely approximates their state in living cells. Ideally, structural biologists employ both X-ray crystallographic structures and NMR structures to deduce the natural structure of a protein *in vivo*.

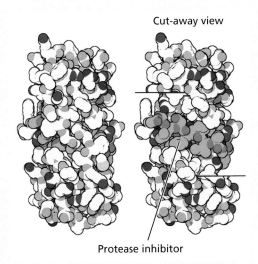

Cut-away view

Protease inhibitor

Figure 3.18 Structure of the HIV protease bound to a protease inhibitor. The protease is a globular enzyme that has an active site obscured from view by two flaps, as on the left. In this view, the van der Waals radii of all the atoms are depicted and the color code shows neutral (white), acidic (red), and basic (blue) regions of the protein. The protease inhibitor is colored green and orange to make it stand out in the images. When the flaps are removed using a computer, as on the right, the green protease inhibitor molecule can be seen binding to the enzyme's active site by mimicking the shape of protein chains that are normally hydrolyzed by the enzyme. (From PDB-101 [http://pdb101.rcsb.org]. Courtesy of David Goodsell. doi: 10.2210/rcsb_pdb/mom_2001_8.)

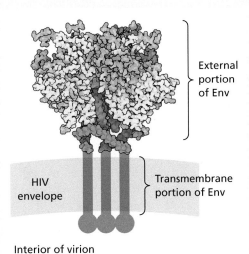

External portion of Env

Transmembrane portion of Env

HIV envelope

Interior of virion

Figure 3.17 Structure of the exterior soluble portion of the HIV envelope protein. The crystal structure of the external soluble portion is shown atop a diagram of the three transmembrane segments. (From PDB-101 [http://pdb101.rcsb.org]. Courtesy of David Goodsell. doi: 10.2210/rcsb_pdb/mom_2014_1.)

The structures of viral proteins are important for antiviral drug design. Most antiviral drugs are small molecules that bind to specific locations on a viral protein and thereby inhibit its function. Computational methods exist to determine whether a small molecule has the potential to bind to a specified crevice in a protein, making it possible to identify therapeutically active candidates *in silico* before proceeding with more extensive experiments designed to determine whether the candidate small molecule inhibitors affect that protein's normal activity. The goal is that these new computerized techniques will increase the pace of drug discovery, which can take 10 or more years from discovery of a molecule that inhibits the action of a viral protein to a clinically useful therapeutic treatment. Experimental tests can determine whether a small molecule binds to the target protein in the expected way (see Figure 3.18).

Virions are at least 10–100 times larger than their protein component parts. Nevertheless, naked virions (lacking a membrane) can also be studied using X-ray crystallography (**Figure 3.19**). The crystallization of tobacco mosaic virus was one of the earliest clues that virus structure is much more simplified than the structure of their host cells, even though viruses can overtake their larger, more complex host cells. Enveloped virions, by contrast, cannot form analyzable crystals so their structures are most commonly examined using cryo-electron microscopy (cryo-EM) or even atomic force microscopy (**Figure 3.20**; see also Technique Box 2.1). During cryo-electron tomography, samples are tilted slightly and then imaged, repeating many times to get a view through many different planes of focus. Then a computer combines the images to create a three-dimensional view of the virus.

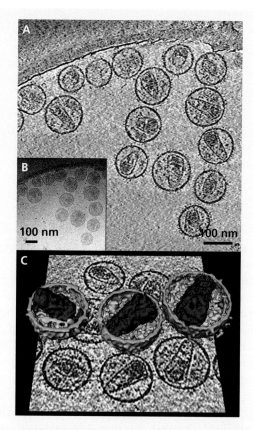

Figure 3.20 Three-dimensional rendering of enveloped virions using cryo-electron tomography. (A, B) One-dimensional slices through HIV virions imaged using cryo-electron tomography. These are used to construct C. (C) The three-dimensional rendering of the virus. The image has been false-colored to highlight the virion core (red), material between the core and envelope (yellow), and the viral envelope (blue). (From Briggs JA et al. 2006. *Structure* 14:15–20. With permission from Elsevier.)

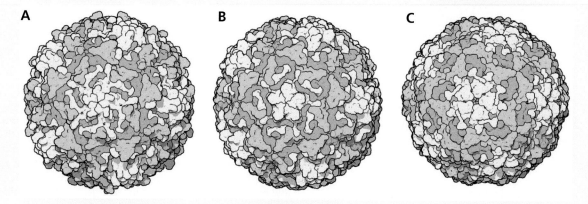

Figure 3.19 The crystal structures for three closely related picornaviruses. (A) Poliovirus. (B) Rhinovirus. (C) Foot-and-mouth disease virus. All three virions are similar in size and symmetry.

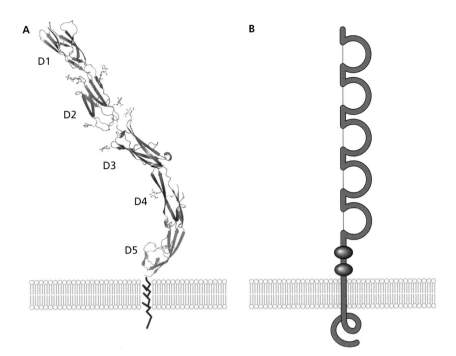

Figure 3.21 ICAM-1 is the cellular receptor for rhinovirus. (A) Crystal structure of the extracellular domains of ICAM-1. (B) Diagram of ICAM-1. (A, From Jin-lab. With permission from Dr. Jin Moonsoo.)

3.14 Several independent lines of evidence indicate that ICAM-1 is the rhinovirus receptor

The receptor for most rhinoviruses, intercellular adhesion molecule 1 (ICAM-1) (**Figure 3.21**), was first discovered by screening monoclonal antibodies and identifying those that could block infection. Next, an affinity chromatography column was loaded with the same antibody attached to a solid substrate and used to purify a 95-kDa protein from permissive host cell extracts. The protein was identified as ICAM-1. Finally, transfection of mouse cells with the human *ICAM-1* gene rendered them susceptible to infection by rhinovirus. The normal function of ICAM-1 is to stimulate an inflammatory response, so the inflammation (itchy red nose) induced by rhinovirus infection probably causes an increase in the expression of ICAM-1 on respiratory cells, making them more susceptible to the virus. This may be an example of viral adaptation to its animal host's physiology.

3.15 Experiments using molecular genetics support the conclusion that ICAM-1 is the rhinovirus receptor

Because immunological, biochemical, and transfection data all pointed to ICAM-1 as the receptor, the rhinovirus–ICAM-1 interaction was investigated in greater detail. Experiments using molecular genetics sought to find the part of ICAM-1 that likely interacts with the virus. **Missense mutations** that substitute one amino acid with a different amino acid should severely disrupt the ICAM-1–rhinovirus interaction only if the substitutions occur in or near the site of attachment, thereby disrupting the intermolecular forces between the virus and ICAM-1 and reducing the avidity of the virus for its receptor. To test this hypothesis, the cDNA encoding *ICAM-1* was mutated in order to make missense mutations and used to transfect nonpermissive host cells (**Figure 3.22**). Comparing the amount of rhinovirus that bound to host cells expressing altered ICAM-1 with the amount bound to cells expressing

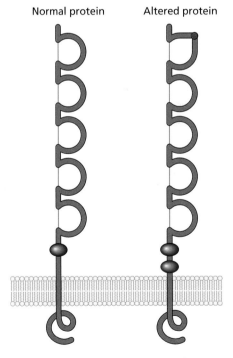

Normal protein   Altered protein

Figure 3.22 Missense mutation in the *ICAM-1* gene results in an ICAM-1 protein with an altered amino acid at the distal tip. The effect of this alteration is portrayed by the diagram on the right in which the distal domain has a change representing the altered shape and chemical properties caused by a missense mutation in the *ICAM-1* gene.

normal ICAM-1 showed that even small alterations in the amino acids near the tip of the ICAM-1 molecule, in the domain most distal to the membrane, caused a drastic loss of rhinovirus attachment. These findings strongly suggest that the distal tip of ICAM-1 binds to rhinovirus, a hypothesis that could be further investigated using structural biology, as explained in Section 3.16.

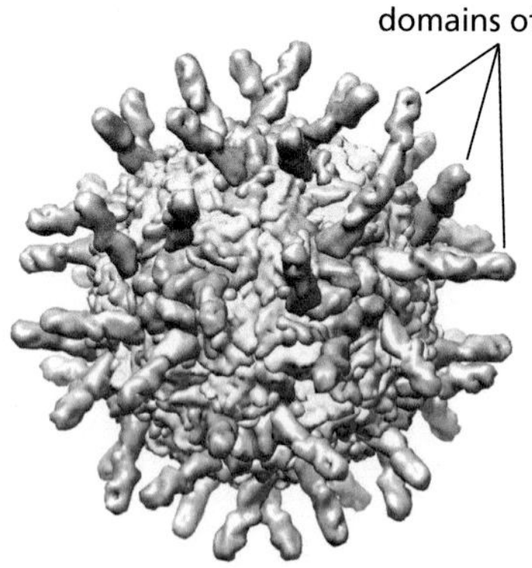

3.16 Structural biology experiments support the conclusion that ICAM-1 is the rhinovirus receptor

Structural biology is the study of the structure of macromolecules, especially proteins, and of multiprotein complexes including virions (see Technique Box 3.2). ICAM-1 is a transmembrane protein, but an ICAM-1 fragment containing the two domains furthest from the membrane has been purified. When mixed with rhinovirus and the mixture is imaged using cryo-EM, the ICAM-1 fragments reach into the canyon of the rhinovirus and bind (Figure 3.23), using many of the amino acids implicated by the genetic study. In this way, structural data also support the conclusion that the distal tip of ICAM-1 serves as the receptor for rhinovirus. This conclusion can be tested even further by examining diversity in the primary amino acid sequence of natural rhinovirus isolates, as described in Section 3.17.

3.17 Bioinformatics comparisons support the conclusion that ICAM-1 is the rhinovirus receptor

The previous structural studies indicate that the ICAM-1 tip binds to the canyon portion of the rhinovirus. Sequence data comparing the rhinovirus capsomers from many viral isolates also support this conclusion (Figure 3.24). For example, there are about 100 immunologically distinct human rhinovirus species, and alignments of the capsomer amino acid sequences show that the amino acids in the base of the canyon, where the ICAM-1 makes very close contact in the cryo-EM structure, are highly conserved among rhinovirus isolates. Two groups of rhinoviruses have divergent

Distal extracellular domains of ICAM-1

Figure 3.23 Crystal structure of two ICAM-1 extracellular domains bound to the surface of rhinovirus. This three-dimensional reconstruction uses the structures for rhinovirus (yellow, green, and purple) and ICAM-1 (gray) to show all the sites that ICAM-1 could bind to the virion. (From Bella J & Rossmann MG, courtesy of RCSB PDB.)

Figure 3.24 Alignment of the section of VP1 that includes the amino acids in the canyon that make contact with ICAM-1 in rhinovirus HRV-14. Amino acids lining the canyon in HRV-14 are in boldface and the corresponding positions in other rhinoviruses are highlighted in yellow. Below the alignment, * indicates that all amino acids at that position are identical, whereas : and . indicate decreasing amounts of chemical similarity between the amino acid side chains.

```
Rhinovirus 1   YMHYDGTETSLESFLGRAACVHVTDIENKLPTRES--------THKEQKLYNDWKINLSS
HRV-14         YMHFNGSETDVECFLGRAACVHVTEIQNKDATGID--------NHREAKLFNDWKINLSS
Rhinovirus 2   YMHFNGSETDVECFLGRAACVHVTEIQNKDATGID--------NHREAKLFNDWKINLSS
Rhinovirus 3   YMHFNGSETDVESFLGRAACVHITEIENKNPADIQ--------NQKEEKLFNDWKINFSS
Rhinovirus 4   YMHFNGSETDVECFLGRAACVHMVKIVNKNPTDIV--------NQKEHLLFNDWKINLSS
Rhinovirus 5   YMHFSGSETTLENFLGRSACVHITEIQNKRPEEFT--SEETAKTHKEQKLFNDWKISLSS
Rhinovirus 6   YMHFTGSETSLENFLGRSACVHITEIKNKLPTEPVMDGNQMRNTHKEQGLFXDWKINLSS
               ***: *:** :* ****:**** :..* **       .::*  *: **:*.:**

Rhinovirus 1   LVQLRRKLDMFTYVRFDSEYTIIATSSQPQDAQFSNTLTVQAMFIPPGAPNPQEWDDYTW
HRV-14         LVQLRKKLELFTYVRFDSEYTILATASQPDSANYSSNLVVQAMYVPPGAPNPKEWDDYTW
Rhinovirus 2   LVQLRKKLELFTYVRFDSEYTILATASQPDSANYSSNLVVQAMYVPPGAPNPKEWDDYTW
Rhinovirus 3   LVQLRKKLELFTYIRFDSEYTILATASQPK-SNYASNLVVQAMYVPPGAPNPKEWDDFTW
Rhinovirus 4   LVQLRKKLELFTYIRFDSEYTILATASQPNDSQYSSNLTVQAMYVPPGAPNPTKWDDYTW
Rhinovirus 5   LVQLRKKLELFTYVRFDSEYTILATASQPDQAQYASNLIVQAMYVPPGAPNPIEWNDYTW
Rhinovirus 6   LVQFRKKLELFTYVRFDSEYTILATASQPNTAQYASNLTVQAMYVPPGAPNPVKWDDYTW
               ***:*:**::***:*********:**:***.  .::::..* ****:*:****** :*:*:**

Rhinovirus 1   QSASNPSIFFNVGKSARFSVPYLGIASAYNNFYDGYSHDDKTTVYGINVLNHMGSIAFRV
HRV-14         QSASNPSVFFKVGDTSRFSVPYVGLASAYNCFYDGYSHDDAETQYGITVLNHMGSMAFRI
Rhinovirus 2   QSASNPSVFFKVGDTSRFSVPYVGLASAYNCFYDGYSHDDAETQYGITVLNHMGSMAFRI
Rhinovirus 3   QSASNPSVFFKVGDTSRFSVPFVGLASAYNCFYDGYSHDDKDTPYGITVLNHMGSIAFRV
Rhinovirus 4   QSASNPSVFFKVGDTARFSVPFVGLASAYNCFYDGYSHDDENTPYGITVLNHMGSMAFRV
Rhinovirus 5   QSASNPSVFFEVGKTARFSVPFTGIASAYNCFYDGYSHDDENTQYGINVLNHMGSIAFRV
Rhinovirus 6   QSASNPSVFFEVGKMARFSVPFIGIASAYNCFYDGYSHDNEDTPYGINVLNHMGSIAFRI
               *******:**:**.  . :*****:  *:***** ******* : *** .*****:***:
```

sequences and these rhinovirus groups use different host cell receptors rather than ICAM-1.

Although the canyon of the rhinovirus serves as the interaction site between the virus and its receptor, note that other naked viruses such as adenoviruses use protrusions or spikes to bind to host cell receptors.

Rhinovirus is a model for how naked viruses in general attach to host cells because it illustrates that both shape and chemical complementarity enable high-avidity interactions between a virus and its receptor. It illustrates the general principle that host receptors are abundant on the target host cells and contribute tremendously to tissue tropism. It exemplifies that virions have more than one attachment site available for receptor binding. ICAM-1 is a model virus receptor because many other viruses use receptors that are similar to ICAM-1 in that it is the most distal portion of the receptor that typically binds to the virion with high avidity. Furthermore, the use of crystallographic and sequence conservation data for the rhinovirus–ICAM-1 interaction are models for applying those approaches to other virus–receptor interactions.

3.18 Influenza serves as a model for attachment by enveloped viruses

Influenza virions are enveloped and contain eight different viral nucleoprotein complexes of single-stranded (−) sense RNA. The virion contains multiple copies of two different spike proteins: **neuraminidase (NA)** and **hemagglutinin (HA)** (**Figure 3.25**). Although NA is particularly important for the release stage of the virus replication cycle, HA is used for the attachment and penetration stages of the replication cycle. Influenza A viruses, the cause of devastating human pandemics, are often named according to the type of immunologically distinctive HA and NA proteins they encode. There are 17 immunologically distinctive HA molecules and 10 different NA proteins. All HA proteins share the same function, but antibodies that react with HA protein type 1 (H1) do not react with H2 proteins, and so on. The same is true of different NA proteins. An influenza A virus might therefore be called H1N1, indicating that it has the first type of HA protein (HA type 1) and the

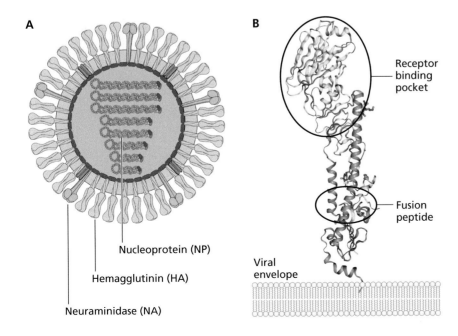

Figure 3.25 Influenza has two spike proteins. (A) Influenza diagram showing the hemagglutinin and neuraminidase spikes on the surface. (B) Hemagglutinin is a trimer of the depicted monomer, which has a receptor binding pocket and a fusion peptide, and crosses the viral envelope. The crystal structure depicts the extracellular components of one monomer. (B, From Wang TT & Palese P 2009. *Nat Struct Mol Biol* 16:233–234. With permission from Springer Nature.)

first type of NA protein (NA type 1). Similarly, H7N3 denotes the presence of HA7 and N3 spikes. These immunological distinctions are very important for influenza vaccination campaigns because a vaccine that protects someone against H7N3, for example, will not protect that person from H1N1 (see **Chapter 16** for more information about vaccines).

3.19   The influenza HA spike protein binds to sialic acids

The influenza HA spike has two component parts: HA2 and HA1. HA2 includes the transmembrane portion, whereas HA1 makes contact with the influenza host receptor, which is the sugar sialic acid (**Figure 3.26**). HA1 can be viewed as a knob at the end of the spike and, like most virions with spike proteins, the distal knob portion mediates attachment to the host cell. The surface of human host cells is a glycocalyx, namely a layer of sugars that are in turn covalently attached to glycoproteins that are associated with the external surface of the plasma membrane (see **Figure 3.26A**). The exact components of the glycocalyx are often specific for different types of tissues. An example is the presence of sialic acids (see **Figure 3.26B**) as carbohydrate components of the glycocalyx of respiratory tissues. The sialic acid sugar is located at the distal end of the oligosaccharide chains that make up the glycocalyx, thus placing them in the outermost surface that makes contact with influenza virus. The terminal sialic acid is connected to the adjacent sugar in variable ways, creating additional structural variation important for virus attachment. On human respiratory cells, for example, the sialic acid is typically attached to an adjacent galactose sugar through an α-2,6 linkage (see **Figure 3.26C**). This linkage has a distinctive shape compared with other types of linkages, such as α-2,3. The linkage between the sialic acid and the next sugar contributes substantially to the HA1 spike–sialic acid interaction.

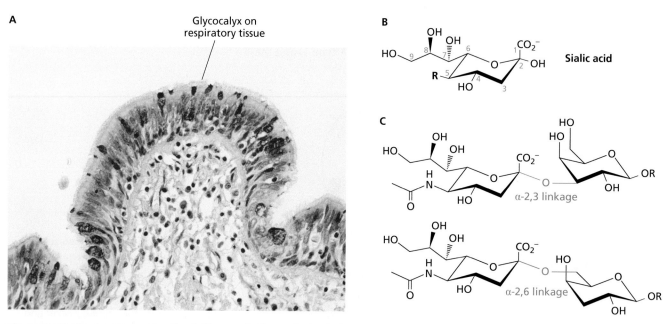

Figure 3.26 The host receptor for influenza is the carbohydrate sialic acid, also known as neuraminic acid. (A) All human cells have a glycocalyx consisting of a variety of carbohydrates, as shown by staining of the sugars on the surface of this human respiratory tissue. The glycocalyx is more than six times as thick as the plasma membrane, and in this case the red stain shows the presence of sialic acid with an α-2,6 connection (typical human respiratory sialic acid). (B) Neuraminic (sialic) acid sugar with an R where different chemical groups result in different members of the sialic acid family. (C) Specific sialic acid connected to the next sugar (galactose or Gal) through an α-2,3 connection or through an α-2,6 connection. (A, Courtesy of John Nicholls.)

The HA1 protein of influenza A viruses that cause human infections binds preferentially to the form of α-2,6-linked sialic acid that is most common on nonciliated human respiratory cells. In contrast, the HA1 protein of influenza A viruses that live in bird populations binds better to α-2,3-linked sialic acids that are common in birds but can also be found along with α-2,6-linked sialic acids on ciliated human respiratory cells. Avian influenza viruses that cross the species barrier and infect humans typically have HA1 knobs that attach better to α-2,6-linked sialic acids than the other avian HA1 knobs do. Avian influenza A cannot undergo person-to-person transmission until the HA1 knob evolves and acquires the right sequence and structure to bind preferentially to human respiratory epithelial α-2,6-linked sialic acids.

Although not covered in detail here, there are many molecular genetic and structural studies that have investigated the HA–sialic acid receptor interaction in just as much detail as that for the rhinovirus–ICAM-1 interaction. Examples of publications addressing the HA–sialic acid interaction are listed in the **Further reading** section.

Influenza thus serves as a model for attachment by enveloped animal viruses because they all use spikes to attach to their receptors. Most animal viruses, such as influenza, interact with glycosylated host receptors because essentially all surface-exposed host proteins are glycosylated. As is the case for other enveloped virus spike proteins, the avidity of the influenza HA spike for its receptor relies upon noncovalent intermolecular forces such as hydrogen bonds, ionic bonds, and London dispersion forces, all of which are enabled by shape complementarity to its receptor. Furthermore, influenza is a model for how virus–receptor interactions heavily influence host range, exemplified by the differences among strains that attach to α-2,6-sialic acid and those that attach to α-2,3-sialic acid.

The study of influenza attachment also has been useful because human tissues are surrounded by an extracellular matrix of polysaccharides that serve as receptors for a growing list of viruses. Many of these receptors are glycosaminoglycans, a type of polysaccharide composed of particular sugar units attached to one another by specific connections, much like different sialic acids. Understanding the use of glycosaminoglycans as receptors has been deeply informed by the HA–sialic acid paradigm.

Keep in mind that viruses are diverse, and there are many examples of enveloped viruses that use their spike proteins to attach to specific host receptor proteins rather than to glycosyl groups attached to proteins, lipids, or extracellular matrix in general. An example is that between HIV and its receptor CD4, addressed in **Section 3.29** when we consider the next stage of the virus replication cycle, the penetration and uncoating stage.

3.20   The second stage of the virus replication cycle includes both penetration and uncoating and, if necessary, transport to the nucleus

Attachment events are sometimes characterized as irreversible because penetration and uncoating ensue automatically, making it impossible for the virion to reform its infectious particle. Penetration is the entry of the virion or subcomponents of the virion into the host cell. For purposes of dividing the virus replication cycle into smaller stages, this book defines the penetration and uncoating stage as including all molecular events that occur between penetration and the production of the first viral protein or mRNA. In the simplest cases, binding to a single receptor molecule leads directly to penetration and uncoating without any other virus–host cell interactions, and the entire uncoating stage is just the physical separation of the viral genome from the capsid. For example, poliovirus forms a pore in the host cell membrane and releases its (+) RNA strand genome directly into the host cytoplasm, after which its genome can be translated (**Figure 3.27A**). In this case, penetration and uncoating cannot be separated and the uncoating stage is brief. The enveloped togaviruses provide a second example of a simple penetration and uncoating stage. Penetration occurs when the virus is internalized into an endocytic vesicle that fuses its envelope with the endocytic membrane and releases its nucleocapsid into the cytoplasm (**Figure 3.27B**). Uncoating then occurs when the nucleocapsid disassembles and the (+) ssRNA genome can then be translated.

In contrast, for many viruses, the penetration and uncoating stage has a longer duration (an hour or more) and includes molecular events in addition to dissociation of the genome from its capsid. In most cases, however, penetration includes internalization of the entire virion (**Figure 3.27C**). It is probably advantageous for the entire virion particle to enter the host cell, because doing so leaves no viral proteins on the surface of the host cell where they might alert the immune system (see **Chapters 14** and **15**). Subsequent uncoating processes are then needed for the virus capsid and genome to escape from the endocytic or phagocytic vesicle.

A virus will be exposed to different conditions and host factors depending on its mode of entry. These various conditions and factors are required to complete uncoating. For example, a virus that enters through a host **endosome** will be exposed to increasingly acidic conditions as the endosome matures through the endocytic vesicle system. This process selects for viruses that use a drop in pH to initiate the uncoating process. For viruses that enter independently of the endosome system, a drop in pH does not trigger uncoating because those modes of entry do not expose internalized particles to increasingly acidic conditions.

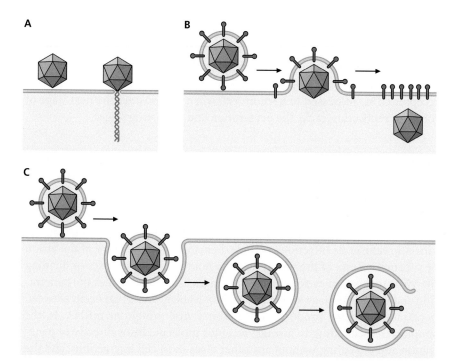

Figure 3.27 Overview of penetration and uncoating. (A) Release of viral genome into the cytoplasm through a pore. (B) Fusion of enveloped virus at the plasma membrane, releasing the nucleocapsid into the cytoplasm and leaving viral spike proteins on the cell surface. (C) Endocytosis of a virus followed by release of the nucleocapsid into the cytoplasm; no viral proteins are left on the cell surface.

3.21 Viruses subvert the two major eukaryotic mechanisms for internalizing particles

Eukaryotic cells internalize particles using several processes including endocytosis and phagocytosis. Endocytosis is the collective term for a variety of ways that cells internalize small particles (<0.2 μm in diameter) in vesicles. Phagocytosis, by contrast, can engulf much larger particles (1–2 μm in diameter). Typically, only cells that have a specialized function in the immune system have the capacity to use phagocytosis. Most virions are smaller than 0.2 μm, and most infect cells through endocytosis.

3.22 Many viruses subvert receptor-mediated endocytosis for penetration

Receptor-mediated endocytosis occurs after a host receptor molecule binds to its target ligand (**Figure 3.28**), which in normal circumstances is a nutrient or growth factor. Viruses that bind to the receptor even though they are not the intended cellular ligand usurp this process. Cells use receptor-mediated endocytosis for many essential functions, for example the acquisition of nutrients such as iron, the regulation of growth factor receptors, and signal transduction. After a ligand in the extracellular media binds to its receptor, the ligand–receptor complex diffuses in the plane of the membrane until it encounters a small indentation where there is abundant **clathrin** on the cytoplasmic face of the indentation (see Figure 3.28). The fibrous clathrin proteins then assemble into a cagelike structure that forms an indentation and ultimately pulls the receptor and its ligand into the cell. Other host proteins required for this activity include actin. Subsequently, the clathrin-coated vesicle loses its clathrin and then the vesicle fuses with an **early endosome**. Early endosomes have slightly acidic internal conditions and are a site for sorting. For example, cells typically sort the receptor molecules away from their ligands and send the receptor back to the cell surface for reuse. Early endosomes then mature, becoming more and more acidic as a motor protein pulls the vesicles along the microtubule cytoskeleton toward the centrosome. Finally, the acidic late endosomes fuse with a lysosome, where the low pH of the interior activates the lysosome's degradative enzymes.

Viruses that enter by receptor-mediated endocytosis and traffic through the endosome system during penetration and uncoating exploit these same

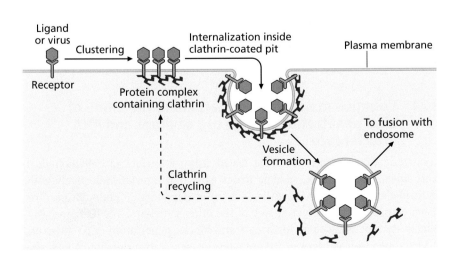

Figure 3.28 Receptor-mediated endocytosis. Binding of several ligands to their receptors causes them to cluster together, after which the clathrin forms a clathrin-coated vesicle that internalizes the receptors. After internalization, the clathrin proteins are recycled. Viruses usurp this process by mimicking ligand binding, resulting in virus internalization.

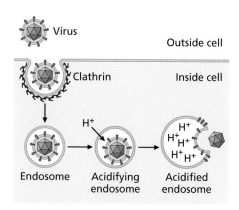

Figure 3.29 Viral entry via receptor-mediated endocytosis. Virus binding to a receptor triggers internalization through clathrin-coated pits. After internalization, the early endosome acidifies, triggering viral uncoating.

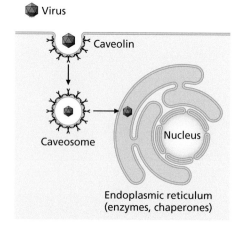

Figure 3.30 Entry via caveolae. After internalization, the virus traffics through the caveosome and from there to the endoplasmic reticulum. Other endocytic caveolar vesicles may instead fuse with an early endosome.

processes (**Figure 3.29**). After internalization, the virus enters the normal endocytic pathway, so that the vesicle acidifies. Acidification triggers the next stage of uncoating, which is typically to release the nucleocapsid or genome into the cytoplasm. Viruses that enter through clathrin-dependent, receptor-mediated endocytosis include vesicular stomatitis virus, influenza A virus, and human rhinovirus 2.

There are alternative forms of endocytosis that do not involve clathrin. For example, there is a pathway that uses the protein **caveolin** instead (**Figure 3.30**); vesicles formed by caveolin require a protein called **dynamin** and do not always become endosomes or fuse with the lysosome. Instead, sometimes caveolin-coated vesicles fuse with an organelle called the **caveosome**, which ultimately releases particles, including viruses, into the endoplasmic reticulum. Here, too, the actin cytoskeleton is involved. Viruses such as the polyomavirus SV40 that enter by the caveolin route are not exposed to low pH but are consequently exposed to the resident proteins in the endoplasmic reticulum, such as the Hsp105 and Hsc70 proteins found exclusively in the membrane of the endoplasmic reticulum. These endoplasmic reticulum proteins are needed for uncoating, comparable to the need for low pH for viruses that enter through receptor-mediated endocytosis.

3.23 Herpesvirus penetrates the cell through phagocytosis

The second major way that cells internalize particles is **phagocytosis** (**Figure 3.31A**), which is generally used by specialist cells, such as macrophages and neutrophils that ingest microbes as part of an immune response. It has long been known that bacterial cells can trigger normally nonphagocytic cells to engulf the bacteria using a process similar to phagocytosis, and it appears that some viruses can do the same. Cultured corneal fibroblasts, for example, provide an *in vitro* model of ocular herpesvirus infections (**Figure 3.31B**), and the virus enters these cells by inducing its own phagocytosis, even though ocular fibroblasts are normally nonphagocytic. Herpesviruses, which have complex tissue tropism, may use other mechanisms to enter epithelial or neuronal cells. For example, herpes simplex viruses and Epstein–Barr virus (formally known as human herpesvirus 1, 2, and 4, respectively) can enter cells by endocytosis or by fusion at the plasma membrane.

Herpesviruses are not alone in using multiple modes of entry into permissive cells and yet the mechanisms by which the same virion can enter multiple cell types using different means are not well understood. When multiple modes of entry occur, every entry pathway might not lead to successful uncoating and infection. In addition, it is uncertain whether the events that occur in cultured laboratory cells are the same as those that occur in living animal hosts. Further technological innovations are likely needed to investigate these modes of entry.

3.24 Common methods for determining the mode of viral penetration include use of drugs and RNA interference

The mode of penetration can be tested using a variety of cell biological techniques. For example, specific drugs exist that interfere selectively with clathrin-dependent endocytosis, caveolin-dependent endocytosis, or phagocytosis without affecting any of the other pathways. The drug nystatin selectively blocks caveolae formation and blocks penetration of SV40 virus. If virus infection is blocked by one class of drugs, but not the others, the

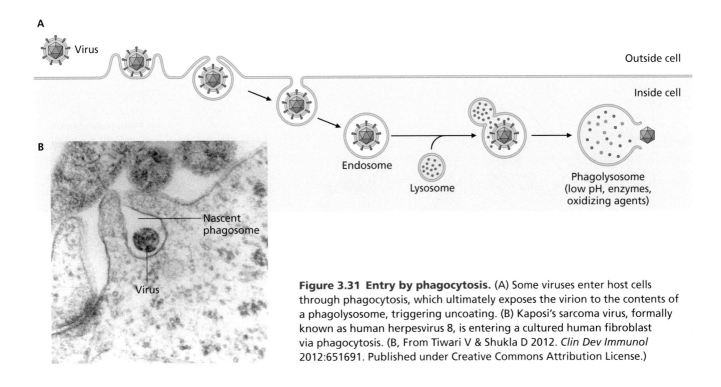

Figure 3.31 Entry by phagocytosis. (A) Some viruses enter host cells through phagocytosis, which ultimately exposes the virion to the contents of a phagolysosome, triggering uncoating. (B) Kaposi's sarcoma virus, formally known as human herpesvirus 8, is entering a cultured human fibroblast via phagocytosis. (B, From Tiwari V & Shukla D 2012. *Clin Dev Immunol* 2012:651691. Published under Creative Commons Attribution License.)

virus is probably using the pathway poisoned by the drug for entry. Virion movement can also be tracked using fluorescence microscopy to attempt to co-localize the virion with other proteins characteristic of one or another mode of entry. For instance, if the virion co-localizes with another protein already known to localize to the caveosome, then the virion probably enters through the caveolin-dependent pathway. Newer approaches to identify the type of entry employ genetic manipulation of host cells. One such technique is **RNA interference** (**RNAi**), which blocks the production of specific proteins, such as dynamin or caveolin. (RNAi is covered more completely in **Chapter 14**.) Interfering with the mRNA encoding a Rab-guanosine triphosphatase (GTPase) protein that controls endosomal vesicle trafficking prevents normal uncoating of poliovirus in human brain microvascular endothelial cells, indicating that poliovirus requires endosomal vesicle trafficking to infect human brain microvascular endothelial cells.

3.25 The virion is a metastable particle primed for uncoating once irreversible attachment and penetration have occurred

Fundamentally, virions have two distinct roles: to protect the virus genome during transmission and to release the virus genome into a host cell. These two roles are somewhat paradoxical, as the ability to protect the genome seems to be antithetical to that of exposing the genome for transcription and replication. The key to this paradox is that virions are **metastable** particles, and intermolecular interactions affect their stability. Thermodynamically, the infectious virion's structure is not the conformation with the lowest possible free energy. However, the energy barrier (the **energy of activation**) between the virion structure and the rearrangement of the virion that would release the genome is high so that it rarely, if ever, occurs until the virion interacts with a host (**Figure 3.32**). In this way, virions are structurally

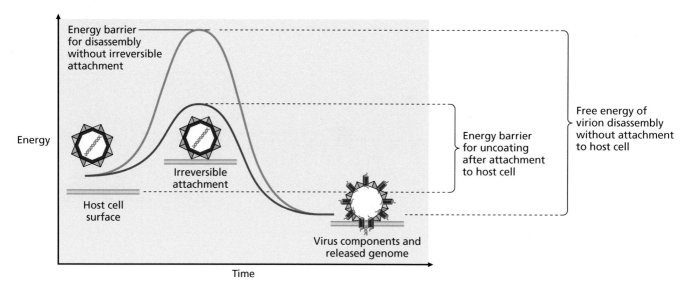

Figure 3.32 Conceptual energy diagram of virion before and after genome release with energy barrier in between. The virion is a metastable particle; interaction with the host cell lowers the energy barrier and the virion subsequently uncoats, releasing its genome. In the absence of attachment, the energy of activation is very high so that most unattached virions tend to remain intact (blue line).

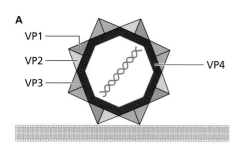

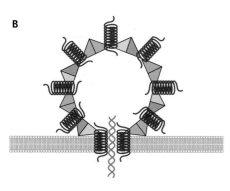

Figure 3.33 Penetration and uncoating of picornaviruses. (A) The capsid of a picornavirus has VP1, VP2, and VP3 on its surface, with VP4 buried in a layer beneath them, not accessible to the surrounding solvent. (B) Penetration is enabled by dramatic protein rearrangements, such that some parts of VP4 become displayed on the surface of the virion particle and subsequently form a pore through the membrane.

biased toward protecting the genome until they engage irreversibly with a host receptor. This situation has similarities to the use of enzymes to catalyze thermodynamically favorable reactions. The oxidation of glucose into carbon dioxide and water is quite favorable under cellular conditions, yet it does not occur without catalysis to overcome the energy of activation.

How does the virion achieve its two roles? The intermolecular interactions among the structural proteins of a virion, and between them and the genome, maintain the virion in its characteristic shape. Virions exist in this metastable state until a host cell triggers them to advance from attachment to the penetration and uncoating stage. Triggers from the host cell include binding to a cellular receptor, a decrease in endosomal pH, proteolytic degradation by a host enzyme, or some combination of these factors. In fact, there is no paradox: instead, the virion's structural properties are ideally suited for both its protective and its cargo delivery roles.

3.26 Picornaviruses are naked viruses that release their genomic contents through pore formation

Naked viruses can enter cells by endocytosis and phagocytosis, but, at some point, the capsid has to disassemble in order to release the genome or the genome has to be extruded from the capsid. The picornaviruses poliovirus and rhinovirus are close relatives that provide models for penetration from the capsids of naked virions (**Figure 3.33**). Poliovirus, which is contracted by ingestion of contaminated food or water, releases its genome at or very near the cell surface, following receptor-mediated endocytosis or perhaps as a consequence of binding to its cellular receptor. Even while it is extracellular, poliovirus may become exposed to the low pH of the human stomach. Rhinovirus, contracted by inhalation and thus not exposed to low pH until after endocytosis, releases its genome after acidification of an endocytic vesicle. In both cases, substantial protein rearrangement apparently occurs, with the VP4 protein, normally buried on the inside of the capsid, becoming

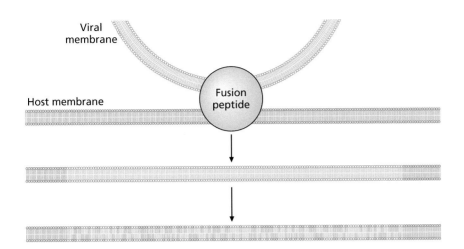

Viral membrane

Host membrane

Fusion peptide

Figure 3.34 Membrane fusion. The blue lipid bilayer fuses with the orange lipid bilayer, catalyzed by a protein complex containing a fusion peptide. Soon after fusion the phospholipids (blue and orange) from both bilayers will mix.

exposed on the surface of the virion. VP4 forms a multimeric pore in the host cell membrane, allowing the (+) ssRNA genome to be released into the cytoplasm.

We have emphasized the picornaviruses here because they were used to discover how naked viruses release their genomes into a state in which they can be expressed. Other naked viruses use different mechanisms; some even remain partially intact during gene expression (see **Chapter 7**). In every case, however, capsid proteins play a key role in maintaining the virus in a meta-stable state and in uncoating once a virus-specific interaction occurs with a host cell, ultimately enabling the genome to be expressed.

3.27   Some enveloped viruses use membrane fusion with the outside surface of the cell for penetration

An enveloped virion is covered by a protein-rich lipid bilayer derived from its former host and contains proteins of both host and viral origin. Penetration and uncoating by enveloped viruses necessarily includes **fusion** of the virion envelope with a host cell membrane (**Figure 3.34**). Fusion can be defined as the two lipid bilayers becoming a single lipid bilayer. Simple collision of lipid bilayers does not result in fusion of those bilayers or else the eukaryotic cell would be unable to maintain the distinctive characteristics of each membrane-bound organelle. Instead, membrane fusion requires catalysis. In the case of enveloped viruses, a component of one of the virus spike proteins, termed the fusion protein, has a region, termed the fusion peptide, that becomes catalytically active following irreversible attachment.

3.28   Vesicle fusion in neuroscience is a model for viral membrane fusion

A well-understood cellular model for membrane fusion is found in neuroscience, because regulated membrane fusion is essential for the transmission of nerve impulses. The terminus of a neuron's axon, where it connects with another cell, is loaded with vesicles full of neurotransmitters (**Figure 3.35**). An electrical signal triggers these vesicles to fuse with the plasma membrane, releasing the neurotransmitters, which diffuse to the next cell and trigger the propagation of the electrical signal through the next neuron. In this case, membrane fusion is catalyzed by proteins called **SNAREs** (soluble *N*-ethylmaleimide-sensitive factor attachment protein receptors). Complex

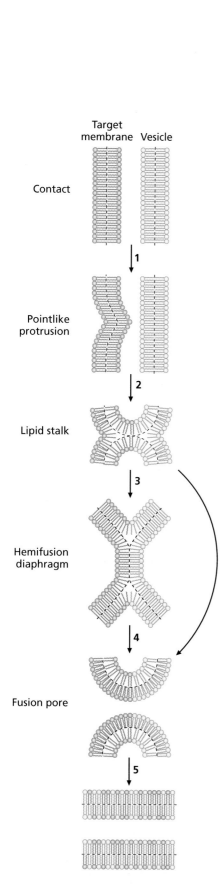

Target membrane — Vesicle

Contact

1

Pointlike protrusion

2

Lipid stalk

3

Hemifusion diaphragm

4

Fusion pore

5

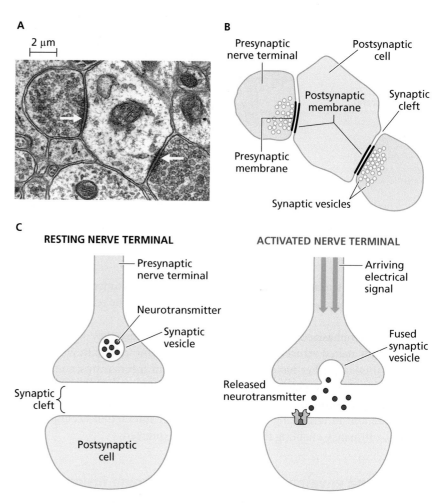

Figure 3.35 Vesicle fusion during transmission of a signal in the nervous system. (A) Neurons connect to their target cell through synapses, indicated by arrows. (B) Diagram of cells in Panel A. (C) A vesicle containing a neurotransmitter fuses with the plasma membrane of the presynaptic cell and releases its contents into the synaptic cleft. (A, Courtesy of Cedric S. Raine.)

formation between multiple SNAREs drives membrane fusion in multiple steps that can best be understood by focusing on the movement of the phospholipids (**Figure 3.36**). After contact, the first step is formation of lipid protrusions. Next, the lipid stalk forms, followed by formation of a hemifusion diaphragm. Next a **fusion pore** forms, ultimately leading to fusion of the lipid bilayers. After fusion, the phospholipids that originated from the two compartments comingle.

Key to this model is that SNAREs have transmembrane domains that anchor the protein to the vesicles that will ultimately fuse, as well as extravesicular

Figure 3.36 Model of membrane fusion as catalyzed by cellular SNARE proteins. Membrane fusion catalyzed by SNARE proteins occurs in multiple steps. Formation of a lipid protrusion (1). Formation of a lipid stalk (2), followed by formation of a hemifusion diaphragm (3), then a fusion pore (4), and finally full fusion of the lipid bilayers (5). (From Chernomordik LV & Kozlov MM 2008. *Nat Struct Mol Biol* 15:675–683. With permission from Springer Nature.)

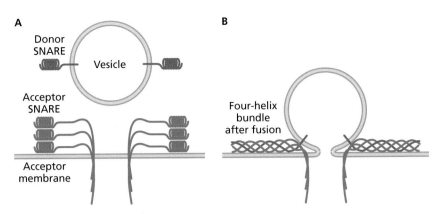

Figure 3.37 Donor and acceptor SNAREs catalyze neuronal membrane fusion. (A) Vesicle and acceptor membrane with extravesicular α-helical domains of the donor SNARE and trimeric acceptor SNARE exaggerated. (B) The donor and acceptor SNARES form a four-helical bundle to catalyze membrane fusion.

domains that assemble to form a bundle of four α-helices (**Figure 3.37**). This bundle is extremely stable; it is held together by many hydrophobic interactions and by highly conserved ionic bonds between the side chains (R groups) of glutamine on three of the helices and the positively charged side chain of an arginine on the fourth helix.

Virus fusion peptides have a similar structure to SNAREs, suggesting a similar mechanism for catalyzing fusion of the virion envelope to a host cell membrane. Virus fusion peptides are part of virus spike proteins with transmembrane components, but they are activated during attachment or penetration. The molecular events that trigger activation of the fusion peptide depend on the specific virus and can include acidification, proteolysis, or other events that occur when the virion is introduced into the endosome, caveosome, endoplasmic reticulum, or phagolysosome. Once triggered, the virus spike proteins undergo significant structural rearrangement that causes one domain of the fusion peptides to insert into the host membrane and other domains to form a SNARE-like helical bundle. In this catalytically active mode, virus fusion peptides catalyze membrane fusion through the same mechanism used by SNAREs. Two particular examples are HIV, in which a cascade of protein–protein interactions causes the fusion peptide to become active, and influenza, in which acidification of an endocytic vesicle containing the virion triggers activity of the fusion peptide (**Figure 3.38**).

3.29   HIV provides a model of membrane fusion triggered by a cascade of protein–protein interactions

Fusion mediated by HIV has been the focus of intensive research directed toward development of anti-HIV medications that block fusion. The HIV fusion peptide is part of the HIV spike, known as **Env**. Env has two parts: **surface (SU)** and **transmembrane (TM)**. These components are also known by their molecular weight, as in **gp120** (SU) and **gp41** (TM). Env is a trimer of SU–TM dimers. In the virion, SU and TM are held together by noncovalent intermolecular forces, which also prevent the catalytic fusion activity associated with gp41 (TM). During penetration, HIV virions fuse with the external surface of the plasma membrane (**Figure 3.39**). This fusion is triggered by a series of protein–protein interactions at the interface between the virion and

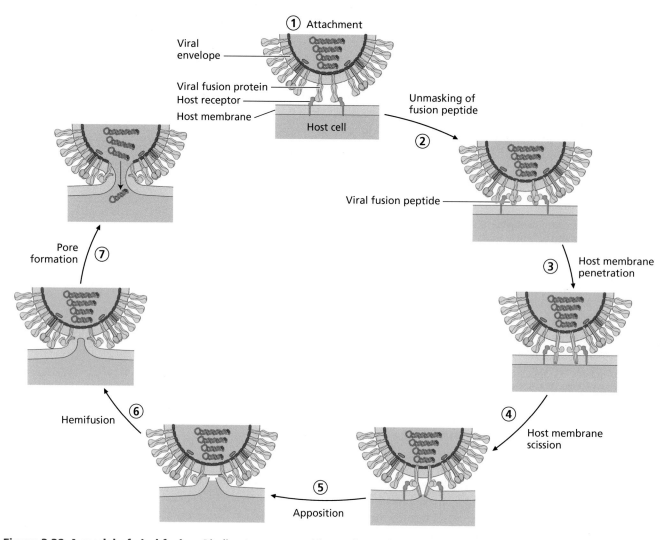

Figure 3.38 A model of viral fusion. Binding to a receptor (1) or a change in pH causes a rearrangement in the viral spike protein, revealing the SNARE-like fusion peptide (2), which is an α-helix with a hydrophobic surface that can penetrate host membrane lipids (3). The viral fusion peptide causes scission of the cell membrane (4), draws the viral envelope into close apposition with the host cell membrane (5), creates a hemifusion structure (6), and then catalyzes pore formation (7) so that the viral contents enter the cytoplasm. (Courtesy of Philippe Le Mercier, ViralZone, © SIB Swiss Institute of Bioinformatics.)

Figure 3.39 Penetration by HIV. After gp120 binds to CD4, the gp120 changes conformation and becomes able to bind to a co-receptor such as CCR5. The gp41 protein rearranges to expose the fusion peptide and insert the distal part of the gp41 protein into the host cell membrane. The gp120 protein then undergoes a dramatic conformational change, which ultimately enables the fusion peptide to catalyze fusion pore formation between the virus interior and the cytoplasm. (Courtesy of Mike Jones. Published under CC BY-SA 3.0.)

the plasma membrane. When SU engages a CD4 host receptor molecule, it triggers a conformational change that initiates SU binding to a second host molecule termed a **co-receptor**. HIV can use a variety of co-receptors (such as CCR5 and CXCR4), depending on the type of host cell. Binding to the co-receptor then triggers a profound rearrangement of the gp120–gp41 complex so that a portion of gp41, known as the fusion peptide, inserts into the plasma membrane, while other domains in gp41 form a helical bundle similar to SNAREs. Pharmaceuticals that block HIV entry target this stage by binding to the extracellular helices of gp41 in a way that prevents the conformational changes required for fusion. Once the virion envelope and the plasma membrane fuse, the HIV nucleocapsid is released into the cytoplasm.

3.30 Influenza provides a model for viral envelope fusion triggered by acidification of an endocytic vesicle

Although HIV provides a counterexample, most fusion peptides become exposed following virion internalization so that the fusion between the virus envelope and a host cell membrane occurs within the cell, leaving no virus antigens on the cell surface. Influenza provides the best-studied model. In this case, the virus is internalized by receptor-mediated endocytosis. Subsequent acidification of the compartment through normal host processes causes a massive rearrangement of the HA spike, making a catalytically active fusion peptide. Once the fusion peptide becomes active, a trimer of fusion peptides works in the same way as the HIV fusion peptides.

There are two other classes of viral fusion proteins. Although the structural and mechanistic details of how they work differ in detail from the Class I fusion catalyzed by influenza and HIV, all three mechanisms catalyze membrane fusion in a similar manner.

3.31 The destination for the virus genome may be the cytoplasm or the nucleus

Because many animal viruses replicate in the cytoplasm, the final event in uncoating is release of the genome into the cytoplasm. Viruses of this type include all RNA viruses except orthomyxoviruses (influenza) and retroviruses (HIV), as well as the poxviruses, despite their dsDNA genomes. Other viruses, however, must express and replicate their genomes in the nucleus. Viruses that express and replicate their genomes in the nucleus have uncoating stages that involve not only dissolution of the nucleocapsid but also trafficking to the nucleus along the cytoskeleton and interacting with nuclear pore complexes.

3.32 Subversion of the cellular cytoskeleton is critical for uncoating

Even for viruses that express and replicate their genomes in the cytoplasm, specific cytoskeletal elements such as microfilaments (actin) are typically required for penetration and uncoating. The role of the cytoskeleton in uncoating is particularly prominent, and better understood, for those viruses that must introduce their genomes into the nucleus. On a molecular scale, a virion that is 100 nm in diameter can have a distance greater than 50,000 nm, or 500 times its own diameter, to travel from the cell periphery to reach the nucleus. To move the same relative distance, an adult person would have to run the length of five soccer fields. As explained in **Chapter 2**, all particles as large as virions diffuse slowly within the cell, where the viscosity can be likened to that of wet sand. Thus, viruses and their component parts must use the cytoskeleton and its associated motor proteins to move to and from the cell periphery as needed.

Microtubules, which are components of the cytoskeleton, are structures exploited by viruses. Microtubules are hollow tubes made up of tubulin dimers; they have chemically and functionally distinctive ends that are arbitrarily named the minus (−) and plus (+) ends (**Figure 3.40A**). The (−) ends are near a single microtubule-organizing center (MTOC; for example, the centrosome) in nondividing cells, with the microtubules projecting either toward the cell periphery or toward the nucleus from there. Dynein motor proteins hydrolyze adenosine triphosphate (ATP) in order to catalyze directed movement, and thereby carry cargo toward the (−) ends (that is, toward the MTOC), whereas kinesin motors carry cargo toward the (+) ends at the cell periphery (**Figure 3.40B**). Viral infections that depend on the microtubule cytoskeleton and its associated motor proteins can be prevented in tissue culture by drugs that selectively disrupt microtubules or their motor proteins. Viruses that must access the nucleus can be carried from the cell periphery to the MTOC when they are still enclosed by an endocytic vesicle or after the genome and any associated nucleocapsid proteins have been released into the cytoplasm. HIV, influenza, and herpesviruses are examples of viruses with a nucleocapsid or nucleocapsid components that must be transported toward the nucleus during uncoating. Once the nucleocapsid reaches the MTOC, motor proteins can subsequently carry it along a different microtubule, toward the nucleus.

The HIV nucleocapsid disassembles to at least some extent before the viral nucleic acids gain access to the nucleus through a nuclear pore. There are currently three competing theories to explain how the process occurs on a mechanistic level. In one model, the capsid disassembles very early and only a few capsomeres remain associated with the genome during transport to the nucleus. In a second model, capsid disassembly occurs more gradually during transport. The third model is that interactions between the capsid and the nuclear pore complex are needed for capsid disassembly. Experimental evidence in favor of each of these options depends in part on the method used to study disassembly.

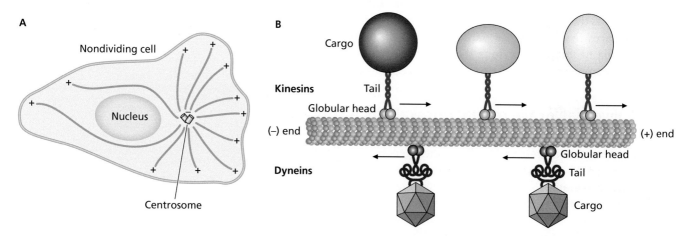

Figure 3.40 Microtubules and their associated motor proteins. (A) Organization of microtubules in an idealized cell, showing (−) ends at a microtubule-organizing center (a centrosome) and (+) ends at the periphery. (B) Kinesins and dyneins carry cargo along a microtubule. Dyneins move cargo such as viruses toward the (−) ends of the microtubule and therefore toward the nucleus.

3.33 Viruses that enter an intact nucleus must manipulate gated nuclear pores

Eukaryotic cells have a complex nuclear envelope defining the boundaries of the nucleus. The nuclear envelope has two lipid bilayers; the outer layer is contiguous with the endoplasmic reticulum (**Figure 3.41**). Macromolecules enter the nucleus through gated structures called nuclear pore complexes, which contain more than 450 different polypeptides. The overall structure of a nuclear pore complex is that of a ring with filaments that extend toward the cytoplasm, and a basket structure that extends into the nucleus. Although small molecules (<60,000 Da) can diffuse through the pores, large molecules such as virus genomes require active transport through the nuclear pore complex.

The normal process cells used to translocate proteins into the nucleus involves the protein **importin**. Importin binds to **cargo** containing a short sequence of amino acids called a **nuclear localization signal** (**Figure 3.42**). This cargo is typically a protein, such as a transcription factor, needed in the nucleus; it could also be a virus nucleocapsid that subverts the normal nuclear import cycle. Subsequently, the importin–cargo complex associates with a molecule of Ran that has guanosine diphosphate (GDP) bound in its regulatory site. The importin–cargo–Ran–GDP complex then translocates through the nuclear pore complex. Once inside the nucleus, a nuclear protein catalyzes the exchange of GDP for GTP in the Ran protein, which causes conformational changes that release the cargo and importin from the complex. The importin, associated with Ran-GTP, is then translocated back to the cytoplasm for reuse. GTP hydrolysis by Ran causes another conformational change, ultimately releasing the importin so that it can be reused.

3.34 Viruses introduce their genomes into the nucleus in a variety of ways

All eukaryotic RNA viruses except orthomyxoviruses (influenza) and retroviruses (HIV) express and replicate their genomes in the cytoplasm, whereas all eukaryotic DNA viruses except poxviruses (vaccinia and variola) express

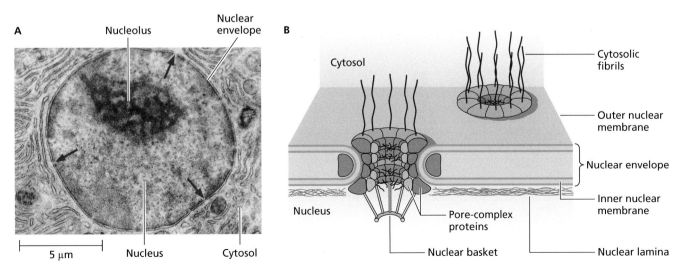

Figure 3.41 Structure of the nucleus. (A) Electron micrograph of a nucleus with pink arrows indicating nuclear pores. (B) Labeled diagram of the nuclear pore complex. (A, From Fawcett DW & Bloom W 1986. *A Textbook of Histology*, 11 ed. Saunders. With permission from Elsevier.)

Figure 3.42 Importin cycle. Proteins with a nuclear localization signal enter the nucleus through an importin-dependent process driven by GTP hydrolysis.

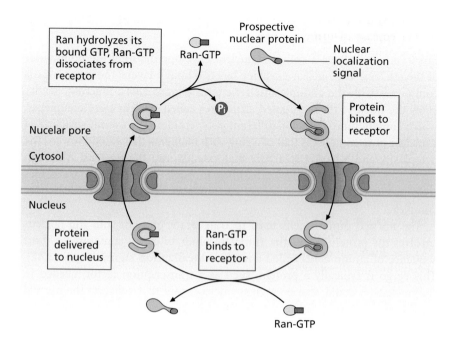

and replicate their genomes in the nucleus (**Figure 3.43**). Herpesviruses and adenoviruses are examples of dsDNA viruses (Baltimore Class I). Herpesvirus is enveloped with a complex amorphous protein containing a layer called the **tegument** between the envelope and the nucleocapsid. Herpesvirus uncoating culminates in the nucleocapsid and some associated tegument proteins docking with a nuclear pore complex. The docking causes a conformational change that releases the genome into the **nucleoplasm** (nucleus interior).

Adenovirus is a naked virus, and uncoating in this case occurs gradually during transport to the nucleus. The last step of the uncoating stage involves

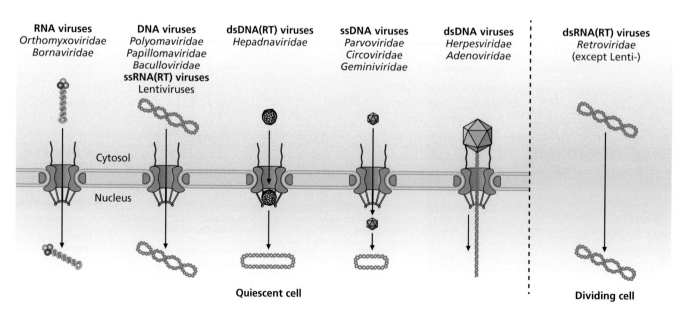

Figure 3.43 Virus entry into the nucleus. Viruses introduce their genomes into the nucleus in several different ways. In a few cases, all or part of the virion itself enters the nucleus along with the genome. (Courtesy of Philippe Le Mercier, ViralZone, © SIB Swiss Institute of Bioinformatics.)

docking with a nuclear pore complex and releasing the viral genome into the nucleoplasm. During uncoating, capsid proteins of tiny (<26 nm) parvoviruses (Baltimore Class II) rearrange, revealing nuclear localization signals that enable transport of the entire capsid into the nucleoplasm, where the genome is released. Naked polyomavirus (Baltimore Class II) virions are disassembled in the endoplasmic reticulum, and the genomes subsequently pass through a nuclear pore complex into the nucleoplasm either after export to the cytoplasm or within the endoplasmic reticulum.

Orthomyxovirus (influenza, Baltimore Class V) (−) ssRNA genomes are bound tightly with proteins, forming distinctive viral ribonucleoprotein (vRNP) complexes. These vRNPs have nuclear localization signals that result in their import across nuclear pore complexes. Retroviruses (Baltimore Class VI) such as HIV have particularly complex uncoating stages, which are discussed in **Chapter 10**. Hepadnavirus (Baltimore Class VII) virions partially enter the nuclear pore complexes, but are too large to pass completely through the nuclear pore complex. Instead, conformational changes cause the capsid to disassemble once it has entered the basket portion of the nuclear pore complex, which releases the virus genome into the nucleoplasm.

3.35 Adenovirus provides a model for uncoating that delivers the viral genome into the nucleus

Adenoviruses have naked virions with prominent spikes, enclosing a linear dsDNA genome that must enter the nucleus for expression and replication. The capsid, which ranges from 70 to 100 nm in diameter not counting the prominent fibers (spikes), is composed of 11 different proteins, some of which are present in hundreds of copies, whereas others are less abundant. The structural complexity and large size of adenovirus make penetration and uncoating among the most elaborate for naked viruses (**Figure 3.44**), but the process has nevertheless been studied in detail. After attachment, adenovirus is internalized through clathrin-coated pits and enters the endosome system. The virus spike proteins dissociate from the rest of the capsid early after internalization. Acidification of the late endosome releases several proteins including protein VI, which is cleaved by a viral protease activated by the chemical conditions in the maturing endosome. Protein VI ultimately causes lysis of the endosomal membrane and the release of the partially disassembled adenovirus capsid into the cytoplasm. In an extended penetration and uncoating stage that can last 40–60 min, the capsid moves along the microtubule cytoskeleton, using dynein motors at 1–3 μm per second, ultimately reaching a nuclear pore complex. As described earlier, docking with the

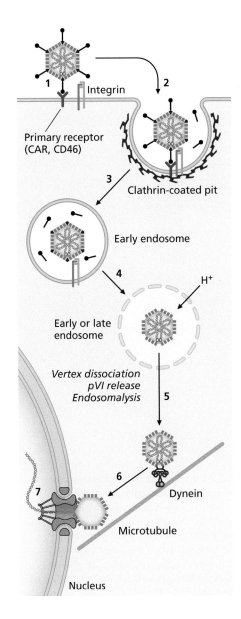

Figure 3.44 The long uncoating stage of adenovirus. The structural complexity of adenovirus coupled with the need to introduce genomic DNA into the nucleus results in an extended uncoating stage. There is irreversible attachment of the virion to its primary receptor (1). Endocytosis follows (2). Several proteins are released from the virion in the early endosome (3). A viral protease activated by conditions in the late endosome releases more proteins, including pVI, from the virion (4). The pVI protein lyses the endosomal membrane (endosomalysis), releasing the partially uncoated capsid into the cytoplasm (5). The capsid is conveyed to the nucleus along a microtubule, using dynein motor proteins (6). After docking with the nuclear pore, the genome is released into the nucleoplasm, completing the uncoating reactions (7). (Adapted from Nemerow GR et al. 2009. *Virology* 384:380–388. With permission from Elsevier.)

nuclear pore complex causes conformational changes in both the capsid and the nuclear pore complex, which together result in the release of virus DNA into the nucleoplasm.

3.36 The unusual uncoating stages of reoviruses and poxviruses leave the virions partially intact in the cytoplasm

Reoviruses and poxviruses are unusual in that the infecting virion remains largely intact inside the host cell for all or some of the gene expression stages. Reoviruses have segmented dsRNA genomes enclosed by three layers of protein (**Figure 3.45**). The innermost protein surrounding the genome segments is called the core or **inner capsid** made up of VP1 and VP3 capsomers. Another capsomer, VP6, coats this layer, forming a double-layered particle, and two more capsomers (VP7 and the VP4 spike) then coat this layer, forming a triple-layered particle. The structure unique to the double-layered particle is sometimes called the **intermediate capsid**, and the structure unique to the triple-layered particle is called the **outer capsid**. These naked viruses are internalized by endocytosis and, following endocytosis, the outer capsid disassembles and the double-layered particle is present in the cytoplasm. The exact order and nature of the changes that result in cytoplasmic localization of the double-layered particle are not clear. It is clear, however, that the double-layered particle is transcriptionally active. The double-layered particle remains intact and serves as the first site of virus mRNA production, when enzymes in the core use the genomic RNA as a template to synthesize, cap, and tail viral mRNAs. Viral mRNA production marks the end of the uncoating stage even though the double-layered particle remains intact for the remainder of the replication cycle.

Another virus that remains partially intact at the conclusion of the uncoating stage is vaccinia virus, a member of the poxviruses, which are enveloped dsDNA viruses with particularly large, complex virions (**Figure 3.46**). Vaccinia's name is derived from its historical importance as the virus first used in intentional vaccination; immunization with vaccinia protects against smallpox, caused by its very close relative variola virus, which has been

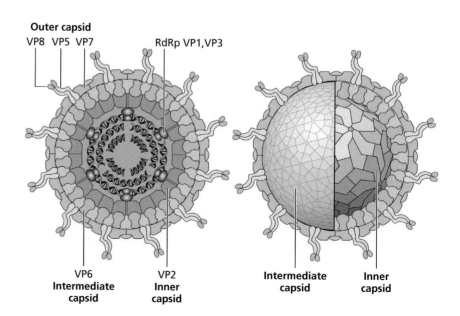

Figure 3.45 Structure of reovirus. Reoviruses have capsids with multiple concentric layers that remain partially intact during gene expression and genome replication. The structural proteins (VP) are labeled, as are the inner, intermediate, and outer capsids. (Courtesy of Philippe Le Mercier, ViralZone, © SIB Swiss Institute of Bioinformatics.)

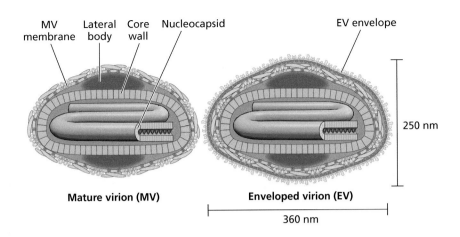

Mature virion (MV)   Enveloped virion (EV)

Figure 3.46 Structure of vaccinia virus. One of the most complex viruses known, the large poxviruses have multiple layers including nucleocapsid, core, and two membranous layers. (Courtesy of Philippe Le Mercier, ViralZone, © SIB Swiss Institute of Bioinformatics.)

eliminated from natural circulation among human populations. Vaccinia virus has two infectious forms, the **external enveloped virion** and the **internal mature virion**. Both forms contain lipids and approximately 100 different proteins, and they are shaped like a sphere flattened into a brick. The external enveloped virion is identical to the internal mature virion in structure except that it is enclosed by several layers of membrane derived from the host Golgi body. The internal mature virion has a lipid bilayer called the mature virion membrane. Internal to this are two structures called the lateral bodies. Further inside, there is a core, also called a palisade layer, and inside this, the nucleocapsid. The innermost layer contains the nucleocapsid and is surrounded by an additional layer called the core.

The uncoating stage for poxviruses includes release of the core into the cytoplasm. The core contains all enzymes necessary for synthesizing mRNA; because they replicate in the cytoplasm, poxviruses cannot rely on host proteins for transcription. The final stages of core disassembly cannot be completed until after new poxvirus mRNA and proteins are synthesized. Thus, the final steps of the uncoating stage overlap with gene expression. After the poxvirus core extrudes capped, tailed early mRNA molecules, those mRNA molecules are translated, and then an additional uncoating event occurs. Some of the proteins encoded by the early mRNA molecules catalyze further disassembly of the core, releasing the DNA genome into the cytoplasm, where it can be copied by viral enzymes and used as a template for expression of other genes (again using viral enzymes for mRNA synthesis, capping, and polyadenylation). Thus, poxvirus uncoating is extremely unusual in that it does not end until after the first round of virus gene expression.

3.37 Viruses that penetrate plant cells face plant-specific barriers to infection

Plant cells are surrounded by a particularly thick cell wall made of inert, non-living materials, such as cellulose and lignin. Plant structures such as leaves and stems are typically surrounded by a waxy **cuticle** layer that protects the plant from dehydration (**Figure 3.47**). Plant viruses must overcome both of these significant physical barriers in order to infect a new cell. A virus cannot diffuse across these structures. Once a virus has established itself in a plant cell, however, its offspring can readily spread to neighboring cells, bypassing the cell wall and cuticle altogether. Plant cells are connected to each other by cytoplasmic channels called plasmodesmata (**Figure 3.48**); a virus can spread through the plasmodesmata and thereby avoid the problem of breaching the

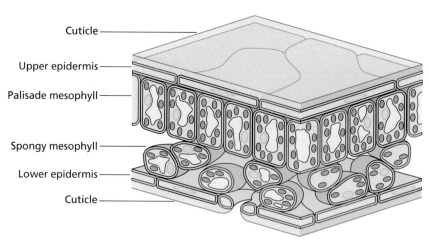

Figure 3.47 The plant cuticle. The cuticle is an abiotic waxy substance that forms a layer atop the cells in a plant, as illustrated in this diagram of the ultrastructure of a typical leaf. Because of the cuticle, viruses that fall on the surface of the leaf cannot reach the living tissue below. (Courtesy of Zephyris. Published under CC BY-SA 3.0.)

Cuticle

Upper epidermis

Palisade mesophyll

Spongy mesophyll

Lower epidermis

Cuticle

Plant cells connected by plasmodesmata

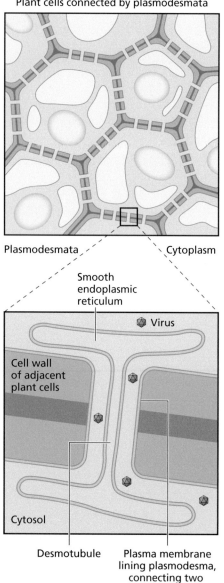

Plasmodesmata

Cytoplasm

Smooth endoplasmic reticulum

Virus

Cell wall of adjacent plant cells

Cytosol

Desmotubule

Plasma membrane lining plasmodesma, connecting two adjacent cells

Figure 3.48 Plasmodesmata. Plasmodesmata are aqueous channels that connect plant cells. They are large enough to allow the passage of some virions and of viral nucleic acids.

cell wall. Plant viruses typically encode movement proteins that make spread through plasmodesmata maximally efficient, perhaps by widening the diameter of the plasmodesmata.

How does a virus get into a plant cell in the first place, given the cuticle and cell wall? The first method is to avoid this problem entirely by being inherited through the generations, by infecting pollen, or by infecting seeds so that the seedlings become infected. Some plants, such as potatoes, have asexual reproduction and can pass a virus vertically on through the generations during asexual reproduction. This method, however, simply delays answering the question of how the virus got into the ancestral cell in the first place. Another common route of entry is through mechanical inoculation. During mechanical inoculation, a grazing animal damages the cuticle and provides an opportunity for viruses to take advantage of the physical breach.

The third method, the most common when considering all known plant viruses, is transmission during feeding of an arthropod. Arthropods are invertebrate animals such as aphids and mites that have mouth parts that pierce through the plant cuticle and cell wall in order to feed on the contents of the plant. They can thereby inject a plant cell cytoplasm directly with virions. Plant viruses may require damage to the actual plasma membrane for transmission. We reach this conclusion because plant viruses don't seem to have specific plant cell receptors for attachment, and they do not appear to enter plant cells through receptor-mediated endocytosis or other processes that result in vesicle formation.

3.38 Plant viruses are often transmitted by biting arthropod vectors

There are five types of transmission by arthropod **vectors**. **Nonpersistent transmission** occurs when a virus adheres to the mouth parts of an arthropod, which then carries the virus to a new host and injects the virus into the plant. This is analogous to touching a doorknob contaminated with rhinovirus and then inoculating yourself with the virus by rubbing your eyes.

Semipersistent transmission is similar, but the virus is found in the insect foregut for many days. During **persistent transmission**, the virus is associated with many tissues in the animal host, where it may remain for long periods of time, even as long as the entire life span of the arthropod. **Circulative persistent transmission** occurs when the virus is transported through the gastrointestinal tissues of the animal vector, ultimately reaching the salivary glands, where it can be introduced into the next plant when the animal feeds. During circulative persistent transmission, the virus is internalized by specific cells in the animal host prior to its ultimate release into the extracellular fluid of the salivary glands.

Propagative persistent transmission is more complex because the virus not only enters the gastrointestinal tract of the animal host but replicates in some of the animal's cells, thus substantially increasing the amount of virus in the arthropod host (**Figure 3.49**). As such, the plant virus is also an animal virus, able to replicate in both plant and animal cells. To attach to, penetrate into, and uncoat inside animal cells, these viruses use the same processes as all animal viruses. The infection of the arthropod cells leads to a dramatic increase in the amount of virus present in the mouth parts of the arthropod, substantially increasing the likelihood of virus transmission when the animal feeds on a new plant. The plant receptors (if any) needed for penetration and uncoating are yet to be characterized.

Propagative persistent transmission is required of all viruses that infect humans through the bite of an arthropod, which includes medically important pathogens such as yellow fever virus. Blood-borne viruses that cannot replicate in an arthropod, such as HIV and Ebola, are not transmitted by arthropods because they cannot replicate in the arthropod and so insect bites do not deliver an infectious dose of these viruses.

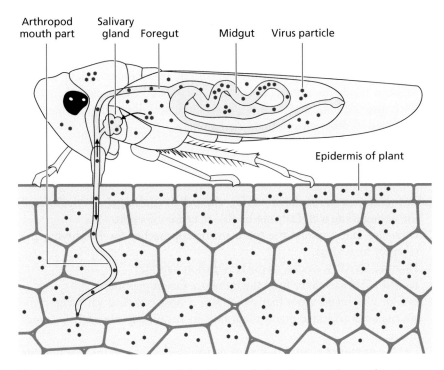

Figure 3.49 Propagative persistent transmission. Propagative persistent transmission of a plant virus is shown using an arthropod. In the animal vector, the virus can replicate in one or more organs and is ultimately passed to the salivary glands so that the infected arthropod transmits the virus to the next plant it bites.

Essential concepts

- Viruses enter the human body through one of six routes, where they gain access to permissive host cells.
- Viruses are selective in their host range and tissue tropism; much of this selectivity can be explained by the attachment stage of the virus life cycle.
- Noncovalent intermolecular forces between virus proteins and host receptor molecules mediate attachment. When present, viruses use spike proteins for attachment. Host receptors are usually proteins, or glycoproteins in the case of animal viruses.
- Host receptors can be identified through genetic, biochemical, or immunological approaches.
- Penetration is the release of the virion or parts of the virion into the interior of a host cell.
- Common ways that viruses penetrate animal cells include endocytosis, phagocytosis, or pore formation that releases the genome directly into the host cell.
- During penetration, enveloped viruses use fusion peptides to catalyze fusion with host membranes. Fusion peptides are components of virus spike proteins that are inactive until activated by a cascade of protein–protein interactions (HIV) or late endosome acidification (influenza).
- For the purposes of dividing viral activities into understandable stages, we define all reactions that occur between initial penetration and the first production of new viral protein or mRNA as belonging to the uncoating stage of the virus life cycle. Uncoating can include release of a nucleocapsid or genome from the intracellular vesicle that formed as a result of penetration and transport of the virion or parts of the virion to a particular subcellular destination such as the nucleus.
- Most DNA viruses must deliver their genomes into the host cell nucleus, typically through interactions with a nuclear pore complex.
- Plant viruses do not attach to specific plant cell receptors because of plant cell walls and cuticles, but instead enter new host cells through mechanical damage or when the mouth parts of an arthropod vector pierce through these layers. During propagative persistent transmission, plant viruses replicate inside the animal cells of their arthropod host species where they likely use the same attachment, penetration, and uncoating mechanisms exhibited by other viruses that infect animals.

Questions

1. List some possible explanations for why a virus entering the body through the wrong route will likely not be able to cause infection.
2. Compare and contrast the three major approaches for identifying the receptor for a virus.
3. How are SNAREs similar to the HIV fusion peptide?
4. How do SNAREs differ from the influenza fusion peptide?
5. For an enveloped virus that penetrates a host cell through phagocytosis, make an educated guess about how its fusion peptide likely becomes active.
6. List the groups of viruses that must deliver their genomes into the nucleus and then design an experiment to determine which depend on importin.
7. Plant viruses that use propagative persistent transmission probably attach to specific receptor molecules in some circumstances, but not in others. Explain.

8. Separation of macromolecules is possible because their properties are different from one another. Classify the four major macromolecules according to their relative mol percent of C, H, N, O, P, and S.
9. Compare and contrast a Southern, northern, and western blot.

Interactive quiz questions

In addition to the questions provided above, this edition has a range of free interactive quiz questions for students to further test their understanding of the chapter material. To access these online questions, please visit the book's website: www.routledge.com/cw/lostroh.

Further reading

Attachment and receptor identification

Li W, Moore MJ & Vasilieva N 2003. Angiotensin-converting enzyme 2 is a functional receptor for the SARS coronavirus. *Nature* 426(6965):450–454.

Liu Y, Hu G, Wang Y, Ren W, Zhao X et al. 2021. Functional and genetic analysis of viral receptor ACE2 orthologs reveals a broad potential host range of SARS-CoV-2. *Proc Natl Acad Sci* 118:e2025373118.

Meertens L, Labeau A, Dejarnac O, Cipriani S, Sinigaglia L et al. 2017. Axl mediates Zika virus entry in human glial cells and modulates innate immune responses. *Cell Reports* 18:324–333.

Raj VS, Mou H & Smits SL 2013. Dipeptidyl peptidase 4 is a functional receptor for the emerging human coronavirus-EMC. *Nature* 495(7440):251–254.

Zamorano Cuervo N & Grandvaux N 2020. ACE2: Evidence of role as entry receptor for SARS-CoV-2 and implications in comorbidities. *Elife* 9:e6139.

Zhou P, Yang XL, Wang XG, Hu B, Zhang L et al. 2020. A pneumonia outbreak associated with a new coronavirus of probable bat origin. *Nature* 579:270–273.

Penetration and uncoating

Gluska S, Zahavi EE, Chein M, Gradus T, Bauer A et al. 2014. Rabies virus hijacks and accelerates the p75NTR retrograde axonal transport machinery. *PLoS Pathogens* 10:e1004348.

Hong Y, Jeong H, Park K, Lee S, Shim JY et al. 2021. STING facilitates nuclear import of herpesvirus genome during infection. *Proc Natl Acad Sci* 118:e2108631118.

Tang Y, Woodward BO, Pastor L, George AM, Petrechko O et al. 2020. Endogenous retroviral envelope syncytin induces HIV-1 spreading and establishes HIV reservoirs in placenta. *Cell Reports* 30:4528–4539.

Zhao X, Zhang G, Liu S, Chen X, Peng R et al. 2019. Human neonatal Fc receptor is the cellular uncoating receptor for enterovirus B. *Cell* 177:1553–1565.

Envelope–membrane fusion

Chen J, Schaller S, Jardetzky TS & Longnecker R 2020. Epstein–Barr virus gH/gL and Kaposi's sarcoma-associated herpesvirus gH/gL bind to different sites on EphA2 to trigger fusion. *J Virol* 94:e01454-20.

Oliver SL, Xing Y, Chen DH, Roh SH & Pintilie GD 2021. The N-terminus of varicella-zoster virus glycoprotein B has a functional role in fusion. *PLoS Pathogens* 17:e1008961.

Structural biology

Piplani S, Singh P, Winkler, DA & Petrovsky N 2021. *In silico* comparison of SARS-CoV-2 spike protein–ACE2 binding affinities across species and implications for virus origin. *Sci Rep* 11:13063.

PDB-101 (Protein Data Bank 101) Molecular explorations through biology and medicine. Visualizing Molecules. http://pdb101.rcsb.org/browse/visualizing-molecules

Gene Expression and Genome Replication in Model Bacteriophages

Virus	Characteristics
T7	Lytic bacteriophage with dsDNA genome; waves of gene expression controlled by phage-encoded RNAP; simplest known DNA replication machine consisting of three phage polypeptides and one subverted host protein; genome replication produces concatamers through ligation and recombination.
λ	Lysogenic bacteriophage; lytic replication regulated by phage-encoded transcription antitermination and repressor proteins; many host proteins required for two-stage DNA replication from a circular template; genome replication produces concatemers through rolling-circle replication.
ΦX174	ssDNA phage that illustrates encoding many proteins in a single small genome using overlapping reading frames with alternative start codons, stop codons, and reading frames.
M13	Model for genome replication of ssDNA phages; three stages of replication via a replicative form intermediate; stages are regulated by phage proteins and result is concatemers of ssDNA genomes.
MS2	Model of (+) ssRNA phage; encodes four proteins; elaborate secondary structure controls translation.
Qβ	Model of replication in (+) ssRNA phages; phage RdRp forms a complex with two subverted host proteins, resulting in antigenome and genome synthesis.

The viruses you will meet in this chapter and the concepts they illustrate

In **Chapters 4–10**, we focus on Stages 3 and 4 of the virus replication cycle, namely the synthesis of early proteins (Stage 3) and the synthesis of late proteins and new viral genomes (Stage 4). The viral genome and the way that genome interacts with its host cell determines how viruses accomplish gene expression and genome replication. In this chapter, we examine bacteriophages with three different types of genomes; in **Chapters 5–10**, we will study gene expression and genome replication in viruses that infect eukaryotic cells. Some features of bacteriophages differ from those of viruses that infect eukaryotes because the host cells handle synthesis and translation of mRNA differently. Consequently, this chapter begins with background discussions of bacterial transcription, mRNA features, and translation. Other features of bacteriophages are similar to those in viruses that infect eukaryotes; these cases will be pointed out to provide a prelude to **Chapters 5–15**.

Bacteriophages are everywhere in research in molecular biology both historically and in the present day. For example, introductory biology courses

DOI: 10.1201/9781003463115-4

often discuss the research of Hershey and Chase, which helped to show that DNA is the genetic material rather than protein. These experiments involved the use of a double-stranded DNA bacteriophage with either radioactive protein or DNA; it was the radioactive DNA (not protein) that entered the host cells to direct the production of new offspring phages. In the present day, we often use phage proteins or nucleic acids as research tools. One example is that a common protein overexpression system relies upon a bacteriophage protein called **T7 RNA polymerase (RNAP)**.

The chapter focuses on specific bacteriophages that infect the model bacterial host *Escherichia coli*: T7, λ, ΦX174, M13, MS2, and Qβ. We begin by discussing two phages with dsDNA genomes, which led to many fundamental discoveries in virology. One of them, phage T7, causes exclusively lytic infections, whereas the other, phage λ, causes both lytic and latent infections. We continue with a consideration of model phages (ΦX174, M13, MS2, and Qβ) with other kinds of genomes such as ssDNA or (+) RNA. In all cases, these phages have been singled out for attention because of their historical importance in the development of molecular biology and virology and because they illustrate general themes in virology, such as regulation of gene expression, mechanisms of encoding multiple proteins in condensed genomes, and the use of a uniquely viral enzyme called RNA-dependent RNA polymerase. We end the chapter with a discussion of the use of phages and phage proteins in molecular biology research laboratories.

4.1 Bacterial host cell transcription is catalyzed by a multisubunit machine that catalyzes initiation, elongation, and termination

In order to express proteins and replicate their genomes, virus nucleic acids and proteins typically interact with the host transcription and translation machinery. In order to learn how bacteriophages use these systems, we will first describe transcription and translation in bacterial hosts.

Both transcription and translation contribute to gene expression. Bacterial transcription (**Figure 4.1**) uses RNA polymerase, which is made up of multiple subunits, termed σ, α, β, and β′. The purpose of the σ subunit is to confer site-specific binding upon the rest of the **holoenzyme** (ααββ′), whereas the holoenzyme has both the **helicase** activity (disrupting hydrogen bonds in double-stranded nucleic acids) and the RNA polymerase activity. After the holoenzyme associated with its **sigma (σ) factor** binds to a promoter, transcription initiation ensues. During elongation, the σ factor dissociates; termination usually involves the formation of a stem–loop structure in the newly synthesized RNA. The holoenzyme and σ factor are then free to reassociate and begin the cycle again.

Bacteria have a limited amount of holoenzyme and can make many different σ factors that all bind to the holoenzyme but target it to recognize different promoter sequences. Promoters are the DNA sequences to which RNA polymerase containing σ factor binds in order to initiate transcription. Therefore, there is competition among σ factors for RNA polymerase holoenzyme; bacteria control the relative abundance of different σ factors as one strategy for gene regulation. For convenience, the base pairs in a promoter are numbered according to their position relative to the first base pair used as a template to synthesize RNA, called +1. Base pairs downstream, in the direction of transcription, are numbered using the plus sign, so that base pair +20 is the 20th one used as a template. Base pairs upstream, opposite the direction of transcription, are numbered using a minus sign. There is no

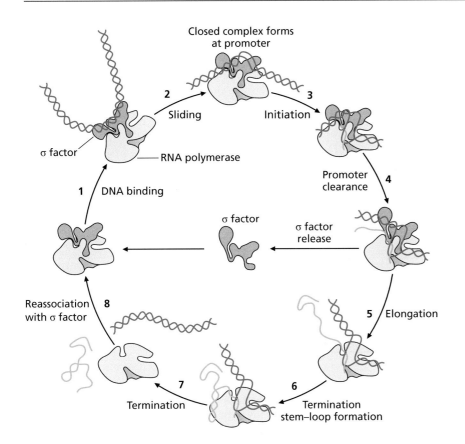

Figure 4.1 The cycle of bacterial transcription. DNA-containing bacteriophages rely on host transcription machinery for at least some of their gene expression. The bacterial RNA polymerase holoenzyme associates with a σ factor in order to bind to DNA (1). The polymerase slides along the DNA until it encounters a promoter sequence, where it binds more tightly, forming the closed complex (2). After transcription initiation (3), the polymerase clears (leaves) the promoter (4) and elongation ensues (5). During elongation, the σ factor dissociates from the complex. Near the end of the mRNA transcript, a region of self-complementarity causes formation of a secondary structure known as the termination stem–loop (6), at which point nucleotide polymerization stops (7). Transcription can be viewed as a cycle because the holoenzyme and σ factor can be used over and over again (8).

base pair assigned to the numeral 0, so −1 is immediately adjacent to +1. RNA polymerase associated with the most common *E. coli* σ factor, σ^{70}, typically binds to promoters with specific sequences at approximately −10 and −35 relative to the start site of transcription. The consensus sequence of 300 known *E. coli* promoters is TTTGCA at −35 and TATAAT at −10 with 17 base pairs of any sequence between them (**Figure 4.2**). The ability of RNA polymerase containing σ to recognize a promoter can also be modified by regulatory DNA-binding proteins that either repress transcription (by binding to a site overlapping the promoter or downstream of the promoter) or activate transcription (by binding upstream and stabilizing the interaction of RNAP to the promoter). Host cells can contain many different σ factors; those controlled by RNA polymerase associated with σ^{70} are called σ^{70}-dependent promoters. Promoters that are dependent on other σ factors have completely different consensus binding sequences.

Formation of a **stem–loop structure** in the RNA terminates transcription in bacteria (**Figure 4.3**). The stem–loop structure forms as the mRNA emerges from the RNA polymerase, thereby putting mechanical stress on the polymerase, which disrupts the normal interaction of the polymerase with the DNA. The NusA protein enhances termination by stem–loop structures. In other cases, transcription is terminated using a combination of such a stem–loop structure along with the **Rho (ρ) protein**, which has helicase activity. The ρ protein increases the frequency of termination when stem–loop termination structures form by scanning along the mRNA through the loop structure using its helicase activity, binding to rho utilization (*rut*) sequences in the RNA, and colliding with the holoenzyme. NusG enhances the termination activity of ρ.

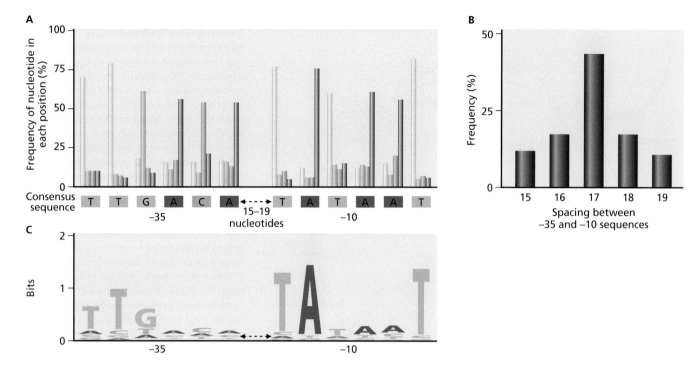

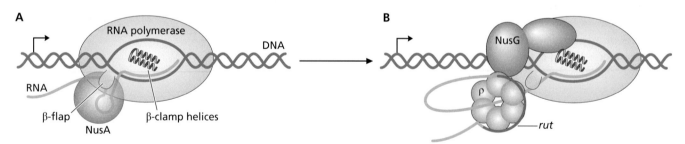

Figure 4.2 Features of a σ⁷⁰-dependent promoter in *E. coli*. Expression of some bacteriophage genes relies on σ⁷⁰-dependent promoters. (A) On the basis of a comparison of 300 promoters, the frequencies of each of the four nucleotides at each position in the promoter are given. The consensus sequence, shown below the graph, reflects the most common nucleotide found at each position in the collection of promoters. These promoters are characterized by two hexameric DNA sequences. These are the –35 sequence and the –10 sequence, named for their approximate location relative to the start point of transcription (designated +1). The sequence of nucleotides between the –35 and –10 hexamers shows no significant similarities among promoters. For convenience, the nucleotide sequence of a single strand of DNA is shown; in reality, promoters are double-stranded DNA. The nucleotides shown in the figure are recognized by σ factor, a subunit of the RNA polymerase holoenzyme. (B) The distribution of spacing between the –35 and –10 hexamers found in *E. coli* promoters. (C) A sequence logo displaying the same information as in panel A. Here, the height of each letter is proportional to the frequency at which that base occurs at that position across a wide variety of promoter sequences.

Figure 4.3 Termination of transcription in bacteria. Some bacteriophage regulatory proteins manipulate transcription termination. (A) Rho-independent termination occurs when a stable RNA stem–loop interacts with the NusA protein and with the β-flap component of RNAP to cause dissociation of the complex. (B) Rho-dependent termination occurs when the rho helicase protein assembles around the RNA, moves toward a sequence called a rho utilization (*rut*) element, and finally collides with the RNAP. The NusG protein interacting with ρ and the β clamp helices region of RNAP enhances rho-dependent termination. (From Santangelo TJ & Artsimovitch I 2011. *Nat Rev Microbiol* 9:319–329. With permission from Springer Nature.)

4.2   Bacterial host cell and bacteriophage mRNA are typically polycistronic

The mRNA produced by bacterial transcription is not processed; it does not have a 5′ cap, a poly(A) tail, or any introns—but it does have at least one **ribosome-binding site** adjacent to a start codon (AUG) (**Figure 4.4**). Bacterial mRNAs can encode more than one protein by having the sequence

```
            lacZ                    lacY                    lacA
        ┌──────────┐          ┌──────────┐          ┌──────────┐
      ★ AUG      UGA        ★ AUG      UAA        ★ AUG      UAG
5' ⓅⓅⓅ                                                              3'
```

Figure 4.4 Features of bacterial mRNA. Bacteriophage and bacterial mRNA, such as this one encoding the three proteins of the *lac* operon, can be polycistronic, with ribosome-binding sites upstream of and adjacent to each start codon (AUG). Stop codons are also indicated (UGA, UAA, or UAG). The 5′ end has three phosphate groups and the 3′ end has a free OH.

for multiple proteins encoded one after the other on the mRNA, each separated from its predecessor by a stop codon, a small amount of noncoding RNA, and another ribosome-binding site–start codon pair. When an mRNA molecule encodes two or more proteins in this way, the mRNA is called **polycistronic**. Bacteriophage mRNA is also typically polycistronic. Polycistronic mRNA may be advantageous because it allows proteins with a common function to be expressed in a physically coordinated way from the same mRNA. For bacteriophages, structural proteins are often encoded by polycistronic operons.

4.3 Transcription and translation in bacterial host cells and bacteriophages are nearly simultaneous because of the proximity of ribosomes and chromosomes

The second step of gene expression is translation, when a protein is synthesized using mRNA to specify the order of amino acids. Because bacterial cells do not have a nucleus enclosing their DNA, bacterial transcription and translation occur at nearly the same time in the cytoplasm and are thereby said to be coupled (**Figure 4.5**). As soon as the mRNA emerges far enough from RNA polymerase for the most 5′ ribosome-binding site to be exposed, the ribosome binds and begins translating. Because transcription and translation are so tightly coupled in both time and physical space, most bacterial gene expression is regulated by controlling the first step in gene expression, the amount of transcription. Thus, bacteriophages are experts at manipulating transcription initiation, elongation, and termination.

4.4 Bacterial translation initiation, elongation, and termination are controlled by translation factors

In all cells, translation occurs in the three stages of initiation, elongation, and termination (**Figure 4.6**). Viruses that interfere with host translation typically do so by affecting initiation. Translation initiation refers to all molecular events needed for the ribosome to assemble around a ribosome-binding site (also known as a **Shine–Dalgarno sequence**) on the mRNA, with the initiation tRNA carrying the amino acid methionine bound to the correct start codon. This ribosome-binding site does not have to be near the 5′ end; polycistronic messages encode many proteins and each of the coding sequences has its own ribosome-binding site. The assembled ribosome has three internal sites that can be occupied by tRNA. They are, from 5′ to 3′ relative to the mRNA, the E, P, and A sites. The assembly of the ribosome around the start codon sets the frame for translation, determining which set of three non-overlapping nucleotides will be used as codons for the rest of the coding sequence.

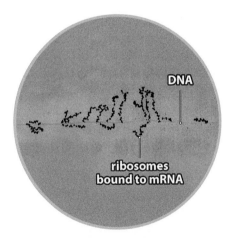

Figure 4.5 Bacterial transcription and translation are coupled in time and space. The DNA in the center is being transcribed by RNAP molecules that are too small to be visualized in this micrograph. RNA emerging from the polymerase is almost immediately bound by ribosomes, which can be seen in the micrograph. (From Miller OL Jr. et al. 1970. *Science* 169:392–395. With permission from American Association for the Advancement of Science.)

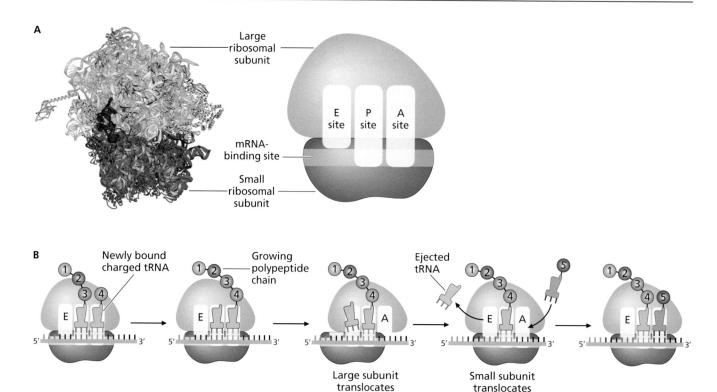

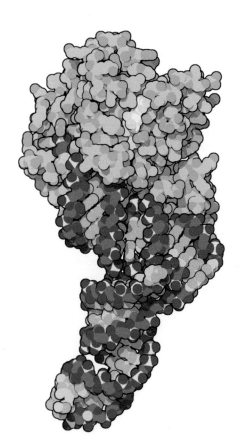

Figure 4.6 Ribosomes and translation. Viruses rely on host ribosomes for translation. (A) Ribosome with A, P, and E sites. (B) During protein synthesis, the addition of new amino acids occurs in a stepwise fashion as tRNAs bind to the codons occupying the P and A sites. (A, From Hu B et al. 2013. *Science* 339:576–579. With permission from American Association for the Advancement of Science.)

Once initiation has occurred, translation elongation can ensue, ultimately followed by translation termination to release new polypeptides. An important aspect of translation termination (cessation) is that it is caused by the ribosome reaching one of three stop codons, which are recognized not by tRNA molecules but by protein **release factors**. Release factors mimic the shape of tRNA molecules; they bind to stop codons inside a ribosome. The binding of a release factor in the A site of the ribosome causes the peptide bound to the tRNA at the P site to be released. In the case of bacterial translation, the ribosome is destabilized but does not always dissociate entirely from the mRNA. Instead, it scans down the mRNA, moving along the mRNA without dissociating, and if it encounters another ribosome-binding site and AUG start codon within 50 nucleotides, translation will reinitiate and a second protein can be synthesized.

Translation requires the action of many cellular proteins in addition to those that actually comprise the ribosome. These proteins are called **initiation factors**, **elongation factors**, and **termination factors**. Viruses manipulate or mimic some of these factors, so it is important to understand their normal cellular function. The elongation factors EF-Tu, EF-T, and EF-G provide an example for how translation factors in general work. During elongation, a GTPase protein called EF-Tu carries the aminoacyl-tRNA (**Figure 4.7**)

Figure 4.7 Bacterial EF-Tu translation factor bound to tRNA. The EF-Tu protein is in blue and the tRNA is in red with the phosphates of the backbone marked in yellow. The yellow molecule bound to EF-Tu is GTP. (From PDB-101 [http://pdb101.rcsb.org]. Courtesy of David Goodsell. doi: 10.2210/rcsb_pdb/mom_2006_9.)

into the free site of the ribosome. If the codon–anticodon match is correct, EF-Tu catalyzes the hydrolysis of GTP, releasing P_i but retaining GDP in its catalytic site. Exit of P_i enables the EF-Tu to release the aminoacyl-tRNA so that the peptide bond can form. After the EF-Tu leaves the ribosome, a second elongation factor, EF-T, stimulates the release of GDP from the EF-Tu so that the EF-Tu can bind to a new molecule of GTP and continue to participate in translation. Finally, the EF-G elongation protein is also needed to catalyze ribosomal translocation at the expense of more GTP.

4.5   Bacteriophages, like all viruses, encode structural and nonstructural proteins

Now that we have described bacterial host cell gene expression and the processes that bacteriophages must exploit in order to induce their host cells to synthesize new viruses, we will introduce the proteins encoded by virus genomes. All viruses, regardless of the host cells they infect, must encode at least one **structural protein**, a protein that is by definition found in the virion; in the simplest cases, structural proteins have no enzymatic activity. A capsomere is the quintessential example of a structural protein. In actual fact, even the simplest of viruses encode several structural proteins. The smallest known naked icosahedral viruses encode at least three capsomeres, whereas enveloped viruses always encode a transmembrane envelope spike protein in addition to encoding at least one capsomere. **Nonstructural proteins** are any proteins that are not found in the virion, as well as enzymes that are packaged into the virion in order to serve a catalytic role upon uncoating. Nonstructural proteins might form a scaffold so that the virion can assemble; interfere with host cell processes so that the virus replication cycle can proceed; catalyze synthesis of viral nucleic acids; process viral proteins, for example by proteolysis; or regulate viral gene expression. For viruses that infect animals, nonstructural proteins might manipulate the immune system, increase spread to new animal hosts, or cause symptoms of illness.

Most viruses express their nonstructural and structural proteins at different times. In general, the nonstructural proteins are expressed as **early proteins**, soon after the virus genome has been uncoated, whereas the structural proteins are expressed as **late proteins**, typically after virus genome replication has begun. The major functions of the nonstructural proteins are to take over the host cell and to cause or enhance the production of viral nucleic acids and proteins, which explains why these proteins are usually expressed early. The structural proteins needed to assemble virions are expressed last, when there are replicated genomes to be encapsidated. Sometimes a nonstructural viral protein with enzymatic activity is expressed late. In that case, the enzyme is typically packaged with the virion and used during penetration and uncoating or during the production of the very first virus nucleic acids. The expression pattern of a viral protein therefore provides a clue as to its general function. Early proteins are almost always enzymes, gene expression regulators, or proteins that interfere with host processes, whereas late proteins are almost always scaffolds, structural proteins, or enzymes needed for penetration and uncoating. There is also a correlation between relative abundance and the general function of a viral protein. Nonstructural enzymes and regulators are expressed at low levels compared with structural proteins. Enzymes and regulatory proteins can be used over and over again, whereas structural proteins must assemble together to form infectious virions.

As is almost always the case in biology, there are exceptions to every rule because evolution proceeds through a series of themes with variations. Viruses, which have very rapid reproduction and have been evolving for a very long time, are particularly diverse, and so it is not surprising that there are exceptions to typical patterns. Although it is common for early and late gene expression to be separated, there are exceptions in viruses with particularly small genomes, defined as those in which almost every nucleotide is used to encode one or more proteins. The genomes in these cases are so condensed that there is no room for regulatory regions that could alter mRNA production. Small viruses usually still control the amount of structural versus nonstructural proteins by manipulating translation in a variety of ways, which can be illustrated using the bacteriophages ΦX174, M13, and MS2. Even here, however, there can be exceptions. An example is provided by the simplest (+) ssRNA viruses that infect animal cells, such as the medically important poliovirus.

4.6   The T7 bacteriophage has naked, complex virions and a large double-stranded DNA genome

Our discussion of bacteriophage gene expression and replication begins with the dsDNA phages, which have large genomes of the same type as their host cells. The replication cycles of dsDNA bacteriophages have many steps in common with each other (**Figure 4.8**), described here to set the stage for the detailed discussions of gene expression and genome replication in T7 and λ phages. After uncoating, the dsDNA genome is used by host enzymes to synthesize the first viral mRNA molecules, which are then translated, resulting in production of early proteins. These early nonstructural proteins interfere with host processes that could stop virus replication and stimulate expression of the next set of virus proteins. This second group often includes nonstructural proteins required for genome replication. Usually, there is a third wave of gene expression that occurs at the same time as genome replication, producing the late proteins, which are usually structural. The new genomes are usually found as **concatemers**, which are DNA molecules with genomes attached covalently to one another, one genome after another. Viral proteins process the concatemers into new individual genomes during maturation (the term bacteriophage specialists use to specify the assembly of virion components).

As an example of a typical Baltimore Class I bacteriophage, this chapter focuses on T7, a member of the family *Podoviridae* in the order *Caudovirales* (**Figure 4.9**). All bacteriophages in the *Caudovirales* order have complex

Figure 4.8 Overview of dsDNA bacteriophage gene expression and genome replication. Uncoating releases the phage genome into the cytoplasm (1). Immediate-early proteins are transcribed (2) and translated (3). Immediate early proteins cause production of delayed early mRNA (4) and proteins (5). Delayed-early proteins result in production of new genomes (6), which is followed by transcription (7) and translation (8) of late proteins. Late proteins and genomes assemble into new phages (9, 10), and the host cell bursts open releasing offspring phages (11). The bacterial chromosome has been omitted from the drawing; the chromosome occupies the vast majority of the cell interior.

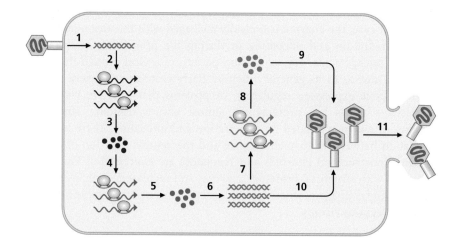

naked virions composed of an icosahedral head, which contains the double-stranded DNA genome, and a helical tail (*cauda* in Latin) that is necessary for attachment and penetration. Podoviruses such as T7 have the shortest tails in the group and have six fibers protruding from the top of the tail, so that the virus resembles the historic American lunar lander of the Cold War's space race. The head is made up of six different structural proteins, and the tail is made of four. The tail fibers are themselves composed of another different structural protein.

Podoviruses such as T7 have large double-stranded DNA genomes typically between 20 and 65 kbp; the genomes usually encode about one protein for every 1,000 bp. Viruses such as T7 that encode more than 20 proteins are considered large. The T7 genome is 40 kbp long with **terminal redundancy**, meaning that the first 160 bp are identical to the last 160 bp. In general, the physical and sequence characteristics at the termini of any linear virus genome provide clues as to the mechanisms of genome replication and **genome packaging**, the stage of maturation when the genome assembles with capsomeres (see **Chapter 11**). Inside the T7 virion, the genome is linear and is packaged in such a way that one particular end always enters the host cell first (**Figure 4.10**).

4.7   Bacteriophage T7 encodes 55 proteins in genes that are physically grouped together by function

Bacteriophage T7 encodes about 55 proteins. Strikingly, genes encoding proteins with similar function are clustered together in the genome. For example, genes that encode enzymes that synthesize, degrade, or modify nucleic acids are clustered together, and the end that enters the host last encodes all of the structural proteins. When genes are functionally clustered together in any virus genome, the implication is that this physical proximity is important for viral gene expression (see Figure 4.10). In the case of T7, genes are clustered together in polycistronic operons, where a single mRNA physically encodes several proteins with similar functional roles, one after the other, thus tightly coordinating their expression.

4.8   Bacteriophage T7 proteins are expressed in three major waves

The **T7 bacteriophage** was originally selected for study because of its remarkable ability to produce hundreds of progeny in less than 30 min. The short time course for its reproduction in combination with the presence of hundreds of proteins per infectious virion indicates that this virus must take

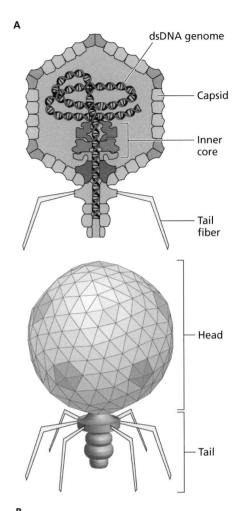

A

dsDNA genome

Capsid

Inner core

Tail fiber

Head

Tail

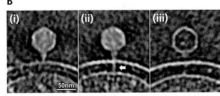

B

(i)   (ii)   (iii)

50nm

Figure 4.9 T7 virion. (A) Labeled diagram of the virion. (B) Electron micrograph of (i) virion attached to host, (ii) injecting host with genome, and (iii) after injection. The arrow in (ii) shows the tail of the phage burrowing through the bacterial cell wall. (A, Courtesy of Philippe Le Mercier, ViralZone, © SIB Swiss Institute of Bioinformatics. B, From Hu B et al. 2013. *Science* 339:576–579. With permission from American Association for the Advancement of Science.)

Class III genes   Class II genes   Class I genes

TR   TR

T7 RNAP

Figure 4.10 T7 genome. Genes in the 40,000-bp, double-stranded linear genome of T7 are grouped by function and by time of expression, with Class III late genes on the left and Class I early genes on the right. The T7 RNA polymerase gene is indicated near the right end, which enters the host cell first. The genome has terminal redundancy (indicated by the initials TR), meaning that the first 160 base pairs are identical to the last 160 base pairs.

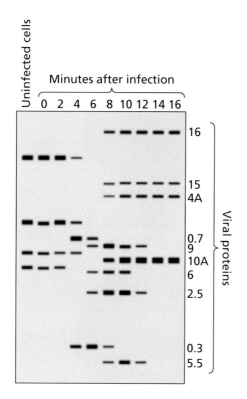

Figure 4.11 Pulse analysis of T7 gene expression. Idealized pulse labeling experiment in which a T7 infection was allowed to proceed. Every 2 min, the cells were pulsed with a radioactive amino acid, thereby making proteins synthesized during the pulse radioactive, and then killed to collect their proteins. After sodium dodecyl sulfate–polyacrylamide gel electrophoresis (SDS-PAGE), the gel was exposed to X-ray film to detect the radioactive bands. The uninfected cells (left lane control) show those proteins normally synthesized by healthy *E. coli* cells. The rest of the gel shows the proteins synthesized during a 2-min pulse starting at the number of minutes postinfection as indicated by the lanes. For example, the lane labeled 4 comes from cells infected for 4 min, pulsed for 2 min, and then killed to collect all their proteins and separate them by electrophoresis. Viral proteins have numerical names as indicated on the right. Three waves of viral gene expression are visible based on the timing of the first time that a protein can be detected. For example, the two proteins expressed during the first wave are 0.7 and 0.3. The 0.7 protein is the viral RNA polymerase.

over the host's gene expression machinery very effectively. One of the oldest experiments to ever use gel electrophoresis to examine protein synthesis used *E. coli* bacteria infected with T7 (**Figure 4.11**). Using a technique called **pulse labeling**, the infected bacteria were provided with a pulse of radioactive ^{35}S-methionine for a few minutes, and then all of the proteins were collected from the cells. Methionine, the amino acid corresponding to the AUG start codon, contains sulfur, so that all proteins synthesized in the presence of this radioactive methionine become radioactive (labeled) during the pulse. Proteins that had already been synthesized prior to the pulse would not become radioactive because the existing amino acids polymerized into a folded protein do not swap with free amino acids in the cytoplasm. The next step of the experiment is to run all of the proteins out on a gel, and finally to use X-ray film to detect exclusively the radioactive proteins.

Two different phenomena may be observed in the 4-min lane (see Figure 4.11). First, many host proteins that were normally synthesized in uninfected cells are being synthesized in smaller amounts, as indicated by the lighter signal from those proteins. Second, the cells have begun to synthesize proteins unique to the infected cells. These include proteins 0.3, 0.7, and 16, named this way for historic reasons. By 6 min postinfection, none of the newly synthesized proteins are found in uninfected cells. At 8 min postinfection, the cells are synthesizing only nine proteins, all of which are unique to infected cells. Further analysis of the proteins found uniquely in infected cells showed that the proteins are all encoded by the T7 genome. These observations and similar ones led to the insight that in general the timing of viral gene expression is important for viral replication.

The gel in Figure 4.11 reveals that there is a group of T7 proteins synthesized very soon after infection, during the 2- to 6-min window. A second group of proteins is synthesized mainly between 6 and 12 min postinfection. A final group of proteins is synthesized after 8 min of infection, and translated very abundantly in cells after 10 min of infection. In virology, each group of genes that shares the timing of their expression is called a **wave**. For T7, then, there are three waves of expression. The proteins and the genes that encode them can be referred to as Class I, Class II, and Class III based on the timing of their expression. This example illustrates that, for some viruses, there can be more than one group of early or late proteins; strictly speaking most virologists use the term late to describe only those waves of gene expression that occur at the same time as or later than the initiation of genome replication.

4.9   The functions of bacteriophage proteins often correlate with the timing of their expression

T7 genes expressed earliest tend to be used to counteract any host antiphage defenses and to bias transcription and translation toward viral genes. Other Class I and Class II genes include those needed for genome replication. The most abundant Class III proteins are structural proteins and other proteins needed for genome encapsidation, or enzymes needed to be introduced into the next host cell along with the genome. These patterns observed with T7 generally hold true for other phages that have distinct waves of gene expression. It is also typical for early proteins to be produced in smaller amounts than late proteins. In most cases, early proteins are expressed using the original infecting genome as the template for transcription, which results in smaller amounts of these proteins being produced. Because these proteins are usually regulatory or enzymatic, they are not needed in abundance. Late proteins, in contrast, are often transcribed from the many copies of newly replicated genomes, because they are needed in such high amounts that many DNA templates are required to manufacture enough of them for the offspring phages to assemble. Once the Class III proteins become abundant and phage heads begin to assemble, the new genomes are packaged into the phage heads, which sequesters the genomes away from the transcription and translation machinery and thereby ends Stage 4 of the virus replication cycle and begins Stage 5, assembly.

4.10   Bacteriophage T7 gene expression is highly regulated at the level of transcription initiation

The three precise waves of T7 gene expression occur because of specific regulatory molecular interactions. The first regulatory events occur because the end of the phage genome encoding the Class I genes always enters the host cell first, which exposes the four Class I promoters to host RNA polymerase. The Class I promoter sequences can be aligned using the +1 nucleotide; the alignment reveals that they have sequences recognized by host RNAP containing σ^{70} (**Figure 4.12**). The ideal binding site for such RNAP is the sequence TATAAT at −10, separated by 17 base pairs from the sequence TTGACA at −36. When the T7 genome enters, there are approximately 2,000 $\alpha\alpha\beta\beta'$ RNAP holoenzymes competing for 650 host σ^{70}-dependent promoters. Under these conditions, the Class I promoters are able to compete well with host cell promoters because the viral ones have nearly ideal −10 and −35 sequences for RNAP-σ^{70} binding. Transcription of the Class I genes provides the mechanical force required to bring the rest of the genome into the host cytoplasm, which means that the final stages of uncoating require some gene expression to occur first. Poxviruses are eukaryotic viruses that similarly require some viral gene expression prior to completion of uncoating.

The Class II promoters have very little sequence similarity to Class I promoters. Instead, they have sequences similar to

A   **Class I promoters transcribed by host RNA polymerase**

Promoter	−40	−30	−20	−10	+1	+10
A1	AAAAGAGTA**TTGACT**TAAAGTCTAA-CCTATAG**GATACT**TACAGCCATACGAGAGG					
A2	AAACAGGTA**TTGACA**ACAAGAAGTAACATGGAG**TAAGAT**ACAAATCGCTAGGTAAC					
A3	ACAAAACGG**TGGACA**ACATGAAGTA-AACACGG**TACGAT**GTACCACATGAAACGAC					
Consensus	A A TtGACa			tA aT		

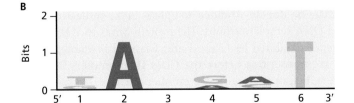

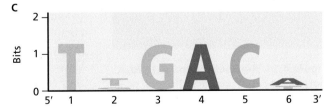

Figure 4.12 The Class I promoters found in bacteriophage T7. Bacteriophage T7 has three classes of promoters. (A) Class I is transcribed by host RNAP containing the σ70 subunit, which targets the polymerase to the −10 and −35 sequences, which have been underlined. (B) Sequence logo for the −10 sequence at the three Class I T7 promoters. Each position consists of a stack of letters corresponding to the nucleotides at that position. The total height of all letters combined indicates the amount of sequence conservation (bits) at that position, and the height of each individual letter indicates its frequency relative to the total number of different nucleotides at that position. (C) Sequence logo for the −35 sequence at the three Class I T7 promoters. (Sequence made using WebLogo [http://weblogo.berkeley.edu/logo.cgi].)

Figure 4.13 The Class II and Class III promoters found in bacteriophage T7. Bacteriophage T7 has three classes of promoters. (A) Both Class II and Class III are transcribed by phage T7 RNAP, which binds to a 23-bp sequence centered at −11.5. (B) Sequence logo for the sequence of Class II promoters. The total height of all letters combined indicates the amount of sequence conservation (bits) at that position while the height of each individual letter indicates its frequency relative to the total number of different nucleotides at that position. Class III promoters are completely identical at the 23-bp motif required for T7 RNAP binding. (Sequence made using WebLogo [http://weblogo.berkeley.edu/logo.cgi].)

A   **Class II promoters transcribed by viral RNA polymerase**

Promoter	−20	−10	+1	+10
Φ1.1A	AACGCCAAAT**CAATACGACTCACTATAGAGGGA**CA			
Φ1.1B	TTCTTCCGGT**TAATACGACTCACTATAGGAGGA**CC			
Φ1.3	GGAC**TGTAATACGACTCAGTATAGGGA**GAAT			
Φ1.5	GA**AGTAATACGACTCACTAAAGGAG**GTAC			
Φ1.6	AGTTAACT**GGTAATACGACTCACTAAAGGAG**ACAC			
Φ2.5	TGGTCACGCT**TAATACGACTCACTATTAGGAGA**AGA			
Φ3.8	AGCACC**TAATTGAACTCACTAAAGGGAGA**CC			
Φ4c	GAAG**CAATCCGACTCACTAAAGAGAGA**GA			
Φ4.3	CGTGGA**TAATTAATACGACTCACTAAAGG**AGACA			
Φ4.7	CCGACTGAGA**CTATTCGACTCACTATAGGAGAT**AT			
Consensus	taATacgactcActataggg aga			

Class III promoters transcribed by viral RNA polymerase

Promoter	−20	−10	+1	+10
Φ6.5	GTCCCTAAAT**TAATACGACTCACTATAGGGAGA**TA			
Φ9	GCCGGGAATT**TAATACGACTCACTATAGGGAGA**CC			
Φ10	ACTTCGAAAT**TAATACGACTCACTATAGGGAGA**CC			
Φ13	GGCTCGAAAT**TAATACGACTCACTATAGGGAGA**AC			
Φ17	GCGTAGGAAA**TAATACGACTCACTATAGGGAGA**GG			
Consensus	TAATACGACTCACTATAGGGAGA			

B

Figure 4.14 The bacteriophage T7 RNA polymerase. The protein is shown in pink, the DNA is in red and blue, and the newly synthesized mRNA is in green.

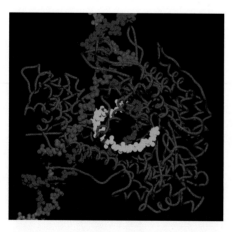

TAATACGACTCACTATAGGGAGA between −19 and +5 of transcription (**Figure 4.13**). Most of the Class II promoters match this consensus sequence at 21/23 positions. The Class III promoters, meanwhile, have this exact sequence with no variations at all. The lack of a typical −10/−35 signature implies that these promoters are not transcribed by the usual host RNAP containing σ⁷⁰. In fact, these promoters are transcribed by an entirely different enzyme known as T7 RNA polymerase (**Figure 4.14**).

The Class I protein 1, expressed almost immediately upon genome penetration of the host cell, is this remarkable T7 RNAP enzyme. In a single polypeptide, it accomplishes all of the functions ascribed to host holoenzyme and the σ factor. T7 RNA polymerase does not require any accessory proteins to bind to DNA, melt the double-stranded template, and synthesize RNA complementary to a DNA template strand. The protein binds to DNA with the consensus sequence exhibited by Class II and Class III promoters. The Class II promoters are transcribed before the Class III promoters because they enter the cell first, before an appreciable number of Class III promoters become available. The third wave of gene expression excludes Class II promoters because the Class III promoters are better binding sites for the T7 RNAP enzyme, so when Class III promoters are available and abundant, particularly after DNA replication results in many available templates, the T7 RNAP uses them preferentially. Class I promoters are no longer transcribed during Class II and Class III expression because of the collective action of Class II genes, which selectively and thoroughly blocks both host and viral RNAP- σ⁷⁰ initiation.

4.11   Bacterial host chromosome replication is regulated by the DnaA protein and occurs via a θ intermediate

In order to appreciate the ways in which bacteriophages such as T7 replicate their double-stranded DNA genomes, it is important to understand the details of bacterial DNA replication. Just as bacteriophage proteins such as T7 RNAP can sometimes substitute completely for host proteins normally needed for transcription, some bacteriophages rely on viral proteins for certain aspects of DNA replication. Early molecular biologists anticipated an elegant DNA replication process because of the obvious simplicity of complementary strands of DNA that could each serve as the template for synthesis of new double-stranded DNA identical in sequence to the first molecule. Detailed research, especially in the model bacterium *E. coli*, however, soon revealed that DNA replication is anything but elegant. Because cellular replication is complex, there are many opportunities for viral proteins to interact with or substitute for specific host replication proteins.

Bacterial chromosomes have 1,000,000–5,000,000 bp of circular double-stranded DNA, and typically encode 1,000–5,000 proteins; the rule of thumb for bacterial chromosomes is that there is approximately one gene for every 1,000 bp of DNA. Replication of these chromosomes is highly regulated; bacteria replicate their chromosomes only when they have perceived that environmental conditions are sufficiently favorable for the cell to have enough energy to double in size and replicate the entire chromosome. A bacterial chromosome has only one **origin of replication (*ori*)**, which has a particular sequence. When the cell senses that environmental conditions are favorable for reproduction, replication is initiated at the origin by a protein called **DnaA**, which binds selectively to certain sequences surrounding the origin. Viral genomes also have origins of replication and must interact with origin binding proteins analogous to DnaA to initiate DNA replication. Bacterial chromosome replication continues until the replication forks meet opposite the origin, at a site called *Ter* for termination (**Figure 4.15**). At the termination site, the Tus proteins allow one replication fork to continue while the other stops, in order to copy the chromosome exactly one time. The result is usually two closed **catenated** circles of double-stranded DNA, resembling two links of a metal chain, which must be resolved into two non-overlapping structures that can be segregated into offspring cells. **Topoisomerase** catalyzes this **decatenation**.

The result of DnaA activity is the assembly of two replication forks at the origin, each of which includes a region of single-stranded DNA (**Figure 4.16A**). Although the massive DNA replication factory is probably stationary so that the DNA being copied moves through it, it is mentally convenient to discuss the direction of replication fork movement. The replication forks move away from the origin around the circular chromosome. The appearance of chromosomes undergoing bidirectional replication and visualized using electron microscopy has led to the name **theta (θ) replication**, because during replication the circular chromosomes resemble the Greek letter θ (see Figure 4.15). One strand of the template DNA on each side is copied continuously on the leading strand, in which the 3′ OH site of nucleotide addition moves in the same direction as the replication fork. The other strand of the template DNA on each side is copied discontinuously through Okazaki fragments on the lagging strand, because as the replication fork moves the 3′ OH of the newly synthesized strand gets further and further away from the replication fork.

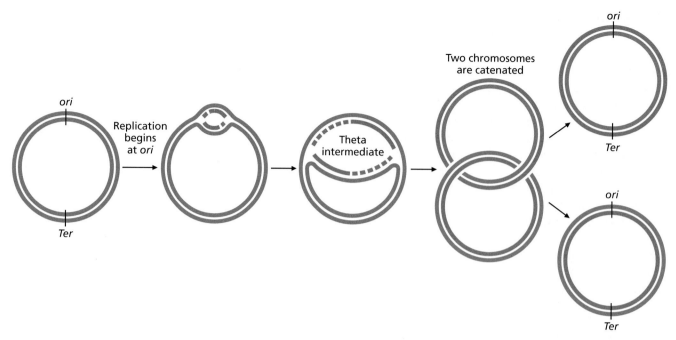

Figure 4.15 Bidirectional theta (θ) replication of a bacterial chromosome. Replication of bacterial chromosomes begins at a sequence called *ori*, and the *ori* sequence is required for replication to occur. As the two replication forks proceed around the template, the partly replicated chromosomes are known as a theta intermediate because the shape is similar to the Greek letter θ. Replication ends at the *Ter* (termination) sequences. The two offspring chromosomes are intertwined or catenated, which cells resolve using dedicated enzymes and the *Ter* sequences, so that the two chromosomes can be segregated into different offspring cells.

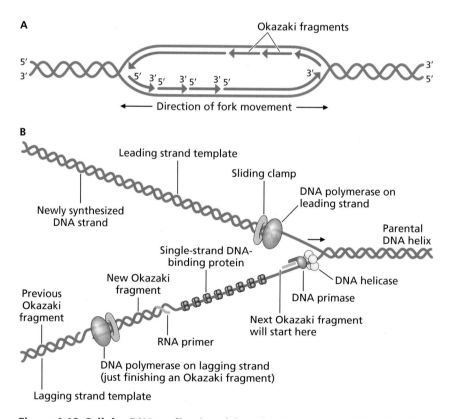

Figure 4.16 Cellular DNA replication. (A) Replication proceeds bidirectionally, with two forks moving in opposite directions. (B) Many proteins are required for bacterial DNA replication. (From Alberts B et al. 2013. *Essential Cell Biology*. Garland Science. With permission from W.W. Norton.)

4.12   Many bacterial proteins are needed to catalyze chromosome replication

Many proteins collaborate to form a replication machine at each fork (**Figure 4.16B**). These proteins include not only a DNA polymerase to copy the DNA, but also helicase, single-strand binding protein, primase, a sliding clamp, a clamp loader, ligase, and topoisomerase (**Table 4.1**). Viral replication sometimes relies upon host proteins for these functions, but in other cases the viruses provide their own proteins to substitute for one or more of these factors. In order to provide a template for DNA polymerase, a helicase enzyme melts the hydrogen bonds holding the template strands together. **Single-strand binding proteins** keep the strands separated long enough for them to be used as a template. **Primase** synthesizes short RNA primers, which are necessary because of the peculiarities of the DNA polymerase III enzyme.

Table 4.1 Proteins needed for bacterial DNA replication.

Protein	Function
Helicase	Breaks hydrogen bonds to form single-stranded templates.
Single-strand binding protein	Prevents template strands from re-annealing.
Primase	Synthesizes short RNA primers.
DNA polymerase III	Copies DNA; has 3′-to-5′ exonuclease editing function as well.
Sliding clamp	Holds DNA polymerase onto the DNA, increasing processivity.
Clamp loader	Loads sliding clamp onto DNA.
DNA polymerase I	Replaces RNA primers with DNA.
Ligase	Synthesizes sugar–phosphate linkages between adjacent Okazaki fragments.
Topoisomerase	Relieves physical stress on DNA from unwinding by helicase.

After these first events, **DNA polymerase III** copies the DNA template to synthesize a complementary strand of DNA. The steps in replication are easiest to understand when considering the leading strand, which is the new DNA synthesized in the same direction as replication fork movement (see Figure 4.16B). The primase enzyme synthesizes a short segment of complementary RNA using a DNA template. This enzyme can bind to a single-stranded template and begins synthesis of nucleic acids *de novo*, without a primer. All DNA polymerases require a primer, so the action of primase is essential for DNA replication to occur. A protein called the **sliding clamp** encircles the DNA template and also binds to DNA polymerase at the replication fork, keeping that DNA polymerase associated with the template for thousands of nucleotides before it dissociates from DNA. The term for the ability to remain associated with a template for a large number of bases is **processivity**. A **clamp loader** is needed to load the clamp onto the DNA in the first place, because the clamp has to move from an open configuration to a closed one that completely encircles the template in order to function. DNA polymerase III is the specific DNA polymerase that then assembles onto the clamp and DNA and copies the DNA. In order to synthesize DNA, DNA polymerase III requires a primer that is hydrogen bonded to a template and that provides a free 3′ OH opposite a single-stranded DNA template. The enzyme cannot synthesize DNA unless these conditions are met. The RNA synthesized by primase serves as the primer for DNA polymerase III to begin synthesizing DNA complementary to the template.

The situation is even more complex on the lagging strand, in which the growing 3′ end of the new DNA is moving in the opposite direction to the replication fork (see Figure 4.16B). The template DNA is drawn out in loops that are folded back on the fork so that the replication factory can copy each segment of template as the replication fork moves. Every time a new Okazaki fragment is synthesized, the clamp loader has to load a new clamp onto the DNA. DNA polymerase III then copies the template until it collides with the previous Okazaki fragment and dissociates from the DNA. After the replication fork has moved some distance away, **DNA polymerase I** binds to the Okazaki fragments and, using **5′ to 3′ exonuclease activity**, degrades the RNA in the RNA–DNA hybrids created by primase. This same DNA polymerase I enzyme then replaces the RNA bases with DNA by extending a free 3′ OH, as all DNA polymerases do. Once all of the RNA has been removed and replaced with DNA, this process still leaves a 3′ OH adjacent to a 5′ phosphate at the junction between one Okazaki fragment and another. No DNA polymerase has the ability to repair this **nick** in the phosphodiester backbone. The nick is instead repaired by **DNA ligase**, which synthesizes the phosphodiester bond at the expense of adenosine triphosphate (ATP) hydrolysis. This same process also occurs on the leading strand in order to remove and replace the much rarer RNA primers. All of the enzymatic activity introduces topological contortions into the DNA; consequently, the enzyme topoisomerase is needed to restore normal supercoiling of both of the replicated chromosomes.

A simplified biochemical system for replication of short segments of DNA (<14 kbp) is called the polymerase chain reaction (PCR). PCR is ubiquitous in biology and biochemistry laboratories (**Technique Box 4.1**). It requires only a single enzyme that is a heat-resistant DNA polymerase, and it results in exponential amplification of DNA between two oligonucleotide primers.

4.13 Although many bacteriophages have linear dsDNA genomes, bacterial hosts cannot replicate the ends of linear DNA

Linear DNA replicated using the typical bidirectional replication fork movement cannot be copied completely because DNA polymerase cannot synthesize DNA *de novo*. Even if RNA primase were to synthesize a primer at the very end of a template molecule, the cell would remove that RNA in the RNA–DNA hybrid as part of its normal DNA repair mechanisms. The result is a gap in the DNA at the end where there is potential template DNA left uncopied (**Figure 4.17**). This phenomenon is called the end-replication

Figure 4.17 The end-replication problem. Because DNA polymerases require a primer and all cells degrade the RNA in RNA–DNA dimers, the ends of linear chromosomes cannot be replicated by DNA polymerase.

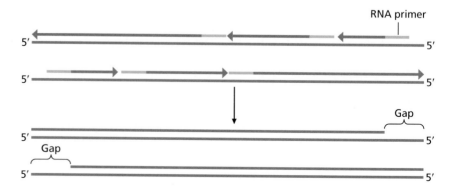

problem, and eukaryotic cells that have linear chromosomes have found an enzymatic solution, telomerase, which extends the ends of linear chromosomes with noncoding DNA. Even if a small amount of this DNA is not copied perfectly from one generation to the next, telomerase ensures that eukaryotic cells do not lose any coding capacity because the repeated telomere DNA does not encode proteins.

Most bacterial chromosomes are circular, thereby avoiding the end-replication problem entirely. Many very abundant bacteriophages, however, have linear rather than circular double-stranded DNA genomes. Bacteriophages with linear dsDNA genomes must have a way of solving the end-replication problem because their hosts do not encode any of their own enzymes that are analogous to eukaryotic telomerase. Both T7 and λ bacteriophages are examples of phages with linear dsDNA genomes; they solve the end-replication problem in different ways, as described in Sections 4.14 and 4.22, respectively.

4.14   T7 bacteriophage genome replication is catalyzed by one of the simplest known replication machines

The T7 genome is linear double-stranded DNA, which implies that it could simply use host enzymes to replicate its genome. Instead, however, it actually encodes its own DNA replication machinery, which has been investigated intensively because it is one of the simplest known DNA replication factories. Only four polypeptides are required to replicate T7 DNA: Class I and Class II proteins **gp2.5**, **gp4**, and **gp5**, and host protein Trx (**Figure 4.18**). The gp4 protein has two enzymatic activities. Its helicase melts the double-stranded template, and its primase activity synthesizes RNA primers. The gp2.5 protein is a single-strand binding protein, whereas gp5 is the DNA polymerase. The host protein Trx is not normally involved in DNA replication but it is an abundant cytoplasmic host protein that the phage replication apparatus co-opts (subverts) for a new purpose. This new purpose is to bind to gp5 and improve the processivity of the phage DNA polymerase, substituting for the clamp normally used by cellular chromosome replication. Host RNase H and ligase activity are also required for T7 genome replication; RNase H degrades the RNA component of RNA–DNA hybrids.

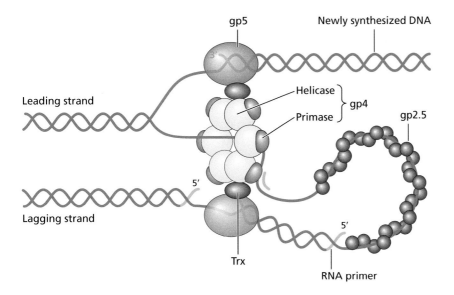

Figure 4.18 The bacteriophage T7 replication machinery. The bacteriophage T7 replication machinery is the simplest known DNA replication complex, composed of only four polypeptides: phages gp2.5, gp4, and gp5, and subverted host protein Trx.

TECHNIQUE BOX 4.1   POLYMERASE CHAIN REACTION

The polymerase chain reaction (PCR) revolutionized biology because it enables a scientist to begin with a very small amount of DNA and obtain a large quantity of a DNA sequence of interest within a few hours. In virology, PCR is used most often to detect viral nucleic acids in a sample, to amplify viral genes in order to use the sequence information for evolutionary comparisons, or to amplify viral genes for cloning. The components of a PCR are based on DNA replication and include deoxyribonucleotide triphosphates (dNTPs), a heat-resistant DNA polymerase that may or may not have its 3′ to 5′ exonuclease editing function intact, a buffer to optimize conditions for the enzyme, and primers (**Figure 4.19**). DNA polymerases require a double-stranded "landing pad" adjacent to a free 3′ OH that is opposite a single strand of template DNA. The primers hybridize to selected regions of the complex DNA template and provide DNA polymerase with these three requirements. PCR results in exponential amplification of the DNA between the primers so that after 25 or 30 rounds of PCR the amplified DNA in a 25-μL reaction is sufficient for it to be visible as a discrete band after agarose gel electrophoresis. Because the primers are so specific and the amplification of DNA is exponential, PCR is more than sensitive enough to detect specific viruses in clinical samples such as urogenital swabs.

Both PCR primers must hybridize to the DNA in order to obtain successful amplification (**Figure 4.20**). The DNA *in between the primers*, however, does not have to have any particular sequence and, in fact, two or more different DNA molecules can be amplified by the same pair of primers provided that the primers hybridize on either side of the sequence to be amplified and their 3′ OH groups face each other. This property of PCR can be very useful for

examining the evolutionary relationships among viruses isolated from different places or at different times or among viruses in a single person with an infection such as human immunodeficiency virus (HIV) or hepatitis. The PCR primers are designed to hybridize to two regions of a target gene that do not vary in one virus compared with another; some research goes into this selection process. These primers can subsequently be used to amplify any DNA between them. For example, PCR has been used in this way to diagnose exactly which strain of the dsDNA human papillomavirus has infected a patient. This information is useful to know because different isolates of the virus are associated with differing risk of developing cervical cancer. The conserved regions of the genome can be identified using sequence alignments and then PCR primers can be designed to amplify the variable DNA between the primer annealing sites. These specially designed primers bind to conserved regions of viral genomes and amplify the less conserved DNA in between them.

PCR is also commonly used to clone viral genes, for example to express these genes in *E. coli* and purify the proteins for further analysis. In such cases, it is common to use a DNA polymerase that has its editing functions intact in order to minimize the chance that PCR will introduce mutations into the DNA. PCR polymerases with editing functions intact are more expensive, slower, and less processive than PCR enzymes lacking the editing function, so it is common to use PCR polymerases with editing activity only when necessary. For eukaryotic DNA viruses, it can be important to clone cDNA instead of genomic DNA because of the presence of introns in pre-mRNA. To start with mRNA, the transcript must first be subjected to reverse transcription using an enzyme that

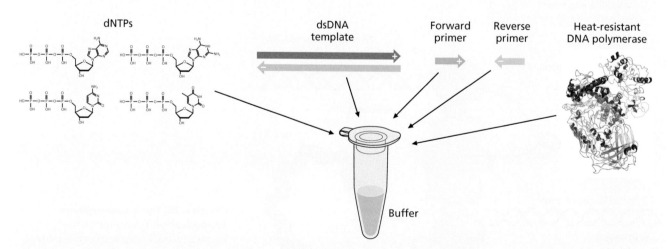

Figure 4.19 Components of a polymerase chain reaction (PCR). The components of a PCR include deoxyribonucleotide triphosphates, template DNA, a forward primer, a reverse primer, heat-resistant DNA polymerase, a buffer with optimal pH, and ions for the heat-resistant polymerase.

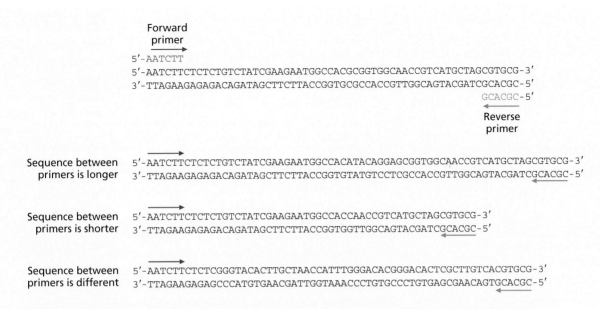

Figure 4.20 PCR amplifies DNA between two primers regardless of the sequence between the primers. The forward primer with the sequence 5'-AATCTT and the reverse primer with the sequence 5'-CGCACG can amplify any DNA between them, whether the sequence is longer, shorter, or just different from the first sequence used to design the PCR.

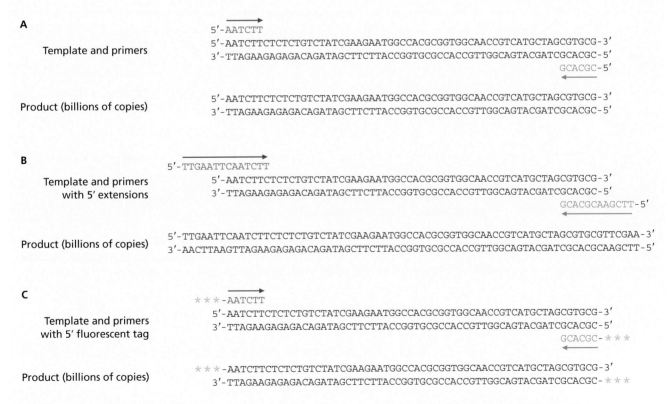

Figure 4.21 Addition of sequences or chemical groups to the 5' end of PCR primers. (A) Normal PCR. (B) PCR using primers that have 5'-sequence extensions. The PCR product is therefore longer, having incorporated those sequences. (C) PCR using primers in which the 5' end has been covalently modified by a fluorescent molecule represented by asterisks. The PCR product is therefore fluorescent.

can make a cDNA copy of mRNA in combination with a poly(T) primer to hybridize to the poly(A) tail found on eukaryotic mRNA transcripts. After that, PCR proceeds as normal.

There are other PCR-related techniques that make molecular cloning easier. For example, almost any sequence can be added to the 5′ end of a primer, which is nowhere near the active site of the DNA polymerase during PCR (**Figure 4.21**). Thus the 5′ end can be modified to carry a convenient restriction enzyme site. This way we can use PCR to more easily clone a gene into any particular plasmid vector even if the gene is not normally flanked by convenient restriction enzyme sites in the natural situation. Other chemical modifications of the 5′ end of the primer are also possible; for example, if the primer has a fluorescent tag at the 5′ end, all the amplified DNA molecules will have that fluorescent tag at their 5′ ends.

The phage cannot replicate the very ends of its genome because all DNA polymerases require a primer. T7 solves its end-replication problem using the direct repeats found at the end of each genome. These direct repeats, when single-stranded, base pair with the single-stranded regions of other replicated genomes. When the ends base pair, they can be ligated together by host ligase (**Figure 4.22**). There can also be extensive homologous recombination among genomes. Because so much replication occurs in a short period of time, the cell fills up with concatemers that are composed of many genomes ligated together. These genomes are subsequently processed into one-unit lengths during the later maturation stage of the virus replication cycle.

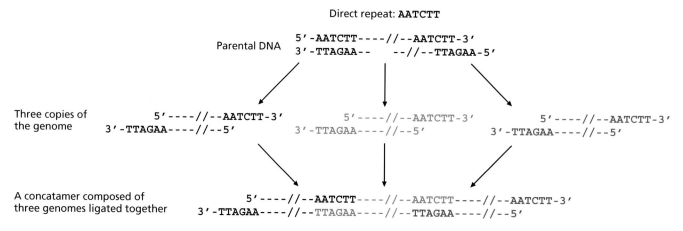

Figure 4.22 Concatemers of bacteriophage T7 genomes. Bacteriophage T7 uses the single-stranded direct repeats at each end to form concatemers catalyzed by host ligase. The direct repeats used here for illustration are much shorter than the real ones. The two forward slashes represent the approximately 40,000 bp that separate the direct repeats in a single genome.

4.15 The λ bacteriophage has naked, complex virions and a large double-stranded DNA genome

The next bacteriophage we consider is λ (**Figure 4.23**). Like T7, bacteriophage λ is also a member of the order *Caudovirales*, but it belongs to the family *Siphoviridae*, which has much longer helical tails than the *Podoviridae*. The large double-stranded DNA genomes of the *Siphoviridae* are between 22 and 70 kbp. As is the case for T7, inside the virion, the genome is linear. Although the genome is mostly double-stranded, at each end there are single-stranded extensions, called **cos** sites, which are complementary to one another and base pair after the genome enters the host cell, forming a circular structure (**Figure 4.24**). The term *cos* stands for *co*hesive ends because the single-stranded extensions at the end of each genome base pair, or cohere, with one another. Lambda gene expression and genome replication can be compared with those processes in T7 in order to get a small taste for the diversity of bacteriophages.

4.16 Bacteriophage λ can cause lytic or long-term infections

Bacteriophage λ encodes 73 proteins, of which 59 have a known function. Approximately 53 of the proteins are essential, which means that they are required for the phage to produce offspring. These essential genes are clustered together physically by function in polycistronic operons (see Figure 4.24). Lambda is called a temperate phage because it has two alternate replication cycles (**Figure 4.25**). During the **lytic cycle**, the bacteriophage infects a host cell and completes the virus life cycle immediately. In the other, termed **lysogeny**, the bacteriophage prophage DNA recombines with the host chromosome and no virions are produced. This choice is called a **developmental switch** because there are only two options: lytic reproduction or lysogeny. There are only two discrete and mutually exclusive outcomes of infection, making control of molecular events analogous to on or off for a lamp. That is, the infecting genome and its interactions with host molecules turn the switch to one option or the other. Lysogeny can persist for many generations of its bacterial host, although it is possible for the phage to get switched back into lytic growth, after which time the prophage DNA is excised from the chromosome and then initiates a lytic cycle. The lytic cycle is the focus of this chapter; lysogeny is addressed in **Chapter 13**.

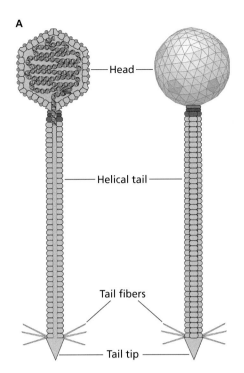

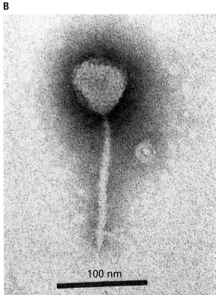

Figure 4.23 The λ virion. (A) Diagram of the λ virion. (B) Electron micrograph of the virion, showing the icosahedral head and long helical tail with tail fibers. (A, Courtesy of Philippe Le Mercier, ViralZone, © SIB Swiss Institute of Bioinformatics. B, Courtesy of H.W. Ackermann.)

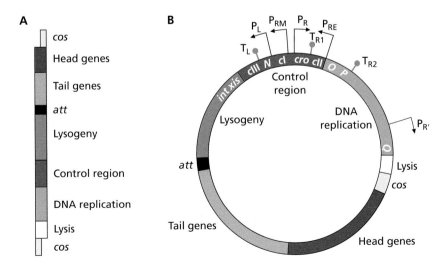

Figure 4.24 The λ genome. The λ genome is linear inside the infectious virion (A), but upon entry into the host cell the *cos* sites base pair and host ligase enzyme repairs the nicks, forming a circular structure resembling a plasmid (B). Some of the important genes, promoters (indicated by black arrows), and terminators (blue lollipops) that control the replication cycle have been indicated. The length of the *cos* and *att* sites has been exaggerated to make them more visible.

Figure 4.25 Temperate phage replication. The λ phage can replicate lytically, producing offspring phages and killing its host cell, or it can exist as a lysogen, which is viral DNA incorporated into the host chromosome without directing synthesis of new phages. The viral DNA is inherited through the generations as part of the chromosome. The phage can exit lysogeny at any time, however, and enter the lytic cycle.

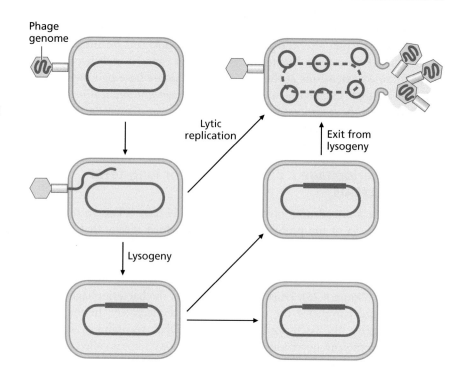

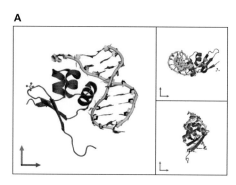

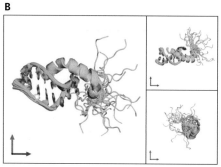

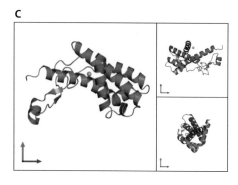

Only 14 of the 53 essential λ genes encode structural proteins. The high number of enzymes and regulatory proteins is needed to control the developmental switch between lytic and lysogenic cycles, to control the timing and abundance of virus proteins during the lytic cycle, and to control exit from lysogeny. The collective ability of nonstructural proteins to control gene expression and to physically manipulate DNA for recombination and other processes made λ one of the darlings of molecular biologists between 1960 and 1990. Because of that intense focus, today we understand most λ nonstructural proteins in incredible detail. In **Sections 4.17–4.21** we will focus on the proteins that regulate three waves of gene expression during lytic replication.

4.17 There are three waves of gene expression during lytic λ replication

Because the complexity of gene expression in λ can be overwhelming, this section focuses only on a few key regulatory events that promote the lytic cycle. **Chapter 13** covers the developmental switch that enables lysogeny, and subsequently the switch that enables the phage to exit lysogeny and re-enter lytic reproduction.

The lytic cycle includes three waves of gene expression: immediate-early, delayed-early, and late. The regulatory proteins required to coordinate these three waves of gene expression in λ virus are **Cro**, **N**, and **Q** (**Figure 4.26**). Although both the T7 and λ viruses have three waves of gene expression,

Figure 4.26 Crystal structures of Cro, N, and Q. (A) The λ Cro protein (pink) bound to DNA (green). (B) Part of the λ N protein (teal) bound to its RNA ligand (green). (C) Major domain of the λ Q protein. (Courtesy of Protein Data Bank in Europe.)

the molecular mechanisms by which they control these waves differ. T7 relies on production of its own RNAP and viral promoters with different sequences from typical host promoters. In contrast, λ relies on viral proteins that manipulate host RNAP. Cro is a DNA-binding protein that is a transcriptional repressor. N and Q are transcription antiterminators that work through two different mechanisms. N is an RNA-binding protein, whereas Q is a DNA-binding protein. Both interact with RNA polymerase in order to prevent the termination of transcription at particular sites in the λ genome.

The regulatory DNA required to coordinate the events of the lytic cycle include promoters (abbreviated P) and terminators (abbreviated T), which are labeled with subscripts to distinguish them from each other (**Figure 4.27**). The DNA with these promoters and terminators is clustered together in the control region along with the DNA encoding the *cro* and *N* genes, and the Q protein is encoded some distance away, to the right of the DNA replication genes *O* and *P*. The subscripts of these promoters and terminators include L or R to indicate whether transcription proceeds to the right or the left relative to the control region drawn with *cro* transcription proceeding from left to right.

4.18 The λ control region is responsible for early gene expression because of its promoters and the Cro and N proteins it encodes

Lytic gene expression can best be understood by focusing on the regulatory features of the λ control region (see Figure 4.27). Immediate-early gene expression occurs before the developmental switch into lytic replication or lysogeny occurs. Host RNA polymerase recognizes P_R and transcribes the DNA encoding the Cro protein, terminating transcription at T_{R1}. Host RNA polymerase also recognizes the promoter driving expression of the N protein, terminating transcription at T_{L1}. Although Cro is a DNA-binding protein, it binds to DNA as a multimer, and it therefore requires the production of many Cro molecules before Cro binds efficiently to DNA. The N protein, even at low levels, can regulate transcription right away, causing delayed-early gene expression to occur. There is also transcription from $P_{R'}$, but it terminates at T_{R4} without encoding any proteins.

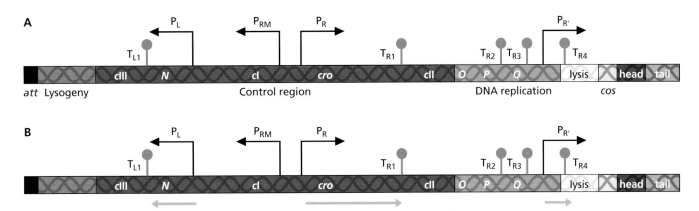

Figure 4.27 Genes, promoters, and terminators that control the lytic cycle in phage λ. (A) Diagram of genome. Although the genome is circular during gene expression, here it is shown as linear to make the relative positions of the genes, promoters, and terminators clearer. The genes, promoters, and terminators are not depicted to scale so that the regulatory DNA and regulatory genes are easier to see. (B) Early gene expression. Green lines indicate mRNA, with the arrowhead indicating the 3′ end.

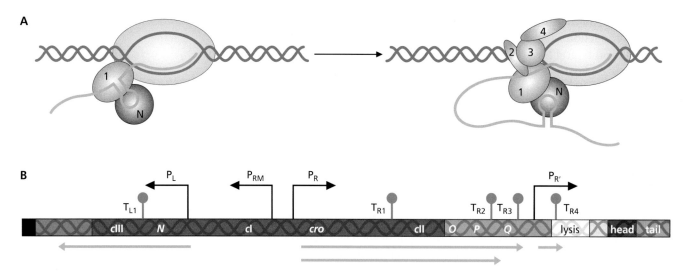

Figure 4.28 Process of antitermination by N and the resulting gene expression. (A) Mechanism of antitermination by λ protein N. The RNA polymerase pauses because of the stem–loop structure in the mRNA. The λ N protein binds to a stem–loop structure in the RNA, recruiting host protein 1 to join the complex. Subsequently, transcription continues and, when subverted host proteins 2, 3, and 4 join the complex, the RNA polymerase becomes highly processive and continues past T_{L1}, T_{R1}, and sometimes T_{R2}. (B) Delayed-early gene expression. The N protein prevents termination at T_{L1}, T_{R1}, and T_{R2}. Green lines indicate mRNA, with the arrowhead indicating the 3′ end. Transcription initiates at P_L and terminates after the head genes (not shown). Transcription initiates at P_R and terminates at T_{R3}. Transcription initiates at $P_{R'}$ and terminates at T_{R4}, which is not affected by the N protein.

4.19 The λ N antitermination protein controls the onset of delayed-early gene expression

Delayed-early gene expression occurs because of the activity of the N anti-terminator (**Figure 4.28A**). In the process of terminating transcription, RNA polymerase stalls at T_{R1} and T_{L1}. The mRNA protruding from the enzyme forms a stem–loop structure that includes a binding site for a complex of subverted host proteins and phage protein N. After the complex binds to the RNA loop, it interacts with the paused RNA polymerase elongation complex and causes transcription to continue. This antitermination results in a polycistronic mRNA initiated at P_R, encoding Cro and the P and O proteins (**Figure 4.28B**), which are needed for replication of phage DNA. N can also relieve a transcription pause at T_{R2}, so that a longer polycistronic mRNA initiated at P_R encodes not only Cro but also the Q antitermination protein. Relief of termination at T_{L1} also occurs, resulting in continued production of N and proteins encoded downstream of N.

4.20 The λ Q antitermination protein and Cro repressor protein control the switch to late gene expression

Late gene expression is regulated by a combination of the Q antitermination protein and high levels of the Cro repressor protein. The Q protein's anti-termination mechanism differs from that of the N protein. Protein Q first binds to target DNA sequences upstream of $P_{R'}$, and from there can join the RNA polymerase complex. Next, a subverted host protein associates with the RNAP–Q complex (**Figure 4.29A**). Together, this form of RNA polymerase does not terminate transcription at T_{R4}, but instead it continues on and tran-scribes a very long polycistronic mRNA, encoding all of the structural pro-teins as well as a few needed for maturation and release (**Figure 4.29B**). At this point, Cro repressor protein levels finally become high enough for Cro

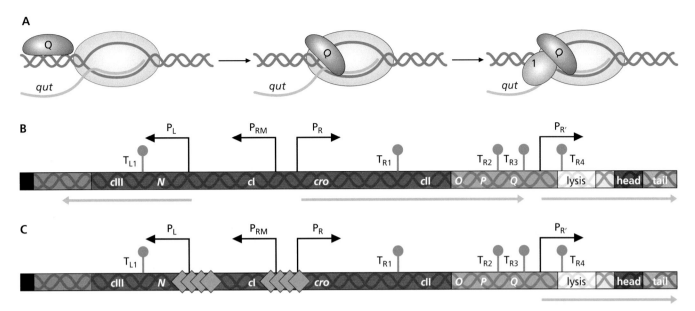

Figure 4.29 Antitermination by Q and late gene expression in bacteriophage λ. (A) Antitermination by Q. Q protein binds to a *qut* sequence immediately upstream of $P_{R'}$. From there, Q binds to the RNA polymerase. Next, a host protein, one of the same ones used by λ N, also joins the RNAP–Q complex. The resulting RNA polymerase complex transcribes through T_{R4}. (B) Late gene expression in bacteriophage λ. Antitermination by Q prevents termination at T_{R4}, leading to production of a polycistronic message encoding not only O, P, and Q, but also the lysis and structural proteins. (C) Very-late gene expression in bacteriophage λ. High levels of Cro protein (gray diamonds) bind to the DNA, repressing P_L and P_R. $P_{R'}$ remains active resulting in abundant expression of structural genes.

to bind to sites in the P_{L1}, and P_R promoters, repressing them (**Figure 4.29C**). This repression turns off the production of new Cro, N, and Q regulatory proteins. The other proteins encoded by the P_{L1} and P_R transcripts, CI, CII, and CIII, are involved in the switch to lysogeny or maintaining the prophage in a lysogenic state. Mutations that knock out the *cro* gene bias the phage toward switching to the lysogenic state by favoring expression of the CI protein, also known as the λ repressor because it represses lytic replication. This phenotype accounts for the name of Cro, which is an acronym for *c*ontrol of *r*epressor and *o*ther things.

4.21   Bacteriophages T7 and λ both have three waves of gene expression but the molecular mechanisms controlling them differ

The regulation of gene expression during the λ lytic phase provides an interesting comparison with that found in phage T7. Phage T7 coordinates its waves of gene expression through the use of a phage-encoded RNA polymerase that recognizes promoters that are distinct from host promoters. In contrast, phage λ proteins control lytic gene expression by preventing transcription termination or by repressing the initiation of transcription.

As is illustrated by both T7 and λ, in general, large dsDNA viruses (Baltimore Class I) have organized waves of early and late gene expression. As is the case with T7 and λ, the molecular mechanisms that control the different waves of infection are idiosyncratic in that they are unique to the type of virus. People interested in transcription regulation often study viral systems because of the variety of different ways that viruses control the production of viral mRNA.

Theta intermediate

Sigma intermediate

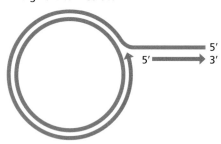

Figure 4.30 Lambda replicates through both theta (θ) and rolling-circle (σ) intermediates. Cells in the midst of a lytic λ replication cycle contain about 50 copies of the λ chromosome synthesized through a θ intermediate (top). Late during the replication cycle, these 50 templates are copied through a σ intermediate (bottom), creating catenated chromosomes that are many genome lengths long.

Figure 4.31 Rolling-circle mechanism of DNA replication. (A) A nick in the double-stranded template frees a 3′ OH. Replication fork proteins helicase, the sliding clamp, and the DNA polymerase assemble at the nick and then copy just one of the strands moving around and around the circle. The old strand peels away and is coated by single-strand binding protein. Newly replicated DNA is shown in red. (B) Later, as the strand of peeling away DNA gets longer, new replication forks assemble using that single-stranded DNA as a template to synthesize the second strand. In this case, replication begins with primase synthesizing an RNA primer. Ligase synthesizes the phosphodiester bond needed to connect the newly synthesized DNA, ultimately creating concatemers that are many genome lengths long.

4.22 Bacteriophage λ genome replication occurs in two stages, through two different intermediates

Cells infected with phage λ during lytic replication contain two forms of λ DNA: circular DNA replicating through a bidirectional θ intermediate, and circular DNA replicating through a **rolling-circle** mechanism (**Figure 4.30**). These different forms of partially replicated λ DNA can be detected by electron microscopy and can be quantified in a variety of ways. The λ chromosome undergoes θ replication for five or six rounds until the cell contains approximately 50 copies of it ($2^5 = 32$; $2^6 = 64$). At that point, there is a switch in which these circular templates are used for rolling-circle, rather than θ, replication. During rolling-circle replication, one strand of the closed circular DNA is used as a template to synthesize a single new double-stranded molecule. Rolling-circle replication requires all of the same proteins required for θ replication. To initiate rolling-circle replication, an enzyme creates a nick in the origin of replication (**Figure 4.31A**). The DNA replication factory assembles at that site, and the DNA is pulled through in a single direction, using one strand as the template to synthesize a single leading strand. The appearance of plasmids replicating using this method has led to the term sigma replication, for the Greek letter σ, as a synonym for rolling-circle replication.

As the copying process proceeds, the factory eventually arrives at the origin. For some plasmids and phages, the result is termination and release of the single-stranded DNA, covalent circularization of the single-stranded DNA, and ultimately synthesis of a second strand of DNA. However, for other plasmids and phages, including λ, instead of stopping once it reaches the origin, replication continues. As a result, the newly synthesized complementary DNA strand peels off the template strand (**Figure 4.31B**). The copying can go on for dozens of cycles before the replication factory stops. As the older nontemplate single strand peels away, it becomes coated by host single-stranded binding proteins, and then a new replication fork assembles and synthesizes the second strand, resulting in long concatemers. As in T7, during maturation the DNA is processed, in this case resulting in phage heads filled with exactly one complete linear genome. Thus, although they replicate their genomes in different ways, both T7 and λ phage genome replication result in the presence of genome concatemers.

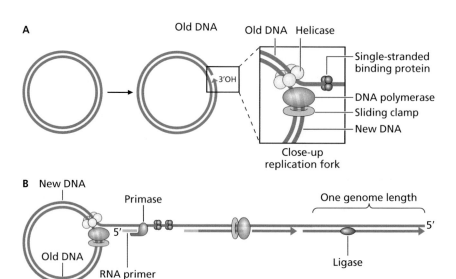

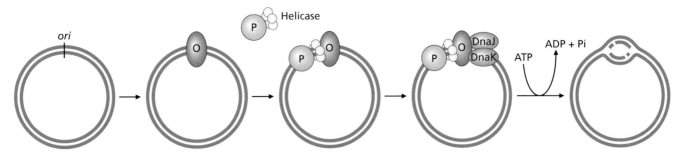

Figure 4.32 Initiation of θ replication in bacteriophage λ. An ordered cascade of protein–DNA and protein–protein interactions controls the initiation of replication of the λ chromosome. The phage O protein initiates the cascade by binding to a specific sequence in *ori*. After that, the phage protein P joins the complex, followed by subverted host proteins DnaJ and DnaK. DnaK hydrolyzes ATP, which results in remodeling of the initiation complex so that DNA replication proteins such as primase can begin replication.

4.23   Lambda genome replication requires phage proteins O and P and many subverted host proteins

The bacteriophage O and P proteins, produced during delayed-early gene expression, are needed to replicate the λ chromosome. Many host DNA replication proteins such as DnaA, helicase, and the chaperones DnaJ and DnaK are also required (**Figure 4.32**). The λ origin of replication (λ_{ori}), a DNA sequence found in the middle of the coding sequence for the O protein, is also required for replication of the λ chromosome. The O protein binds to the origin, whereas the P protein associates with the host helicase protein. Once the O protein has bound to the origin, the P–helicase complex loads onto the origin of replication. At this point, the chaperones DnaJ and DnaK are subverted to join the complex, after which ATP hydrolysis catalyzed by DnaK and stimulated by DnaJ remodels the complex in such a way that all of the other necessary replication proteins can bind to the origin–protein complex. After that, DNA replication follows, using host proteins for all replication functions other than initiation at *ori*. The host protein DnaA is also needed for normal θ replication, although the mechanism for this requirement is not clear. Evidence supporting this view includes the observation that host cells depleted of DnaA protein have reduced levels of bidirectional θ replication of λ chromosomes. Virologists do not agree on the molecular interactions that cause this effect and it may be complicated.

4.24   The abundance of host DnaA protein relative to the amount of phage DNA controls the switch to rolling-circle replication

The phage proteins O and P have been known as replication proteins for many years, but the mechanism controlling the switch from θ replication to rolling-circle replication is still being investigated. Rolling-circle replication uses the same origin sequences and also requires O, P, DnaJ, and DnaK. Initiation of rolling-circle replication occurs when θ replication is initiated, but only one fork proceeds around the chromosome. Once the replication fork reaches the *ori*, it continues and rolling-circle replication ensues. Recent experiments suggest that the host DnaA protein might be responsible for the switch (**Figure 4.33**). As described in **Section 4.11**, DnaA is a DNA-binding protein that is normally required for recognition of host chromosome *oriC* sequences. Although it binds to DNA near the host chromosome origin,

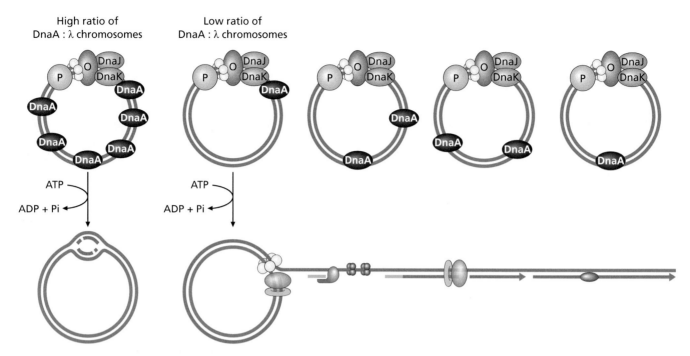

Figure 4.33 Model for DnaA depletion governing a switch to rolling-circle replication. When the ratio of host DnaA protein to λ chromosomes is high, the chromosomes undergo θ replication. When there are about 50 copies of the λ chromosome in the cytoplasm, the ratio of DnaA protein to λ chromosomes is low. The result is that, instead of initiating θ replication, the proteins can initiate only one fork, resulting in σ replication.

there are also 240 strong DnaA binding sites throughout the *E. coli* chromosome and 11 such sites in the λ chromosome. The balance of the number of DnaA molecules and DnaA binding sites is important for host chromosome replication and is important for the switch from θ to σ replication in bacteriophage λ. When λ chromosomal DNA reaches 50 copies per cell, the cytoplasm contains 550 more strong DnaA binding sites than are found in uninfected cells. Host cells contain limiting amounts of DnaA, approximately 200 copies per cell. When the number of new λ chromosomes reaches approximately 50, DnaA binds to the 550 sites in the λ chromosomes and is thus depleted from binding to the host chromosome. At that point, DnaA can no longer carry out its function of promoting θ replication. Instead, the next time O, P, and the host proteins form a functional initiation complex at λ$_{ori}$, θ replication proceeds unidirectionally and becomes rolling-circle replication when the replication complex reaches λ$_{ori}$.

4.25 There are billions of other bacteriophages that regulate gene expression in various ways

There are many additional described dsDNA bacteriophages, perhaps because our procedures for detecting phages are biased in favor of phages with these cell-like genomes. In general, bacteriophages with dsDNA genomes control transcription initiation or termination to coordinate gene expression, but there are many mechanisms of doing so. For example, bacteriophage T4 encodes proteins that alter host RNA polymerase without replacing it completely, resulting in the ordered expression of genes from promoters with different sequences without encoding its own RNA polymerase as T7 does. There are many other variations on the theme of controlling transcription, and the hundreds of known double-stranded phage genomes undoubtedly encode as yet unknown regulatory proteins with distinctive mechanisms.

4.26 Some bacteriophages have ssDNA, dsDNA, or (+) ssRNA genomes

Of the other types of bacteriophages, approximately two dozen have been studied in enough detail to understand their gene expression and genome replication. These include some Class II, Class III, and Class IV viruses (**Figure 4.34**). So far there are no described bacteriophages assigned to Classes V ([−] RNA), VI (retroviruses), or VII (reversiviruses), but we don't know whether we have simply failed to detect them or whether they do not exist. Rather than examine other bacteriophages in as much detail as we did for T7 and λ, Sections 4.26–4.35 describe a few aspects of gene expression and genome replication in four of these phages, ΦX174, **M13**, **MS2**, and **Qβ**. We use ΦX174 to examine how viruses can encode many proteins in a small genome, M13 to examine one mechanism to replicate a single-stranded DNA genome, MS2 to show how a small (+) RNA virus is expressed and controls the relative amounts of proteins expressed from its genome, and Qβ to show how (+) RNA genomes can be copied to provide genomes for offspring phages.

4.27 The replication cycles of ssDNA bacteriophages always include formation of a double-stranded replicative form

The replication cycles of ssDNA bacteriophages have some common features (**Figure 4.35**). After uncoating, the single-stranded circular DNA genome is converted to a dsDNA molecule using the secondary structure in the ssDNA molecule as an origin of replication for assembly of

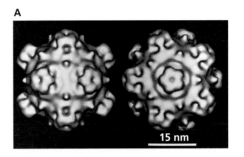

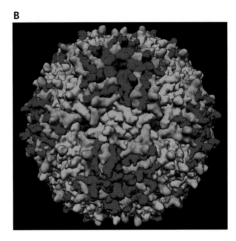

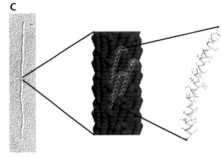

Figure 4.34 Phages with ssDNA or (+) ssRNA genomes. (A) Cryo-electron microscopy reconstruction of icosahedral ΦX174 phage (Class II). (B) Crystal structure of icosahedral MS2 phage, which is 27 nm in diameter (Class IV). (C) Close-up of the middle of the structure of the filamentous fd phage (Class II), which is 800 nm long and 7 nm in diameter. (A, From Olson NH et al. 1992. *J Struct Biol* 108:168–175. With permission from Elsevier. B, UCSF Chimera image from PDB 2MS2, courtesy of Dr. Jean-Yves Sgro, UW-Madison. C, From Marvin DA 2017. *Prog Biophys Mol Bio*. doi: 10.1016/j.pbiomolbio.2017.04.005. With permission from Elsevier.)

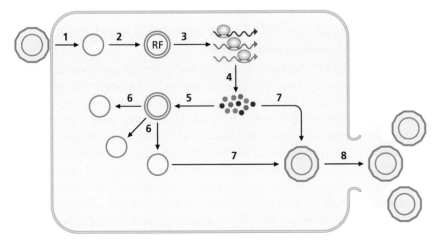

Figure 4.35 Overview of gene expression and genome replication in ssDNA phages. Uncoating releases the circular single-stranded DNA genome into the cytoplasm (1). Host enzymes copy the viral genome making the double-stranded replicative form (RF) (2). Host enzymes transcribe the DNA (3), resulting in translation of viral proteins (4). Transcription and translation are simultaneous, but the figure depicts them separately for clarity. Viral and host proteins use the RF (5) to synthesize new genomes (6). Viral proteins form empty procapsids that are filled with new ssDNA genomes (7). The offspring virions exit the cell by lysis (8).

host enzymes that carry out synthesis of the second strand. The double-stranded version of the virus genome is called the **replicative form**. Host enzymes transcribe the phage genes; unlike the dsDNA phages, the timing of gene expression is not under tight control, so there are no early or late proteins. This collapsing of gene expression into a single stage of a viral replication cycle is common for viruses with small genomes, which can be defined as those in which most nucleotides encode protein, leaving little to no room for regulatory sequences. These phages still control the relative amounts of different proteins, however, so that the structural proteins become much more abundant than the nonstructural ones. Control over protein amounts arises as a result of both transcription and translation. The phages encode one or more proteins needed to use the replicative forms to synthesize new ssDNA genomes, which are subsequently processed and packaged during maturation.

4.28 Bacteriophage ΦX174 is of historical importance

There are several known Class I bacteriophages with single-stranded DNA genomes, including M13 and ΦX174. M13 is commonly used in biotechnology applications, whereas ΦX174 (pronounced phi ex 174) has been studied extensively to understand replication of single-stranded DNA phages. Phage ΦX174 is also important for historic reasons. For example, before the structure of DNA was known, Erwin Chargaff found that, in most organisms, the percent of each type of nucleotide in their DNA always occurred in a predictable pattern, with %A = %T and %G = %C. These proportions inspired the idea that perhaps the structure of DNA included A–T and G–C pairwise interactions. However, ΦX174 was strikingly different: %A was not equal to %T and %G was not equal to %C, ultimately indicating that the genome is single stranded. In 1977, the ΦX174 genome of 5,387 bases was also the first completely sequenced DNA molecule, leading to Frederick Sanger's second Nobel Prize.

4.29 Bacteriophage ΦX174 has extremely overlapping protein-coding sequences

Bacteriophage ΦX174 manages to encode six structural and five nonstructural proteins in very little genomic information. Dividing the ΦX1744 genome length by the 11 coding sequences leads to the expectation of an average gene length of 489 bases, which would make the proteins, on average, only 163 amino acids long; this is much shorter than the length of most host proteins. The average size of the 11 phage proteins is actually known to be 211 amino acids, and the total number of amino acids in all 11 proteins is 2,327. This math is curious, as it would take a minimum of 2,327 amino acids × 3 bases per amino acid, or 6,981 bases, in a conventional cellular genome to encode so many amino acids, not counting any bases needed to form promoters and other regulatory sequences.

The phage therefore raises the question of how it can encode so many proteins in so few bases of DNA. The answer is that 7 of the 11 coding sequences overlap, in some cases to an extreme degree (**Figure 4.36**). For example, the coding sequences for the A, A*, and B proteins entirely overlap each other, and the coding sequences for C, K, D, and E also overlap with at least one other coding sequence. A and A* have the same reading frames but have two different translation start sites. The B coding sequence is actually in a

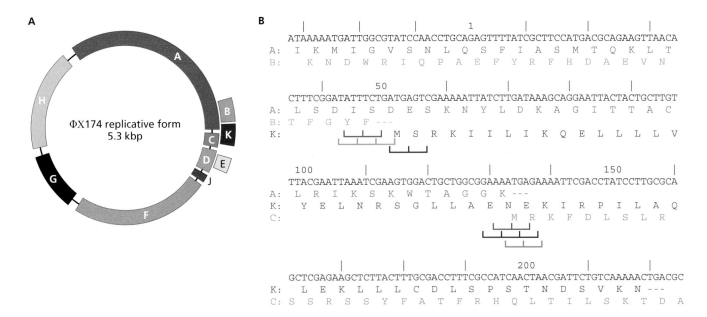

Figure 4.36 Genes encoded by the ΦX174 replicative form. (A) The map of ΦX174 is circular and genes are indicated by letters on various colored blocks; gene A* overlaps completely with A but uses a different start codon and so has not been shown separately. Note that the coding sequences for A, B, and K; K and C; and D and E overlap. (B) Close-up on sequences encoding A, B, K, and C. The top row shows the nucleotide count with nucleotides 1, 50, 100, 150, and 200 noted in numbers and every other 10-nucleotide interval marked by a black vertical line. The one-letter amino acid codes for A, B, K, and C are shown; A is in pink, B is in green, K is in purple, and C is in blue. Areas of particularly intense overlap are indicated by color-coded brackets marking the frames for the different proteins where they overlap.

different reading frame from A and A*. The K protein is translated in a different frame from either the A or C proteins, with which its coding sequence overlaps, and the E protein is translated from the same sequence as the D protein, but again using a different reading frame with a different translation initiation signal. Because of this overlap, genetic changes at a single nucleotide could potentially change the amino acids in two or three proteins at a time, meaning that selective pressure on the function of multiple proteins affects the evolution of the genome sequence. Nevertheless, overlapping genes are found in many groups of viruses, so sequencing and *in silico* (computerized) translation of virus genomes must be interpreted cautiously as virus genomes might encode overlapping proteins. The smaller a virus's genome, the more likely it is to encode proteins using overlapping sequences.

4.30 Bacteriophage ΦX174 proteins are expressed in different amounts

The relative timing of the expression of the 11 ΦX174 proteins is much less important for phage replication than is controlling their relative amounts. The six structural and scaffold proteins are needed in the greatest abundance, with 240 copies per virion for scaffold protein D, 60 copies each for proteins B, F, G, and J, and 12 copies of H per virion (**Figure 4.37**). The five nonstructural proteins are needed in much smaller amounts. The relative abundance of all of the proteins produced results from a combination of the strength of the virus promoters and terminators and the way that overlapping coding sequences affect translation of those proteins.

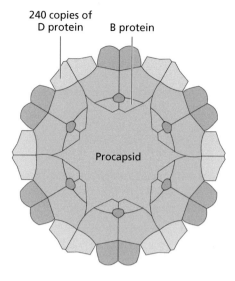

240 copies of D protein   B protein

Procapsid

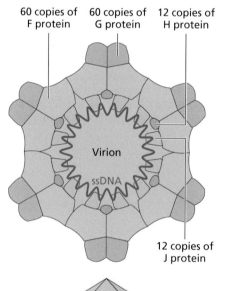

60 copies of F protein   60 copies of G protein   12 copies of H protein

Virion

ssDNA

12 copies of J protein

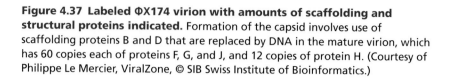

Figure 4.37 Labeled ΦX174 virion with amounts of scaffolding and structural proteins indicated. Formation of the capsid involves use of scaffolding proteins B and D that are replaced by DNA in the mature virion, which has 60 copies each of proteins F, G, and J, and 12 copies of protein H. (Courtesy of Philippe Le Mercier, ViralZone, © SIB Swiss Institute of Bioinformatics.)

4.31   A combination of mRNA levels and differential translation accounts for levels of bacteriophage ΦX174 protein expression

The transcribed phage genome is a double-stranded DNA replicative form, which includes three promoters, P_A, P_B, and P_D, and four transcription terminators, T_J, T_F, T_G, and T_H. The three promoters and four terminators combined result in multiple mRNA molecules (**Figure 4.38**). The longest, initiated by a promoter upstream of the A coding sequence, is transcribed from the entire genome, but it is very unstable, which contributes to keeping levels of the nonstructural A and A* proteins low. The polycistronic mRNA molecules initiated at P_B can encode proteins B, K, C, D, E, J, F, G, and H. The B protein, then, is mostly translated from this mRNA rather than from the unstable mRNA initiated at P_A, so that the overlapping of the sequence encoding the structural B protein (60 copies per virion) with that of A and A* is not that important for gene expression. The nonstructural protein K is translated less frequently than B because ribosomes that initiate at B continue to translate B, and K is translated only by ribosomes that, much more rarely, bind to the ribosome-binding site upstream of K. This ribosome-binding site is often physically occluded by the polyribosomes translating the B protein.

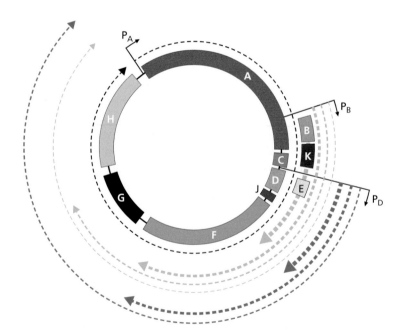

Figure 4.38 Eight mRNAs produced by transcribing the ΦX174 replicative form. The map of ΦX174 is circular and genes are indicated by letters on various colored blocks. Three promoters are indicated as bent arrows. The one mRNA that originates at P_A is a black dashed line. Pale green dashed lines indicate mRNA transcripts that initiate at P_B, whereas dark green dashed lines indicate mRNA transcripts that initiate at P_D. The relative thickness of the dashed lines indicates the relative abundance of the different mRNAs.

The polycistronic mRNA molecules initiated at P_D can encode D, E, J, F, G, and H. The D scaffold protein (240 copies per virion) is the most abundant, whereas the nonstructural E protein is found in very low amounts in infected cells. The D and E proteins are always encoded on the same mRNA, with D coding sequence upstream of that of E. Translation of E from these messages is as similarly rare as translation of K, because E can be translated only by ribosomes that initiate adjacent to the E start codon instead of at the much stronger, and out-of-frame (relative to D), ribosome-binding site for D.

The abundant structural proteins D, B, F, G, J (60 copies each per virion), and H (12 copies per virion) are encoded on every mRNA transcribed from the phage genome. Most of them have coding sequences that do not overlap with any other coding sequences. Although two (B, D) have coding sequences that overlap with those of other proteins, those coding sequences are always found upstream of the gene with which they overlap, resulting in much greater translation compared with the nonstructural proteins with which their coding sequences overlap.

It is notable that this tiny bacteriophage does not encode any DNA-binding proteins that regulate transcription or any RNA or DNA polymerases. Instead, the phage relies on host enzymes for these functions. This exemplifies a common trend in virology: viruses with small genomes have very few regulatory proteins or nucleotide polymerase enzymes, whereas viruses with large genomes encode gene regulatory networks and proteins including enzymes used for gene expression and genome replication. Describing a virus as having a small or large genome is of course relative, but in general small genomes encode fewer than 10 different proteins and have few if any regulatory sequences, whereas large viruses encode more than 20 different proteins and have at least a few regulatory sequences. The presence of extensively overlapping open reading frames is a signature feature of viruses with particularly small genomes (fewer than 3,000–5,000 nucleotides, or base pairs, for double-stranded genomes). A consequence of the lack of regulatory sequences is that these viruses control the relative abundance of nonstructural and structural proteins without separating the timing of their production into early and late phases. That is, for small viruses there are no distinctive waves of early and late gene expression, though it is common for there to be a switch away from mRNA production and toward production and packaging of new genomes. As illustrated by ΦX174, they may still exert control over relative protein abundance.

4.32   Bacteriophage M13 genome replication is catalyzed by host proteins and occurs via a replicative form

Whereas ΦX174 is a model for gene expression by ssDNA phages, phage M13 is a model for genome replication by ssDNA phages. The M13 phage has a (+) orientation linear 6.4-kilobase genome; upon entry into the host cytoplasm, host enzymes convert this genome into a circular dsDNA replicative form (**Figure 4.39**). New DNA synthesis proceeds in three general steps. Step one has already been described and consists of the infecting ssDNA genome being converted into the replicative form (**Figure 4.40A**). As we discussed earlier, however, all host DNA polymerase enzymes require a primer, which is not present in the infecting genome. The need for a primer raises the question of how the phage can accomplish stage one of genome replication. The first step is that, upon infection, most of the virus genome is coated by host single-stranded binding protein, the same one used in host replication.

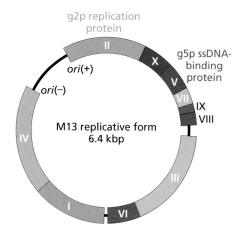

Figure 4.39 The M13 replicative form. The (–) DNA strand is synthesized beginning at *ori* (–) and new (+) DNA is synthesized beginning at *ori* (–). Protein-coding sequences are shown.

A small region of the genome is left uncoated; this part of the genome forms an unusual secondary structure by folding up on itself. This secondary structure can be bound by host RNAP. The RNAP then synthesizes a short RNA primer and dissociates from the DNA. At that point, the normal host proteins can assemble around the primer and then synthesize the (–) strand, using DNA polymerase III, DNA polymerase I, and ligase, among other host enzymes.

After the formation of the first replicative form and after viral gene expression, step two of DNA synthesis begins (**Figure 4.40B**). The purpose of step two is to increase the number of replicative forms in the cytoplasm. The replicative forms are used for gene expression and later will be used to synthesize new genomes. In order to produce more replicative forms, the viral g2p protein nicks the replicative form DNA at the (+) sense origin of replication, meaning that the enzyme hydrolyzes one sugar–phosphate bond on a single strand in the *ori* (+) sequence. This nicking is required for new (+) strands to be synthesized by rolling-circle replication using host DNA replication machinery. These (+) strands are converted into new replicative form molecules in the same way that the infecting genome was, and gene expression continues from these more abundant templates.

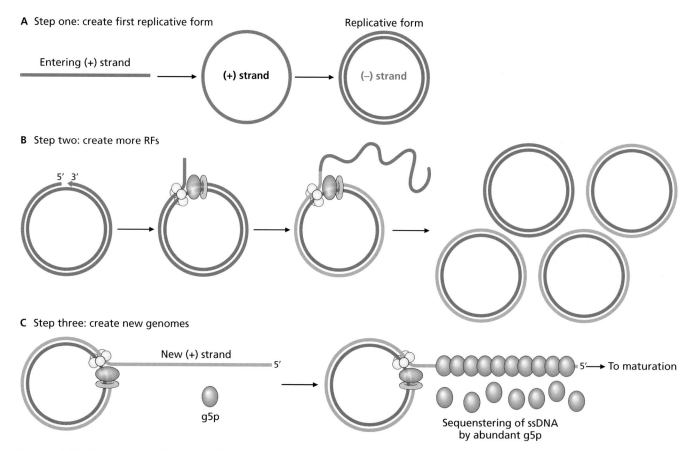

Figure 4.40 Three steps of DNA synthesis in bacteriophage M13. (A) Entering (+) ssDNA is linear (dark blue) and during step one it circularizes and serves as a template to synthesize the complementary (–) strand (in red), all catalyzed by host enzymes. (B) In step two, the original replicative form (RF) is used to synthesize new (+) strands by a rolling-circle mechanism. New (+) DNA is processed into single-stranded circular DNA (not shown) and then copied to give rise to new replicative forms. New (+) DNA is light blue. (C) In the third step, the abundant replicative forms are used to synthesize more (+) DNA, still using a rolling-circle mechanism, but because of viral gene expression, there are high levels of the g5p phage protein. This single-strand DNA-binding protein coats the new (+) strands, preventing them from being converted into new replicative forms and targeting them for processing into linear genomes and packaging during assembly.

At some point, a threshold of g5p protein will have accumulated in the cytoplasm. This accumulation of g5p late in infection triggers step three of DNA synthesis (**Figure 4.40C**), which is the use of replicative form DNA to make new linear (+) ssDNA genomes. The switch occurs when g5p proteins reach a threshold. At that point, g5p proteins are sufficiently abundant that while new (+) ssDNA is synthesized, the g5p protein binds to the nascent ssDNA and coats it. The effect is to sequester this new DNA away from the proteins that would otherwise convert the newly synthesized DNA into another replicative form. The phage is now ready to enter the next stage of the virus replication cycle, assembly.

4.33   Bacteriophage MS2 is a (+) ssRNA virus that encodes four proteins

In **Sections 4.33–4.35**, we will learn about gene expression and genome replication in a virus that has an RNA genome made of a different type of macromolecule than the genome of its host cell. MS2 is a small icosahedral bacteriophage with 180 copies of the capsomere (known as coat protein) and a single asymmetric vertex occupied by the **maturation protein**, which is also bound to the linear (+) ssRNA genome within. The genome is short, only 3,569 bases long, and encodes only four proteins: maturation protein, coat protein, lysis protein, and an **RNA-dependent RNA polymerase**, typically abbreviated **RdRp**. Because no cellular host encodes its own RdRp, all viruses (including phages and eukaryotic viruses) with RNA genomes must encode their own RdRp to carry out gene expression and genome replication. To date, there is no known example in which a cell uses RdRp for its own survival or reproduction; all known RdRps, including telomerase, are viral in origin.

Except in retroviruses, (+) ssRNA virus genomes are used directly as mRNA when they enter the host cytoplasm at the conclusion of uncoating, because they are in the same "sense" as mRNA. There are several common events in the replication cycles of the simplest (+) ssRNA virus genomes, including phage MS2 (**Figure 4.41**). Following the penetration and uncoating stage, the genome is translated, resulting in production of the RdRp and other proteins. Next, the RdRp uses the genome as a template to synthesize a negative strand copy of the genome, also known as an antigenome. Subsequently, the RdRp uses the antigenome as a template to synthesize more (+) ssRNA molecules, which could in principle be used either as mRNA or

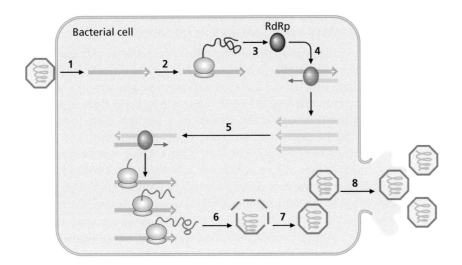

Figure 4.41 Events in the replication of (+) ssRNA bacteriophage genomes. Following uncoating (1) the linear (+) genome enters the cytoplasm and is translated (2), making viral proteins including the RdRp (3), also known as the replicase. The replicase copies the genome template to make many (−) strand antigenomes (4). The replicase copies the antigenomes to make more (+) strands (5), which are translated (6) simultaneously by ribosomes. (7) Viral genomes and proteins assemble into offspring phages. (8) The phages escape by host lysis. Gray arrows indicate the direction of RdRp. The arrowheads are the 3′ ends of the nucleic acids.

Figure 4.42 Map of the (+) RNA bacteriophage MS2 coding sequences. The genome encodes the A protein, the coat protein, the lysis protein, and the RNA-dependent RNA polymerase (RdRp), also known as replicase. The coding sequence for the lysis protein overlaps that of the coat protein and the RdRp.

as new genomes packaged into assembling offspring virions. Early during an infection, the (+) RNA is used preferentially as mRNA; later during infection, when the coat protein has accumulated above a certain threshold, the (+) RNA is packaged into new virions. After packaging and maturation, the offspring phages are released from the host cell.

Because a virus must have a capsomere, the simplest possible RNA virus genome is just a single strand of (+) ssRNA encoding an RdRp molecule and a capsomere. Even in the simplest known (+) ssRNA viruses, however, the situation is more complicated. For example, MS2 encodes four proteins, which include a lysis protein, an RdRp, a coat protein, and a maturation protein (also known as protein A). A linear map of the genome reveals that the lysis protein coding sequence overlaps those of both the coat protein and RdRp (**Figure 4.42**); as we have seen with small ssDNA phages, the overlap of an enzyme's coding sequence with that of the most abundant viral protein (the coat protein) indicates that differential translation will likely affect their relative abundance.

The following experimental results indicate unexpected complexity in the regulation of MS2 protein expression. The experiment begins with an *in vitro* translation suspension made up of bacterial ribosomes, translation factors, and charged tRNAs, but lacking any mRNA. After a purified MS2 genome was added, the result was only a single protein. Although at first it might seem as if the protein is likely the maturation protein because it is encoded closest to the 5′ end of the linear genome, the single protein synthesized is the coat protein, which is encoded near the middle of the genome. Further experimentation revealed that physically fragmenting the genome prior to mixing it with the *in vitro* translation mix resulted in production of all four proteins. The difference depends on the way MS2 controls the relative amounts of protein produced during infection, as explained in **Section 4.34**.

4.34 Bacteriophage MS2 protein abundance is controlled by secondary structure in the genome

Every new MS2 virion needs 180 coat proteins, one genome, and one maturation (A) protein. The RdRp and lysis enzymes are also needed in small amounts, more than one but far less than 180 per new genome. The phage, having a genome composed of (+) RNA, cannot rely upon transcriptional promoters or terminators to alter the amount of gene expression. Instead, it must rely on control of translation. It does so in several different ways. First, the secondary structure of the genome itself plays a major role. Second, the overlapping coding sequences ensure that the lysis protein is translated infrequently compared with the coat protein. Third, an unusual start codon also restricts expression of the maturation protein to one per genome. The critical roles of secondary structure and overlapping coding sequences are typical of many small (+) ssRNA viruses, including those that infect animals and plants. We will explore this situation in MS2 in more detail.

The secondary structure of the genome explains why coat protein is translated preferentially in an *in vitro* translation system containing viral genomic RNA. Unlike DNA, which is a double helix under cellular conditions, RNA can form more complex structures by base pairing within the same molecule or to different molecules. **RNA secondary structure** typically consists of stem–loop structures, in which some of the RNA is folded and base paired with other nucleotides in the same strand, making a helical stem, and where the RNA in between the elements forming the stem are not base paired with each other and form a loop (**Figure 4.43A**). **RNA tertiary structure** arises from interactions between different stem–loop structures (**Figure 4.43B,C**); the most typical interaction is base pairing between a loop at one site and another loop, which could be hundreds or even thousands of bases away in **primary sequence** (nucleotide sequence) of the RNA molecule. A familiar ssRNA molecule with secondary structure is tRNA (**Figure 4.44**). Like tRNA, the genome of MS2 is also folded into a structure, and this structure controls translation of most of the proteins. This structure explains the finding that fragmenting the genome prior to adding it to the *in vitro* translation system allows for production of all four proteins, because fragmented genomes cannot assume the same structure as the original molecule.

Under cellular conditions, the genome of MS2 and its relatives is folded into a complex tertiary structure. The secondary structures surrounding the start and stop codons for the sequences encoding the four protein is particularly extensive (**Figure 4.45**). The start codon for the sequence encoding coat protein is found at the end of a stem–loop structure, in the loop. The start codon for the replicase protein, in contrast, is buried in a long stem that participates in many base-pairing interactions. The start codon for the sequence encoding the lysis protein is found in the coding sequence for the coat protein, and it occurs following and nearly adjacent to two out-of-frame stop codons in the coat protein. Finally, the start codon for the maturation (A)

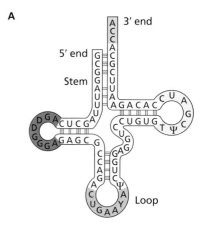

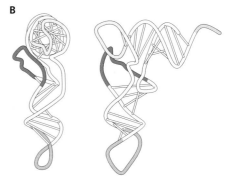

Figure 4.44 The extensive secondary structure of tRNA. (A) Diagram of tRNA emphasizing the presence of stem–loop secondary structure. There are four stems in the image (bases paired through red rungs) and three colored loops. (B) Three-dimensional model of tRNA structure, shown in two different orientations, illustrating helical secondary structure in the stem portion of the stem–loops. (From Alberts B et al. 2013. *Essential Cell Biology*. Garland Science. With permission from W.W. Norton.)

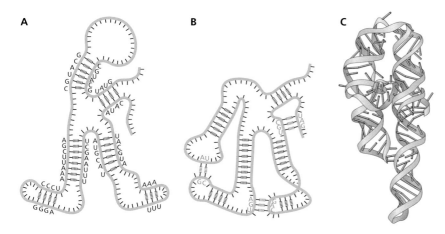

Figure 4.43 RNA can form secondary and tertiary structures. (A) A diagram of a hypothetical folded RNA structure showing conventional (G–C and A–U) base-pair interactions that result in base-paired stem regions (red rungs) and loops (unpaired bubbles). (B) Tertiary structure occurs when bases that are part of loops pair with other bases that are distant from one another in the primary structure and can involve conventional base-pair interactions or nonconventional interactions (A–G and C–U). (C) Structure of an actual RNA molecule that is involved in RNA splicing. This RNA has a considerable secondary and tertiary structure. (From Alberts B et al. 2013. *Essential Cell Biology*. Garland Science. With permission from W.W. Norton.)

A

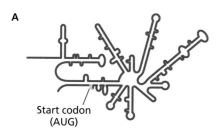

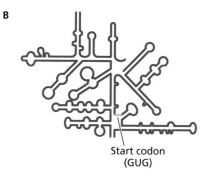

Start codon
(AUG)

B

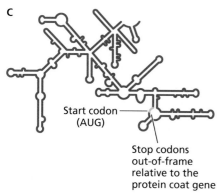

Start codon
(GUG)

C

Start codon
(AUG)

Stop codons
out-of-frame
relative to the
protein coat gene

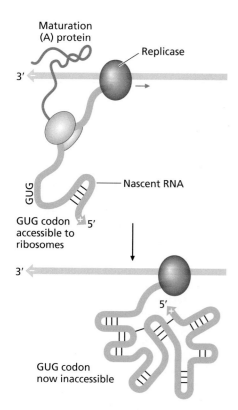

Maturation
(A) protein

Replicase

3′

GUG

Nascent RNA

GUG codon
accessible to
ribosomes

5′

3′

5′

GUG codon
now inaccessible

Figure 4.45 Secondary structure surrounding the start and stop codons for three of the MS2 genes. (A) The start codon for the abundant coat protein is at the top of an exposed loop, whereas the start codon for the low-abundance RdRp is buried in secondary structure; surrounding stem–loops are also likely to participate in tertiary interactions. (B) The alternative GUG start codon for the maturation (A) protein is obscured by secondary structure; surrounding stem–loops are also likely to participate in tertiary interactions. (C) The start codon for the lysis protein is immediately adjacent to and downstream of two out-of-frame stop codons in the coat protein-coding sequence. (From Voyles BV 2001. *The Biology of Viruses*, 2nd ed. McGraw-Hill. Courtesy of © McGraw-Hill Education.)

protein, although closest to the 5′ end of the (+) ssRNA, is buried in extensive stem–loop structures that probably participate in tertiary interactions as well.

The best model for expression of proteins from the MS2 genome is thus as follows. After uncoating, ribosomes initiate at the coat protein AUG, which is the only start codon available in the tertiary structure of the genome. As the ribosome progresses through the gene, this necessarily disrupts the tertiary and secondary structures it encounters. Relaxation of these structures allows ribosomes to either initiate independently at the replicase start codon or to continue from the coat protein sequence, release the coat protein, and then reinitiate at the replicase start codon without dissociating completely from the mRNA. In a process that is not well understood, premature termination of coat protein translation at the two out-of-frame stop codons is required for translation of the lysis protein. The idea is that the ribosome releases the truncated coat protein and then reinitiates at the lysis protein sequence. These events can only occur if the ribosome experiences a frameshift (a shift in reading frame) during translation of the coat protein, but the regulation of this frameshift and its precise nature are not known. The frameshift event must be relatively rare, because the lysis protein is needed in small amounts, much fewer than 180 coat proteins per offspring virion. Frameshifts required to achieve expression of low abundance viral proteins are common in both RNA bacteriophages and eukaryotic RNA viruses.

The replicase enzyme is also needed in small amounts relative to the coat protein. This balance of proteins is accomplished by **translational repression**, in which high levels of the coat protein prevent translation of the replicase. Coat proteins are able to bind to a secondary structure in (+) ssRNA genomes near the start site of the replicase-coding sequence thereby preventing translation of the replicase enzyme when concentrations of coat protein are high in the maturation stage of replication.

The regulatory mechanism that results in one maturation protein per new genome involves not only secondary and tertiary structure burying the start codon, but also an **alternative start codon**, GUG. The tertiary structure obscuring the start codon is so extensive that the maturation protein cannot be expressed from an intact complete genome. Instead the maturation protein is translated exclusively from nascent genomes as they emerge from the RdRp, before the spontaneous secondary and tertiary folding reactions obscure the ribosome-binding site and start codon (**Figure 4.46**).

Figure 4.46 Replicase and nascent translation of maturation protein. The maturation (A) protein can only be translated from nascent (+) RNA because the GUG start codon is only accessible before the RNA genome folds into its characteristic secondary and tertiary structures. The nascent (+) RNA protrudes from the replicase and an average of one ribosome translates the maturation (A) protein before the genome folds up.

Additionally, the start codon for the maturation coding sequence is not the canonical AUG but is instead GUG, which results in approximately 10 times less translation than the canonical AUG. The use of GUG as a start codon in the case of the maturation protein likely ensures that only one or two ribosomes successfully initiates translation before the nascent transcript folds too much for any additional ribosomes to have access to the (+) ssRNA, resulting in an average of one maturation protein per new virion.

4.35   Bacteriophage RdRp enzymes subvert abundant host proteins to create an efficient replicase complex

Although the phage replicase protein has RdRp activity, it cannot work efficiently without subverting proteins from its host. This situation is best understood in the case of Qβ, a phage very similar to MS2. During a Qβ infection, the phage RdRp forms a complex with three host proteins: EF-Tu, EF-Ts, and S1 (**Figure 4.47**). EF-Tu and EF-Ts are cellular translation factors and S1 is a ribosomal protein, but Qβ subverts these proteins for use during viral RNA synthesis. Because virus genome lengths are limited by selective pressure in favor of rapid replication, by the need to fit within the capsid, and in favor of large burst sizes, viral genomes are under pressure to remain small. Therefore, viral RdRps almost always subvert host proteins to improve the activity of the viral RdRp, using abundant proteins encoded by the large host genome. Functions associated with subverting host proteins that interact with viral RdRps include improving the RdRp's processivity, improving its ability to copy templates with extensive secondary or tertiary structures, or altering its activity in favor of the production of either antigenomes, genomes, or subgenomic mRNA, of the same sense as the genomes but shorter. Subgenomic mRNAs are not found in Qβ or its relatives, but there are some eukaryotic ssRNA viruses that express subgenomic mRNA molecules (see **Chapter 5**). The normal cellular function of these subverted host proteins is most often translation for both RNA bacteriophages and animal viruses. This ability likely reflects not the original functions of the translation proteins but rather their abundance in the cytoplasm, which makes them available to infecting viral genomes.

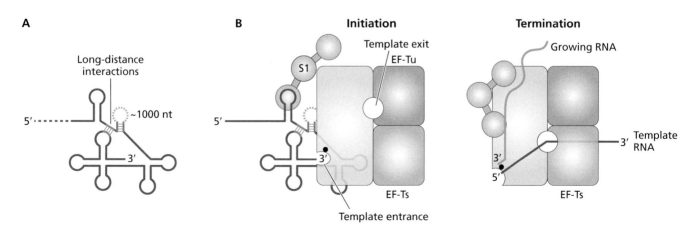

Figure 4.47 The Qβ RdRp complex. (A) Diagram of the Qβ genome during replication. (B) The phage RdRp (yellow) forms a complex with three host proteins: EF-Tu (red), EF-Ts (blue), and S1 (green). The model in the diagram is based on the crystal structures of the proteins. During initiation, shown on the left, the secondary structure of the genome is critical for complex formation. A conformational change in the S1 protein is critical for termination, as shown on the right. (From Tomita K 2014. *Int J Mol Sci* 15:15,552–15,570. Courtesy of Kozo Tomita. Published under CC BY 3.0.)

The molecular biology of the trinary complex consisting of the Qβ phage RdRp, EF-Tu, and EF-T is well understood and supported by structural, biochemical, and genetic evidence. Both host translation factors join the RdRp in a complex needed for initiation of RNA synthesis as well as elongation. Although the RdRp provides the catalytic activity, the EF-Tu protein forms an exit channel used by the growing new RNA and also plays a crucial role in separating the newly synthesized RNA from the template to which it is base paired. Studying the RdRp and its interactions with EF-Tu and EF-T required cloning and expressing these proteins individually in order to reconstitute an *in vitro* system that could be manipulated to study the complex. Cloning viral genes or even genomes in order to express wild-type or altered versions of them is a key experimental approach in molecular and cellular virology (**Technique Box 4.2**).

Study of the bacteriophage Qβ genome replication proteins *in vitro* and during an infection reveals that the RdRp complex amplifies genomes and antigenomes exponentially. Both the genomes and the antigenomes remain single stranded in that, although they likely form secondary and tertiary structures with nucleotides in their same strand, there is no completely double-stranded RNA template equivalent to one genome bound to one antigenome through a double-helical structure. Although such double-stranded RNA would seem to be produced as a matter of course during RNA synthesis and would be very stable, many experiments have demonstrated that dsRNA does not serve as a template for RNA synthesis by the MS2 RdRp complex. The crystal structure of EF-Tu, EF-Ts, and RdRp shows that the EF-Tu subunit forms a wedge near the enzyme's active site, which pulls apart the nascent RNA from its template. The structure also reveals that the exit tunnel for nascent RNA is at a distance from the exit tunnel for template RNA. Thus, the replicase complex is able to amplify single-stranded RNA and give rise to single-stranded RNA molecules which can then fold independently, a capacity that is critical for the function of new (+) strands. Whether secondary structure in the antigenomes is also important remains an open question requiring further research, as is whether the phage switches from synthesizing antigenomes to synthesizing new genomes late in infection in order to prepare for maturation.

As the phage examples including Qβ genome replication illustrate, it is very common for viruses to borrow host proteins and subvert them for their own purposes. This general principle is as true of animal and plant viruses as it is of bacteriophages.

4.36 Bacteriophage proteins are common laboratory tools

Bacteriophage proteins are common laboratory tools in most molecular biology labs. A ubiquitous example is the use of RNA polymerase T7, which is often incorporated into schemes for overexpressing proteins in order to purify them (**Figure 4.48**). Many other phage proteins are useful in

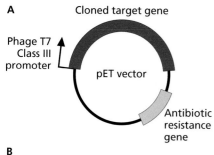

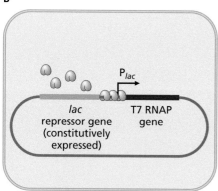

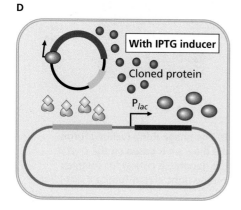

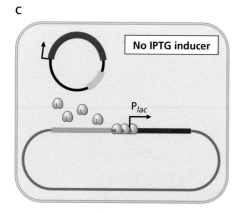

Figure 4.48 The pET expression system using T7 RNAP. (A) The target gene is cloned under the control of a bacteriophage T7 Class III promoter, which can be transcribed only by T7 RNA polymerase. (B) The host cell used for expression has a genetically engineered chromosome so that the *lac* repressor is expressed constitutively, whereas the T7 RNA polymerase gene is expressed under the control of the lactose promoter. (C) In the absence of the chemical IPTG, the *lac* repressor binds to P_{lac} and represses expression of the T7 RNA polymerase. The cloned gene in the pET vector is not expressed. (D) In the presence of IPTG, the *lac* repressor changes shape and cannot bind to DNA. P_{lac} is therefore derepressed and the T7 RNAP is expressed. The T7 RNAP then transcribes the cloned gene, leading to high levels of the protein encoded by that gene.

traditional molecular cloning using restriction enzymes and plasmids. For example, bacteriophage **T4 ligase** is the enzyme used to seal the nicks in a vector's backbone once it has hybridized to the sticky ends of the target gene. Target genes cut with a restriction enzyme have a 5′ phosphate group, which is required for the T4 ligase reaction to occur. The ligases from phages T3 and T7 have also been commercialized for similar purposes. Another useful phage enzyme is **T4 polynucleotide kinase** enzyme, which is typically used to phosphorylate PCR products before cloning them.

TECHNIQUE BOX 4.2   MANIPULATION OF CLONED DNA

Molecular cloning of a viral gene or genome enables many downstream applications such as sequencing of the DNA, targeted mutagenesis of the cloned DNA, and overexpression of the protein encoded by the cloned gene (**Figure 4.49**). Purified proteins can then be subjected to biochemical and structural analysis. Another advantage of cloning is that bacteria containing a plasmid can be cryogenically preserved indefinitely, thus providing a long-term source

of the cloned DNA. Viral genes can be cloned into a dizzying array of vectors (**Figure 4.50**). Some vectors are optimized for expressing genes in *E. coli* or insect cells for biochemical purification, whereas others are optimized for expressing a gene in mammalian cells, typically to study the gene's function in those cells. Some features commonly found in today's cloning vectors include a promoter that can be controlled by manipulating the cell's environment

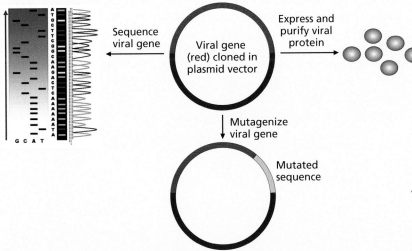

Figure 4.49 Purpose of cloning genes and genomes. Cloning a gene or genome makes it easier to sequence it, mutagenize it, or express and purify viral proteins from the cloned DNA. (Courtesy of Abizar. Published under CC BY-SA 3.0.)

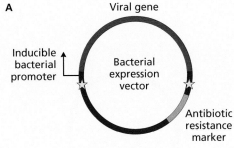

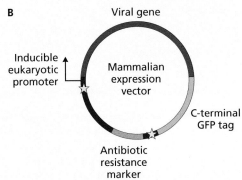

Figure 4.50 Examples of cloning and expression vectors. (A) A typical bacterial expression vector in which a viral gene is under the control of an inducible promoter. The cloned gene is flanked by primer binding sites (indicated by stars) convenient for sequencing, and the vector has an antibiotic resistance gene conferring resistance to an antibiotic such as ampicillin, kanamycin, tetracycline, or chloramphenicol. (B) The mammalian expression vector is designed to add a C-terminal green fluorescent protein (GFP) tag to the cloned protein and includes an inducible eukaryotic promoter, convenient primer binding sites (stars), and an antibiotic resistance gene conferring resistance to an antibiotic such as neomycin. Selection in bacterial cells compared with eukaryotic cells requires use of different antibiotics.

(for example by shifting the cells to a higher temperature or making the sugar arabinose available to cells), epitope tags or fluorescent protein sequences to make chimeras convenient for biochemical or microscopic applications, and, always, a marker that enables selective propagation of the cells containing the plasmid. Cloning vectors also typically include convenient primer annealing sites flanking the cloned gene, so that the inserted DNA can be sequenced from both ends.

Some viral genomes are too large to be cloned into a traditional plasmid and are not reliable for cloning DNA that is larger than 10,000–15,000 bp. Instead, vectors derived from phages such as λ can be used for cloning larger segments of viral DNA or cDNA. An example is the use of a **cosmid** (**Figure 4.51**). Cosmids are bacterial plasmids that have an origin of replication and a selectable marker for maintenance in *E. coli* host cells and also have two *cos* sites from λ cloned on either side of a polycloning site. Large (25–37 kbp) DNA segments can be ligated into the polycloning site, and then the ligation mix is combined *in vitro* with a λ phage packaging system that contains empty virus heads, tails, and the viral packaging enzyme that fills the heads with DNA between two *cos* sites. The filled heads and tails assemble into an infectious virus particle that can attach to a new *E. coli* host and introduce the cloned DNA into the new host cell, where it will persist in the cytoplasm as a plasmid.

The first entire cloned viral cDNA was reverse transcribed from poliovirus, which normally has a positive-strand RNA genome. Unexpectedly, plasmid encoding the viral cDNA could be introduced into host cells that normally support the replication of poliovirus, and the cells transfected with the plasmid made infectious poliovirus. Other viruses can be manipulated in the same manner; in most cases, including that of poliovirus, it is not clear mechanistically how forcing DNA into the cytoplasm of a host cell results in virus production. Being able to do so is so experimentally valuable that we use it without worrying too much about why it works. For example, it is possible to introduce any mutation imaginable into cloned DNA; we can create mutant cDNA clones to evaluate whether deletion of a certain viral gene, or even a small alteration to the coding sequence for a certain viral protein, will alter the course of a viral infection initiated by the cloned DNA or cDNA (**Figure 4.52**). Molecular manipulation of DNA, RNA, and protein sequences, in combination with microscopy, biochemistry, and bioinformatics, is likely to remain one of the most important pillars of virological research techniques for the foreseeable future.

Another common phage genome that has been manipulated for molecular cloning is that of M13. Sometimes experiments require the use of single-stranded DNA, in which case we can clone the desired DNA into a vector derived from the double-stranded replicative form of the M13 (ssDNA) phage genome. Then when the DNA is introduced into a host cell, the M13 genome replication proteins encoded by the vector synthesize ssDNA corresponding to the cloned DNA.

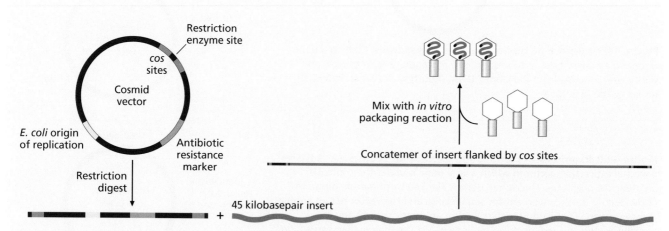

Figure 4.51 Cloning large inserts using λ cosmids. The starting vector has a bacterial origin of replication and antibiotic resistance gene for maintenance in *E. coli* as a plasmid. The vector and insert can then be digested with the same restriction enzyme and then ligated together, forming concatemers of insert separated by the linearized vector, which in effect flanks the insert with *cos* sites. Upon mixing with an *in vitro* reaction containing phage maturation enzymes, heads, and tails, the DNA including the *cos* sites will be packaged into infectious λ phages.

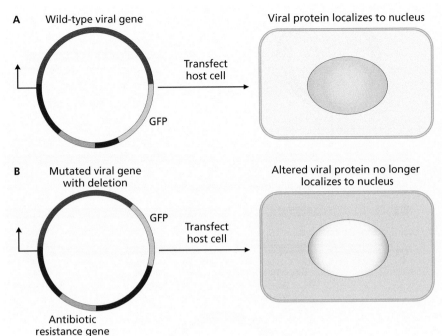

Figure 4.52 Testing the function of a viral protein through manipulation of a cloned viral gene. (A) Tagging the wild-type gene with GFP results in a green nucleus, indicating that the normal viral protein localizes to the nucleus. (B) A mutant version of the gene with DNA encoding just the N-terminal domain of the viral gene localizes instead to the cytoplasm. This experiment suggests that the C-terminal region of the viral protein is needed for nuclear localization.

Although molecular cloning with restriction enzymes and ligase revolutionized biology in the twentieth century, using restriction enzymes and ligase for cloning has many drawbacks. For example, it is laborious, can take many days, and can be inefficient. In contrast, newer recombination-based cloning techniques are altering the way that labs perform routine cloning. These techniques rely on bacteriophage proteins and are so much faster and more efficient that they are replacing the use of restriction enzymes. One of the most rapid and flexible methods, called **Gibson cloning**, uses a combination of phage and other enzymes and enables us to join any two fragments of DNA with at least 20 bases of overlap in sequence at one end (**Figure 4.53**). After PCR is used so that the destination vector and target sequence have these overlaps, a combination of phage and bacterial proteins recombines the DNA molecules where they overlap. A tremendous advantage to this method of cloning is that it is independent of the sequence at the desired site of insertion in the plasmid and independent of the sequence of the insert, so that it is no longer necessary to find just the right combination of restriction enzymes. A single reaction can also be used to connect multiple DNA sequences together. Gibson cloning has been used not only for cloning single inserts but also to assemble very large DNA molecules using synthetic DNA or PCR products. An example is the assembly of cDNA corresponding to an RNA viral genome as large as 15,000 bp. It could be used to assemble much larger genomes, such as those of herpesviruses or poxviruses (which can be more than 200,000 bp).

A second very common recombinational cloning strategy that has also been commercialized relies upon temperate bacteriophage enzymes that catalyze recombination between short segments of specific target DNA sequences. One of the first systems that was commercialized relies upon the phage λ protein Int, which is a site-specific recombinase used to establish and reverse lysogeny (**Figure 4.54**; see **Chapter 13**). To establish lysogeny, the enzyme, in combination with subverted host protein IHF, catalyzes recombination between one specific sequence (*attB*) in the host chromosome and another similar sequence (*attP*) in the phage genome. The result is that the

Figure 4.53 Gibson cloning. (A) A plasmid vector is linearized and amplified using PCR with primers facing away from each other. One of the primer annealing sites (orange) is much closer to the antibiotic resistance marker (purple). (B) The target gene is amplified using primers with 20 base-pair extensions at their 5′ ends; the extensions have the same sequence as the DNA bound by the primers that linearized and amplified the plasmid. (C) The two PCR products are mixed together with a phage exonuclease, which degrades one strand of double-stranded DNA, leaving 3′ single-stranded DNA at each end. The single-stranded "A" DNA is complementary to the single-stranded "a" DNA; the orange "B" and "b" sequences are also complementary to each other. (D) The temperature is raised to 75°C, which allows the overlapping DNA to anneal and at the same time inactivates the exonuclease by denaturing it. (E) Heat-resistant cellular DNA polymerase and ligase added to the reaction fill in the gaps and seal the nicks, resulting in closed circular plasmid DNA without the use of restriction enzymes.

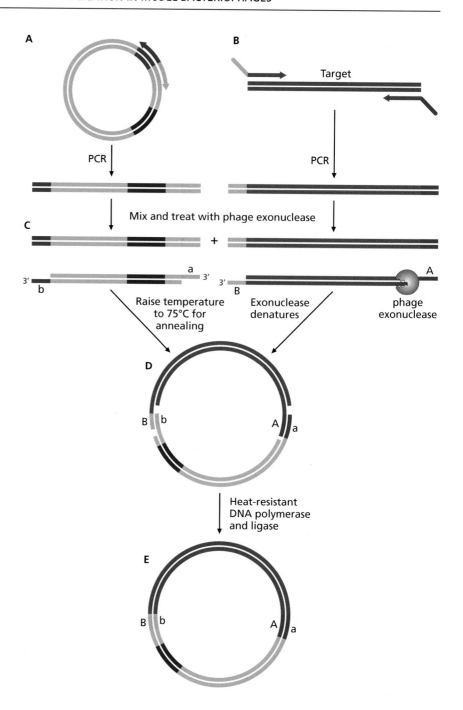

phage DNA inserts into the chromosome and is flanked by new sequences, which are called *attL* and *attR*. To initiate a lytic infection starting with a lysogen, the phage proteins Int and Xis and host IHF are needed to catalyze the opposite reaction, in which they recombine the *attL* and *attR* sequences, producing a circularized λ genome.

This natural λ recombination system has been adapted for cloning (**Figure 4.55**). Cloning vectors with the recombination target *attP* are first created; in these vectors, the *attP* sites surround a stuffer gene that encodes an indicator protein like GFP. Next the target sequence to be cloned is amplified with primers that add the necessary *attB* sequences on either side of the target. The λ Int recombination enzyme is very specific so that the vector and PCR product are mixed *in vitro* with Int and IHF protein, *attP* and *attB* are recombined, resulting in directional cloning of the PCR product. The flanking sites

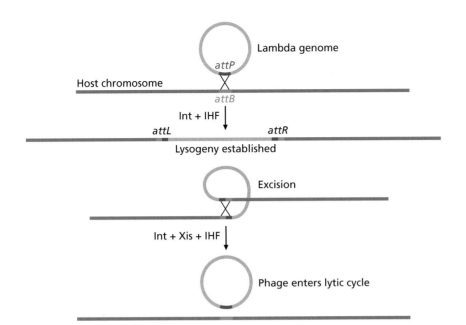

Figure 4.54 Recombination catalyzed by bacteriophage λ proteins. To establish lysogeny, expression of the phage Int protein causes recombination between a short sequence in the phage genome called *attP* and a different short sequence in the bacterial chromosome called *attB*. The host protein IHF is also required for this recombination. The result is that the prophage DNA enters the chromosome flanked by *attL* and *attR* sequences. To exit lysogeny, the phage proteins Int and Xis, along with the host protein IHF, catalyze the reverse reaction, recombining *attL* and *attR* so that the phage genome reforms. The phage genome is then used to initiate lytic replication.

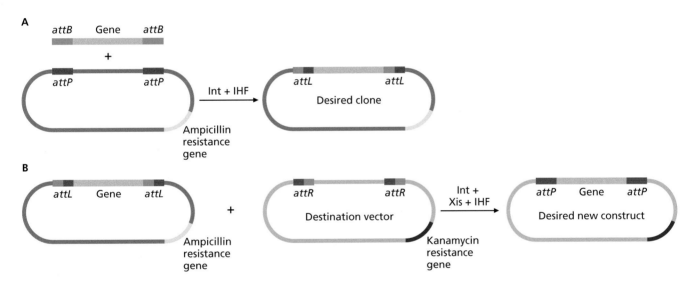

Figure 4.55 Use of λ recombination proteins in cloning. (A) To clone a gene using λ recombination proteins, the gene is amplified with primers that add *attB* sites to both ends of the gene. After mixing with a vector that has *attP* sites, the λ Int and host IHF proteins will recombine the DNA, cloning the gene of interest. (B) The first clone can subsequently be used to move the gene into a variety of new destination vectors that contain *attR* sites and a different antibiotic resistance marker. Two λ proteins (Xis and Int) and the host IHF protein are necessary to recombine the *attL* sites surrounding the clone with the *attR* sites in the new destination vector.

where recombination occurred are now flanked by *attL*. This reaction occurs with a higher percentage of successful cloning than is usually achieved by cloning with restriction enzymes. Once cloned into a plasmid and flanked by *attL* sites, the gene can easily be moved into any different vector that has *attR* sites, using Int, Xis, and IHF in the reaction. The biotech industry has commercialized this system, providing labs with a variety of vectors that can be used to clone a gene for bacterial expression, mutagenesis, or expression in a variety of eukaryotic cells.

Phage proteins are also used in applications other than molecular cloning. For example, the Φ29 phage DNA polymerase protein is highly processive,

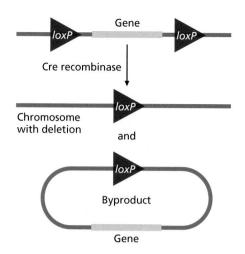

Figure 4.56 The Cre–*lox* system can be used to delete a gene. A chromosome has been engineered so that *loxP* sites flank the gene of interest. When the Cre recombinase is expressed in the eukaryotic cells, it catalyzes recombination, resulting in deletion of the gene from the chromosome and a circular byproduct that will be degraded. The result is that the target gene has been knocked out.

meaning that it is able to copy many thousands of bases before it dissociates from the template. This property makes it possible to use the enzyme in whole genome amplification (WGA). In this procedure, the genome from a single cell is used as a template to make enough DNA to determine that cell's genome sequence with a high level of confidence because the enzyme also has very high **fidelity**. Consequently, the DNA synthesized by the enzyme is a faithful copy of the original with a low misincorporation rate. WGA is typically used to generate enough DNA to determine the sequence of the amplified DNA. It can also be used in combination with other techniques to clone viral genomes or cDNA representing RNA genomes.

Another indispensable contemporary use of phage proteins is to catalyze site-specific recombination in genetically engineered model organisms. This recombination can be useful in various ways. Here we will consider using this technology to knock out a gene. One virus-based system used for this purpose is based on the bacteriophage P1 recombinase protein Cre, which catalyzes recombination between specific 34 base-pair sequences of DNA called *loxP* sites (**Figure 4.56**), similar to the way that the λ Int protein and host IHF protein catalyze recombination between *attP* and *attB* sites. Mice can be genetically engineered to contain *loxP* sites flanking any exon of interest and to express the Cre recombinase during a specific stage of development *in utero*, or in a specific tissue or organ (**Figure 4.57**), such as the liver, skeletal muscle, or brain. Expression of the P1 phage Cre recombinase causes deletion of the exon flanked by *loxP* sites, typically resulting in a null mutation that inactivates the protein encoded by the target gene. Genetically engineered mice are important animal hosts used to study virus replication and pathogenesis *in vivo*. Any recombinase system that is effective for genetically engineering host DNA can also be used to genetically engineer viral DNA or cDNA. For this application, the Cre–*lox* system is preferable to the

Figure 4.57 Use of Cre–*lox* system to make a knockout in a mouse. The parental female mouse (left) has been engineered so that an exon is flanked by *loxP* sites. Although the mouse is diploid, only one copy of the chromosome has been depicted. This mouse can be mated to a male (right) in which the P1 phage Cre recombinase protein is expressed only in skeletal muscle tissue. As indicated by the picture of a muscle cell under each parent, their muscles are normal. When the two mice are mated, their offspring will have the exon deleted specifically in skeletal muscle because of expression of Cre in that tissue, although the exon remains intact in all other cells of the body. If the knockout mouse has abnormal muscle tissue, the normal gene likely plays a role in muscle structure or function.

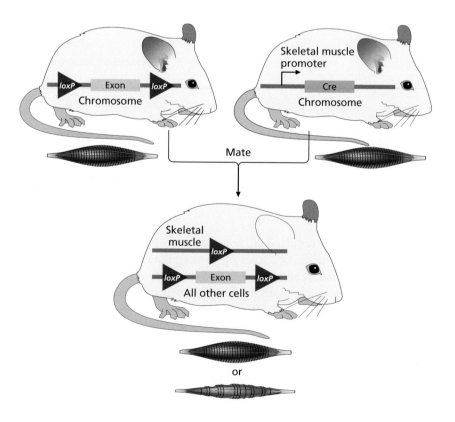

λ Int system because it requires expression of just one protein (Cre) and uses shorter recombination sites (*loxP*).

Some phage-based technologies have also been developed for studying protein–protein interactions. A key innovation was to create recombinant filamentous phages with genomes that encode protein fusions with the coat protein. The result is that the phage particles display the desired folded protein on their surfaces. Phage display is especially valuable when combined with experimental evolution as follows. A mutagenized library of the gene encoding the protein of interest is recombined with the phage genome, resulting in infectious phages that have a novel protein fusion. The collection of phages is then subjected to an artificial selection such as the ability to bind to a desired target. The phages that bind to the desired target can be further purified and then used in subsequent rounds of mutagenesis, expression, and screening to isolate proteins with optimized properties. This work was rewarded with the 2018 Nobel Prize in Chemistry.

Essential concepts

- Bacterial transcription and translation are physically and temporally coupled; therefore, dsDNA bacteriophages typically regulate transcription (but not translation) in order to regulate gene expression.
- Bacteriophages use polycistronic mRNAs, thus mimicking host molecules.
- Bacteriophage T7 serves as a model for the expression of early mRNA and proteins (Stage 3) and late mRNA and protein expression and genome replication (Stage 4).
- Pulse-chase technology can reveal temporal waves of viral gene expression in which the timing of expression is correlated with the function of the proteins expressed at that time. For example, regulatory proteins and enzymes are often expressed early, whereas structural proteins are expressed late.
- Phages with dsDNA genomes can regulate either transcription initiation or termination, as exemplified by T7 and λ, respectively.
- Phages with dsDNA genomes typically replicate by forming concatemers; concatemerization is a mechanism for solving the problem of replicating linear genomes completely.
- Phages with small ssDNA genomes, such as ΦX174 and M13, typically encode proteins using overlapping sequences that maximize the use of a small number of nucleotides to encode many proteins.
- Both ssDNA and ssRNA phages replicate through a double-stranded replicative form intermediate.
- All (+) ssRNA viruses encode an RNA-dependent RNA polymerase because there are no host enzymes that catalyze polymerization of RNA using an RNA template.
- In (+) RNA phages, the relative abundance of viral proteins is controlled by the genome's secondary structure.
- Bacteriophages and their proteins are typical laboratory research tools essential for contemporary research.

Questions

1. List examples from this chapter of the process of co-optation, also known as subversion, in which a virus uses a host protein for a purpose that promotes viral replication, usually using that host protein differently from the way the host uses the same protein.

2. Compare and contrast the regulation of gene expression in T7 and λ. List as many similarities and differences as possible.

3. Explain the differences and similarities between θ and rolling-circle (bidirectional and unidirectional) DNA replication.

4. Provide examples of overlapping genes in at least two phages and explain how they are expressed even though they overlap.

5. Would you expect that %A = %U for the RNA phages discussed? Why or why not?

6. Consider the λ genome. Design primers that would amplify the DNA when the phage was inside a host in the midst of gene expression yet would *not* amplify the DNA when the genome was isolated directly from phages.

7. Explain the different roles of the proteins needed for normal bacterial transcription.

8. How is T7 RNAP different from host RNAP?

9. Explain the different roles of the proteins needed for normal bacterial translation elongation.

10. List the host proteins required for T7, λ, M13, and MS2 genome replication. Considering these examples, do you note any trends or associations?

Interactive quiz questions

In addition to the questions provided above, this edition has a range of free interactive quiz questions for students to further test their understanding of the chapter material. To access these online questions, please visit the book's website: www.routledge.com/cw/lostroh.

Further reading

Bacteriophage T7

Geertsema HJ, Kulczyk AW, Richardson CC & van Oijen AM 2014. Single-molecule studies of polymerase dynamics and stoichiometry at the bacteriophage T7 replication machinery. *Proc Natl Acad Sci USA* 111:4073–4078.

Koh HR, Roy R, Sorokina M, Tang GQ, Nandakumar D et al. 2018. Correlating transcription initiation and conformational changes by a single-subunit RNA polymerase with near base-pair resolution. *Mol Cell* 70:695–706.

Sun B, Pandey M & Inman JT 2015. T7 replisome directly overcomes DNA damage. *Nat Commun* 6:10260 (doi: 10.1038/ncomms10260).

Bacteriophage λ

Conant CR, Goodarzi JP, Weitzel SE & von Hippel PH 2008. The antitermination activity of bacteriophage lambda N protein is controlled by the kinetics of an RNA-looping-facilitated interaction with the transcription complex. *J Mol Biol* 384:87–108.

Echols GE 2001. *Operators and Promoters: The Story of Molecular Biology and Its Creators* (Gross CA ed.). University of California Press.

Lewis DE, Gussin GN & Adhya S 2016. New insights into the phage genetic switch: effects of bacteriophage lambda operator mutations on DNA looping and regulation of P_R, P_L, and P_{RM}. *J Mol Biol* 428:4438–4456.

Liu X, Jiang H, Gu Z, & Roberts JW 2013. High-resolution view of bacteriophage lambda gene expression by ribosome profiling. *Proc Natl Acad Sci* 110:11928–11933.

Bacteriophage ΦX174

Sanger F 1980. Determination of nucleotide sequences in DNA. http://www.nobelprize.org/nobel_prizes/chemistry/laureates/1980/sanger-lecture.html

Sanger F & Rose J 2001. Interview with Frederick Sanger. www.nobelprize.org/prizes/chemistry/1958/sanger/interview/

Zhao LY, Stanick AD & Brown CF 2012. Differential transcription of bacteriophage ΦX174 genes at 37°C and 42°C. *PLOS One* 7:e35909.

Biotechnology applications of phage biology

Gibson DG, Young L, & Chuang RY 2009. Enzymatic assembly of DNA molecules up to several hundred kilobases. *Nat Methods* 6:343–345.

Jia H, Yue X, & Lazartigues E 2020. ACE2 mouse models: a toolbox for cardiovascular and pulmonary research. *Nat Comm* 11:5165.

Katzen F 2007. Gateway® recombinational cloning: a biological operating system. *Expert Opin Drug Discov* 2:571–589.

Xu H, Li L, Deng B, Hong W, Li R et al. 2022. Construction of a T7 phage display nanobody library for bio-panning and identification of chicken dendritic cell-specific binding nanobodies. *Sci Rep* 12:12122.

Gene Expression and Genome Replication in the Positive-Strand RNA Viruses

Virus	Characteristics
Poliovirus	Picornavirus; genome has covalently attached 5′ VPg protein and 3′ poly(A) tail; encodes one polyprotein processed into individual proteins by viral proteases. Viral RdRp uses protein primer.
Hepatitis C virus	Flavivirus; genome lacks a 5′ cap, has secondary structure at 3′ end, and lacks a poly(A) tail; it encodes one polyprotein, most of which must be translated on rough endoplasmic reticulum (ER) because of transmembrane segments.
Sindbis virus	Togavirus; genome has 5′ cap and 3′ poly(A) tail; synthesizes two (+) strand RNAs and encodes multiple polyproteins; uses suppression of translation termination and ribosome frame-shifting during translation.
SARS-CoV-2	Coronavirus; encodes two polyproteins and multiple other proteins; synthesizes (–) strands in replicative form through discontinuous method; uses leaky scanning during translation; very large genome possible because of proofreading associated with ExoN nonstructural protein. Cause of the COVID-19 pandemic.

The viruses you will meet in this chapter and the concepts they illustrate

Chapters 5–10 address gene expression and genome replication in viruses that infect animals and plants with an emphasis on viruses that infect humans. We start this chapter with the Class IV (+) strand RNA viruses, which are the most abundant known viruses that infect animals and plants. They may really be the most abundant eukaryotic viruses in nature or they just might be easy for us to detect. One of the most well-known Class IV viruses is poliovirus, which has been studied intensively from the beginning of molecular and cellular virology as a discipline. The COVID-19 pandemic has intensified global interest in coronaviruses, which are also Class IV viruses.

The themes introduced in **Chapter 4** about bacteriophage gene expression and genome replication are expanded when we consider the Class IV viruses of animals and plants. For example, all the viruses in this chapter have genomes in which secondary structures such as stem–loops or cloverleafs affect translation or RNA synthesis. Class IV viruses that infect plants and animals differ from those that infect bacteria in that, generally speaking, eukaryotic mRNA is monocistronic. Monocistronic mRNA permits translation of just one protein per mRNA because eukaryotic mRNA does not have Shine–Dalgarno sequences that facilitate ribosome assembly around

DOI: 10.1201/9781003463115-5

internal start codons. Instead, eukaryotic translation initiation depends on the 5′ methylated cap and the poly(A) tail, features not found in bacteria and that result in translation of whatever protein is encoded closest to the 5′ end of the mRNA. Therefore, all the animal viruses in this chapter have evolved interesting mechanisms to express more than one protein from a single infecting (+) strand RNA genome. Some of the Class IV plant viruses have solved this problem by being multipartite, that is, having more than one genome segment, but that is not the case for the four animal virus families described in this chapter.

Nucleic acid synthesis is also an issue for all Class IV viruses because host enzymes do not synthesize RNA using an RNA template; instead, a virus-encoded RNA-dependent RNA polymerase (RdRp) is required. Because the viral genome can be translated after uncoating, Class IV viruses do not package RdRp into their virions. We will examine the Class IV poliovirus RdRp in the most detail. Keep in mind that the mechanisms of mRNA synthesis and genome synthesis must produce the 5′ and 3′ terminal features necessary for translation, replication, or both.

The chapter starts with the simpler Class IV animal viruses, namely, picornaviruses and flaviviruses. Their gene expression and genome replication are similar, with the exception that flaviviruses must translate transmembrane proteins because, unlike picornaviruses, they are enveloped. These two families are so simple that they do not have early and late gene expression, and they do not produce a tremendous excess of structural proteins compared with nonstructural proteins. In contrast, the togaviruses are more complex in that they have both early and late gene expression and produce nonstructural proteins in great abundance during late gene expression. They also exhibit common viral mechanisms for encoding different proteins with overlapping mRNA sequences. The coronaviruses are introduced last. Their genomes are five times larger than those of picornaviruses and they have the most complex gene expression and genome replication of all the viruses discussed in this chapter. They have a unique mechanism by which they produce new RNA. Coronaviruses exhibit not only early and late gene expression and abundant production of nonstructural proteins, but also something quite unexpected in the RNA world: a novel editing mechanism that corrects misincorporations during RNA synthesis. The chapter concludes with a brief consideration of Class IV plant viruses, emphasizing aspects of their gene expression and genome replication that are distinctive compared with the animal virus families described in Sections 5.1–5.29.

5.1 Class IV virus replication cycles have common gene expression and genome replication strategies

The four families of (+) strand RNA animal viruses share a general scheme for ensuring the production of RdRp and other nonstructural proteins, the production of structural proteins, the synthesis of antigenomes, and the synthesis of new genomes (**Figure 5.1**). After uncoating, the (+) strand RNA is used as mRNA and virus proteins are translated, including the RdRp. The RdRp assembles with both virus and host proteins on a specific membrane, and together the proteins and viral nucleic acids result in formation of a **virus replication complex (VRC)**. Within the VRC, the (+) strand genome is used as a template to synthesize a full-length antigenome, which remains hydrogen bonded to the (+) strand genome. Cells do not normally contain long (>1 kb) cytoplasmic dsRNA; in fact, animal cells have innate immune mechanisms for recognizing dsRNA, which trigger an antiviral response

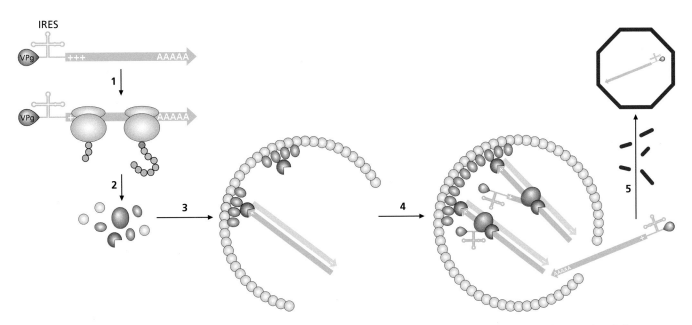

Figure 5.1 Overview of events during gene expression and genome replication in (+) strand RNA viruses that infect eukaryotes. After uncoating, the internal ribosome entry site (IRES) enables the genome to be translated (1), making a polyprotein that is processed into individual proteins (2). The proteins go on to form virus replication compartments (3) in which double-stranded replicative forms are used to make mRNA and new genomes (4), which are ultimately used to make new infectious virions (5).

(see **Chapter 14**). The VRCs are probably also important for sequestering the replicative form away from cytoplasmic host antiviral detection systems. The viral dsRNA intermediate, called the replicative form, is then used as the template for synthesis of many more (+) strand RNA molecules, which can be used as mRNA or can become new genomes during the assembly stage.

5.2   Terminal features of eukaryotic mRNA are essential for translation

The viruses in this chapter have genomes that, upon uncoating, are recognized and translated by the host's ribosomes. Eukaryotic ribosomes recognize and bind to specific features of mRNA that are unique to eukaryotes. For example, eukaryotic mRNA has a covalent linkage to 7-methylguanylate, also known as a cap, at the 5′ end (**Figure 5.2**). Following this cap, there can be hundreds of untranslated nucleotides, called the 5′ untranslated region (UTR). Eukaryotic mRNA typically encodes a single protein, so that there is a single start codon, followed by a coding sequence and a single stop codon. After the stop codon, there is a 3′ UTR, followed finally by a poly(A) tail of 200–300 As at the 3′ end. Unlike bacterial mRNA, there is no specific ribosome-binding site sequence surrounding a start codon; instead both the 5′ cap and the 3′ poly(A) tail are essential for translation initiation. Both the 5′ UTR and the 3′ UTR can play regulatory roles that affect gene expression, which some viruses also subvert for translation.

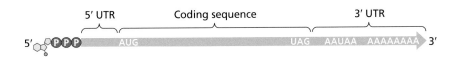

Figure 5.2 Eukaryotic mRNA. Eukaryotic mRNA has a methylated cap at the 5′ end, a 5′ UTR, a single coding sequence, a 3′ UTR, a polyadenylation sequence (AAUAA), and a polyadenylated tail at the 3′ end.

Figure 5.3 Eukaryotic mRNA is always bound by proteins, forming mRNP. Before translation, eukaryotic mRNA is bound to translation factors and the poly(A)-binding protein.

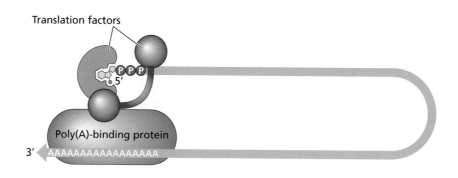

Features of the mRNA alone are not sufficient for translation initiation; instead, eukaryotic mRNAs exist as **messenger ribonucleoprotein complexes (mRNPs) (Figure 5.3)** in which the protein components of mRNPs have dramatic effects on translation. A particular collection of **eukaryotic translation initiation factor** proteins is needed for translation initiation to occur. Viral (+) strand RNA must mimic this mRNP in order to be translated. Class IV (+) strand ssRNA virus genomes must have either the same terminal features as eukaryotic mRNA or a mechanism for bypassing the use of these terminal features during translation.

5.3   Monopartite Class IV (+) strand RNA viruses express multiple proteins from a single genome

Monopartite Class IV (+) strand RNA viruses such as picornaviruses, flaviviruses, togaviruses, and coronaviruses have just one genome segment that host cells translate immediately after uncoating. Immediate translation is critical for viral replication because it results in synthesis of the viral RdRp. In turn, the RdRp subsequently synthesizes the replicative forms and viral mRNA. Eukaryotic mRNA typically encodes just one protein, however, whereas (+) strand RNA virus genomes must encode at least two proteins: a capsomer and an RdRp. The Class IV viruses have a common solution to this problem: the genome encodes a **polyprotein** that is **proteolytically processed** to release many individual proteins including the capsomers and an RdRp.

5.4   Picornaviruses are models for the simplest (+) strand RNA viruses

Picornaviruses are some of the simplest viruses known to infect humans, in terms of their genome size and the number of viral proteins they express. Despite their simplicity, they can cause serious human infections. For example, poliomyelitis, commonly known as polio, is caused by the picornavirus **poliovirus** (order *Picornavirales*, family *Picornaviridae*, genus *Enterovirus C*). Polio begins as a gastrointestinal infection, but sometimes the virus moves from the gastrointestinal tract into the nervous system and can subsequently cause paralysis. If the paralysis affects the respiratory system, the infected person will probably die without artificial respiration provided in a hospital. Paralysis in other parts of the body can lead to long-term problems, such as difficulty walking or an inability to use a particular limb. The World Health Organization is in the midst of a campaign to eliminate poliovirus completely. In wealthier countries, polio is viewed as a scourge of the past—someone might have a grandparent or great aunt who walks with a cane because of a childhood polio infection, but likely does not know anyone their own age who has had a polio infection. Within living memory, however, there has been active polio in Southeast Asia, East Africa, West Africa, and North Africa. If the eradication campaign succeeds, poliovirus will be the third virus eliminated through vaccination, joining smallpox and rinderpest (a disease of cattle, not humans).

A second enterovirus, rhinovirus, will likely be around for many more human generations. Fortunately, it is responsible for the common cold rather than for a life-threatening infection. Studies of rhinovirus and poliovirus have provided most of the insights into picornavirus gene expression and genome replication. **Sections 5.5–5.11** focus on poliovirus as the specific example, but the strategies of this virus for gene expression and genome replication can be generalized to other picornaviruses. Our detailed knowledge of the mechanisms for gene expression and genome replication arose from investment in basic science related to human health that was not focused on clinical application. This detailed molecular information is clinically important, however, in many ways. Some examples include the design or discovery of antiviral medicines that target virus gene expression or genome replication, along with prediction of virus evolution in response to use of those drugs. Although this book is not about the clinical manifestations of viral infections, it does provide the foundations for understanding infectious diseases on a molecular and cellular level, which in turn is likely to yield clinical benefits.

The name of the picornaviruses derives from their size (tiny, or pico) and the nucleic acid found in the virions (RNA) (**Figure 5.4**). All picornaviruses have simple naked icosahedral virions containing single-stranded

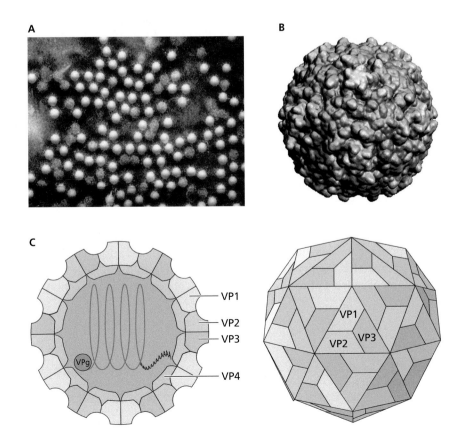

Figure 5.4 Picornavirus structure.
(A) Electron micrograph of poliovirus. These viruses are approximately 35 nm in diameter. (B) Crystal structure of poliovirus, a picornavirus. (C) Proteins in the virion include structural capsomeres VP1, VP2, VP3, and VP4 and one copy of the VPg protein covalently attached to the 5′ end of the genome. (A, Courtesy of Dr. Graham Beards. Published under CC BY-SA 4.0. C, Courtesy of Philippe Le Mercier, ViralZone, © SIB Swiss Institute of Bioinformatics.)

Figure 5.5 Picornavirus genome. Important elements of the single-stranded (+) sense RNA genome are indicated, such as the covalently attached terminal VPg protein, the secondary structures needed for translation, and the polyprotein separated into polyproteins P1, P2, and P3. These are subsequently digested into the smaller proteins named in the boxes, such as VP4 or 2A.

linear (+) sense RNA. There are four different capsomers in the virion. Picornaviruses have small genomes typically between 7 and 8.8 kb. The 5′ end of their genomes is not capped with 7-methylguanylate, as might be expected for a (+) strand genome that the host cell will translate. Instead, the 5′ end is covalently bound to the **VPg protein**, and the RNA in the 5′ UTR has extensive secondary structure (**Figure 5.5**). The 3′ UTR also contains secondary structure and the 3′ end of the genome is polyadenylated.

5.5 Class IV viruses such as poliovirus encode one or more polyproteins

Cells infected with a picornavirus typically contain a few viral structural proteins and seven viral nonstructural proteins, which are mostly enzymes. Yet, typical animal mRNA encodes just one protein, and poliovirus has only one genome segment. How does poliovirus accomplish this feat? We can use a computer to examine the sequence of the poliovirus genome and translate its genome *in silico*, using software to identify start codons and stop codons. In the case of picornaviruses, the software predicts that the genome does not encode multiple proteins. Instead, the entire genome should be translated into a single, very large protein with more than 2,200 amino acids (>200,000 Da; **Figure 5.6**). This example illustrates that, although software analysis of sequences is essential, in the case of viruses especially, software predictions often do not tell the whole story.

The contradiction between the software prediction and the existence of many viral proteins must be resolved with experimentation. In this case, a useful experimental approach is **pulse-chase analysis** with radioactive ^{35}**S-methionine**. In this technique, cells infected by a virus are **pulsed** with short exposure to radioactive ^{35}S-methionine, which they import and incorporate into any proteins being actively translated during the pulse. The pulse selectively labels proteins that are being synthesized during the pulse because only ribosomes active during the pulse use the radioactive ^{35}S-methionine. The thousands of already extant proteins in the cell do not become radioactive because amino acids already incorporated into proteins do not spontaneously exchange with the radioactive amino acids in the cytoplasm. Radioactive ^{35}S-methionine can be detected with X-ray film or with an instrument called a **phosphorimager**, which also allows quantification of

Figure 5.6 A single start codon and stop codon are found in the picornavirus genome. The picornavirus genome has one start codon and one stop codon separated by about 2,200 codons that specify amino acids. (IRES, internal ribosome entry site.)

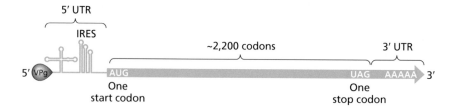

the amount of radioactivity over many orders of magnitude. Infected cells pulsed with radioactive methionine can be lysed immediately following the pulse in order to observe the proteins labeled during the pulse. In the case of a picornavirus, the result of such a procedure would reveal many proteins, including some very large ones (**Figure 5.7**). Alternatively, after the pulse, the infected cells can be exposed to a large excess of nonradioactive methionine. This procedure, called a **chase**, prevents proteins synthesized during the chase from being labeled because the concentration of radioactive methionine in the cytoplasm plunges by several orders of magnitude. The result is that proteins synthesized during the pulse are radioactive, and if they change size during the chase, that change can be detected using techniques to detect radioactivity while observing the size of the labeled proteins.

In the case of picornaviruses, the chase reveals that large proteins get shorter during the chase. For example, in Figure 5.7, compare the 15-min chase with that of the 240-min chase. Large radioactive proteins are more abundant in the 15-min chase, and there are few small radioactive proteins. After the 240-min chase, however, the amount of radioactivity in the large protein bands has decreased significantly, and several smaller radioactive proteins can be observed. These small proteins must have been synthesized on ribosomes 240 min ago, because that is the only way that they could have incorporated the radioactive label. The chase reveals that the small proteins originated from the large proteins. Poliovirus genomes *both* direct the synthesis of a single, large polyprotein *and* cause the appearance of many shorter, individual proteins in an infected cell.

Pulse-chase analysis remains important in virology, although use of radioactivity in research has fallen out of favor as unnecessarily hazardous and not especially environmentally friendly. One of the most common pulse-chase strategies in use today examines viral RNA rather than protein. In this case, the cells are pulsed with an altered RNA nucleotide called bromouridine triphosphate (BrUTP). This nucleotide is incorporated into RNA molecules in place of normal UTP, and it can subsequently be detected by antibodies that bind to the BrU epitope (**Technique Box 5.1**).

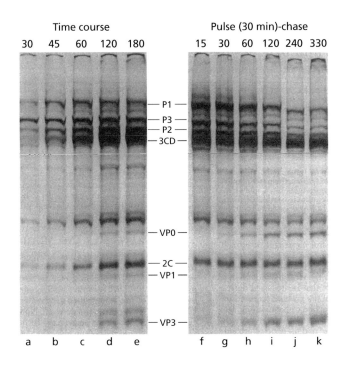

Figure 5.7 Pulse-chase analysis of picornavirus gene expression. This autoradiograph shows the results of an *in vitro* pulse-chase experiment in which an *in vitro* translation system was mixed with purified picornavirus genomes. The left segment shows the result of continuous ^{35}S-methionine labeling that began when the genome was added to the *in vitro* translation system and allowed to continue for the number of minutes indicated above each lane before protein synthesis was stopped and the proteins were collected and separated using sodium dodecyl sulfate–polyacrylamide gel electrophoresis (SDS-PAGE). The right segment indicates the result of a pulse-chase experiment in which the ^{35}S-methionine pulse was carried out for 30 min and then the radioactive methionine was chased with cold methionine for the number of minutes indicated above each lane before protein synthesis was stopped and the proteins were collected and separated using SDS-PAGE. Radioactive proteins were then detected by exposing the gel to X-ray film.

TECHNIQUE BOX 5.1   IMMUNOBLOTTING AND IMMUNOSTAINING

Antibodies are versatile research tools. They are proteins produced by vertebrate B cells as part of an adaptive immune response. They are typically diagrammed using a Y shape to emphasize their constant regions at the base of the Y and their **epitope-binding sites** at the end of each arm on the Y (**Figure 5.8**). An **epitope** is a small part of a foreign molecule that cells of the adaptive immune system can recognize. Our focus here is on epitopes that are parts of proteins. **Conformational epitopes** are those that form a specific shape bound by the antibody and are composed of amino acids that are not necessarily contiguous in the primary sequence of the protein antigen. **Linear epitopes** are those in which the antibody binds to amino acids that are contiguous in the primary sequence of a protein antigen. An antibody's epitope-binding sites are identical to each other and bind to their epitope with high selectivity. For example, an antibody that binds to a 15-amino acid linear epitope will interact with a similar epitope in which 1 of the 15 amino acids differs. But, the affinity would be many times lower and might even be below the limit of detection in a typical experiment using that antibody.

One use of antibodies to analyze protein samples is **immunoblotting**, also known as **western blotting** (**Figure 5.9**). Because antibodies are so specific, they can bind to just one protein antigen in a complex sample such as proteins collected from the cytoplasm of a cell infected by a virus. In western blotting, proteins in a complex sample are first separated by sodium dodecyl sulfate–polyacrylamide gel electrophoresis (SDS-PAGE). Then the separated proteins are transferred to a supportive solid substrate called a **membrane**, which is much less fragile than the polyacrylamide gel. The membrane can be physically manipulated in subsequent steps without tearing

or falling apart. The transfer process conserves the location of the proteins relative to one another and can be visualized as a mirror image of the gel. The membrane is subsequently incubated with an antibody known to bind to an epitope on a particular protein antigen in order to determine whether that protein is present in the sample. There are several procedures for detecting where this **primary antibody** binds to the membrane. For example, the primary antibodies themselves might be tagged with a fluorescent dye that permits detection of where they have bound. Older technologies employ **secondary antibodies** that bind to the **constant region** of the primary antibodies. The secondary antibody is typically tagged with a dye or

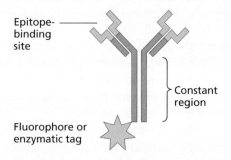

Figure 5.8 Structure of an antibody. The simplest antibodies consist of four polypeptides held together with disulfide bridges (not shown) to make a very large Y-shaped molecule that has two identical epitope-binding sites. A portion of the antibody is called the constant region because, unlike the epitope-binding sites, it is the same in all antibodies of the same type, from the same species. This is an example of an artificially tagged antibody with a fluorophore or enzymatic tag (green star) that will make it possible to detect where the antibody has bound to a sample.

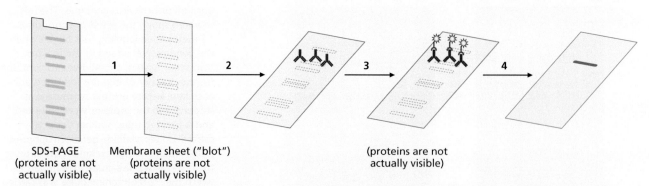

Figure 5.9 Immunoblotting (western blotting). After proteins have been separated by SDS-PAGE, they are transferred to a membrane (blot) using an electrical current (1) instead of staining them. The blot preserves the relative positions of the proteins that were separated by size during SDS-PAGE. The blot is immersed in a solution of primary antibodies that bind to an epitope unique to the protein of interest (2). Excess primary antibodies are washed away and then secondary antibodies that bind to the constant region of the primary antibodies are added (3). The secondary antibodies are also tagged, making them easy to detect. Common tags include a fluorophore or an enzyme that reacts with a clear substrate to make a colored product that sticks to the membrane at the position of the protein of interest (4).

attached to an enzyme that converts a clear substrate to a colored product that precipitates out on the membrane. Use of secondary antibodies amplifies the signal from the primary antibodies bound to the membrane.

An application that combines microscopy with the use of antibodies is **immunostaining**. In immunostaining, fixed, permeabilized cells are stained with an antibody already known to bind to a protein of interest. The location of the protein can then be imaged using fluorescence microscopy (if the antibody is tagged with a fluorophore) or electron microscopy (if the antibody is conjugated to gold nanoparticles). Staining of tissue sections using antibodies is called **immunohistochemistry**, which can reveal

the location of viruses or virus proteins in a complex sample such as a biopsy.

Immunohistochemistry remains important for determining the localization of proteins in research as well because localization patterns based on the analysis of green fluorescent protein (GFP) chimeras can be misleading if the large size of GFP affects the folding, activity, or localization of its partner in the chimera. In practice, we have greater confidence in the localization of a certain viral protein when the results using GFP chimeras in living cells and those obtained through immunostaining of dead, fixed cells are in agreement.

5.6   Class IV viruses such as poliovirus use proteolysis to release small proteins from viral polyproteins

All Class IV viruses synthesize large polyproteins that are subsequently processed into smaller proteins. In the case of poliovirus, the very large polyprotein is not visible in the pulse-chase experiments because it is subjected to proteolytic cleavage *during* synthesis. This phenomenon can be demonstrated by adding protease inhibitors to the *in vitro* translation mix, which results in the accumulation of the gigantic polypeptide encoded by the genome. In order to understand proteolysis of the polypeptide, it is important to know that all proteins begin to fold during translation, when the **nascent** polypeptide is still being synthesized. One of the segments of the picornavirus polyprotein folds into protein 2A^{pro}, which is a **site-specific protease** (as indicated by the superscript; **Figure 5.10**). Site-specific proteases recognize a certain amino acid sequence by binding to that sequence in their active sites, and they subsequently catalyze hydrolysis of the peptide backbone at that sequence, analogous to the sequence specificity of a restriction enzyme. The 2A^{pro} protease hydrolyzes the polyprotein at a particular amino acid sequence found between the 1D and 2A^{pro} amino acid sequences, releasing the P1 polyprotein from the nascent polypeptide. Similarly, the 3C^{pro} protease folds during translation and subsequently hydrolyzes the nascent polyprotein between the 2C and 3A amino acid sequences, releasing the P2 polyprotein. The P3 polyprotein is then produced as a matter of course.

The polyproteins are then subjected to further proteolysis after they have been released from the ribosome; these reactions are called **maturation**

> ▶ To see an animation showing protease activity on nascent proteins in polio, please visit the book's website: www.routledge.com/cw/lostroh.

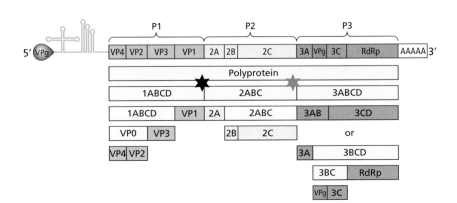

Figure 5.10 Translation and proteolytic digestion of a picornavirus polyprotein. The active protein subunits are shown. Nascent cleavages catalyzed by 2A^{pro} or 3C^{pro} are marked with a star (black = 2A^{pro}; gray = 3C^{pro}). All other proteolytic events are maturation cleavages that are typically catalyzed by 3CD^{pro} or 3C^{pro}. The 13 active proteins are VP4, VP2, VP3, VP1, 2A^{pro}, 2B, 2C, 3AB, 3CD^{pro}, 3A, VPg, 3C^{pro}, and the RdRP (3D).

cleavages because they occur after the polyproteins have been released from the ribosome. The most common protein catalyzing the maturation cleavages is **3CD^{pro}**. It processes **P1 polyprotein** into structural proteins VP0, VP3, and VP1. Cleavage of VP0 into VP4 and VP2 occurs during maturation of the virion and is catalyzed by VP0 itself. The viral 3CD^{pro} also proteolyses **P2 polyprotein**, releasing nonstructural proteins 2A^{pro}, 2B, and 2C. These P2-derived proteins are important for preventing the host cell from stopping the replication cycle, for example by blocking translation of host mRNA. The proteolysis of **P3 polyprotein** can occur in two different ways. The result of one pathway is production of 3AB and 3CD^{pro} without further proteolysis. In fact, this pathway occurs most of the time. More rarely, 3CD^{pro} processes the P3 polypeptide completely, releasing 3A, 3B, 3C, and 3D as separate proteins. The six P3 proteins are needed for synthesis and folding of viral RNAs.

As a consequence of encoding a single polyprotein, picornaviruses do not have distinct waves of early and late gene expression. Furthermore, most individual proteins are produced in equimolar amounts through proteolytic digestion of the polyprotein (VP1 = VP2 = VP3 = VP4 = 2A^{pro} = 2B = 2C). The only variation is that the amounts of 3AB and 3CD^{pro} are higher than the amounts of 3A, 3B, 3C, and 3D. The ratio of structural capsomers to nonstructural proteins is less than one, indicating that the structural proteins are actually less abundant than the nonstructural ones. Although it would seem logical to have different early and late gene expression and to have an excess of structural proteins compared with nonstructural proteins, clearly the picornaviruses are evolutionarily successful as they are. Actually, this is an instance of a common theme in virology: for every rule, there are exceptions. Exceptions arise because of the tremendous evolutionary potential of viruses, which reproduce so quickly and often exhibit higher mutation rates than their cellular hosts (**Chapter 17**), resulting in tremendous genetic diversity.

5.7 Translation of Class IV virus genomes occurs despite the lack of a 5′ cap

In general, features at the end of a viral genome reflect its mechanism of synthesis; for (+) strand RNA viruses, the features are also critical for the mechanism of translation initiation. In the case of picornaviruses, the genomes have a covalently attached VPg protein at their 5′ ends and a poly(A) tail at their 3′ end. Translation of host mRNA depends on the 5′ cap on the mRNA, raising the question of how picornavirus genomes are translated without this molecular feature (see Figure 5.3). Picornavirus genomes can be translated despite their lack of a 5′ cap because of the 5′ UTR of the virus genome, which folds into a structure known as an **internal ribosome entry site (IRES)** (**Figure 5.11**).

The discovery of internal ribosome entry in eukaryotes was unexpected because internal ribosome entry (downstream of the 5′ end) was thought to be a feature restricted to bacteria and archaea. The viral IRES was discovered in part using an *in vitro* translation system made from the cytoplasmic contents of host cells. The strategy was to incubate different controls and test mRNA constructs in the *in vitro* translation system, then measure the amount of reporter proteins translated from each mRNA using SDS-PAGE. These experiments revealed that the 5′ UTR allows translation to begin despite the lack of a 5′ cap on the viral RNA.

Just as cap-dependent translation of normal host mRNA requires a series of assembly reactions so that a translation initiation complex can form, so does IRES-dependent translation. Through a cascade of RNA–protein and

0   1
Sequence
conservation

Figure 5.11 Picornavirus IRES. The most likely folding of the IRES. A rainbow color scale indicates the evolutionary conservation across the picornaviruses; the least conserved sequences are purple and the most conserved are red.

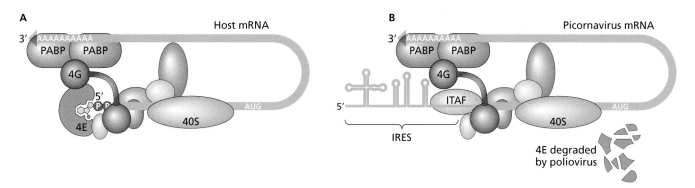

Figure 5.12 Cap-dependent initiation complex compared with picornavirus initiation complex. (A) Normal initiation of host mRNA involves the eIF4E protein binding to the 5′ cap and forming a complex with PABP. Other initiation factors are involved (shown in different colors). The small subunit of the ribosome (40S) is closest to the AUG start codon. (B) For initiation of the translation of the picornavirus genome, the host ITAF protein binds to the IRES and substitutes for eIF4E. The IRES is a complex stem–loop structure in the 5′ UTR. The terminal VPg protein was removed from the 5′ end of the genome by a host enzyme. Meanwhile, poliovirus proteolytically degrades eIF4E, thus preventing cap-dependent translation of host mRNA.

protein–protein interactions, a 48S initiation complex forms around the IRES (**Figure 5.12**). Formation of the 48S initiation complex does not require eukaryotic initiation factor 4E (eIF4E), which is the protein that normally binds to the cap and initiates formation of mRNP that is ready for translation. For IRES-dependent translation, all of the other eukaryotic translation initiation factors and the PABP-coated poly(A) tail are still required. The resulting circularized initiation 48S complex is similar to that formed by normal host mRNA. Meanwhile, poliovirus proteases 2A^{pro} and 3CD^{pro} proteolytically degrade two host translation initiation factors needed for cap-dependent host mRNA translation, including eIF4E, thus blocking translation of host mRNA. Interfering with host translation is beneficial because it makes more tRNA and amino acids available for viral protein synthesis, and it also interferes with innate cellular antiviral responses (see **Chapters 12** and **14**).

5.8 Class IV virus genome replication occurs inside a virus replication compartment

Virus replication compartments (**VRCs**) are induced by many if not most viruses that infect animals and plants; picornaviruses are no exception. VRCs are assembled by nonstructural viral proteins, viral genomes, host lipids, and host proteins, all of which vary from virus to virus. They typically appear as invaginations of membranes associated with specific organelles, such as mitochondria, **endoplasmic reticulum** (ER), the Golgi apparatus, autophagosomes, lysosomes, peroxisomes, or the plasma membrane. They can appear bulbous, with a narrow neck that connects the interior of the VRC to the rest of the cytoplasm and presumably serves as the exit point for viral mRNA. For some viruses, the VRCs are interconnected by a membranous web, and some VRCs are bounded by single membranes early in infection, and double membranes later in infection (**Figure 5.13**).

In the case of poliovirus, the VRCs are closely associated with the Golgi apparatus. The nonstructural viral proteins required for VRC formation and organization include 2C^{ATPase} and the 3A protein. The set of host proteins required for VRC formation includes enzymes that modify lipids, leading to the creation of a novel lipid profile in the VRC.

Figure 5.13 Virus replication compartments (VRCs). Electron tomography was used to create this three-dimensional model for a (+) RNA virus replication complex. These particular VRCs have both an outer membrane (gold) and an inner membrane (silver). The outer membrane is connected to the endoplasmic reticulum (bronze). Arrows I, II, and III indicate sites where the outer membranes are connected to those of other VRCs or to the endoplasmic reticulum. The inset elecron micrograph views show the tomographic slices that revealed the connections indicated by arrows I, II, and III. (From Knoops K et al. 2008. *PLOS Biol* 6:e226. doi: 10.1371/journal.pbio.0060226. Published under CC BY 4.0.)

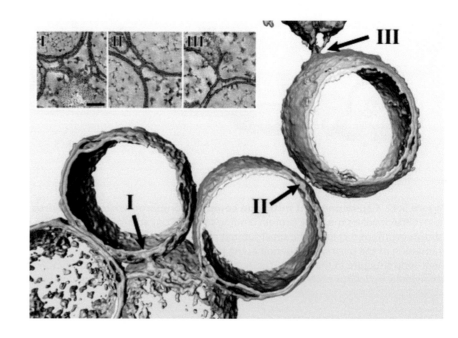

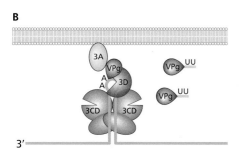

Figure 5.14 The enzymatic activities of the picornavirus 3D protein. (A) 3D^{pol} catalyzes RNA synthesis; the black arrow indicates the movement of the polymerase. (B) 3D^{pol} can also catalyze uridylylation of VPg (also known as 3B). It does so when it is part of a complex with a structure in the (+) RNA (stem–loop) and with 3CD^{pro}. A pair of adenines in the stem–loop structure serves as the template for uridylylation. The 3AB substrate for uridylylation is associated with the membrane of a VRC through the 3A protein.

5.9 The picornavirus 3D^{pol} is an RdRp and synthesizes a protein-based primer

Inside the poliovirus VRC, the **3D^{pol}** protein is an RdRp (**Figure 5.14**). Like most enzymes used for genome replication, 3D^{pol} requires a primer. In this case, 3D^{pol} synthesizes the primer by binding to the VPg protein (3B) and **uridylylating** it, adding two uridylyl groups to the OH side chain of a particular tyrosine amino acid in the VPg (**Figure 5.15**). Whether 3D^{pol} acts as a polymerase or as a primer-synthesis enzyme depends on the context of the other proteins interacting with it. Uridylylation activity occurs exclusively when 3D^{pol} is part of a complex made up of 3D^{pol}, 3CD, and certain secondary structures in the (+) RNA template. Although cells do not use proteins such as VPg (3B) to prime nucleic acid synthesis, many viruses do. Another example is that of the adenoviruses (dsDNA; see **Chapter 8**). It is typical for viral proteins to have multiple functions that depend on their interacting partners; in this way, a single gene can encode one protein that catalyzes several reactions. Multifunctional proteins enable viral genomes to remain small, which is necessary for packaging into icosahedral capsids and for the most rapid genome replication possible.

5.10 Structural features of the viral genome are essential for replication of Class IV viral genomes

As is the case for all (+) strand ssRNA viruses, many structural elements in the picornavirus genome are required for synthesis of the antigenome. For example, a **cloverleaf** structure near the 5′ end is required, as is a stem–loop structure within the coding sequence, called the **cis-responsive RNA element** or **CRE**. The word *cis* indicates that a sequence in DNA, RNA, or protein acts only on other sequences within the same molecule, so the CRE is needed to copy the very RNA molecule that contains it. Two more *cis*-active RNA elements in the genome needed for replication are stem–loop structures in the 3′ UTR and the poly(A) tail (**Figure 5.16**). **Tertiary interactions** between one of the RNA loops near the 5′ end and one of the loops in the 3′

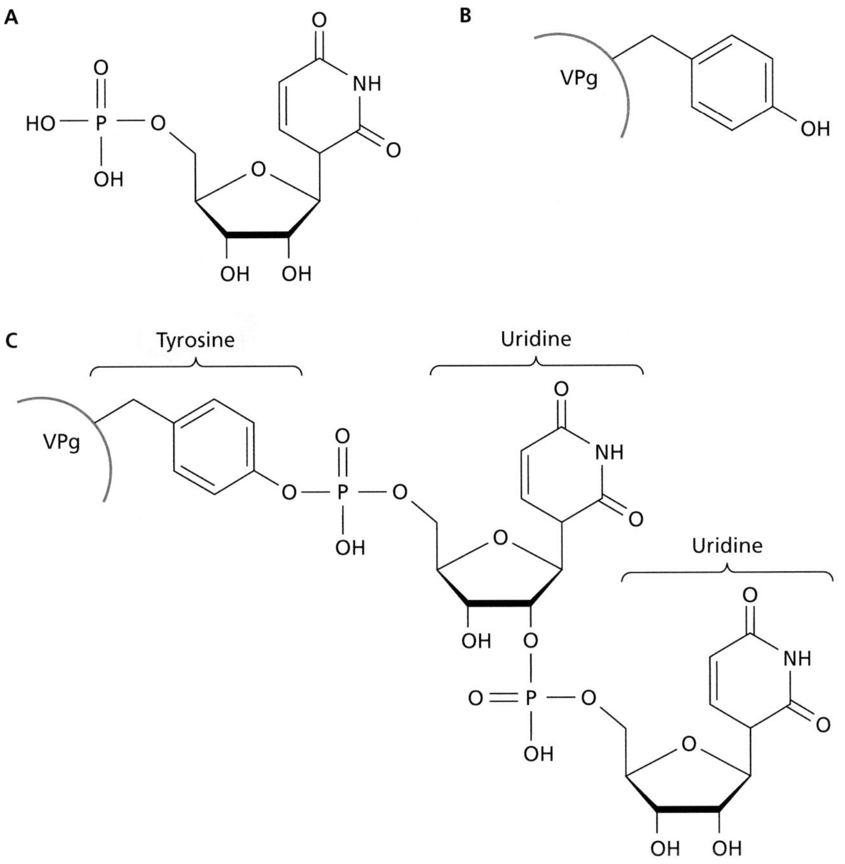

A

B

C

Tyrosine

Uridine

VPg

Uridine

Figure 5.15 Uridylylation of VPg by the picornavirus 3D^pol protein. (A) Uridine monophosphate. (B) VPg tyrosine side chain prior to uridylylation. (C) VPg tyrosine side chain after uridylylation.

structure are also required for genome replication; this hydrogen bonding between loops is sometimes described as kissing.

5.11 Picornavirus genome replication occurs in four phases

Genome replication in (+) strand ssRNA viruses uses a double-stranded replicative form as the template and ultimately produces many full-length genomes with the same terminal features as the infecting uncoated genome (**Figure 5.17**). Any hypothesis to explain viral genome replication must take the terminal features of the genome into account. In the case of polioviruses, the terminal features include a terminal covalently attached protein (VPg) and a poly(A) tail. Genomic RNA is used as a template to synthesize (−) RNA, which remains associated with the (+) strand, resulting in a double-stranded replicative form with VPg proteins covalently attached to the 5′ ends of both strands. The replicative form is used to synthesize many (+) strands that have the same features as the infecting genome.

Poliovirus genome replication is understood in some detail. In the first phase of genome replication, various protein complexes have assembled onto the CRE and 5′ cloverleaf, forming a **viral ribonucleoprotein complex** or **vRNP** (**Figure 5.18**). These proteins include a complex of 3AB, 3CD^pro, 3D^pol, and the host protein **PCBP**. As a component of this complex, 3D^pol synthesizes a primer by uridylylating the VPg (3B) protein (**Figure 5.19**). Phase two is synthesis of the negative strand. When it is not in the primer-synthesizing complex, the 3D^pol protein uses VPg-pU-pU as a primer and subsequently

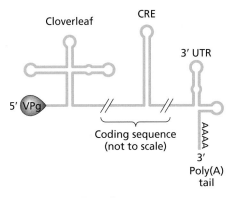

Figure 5.16 *Cis*-acting RNA sequences and structures important for replication of picornavirus RNA. The cloverleaf near the 5′ end, the CRE within the coding sequence, stem–loop structures in the 3′ UTR, and the poly(A) tail are all critical for replication of picornavirus RNA. The coding sequence between the double slashes and indicated by the bracket has been severely truncated in order to highlight instead the *cis*-acting RNA structures. Tertiary interactions among the RNA loops have also been omitted for clarity.

Figure 5.17 Overview of replication in picornaviruses with a focus on the nucleic acids. The infecting genome is a single strand of (+) RNA that contains secondary structures important for translation (the IRES) and for genome replication (the cloverleaf, the CRE, and the structures in the 3' UTR). This RNA is used as a template to synthesize (–) RNA, resulting in a double-stranded replicative form. The replicative form is used to synthesize many (+) strands.

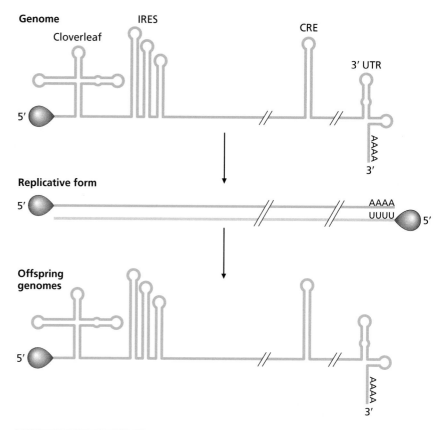

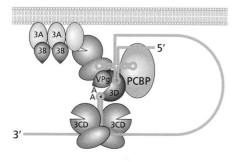

Figure 5.18 Picornavirus vRNP during initiation of RNA replication. Inside a VRC, the CRE portion of the genome assembles with many proteins such as 3AB, 3CD^pro, 3D^pol, VPg/3B, and the host protein PCBP. The cloverleaf in the 5' UTR of the RNA is also part of the complex.

Phase	Description	Template	Product
One	Synthesis of VPg-pU-pU primer	Two adjacent As in CRE loop	VPg-pU-pU primer  $\quad$ VPg^{UU}
Two	Synthesis of (–) strand	(+) strand genome	Replicative form
Three	Cessation of VPg-pU-pU production	None	None
Four	Synthesis of offspring (+) genomes	(–) strand in replicative form	New (+) genomes 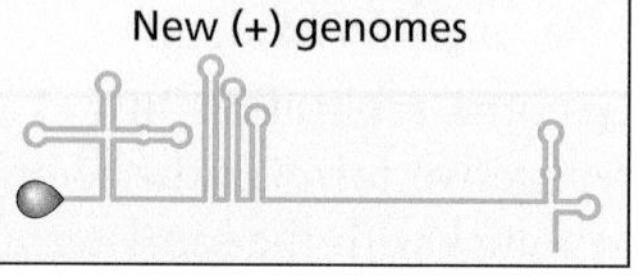

Figure 5.19 Summary of four phases in the replication of picornavirus genomes. During phase one, 3D^pol uses two adjacent As in the CRE and the 3B protein to produce uridylylated VPg proteins. During phase two, 3D^pol uses the uridylylated VPg primer and the genome as a template to synthesize negative strands. During phase three, synthesis of new uridylylated VPg primers ceases because the CRE structure is no longer present in the VRC, which contains instead double-stranded replicative forms. During phase four, 3D^pol synthesizes new (+) offspring genomes.

synthesizes an antigenome, which remains base paired to the genome over its whole length. The base-paired molecule is the replicative form.

Phases one and two of picornavirus replication overlap temporally so that VPg uridylylation is ongoing during antigenome synthesis (**Figure 5.20**). In the third phase of picornavirus genome replication, uridylylation of VPg stops because the synthesis of the antigenome destroys the structure of the CRE. In the absence of the CRE, there is no VPg uridylylation. The 3D^pol

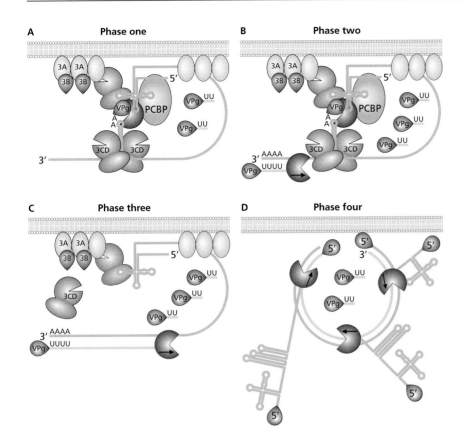

Figure 5.20 The interior of a VRC with RNA and protein elements indicated during the four phases of genome replication. (A) Phase one, synthesis of uridylylated VPg primers. (B) Phase two, synthesis of the complementary (−) strand (light green). (C) Phase three, synthesis of uridylylated VPg primers now ceases because the CRE is no longer present. (D) Phase four, synthesis of new (+) genomes.

protein reaches the end of its template and dissociates. The VRC is now full of double-stranded replicative forms and uridylylated VPg-pU-pU primers so that phase four can begin. In phase four, the 3D^{pol} protein uses the VPg-pU-pU primers for (+) strand synthesis. Many 3D^{pol} proteins use the same template multiple times and the offspring single-stranded (+) orientation genomes are released into the cytoplasm. The use of VPg-pU-pU as a primer explains why the 5′ ends of the genomes and antigenomes always have the VPg protein covalently attached. The existence of the poly(A) tail in the infecting genome ensures the presence of a poly(U) tract in the antigenome. This poly(U) tract is copied to create the poly(A) tail found in the genome—normal host mRNA obtains its poly(A) tail in the nucleus instead through an entirely different mechanism.

Sections 5.12 and 5.13 describe another medically important family of slightly larger Class IV viruses, the **flaviviruses**.

5.12   Flaviviruses are models for simple enveloped (+) strand RNA viruses

Viruses in the family *Flaviviridae* have enveloped, spherical virions with prominent envelope proteins projecting from their surfaces arranged as an icosahedron (**Figure 5.21**). A single capsomere, called the core, forms the icosahedral center and two different transmembrane envelope proteins, E1 and E2, form the spikes. Flaviviruses have linear (+) strand RNA genomes that are 10–12 kb in length, somewhat longer than the genome of picornaviruses. They are named for the Latin word *flavus* (yellow) because the model species, yellow fever virus, causes jaundice, including yellowing of the eyes. Other flaviviruses are hepatitis C virus, West Nile virus, and dengue fever

Figure 5.21 Hepatitis C virion. Electron micrograph of a hepatitis C virus (HCV), a spherical enveloped flavivirus. (From Lindenbach BD & Rice CM 2005. *Nature* 436:933–938. With permission from Springer Nature.)

virus. **Sections 5.13** and **5.14** focus on **hepatitis C virus** (**HCV**) because as many as 3% of people globally are infected with it, and it can cause debilitating and even fatal interference with the function of the liver. The World Health Organization estimates that 4 million new HCV infections occur every year, primarily through IV drug use and contaminated blood products. Flaviviruses share the same overall genome expression and replication strategies as the other Class IV animal viruses covered in this chapter (see Figure 5.1), and thorough understanding of the molecular biology of HCV gene expression and genome replication has led to the development of anti-HCV medicines, discussed in **Chapter 16**.

5.13   The linear (+) strand RNA flavivirus genomes have unusual termini

Flavivirus genomes are similar to those of the picornaviruses. Whereas some flavivirus genomes have a terminal covalently linked VPg protein at the 5′ end, others have a 5′ cap similar to host mRNA. HCV has neither a cap nor a VPg protein at the 5′ end of its genome; instead, it has a 5′ nucleotide with a triphosphate group. Flavivirus genomes also lack the poly(A) tail found on normal eukaryotic mRNA and in picornavirus genomes. Instead, the viral 3′ UTR folds into a tertiary structure consisting of several stem–loop structures (**Figure 5.22**). The HCV genome encodes one large polyprotein, similar to the genome of picornaviruses. Viral and cellular proteins cleave the polyprotein into smaller functional proteins, and the cleavage products closer to the N-terminus are structural proteins. The cleavage products closer to the C-terminus are the nonstructural proteins.

5.14   Enveloped HCV encodes 10 proteins including several with transmembrane segments

HCV has three structural proteins: C or **core** (capsid), **E1**, and **E2** (two different envelope proteins; **Figure 5.23**). The other proteins are nonstructural and include two proteases (**NS2** and **NS3**) that cleave the viral polypeptide

Figure 5.22 Hepatitis C virus genome. The genome has significant secondary structure at both the 5′ and 3′ ends, and it encodes a single polyprotein that is processed into structural proteins and seven nonstructural proteins.

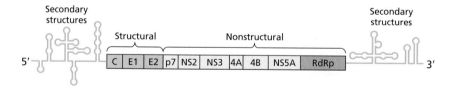

Figure 5.23 The HCV virion. The virion has three structural proteins: the envelope proteins E1 and E2 and the internal capsomere, called core in HCV. (Courtesy of Philippe Le Mercier, ViralZone, © SIB Swiss Institute of Bioinformatics.)

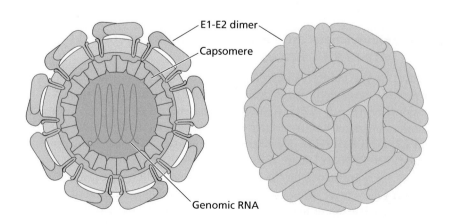

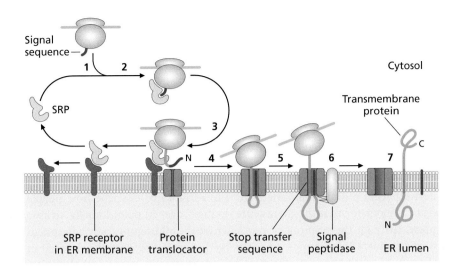

Figure 5.24 The signal recognition particle (SRP) system for translation of viral transmembrane proteins. As soon as the signal sequence, also known as a start-transfer sequence, emerges from the active ribosome (1), the SRP binds to it, causing a translational pause (2). The SRP and ribosome then dock with the protein translocator or translocon in the ER membrane (3), relieving the translation pause (4). Next, the SRP component is released to be used again. The protein is then synthesized directly into the lumen of the ER. When a stop-transfer sequence is reached (5), extrusion into the ER stops. The signal peptidase cleaves the signal peptide from the protein (6) and the protein is released from the translocon. In the mature protein, the stop-transfer sequence is a transmembrane segment (7). Combinations of start-transfer and stop-transfer sequences result in multipass transmembrane proteins such as HCV p7. (From Alberts B et al. 2014. *Molecular Biology of the Cell*, 6th ed. Garland Science. With permission from W.W. Norton.)

and an enzyme with RdRp activity (**NS5B**). The translation and cleavage of the HCV polyprotein is similar to that of picornaviruses but is more complex because seven proteins with transmembrane segments must be synthesized.

In eukaryotes, ribosomes bound to the rough endoplasmic reticulum (ER) translate transmembrane proteins such as HCV E1, E2, p7, NS2, NS4A, NS4B, and the HCV RdRp (also known as NS5B) (**Figure 5.24**). Ribosomal docking with the ER occurs through the **signal recognition particle** (**SRP**) system. This system relies upon translation of an N-terminal **signal sequence**. When this sequence of amino acids emerges from the ribosome on the nascent polypeptide, it binds to the SRP, causing translation elongation to pause temporarily. The SRP then docks with a protein-conducting channel in the ER, known as the **translocon**. The translation arrest is relieved by these interactions, so that protein synthesis then continues. The nascent protein folds in the membrane of the ER if it has transmembrane segments. Soluble portions of transmembrane proteins often fold assisted by **chaperones**, proteins that consume ATP to catalyze rapid protein folding. An ER protein known as the **signal peptidase** processes the proteins, removing their signal sequences so that signal sequences are not found in the mature folded proteins.

In the case of HCV, the cytoplasmic core protein is translated first, when the ribosome is still free in the cytoplasm. After translation of the core protein, a signal peptide is translated, triggering the SRP system. The rest of the viral proteins are then translated by ribosomes docked to the rough ER. The cellular signal peptidase enzyme in the ER makes several of the proteolytic cleavages, releasing E1, E2, and p7 from the polyprotein. The NS2 protease releases NS2 by cleaving between NS2 and NS3, and the NS3 protease along with its cofactor NS4A makes the proteolytic cuts that release NS3, NS4A, NS4B, NS5A, and NS5B from the polyprotein (**Figure 5.25**).

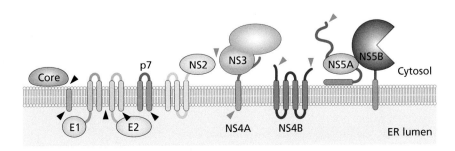

Figure 5.25 Synthesis and processing of HCV polyprotein into 10 proteins, some of which have transmembrane segments. Black arrowheads indicate sites of proteolytic processing catalyzed by host enzymes and blue arrowheads indicate processing catalyzed by viral proteases. (From Lindenbach BD & Rice CM 2005. *Nature* 436:933–938. doi: 10.1038/nature04077. With permission from Springer Nature.)

A

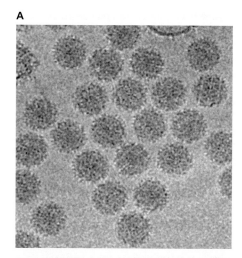

B

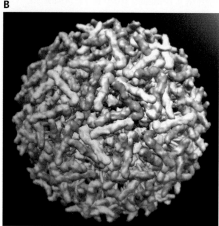

Figure 5.26 Sindbis virus. (A) Electron micrograph of a Sindbis virus (SINV) showing its spherical shape. (B) Structure solved by cryo-electron microscopy. The virions are approximately 65 nm in diameter. (A, Courtesy of Tuli Mukhopadhyay lab, Indiana University. B, Courtesy of Jean-Yves Sgro.)

5.15 Togaviruses are small enveloped viruses with replication cycles more complex than those of the flaviviruses

Viruses in the family *Togaviridae* have enveloped, icosahedral virions with prominent spikes protruding from their surfaces (**Figure 5.26**). Medically important **togaviruses** include rubella virus and chikungunya virus. Although nearly universal vaccination has eradicated rubella from the Western hemisphere, chikungunya is an emerging infectious disease that is expected to continue its explosive spread around the globe. Its symptoms include severe pain. In fact, the word *chikungunya* means to bend up (derived from a Tanzanian language, Kimakonde) to describe how people with a serious form of the disease are doubled over in pain. Rubella virus is spread through the respiratory route, whereas chikungunya virus is spread by mosquitos. Chikungunya virus emerged in west and central sub-Saharan Africa but is now spreading through Asia, Australia, Europe, South America, Central America, and the southern parts of North America.

Togavirus molecular and cellular biology has been studied closely for much longer than that of the flaviviruses, historically through focus on **Sindbis virus**. Sindbis viruses have a single capsomer protein in the icosahedral core, and three envelope proteins. The spikes are dimers of E1 and E2; E3 stabilizes the spikes. The proteins TF and 6K can also be found in the virus envelope in small amounts. The genomes have features of eukaryotic mRNA, such as a normal 5′ cap and a poly(A) tail (**Figure 5.27**).

The 9- to 12-kb togavirus genomes are similar in length to flavivirus genomes. Despite this similarity in genome size, the togaviruses have far more complex gene expression and genome replication strategies. For example, togaviruses synthesize three viral nucleic acids rather than two and synthesize four unprocessed polyproteins rather than one. Togaviruses also have distinct early and late phases of gene expression and nucleic acid

Figure 5.27 Sindbis virus genome. The Sindbis virus genome has a 5′ methylated cap, nonstructural and structural polyproteins, and polyadenylation at the 3′ end.

Figure 5.28 Togavirus RNA. Togaviruses such as Sindbis virus direct synthesis of three RNAs. (A) Full-length (+) strands that have a 5′ methylated cap and poly(A) tail. (B) Full-length (−) antigenomes (light green) that are part of the double-stranded replicative form. (C) Subgenomic (+) RNA with a 5′ methylated cap and a poly(A) tail, encoding the structural polyproteins.

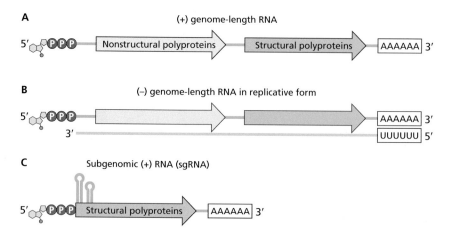

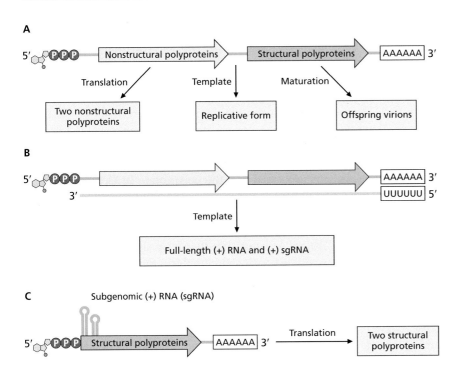

A

B

C

Subgenomic (+) RNA (sgRNA)

Figure 5.29 Purposes for the viral nucleic acids made during togavirus infections. (A) Full-length (+) RNA is used as mRNA to translate the two nonstructural polypeptides, as a template for synthesis of replicative forms, and late during infection as genomes packaged into offspring virions. (B) Replicative forms are used to synthesize full-length and subgenomic (+) RNA. (C) Subgenomic (+) RNA is used as mRNA to translate the two structural polypeptides.

synthesis and express structural proteins much more abundantly than nonstructural proteins.

Cells infected with togavirus contain three different viral RNA molecules: genomic-length (+) strands, full-length (–) antigenomes, and a **subgenomic (sg)** (+) RNA equivalent to the 3′ third of the genome (**Figure 5.28**). Both the full-length (+) and sgRNA have a 5′ cap and a poly(A) tail; the antigenome is not capped. The antigenome is found as part of the dsRNA replicative form inside VRCs. Synthesis of antigenomes occurs exclusively during early gene expression, whereas synthesis of the two (+) orientation RNAs occurs throughout infection.

The three different viral nucleic acids serve different purposes. The genome-length (+) strand RNAs can be used in three ways. First, they can be translated to produce the two nonstructural polyproteins (**Figure 5.29**). Second, they can serve as templates for synthesis of antigenomes and thereby formation of replicative forms. Third, toward the end of the virus replication cycle, they can be packaged to serve as genomes for offspring virions. The purpose of the replicative form is to be transcribed into the two different (+) strand RNAs. The purpose of the sgRNA is to be translated into two slightly different structural polyproteins. Nonstructural proteins are translated exclusively from the longest (+) strand, whereas structural proteins are translated exclusively from the sgRNA.

5.16   Four different togavirus polyproteins are found inside infected cells

As mentioned earlier, another aspect of togaviruses that makes them more complex than flaviviruses is that togaviruses produce not one but four different polyproteins. There are two structural polyproteins that differ near their C-termini (**Figure 5.30**). The longer one is produced about 85% of the time. The N-terminal regions of both are processed into the capsid protein, the E3 envelope protein, and the E2 envelope protein. The C-terminal region of

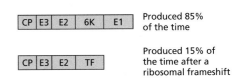

Produced 85% of the time

Produced 15% of the time after a ribosomal frameshift

Figure 5.30 Togavirus structural polyproteins. The longer polyprotein is processed into the coat protein, three envelope proteins (E1–E3), and 6K. It is produced from translating the sgRNA about 85% of the time. The shorter polyprotein is processed into the coat protein, two envelope proteins, and the TF protein. The shorter polyprotein requires a programmed ribosomal frameshift in order to be synthesized and is produced from translating the sgRNA about 15% of the time.

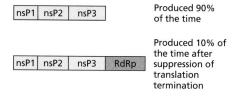

Produced 90% of the time

Produced 10% of the time after suppression of translation termination

Figure 5.31 Togavirus nonstructural polyproteins. The shorter polyprotein is processed into nsP1, nsP2, and nsP3. It is produced from translating the genome-length RNA about 90% of the time. The longer polyprotein is processed into nsP1, nsP2, nsP3, and RdRp. The longer polyprotein requires a programmed suppression of translation termination in order to be synthesized and is produced from translating the genome-length RNA about 10% of the time.

the longer structural polyprotein is processed into the 6K and E1 envelope proteins, and the C-terminal region of the shorter structural polyprotein is processed to release the envelope transframe (TF) protein. Although derived from the structural polyprotein, 6K and E3 are not necessarily found in all togavirus virions and may be important for assembly or release.

Similarly, there are two nonstructural (ns) polyproteins that differ in their C-terminal regions (**Figure 5.31**). The shorter one is produced about 90% of the time and early during infection it is processed into nsP1, the P23 polyprotein, nsP2, and nsP3. Later it is processed into the polyprotein P12 and nsP3. The longer one is processed into the polyprotein P123, nsP4, the P23 polyprotein, nsP2, and nsP3. Later it is processed into the polyproteins P12 and P34. A protease active site can be found within the sequence corresponding to nsP2, so that any polyproteins containing that sequence also have a protease active site. The nsP4 protein has the active site for RNA synthesis. The nsP4 protein is part of two different complexes that determine its activity. Complex 1 is the RdRp that synthesizes antigenomes, whereas complex 2 is the RdRp that synthesizes (+) strand genomes and subgenomic RNA. Only complex 2 has capping and polyadenylation activity; consequently, only (+) sense togavirus RNA has a 5′ cap and poly(A) tail. Complex 1 is active early during infection, whereas complex 2 dominates at later time points.

5.17 Different molecular events predominate early and late during togavirus infection

Early during a togavirus infection, translation of the infecting genome results in production of nonstructural proteins that cause VRC formation; the VRCs then fill with replicative forms (**Figure 5.32**). Throughout infection, both species of (+) strand RNA are synthesized using the replicative forms as a template. Early during infection, the long (+) strand RNA is translated to produce the two nonstructural polyproteins and the sgRNA is translated to produce the structural polyproteins. Late during infection, creation of new VRCs and replicative forms stops. Although the long (+) strand RNA is still produced, it is no longer translated. Instead it is targeted for assembly. The short sgRNA continues to be translated, resulting in ongoing production of all the nonstructural proteins. Because the sgRNA is translated throughout infection, the structural proteins become much more abundant than the nonstructural proteins.

5.18 Translation of togavirus sgRNA requires use of the downstream hairpin loop

Regulation of translation is critically important for causing the switch to late gene expression. Both the sgRNA and the long (+) strand RNA have a 5′ cap and poly(A) tail just as normal host mRNA does. During late virus gene expression, however, cap-dependent translation initiation is blocked as part of an antiviral cellular defense that depletes the cytoplasm of functional eukaryotic translation initiation factor 2 (see also **Chapter 14**). This situation raises the question of how the togavirus sgRNA can be translated. The solution involves an extensive stem–loop structure called the **downstream hairpin loop** or **DLP** (**Figure 5.33**). The DLP is unique to sgRNA and forms adjacent to and downstream of the start codon for the structural polyprotein. The DLP is needed to allow translation when cap-dependent translation is suppressed. The DLP replaces eukaryotic translation initiation factor 2 in the initiation complexes so that the sgRNA can be translated.

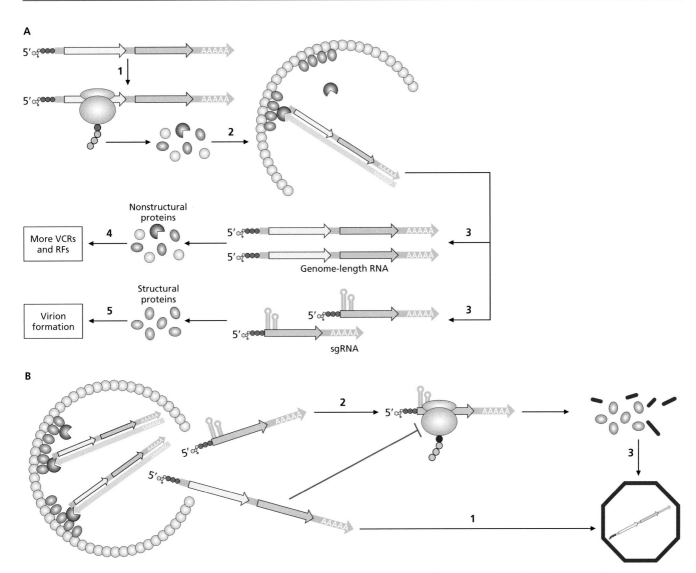

Figure 5.32 Timing of events during togavirus replication. (A) Early during infection, translation of the genome (1) results in production of nonstructural proteins, used for VRC formation and the creation of double-stranded replicative forms (RFs) (2). The RFs are used as templates for synthesis of both genome-length and subgenomic (+) RNA (3). Translation of genome-length RNA results in formation of more VRCs and RFs (4), whereas translation of sgRNA results in the production of structural proteins (5) that will be used later for virion formation during maturation. (B) Late during infection, genome-length RNA is no longer translated because cap-dependent translation has been blocked by the viral nonstructural proteins; it is instead targeted for packaging into new virions (1). The sgRNA continues to be translated because of a cap-independent mechanism that depends upon RNA secondary structure, resulting in the accumulation of many structural proteins (2) that are used to create new virions (3).

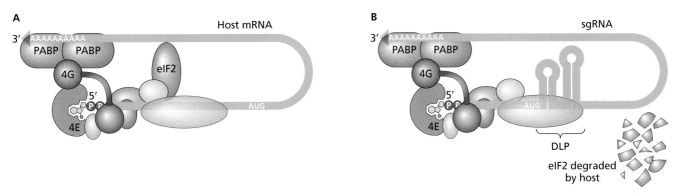

Figure 5.33 The downstream hairpin loop of togavirus sgRNA. (A) Initiation factors including eIF2 assembled on normal host mRNA for cap-dependent translation initiation. (B) The viral DLP compensates for the loss of eIF2, making translation initiation of sgRNA possible.

The use of the DLP is an example of a virus overcoming a host antiviral defense (see **Chapter 14**). In this case, late during infection, a host antiviral defense shuts down cap-dependent translation, ultimately causing cell death; the idea is that the host sacrifices the infected cell in order to try to protect the whole animal from the infection. The 5′ cap on the sgRNA is used for translation early during infection, before the host antiviral response occurs. The DLP is used after the host has ramped up its antiviral defenses—ironically, the virus subverts the antiviral defenses in order to complete its replication cycle.

5.19 Suppression of translation termination is necessary for production of the nonstructural P1234 Sindbis virus polyprotein

Unusual translation events often enable viruses to encode multiple proteins using overlapping mRNA sequences. Sindbis virus uses two such strategies, one of which is called **suppression of translation termination** (**Figure 5.34**). Suppression of translation termination is used to produce the less abundant nonstructural polyprotein P1234 from sequences that overlap with the ones encoding the more abundant P123 polyprotein. In both cases, translation begins at a typical start codon approximately 60 nucleotides from the 5′ end of the RNA. When the ribosome encounters a UGA stop codon at the end of the sequence encoding nsP3, 90% of the time translation termination occurs, releasing the P123 polyprotein. Translation termination is mediated by the usual host translation release factor binding to the UGA stop codon. But, 10% of the time, a tRNA with an anticodon that can hydrogen bond with UGA allows translation to continue, so that the result is the P1234 polyprotein. This suppression of the translation termination process depends not only on the stop codon but also on the C nucleotide immediately adjacent to and downstream of the stop codon in the RNA.

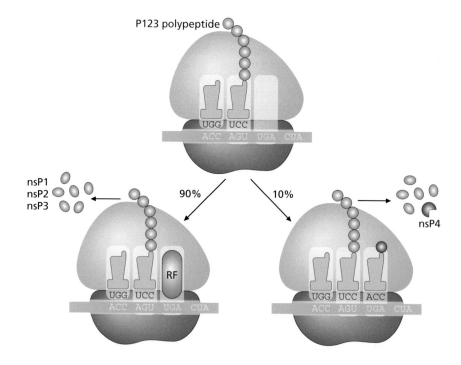

Figure 5.34 Suppression of translation termination in Sindbis virus. Translation termination depends on the UGA and CUA sequences in the mRNA. These critical sequences are in boldface text, highlighted yellow. When the ribosome reaches the stop codon (UGA) for the P123 polyprotein, 90% of the time a release factor (RF) catalyzes translation termination and the polyprotein is released, ultimately producing nsP1, nsP2, and nsP3. But, 10% of the time, a tRNA with the anticodon ACC base pairs to the stop codon and translation continues, ultimately producing nsP1, nsP2, nsP3, and the nsP4 protein.

Many viruses employ suppression of translation termination to encode more than one protein with the same genetic information. Doing so is an example of how viruses with small genomes use their coding capacity more efficiently than cells do, a concept that was introduced using the bacteriophages ΦX174 and MS2 as examples (see **Chapter 4**). In the case of togaviruses, the nsP4 protein is needed in smaller amounts than the rest of the nonstructural proteins; the virus uses suppression of translation termination to keep levels of nsP4 synthesis relatively low, which has the benefit of making more amino acids and ribosomes available for translating proteins that are needed in greater abundance.

5.20 Sindbis virus uses an unusual mechanism to encode the TF protein

Translation of sgRNA is also regulated so that the shorter polyprotein (C/E3/E2/TF) is produced less frequently; the shorter one allows production of the rarer TF protein, whereas the longer one allows production of the more abundant 6K and E1 proteins. The sequence of the sgRNA indicates that it includes a single open reading frame that encodes the longer polyprotein that can be processed into individual proteins C, E3, E2, 6K, and E1. C is produced and released from the polypeptide through a *cis* proteolytic cleavage mediated by C itself. Following autoproteolysis, a signal sequence is exposed, targeting the ribosome to the rough ER so that the rest of the proteins can be inserted co-translationally into the membrane. The rest of the cleavages are catalyzed by host proteases.

However, cells infected with togavirus produce another transmembrane protein, called TF, from sgRNA, again as part of a polyprotein. This polyprotein includes the sequences for C, E3, E2, and TF, but terminates after TF and therefore lacks the 6K and E1 sequences. Oddly, the TF primary sequence is not found when a computer translates the sgRNA. This situation raises the question of how TF protein is translated at all.

A clue as to the way that TF is translated is that the N-terminal amino acids of TF are identical to those of 6K, but the C-terminal amino acids of TF are unique (**Figure 5.35**). Examination of the genome reveals that the

A

```
6K   ETFTETMSYLWSNSQPFFWVQLCIPLAAFIVLMRCCSCCLPFLVVAGAYLAKVDA
TF   ETFTETMSYLWSNSQPFFWVQLCIPLAAFIVLMRCCSCCLPFLSGCRRLPGEGRRLRTCDHCSKCATDTV
     *****************************************
```

B

```
6K    P    F    L    V    V
      CCU  UUU  UUA  GUG  GUU

TF    P    F    L    S    G    C
      CCU  UUU  UUA  AGU  GGU  UGC
```

Figure 5.35 Alignment of 6K and TF sequences. (A) The amino acid sequences of 6K and TF have been aligned with asterisks to indicate positions of identity. The N-terminal and central regions are identical, whereas the C-termini differ from each other. (B) The codons and one-letter amino acid codes at the boundary between the identical and different sequences in proteins 6K and TF. The red UUA codon specifies the last identical amino acid, leucine. The A in this codon is used twice when translating the TF sequence, in a process called a programmed −1 ribosome frameshift.

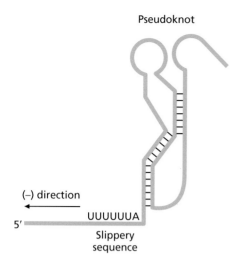

Figure 5.36 Slippery sequence and pseudoknot. Sindbis virus RNA has a slippery sequence at the bottom of an RNA pseudoknot in the (+) sgRNA that encodes the structural polyproteins. The lines show base pairing between different elements of the pseudoknot. The slippery sequence and pseudoknot cause a programmed –1 frameshift that results in translation of the TF protein 15% of the time.

codons representing the C-terminal amino acids of TF are present in the genome and are found in the same sequence as that used to translate the 6K protein, except that the reading frame for TF is shifted one nucleotide in the **–1 direction**. Bioinformatic comparisons of many togavirus genomes revealed a conserved UUUUUUA **slippery sequence** within the site of the frameshift; further analysis indicated that this motif is part of a **pseudoknot** stem–loop structure and occurs at the base of the stem (**Figure 5.36**).

5.21   A programmed –1 ribosome frameshift is needed to produce the togavirus TF protein

The mechanism that enables translation of TF is called a **programmed –1 ribosomal frameshift** (**Figure 5.37**) because it occurs with regular frequency (and so is programmed) and enables the ribosome to shift the reading frame by one nucleotide in the upstream direction. Programmed ribosomal frameshifts are common in viruses. In this case, approximately 15% of the time the pseudoknot causes the ribosome to pause during translation elongation. Within the paused ribosome, the first U in the slippery sequence is the last nucleotide in the E site of the ribosome, and tRNAs are bound to the adjacent codons UUU and UUA occupying the P and A sites, respectively. The ribosome pauses 85% of the time but then continues as normal, so that the frame for translation is xxU-UUU-UUA within the slippery sequence, where x represents the two nucleotides upstream of the slippery sequence. The UUU codon specifies phenylalanine and the UUA codon specifies leucine, carried by tRNAs with 3′AAG and 3′AAU anticodons, respectively. These are the only two possible tRNAs that can translate these particular codons. The programmed –1 frameshift takes advantage of the wobble property of tRNAs, in which the third position in the codon does not have to hydrogen bond to the anticodon. In the paused ribosome on the togavirus sgRNA, there is hydrogen bonding between five of the six bases in the codon–anticodon pairs. The UUUUUUA sequence is described as slippery because the UA and AU base pairs between the tRNA anticodons and their codons have only two hydrogen bonds each. As a result of the slippery sequence and its particular base pairing with cognate tRNAs, about 15% of the time the paused ribosome shifts backward one nucleotide (–1) before continuing translation (see Figure 5.37). The pseudoknot is also essential for causing the frameshift.

After the –1 frameshift, the frame for translation has shifted so that the frame in the slippery sequence has changed from xxU-UUU-UUA to UUU-UUU-Axx. The codons in the P and A sites are now UUU and UUU, respectively. Despite this shifting, the tRNAs that had already been occupying the P and A sites remain base paired to the codons in the P and A sites. This base pairing is possible because of the wobble allowed at the third position of each codon. The ribosome then catalyzes peptide bond formation between the phenylalanine and leucine. When the ribosome next moves forward during a normal elongation reaction, the A at the end of the slippery sequence now serves as the first nucleotide in the codon AGU. Thus, the amino acids phenylalanine and leucine get incorporated into the polypeptide but the next amino acid is not phenylalanine but serine (AGU). The ribosome then continues, disrupting the secondary structure that caused the pause and frameshift in the first place. The TF sequence ends with a stop codon so that the E1 sequence is not part of the polyprotein. Because the N-terminal sequences of TF are the same as those for 6K, the same host protease that normally releases 6K from the polypeptide similarly acts on the E3E2^TF bond.

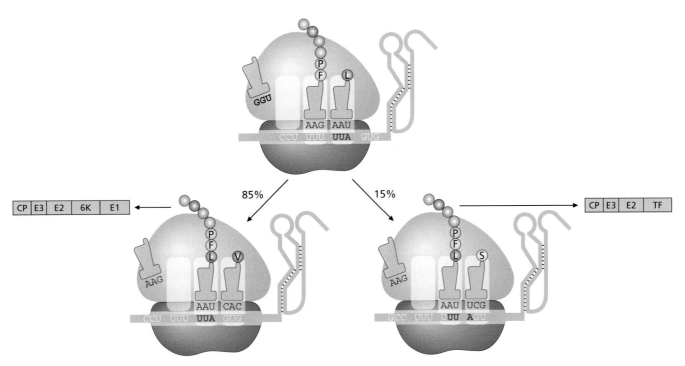

Figure 5.37 Programmed −1 ribosome frameshift. There are sites for three codon–anticodon interactions within the ribosome. During a programmed −1 frameshift, a slippery sequence such as UUUUUUA has an A that is used twice, once as the last nucleotide of a codon (UUA in red in this example) and a second time as the first nucleotide of the next codon (AGU in Sindbis virus). The pseudoknot is essential for the frameshift to occur, at a frequency of about 15%, making the polyprotein encoding TF instead of the polyprotein encoding 6K.

5.22   The picornaviruses, flaviviruses, and togaviruses illustrate many common properties among (+) strand RNA viruses

Because picornaviruses, flaviviruses, and togaviruses all have short (+) strand RNA genomes and use a polyprotein strategy for making multiple proteins from only one or two template mRNAs, it is useful to compare them. All of them carry out replication within a VRC that is created through the actions of nonstructural proteins that manipulate host proteins and membranes. Because all three viruses have RNA genomes that can be translated immediately following uncoating, they encode an RdRp but do not have to carry a copy of that RdRp into the cell. Instead, the RdRp is always one of the first proteins translated in its mature form in order to begin synthesizing the first antigenome. Antigenomes remain bound to their templates, so that the VRCs contain dsRNA, known as the replicative form, which is the template for synthesis of new (+) strands. The picornaviruses are the simplest because they encode only soluble proteins that do not require synthesis on ribosomes docked with the ER and because proteolysis of the viral polypeptide is carried out exclusively by viral proteases. The flaviviruses are somewhat more complex because part of the genome is translated on cytosolic ribosomes, but following a proteolysis event that reveals a signal peptide, the rest of the proteins are synthesized on ribosomes docked to the ER. Moreover, proteolytic digestion of flavivirus polyproteins requires both viral and host proteases. The togaviruses are the most complex of the three because they synthesize two types of (+) strand RNA molecules that can each be translated into two different polyproteins. Two different polyproteins arise either through suppression of translation termination mechanism or through a programmed

(−1) ribosome frameshift. Additionally, unlike the other two viruses, toga-viruses have distinctive early and late gene expression stages, which are dominated by the production of nonstructural proteins and structural proteins, respectively. The genomes are also organized differently; the picornaviruses and flaviviruses encode their nonstructural proteins nearer the 3′ end of the genome, whereas the togaviruses encode their nonstructural proteins near the 5′ end of the genome. Finally, all three use secondary structure in the (+) strand RNA for the regulation of translation, genome replication, or both.

5.23 Coronaviruses have long (+) strand RNA genomes and novel mechanisms of gene expression and genome replication

Picornaviruses, flaviviruses, and togaviruses share many characteristics, from their small genome length to their use of polyproteins. The last group of Class IV viruses considered in this chapter, the **coronaviruses**, is a bit different. Coronaviruses encodes two polyproteins and many individual proteins. Coronavirus genomes are three to five times the length of the picornavirus, flavivirus, and togavirus genomes. Although both togaviruses and corona-viruses produce sgRNAs, the coronavirus mechanism for doing so differs completely from that of the togaviruses. Because of their odd gene expression and genome replication mechanisms, along with their apparent lack of association with deadly human diseases, coronaviruses were once considered a bit obscure. In 2002–2003, however, the severe acute respiratory syndrome, or SARS, outbreak that occurred was caused by a coronavirus, now known as SARS-CoV. The outbreak affected more than 8,000 people, of which 9% died even with excellent medical care. Many of those affected early in the outbreak were healthcare providers who got infected merely by examining patients in order to make a diagnosis. Because of that SARS epidemic, interest in the molecular biology of coronaviruses increased. A new coronavirus disease, Middle East respiratory syndrome coronavirus (MERS-CoV), emerged in 2012; 50% of patients infected with MERS-CoV have died. In 2019, the SARS-CoV-2 virus emerged, causing COVID-19. The global COVID-19 pandemic has caused research on coronaviruses to surge. Studies of the molecular biology of coronaviruses will likely reveal additional viral proteins that could be good targets for new antiviral drugs (see **Chapter 16**).

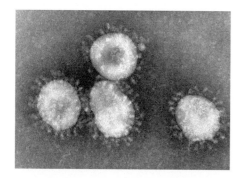

Figure 5.38 Severe acute respiratory syndrome coronavirus (SARS-CoV) virion. Electron micrograph of a SARS-CoV virion showing its spherical morphology crowned with prominent spikes. Virions are approximately 120 nm in diameter. (CDC/Dr. Fred Murphy.)

5.24 Coronaviruses have enveloped spherical virions and encode conserved and species-specific accessory proteins

Coronaviruses (family *Coronaviridae*, order *Nidovirales*) have enveloped spherical virions with striking spikes akin to a crown (corona) projecting from their surfaces (**Figure 5.38**). We will focus on the medically important **SARS coronavirus 2 (SARS-CoV-2)**. Coronavirus virions are spherical overall, but the RNA genome inside is arranged in a flexible helical structure, closely bound to a nucleocapsid protein, N. Coronavirus genomes are the largest of any (+) strand RNA viruses at 27–32 kb in length, which is between four and five times longer than those of the picornaviruses. Coronavirus genomes have the same terminal features as eukaryotic mRNA: a 5′ cap and a poly(A) tail (**Figure 5.39**). Both the 5′ and 3′ ends of the genome contain sequences and secondary structures that are important for synthesizing RNA using the genome as a template. The genome encodes two large polyproteins and many other individual proteins. The polyproteins are encoded at the 5′

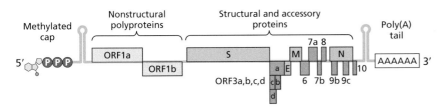

Figure 5.39 SARS-CoV genome. Diagram of the virus genome showing the 5′ methylated cap, secondary structure in the 5′ UTR, the RNA encoding two nonstructural polyproteins, the RNA encoding the structural (blue) and accessory proteins (orange), secondary structure in the 3′ UTR, and the poly(A) tail at the 3′ end. (ORF, open reading frame.)

end of the genome, and there is a shorter polyprotein (**pp1a**) and a longer polyprotein (**pp1ab**) produced using the same nucleotides for most of their length. The structural proteins are translated one at a time and do not arise by proteolysis; they are encoded near the 5′ end of subgenomic viral mRNAs that have a 5′ cap and poly(A) tail.

The coronavirus nonstructural proteins called ORF1a and ORF1b are encoded by two open reading frames (ORFs) in the third of the genome at the 5′ end. ORF1a encodes a shorter polyprotein, pp1a, which is processed into nonstructural proteins 1–11. ORF1b is not translated on its own but is expressed following a ribosomal −1 frameshift at the boundary between ORF1a and ORF1b, so that the cells produce an even longer polyprotein, pp1ab, encoding nonstructural proteins 1–16 (**Figure 5.40**). Viral proteases process the polyproteins to release individual nonstructural proteins. Functions for the nonstructural proteins include not only RNA synthesis but also formation of a membranous VRC.

The two-thirds of the genome at the 3′ end encode the four structural proteins S, E, M, and N (**Figure 5.41**). S is the **spike protein** and **E** and **M** are transmembrane proteins found in the virion envelope. The M protein is more abundant in the virion than the E protein; both are important for the assembly and release stages of the virus replication cycle. The N or **nucleocapsid** protein forms a helical complex with the (+) strand RNA genome and is important for multiple stages of the virus life cycle, including not only assembly and release but also gene expression and genome replication. N is the most abundant structural protein. In addition to the four structural proteins, coronaviruses encode **accessory proteins**. Accessory proteins are encoded surrounding the S, E, M, and N genes in the last two-thirds of the genome. They are often dispensable for replication in tissue culture, but they

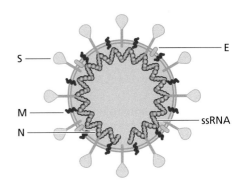

Figure 5.41 Coronavirus structural proteins. The enveloped virus has spikes composed of the S protein, the two transmembrane proteins E and M, and the N nucleocapsid protein that is bound to the genome forming a helical conformation inside the envelope. (From Perlman S & Netland J 2009. *Nat Rev Microbiol* 7:439–450. doi: 10.1038/nrmicro2147. With permission from Springer Nature.)

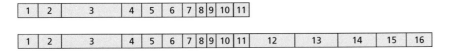

Figure 5.40 Coronavirus nonstructural polyproteins. The first polyprotein, translated from ORF1a, encodes nonstructural proteins 1–11. The second polyprotein, translated from ORF1a and ORF1b, encodes nonstructural proteins 12–16 after a −1 ribosomal frameshift at the boundary of ORF1a and ORF1b.

are essential for infecting an animal, which suggests that they interact with the animal's immune system. SARS-CoV-2 encodes 11 accessory proteins: 3a, 3b, 3c, 3d, 6, 7a, 7b, 8, 9b, 9c, and 10.

5.25 Coronaviruses express a nested set of sgRNAs with leader and transcription regulating sequences

Cells infected by a coronavirus produce at least 10 different RNAs of viral origin, not counting those encoding accessory proteins (**Figure 5.42**). Cells infected with coronaviruses contain full-length (+) strand genomes, used to translate polyproteins 1a or 1ab, ultimately processed into nonstructural proteins. They also contain a minimum of four more different (+) sense sgRNAs, which are used to translate the S, E, M, and N proteins. Infected cells also contain five double-stranded replicative forms corresponding to these five (+) sense RNAs. Cells infected by coronaviruses that also encode accessory proteins contain more viral RNAs because each of the accessory proteins is also encoded by its own (+) sense sgRNAs, which in turn arise from replicative forms containing one (−) sgRNA and one (+) sense sgRNA. All of the (+) sense RNAs are capped and polyadenylated, but the (−) strands are not. Because the replication of the virus occurs in the cytoplasm, not the nucleus, capping and tailing are catalyzed by viral nonstructural proteins.

The (+) strand sgRNA molecules have two curious properties that reflect the mechanism by which they are synthesized. First, the (+) sense sgRNAs form a **nested set**, in which the sequence of the smallest (+) strand sgRNA is found in the next largest (see Figure 5.42). The sequences in this next largest (+) sense sgRNA are found in the one larger than it, and so on, like nested Russian matryoshka dolls. Second, they all contain identical sequences at

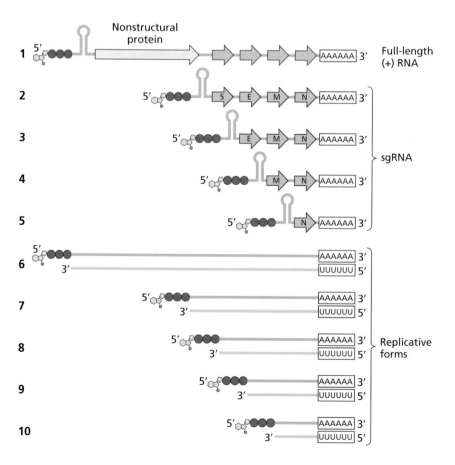

Figure 5.42 Ten coronavirus RNAs found in infected cells. Cells infected with a coronavirus contain a minimum of 10 viral RNAs. Five of them are a nested set of single-stranded (+) RNAs. The longest of these corresponds to the full-length genome. The smallest encodes just one protein (N). They are referred to as a nested set because the sequence of the smallest one is found in the next longest one, and so on. Infected cells also contain five double-stranded replicative forms that serve as the templates to produce the (+) RNAs. If the coronavirus has any accessory genes, infected cells will contain more (+) and replicative forms corresponding to those genes (not shown).

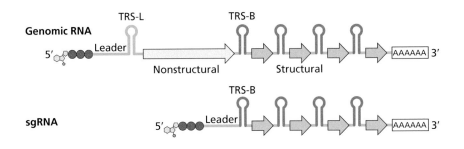

Figure 5.43 Leader sequence and TRS in the genome and sgRNA. Sequences that are critical for RNA synthesis have been exaggerated compared with the coding sequences. The genome begins with a unique leader sequence in the 5′ UTR followed by the leader TRS (TRS-L). Every structural gene (and accessory gene) is preceded by its own body TRS (TRS-B). Each (+) sgRNA has a 5′ sequence identical to the genome leader and their corresponding TRS-B.

their 5′ ends, consisting of a **leader sequence** and a **transcription regulating sequence** (**TRS**) (**Figure 5.43**). Even though it appears in all of the sgRNAs, the leader sequence is found only once in the genome, whereas the TRS is found in many places, adjacent to and downstream of the leader in the 5′ end of the genome and also at the boundary between most of the coding sequences in the 3′ end of the genome. The TRS near the leader is called **TRS-L** and those closer to the coding sequences are called **TRS-B** (for TRS-body). Although there is a consensus sequence for all of the TRSs, some of them have a few variant nucleotides that do not match the consensus perfectly. The TRSs are found as components of secondary structures such as stem–loops.

5.26 Coronaviruses use a discontinuous mechanism for synthesis of replicative forms

Production of (−) sgRNA in coronaviruses occurs through a **discontinuous mechanism**, often drawn as a jump (**Figure 5.44**). In a discontinuous

 To see an animation showing discontinuous minus strand synthesis in coronaviruses, please visit the book's website: www.routledge.com/cw/lostroh.

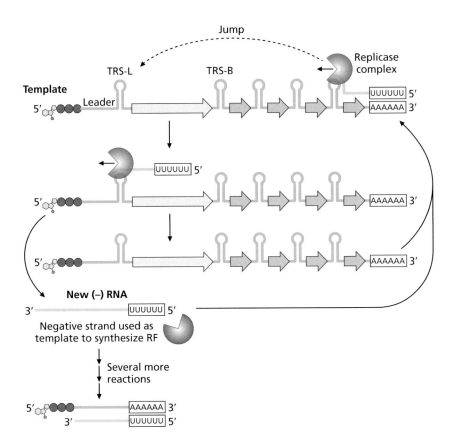

Figure 5.44 Discontinuous synthesis of coronavirus (−) sgRNA. The genome is used as a template by the RdRp replicase complex. Upon reaching a TRS-B template sequence, sometimes the RdRp jumps and transfers to the TRS-L sequence, which it then uses as a template to finish synthesizing (−) RNA. The (−) RNA therefore has sequences complementary to the TRS and also, at the 3′ end, a sequence complementary to the leader. The result is a (−) strand that will subsequently be used to synthesize one of the replicative forms. This figure shows the process of synthesizing the shortest (−) strand, which will ultimately be part of the shortest replicative form (RF).

mechanism, the polymerase skips over some bases in the template molecule. Reverse transcription is another example of discontinuous viral nucleic acid synthesis. In this case, the coronavirus genome is used by the RdRp replicase complex as a template to synthesize a (−) orientation strand. When the replicase reaches the sequence of a TRS-B, sometimes the nascent (−) RNA stops base pairing with the TRS-B and hydrogen bonds instead with TRS-L near the 5′ end of the genome (template). The replicase then finishes copying the 5′ end genome, continuing on from the TRS-L. In this way, every (−) sgRNA molecule has the same 3′ ends, creating a nested set of (−) RNA templates. The (−) strands are soon converted into double-stranded replicative forms, which are subsequently used to synthesize a large number of (+) strand sgRNA molecules. These molecules form a nested set because they are exactly complementary to the (−) sense sgRNA templates.

5.27   Most coronavirus sgRNA is translated into a single protein

Translation of genome-length (+) strand RNA in coronaviruses was discussed in Section 5.24 to explain the origin of the nonstructural polyproteins 1a and 1ab. There are nine SARS-CoV sgRNAs used to encode the other proteins. The structural proteins S, E, M, and N and accessory protein 6 are each translated from just one of the (+) strand sgRNAs. Ribosomes translate the first ORF in any given sgRNA and then dissociate from the sgRNA. For example, although the sgRNA with the S coding sequence also has the sequences for all downstream proteins such as E, M, and N, the only protein synthesized using that RNA as a template is the S protein. There are no internal ribosome entry sites that would allow for translation of downstream genes.

5.28   Coronaviruses use a leaky scanning mechanism to synthesize proteins from overlapping sequences

The other four sgRNAs encode two or more proteins each, employing a combination of mechanisms. The ORFs for 3a and 3b overlap with each other on the same sgRNA, as do the sequences for 8a and 8b. The coding sequence for accessory protein 9b completely overlaps the coding sequence for the N protein (**Figure 5.45**). The mechanism for translating the 9b protein has been investigated in some detail, and probably accounts for translation of the second ORF in the other sgRNAs used to synthesize two proteins. In the case of the N and 9b proteins, there are seven bases separating the AUG start codon for the N protein and the start codon for the 9b protein. The sequence surrounding the start codon has a strong influence on the efficiency of translation, and the optimal sequence for translation is called the **Kozak sequence**. The Kozak sequence surrounding the N start codon is suboptimal, whereas the Kozak sequence surrounding the start codon for the 9b protein is closer to the ideal. In this circumstance, the 48S preinitiation complex containing the small subunit of the ribosome sometimes scans past the first AUG to initiate translation at the downstream ORF instead. This mechanism is found in many viruses and is called **leaky scanning**, to indicate that the first Kozak sequence and AUG are leaky, so that the 48S preinitiation complex sometimes leaks past it without initiating translation at the first AUG (**Figure 5.46**).

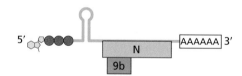

Figure 5.45 The N and 9b coding sequences overlap on a single sgRNA. The smallest sgRNA has a 5′ cap and 3′ poly(A) tail, and encodes both the N and 9b proteins. The 9b coding sequence completely overlaps with that of the N protein and is in a different frame.

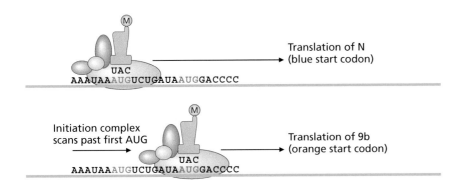

Figure 5.46 A leaky scanning mechanism allows translation of 9b from the same sgRNA as that encoding N. Sometimes the translation initiation complex, which formed at the 5′ cap, stabilizes around the first AUG, resulting in translation of the N protein. At other times, the initiation complex physically scans past the first AUG and instead stabilizes around the downstream AUG, resulting in translation of the 9b protein.

5.29 Coronaviruses proofread RNA during synthesis

Most groups of (+) strand RNA Class IV viruses have small genomes, around 10,000 bases in length. This limit on genome length may arise from the properties of viral RdRps, which, unlike the host DNA polymerase used for replication, have no **proofreading** function (**Figure 5.47**; see also **Chapter 17**). In molecular biology, proofreading means removing a base that has just been covalently added to a growing chain of nucleic acids because the base does not hydrogen bond properly with the template. For example, if the template has a G and the polymerase enzyme adds an A instead of a C, that is a **misincorporation**. Proofreading removes nucleotides in a 3′–5′ direction, the opposite of polymerization. The proofreading enzyme is said to have **exonuclease** activity because exonucleases degrade nucleic acids starting at one end. Removal of the misincorporation leaves behind a free 3′ OH so that the enzyme can make a second attempt to incorporate the correct nucleotide and continue nucleic acid synthesis.

Viral RdRp enzymes misincorporate the wrong base approximately 1 in 10,000 times, with a range of 1 in 1,000 to 1 in 100,000, depending on the RdRp. These values contrast with 1 in 10,000,000 misincorporations for host DNA polymerase with proofreading functions. The lack of proofreading by RdRp in combination with its typical misincorporation rate means that most of the thousands of offspring virions in an infection will have at least one mutation. This high rate of mutagenesis may limit the length of Class IV virus genomes because longer genomes would have even more mutations per offspring virion, an increasing proportion of which would result in a genome that could no longer direct a complete virus replication cycle. For example, the picornavirus RdRp misincorporates the wrong base approximately once for every 2,000 RNA template bases copied. At this rate, on average, the 7.2- to 8.5-kb genomes of offspring picornaviruses have three or four mutations relative to the template from which they were synthesized. A similarly high misincorporation rate in coronaviruses would result in 15 mutations in each offspring virion, and that high amount would probably result in at least one mutation that would be detrimental to viral replication in the next generation. This situation led to a desire to understand how coronaviruses can faithfully reproduce such large genomes, with the surprising finding that coronaviruses have at least one mechanism for removing and replacing misincorporated bases. This finding alters our understanding of the capabilities of RdRp complexes, which were always thought to lack proofreading.

Interest in the possibility of proofreading arose due to the length of the coronavirus genome, and the predicted protein domains and motifs in the nonstructural proteins led to a focus on a particular nonstructural protein,

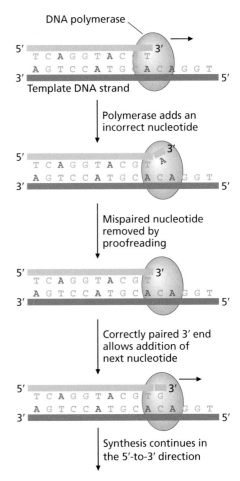

Figure 5.47 Polymerases with proofreading activity. If an incorrect nucleotide is added to a growing strand, the polymerase cleaves it from the strand and replaces it with the correct nucleotide before continuing.

nsp14 (ExoN). Because of a DE-D-D amino acid sequence motif, ExoN was predicted to have 3′–5′ exonuclease activity, which is a hallmark of proofreading enzymes known to be active during DNA replication.

There are several findings in favor of ExoN participating in proofreading during synthesis of coronavirus RNA. First, purified ExoN does have exonuclease activity. Further, the ExoN DE-D-D motif is essential for catalysis. Replacing the aspartic acids (D) or glutamic acid (E) with an alanine resulted in reduced or abolished exonuclease activity in biochemical tests and higher rates of misincorporation among offspring virions (**Figure 5.48**). Further biochemical testing has shown that the ExoN protein removes mismatched bases from the 3′ end of base-paired RNA and that this activity is stimulated by the coronavirus nsp10 protein.

Taken together with the predicted functional domains and motifs of the other nonstructural proteins, a model for the coronavirus RNA replication machine has emerged (**Figure 5.49**). According to the model, a ring formed by **nsp7** and **nsp8** serves as a sliding clamp to keep the **nsp12/14/10 complex** attached to the RNA template. The nsp8 protein is also a primase that synthesizes a short sequence of RNA that provides a primer for the main RdRp to extend. The catalytic RdRp active site resides in the nsp12 protein. When the RdRp misincorporates a nucleotide, ExoN removes the mismatch and the RdRp has a second chance to incorporate the correct nucleotide. The **nsp10** protein stimulates the exonuclease activity of ExoN and the **nsp9** protein may function as a single-stranded RNA binding protein. The nsp9 protein

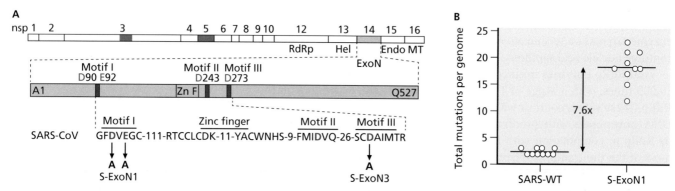

Figure 5.48 Amino acid substitutions in the DE-D-D motif of ExoN cause an increase in mutations per genome. (A) Close-up of ExoN (nsp14) in the SARS-CoV genome. The ExoN primary sequence shows the DE-D-D motif and the position of alanine substitutions in the S-ExoN1 mutant. (B) Virus with the S-ExoN1 mutations gives rise to offspring with more mutations per genome (S-ExoN1) than wild-type virus (SARS-WT). (From Eckerle LD et al. 2010. *PLOS Pathog* 6:e1000896. doi: 10.1371/journal.ppat.1000896. Published under CC BY 4.0.)

Figure 5.49 Model for the coronavirus replication machinery. A model of how nsp7, nsp8, nsp9, nsp10, nsp12, and nsp14 (ExoN) could assemble on viral dsRNA and interact with the putative multisubunit CoV polymerase complex. The nsp14 (ExoN) subunit has proofreading activity. (From Smith EC & Denison MR 2013. *PLOS Pathog* 9:e1003760. doi: 10.1371/journal. ppat.1003760. Published under CC BY 3.0.)

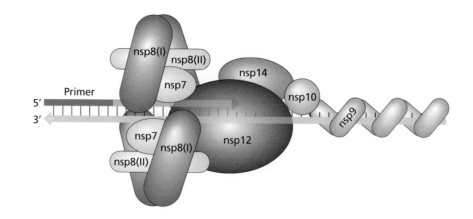

prevents secondary structure in the template so that the complex containing RdRp can move rapidly down the template without dissociating. The nsp13 protein is a helicase needed for copying double-stranded replicative forms.

The model raises additional questions. Three proteins, nsp13, nsp14, and nsp16, are probably needed for synthesizing and attaching the 5′ cap to (+) sense RNAs, but it is not known how coronavirus exclusively caps (+) strands without modifying (−) strands. Nsp15 has endonuclease activity that degrades RNA and is essential for normal genome replication, but its function is not understood beyond these observations. Furthermore, the model does not explain how the machine switches from (−) to (+) strand synthesis, or selects or bypasses various TRS-B sequences during (−) synthesis. Nevertheless, the model provides a starting place to design experiments that will further the understanding of coronavirus RNA synthesis. Finally, coronaviruses are recombinogenic. While discontinuous synthesis of replicative forms is hypothesized to be crucial for that recombination, coronavirus recombination has never been observed directly in the lab. Recombinant SARS-CoV-2 viruses are an increasing problem in the COVID-19 pandemic.

5.30 Plants can also be infected by Class IV RNA viruses

Animals are not the only multicellular eukaryotes infected by Class IV (+) strand RNA viruses. There are many Class IV (+) strand RNA viruses that infect plants. Based on the features of their genomes, most use gene expression and genome replication strategies similar to those in the picornaviruses, flaviviruses, and togaviruses that infect animals. For example, there are picornaviruses that infect plants (families *Secoviridae* and *Marnaviridae*). Other Class IV plant viruses, such as those of the potyvirus (family *Potyviridae*, order *Tymovirales*) and **tobacco mosaic virus** (TMV; family *Virgaviridae*, order *Tymovirales*) have naked, helical virions (**Figure 5.50**). Potyvirus genomes have terminal features similar to those of picornaviruses while the TMV genome has a 5′ cap and a 3′ UTR with unusual secondary structure (**Figure 5.51**).

The TMV genome differs from those we discussed earlier in the chapter; although it has a 5′ methylated cap instead of a VPg, its 3′ UTR is folded into a structure resembling a tRNA, so that the 3′ UTR is called a **tRNA-like sequence** (**TLS**; **Figure 5.52**). TLSs are so far known only in plant viruses. The **molecular mimicry** is so good that cellular enzymes bind to the TLS and aminoacylate it, covalently attaching a histidine as though the TLS were an actual tRNA. Other viruses have somewhat different TLSs to which cellular enzymes attach valine or tyrosine amino acids. Aminoacylation of the TLS

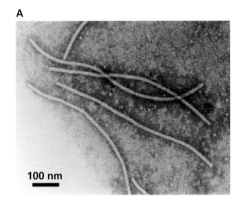

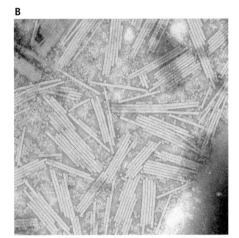

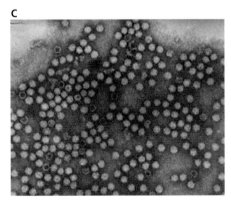

Figure 5.50 (+)-Strand RNA viruses of plants. (A) Electron micrograph of a potyvirus showing its flexible helical morphology. (B) Crystal structure of TMV. The helical virions are about 300 nm long and 18 nm in diameter. (C) Electron micrograph of cowpea mosaic viruses, which are about 28 nm in diameter.

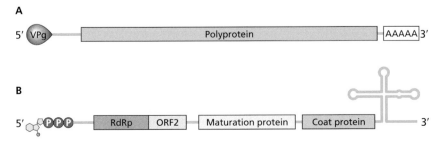

Figure 5.51 (+)-Strand RNA genomes in plant viruses. (A) Potyvirus genome with 5′ terminal VPg protein and 3′ poly(A) tail. (B) TMV genome with 5′ cap and 3′ untranslated region resembling tRNA.

A

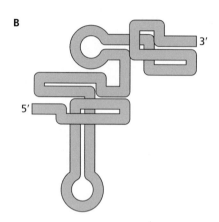

B

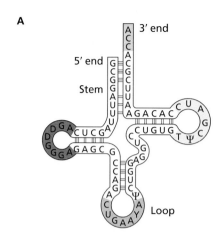

Figure 5.52 The tRNA-like sequence at the 3′ end of the TMV genome. (A) Structure of tRNA. (B) Structure of the TLS. (From Alberts B et al. 2013. *Essential Cell Biology*. Garland Science. With permission from W.W. Norton.)

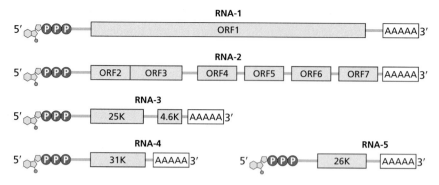

Figure 5.53 Multipartite plant picornaviruses. Benyviruses have five genome segments, each with a 5′ cap and a 3′ poly(A) tail and each encapsidated into its own virion.

has been shown to be important for translation of one virus, and for synthesis of RNA in another. The role of aminoacylation of the TMV TLS has not been investigated experimentally. It is also possible that the TLS has a function in gene expression or genome replication that is independent of its ability to be aminoacylated.

Another variation on Class IV genome structure found mainly in plant viruses is that some are **multipartite**, meaning that instead of having just one genomic molecule, their genomes consist of two or more different RNA molecules (**Figure 5.53**). Furthermore, each genome segment is encapsidated separately into its own virion, so that a cell must be simultaneously co-infected with all of the different particles in order to carry out a full virus replication cycle. An extreme example with five different genome segments, each capped and tailed like eukaryotic mRNA and enclosed in its own separate virion, is the **beet necrotic yellow vein virus** (genus *Benyvirus*). The genome segments are named RNA-1 to RNA-5, from longest to shortest. All five have a 5′ cap and poly(A) tail. RNA-1 encodes the RdRp, which is part of a polyprotein similar to that in picornaviruses. RNA-2 resembles the 3′ end of a coronavirus, in that it encodes two different proteins at its 5′ end through a ribosomal frameshift and it causes production of a nested set of sgRNAs in which only the most 5′ ORF is translated. One of these sgRNAs can be translated into either of two proteins using a leaky scanning mechanism. RNA-3 encodes two proteins, which are translated either from the full-length genome or from an sgRNA. RNA-4 and RNA-5 each encode just one protein. The similarities with picornaviruses and coronaviruses suggest hypotheses for the mechanisms of gene expression and genome replication in these viruses, but so far very little is known about the mechanisms.

5.31 Comparing Class IV viruses reveals common themes with variations

Class IV viruses have genomes that are translated following uncoating, resulting in production of an RdRp enzyme that synthesizes viral RNAs. In all cases, the single RdRp enzyme is part of a multiprotein machine that actually carries out RNA synthesis. Viral RNA occurs in membranous VRCs, where the early production of double-stranded replicative forms leads to synthesis of many (+) orientation molecules. The regulatory mechanism that leads to selective synthesis of (−) or (+) strands is not clear in every case, but because of the picornavirus and togavirus examples, it likely involves the ordered proteolysis of a polyprotein that contains the RdRp. Before the accumulation of structural proteins, the full-length (+) sense molecules are used as mRNA, but after the accumulation of structural proteins, they are packaged as new genomes. All Class IV viruses have distinctive genome features that are necessary for their particular mechanisms of gene expression and genome replication. Examples include the picornavirus IRES needed for translation; the togavirus slippery sequence, pseudoknot, and DLP; and the coronavirus TRS-L and TRS-B sequences. Some of these viruses use a single mRNA to encode more than one protein; examples include the major versus minor proteolysis pathways in picornaviruses, suppression of translation termination in togaviruses, programmed −1 ribosome frameshifting in the togaviruses and coronaviruses, and leaky scanning, also in the coronaviruses. Coronaviruses are unusual because their genomes are many times longer than those of other Class IV groups, which is likely possible because of the unique existence of proofreading during RNA synthesis.

Essential concepts

- Class IV viruses replicate in membranous virus replication complexes, or VRCs, made up of virus proteins, host proteins, and host membranes. VRCs sequester double-stranded replicative forms of RNA viruses so that cytoplasmic surveillance by innate immune mechanisms does not detect the virus.
- All viruses with RNA genomes must encode an RNA-dependent RNA polymerase, or RdRp, because no cellular enzymes use an RNA template to synthesize complementary RNA. Class IV virus genomes are translated immediately upon uncoating and one of the first proteins translated is always the RdRp.
- Picornaviruses are the smallest Class IV viruses that infect animals and encode a polyprotein that is processed into smaller component parts, which include both nonstructural and structural proteins.
- In general, the features at the end of a virus genome, such as a 5′ cap, a covalently linked VPg protein, or poly(A) tail, provide clues as to the mechanisms of gene expression and genome replication.
- Picornaviruses rely upon the IRES in their 5′ UTR for efficient translation of their polyprotein.
- Picornavirus RNA synthesis occurs in four phases that depend on particular secondary structures in the template RNAs and on the catalytic activities of several different virus proteins.
- Flaviviruses such as hepatitis C virus encode a longer polyprotein than that of picornaviruses. The polyprotein includes transmembrane segments because the viruses are enveloped.
- Togaviruses, although similar to flaviviruses in genome length, are more complex because they encode four different polyproteins and have distinct phases of early and late gene expression and produce structural proteins much more abundantly than nonstructural proteins.
- Class IV viruses use strategies such as suppression of translation termination, programmed ribosome frameshifts, and leaky scanning to encode multiple proteins with some overlapping sequences.
- Coronavirus genomes are four to five times longer than those of picornaviruses and encode two nonstructural polyproteins and multiple structural proteins.

- Coronaviruses produce many (+) sense sgRNAs that form a nested set in which the sequence of longer sgRNAs includes the sequences of the next smaller sgRNA, and so on.
- Coronaviruses synthesize (−) strand templates by a discontinuous mechanism that relies upon the TRSs and secondary structure in the template (+) strand RNA.
- Unexpectedly for an RNA virus, coronaviruses can detect, remove, and replace misincorporated bases during genome replication, which is likely essential for maintaining such a long genome.
- Class IV viruses of plants are generally similar to those that infect animals. Unique features of certain Class IV plant viruses include tRNA-like sequences in tobacco mosaic virus and multipartite, individually encapsidated genomes in other viruses.

Questions

1. Why is the RdRp always one of the first proteins translated in any infection by a (+) strand RNA (Class IV) virus?
2. What is an IRES and what is its purpose?
3. Write the nucleic acid sequence for two viral proteins in which the second one is synthesized using a −1 frameshift.
4. What are the shared features of picornaviruses, flaviviruses, and togaviruses?
5. Hypothesize as to the role of the secondary structure in the 3′ UTR of hepatitis C virus. Justify your hypothesis.
6. Compare and contrast picornaviruses with coronaviruses.
7. Provide three examples of unusual structural features in the genomic or subgenomic (+) sense RNAs of Class IV viruses and where possible explain their functions.
8. How does leaky scanning differ from suppression of translation termination?
9. Good antiviral drugs block the activity of a viral enzyme that is needed for the replication cycle and that is unique to the virus, so that the drug selectively binds to viral protein without affecting any human ones. List at least five viral proteins from this chapter that could serve as good antiviral drug targets and justify your choices.
10. Why must enveloped viruses use the SRP system?
11. In general, in any pulse-chase analysis, what is the major experimental goal (enabled by the chase)?
12. What is the purpose of the secondary antibody in a western blot?
13. What are the advantages to separating proteins by SDS-PAGE prior to western blotting rather than spotting all of the proteins in a sample directly onto a blot without doing SDS-PAGE first?

Interactive quiz questions

In addition to the questions provided above, this edition has a range of free interactive quiz questions for students to further test their understanding of the chapter material. To access these online questions, please visit the book's website: www.routledge.com/cw/lostroh.

Further reading

Picornaviruses

Jang SK, Kräusslich HG & Nicklin MJ 1988. A segment of the 5′ nontranslated region of encephalomyocarditis virus RNA directs internal entry of ribosomes during *in vitro* translation. *J Virol* 62:2636–2643.

Oshinksy DM 2005. *Polio: An American Story*, 1st ed. Oxford University Press.

Pelletier J & Sonenberg N 1988. Internal initiation of translation of eukaryotic mRNA directed by a sequence derived from poliovirus RNA. *Nature* 334:320–325.

Twu WI, Lee JY, Kim H, Prasad V, Cerikan B, et al. 2021. Contribution of autophagy machinery factors to HCV and SARS-CoV-2 replication organelle formation. *Cell Reports* 37:110049.

Flaviviruses

Appleby TC, Perry JK, Murakami E, Barauskas O, Feng J et al. 2015. Structural basis for RNA replication by the hepatitis C virus polymerase. *Science* 347:771–775.

Crosby MC 2006. *The American Plague*, 1st ed. Penguin Group.

Evans AS, Lennemann NJ & Coyne CB 2020. BPIFB3 regulates endoplasmic reticulum morphology to facilitate flavivirus replication. *J Virol* 94:e00029-20.

Zhuang X, Magri A, Hill M, Lai AG, Kumar A et al. 2019. The circadian clock components BMAL1 and REV-ERBα regulate flavivirus replication. *Nat Comm* 10:377.

Togaviruses

Hellström K, Kallio K, Meriläinen HM, Jokitalo E & Ahola T 2016. Ability of minus strands and modified plus strands to act as templates in Semliki Forest virus RNA replication. *J Gen Virol* 97:1395–1407.

Junjhon J, Pennington JG, Edwards TJ, Perera R, Lanman J & Kuhn RJ 2014. Ultrastructural characterization and three-dimensional architecture of replication sites in dengue virus-infected mosquito cells. *J Virol* 88:4687–4697.

Lello LS, Bartholomeeusen K, Wang S, Coppens S, Fragkoudis R et al. 2021. nsP4 is a major determinant of alphavirus replicase activity and template selectivity. *J Virol* 95:e00355-21.

Coronaviruses

Li J, Lai S, Gao GF & Shi W 2021. The emergence, genomic diversity and global spread of SARS-CoV-2. *Nature* 600:408–418.

Shannon A, Selisko B, Le NTT, Huchting J, Touret F et al. 2020. Rapid incorporation of Favipiravir by the fast and permissive viral RNA polymerase complex results in SARS-CoV-2 lethal mutagenesis. *Nat Comm* 11:4682.

Yan L, Yang Y, Li M, Zhang Y Zheng L et al. 2021. Coupling of N7-methyltransferase and 3′-5′ exoribonuclease with SARS-CoV-2 polymerase reveals mechanisms for capping and proofreading. *Cell* 184:3474–3485.

Gene Expression and Genome Replication in the Negative-Strand RNA Viruses

6

Virus	Characteristics
Rabies virus	Exemplifies common features of all mononegaviruses.
Measles virus	Mononegavirus that uses RNA editing for gene expression.
Ebola virus	Mononegavirus with unique transcription factor.
Influenza A virus	(–) RNA virus with segmented genome and cap snatching during gene expression.
Lassa fever virus	Virus with segmented ambisense genome.

The viruses you will meet in this chapter and the concepts they illustrate

Class V (–) RNA viruses are interesting on many levels, from the pandemics one of them causes to the molecular biology of their unusual enzymes. For example, one of the viruses in this group is known for cap snatching, a process in which a viral enzyme beheads nuclear mRNA and then uses the snatched RNA as a primer. A second example is the very large replicase protein found in rabies and its close relations (the mononegaviruses); these replicases are more than 2,000 amino acids long and have many enzymatic activities assigned to distinct domains. Although there are known orders of Class V viruses, all Class V viruses share certain features. For example, because the genomes themselves cannot be translated, each virion must contain at least one functional viral replicase protein that enters with the genome and synthesizes mRNA starting from a protein-coated (–) RNA genome template. The viral proteins translated from viral mRNA then continue the process of taking over the host cell and causing the production of offspring virions, and here we see some variation in the mechanisms used to synthesize mRNA, antigenomes, and genomes.

6.1 Studies of two historically infamous Class V viruses, rabies and influenza, were instrumental in the development of molecular and cellular virology

Two negative-strand RNA viruses, **rabies virus** and **influenza virus**, stand out for their historical association with virology as it emerged as a field distinct from bacteriology and became closely allied with immunology. Rabies virus figured prominently in the intellectual biography of Louis Pasteur, a nineteenth century microbiologist who experimented with **attenuation** of lethal viruses by propagating rabies virus isolated from dogs in monkeys. As

DOI: 10.1201/9781003463115-6

the virus adapted to the monkeys, it became less lethal to dogs, that is, it became attenuated. Ultimately, these experiments led Émile Roux and Louis Pasteur to use desiccated rabbit tissue containing rabies virus to inoculate people who had been bitten by rabid animals, thereby saving these people from certain death. Pasteur's work helped establish France, led by the Institut Pasteur, as the global leader in immunology research and applied immunological-based treatments.

Some decades later, the 1918–1919 global influenza pandemic occurred, which had a profound effect on virology as a discipline. At that time, influenza had long been recognized as a dangerous disease with pandemic potential, described in historical documents since the age of Hippocrates (circa 400 BCE). Of the total human population in 1918–1919 (about 2 billion), 25% caught influenza during the pandemic. As many as 50 million died, more than the soldiers and civilians who died as a consequence of World War I (1914–1918). The causative agent was not known, and the great powers of the day raced to find the infectious microbe in hopes of developing some kind of treatment or immunization. Viruses, however, had still not been definitively described as noncellular infectious agents important in human disease, and could not be observed directly with the best microscopy of the day. A bacterium isolated from many patient lung samples was named *Haemophilus influenzae* because it was thought to be the causative agent. It is now known that this species is simply one of the typical secondary respiratory infectious agents that can follow in the wake of the tissue destruction caused by influenza. Influenza virus was finally isolated in the 1930s and was therefore part of molecular and cellular virology at the very beginning, when technologies such as electron microscopy were first making it possible to observe viruses directly. The development of techniques to propagate influenza in the lab led to discoveries enabling the propagation of many more animal viruses.

The historic importance of rabies virus and influenza virus in the development of both immunology and virology means that we have a solid understanding of their molecular biology, which in turn has led to better understanding of their close relatives. In the twenty-first century we are on the verge of many new pharmaceuticals that target the viral enzymes required for the (−) RNA viruses to complete their life cycles. This era is also a period of rapid vaccine development, including a vaccine to protect at-risk populations such as healthcare workers from another disease caused by a (−) RNA virus, namely Ebola virus hemorrhagic fever. These advances in applied virology would not have been possible without the basic research revealing how the viruses work at the cellular and molecular levels.

6.2 The mononegavirus replication cycle includes primary and secondary transcription catalyzed by the viral RdRp

The first group of viruses considered in detail in this chapter is the order *Mononegavirales*; unlike cellular organisms that can all be assigned to higher taxonomic classifications such as orders, it is unusual for virus families to group into orders. The replication cycles of viruses belonging to this order share many features (**Figure 6.1**). After uncoating, the (−) RNA remains associated with **nucleocapsid proteins** (**NPs**) and with viral proteins including an RNA-dependent RNA polymerase (RdRp) and associated factors necessary for RNA synthesis. The genome and its associated nucleocapsid proteins can be referred to as a viral ribonucleoprotein complex (**vRNP**). The RdRp becomes active as a **transcriptase** and uses the vRNP as a template to synthesize multiple different viral mRNAs, which have methylated caps and polyadenylated tails, all created by viral enzymes. This process is referred to as **primary transcription** because it uses the infecting genome as a template. During primary transcription, mRNA-encoding viral proteins become abundant and all of the viral proteins, including NP, are synthesized. When the levels of NP in particular become high enough, the virus switches from primary transcription to production of complete (+) antigenomes. This switch requires use of newly synthesized viral enzymes, including new RdRp molecules. Genome synthesis requires high levels of NP because the nascent RNA must be packaged into ribonucleoproteins (RNPs) during synthesis. The (+) antigenomes are very different from mRNA because the antigenomes are a copy of the entire genome and lack 5′ caps and 3′ polyadenylation.

The complex containing RdRp that creates antigenomes is called the **replicase**, to distinguish it from the transcriptase that produces mRNA. It might contain exactly the same protein subunits as the transcriptase, however, so that it is the availability of NP that determines whether mRNA or antigenomes are synthesized from (−) sense templates. The molecular features of the replicase compared with the transcriptase are not completely understood. The replicase also synthesizes new genomes using the antigenome vRNPs as a template.

Packaging of nascent RNA into RNP consumes the NPs until levels of free NP (not bound to RNA) fall so much that **secondary transcription** begins, using the many new copies of the genome as templates to produce a second wave of abundant viral mRNA. Once again, the accumulation of viral NPs

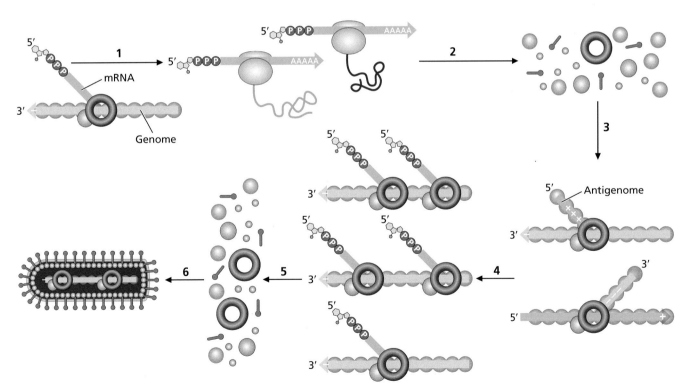

Figure 6.1 Overview of the mononegavirus replication cycle. After uncoating, the genomic vRNP (genome and nucleocapsid proteins) is used as a template for primary transcription (1), resulting in production of viral mRNA and proteins (2). Viral mRNAs have 5′ caps and poly(A) tails. When levels of NPs reach a threshold, the virus makes antigenomes and subsequently new genomes (3), both bound to NPs. When levels of free NPs drop below a threshold, the new genomes are next used as templates in secondary transcription, producing very abundant viral mRNA (4) and proteins (5). When NP levels again reach a threshold, the virus synthesizes antigenomes and new genomes, which are packaged into new virions during maturation (6).

ultimately biases the system toward production of full-length genomes and antigenomes rather than mRNA, leading to a switch from secondary transcription to production of antigenomes and genomes, which concludes the genome synthesis stage of the replication cycle.

6.3 Rhabdoviruses have linear (−) RNA genomes and encode five proteins

Rhabdoviruses such as rabies virus and vesicular stomatitis virus (family *Rhabdoviridae*) are two of the simplest Class V viruses in that they encode only five proteins, all of which are found in all other mononegaviruses. The name of the rhabdoviruses refers to their rod-shaped virions, derived from a Greek word for rod; in the case of rabies virus, the rod shape is curved at one end resulting in a distinctive bullet-like morphology (**Figure 6.2**). The virions are enveloped and covered by a single abundant glycoprotein spike (**G**).

Rhabdovirus genomes are typically 11–15 kb long and include **leader** (**le**) and **trailer** (**tr**) **sequences** that do not encode proteins but are partially complementary to one another (**Figure 6.3**). The le and tr are *cis*-acting sequences important for transcription, replication, and assembly. The protein-coding sequences are separated by conserved sequences called **E** (**end**), **I** (**intergenic**), and **S** (**start**). Inside the virion, the genome is bound to **nucleocapsid** (**N**) **proteins**, forming an overall helical structure. Several enzymatic and regulatory proteins are also found in smaller amounts but in close association with the genome. These include the L and P proteins, which are components of the RdRp complex. Although rabies virus is of obvious clinical interest, many experiments performed to investigate the molecular and cellular biology of the simplest rhabdoviruses used vesicular stomatitis virus because it is safer and easier to handle in a laboratory setting.

All rhabdoviruses have genomes that encode at least the following five proteins (see Figure 6.3). Starting with the protein encoded at the 3′ end and moving toward that encoded at the 5′ end, the proteins are N, P, M, G, and L. The N, M, and G proteins are abundant in the virion and do not have

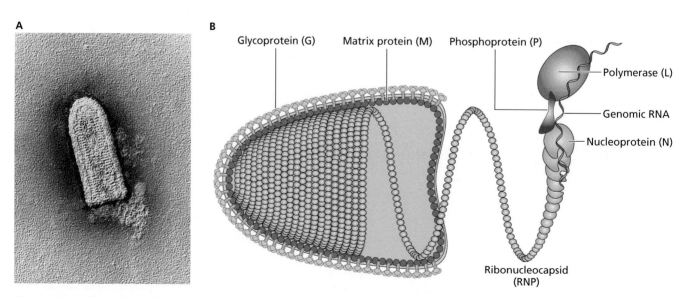

Figure 6.2 Rabies virion. (A) Electron micrograph of rabies virions. The viruses are bullet-shaped, packed full of vRNP (yellow), and have abundant spikes (red). Each virion is about 180 nm long and 75 nm wide. (B) Diagram of rabies virus showing the five proteins: G, M, N, P, and L. (A, Eye of Science/Science Photo Library. B, Courtesy of Philippe Le Mercier, ViralZone, © SIB Swiss Institute of Bioinformatics.)

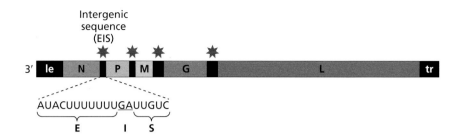

Figure 6.3 Rabies virus genome. The rabies virus genome consists of a 3′ leader, the coding sequences for the N, P, M, G, and L proteins, the intergenic sequences needed for start–stop transcription, and the 5′ trailer.

enzymatic activity. The N or nucleocapsid protein encapsidates the genome, resulting in its helical shape. The **matrix protein** (**M**) is needed for budding, and the G or glycoprotein is a transmembrane spike needed for attachment, fusion, and penetration. The **P** and **L proteins** form the RdRp that, in addition to its enzymatic role, is also a structural component of the virions. The P protein can be phosphorylated and forms a complex with the L or large protein. It also forms a complex with the N protein; this complex is important for controlling the switch from mRNA production to genome replication.

The mononegavirus L protein, which can be as large as 250,000 Da, has multiple domains and functions including RNA-dependent RNA polymerization, synthesis of methylated caps, and capping and polyadenylation of mRNAs (**Figure 6.4**). Sequence comparisons and electron microscopy have been used to examine the structure of the L protein. Amino acid sequence comparisons across all known mononegavirus L proteins revealed six highly conserved regions, numbered I–VI. Various genetic and biochemical experiments in combination with electron microscopy indicated that these conserved regions likely fold into discrete domains with specific functions connected to one another by flexible hinges. Functions have been associated with four of the six conserved regions as follows. The function of region I is probably to allow L to associate with P. Region III contains conserved motifs found in all RNA polymerases, so likely contains the active site for RNA synthesis. Regions V and VI are responsible for capping the mRNAs during transcription.

6.4 Rhabdoviruses produce five mRNAs with 5′ caps and polyadenylated 3′ tails through a start–stop mechanism

Rabies virus genomes remain cytoplasmic for the entire replication cycle. Cells infected with a rhabdovirus contain a minimum of five viral mRNA molecules, which have typical eukaryotic 5′ methylated caps and polyadenylated tails and therefore can be recognized by the normal host translation machinery. Interestingly, the relative amounts of each mRNA correlate with the order in which the gene appears in the genome, with N mRNA being most abundant, followed by P, M, G, and L. Each mRNA is about two-thirds as abundant as the previous one. The noncoding boundaries between the genes are responsible for this pattern and are called E, I, and S to indicate that these regions of the template serve as the end of the template for one mRNA, the intergenic region between genes, and the start of the template for the next mRNA. Conserved sequences in the E and S regions are responsible for causing polyadenylation and release of mRNA and initiation of synthesis of the next mRNA, respectively.

This pattern of mRNA production in which each sequential gene produces less mRNA than the one preceding it must have a molecular basis; the

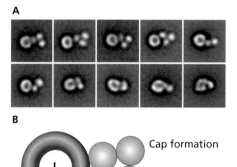

Figure 6.4 Mononegavirus large (L) protein. (A) Electron micrograph of rhabdovirus L protein showing its hollow ring structure, with three spherical domains associated with one side. The position of the three spherical domains changes during synthesis of different types of nucleic acids. (B) Structure of the L protein. The ring structure has the polymerization activity, whereas the three spherical domains off to one side are responsible for synthesizing a methylated cap and attaching it to nascent mRNA. (From Rahmeh AA et al. 2010. *PNAS* 107:20,075–20,080. doi: 10.1073/pnas.1013559107. With permission from Amal Rahmeh.)

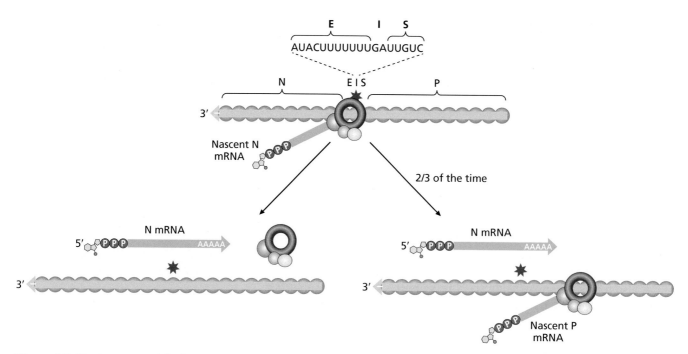

Figure 6.5 Start–stop model of mononegavirus transcription. Only the N and P genes with the intergenic region consisting of E, I, and S sequences are shown. Upon reaching the intergenic E (end) sequence, the polymerase releases the mRNA. Approximately two-thirds of the time, the polymerase then scans past the I (intergenic) sequence and initiates transcription again at the S (start) sequence, leading to transcription of the downstream gene (in this case, the P gene).

currently accepted explanation is called the **start–stop model** (**Figure 6.5**). The transcriptase consists of viral L and P proteins in combination with at least two host proteins called eEF1-α and Hsp60. To make mRNA, the transcriptase enzyme complex begins at a promoter near, but not directly at, the 3′ end of the genome, using genomes that are encapsidated by N protein as the template. After transcribing the leader sequence, that RNA is released and then the complex transcribes the first gene, which is always the N gene encoding the protein that encapsidates the genome. Upon reaching the intergenic E (end) sequence, the polymerase releases the mRNA. Approximately two-thirds of the time, the polymerase then scans past the I (intergenic) sequence and initiates transcription again at the S (start) sequence. This frequency explains why each mRNA is one-third less abundant than the one synthesized from an upstream gene. The intergenic region is required for the transcriptase to reinitiate transcription, which can be demonstrated by deleting or altering it through mutation, using a genetically engineered host-cell system. Similarly, it can be experimentally demonstrated that the S sequence, which is transcribed, is required for proper capping and initiation of the next mRNA.

Mononegavirus mRNA is polyadenylated through a mechanism known as **reiterative transcription** (**Figure 6.6**). When the transcriptase arrives at the E (end) sequence, it encounters seven U nucleotides in a row. These Us,

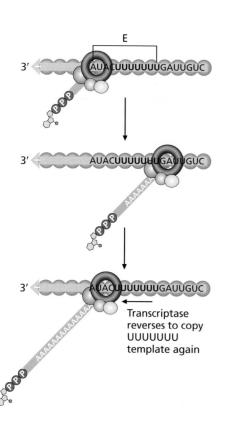

Transcriptase reverses to copy UUUUUUU template again

Figure 6.6 Reiterative transcription to create a poly(A) tail. When the transcriptase arrives at the E (end) sequence, it encounters seven U nucleotides in a row. These U nucleotides are used as a template to add A nucleotides to the 3′ end of the nascent mRNA and then, instead of continuing, the transcriptase moves backward to copy the U sequence again in a reiterative process that will add many As to the 3′ end of the mRNA.

in the context of the rest of the E, I, and S sequences, cause the transcriptase to engage in reiterative transcription, repeatedly using these Us as a template to add many A nucleotides (a poly[A] tail) to the 3′ end of the newly synthesized mRNA.

Many experiments support various aspects of the start–stop model; we describe only a single example illustrating the role of gene order. To test whether the order of genes in the mononegaviruses determines the relative abundance of their expression, investigators constructed a set of mixed-up genomes with the genes in an atypical order. The set consisted of every possible rearrangement of the central P, M, and G proteins; for example, two of the test genomes had the gene order N, M, P, G, L and another had the gene order N, M, G, P, L. After using genetically engineered host cells to package these altered genomes into virions, host cells were infected with these mutant virions. Afterward, the quantities of specific M, P, and G mRNA and protein were measured. The amount of protein and mRNA produced by these manipulations mirrored the order of the genes in the artificial genomes. Experiments like these provide support for the conclusion that gene order is critical for determining the relative amounts of different viral proteins.

6.5 Rhabdovirus genome replication occurs through the use of a complete antigenome cRNP as a template

In order to replicate its genome, a rhabdovirus must first synthesize a complete antigenome. Synthesis of mRNA is very different from synthesis of antigenomes, even though both result in synthesis of (+) sense RNA (**Figure 6.7**). For example, production of mRNA requires a particular start site upstream of the N gene and includes reiterative transcription to create poly(A) tails; moreover, the resulting viral mRNA is not encapsidated. In contrast, synthesis of an antigenome requires that the very first nucleotide be used as a template and cannot include any reiterative copying. Furthermore, new antigenomes are completely encapsidated during synthesis, with the nucleotides coated by N protein as soon as there is space for the N protein to bind to the RNA. The transcriptase that produces mRNA and the replicase that produces antigenomes and genomes both contain viral proteins L

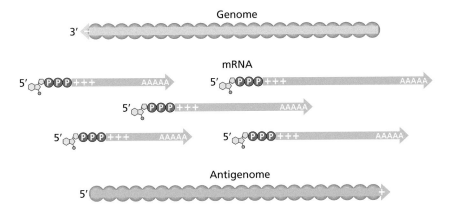

Figure 6.7 Comparing rhabdovirus mRNAs and antigenomes. Both mRNA and antigenomes are (+) sense, but mRNA has a 5′ cap and a poly(A) tail and is much shorter than the entire genome. Antigenomes are in a complex with N protein and are exactly the same length as genomes, so that genomes can be synthesized by copying the antigenome template.

and P, but the replicase does not include host proteins eEF1-α or Hsp60 (**Figure 6.8**). Distinguishing the components of the transcriptase compared with the replicase often takes advantage of classic experimental strategies for discovering protein–protein interactions, such as two-hybrid and co-immunoprecipitation analysis (**Technique Box 6.1**).

Figure 6.8 Rhabdovirus replicase compared with rhabdovirus transcriptase. The rhabdovirus replicase consists of viral proteins L and P; whether it contains host proteins is unknown. It uses as its template (–) vRNP and synthesizes (+) copy RNP (cRNP). The transcriptase consists of viral proteins L and P and host proteins eEF-1α and Hsp60. It uses as its template (–) vRNP and synthesizes capped and tailed mRNA.

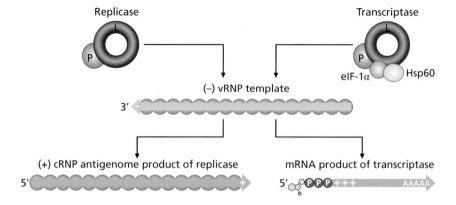

TECHNIQUE BOX 6.1   DETECTION OF PROTEIN–PROTEIN INTERACTIONS, MULTIPROTEIN COMPLEXES, AND PROTEIN–DNA INTERACTIONS

There are two common techniques for studying protein–protein interactions: two-hybrid assays and co-immunoprecipitation. The two-hybrid assay detects interactions *in vivo* and relies on molecular manipulation of a protein that can activate transcription (**Figure 6.9**). Such proteins have two domains, namely a DNA-binding domain and a transcription-activation domain. In a two-hybrid system, these two domains are separated and expressed separately. In this separated state, they can no longer activate transcription because the DNA-binding domain and activation domain are no longer able to associate with one

another as they had when the intact protein connected the two parts through covalent bonds. The two candidate proteins that might interact are known as the bait and the prey. The bait gene is cloned so that it forms a chimera with the DNA-binding domain, and the prey gene is cloned so that it forms a chimera with the activation domain. If the bait and prey interact they will bring the DNA-binding domain and the activation domain in the two chimeras into proximity, after which they will activate a reporter gene, such as one encoding green fluorescent protein (GFP).

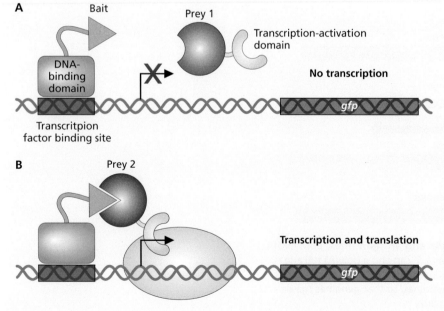

Figure 6.9 Two-hybrid assay. A two-hybrid assay detects whether bait and prey proteins interact *in vivo*. The bait is fused to a DNA-binding domain, and the prey is fused to a transcription-activation domain. (A) If the bait and prey do not interact, there is no activation of a reporter gene such as for green fluorescent protein (GFP). (B) If the bait and prey do interact, the reporter gene is activated and GFP levels can be quantified.

Co-immunoprecipitation is another method to study protein–protein interactions (**Figure 6.10**). The technique is based on the property that an antibody covalently bound to a large particle such as an agarose bead will readily precipitate out of solution when spun in a centrifuge, leaving behind all smaller particles such as most of the proteins. Then it is possible to make a cell lysate in which the cell membrane is not intact and the cell contents are available to the antibody. The antibody will then bind to its antigen in the lysate, and when the mixture is centrifuged, the antibody and its antigen will precipitate out of solution, forming a pellet that can be washed and subsequently analyzed, for example by using sodium dodecyl sulfate–polyacrylamide gel electrophoresis (SDS-PAGE). Most of the rest of the proteins that do not bind to the antibody remain in solution, with one important exception. Any proteins that themselves bind tightly to the antigen will also remain bound to the antigen, which is itself bound to the antibody and which in turn is covalently attached to the bead. The antigen can be thought of as bait that entraps any proteins (prey) that bind to that bait. When the bead–antibody–antigen complex is precipitated, the prey proteins co-precipitate with the bead–antibody–antigen bait, even though the antibody did not itself bind to the prey. Instead, the antibody indirectly caused them to precipitate out of solution through an interaction with its antigen; hence, this procedure is called co-immunoprecipitation. After washing, an elution step that disrupts noncovalent intermolecular molecular forces, such as hydrogen bonds between the antibody, bait, and prey, allow the prey molecules to be recovered and then identified.

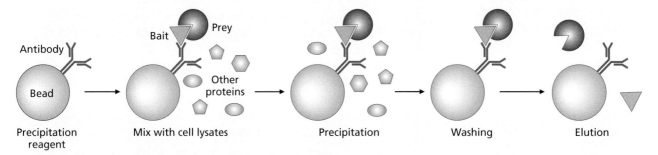

Figure 6.10 Co-immunoprecipitation. Antibodies that bind to a bait protein are themselves covalently linked to a very large substrate such as agarose beads. When the antibody–bead complexes are incubated with cell lysates, prey proteins that normally interact with the bait form a large complex consisting of a bead, its covalently attached antibodies, the antigen (bait), and the prey. Precipitation selectively removes this complex from the solution. After washing, the prey can be recovered using an elution procedure.

The replicase complex initiates RNA synthesis at the very 3′ end of both genomes and antigenomes. The switch from L and P being part of a transcriptase complex to being part of a replicase complex is not understood but it is clear that the switch involves the abundance of free N protein not bound to RNA. As the infection proceeds, the N protein becomes more abundant. At some point, it reaches a threshold level and becomes so abundant that it can encapsidate newly synthesized RNA. This encapsidation of the newly synthesized RNA causes the replicase to ignore the E, I, and S sequences and thus to skip over the sites that would otherwise cause reiterative transcription, leading to the synthesis of faithful full-length antigenomes. These antigenomes are subsequently used to synthesize full-length genomes, likely using the same replicase complex in combination with abundant N protein to coat the newly synthesized (–) nucleic acids.

6.6 The paramyxoviruses are mononegaviruses that use RNA editing for gene expression

The **paramyxoviruses** are a second group of mononegaviruses, with a few unique characteristics compared with rhabdoviruses. The paramyxoviruses include one of the most infectious viruses known, **measles virus** (MeV;

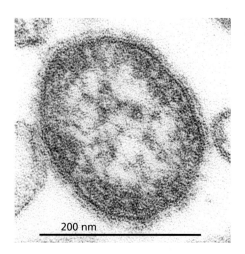

Figure 6.11 Electron micrograph of measles virus. This is a thin-section transmission electron micrograph of measles virus. (CDC/Courtesy of Cynthia S. Goldsmith & William Bellini.)

Figure 6.11). A typical person infected with measles in an unvaccinated population will spread the virus to 12–18 more people. Other medically important paramyxoviruses include respiratory syncytial virus and mumps virus. Agriculturally important paramyxoviruses include Newcastle disease virus, which infects poultry. The cattle disease rinderpest, also caused by a paramyxovirus, was once an important problem in agriculture, but it joins smallpox in being the second infectious animal disease that has ever been globally eradicated by intentional human intervention.

Paramyxoviruses such as MeV have enveloped, spherical, or irregularly shaped virions (see Figure 6.11) with two different types of protein spikes. Like all mononegaviruses, they have (–) strand RNA genomes; in this case, the genomes are 15.1–18.2 kb long (**Figure 6.12**), slightly longer than those of the rhabdoviruses. The mechanisms of primary transcription, genome replication, and secondary transcription are the same in the paramyxoviruses and the rhabdoviruses.

Different paramyxoviruses direct synthesis of 8–11 proteins. In the case of MeV, the proteins (from 3′ to 5′ in the genome) are N, **P/V/C**, M, **F**, **H**, and L (see Figure 6.12). The gene order is conserved with that in rhabdoviruses (N, P, M, G, and L), except that instead of a single glycoprotein spike (G), the paramyxoviruses have two spikes called F and HN. The P/V/C proteins also warrant explanation, as V and C were not found in the rhabdoviruses. V may be important for pathogenesis by interfering with cellular immune responses or may be important for regulating RNA replication, or both. C protein also regulates viral RNA production and interferes with cellular immune responses, and it is additionally important for new virions to exit the host cell. The P/V/C proteins are written with slashes between them to indicate that they are encoded by significantly overlapping sequences in the genome. Their synthesis arises through two different unusual mechanisms that could not be predicted from the sequence of the genome alone.

In MeV, the P protein is most often produced from the mRNA transcribed from the P/V/C region. Sometimes, however, an alternative start codon is used, producing the C protein instead (**Figure 6.13**). The use of this alternative start codon is called leaky scanning (see **Chapter 5**). During leaky scanning, the ribosome ignores the first AUG, where it ought to have begun, and instead scans until it reaches a downstream AUG and begins translation at this later, more 3′ position. Still another protein, V, can be produced from the genomic information in the P/V/C region, but this time the mechanism involves production of a novel mRNA through **RNA editing**. RNA editing is the addition of one or two untemplated nucleotides to a specific sequence in an mRNA, in this specific case a single C (**Figure 6.14**). RNA editing is remarkable as it involves the inherited insertion of specific nucleotides at specific sites in an mRNA molecule so that the mRNA does not hybridize perfectly with its template but nevertheless has a single consistent coding sequence. The sequence is not inherited directly in the DNA template but

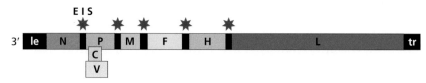

Figure 6.12 Measles virus genome. The measles virus genome has a short 3′ leader (le), coding sequences separated by E I S start–stop signals, and a 5′ trailer (tr). P, V, and C sequences overlap significantly.

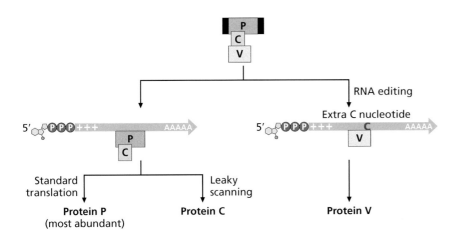

Figure 6.13 Three proteins are encoded by the P/V/C region of the measles virus genome. Two different mRNAs are produced from the P/V/C region of the measles virus genome. One of them can be translated in two different ways through a leaky scanning mechanism, giving rise to P or C proteins. The other, which has an extra C nucleotide relative to the first one, can be translated into the V protein through RNA editing.

```
3' ucgugaaggcucuguggguaauuuuucccgugucugcgcucuaaccggaguaaaccuugc    Genome sequence for
                                                                   P and V proteins

5' agcacuuccgagacacccauuaaaaag ggcaca gacgcgagauuggccucauuuggaacg   mRNA sequence for
    S   T   S   E   T   P   I   K   K   G   T   D   A   R   L   A   S   F   G   T     P protein

5' agcacttccgagacacccattaaaaag ggcCac agacgcgagattggcctcatttggaac   mRNA sequence for
    S   T   S   E   T   P   I   K   K   G   H   R   R   E   I   G   L   I   W   N     V protein
```

Figure 6.14 RNA editing. RNA editing occurs when one or two nontemplated nucleotides are added to an mRNA, resulting in a frameshift. In the case of measles virus, the addition of an untemplated C (bold, uppercase, red) causes translation of the V protein instead of the C protein. The last codon the two mRNAs have in common and the first codon that is different between the two are shown (bold, underlined). The amino acids unique to P are shown in blue and those unique to V are in red. The figure shows just a close-up of the most relevant sequences rather than the entire coding sequence for either protein.

rather is inherited through the protein–RNA interactions that result in a specific editing event. RNA editing changes the frame for translation, so that starting at the codon containing the inserted nucleotide, the protein sequence differs from the protein encoded by the unedited mRNA. Phenomena such as RNA editing make it difficult to predict the comprehensive set of proteins encoded by a virus from the genome sequence alone.

6.7 Filoviruses are filamentous mononegaviruses that encode seven to nine proteins

The last group of mononegaviruses considered here are the **filoviruses**. In humans, filoviruses such as **Ebola virus** (EBOV) and Marburg virus cause deadly hemorrhagic fevers for which there are limited therapeutic interventions, and so must be studied under the strictest containment conditions, called **Biosafety Level 4**, or BSL-4. Virologists working under BSL-4 conditions work in positive-pressure garments that cover them from head to toe, provide pressurized air from an outside source, and resemble spacesuits (**Figure 6.15**). It is possible, however, to study components of filovirus biology under more normal laboratory conditions. For example, the EBOV spike protein, and therefore EBOV attachment and penetration, can be studied by creating a recombinant rhabdovirus that expresses the EBOV spike instead of its normal G-spike glycoprotein. In order to study transcription in EBOV, we

Figure 6.15 Scientists about to enter a BSL-4 facility. These employees of the Centers for Disease Control and Prevention (CDC), US, are putting on their positive-pressure suits in order to enter a BSL-4 laboratory. The person on the left is in the process of attaching the hose that will provide him with uncontaminated air and create positive pressure in the suit, so that microbes in the BSL-4 facility cannot be inhaled. (CDC/Dr. Scott Smith. Courtesy of James Gathany.)

can use minigenomes that include the start–stop signals and a reporter gene but that do not encode any other viral genes. In these instances, necessary virus proteins are supplied to the host cell using the tools of genetic engineering such as transfection of host cells with plasmid DNA. Experiments on these modified viruses do not have to be done under BSL-4 conditions because the virus is not viable.

Filoviruses have long (>500 nm) enveloped, filamentous virions (**Figure 6.16**) with one type of protein spike. Like all mononegaviruses, they have minus (−) strand RNA genomes; in this case, the genomes are 18.9–19.1 kb long (**Figure 6.17**), slightly longer than those of the paramyxoviruses.

Filoviruses encode the following seven proteins, in order starting from the 3′ end of the genome: NP (nucleoprotein for encapsidating RNA), **VP35** (part of the polymerase complex and functionally equivalent to the P protein), **VP40** (matrix protein), **GP** (glycoprotein spike), **VP30** (transcription activator), **VP24** (second matrix protein), and L (RdRp). The gene order is still mostly conserved with that of the other mononegaviruses (N, P, M, G, L). The eighth protein, when present, is a soluble form of GP (**sGP**) in which the C-terminal transmembrane segment is missing. The ninth protein, when present, is an even shorter version of sGP that has been named **ssGP** (**Figure 6.18**). In EBOV, the sGP protein is directly encoded by the genome, and the full-length GP and the small ssGP proteins arise by RNA editing. As described in **Section 6.6**, RNA editing is the programmed addition of one (GP) or two nucleotides (ssGP) to the mRNA at a particular sequence, when

To see an animation showing RNA editing in Ebola virus, please visit the book's website: www.routledge.com/cw/lostroh.

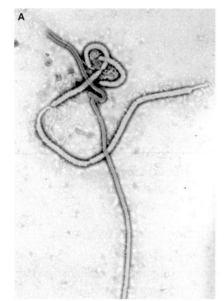
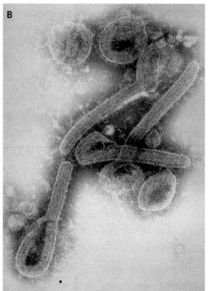

Figure 6.16 Filoviruses. (A) Electron micrograph of Ebola virus; the virus is approximately 60–80 nm in diameter and about 1,000 nm long, although the length of filaments is variable. (B) Electron micrograph of Marburg virus, which is similar in size to Ebola virus. Both viruses have been magnified approximately 100,000 times. (A, CDC/Dr. Frederick A. Murphy. B, CDC/Dr. Frederick A. Murphy; J. Nakano.)

Figure 6.17 Ebola virus genome. The Ebola virus genome has a short 3′ leader, protein coding sequences, E I S start–stop sequences, and a 5′ trailer sequence.

Identical N-terminal sequence

```
sGP    M---//---WETKKTSLEKFAVKSCLSQLYQTEPKTSVVRVRRELLPTQGPTQQLKT
GP     M---//---WETKKNLTRKIRSEELSFTVVSNGAKNISGQSPARTSSDPGTNTTTEDHKIMASENSSAMVQVHSQG...
ssGP   M---//---WETKKKPH
```

Figure 6.18 Ebola virus GP, ssGP, and sGP proteins. The proteins have been aligned in order to demonstrate that they have identical N-terminal regions but different C-terminal sequences because of RNA editing. The GP protein continues on for many amino acids after those depicted here.

the template RNA is not complementary to those nucleotides. In EBOV, addition of a single A occurs 25% of the time and addition of two As occurs 5% of the time, at a particular stretch of the genome where the template sequence is 3′ UUUUUU. This template is copied accurately 70% of the time and only seven consecutive As appear in the mRNA. But, 25% of the time, this template results in eight consecutive As in the mRNA. Similarly, 5% of the time, this template results in nine consecutive As even though there are only seven Us in the template.

Why use RNA editing to make a key abundant structural protein (GP) for the virion? To answer this question, scientists engineered an EBOV that expresses GP 100% of the time because they inserted an extra U into the template. This mutant virus is attenuated in cell culture, apparently because it kills its host cell so fast that the virus cannot complete its replication cycle. Thus, the RNA editing mechanism in wild-type EBOV restricts the levels of GP so that the virus has time to produce offspring before killing the host cell.

6.8   The filovirus VP30 protein, not found in other mononegaviruses, is required for transcription

Like other mononegaviruses, primary transcription occurs following uncoating, which releases vRNP into the cytoplasm. The vRNP is already associated with transcriptase, which in filoviruses includes not only L and VP35, the equivalent of the phosphoprotein in other mononegaviruses, but also VP30, which is unique to filoviruses and is considered to be a transcription factor. Transcription begins at a promoter near the 3′ end of the genome; transcription of the first nucleotides results in a nascent mRNA molecule with a stem–loop structure. This stem–loop structure impedes further transcription until the block is somehow relieved by the VP30 protein. The start–stop E, I, and S sequences between filovirus genes have been defined through sequence conservation and mutagenesis studies. For example, the transcriptional end (E) site is 3′ UAAUUCUUUUUU. The VP30 protein is not thought to be part of the VP35-L replicase complex that synthesizes antigenomes and genomes. As in the rest of the mononegaviruses, the replicase complex has different activities compared with the transcriptase complex; for example, although the replicase synthesizes a complete faithful copy of its templates, the transcriptase uses reiterative transcription at particular tracts of template Us to synthesize poly(A) tails.

6.9   Influenza is an example of an orthomyxovirus

The last group of (−) RNA viruses to consider, the **orthomyxoviruses**, are very different from the mononegaviruses. For example, they do not have the same homologous proteins and they have segmented genomes. The best understood orthomyxovirus is influenza, which is among the most important

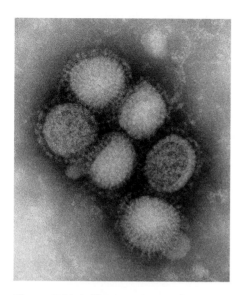

Figure 6.19 Influenza virus. Colorized electron micrograph of spherical influenza viruses showing their prominent spikes. The virions are about 100 nm in diameter.

viruses in the development of molecular and cellular virology as a discipline and remains an important pathogen to this day. In particular, influenza A is responsible for flu pandemics.

Orthomyxoviruses have spherical or filamentous enveloped virions with prominent spikes (**Figure 6.19**). Although they look the same in an electron micrograph, there are two distinct types of spikes, one made of **hemagglutinin (HA)** protein and the other made of **neuraminidase (NA)** protein (see Figure 3.25). Like all orthomyxoviruses, influenza A has a (−) strand RNA genome, but in this case the genome is **segmented** (**Figure 6.20**). That is, inside each virion there are eight helical nucleocapsids, each different from the others. Each of the segments of the (−) RNA genome is bound tightly to many molecules of NP and to one molecule each of proteins PA, PB1, and PB2 (see Figure 2.4). Each of the individual segments can encode one to four proteins and varies from about 890 to 2,400 bases long. Orthomyxoviruses can have between six and eight genome segments, with a total genome size of 10–15 kb of RNA. Although the 3′ and 5′ ends are unmodified RNA, there are secondary structures in the 3′ and 5′ untranslated regions (UTRs) and a short poly(U) tract in the 5′ UTR.

6.10 Of the 17 influenza A proteins, 9 are found in the virion

Structural influenza A proteins include the HA and NA surface glycoproteins, the NP nucleoprotein that makes up most of the vRNP, the **M1** matrix protein, the **M2** ion channel, and the **nuclear export protein (NEP)** (**Figure 6.21**). Three of the structural proteins in the virion are components of the RdRp: **PA**, **PB1**, and **PB2**. Nonstructural proteins NS1 and PB1-F2 interfere with the host innate immune system and PA-X reduces host mRNA expression. Some of the most recently discovered proteins, such as PB1-N40, have no agreed-upon function.

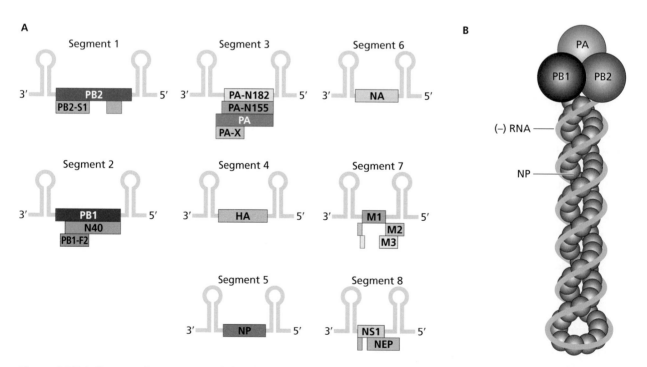

Figure 6.20 Influenza virus genome. (A) Eight genome segments of influenza A, which are numbered from largest to smallest. Both of the 3′ and 5′ UTRs contain important secondary structures. Some of the proteins are produced by alternative splicing, so their coding regions are split in the diagram. (B) Diagram of the genomic vRNP.

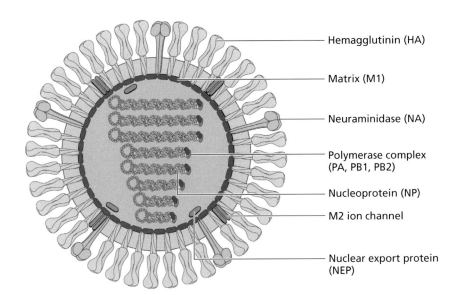

Hemagglutinin (HA)

Matrix (M1)

Neuraminidase (NA)

Polymerase complex
(PA, PB1, PB2)

Nucleoprotein (NP)

M2 ion channel

Nuclear export protein
(NEP)

Figure 6.21 The influenza virion. In addition to the eight genome vRNPs, the virion contains proteins NP, M1, M2, NEP, PA, PB1, and PB2. (Courtesy of Philippe Le Mercier, ViralZone, © SIB Swiss Institute of Bioinformatics.)

Although the most abundant proteins in the virion are encoded by a dedicated mRNA, most of the proteins are encoded by overlapping sequences. For example, the PB1-F2 protein is synthesized from the single PB1 mRNA as a consequence of a programmed +1 ribosomal frameshift (see **Section 5.21**). The PA-X protein is also produced as a consequence of a programmed +1 ribosomal frameshift, this time from the PA mRNA. Proteins M2 and NEP arise from alternatively spliced mRNA originating from the M and NS genes, respectively (for a discussion on splicing, see **Section 8.4**).

The remainder of this chapter focuses on the four viral proteins directly involved in transcription and genome replication: NP, PA, PB1, and PB2.

6.11   Orthomyxovirus nucleic acid synthesis occurs in the host cell nucleus, not in the cytoplasm

Figure 6.22 provides an overview of the replication cycle of orthomyxoviruses, such as influenza A. The last step of uncoating is to release the vRNP segments through nuclear pores into the nucleus instead of the cytoplasm. The PA, PB1, and PB2 heterotrimer serves as the viral RdRp. Viral mRNA synthesis proceeds and the mRNA is then exported to the cytoplasm. To make new genomes, the NP, PA, PB1, and PB2 proteins must be translocated back into the nucleus, along with the M1 and NEP proteins, which are needed for exporting new vRNP genomes out of the nucleus. Late in infection, there is a switch from production of mRNA to production of antigenomes and genomes, both of which are encapsidated by NP. Antigenomes in a complex with NP are called **cRNP** for copy RNP. The new vRNPs are subsequently exported to the cytoplasm for assembly and maturation, and the new virions exit the host cell through budding.

6.12   The first step of transcription by influenza virus is cap snatching

Influenza A produces 10 mRNAs with typical eukaryotic 5′ methylated caps, yet the virus does not encode any enzymes that can synthesize such a cap. Instead, the first step in mRNA synthesis is that the cap-binding motif in the PB2 protein found in the vRNP binds to a host mRNA before that mRNA can

Figure 6.22 Overview of gene expression and genome replication in the orthomyxoviruses. The vRNP genome segments are released into the nucleus (1). Transcription results in viral mRNA (2), which is exported from the nucleus (3) and translated (4). Some of the proteins are translocated back into the nucleus (5), where the genome is then used to synthesize many cRNP antigenomes (6) and new genomes (7). These genomes can make more viral mRNA (8). Viral proteins needed for synthesis of nucleic acids enter the nucleus to catalyze nucleic acid synthesis or to join with the new genomes (9, 10) to make vRNPs that are exported from the nucleus. Viral proteins and genomes come together during maturation (11).

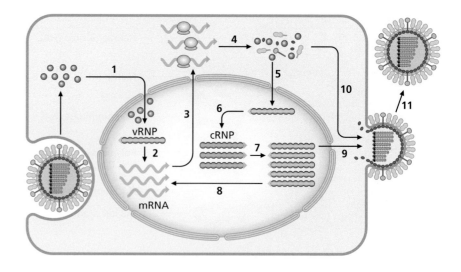

 To see an animation showing influenza virus cap snatching, please visit the book's website: www.routledge.com/cw/lostroh.

leave the nucleus. After PB2 binds to a host mRNA, the PA protein cleaves the mRNA between 9 and 15 nucleotides downstream of the cap; this process is designated as **cap snatching** (**Figure 6.23**). This cleaved mRNA subsequently serves as a primer for synthesis of mRNA that is complementary to the RNA in the vRNP template. Every cap comes from an mRNA synthesized by the host, yet the actual coding sequence of the viral mRNA is synthesized by the viral RdRp. In general, during influenza virus transcription the PA–PB1–PB2 replicase complex acts in *cis*, using as a template the specific vRNP with which the enzymes were physically associated before transcription began.

Synthesis of mRNA begins at the 3′ end of the template; it is catalyzed by the PB1 subunit of the transcriptase. Approximately 17–21 nucleotides from

Figure 6.23 Synthesis of mRNA by influenza A virus. The process begins with cleavage of nuclear host mRNA known as cap snatching (1) because it provides a primer that has a 5′ cap for viral nucleic acid synthesis. Next, the polymerase acts in *cis* to elongate the capped primer, while remaining strongly attached to the 5′ end of the template genomic RNA (2). Because the polymerase is associated so tightly with the 5′ end of the template, reiterative transcription (3) occurs when the polymerase reaches a series of Us near the 5′ end of the template, which creates a poly(A) tail. The mRNA is then released and the vRNP most likely returns to its original configuration (4), ready to begin transcription again.

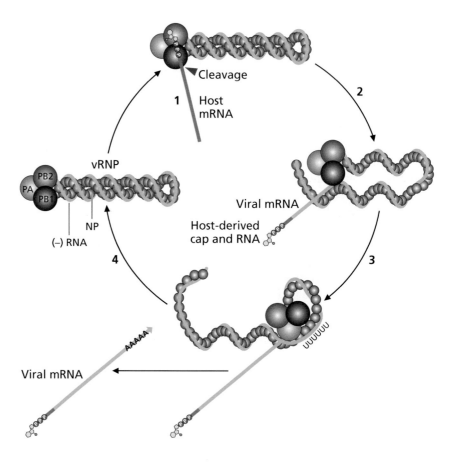

the 5′ end of the template of the polymerase complex encounters a run of five Us in the template prior to secondary structure in the 5′ UTR. Because the 5′ end of the template remains tightly associated with the polymerase complex, there is steric hindrance that prevents the polymerase from clearing all the way through the template at this sequence. Instead, the polymerase synthesizes a poly(A) tail by copying the 3′-UUUUU template dozens of time—a process called reiterative transcription. After polyadenylation, the mRNA dissociates from the polymerase and is subsequently exported to the cytoplasm for translation. What happens to the polymerase next is not clear, but perhaps release of the mRNA causes it to return to its original configuration, ready to begin mRNA synthesis again. In addition to providing the primer needed for transcription, cap snatching also prevents expression of host proteins because decapitated host mRNA can be neither exported from the nucleus nor translated, even if it were to reach the cytoplasm.

6.13   An influenza cRNP intermediate is used as the template for genome replication

As introduced earlier, genome replication in influenza A virus occurs through a (+) strand intermediate called cRNP. As the name implies, the cRNP consists of (+) antigenomes bound to NP. The cRNP differs from mRNA in other ways, even though both are (+) strands (**Figure 6.24**). For example, the 5′ end of mRNA is composed of host sequences (from the snatched cap), and the 3′ end of the mRNA is polyadenylated and lacks nucleotides complementary to the very 5′ end of the vRNP template. In contrast, the cRNP is a faithful copy of all of the vRNP sequences, including the very 3′ and 5′ termini, and it lacks any polyadenylation.

The initiation of transcription through cap snatching is mechanistically distinct from the initiation of replication to synthesize cRNP. For example, synthesis of cRNP begins in a completely different manner, ***de novo***, without the benefit of a primer. The synthesis of cRNP occurs early during infection and for a short duration relative to the rest of the replication cycle. Moreover, at the end of cRNP synthesis, a PA, PB1, and PB2 complex remains associated with the cRNP. Beyond this information, the synthesis of cRNP is poorly understood.

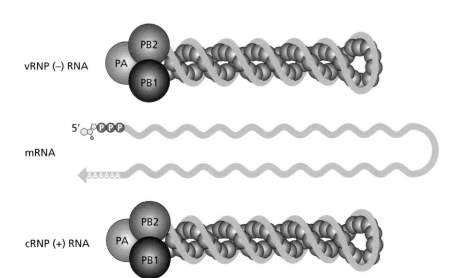

Figure 6.24 Comparison of influenza A mRNA and cRNP. mRNA has a normal 5′ cap, derived from host mRNA, and a poly(A) tail. It is therefore longer than the vRNP that served as a template for its synthesis. mRNA is not bound to NP. The cRNP is not capped and does not have a poly(A) tail, and it is exactly the same length (in nucleotides) as the vRNP. It is also bound to NP.

Once cRNP has been made, new vRNP genome segments can be synthesized using the cRNP as a template. A current model proposes that genome replication differs from transcription in several ways. First, the PA, PB1, and PB2 viral enzymes are part of a genome replicase complex that includes a fourth viral protein, NEP. The replicase complex probably also contains at least one distinctive host protein not found in the transcriptase. In the model, the replicase complex binds to cRNP templates in *trans* and initiates genome replication (**Figure 6.25**). In this context, in *trans* indicates that the replicase uses as its template a cRNP molecule with which it was not physically associated prior to beginning replication. Instead of starting at the very end of the template, the replicase uses a discontinuous mechanism. It first copies nucleotides 4 and 5, synthesizing an AG dinucleotide. After that, the complex shifts back to the very 3′ end of the template, skipping over nucleotide 3 so that the UC dinucleotide at the 3′ end of the (+) template base pairs with the newly synthesized 5′ AG on the (−) RNA. From there, RNA synthesis is continuous with new RNA being encapsidated by NP as soon as the RNA emerges from the replicase. When the active replicase reaches the end of the template, it displaces the resident PA–PB1–PB2 complex that was left over from cRNP synthesis. The end result is multiple vRNP molecules that can be used to synthesize more mRNA or can be exported from the nucleus to be packaged into new virions. The PA–PB1–PB2 complex may reassociate with the template (+) RNA to restore the cRNP structure.

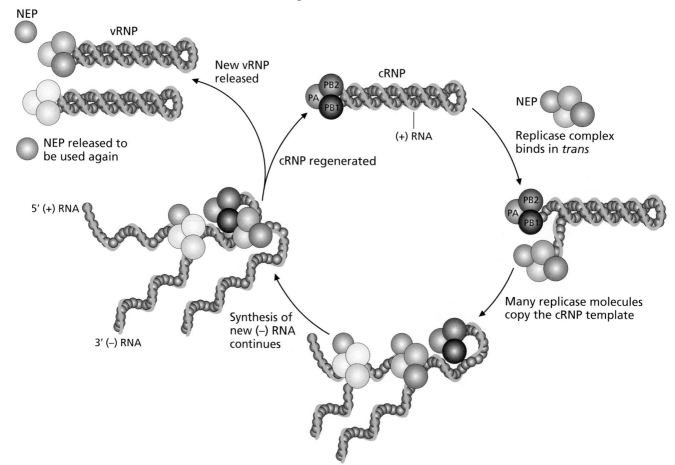

Figure 6.25 Synthesis of new influenza virus genomes. Synthesis of vRNP likely occurs in *trans*. Initial synthesis of an AG dinucleotide followed by a shift backward, to copy the whole genome. Multiple replicase molecules copying a single vRNP. At the end, there are many vRNPs produced for each cRNP template. The original cRNP template may return to its original state, as depicted here. The NEP may dissociate so that the new vRNPs are identical in composition to the original infecting vRNPs.

6.14 Arenavirus RNA genomes are ambisense

The **arenaviruses** include Lassa fever virus, which causes a serious hemorrhagic fever with up to 20% mortality in West Africa. Arenaviruses are included in this chapter because each of their two genome segments has some protein-coding information that is complementary to mRNA, like a negative-strand virus. But, on the other hand, each of their two genome segments has other protein-coding information that is identical to mRNA, like a positive-strand virus, resulting in their designation as **ambisense** viruses. Like negative-strand viruses, arenavirus genomes cannot be translated directly and instead must first be copied into mRNA by a viral RdRp that is carried with the genome into the host cell. This is another reason that they are most often grouped with the negative-strand viruses in textbooks.

Arenaviruses have spherical enveloped virions with helical nucleocapsids inside (**Figure 6.26**), similar to the paramyxoviruses and orthomyxoviruses. The viral spikes are made of proteins GP1 and GP2, and the matrix protein is called Z. Like the orthomyxoviruses, the genome is segmented, in this case into two segments designated L for large and S for small (**Figure 6.27**). The RNA is tightly complexed with abundant NP and also associated with lower amounts of the L protein, also known as the RdRp. The genome segments each have an internal stem–loop structure called the intergenic region, which separates the nucleic acids that encode two different proteins on each

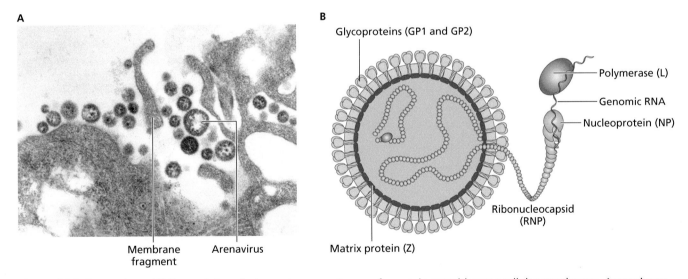

A

Membrane fragment   Arenavirus

B

Glycoproteins (GP1 and GP2)

Polymerase (L)

Genomic RNA

Nucleoprotein (NP)

Ribonucleocapsid (RNP)

Matrix protein (Z)

Figure 6.26 Arenavirus. (A) Transmission electron microscope image of arenaviruses with some cellular membranes. Arenaviruses vary from 60 to 300 nm in diameter. (B) An arenavirus has GP1 and GP2 spike proteins, a matrix protein (Z), and L (RdRp) and NP (nucleoprotein) proteins that are tightly associated with the genomic RNA. (A, CDC/C. S. Goldsmith. B, Philippe Le Mercier, ViralZone, © SIB Swiss Institute of Bioinformatics.)

Figure 6.27 The segmented genome of the arenaviruses includes an L segment and an S segment. The segments have two sections separated by an intergenic region that folds into a stable hairpin. Each genome segment encodes two proteins, one on either side of the intergenic region. Although the Z and GPC sequences in the genome are identical to those in the mRNA encoding those proteins, unlike (+) sense Class IV viruses, the RNA in the genome will not be used as a template for translation, so the viral genes are not drawn as arrows.

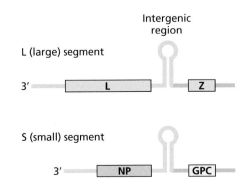

Intergenic region

L (large) segment

3'   L   Z

S (small) segment

3'   NP   GPC

genome segment. The virions also contain ribosomes derived from the previous host cell; the functional significance, if any, of the ribosomes is unknown.

6.15 Expression of the four arenavirus proteins reflects the ambisense nature of the genome

The arenavirus genome encodes four proteins. The RdRp protein (L) is encoded on the large (L) genome segment, as is the matrix protein Z. The S (small) genome segment encodes the NP as well as GPC, the precursor to the mature glycoproteins that will comprise the spike. During maturation, the GPC polyprotein will be cleaved by a host protease, releasing the mature GP1 and GP2 components of the spike. The viral genes are arranged in an unusual way; the L genome segment provides a good example and the arrangement is similar for the S genome segment. The genome is typically drawn from 3′ to 5′ and the first mRNA produced from the L segment is the L (RdRp) mRNA, which is complementary to most of the 3′ end of the genome. This situation is the same as that for negative-strand viruses. Following the L coding sequence, there is an intergenic region that likely forms a secondary structure such as one or more stem–loops, and on the other side of the intergenic region there are sequences that are identical to the Z mRNA sequence, running from 5′ to 3′. These sequences are never recognized as mRNA; instead, expression of Z requires several steps, as explained next.

For ambisense viruses, both the genome and the antigenome segments are used as templates to synthesize mRNA. After uncoating introduces the genome segments and viral RdRp into the cytoplasm, the viral RdRp first uses

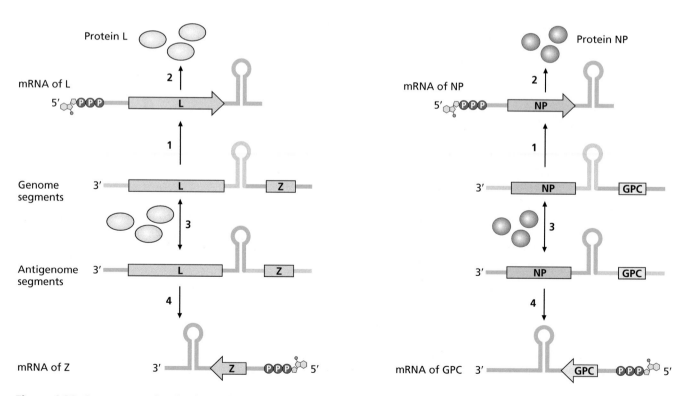

Figure 6.28 Gene expression in the ambisense arenaviruses. RdRp introduced into the cell by viral uncoating first produces capped mRNAs from the genome segments, encoding L and NPs (1). mRNA of L and NP are translated into the L (RdRp) and NPs (2). Accumulation of NP likely allows synthesis of full-length antigenomes and genomes by RdRp (3). The antigenomes are then used to synthesize Z and GPC mRNAs (4). Although the genomes have nucleic acids corresponding to the sequence of Z and GPC mRNA, the genomes are never used as templates for translation.

the 3′ end of each genome segment to synthesize capped mRNA encoding the L and NP proteins (**Figure 6.28**); the RdRp can only copy RNA templates that are encapsidated into a vRNP. The 5′ caps on the mRNA may originate from a cytoplasmic cap-snatching mechanism. The secondary structures formed by the intergenic region sequences in the genome are responsible for termination of transcription and the viral mRNA is not polyadenylated. An accumulation of NP translated from one of the first two mRNAs synthesized is probably needed for the RdRp to switch from mRNA synthesis to synthesizing complete antigenomes, which are most likely coated with NP over their entire length, forming cRNP. The RdRp subsequently uses the antigenomes as a template to synthesize capped mRNA, encoding the Z and GPC proteins, and again terminating in the intergenic region and producing mRNA molecules that are not polyadenylated. At some point, viral nucleic acid synthesis shifts primarily to synthesis of new genomes but the mechanism underlying this shift is unknown. It is likely that there are both sequences and RNA secondary structural elements that are required for proper initiation of synthesis of mRNA, antigenomes, and genomes, but these details are not yet understood.

Essential concepts

- The two major groups of (−) RNA viruses that affect humans and agriculturally important animals are the mononegaviruses and orthomyxoviruses. Both have genomes that are obligatorily associated with proteins in vRNP complexes, and they must introduce a viral RdRp into the host cell during uncoating because the genome cannot be translated.
- The mononegaviruses and the orthomyxoviruses are very different in that the mononegaviruses have one genome segment and replicate entirely in the cytoplasm, whereas the orthomyxoviruses have multiple genome segments, and viral gene expression and genome replication occur in the nucleus.
- The RdRp for both mononegaviruses and orthomyxoviruses requires RNA encapsidated by an abundant dedicated viral protein for a template and cannot copy naked RNA.
- The mononegavirus L protein has many domains and is responsible for all viral RNA synthesis and for capping and tailing viral mRNA. The orthomyxovirus PA, PB1, and PB2 proteins are components of both the viral transcriptase and replicase, but the virus uses an unusual cap-snatching mechanism during mRNA synthesis.
- Mononegavirus and orthomyxovirus genome replication occurs through a vRNP (+) orientation antigenome intermediate.
- Mononegavirus and orthomyxovirus antigenomes and mRNA are both (+) sense but otherwise differ in both structure and sequence compared with viral mRNAs.
- The mononegavirus group shares conserved genes, conserved gene order, and similar expression and replication strategies. Their RdRp uses a start–stop mechanism that results in the genes being expressed in amounts that correlate with their position in the template.
- Filoviruses, although they are mononegaviruses, have a unique transcription protein that makes their gene expression somewhat different from that of other mononegaviruses.
- Influenza A, a model orthomyxovirus, uses a different combination of proteins in the viral transcriptase compared with the viral replicase.

- Arenaviruses are segmented ambisense viruses in which each of two segments is used as a template to synthesize a subgenomic mRNA and the genomes are also used as templates to synthesize antigenomes. Because these antigenomes subsequently are used to synthesize additional subgenomic mRNAs, the genome is designated as ambisense rather than strictly (−) or (+) sense.

Questions

1. If you could use rational drug design to find an antiviral targeting any virus in this chapter, what viral protein would you target, and why? What if you had much more funding and could choose three targets?
2. Identify the purpose of every rhabdovirus protein.
3. List the filovirus proteins that correspond to each of the functions of the rhabdovirus proteins.
4. Compare and contrast rhabdovirus with influenza A.
5. How is cRNP similar to influenza mRNA? How is it different?
6. What similarities between arenaviruses and negative-strand viruses often lead to them being discussed together?
7. How does reiterative transcription, which is used to create a poly(A) tail, differ from the way host cells create poly(A) tails?
8. How is filovirus synthesis of mRNA similar to that in rhabdovirus?
9. Would you expect mononegaviruses to interfere with host mRNA translation initiation? Why or why not?
10. What is the purpose of each of the transcription and replication proteins in influenza A?
11. Compare and contrast the RdRp of rhabdovirus with that of influenza.
12. Draw a two-hybrid system in which two different influenza A proteins are able to positively interact. Label the influenza A proteins.
13. Using an antibody against the L protein, you immunoprecipitate L and any associated proteins from a lysate made from cells infected with filovirus. List all of the proteins you expect to co-immunoprecipitate with L.
14. How is rabies virus similar to any of the (+) RNA viruses in **Chapter 5**?

Interactive quiz questions

In addition to the questions provided above, this edition has a range of free interactive quiz questions for students to further test their understanding of the chapter material. To access these online questions, please visit the book's website: www.routledge.com/cw/lostroh.

Further reading

Rabies

Barr JN, Whelan SP & Wertz GW 2002. Transcriptional control of the RNA-dependent RNA polymerase of vesicular stomatitis virus. *Biochim Biophys Acta* 1577:337–353.

Finke S & Conzelman KK 2005. Replication strategies of rabies virus. *Virus Res* 111:120–131.

Gould JR, Qiu S, Shang Q, Ogino T, Prevelige Jr PE et al. 2020. The connector domain of vesicular stomatitis virus large protein interacts with the viral phosphoprotein. *J Virol* 94:e01729-19.

Qanungo KR, Shaji D, Mathur M & Banerjee AK 2004. Two RNA polymerase complexes from vesicular stomatitis virus-infected cells that carry out transcription and replication of genome RNA. *Proc Natl Acad Sci USA* 101:5952–5957.

Wasik B & Murphy M 2012. *Rabid: A Cultural History of the World's Most Diabolical Virus.* Viking Penguin.

http://www.radiolab.org/story/312245-rodney-versus-death/ (Podcast).

Measles

Corum J, Keller J, Park H & Tse A 2019. Facts about the measles outbreak. http://www.nytimes.com/interactive/2015/02/02/us/measles-facts.html

Houben K, Marion D, Tarbouriech N, Ruigrok RW & Blanchard L 2007. Interaction of the C-terminal domains of Sendai virus N and P proteins: Comparison of polymerase-nucleocapsid interactions within the paramyxovirus family. *J Virol* 81:6807–6816.

LaFrance A 2015. The new measles. http://www.theatlantic.com/health/archive/2015/01/the-new-measles/384738/

Filoviruses

Farmer P 2020. *Fevers, Feuds and Diamonds.* Farrar, Straus and Giroux.

Muehlberger E 2007. Filovirus replication and transcription. *Future Virol* 2:205–215.

Quammen D 2014. *Ebola: The Natural and Human History of a Deadly Virus.* W.W. Norton.

http://www.bbc.com/news/world-africa-28754546

http://www.cdc.gov/vhf/ebola/outbreaks/2014-west-africa/

Influenza virus

Dias A, Bouvier D & Crepin T 2009. The cap-snatching endonuclease of influenza virus polymerase resides in the PA subunit. *Nature* 458:914–918.

Eisfeld AJ, Neumann G & Kawaoka Y 2015. At the centre: Influenza A virus ribonucleoproteins. *Nat Rev Microbiol* 13:28–41.

Fan H, Walker AP, Carrique L, Keown JR, Serna Martin I et al. 2019. Structures of influenza A virus RNA polymerase offer insight into viral genome replication. *Nature* 573:287–290.

Plotch SJ, Bouloy M, Ulmanen I & Krug RM 1981. A unique cap-dependent influenza virion endonuclease cleaves capped RNAs to generate the primers that initiate viral RNA transcription. *Cell* 23:8847–8858.

Arenaviruses

Holm T, Kopicki JD, Busch C, Olschewski S, Rosenthal M et al. 2018. Biochemical and structural studies reveal differences and commonalities among cap-snatching endonucleases from segmented negative-strand RNA viruses. *J Biol Chem* 201:19686–19698.

Gene Expression and Genome Replication in the Double-Stranded RNA Viruses

Virus	Characteristics
Rotavirus	Model for gene expression and genome replication in dsRNA viruses that infect eukaryotic cells.

The viruses you will meet in this chapter and the concepts they illustrate

Although there are several orders of Baltimore Class III double-stranded RNA viruses, in this chapter we investigate the well-characterized family *Reoviridae* because it has been subjected to the most research on viral nucleic acid synthesis. These viruses were first discovered around 1959, when the famous poliovirus vaccine researcher Albert B. Sabin named them **reoviruses**, as an acronym for respiratory, enteric, and **orphan viruses**. Orphan viruses are those that have been discovered but have not been associated with any diseases. For many years, therefore, reoviruses could be isolated from respiratory and enteric human samples but were not considered pathogenic. Since that time, pathogenic genera including **rotaviruses** have been discovered. Rotaviruses are responsible for as many as half of all cases of infectious diarrhea in young children and infants, and they can be deadly. In high-consumption countries, deaths from rotavirus are prevented primarily through immunization, whereas in low-consumption countries, deaths from rotavirus are prevented primarily through treating the symptoms with oral rehydration therapy.

In addition to being interesting because of their clinical significance, the *Reoviridae* are intriguing on a molecular and cellular level. For example, they have double-stranded, segmented RNA genomes. Additionally, gene expression and genome replication in *Reoviridae* are catalyzed by partially intact virions. The structural proteins that make up the inner components of the virion are also the very enzymes used in transcription and genome replication. These viruses thus blur the lines between structural and nonstructural viral proteins. Confining the dsRNA genome segments inside a capsidlike structure is likely critical for their replication because eukaryotic host cells have well-developed innate immune responses that detect and respond to dsRNA as a marker of RNA virus infection (see **Chapter 14**). In this chapter, we will use the pathogen rotavirus A as an example of reoviruses in general.

DOI: 10.1201/9781003463115-7

7.1 The rotavirus replication cycle includes primary transcription, genome replication, and secondary transcription inside partially intact capsids in the host cytoplasm

Figure 7.1 is an overview of gene expression and genome replication in the rotaviruses. Rotaviruses have three concentric icosahedral capsids. The final uncoating step leaves a **double-layered particle** (**DLP**), with two concentric capsids, intact; this DLP serves as the site of viral nucleic acid synthesis in the cytoplasm of infected cells. During primary transcription, the infecting DLP synthesizes 9–12 different capped, polyadenylated mRNA molecules, depending on the species. Each different mRNA is extruded from a particular dedicated vertex of the DLP. These mRNAs direct synthesis of proteins, which together with host factors result in formation of **viroplasm**, where new DLPs enclosing viral mRNA initiate synthesis of complementary (−) RNA. The result is new, transcriptionally active DLPs enclosing dsRNA genomes. Secondary transcription from these DLPs results in an exponential increase in viral RNA synthesis with mRNA outnumbering dsRNA. Viral RNA synthesis stops when a particular nonstructural protein binds to the DLPs and targets them for further encapsidation, assembly, and release from the host cell.

7.2 Rotavirus A has a naked capsid with 3 protein layers enclosing 11 segments of dsRNA

Reoviruses such as rotavirus are naked spherical virions with many distinctive layers (**Figure 7.2**). The mature virion released from host cells has three concentric capsids. The outer layer consists of proteins VP7 (a subunit of a trimeric capsomere) and VP4, which is a subunit of the trimeric spike used for attachment. Interaction with host proteases during attachment and penetration, or protease treatment, releases the outer proteins, leaving

Figure 7.1 Overview of gene expression and genome replication in rotaviruses. After uncoating, the cytoplasmic DLP catalyzes primary transcription, in which capped mRNA leaves each vertex of the infecting particle (1). Translation of these proteins causes formation of viroplasm, in which mRNA becomes enclosed by new DLPs (2). Inside the new DLPs, copying of the (+) templates results in formation of DLPs containing dsRNA genomes (3). The new DLPs cause an exponential increase in viral mRNA and protein (4) and, after a few hours, the virus switches from gene expression and genome replication to the assembly phase of the virus replication cycle (5).

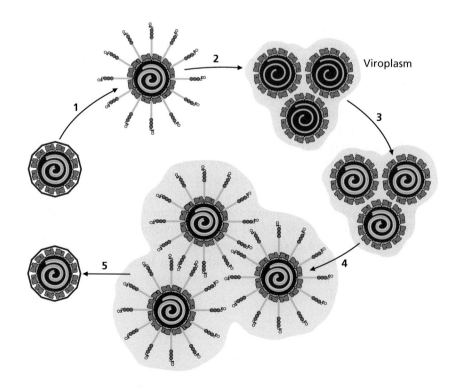

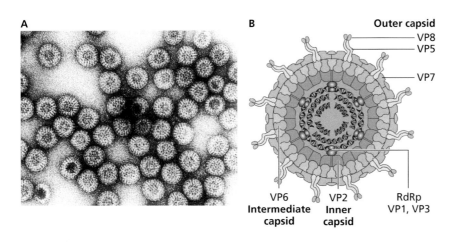

Figure 7.2 Rotavirus. (A) Transmission electron micrograph of spherical rotaviruses that are about 80 nm in diameter. Some of the particles resemble wheels, which was the reason for the prefix *rota* (Latin for wheel). (B) Rotavirus structural proteins enclosing the segmented dsRNA genome. (RdRp, RNA-dependent RNA polymerase.) (A, CDC/Dr. Erskine L. Palmer. B, Courtesy of Philippe Le Mercier, ViralZone, © SIB Swiss Institute of Bioinformatics.)

behind a DLP that has two concentric capsids. The outer layer of the DLP is the VP6 capsomere protein. The DLP is the intracellular form of rotavirus that has the capacity to synthesize nucleic acids. It is also possible to treat DLPs *in vitro* with additional chemical reagents that cause the release of the VP6 layer, resulting in formation of a single-layered core particle with protein VP2 as the outer layer. This core particle is not transcriptionally active in rotavirus but might be similar in structure to the first core structure that forms surrounding (+) RNA during formation of new DLPs. The other proteins found in the core are VP1 and VP3, which form an internal structure that projects toward the center of the particle and has been described as shaped like a flower.

The rotavirus A genome is composed of 11 different segments of double-stranded RNA (**Figure 7.3**). All of the (+) strands have a 5′ cap; neither of the two strands is polyadenylated. Each of the 11 individual genome segments can encode one or two proteins and varies from about 650 to 3,400 bp. The segments are numbered according to size, with segment 1 being the largest, similar to the numbering of influenza A ssRNA genome segments. Reoviruses in general can have between 10 and 12 segments, numbered from largest to smallest, with a total of 18–35 kb of RNA. The RNA inside the virion is highly ordered and tightly associated with proteins including the VP1/VP3 flower structure.

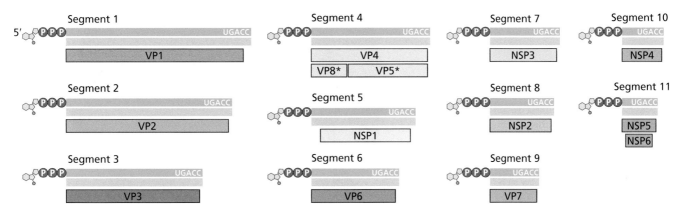

Figure 7.3 Rotavirus genome. There are 11 genome segments numbered from largest (3,302 bp) to smallest (667 bp), for a total of 18,550 bp. The (+) strands have methylated caps and the sequence UGACC at the 3′ end. VP5* and VP8* arise from proteolytic processing of VP4, and NSP5 and NSP6 are encoded by overlapping sequences using a leaky scanning mechanism.

7.3   Rotavirus A encodes 13 proteins

Rotavirus A encodes 13 polypeptides. These include seven structural proteins found in virions: VP1, VP2, VP3, VP5*, VP6, VP7, and VP8*. The other six are nonstructural proteins: NSP1, NSP2, NSP3, NSP4, NSP5, and NSP6. Of these 13 proteins, 11 are encoded on a dedicated mRNA (see Figure 7.3). The exceptions are NSP5 and NSP6, both of which are encoded on the smallest genome segment. The two coding sequences overlap significantly with NSP6 translation enabled by a leaky scanning mechanism (see **Chapter 5**). The VP5* and VP8* proteins are derived from proteolytic digestion of VP4. The nonstructural proteins that play a prominent role in gene expression and genome replication are NSP3, needed for translation, and NSP2 and NSP5, which form the viroplasm.

7.4   Synthesis of rotavirus nucleic acids occurs in a fenestrated double-layered particle

Synthesis of rotavirus nucleic acids occurs inside DLPs. These DLPs are therefore necessarily **fenestrated**. That is, they are pierced by passages that allow the entry of substrates such as nucleotide triphosphates (NTPs) and exit of nascent mRNA (**Figure 7.4**). Synthesis of RNA is catalyzed by proteins that play a major role in the structure of the naked virions: namely by the **VP1/VP3 flower complex** (**Figure 7.5**). The flower complex is found at each vertex; from some perspectives, it resembles a flower in which the petals extend toward each vertex. There is a tunnel that channels template (−) RNA to the flower complex, where new (+) RNA synthesis occurs, thanks in part to a second tunnel that allows entry of substrate NTPs. There is a third tunnel allowing the capped mRNA to leave the vertex. Finally, a fourth tunnel allows exit of the template (−) RNA, which then reassociates with its complementary (+) RNA, allowing the system to begin again and synthesize another mRNA.

The VP1/VP3 flower complex has two different enzymatic activities. First, it is a transcriptase that synthesizes capped mRNA from a dsRNA template

Figure 7.4 The fenestrated double-layered particle. VP6 is the outermost protein, and VP2 lies underneath. Also shown are the 11 double-stranded RNA segments. Every vertex of the rotavirus is fenestrated, allowing for entry of NTPs and exit of newly synthesized mRNA between the viroplasm and DLP interior. The proteins VP1 and VP3 are also found at each vertex but are not included in this drawing.

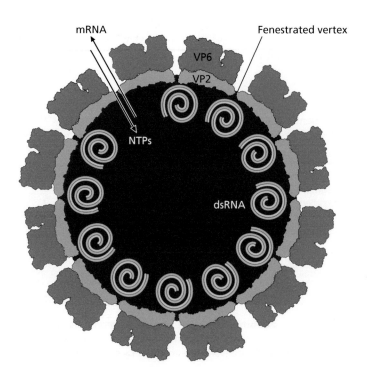

A

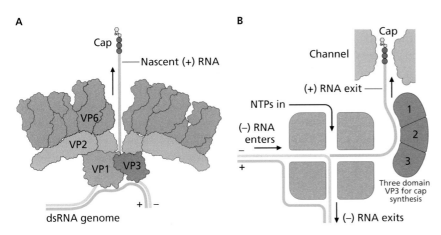

B

Figure 7.5 VP1/VP3 flower proteins in relation to the rest of the rotavirus virion. (A) VP1 and VP3 are found at each fenestrated vertex underneath the VP2 layer, which in turn is covered by a layer of VP6. The size and abundance of the proteins shown here are based on the experimentally determined structural biology of rotavirus. (B) The proposed route of nascent mRNA through the VP1 and VP3 proteins. The VP1 protein (pink) is thought to have four different tunnels. One is an entryway for the dsRNA genome template, and another serves as the site for NTPs to enter VP1. The third is the exit tunnel for newly synthesized (+) RNA, and the fourth is the exit tunnel for the (–) RNA strand of the template. VP1 has an active site for (+) RNA synthesis; (+) RNA exits through the tunnel that brings it in proximity to the three domains of the VP3 protein (purple). The VP3 proteins are necessary for cap synthesis and for attaching the cap to the nascent RNA. When the new (+) RNA exits the particle, the cap exits first. (Adapted from Trask SD et al. 2012. *Nat Rev Microbiol* 10:165–177. With permission from Springer Nature.)

and allows the nascent mRNA to exit the fenestrated DLP through one of the passages found at each of the 12 vertices. Each of the 11 different genome segments is associated with 1 of the 12 VP1–VP3 complexes in the DLP, and so a different mRNA is extruded from each vertex (**Figure 7.6**). One of the vertices may not be transcriptionally active since there are only 11, rather than 12, genome segments. During synthesis of mRNA, the VP3 component is responsible for synthesizing and adding a methylated 5′ cap while VP1 has RNA-dependent RNA polymerase activity.

The second activity of the VP1–VP3 complex is genome replication inside a newly formed DLP, where it synthesizes (–) RNA that is not capped and does not exit the DLP but instead remains associated with the (+) RNA template and with the inner core proteins. It is not known why the capping activity of VP3 is not active during synthesis of (–) RNA. The result of this second activity is formation of a transcriptionally active DLP containing the dsRNA templates needed for mRNA synthesis. This DLP might be used in secondary transcription to dramatically increase the number of viral nucleic acids and proteins in the host cell or, later during infection, might instead associate with other nonstructural and structural proteins to be packaged into a mature, enzymatically inert virion.

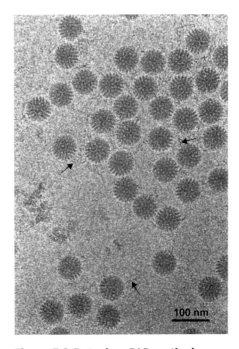

Figure 7.6 Rotavirus DLPs actively transcribing. This cryo-electron micrograph shows mRNA emerging from the vertices of actively transcribing rotaviruses; arrows indicate RNA emerging from the DLPs. (From Lawton JA et al. 1997. *Nat Struct Mol Biol* 4:118–121. With permission from Springer Nature.)

7.5 Translation of rotavirus mRNA requires NSP3 and occurs in viroplasm formed by NSP2 and NSP5

Rotavirus mRNA has the following features that affect translation of that mRNA. The structure of the capped mRNA is likely to be circular because the 5′ and 3′ ends are complementary, with the exception of a single-stranded tail at the 3′ end. Unlike cellular mRNA, the 3′ end is not polyadenylated; instead, each mRNA has a conserved UGACC sequence at the 3′ end. As discussed in **Chapter 5**, translation of cellular mRNA depends upon both the 5′ cap and the poly(A) tail because of translation factors that bind to the cap (eIF4E, eIF4G, and eIF4A) or to the poly(A) tail (PABP). Rotavirus must bypass the requirement for PABP binding because its mRNA does not have a poly(A) tail. The solution is that the viral NSP3 protein binds to a UGACC sequence near the 3′ end of the viral mRNA and substitutes for the PABP, thus interacting with the eIF4 complex and allowing translation initiation to begin efficiently (**Figure 7.7**).

Translation of primary transcripts allows for formation of viroplasm, the name for the cytoplasmic site of viral dsRNA synthesis (genome replication), assembly of new DLPs, and secondary transcription. Rotavirus viroplasm appears as an electron-dense inclusion in electron micrographs and is not bounded by any membranes (**Figure 7.8**). Viroplasm formation occurs within hours of infection and small viroplasms fuse to create fewer, but larger, viroplasms as the infection proceeds.

Viral **NSP2** and **NSP5** proteins are responsible for viroplasm formation. This phenomenon was discovered through a series of experiments, such as the following. When NSP2 is expressed alone in host cells, it forms a diffuse pattern in the cytoplasm as detected by immunofluorescence microscopy. NSP5 expressed alone appears in a similar diffuse pattern. In contrast, when NSP2 and NSP5 are co-expressed in host cells, they form discrete viroplasmlike structures that in appearance closely resemble authentic viroplasm during infection. To investigate the structural features of the NSP5 protein that are required for viroplasm formation, host cells expressing NSP2 and altered versions of NSP5 were examined. Removal of portions of the N- or C-terminal amino acids prevented the formation of viroplasmlike structures, as did most other tested alterations of the NSP5 protein. The only exception was an alteration that removed 46 internal amino acids, which suggests that this portion of NSP5 is not needed for viroplasm formation. Direct interactions between NSP2 and NSP5 have been demonstrated using co-immunoprecipitation with monoclonal antibodies directed against NSP2 (see **Chapter 6**).

The NSP2 protein is the most abundant protein in the viroplasm and likely plays several roles during RNA synthesis, perhaps including assembly of the RNA with VP1 and other structural proteins. NSP6, also present in viroplasm, may regulate the assembly of NSP5 into large decameric complexes, which in turn regulate viroplasm formation.

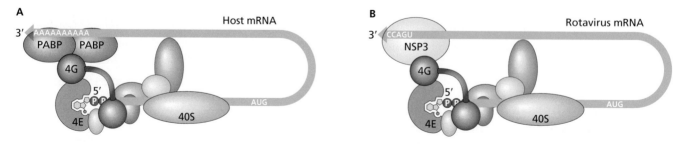

Figure 7.7 Rotavirus mRNA compared with host mRNA, ready for translation to initiate. (A) Translation initiation complex on host mRNA. (B) Translation initiation complex on rotavirus mRNA, in which NSP3 substitutes for PABP by binding to the conserved UGACC sequence near the 3′ end of the viral mRNA.

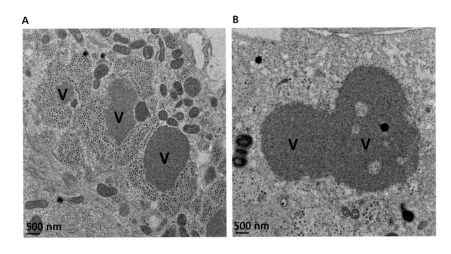

Figure 7.8 Rotavirus viroplasm. (A) During the earlier phases of infection viroplasms (V) are separate. (B) Older viroplasms fuse with one another. (From Eichwald C et al. 2012. *PLOS* 7:e47947. With permission from Public Library of Science. Published under Creative Commons Attribution License.)

7.6 Rotavirus genome replication precedes secondary transcription

Virus genome replication is required before secondary transcription can occur because dsRNA templates are required for synthesis of new mRNA. Furthermore, those new dsRNA templates must be found assembled into new DLPs, where the geometry and enzymatic activity create the perfect conditions for synthesis of capped mRNA. Formation of new DLPs requires an accumulation of VP1, VP3, and (+) mRNA, which together form a complex that targets the (+) RNA to become part of a new genome. VP2 then joins this complex, creating catalytically active cores that synthesize the complementary (−) RNA, which remains associated with the proteins and its template. There are specific sequences in the 3′ end of the (+) template that are required for synthesis of the complementary strand. Each core synthesizes only one (−) strand per template and the (−) strands are not capped, even though VP3 is present. The assortment mechanism that ensures that each core contains one and only one of each genome segment is not clear. Finally, VP6 joins the core, forming DLPs containing 11 different dsRNAs, which can be used for secondary transcription or, later in infection, for assembly.

Secondary transcription occurs in newly formed DLPs enabled by mRNA translation and viroplasm formation. It uses the same mechanism as that described for primary transcription. It is thought that a transition from secondary transcription to assembly occurs when NSP4 interacts with the DLP and targets it for further packaging, thereby turning its enzymatic activity off.

Essential concepts

- Rotaviruses have three-layered naked capsids surrounding a segmented dsRNA genome.
- The double-layered particle is the form of rotavirus that is transcriptionally active in the cytoplasm of a host cell.
- Many individual components of the double-layered particle are enzymes needed for nucleic acid synthesis and also structural components of the virion. VP1 is an example of a rotavirus protein that is structural and has enzymatic activity.
- The VP1/VP3 flower complex synthesizes the 5′ cap structure and the mRNA using one of the dsRNA molecules as a template.

- The mRNA does not have a poly(A) tail; instead, a viral nonstructural protein binds to a conserved sequence in the 3′ end of each mRNA and thereby substitutes for the poly(A) binding protein during translation initiation.

Questions

1. Compare and contrast influenza with rotavirus.
2. In rotavirus, what is primary transcription? Compare it to secondary transcription.
3. In what ways are VP1 and VP3 like structural proteins of other viruses?
4. In what ways are VP1 and VP3 like nonstructural proteins of other viruses?
5. How does rotavirus translate its mRNA when that mRNA lacks a poly(A) tail and the PABP is required for normal mRNA translation?

 Interactive quiz questions

In addition to the questions provided above, this edition has a range of free interactive quiz questions for students to further test their understanding of the chapter material. To access these online questions, please visit the book's website: www.routledge.com/cw/lostroh.

Further reading

Desselberger U 2014. Rotaviruses. *Virus Res* 190:75–96.

Geiger F, Acker J, Papa G, Wang X, Arter WE et al. 2021. Liquid–liquid phase separation underpins the formation of replication factories in rotaviruses. *EMBO J* 40(21):e107711.

Gillies S , Bullivant S & Bellamy AR 1971. Viral RNA polymerases: Electron microscopy of rotavirus reaction cores. *Science* 174:694–696.

Lawton JA, Estes MK & Prasad BVV 1997. Three-dimensional visualization of mRNA release from actively transcribing rotavirus particles. *Nat Struct Biol* 4:118–121.

Pan M, Alvarez-Cabrera AL, Kang JS, Wang L, Fan C & Zhou ZH 2021. Asymmetric reconstruction of mammalian reovirus reveals interactions among RNA, transcriptional factor μ2 and capsid proteins. *Nature Comm* 12(1):1–6.

Prasad BV, Rothnagal R & Zeng CQ 1996. Visualization of ordered genomic RNA and localization of transcriptional complexes in rotavirus. *Nature* 382:471–473.

Silvestri LS, Taraporewala ZF & Patton JT 2004. Rotavirus replication: Plus-sense templates for double-stranded RNA synthesis are made in viroplasms. *J Virol* 78:7763–7774.

Tao Y, Farsetta DL, Nibert ML & Harrison SC 2002. RNA synthesis in a cage: Structural studies of rotavirus polymerase I3. *Cell* 111:733–745.

Gene Expression and Genome Replication in the Double-Stranded DNA Viruses

The viruses you will meet in this chapter and the concepts they illustrate

Virus	Characteristics
Simian vacuolating virus 40 (SV40)	Illustrates manipulation of host cell cycle and apoptosis, viral gene expression, and viral genome replication in a dsDNA virus with a particularly small genome; led to discovery of cellular p53 protein.
Human papillomavirus	Genome size similar to SV40; gene expression and genome replication are linked to host cell differentiation in a stratified epithelium.
Human adenovirus C	Linear dsDNA genome with covalently attached protein at 5′ end used to avoid the end-replication problem; led to discovery of mRNA splicing.
Herpes simplex virus 1	Very large linear dsDNA genome that circularizes prior to genome replication; cascade of viral gene expression triggered by many viral regulatory proteins. The International Committee on Virus Taxonomy recently renamed this virus human herpesvirus 1.
Vaccinia	Encodes its own RNA polymerase and DNA replication proteins in order to replicate in the cytoplasm, not the nucleus. Has multiple waves of gene expression and self-priming mechanism of genome replication.

Baltimore Class I viruses have dsDNA genomes. Class I viruses that infect animals are diverse, ranging from the small polyomaviruses and papillomaviruses to the comparatively much larger adenoviruses and herpesviruses. These viruses replicate in the nucleus and thus take full advantage of their host's normal transcription machinery. They encounter problems when it comes to replicating their genomes because their host cells are usually terminally differentiated. Terminally differentiated cells only rarely undergo mitosis and almost never contain DNA replication proteins. Components of the DNA synthesis machinery are therefore not present when one of these viruses first infects the host cell. Thus, most Class I animal viruses encode proteins that force their host cells into S phase, consequently making host replication proteins available for viral genome replication. Poxviruses, on the other hand, are unique Class I viruses because they have particularly large dsDNA genomes that are expressed and replicated entirely in the cytoplasm

DOI: 10.1201/9781003463115-8

Figure 8.1 Outcomes of infection by DNA viruses. Infection by a DNA virus can result in a lytic infection, cellular transformation, or a latent infection.

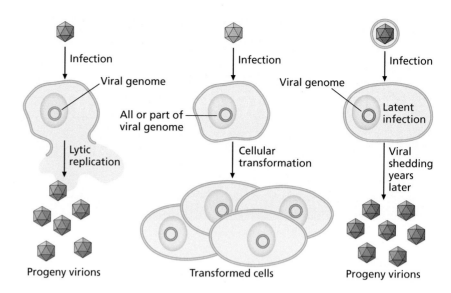

of their eukaryotic hosts. Poxviruses therefore use virally encoded enzymes to catalyze both transcription and genome replication. Interest in poxviruses has been stimulated by the 2022 monkeypox outbreak.

8.1 DNA viruses can cause productive lytic infections, cellular transformation, or latent infections

Infection by most groups of DNA viruses can result in three distinct outcomes (**Figure 8.1**). The outcome is determined by the interaction of the specific virus and host cell; the same virus can have different outcomes when it infects different host cells. The first possible result is a productive infection (also known as a lytic infection) in which the host cell dies as a result of producing progeny virions. The second possible outcome is **cellular transformation**, during which the host cell begins to be transformed into a cancer cell. The cell becomes less responsive to physiological cues that are meant to regulate cell growth, differentiation, and death. Transformation can be a dead end for the virus, if no progeny virions are produced. This chapter focuses on the productive infections, and **Chapter 13** covers transformation. The third outcome of infection of an animal cell with a Class I virus is latent infection, in which the viral genome is mostly quiescent and no virions are produced for a period of time, such as years, that is much longer than the duration of a lytic cycle; this possibility is also discussed in **Chapter 13**.

8.2 Most Class I animal viruses rely on host transcription machinery for gene expression

Transcription by most Class I animal viruses is catalyzed by host enzymes and therefore we begin with a description of eukaryotic transcription. (See **Chapter 5** for foundational information about eukaryotic translation.) Eukaryotic transcription occurs in the nucleus and is catalyzed by a molecular machine called RNA polymerase. Different forms of the polymerase are specialized for producing mRNA, rRNA, and tRNA. This chapter focuses on mRNA because our main concern is how viruses usurp host transcription and translation machinery in order to produce viral proteins. RNA polymerase II, often simply known as Pol II, catalyzes mRNA synthesis.

8.3   Eukaryotic transcription is affected by the state of the chromatin

In eukaryotes, the DNA is packed around nucleosomes forming chromatin, which when condensed can interfere with transcription. The basic unit of chromatin is the nucleosome, which is composed of two copies each of four histone proteins; approximately 147 bp (depending on the species) of eukaryotic DNA are wrapped around each histone so that in electron micrographs the chromatin resembles beads (of nucleosomes) on a string (of DNA; **Figure 8.2**). The histones are packed together by a fifth histone protein that binds to the linker DNA between nucleosomes. When histones are packed together tightly, this heterochromatin cannot be transcribed; heterochromatin silences transcription. In contrast, when histones are associated more loosely with the DNA in the form of euchromatin, the transcription machinery can access the DNA. Covalent modification of histones alters the chromatin. **Histone acetyl transferase** enzymes add negatively charged acetyl groups to the histones, which makes them move apart because of the repelling charges; the result is euchromatin (**Figure 8.3**). **Histone deacetylation enzymes** remove acetyl groups and thus enable close nucleosome packing characteristic of transcription-silencing heterochromatin.

Before Pol II can initiate transcription, the chromatin has to be remodeled so that the promoter DNA is available (**Figure 8.4**). Next, transcription initiation requires assembly of a complex array of basal transcription factors such as TATA binding protein and transcription factor IID. Because the supply of these factors is limited, they cannot assemble on all promoters simultaneously. This situation provides an opportunity for regulating gene expression by using different promoter sequences that have different affinities for the protein transcription factors. The major component that allows Pol II

A

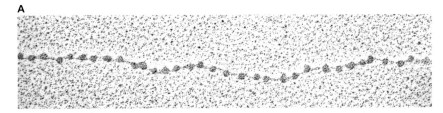

B

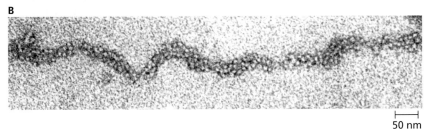

├──┤
50 nm

Figure 8.2 Chromatin structure. (A) The basic unit of chromatin is the nucleosome, which is made up of an octamer of four pairs of histone proteins. Chromatin looks like beads on a string, where the beads are the nucleosomes and the string is the DNA. (B) The nucleosomes can also be packed tightly together. (A, Courtesy of Barbara Hamkalo. B, Courtesy of Victoria Foe.)

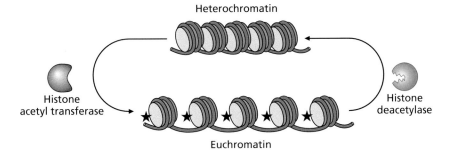

Figure 8.3 Histone acetyl transferases and histone deacetylases. Acetyl groups are represented as stars. Histone acetyl transferase enzymes cause the formation of more relaxed euchromatin. Histone deacetylases cause the formation of more condensed heterochromatin.

Figure 8.4 Chromatin remodeling.
Nucleosomes have to be removed from
the promoter in order for the transcription
machinery to assemble.

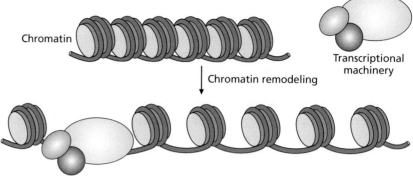

Figure 8.5 Mediator. The mediator is a
very large component of the transcription
machinery that integrates activating and
repressing signals from many different
proteins in order to control the initiation
of transcription by Pol II.

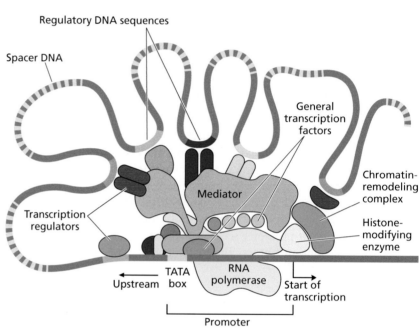

to respond to multiple different transcription regulators at the same time is
called the mediator of the transcription complex (mediator) and is composed
of at least 26 proteins (4,000 kDa; **Figure 8.5**). Ultimately, the transcription
initiation machinery responds to many transcription regulators at the same
time, modulating the amount of transcription initiation in response to multiple inputs including signals from enhancers, activators, and repressors.
Mediators can also integrate input from both cellular and viral transcription
factors.

8.4 Eukaryotic capping, splicing, and polyadenylation occur co-transcriptionally

During transcription elongation, Pol II is in a complex with capping, splicing, and polyadenylation enzymes. As the pre-mRNA is transcribed, it is
capped at the 5′ end by a covalent linkage to 7-methylguanylate. Eukaryotic
primary transcripts contain introns that must be spliced out in order for
the mRNA to encode protein; splicing occurs at the same time as transcription and is accomplished by spliceosome complexes that contain an array
of different protein and RNA components (**Figure 8.6**). The **constitutive
splicing machinery** is constantly active and removes most introns. Typical

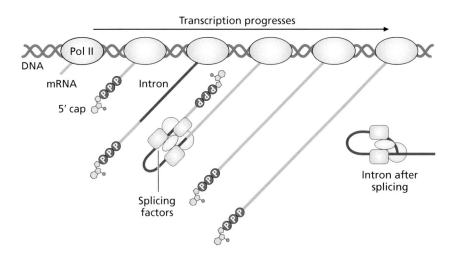

Figure 8.6 Capping and splicing are co-transcriptional. As the pre-mRNA emerges from the transcription machinery, capping and splicing factors modify the RNA during transcription elongation.

introns have several *cis*-acting sequences, such as a GU at their 5′ end (near the upstream exon) and a sequence resembling YUNAY, where Y stands for either pyrimidine (U or C) and N is any ribonucleotide, toward the 3′ end of the intron (**Figure 8.7**). The A in YUNAY is called the branch point A, which participates directly in splicing. This consensus sequence is followed by a polypyrimidine tract, which is a stretch of 8′ pyrimidines. Following the polypyrimidine tract, there is the sequence AG at the very 3′ end of the intron, adjacent to the downstream exon.

The *cis*-acting sequences in the intron interact with the **spliceosome**, a complex of *trans*-acting proteins and RNA molecules (**Figure 8.8**). Some of the component parts are called **snRNPs** (small nuclear ribonucleoprotein complexes), including U1, U2, U4, U5, and U6. Other components include 50–100 different polypeptides. This section focuses on the activities of the snRNPs because they are critical for catalysis. They assemble on the RNA in a cascade with an obligatory order that reflects the protein–RNA and protein–protein interactions required for each step to occur. The U1 component of the spliceosome binds to the 5′ GU of the intron. The U2 snRNP binds to the branch point A and also binds to U1. Another *trans*-acting factor that

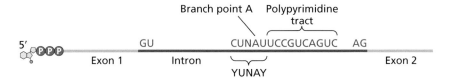

Figure 8.7 *Cis*-acting splicing sequences found in introns. Introns have GU at their 5′ end, but near the 3′ end there is a sequence resembling YUNAY, where Y stands for either pyrimidine (U or C) and N is any ribonucleotide, and the A is the branch point A where the lariat will form. The 3′ YUNAY sequence is followed by an 8–12-bp polypyrimidine tract. Following the polypyrimidine tract, there is the sequence AG at the very 3′ end of the intron.

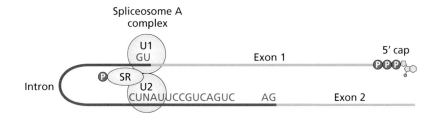

Figure 8.8 Early steps of spliceosome assembly. The U1 component of the spliceosome binds to the 5′ GU of the intron. The U2 snRNP binds to the branch point A and also binds to U1. An SR protein binds to the U1 and U2 snRNPs, forming the spliceosome A complex.

binds near the 3′ end of the upstream exon is an **SR protein**. The proteins are named SR because they have a serine (S)-rich and an arginine (R)-rich domain. They are thought to be involved in almost all splicing and must be phosphorylated before they bind to sequences found in exons. They also bind to the snRNPs.

At this point, the entire complex is called the spliceosome A complex. The spliceosome A complex determines which introns get removed from a transcript. Subsequently, a complex of U4, U5, and U6 joins the spliceosome A complex and U1 and U4 dissociate. Catalysis to remove the intron and attach the exons together requires dephosphorylation of the SR protein and proceeds as shown in **Figure 8.9**. The spliceosome catalyzes the cleavage of the intron at the 5′ end at the junction of the upstream exon and the intron. The spliceosome then synthesizes a covalent bond between the branch site A and the free end of the intron, forming a closed loop. The last step is cleavage of the 3′ end of the intron accompanied by joining the exons with a normal

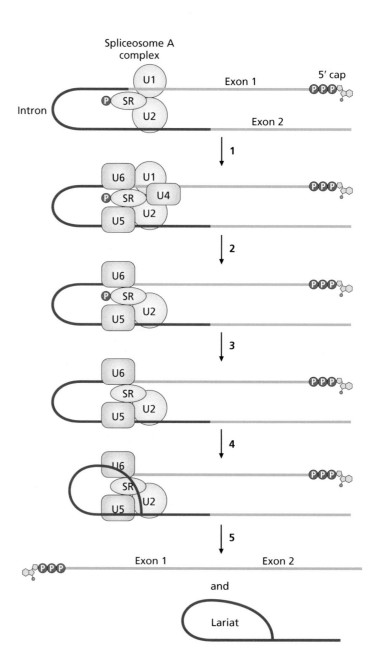

Figure 8.9 The process of intron removal. After the spliceosome A complex forms, snRNPs U4, U5, and U6 join (1). Next, U1 and U4 dissociate from the increasingly large complex (2). After dephosphorylation of the SR protein (3), the spliceosome cleaves the intron at the 5′ end and synthesizes a covalent bond between the branch site A and the free end of the intron, forming a closed loop (4). Finally, spliceosome cleaves of the 3′ end of the intron and joins the exons with a normal phosphodiester bond, releasing the intron lariat (5). The relative size of the intron compared with that of the exons has been exaggerated to focus on the intron.

phosphodiester bond. The released intron is referred to as a lariat because of its physical resemblance to a rodeo rope.

After splicing, each cellular mRNA typically encodes just one protein. Termination of transcription in eukaryotes is less well understood but it can influence gene expression as well. It is clear that eukaryotic pre-mRNAs contain core polyadenylation signals (the consensus sequence is AAUAAA), which are recognized by specific proteins that subsequently cleave the RNA and then polyadenylate it, resulting in a string of 200–300 As at the 3′ end. There can be alternative polyadenylation sites that have a profound effect on gene expression by determining the sequence of the 3′ untranslated region (UTR). Mature mRNA in the nucleus is part of a messenger ribonucleoprotein (mRNP) complex that is competent for export to the cytoplasm. Export through the nuclear pores can also be regulated and is subjected to quality control to ensure that mRNPs that leave the nucleus can be translated. The absence of any normal component of the mRNP typically results in degradation of the mRNA before it leaves the nucleus.

8.5 Polyomaviruses are small DNA viruses with early and late gene expression

Polyomaviruses are models for viruses with some of the shortest known genomes that can replicate in animal cells. For most of the past 50 years, polyomaviruses were not associated with any important human diseases. But the availability of better sequencing and culturing methods has led to the discovery of polyomaviruses associated with post-organ transplant kidney malfunction and with Merkel cell carcinoma. Polyomavirus research is driven in part because knowledge of the molecular biology of polyomaviruses could lead to treatments or preventative vaccines, but the prototype polyomavirus is **simian vacuolating virus 40 (SV40)**, which is not associated with any human disease.

In terms of both genome and virion size, the polyomaviruses are the smallest dsDNA viruses that infect animal cells. They have naked spherical virions that are only 50 nm in diameter (**Figure 8.10**); in contrast, the rotaviruses discussed in **Chapter 7** are about 80 nm in diameter, and the rabies viruses discussed in **Chapter 6** are 75 nm wide and 180 nm long. The polyomavirus genome must be small in order to fit inside such a small capsid. The small (4- to 5-kbp) genome is composed of circular dsDNA packaged into nucleosomes (**Figure 8.11**). The circular genome allows the virus to avoid the end-replication problem facing linear cellular DNA that replicates using a mechanism reliant upon RNA primers.

After polyomavirus attachment and penetration, uncoating results in the introduction of the viral genome with its 24 incomplete nucleosomes into the nucleus (**Figure 8.12**). After nuclear histone H1 joins the dsDNA genome, the nucleosomes are the same as those in host chromatin and thus the genome becomes ready for transcription mediated by host Pol II. The early viral transcripts result in production of viral proteins that are regulatory rather than structural. The regulatory proteins manipulate the host cell cycle, regulate viral gene expression, and participate in viral DNA replication. Late transcription does not occur until genome replication has begun; late genes encode structural proteins needed to complete the replication cycle. The virus also expresses a regulatory RNA during late gene expression, which helps shut down production of early proteins. Late proteins are much more abundant than early ones, in part because of the availability of newly replicated templates that can be used for transcription. Ultimately, new virion assembly occurs in nuclear viral factories and the mature virions exit the host cell through lysis.

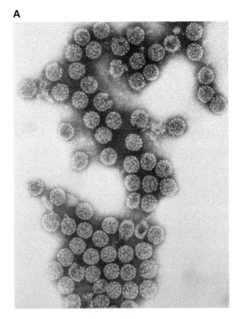

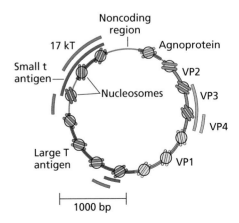

Figure 8.10 SV40. (A) Transmission electron micrograph of SV40. (B) Crystal structure of SV40 virions. Virions are 50 nm in diameter. (A, CDC/Dr. Erskine Palmer. B, Courtesy of deposition authors Stehle T, Gamblin SJ & Harrison SC, and PDB.)

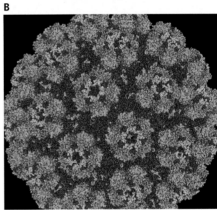

Figure 8.11 SV40 genome. The SV40 genome with nucleosomes. Each protein-coding gene is indicated by a different color and is labeled.

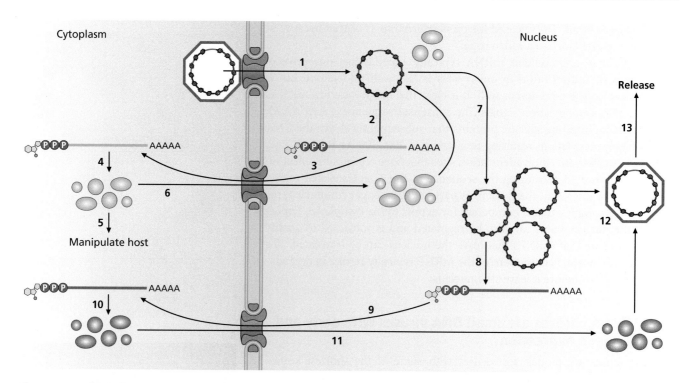

Figure 8.12 The polyomavirus replication cycle. Uncoating releases the genome with nucleosomes into the nucleus (1). Early mRNAs are synthesized by host Pol II (2). They exit the nucleus (3) and are translated (4), making early proteins that manipulate the host cell (5). Early proteins enter the nucleus (6) and enable genome replication (7). Late mRNAs are produced by host Pol II (8). They exit the nucleus (9) and are translated (10). Late proteins enter the nucleus (11), where they assemble with new genomes (12) and are ultimately released through host cell lysis (13).

8.6   The SV40 polyomavirus encodes seven proteins in only 5,243 bp of DNA

The prototype polyomavirus is SV40 because it has been investigated in the most molecular detail. Although the genome has only 5,243 bp, the DNA encodes seven proteins. Here we consider the size of all the proteins to best appreciate the economy with which SV40 uses its genome. SV40 capsids are composed of three proteins: VP1 (364 amino acids), VP2 (352 amino acids), and VP3 (234 amino acids). Each capsid has 360 copies of VP1 and 60 copies each of VP2 and VP3. The VP4 protein (125 amino acids) is expressed at the same time as the other VPs, but it is not a structural protein. Instead, VP4 is a late-acting nonstructural protein required for virion assembly and release. The virus encodes four additional nonstructural proteins that are essential for completing its life cycle: **large T antigen** (708 amino acids), **small t antigen** (174 amino acids), 17 kT (135 amino acids), and agnoprotein (62 amino acids).

Large T antigen plays many regulatory roles during infection, from biasing transcription toward late gene expression to DNA replication. Both large T antigen and small t antigen are required to drive the cell out of G_0 and pass the G_1/S cell cycle checkpoint. The N-terminal regions of the large T antigen and small t antigen are identical, but they have very different middle and C-terminal sequences, conferring different molecular functions on the two regulatory proteins. Agnoprotein has also been implicated as a regulatory protein that affects many stages of the virus replication cycle. Adding up the amino acids in each protein encoded by the genome results in 2,156 amino acids, which must be encoded by three times that many nucleotides.

A calculation demonstrates that the proteins should be encoded by 6,468 bp because (3 × 2,156 = 6,468). Nevertheless, the genome has fewer than 5,300 bp. That is, the tiny virus must use at least some of its nucleotides to encode more proteins than the apparent coding capacity of the genome. How the virus does so is discussed in **Sections 8.7** and **8.8**.

8.7 The synthesis of mRNA in SV40 is controlled by the noncoding control region

Polyomaviruses express their genomes in the nucleus, where host factors needed for mRNA production are abundant. The genome has two promoters facing in opposite directions; the promoters are part of the **noncoding control region** (**NCCR**) (**Figure 8.13**). The NCCR is free of nucleosomes. The NCCR plays a crucial role in gene expression and regulation, suggested by its 414-bp length, which is 8% of the virus's tiny genome, leaving only 4,829 bp to encode the seven viral proteins. By convention, the viral genome is depicted so that the counterclockwise promoter is the one that is active during early gene expression and the clockwise promoter is active during late gene expression.

Specific sequences in the NCCR and certain host proteins are required for early viral gene expression. In the NCCR, regulatory DNA includes AT-rich sequences, a 21-bp repeat region containing two copies of a GC-rich sequence, and a single 72-bp repeat. The AT-rich region serves as a TATA box for initiating the assembly of the Pol II transcription complex. Repeated elements in any promoter are usually sites for protein binding, and these are no exception. The repetitive sites in the NCCR bind to ubiquitous host transcription factors such as SP1, which binds to the GC-rich sequences in the 21-bp repeat region (see Figure 8.13). The 72-bp repeat functions as a binding site for host proteins that enhance transcription.

Early transcription proceeds unidirectionally, terminating approximately halfway around the circular genome. **Alternative splicing** is the major mechanism by which SV40 encodes so many different proteins using a small amount of DNA, resulting in three different mature early mRNAs that encode either small t antigen, large T antigen, or the 17-kT protein (**Figure 8.14A**). The mechanism of alternative splicing is explained in **Sections 8.24** and **8.25**. The mRNAs leave the nucleus as normal mRNP particles, which are translated in the usual way and produce the regulatory T antigens. T antigens have nuclear localization signals that cause them to be imported into the nucleus, where they affect gene expression and genome replication. The large T antigen, for example, binds to sequences in the NCCR and thereby represses early mRNA production.

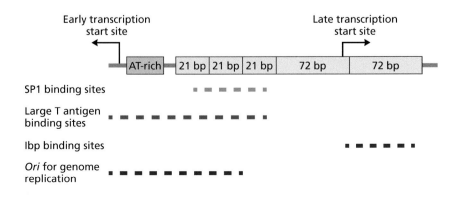

Figure 8.13 The NCCR of SV40 includes many *cis*-acting regulatory sites. The early promoter drives transcription to the left, and the late promoter drives transcription to the right. The NCCR has an AT-rich region, three 21-bp sequences, and two 72-bp sequences, all of which are multifunctional in that they interact with a variety of different proteins depending on the stage of the virus replication cycle. The repeated sequences are sites for host SP1 transcription factors, the viral large T antigen, host initiator-binding protein (Ibp) binding sites, and the origin of DNA replication.

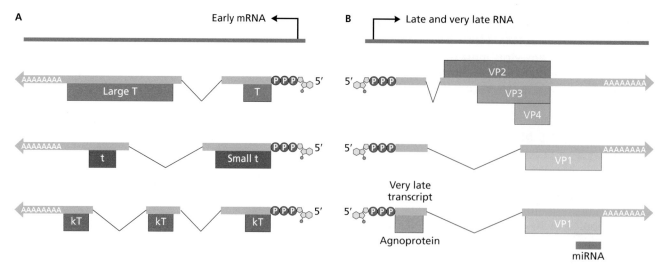

Figure 8.14 Early and late transcripts. (A) Map of early transcripts. Each mRNA is represented by a green line. Thin black lines connect exons. (B) Map of late and very late transcripts. Each mRNA is represented by a green line. Thin black lines connect exons. The miRNA is a small regulatory RNA that is expressed late during infection.

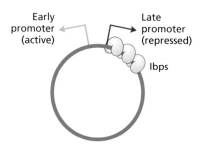

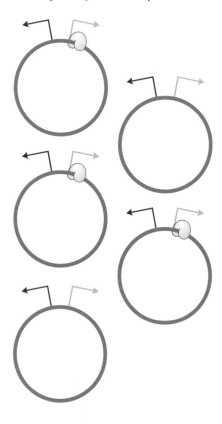

8.8 Late SV40 transcription is regulated by both host and viral proteins

Although early transcripts are found at only hundreds of copies per cell, late transcripts (**Figure 8.14B**) are produced in hundreds of thousands of copies per cell. These transcripts initiate from a second promoter in the NCCR, but this time transcription is directed to proceed clockwise around the genome. Several mechanisms trigger the switch to late transcription. First, late transcription is typically repressed by several copies of different host proteins, collectively called **initiator-binding proteins** (Ibps), binding to the NCCR (**Figure 8.15**). The Ibps remain at constant levels in the nucleus. Consequently, after the viral genome has been replicated many times, the Ibps distribute themselves among the many copies of the NCCR instead. The result is alleviation of Ibp-mediated repression of late transcription, also known as derepression. Second, accumulation of large T antigen in the nucleus stimulates late transcription through a less well-characterized mechanism and represses early transcription by binding to parts of the NCCR to sequences called box I, II, and III (**Figure 8.16**).

Late transcription terminates approximately halfway around the genome and produces two alternatively spliced mRNAs that together encode four proteins: VP1, VP2, VP3, and VP4 (see Figure 8.14B). The ratio of VP1 to VP2 and VP1 to VP3 in the mature capsids is 6:1, so VP1 is needed in much greater abundance than the other two. The VP1 protein is translated from its own dedicated mRNA, as is typical for the most abundant viral protein in many viruses (for example, see influenza virus in **Chapter 6**). In contrast, the less abundant late proteins are translated from overlapping open reading

Figure 8.15 Initiator-binding proteins in polyomavirus gene expression. (A) Cellular DNA-binding proteins, collectively termed initiator-binding proteins (Ibps), bind to the NCCR and repress late transcription. The Ibps remain at constant levels in the nucleus. (B) Consequently, after viral genome replication substantially increases the amount of viral DNA in the nucleus, the Ibps distribute themselves among the many copies of the NCCR. In this configuration with a small number of Ibps per genome, late gene expression is no longer repressed. Transcriptionally active promoters are green while inactive promoters are red.

Figure 8.16 Large T antigen binds to the NCCR, repressing early transcription and activating late transcription. (A) Early during infection, Sp1 and other host transcription factors activate early transcription, and Ibps repress late transcription. (B) Late during infection, accumulation of large T antigen in the nucleus results in the protein binding to the NCCR. The effect is to repress early transcription and activate late transcription. Transcriptionally active promoters are green while inactive promoters are red.

A   Early gene expression

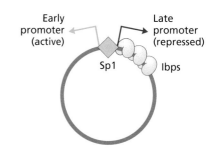

B   Late gene expression

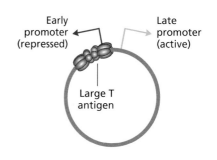

frames on the same mRNA molecule, which reduces their abundance relative to that of VP1. The first AUG is used to begin translation of the VP2. The VP3 protein is translated in the same open reading frame as VP2 but uses an alternative AUG start codon that occurs further toward the 3′ end of the mRNA, so that VP3 overlaps VP2 entirely but lacks some of the amino acids found at the N-terminal end of VP2 (**Figure 8.17**). The mechanism of translating VP3 appears to be through an internal ribosome entry site (see **Chapter 5**). VP4 is also translated from this same mRNA using a more distal alternative start codon, by an unknown mechanism. The use of alternative start codons in combination with alternative splicing therefore accounts for six of the seven SV40 proteins. The seventh, agnoprotein, is translated from a **very late transcript** that is synthesized exclusively during assembly after the VP proteins have accumulated (see Figure 8.14B). This very late transcript begins upstream of the late transcripts, and agnoprotein is encoded near its 5′ end by the mRNA sequence that is unique to the very late transcript.

The late transcript also includes a regulatory RNA called a microRNA or **miRNA**. In general, miRNA molecules associate with cellular proteins that

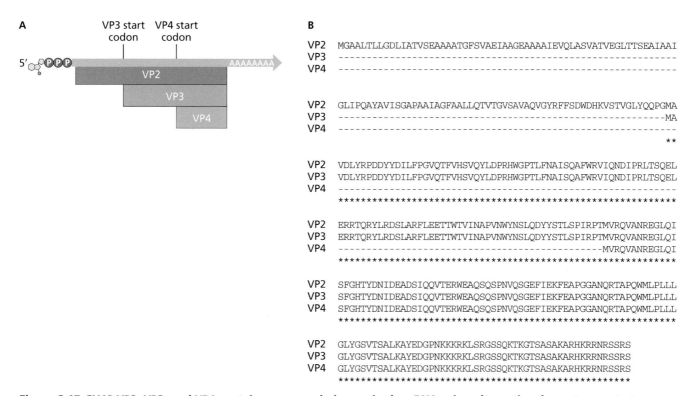

Figure 8.17 SV40 VP2, VP3, and VP4 proteins are encoded on a single mRNA using alternative downstream start codons. (A) The most 5′ start codon is used to begin translation of the VP2. The VP3 protein is translated in the same open reading frame and so is VP4, but they begin at alternative start codons that are further downstream. (B) Alignment of SV40 VP2, VP3, and VP4 proteins. Dashes under the alignment indicate a lack of identity, and asterisks indicate that the amino acids at that position are identical. As can be seen from the alignment, VP2 contains all of the amino acids that are also found in VP3 and VP4, but VP3 and VP4 are initiated at later start codons.

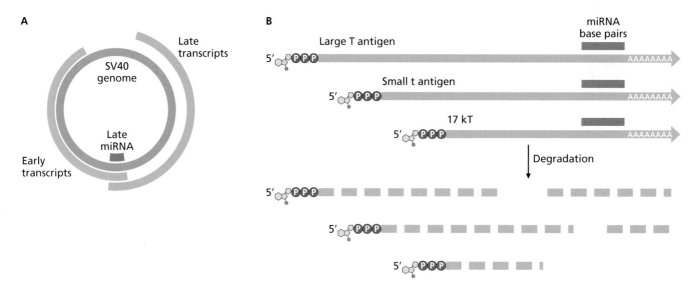

Figure 8.18 SV40 regulatory miRNA. (A) Map of the position of the DNA encoding the regulatory miRNA. The miRNA is complementary to the early transcript. (B) During late gene expression, a late miRNA is produced. The miRNA binds to early mRNA leading to its degradation so that translation of early proteins is reduced.

guide the RNA to bind to complementary mRNA, resulting in a miRNA–mRNA duplex that is subsequently degraded. In this case, the SV40 miRNA is complementary to the early transcript (**Figure 8.18**). Mutations that selectively delete the miRNA without affecting any protein-encoding sequences result in higher levels of large T and small t antigen in infected cells. Downregulation of the large T antigens is probably important during pathogenesis for evading detection by the host immune system. Viral use of miRNAs for their own purposes is ironic, given that the host proteins that participate in miRNA processing, targeting, and degradation of complementary mRNA are thought to have evolved as an antiviral response. In this case, which might be regarded as an ultimate form of subversion, SV40 uses a host antiviral response to regulate its own life cycle and to evade an immune response.

8.9 Most Baltimore Class I viruses including polyomaviruses manipulate the eukaryotic cell cycle

In general, genome replication among the Class I DNA viruses, including SV40, relies heavily on the eukaryotic cell cycle, which has four stages: M, G_1, S, and G_2 (**Figure 8.19**). The stages and regulation of the cell cycle are presented here as background for understanding why the nonstructural polyomaviruses target cell cycle proteins. In the cell cycle, M stands for mitosis and is followed by G_1 or the first gap phase. The G_1 phase is followed by S, which stands for DNA synthesis. DNA replication proteins are present exclusively during the S phase, which lasts about one-third of the lifetime of an actively cycling cell. Between the S phase and mitosis there is another gap, G_2. Many viruses infect cells that are **terminally differentiated** such as macrophages or neurons. Terminally differentiated cells are highly specialized and do not typically undergo mitosis; they have exited the cycle, passing from G_1 into a resting phase called **G_0** instead of continuing on to S phase. The nuclei of cells in G_0 lack most DNA replication proteins. Some differentiated cells remain in G_0 for the rest of their lives, whereas others can respond to signals

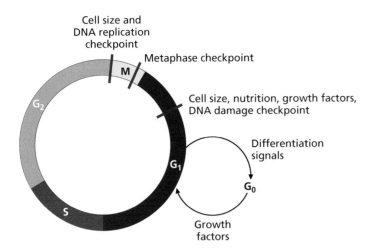

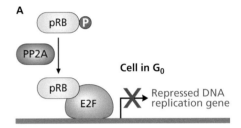

Figure 8.19 Overview of the eukaryotic cell cycle. The eukaryotic cell cycle has four stages: M, G_1, S, and G_2. Progression through the cell cycle is controlled by three checkpoints where cells can exit the cell cycle and undergo apoptosis. Differentiated cells are in G_0, which branches off from G_1. Some cells in G_0 can be stimulated by growth factors to re-enter G_1 and begin cycling.

such as growth factors that induce the cell to exit G_0 and re-enter the cycle at the G_1 stage. They will then progress through G_1, S, G_2, and M.

Progression through the cell cycle is not automatic. Instead, it is controlled by three **checkpoints** (see Figure 8.19). At each checkpoint, the cell assesses whether completion of the cell cycle without any mutations is likely and then the cell either continues through the cell cycle or, if there is something irreparably wrong, exits the cell cycle. Following this exit, the cell typically undergoes **apoptosis**, in which the cell enters a pathway that leads to programmed cell death (see **Chapter 13**). Apoptosis is a natural pathway used during development and during immune responses. It is also a defense to protect the whole animal from uncontrolled cell proliferation that is a hallmark of cancer, and to protect the whole animal from a virus attempting to complete its replication cycle. The checkpoint during G_1 is particularly important for DNA virus replication and therefore merits examination in some detail. In the absence of a viral infection, the G_1 checkpoint is controlled by **growth factors** that can stimulate the cell to enter S phase; many host cells susceptible to viral infection are differentiated and thus exist in G_0 and therefore have not passed the G_1 checkpoint (see Figure 8.19). DNA viruses that replicate in the nucleus typically affect the cell cycle by causing a differentiated host cell to pass the G_1 checkpoint in the absence of external growth factor stimulation.

The molecular mechanism by which a growth factor stimulates entry into S phase is known in detail (**Figure 8.20**). The following description emphasizes cell cycle proteins that are targeted by dsDNA viruses. Cells in G_0 contain a complex of the protein **pRB** bound to **E2F transcription factors**; the pRB–E2F complex represses S-phase promoters that would otherwise express proteins needed for DNA replication. The pRB protein can be phosphorylated, which changes its conformation and disrupts its binding to E2F transcription factors, which in turn derepresses S-phase promoters. A failsafe measure, the **PP2A** protein complex, is also active during G_0 in order to dephosphorylate any pRB that becomes phosphorylated; PP2A thereby helps maintain repression of S-phase promoters.

The body normally controls this progression with extracellular growth factors. After an extracellular growth factor binds to its cellular receptor in the plasma membrane, the cell increases expression of a protein called **cyclin D** (see Figure 8.20). Cyclin D associates with one of two **cyclin-dependent kinases** (**Cdks**). Kinases by definition are enzymes that phosphorylate

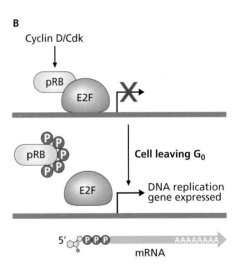

Figure 8.20 Cell cycle stimulation by a growth factor. (A) Cells in G_0 contain a complex of pRB–E2F that represses S-phase promoters. The PP2A protein dephosphorylates any pRB that becomes phosphorylated. (B) Cells are stimulated to leave G_0 when a growth factor increases levels of nuclear cyclin D. Cyclin D/Cdk phosphorylates pRB extensively, causing it to dissociate from E2F, thus derepressing S-phase promoters, leading to transcription of mRNA encoding DNA replication protein.

proteins and together the cyclin/Cdk complex phosphorylates pRB. Cells require not only phosphorylation of pRB by cyclin/Cdk, but also a lack of dephosphorylation of pRB by PP2A to result in enough phosphorylated pRB that the S-phase promoters become active. Viruses that cause artificial phosphorylation of pRB must therefore also prevent the activity of PP2A. Robust phosphorylation of pRB ultimately causes the E2F transcription factors to dissociate from pRB, allowing derepression of genes needed in S phase, including genes that encode DNA replication proteins.

8.10 Most Class I viruses prevent or delay cellular apoptosis

The pRB system also interacts with other signal transduction proteins. For example, as part of the body's natural defense against cell transformation that can lead to cancer, cells have a mechanism to detect and respond to DNA damage. Normal, uninfected cells will exit the cell cycle and undergo apoptosis if DNA damage persists. A key protein regulator of apoptosis in response to DNA damage is **p53** (**Figure 8.21**). The stress caused when a virus forces a cell into S phase resembles DNA damage stress enough that it can trigger the p53 system. In addition to promoting apoptosis, p53 feeds into the pRB signal transduction network by interfering with cyclin/Cdk-mediated phosphorylation of pRB. DNA viruses therefore usually block p53 in order to prevent apoptosis and to enable phosphorylation of pRB. Preventing apoptosis provides the virus with sufficient time to complete its replication cycle, and enabling phosphorylation of pRB forces the host cell past the G_1 checkpoint.

8.11 SV40 forces the host cell to express S-phase genes and uses large T antigen and host proteins for genome replication

In SV40 in particular, the large T and small t antigens have multiple functions, including manipulation of the cell cycle and of apoptotic pathways (**Figure 8.22**). Host cells susceptible to infection by SV40 are typically in G_0 prior to infection, which means that they do not contain nuclear DNA

Figure 8.21 The p53 protein promotes cell cycle arrest and apoptosis. Normal, uninfected cells will exit the cell cycle and undergo apoptosis if the p53 system is activated. Both DNA damage and viral infections activate p53. When p53 is activated, it becomes phosphorylated, which has two major effects. First, phosphorylated p53 stimulates cells to undergo apoptosis. Stimulatory interactions are indicated by black arrows. Second, phosphorylated p53 prevents the p53 protein from becoming phosphorylated. Inhibitory interactions are indicated by red lines. The effect on p53 ensures that, even if apoptosis does not occur, the cells remain in G_0 when they do not express DNA replication proteins. Some viruses encode proteins that prevent the p53 system from triggering apoptosis or from maintaining the cells in G_0.

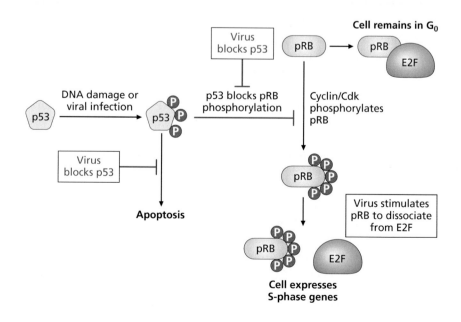

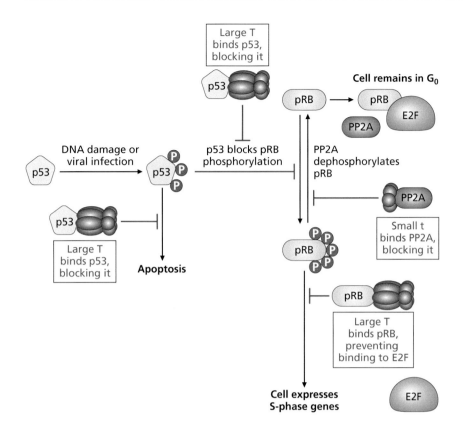

Figure 8.22 The SV40 large T and small t antigens affect the cell cycle and apoptosis. The black arrows indicate stimulatory effects of protein interactions. Red lines indicate the opposite, namely, inhibitory effects. Large T antigen interacts with pRB, preventing pRB from interacting with E2F transcription factors and thus derepressing S-phase genes. The large T antigen also interacts with p53, which prevents p53 from stimulating apoptosis. The small t antigen interacts with the PP2A (phosphatase) complex and inhibits its activity, thereby preventing repression of S-phase genes.

replication proteins. The virus must force the cell to express S-phase genes in order to replicate its genome. These changes in gene expression are accomplished by the early nonstructural proteins large T antigen and small t antigen. Large T antigen interacts with pRB, preventing pRB from interacting with E2F transcription factors. Consequently, the S-phase genes are derepressed. The large T antigen also interacts with p53; this interaction prevents p53 from promoting apoptosis. The small t antigen interacts with the PP2A (phosphatase) complex and inhibits its activity, which would otherwise dephosphorylate pRB. The small t antigen therefore further biases the system toward expression of S-phase genes.

The p53 protein, which is required for normal animal cell development and differentiation, was actually discovered through its interaction with large T antigen using co-immunoprecipitation (**Figure 8.23**). The p53 protein was the most abundant protein that bound to T-antigen bait. Its name reflects its apparent molecular weight, 53,000 Da, on a gel. Since its discovery more than 30 years ago, we have learned that the p53 protein is altered and thus inactive or incorrectly active in most human cancers. The discovery of p53 through its interactions with large T antigen is a telling example of the impact that research in virology can have on molecular and cellular biology in general and on medicine beyond infectious disease.

8.12   SV40 genome replication requires viral and host proteins to form active DNA replication forks

Replication of SV40 DNA is arguably the best understood eukaryotic replicon to date. Many aspects of SV40 genome replication are similar to those of host DNA replication; the unique aspects of viral replication involve large T antigen, which in addition to its regulatory roles plays a direct enzymatic role

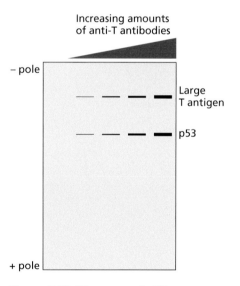

Figure 8.23 Discovery of p53 through co-immunoprecipitation with large T antigen. Cells were fed radioactive methionine to label all newly synthesized proteins. Extracts were immunoprecipitated with negative control antibodies (left lane) or with increasing amounts of antibodies that bind to large T antigen. After precipitation, protein complexes were separated by SDS-PAGE, revealing that p53 co-immunoprecipitates with large T antigen. The radioactive proteins were detected by exposing the gel to X-ray film.

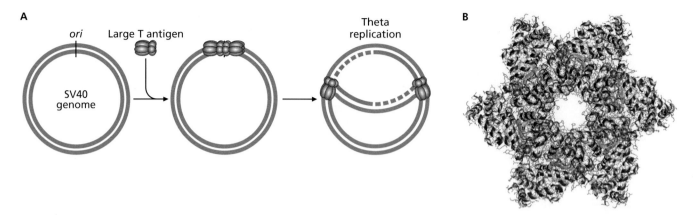

Figure 8.24 SV40 large T antigen and viral genome replication. (A) Large T antigen binds to *ori* using helicase activity to form two replication forks that will proceed bidirectionally around the genome in θ replication. (B) Crystal structure of a hexamer is composed of large T antigen helicase domains and six Zn²⁺ atoms (green). (Deposition authors Li D et al. Image courtesy of Astrojan. Published under CC BY 4.0.)

during DNA replication. DNA replication in SV40 always begins at the same sequence of DNA, which is called the **origin of DNA replication** (*ori*), present in the NCCR. Six large T antigen proteins form a hexamer and two hexamers bind *ori* in a head-to-head arrangement (**Figure 8.24**). The hexamers have helicase activity that facilitates melting of the DNA at the origin. DNA replication then proceeds bidirectionally, with each large T antigen hexamer progressing in opposite directions around the template in a process known as θ replication because of the appearance of partially replicated molecules in electron micrographs (see **Chapter 4**).

The proteins at the replication fork include many host factors in addition to the viral large T antigen (**Figure 8.25**). Replication protein A is found at replication forks, where its primary function is to bind to single-stranded DNA to maintain single-stranded template DNA at the replication fork. The leading strand is copied by DNA Pol δ, which is loaded onto the DNA and clamped into position by replication fork protein C and PCNA protein. The lagging strand has a complex containing Pol α and **primase**, which synthesizes the RNA primers required to provide a double-stranded binding site and free 3′ OH for the other polymerases to extend. The host DNA replication proteins needed for SV40 genome replication also include FEN1 and ligase, which form their own complex needed to remove the RNA components of Okazaki fragments, replace them with DNA, and seal the nicks in the new DNA. Replication of cellular DNA is thought to proceed in a similar fashion but with several host proteins replacing the large T antigen.

8.13 The papillomavirus replication cycle is tied closely to the differentiation status of its host cell

Papillomaviruses are important models for replication of small circular dsDNA viruses, but they are also important causes of human disease. The papillomavirus lytic reproduction cycle causes warts, including genital warts; additionally, human papillomaviruses are the leading cause of cervical cancer (see **Chapter 13**). The gynecological exam procedure known as a pap smear is used to detect papillomavirus activity in cervical tissue in order to diagnose cervical cancer; vaccines such as Gardasil prevent the most common forms of cervical cancer by immunizing the population (including males) against **human papillomavirus** (**HPV**) (see **Chapter 16**).

A   **SV40 replication fork**

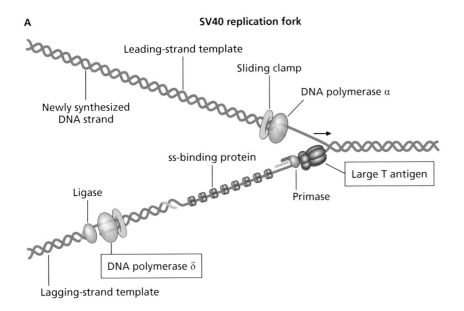

B   **Cellular replication fork**

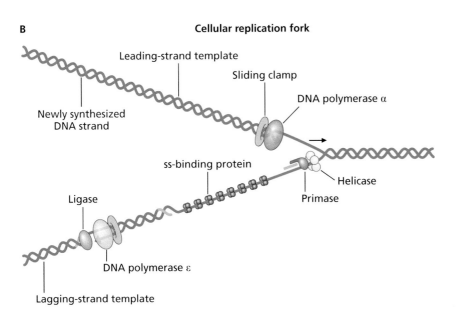

Figure 8.25 Replication fork proteins for SV40 replication compared with cellular DNA replication. (A) The SV40 large T antigen acts as a helicase during viral genome replication. Cellular DNA polymerase α is used on the leading strand while DNA polymerase δ copies the lagging strand template. (B) During normal cellular DNA replication, host helicase substitutes for large T antigen and DNA polymerase ε is used on the lagging strand instead of DNA polymerase δ.

A

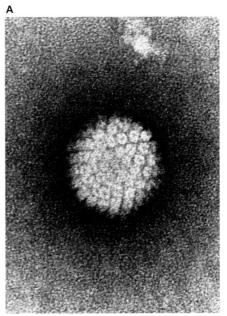

B

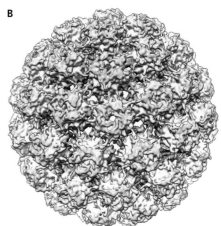

Figure 8.26 Papillomavirus. (A) Electron micrograph of HPV, which is 60 nm in diameter. (B) False-color crystal structure of a papillomavirus emphasizing the symmetry of the capsomers. (A, Courtesy of NIAID-NIH Laboratory of Tumor Virus Biology. B, Courtesy of Vossman. Published under CC BY-SA 3.0.)

Papillomaviruses such as HPV (genus *Alphapapillomavirus*) have naked spherical virions that are 60 nm in diameter (**Figure 8.26**). The capsid is composed of a single type of capsomer, and the small circular genome inside the virion is 6–8.5 kbp of circular dsDNA arranged in nucleosomes (**Figure 8.27**). Although the average coding capacity needed to encode a protein is about 1 kbp, papillomaviruses encode at least 12 proteins. The tiny size of the papillomavirus capsid exerts selective pressure in favor of a genome small enough to be packaged into the virion.

The papillomavirus replication cycle is very similar to that of polyomavirus. After uncoating, which introduces the viral genome into the nucleus, host Pol II transcribes viral early mRNAs. Early gene expression results in production of nonstructural proteins that manipulate the cell cycle, regulate viral gene expression, and ultimately initiate viral genome replication. Late gene expression occurs after genome replication has begun; late genes encode structural proteins needed to complete the reproductive cycle.

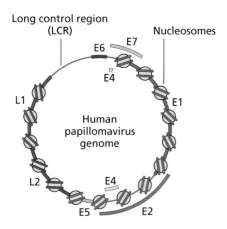

Figure 8.27 Human papillomavirus genome. This map of the HPV genome with nucleosomes has been simplified so that it does not show all of the proteins that result from alternative splicing. E proteins are expressed early, whereas L proteins are expressed late.

Unlike polyomavirus, an important aspect of the papillomavirus replication cycle is that it is closely tied to the differentiation status of its host cells, despite the fact that polyomaviruses also manipulate the cell cycle. Papillomaviruses obligatorily infect stratified squamous epithelia, such as that found in human skin (**Figure 8.28**). Cells in each layer are uniquely differentiated in order to serve the unique functions of each layer. The first layer of basal stem cells is located atop a basement membrane connected to the underlying dermis, which includes fibroblasts and blood vessels. Cells in the basal zone are actively dividing and give rise to cells that progress up through the epithelium toward the exterior of the body. Such cells undergo differentiation along the way, giving rise to the different cell types that provide characteristics such as elasticity and resistance to shearing forces to the epithelial tissue. The differentiated cells use unique groups of transcription factors to maintain the unique array of gene expression that causes their specialization. Atop the basal stem cells are midzone cells, which continue to be pushed up through the epithelium as they undergo differentiation to form the upper level of keratinized dead cells known as a stratum corneum. The stratum corneum protects animals against skin infections much the same way that the cuticle protects plants from viruses.

Papillomavirus can only infect the living basal cell layer, meaning that an injury is necessary to allow the tiny virion to bypass the upper layers of the epithelium. The expression of immediate-early, early, and late proteins is then linked to the layer of the epithelium occupied by the infected host cell (**Figure 8.29**). For example, a pair of immediate-early proteins is expressed in the basal zone of the epithelium. Immediate-early and early proteins are expressed in cells of the lower midzone cells. In the upper midzone cells, there is a switch to late viral protein production and to vegetative DNA replication, resulting in hundreds of copies of the genome per cell (see **Section 8.17**). Virion assembly occurs in the upper midzone cells so that when the cells enter the granular layers (in which they contain increasing amounts of keratin), the cells are full of mature virions. These cells then become the stratum corneum, and normal everyday sloughing of the stratum corneum then spreads the virions to the next person. A similar replication cycle occurs when HPV infects the stratified epithelium of the cervix (which connects the vagina to the uterus), except that the cervical epithelium does not have a dead stratum corneum.

Figure 8.28 Stratified epithelium. Starting with the cells closest to the interior of the human body (proximal cells), the first layer is that of the basal stem cells. Cells in the basal zone are actively dividing to give rise to the rest of the epithelium. Atop the basal stem cells are midzone cells. The upper level formed by dead keratinized cells is the stratum corneum.

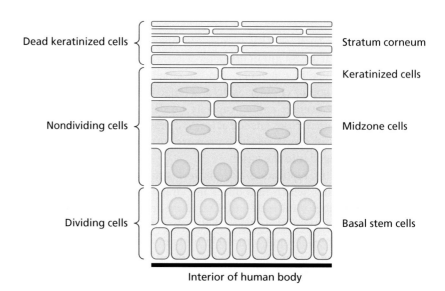

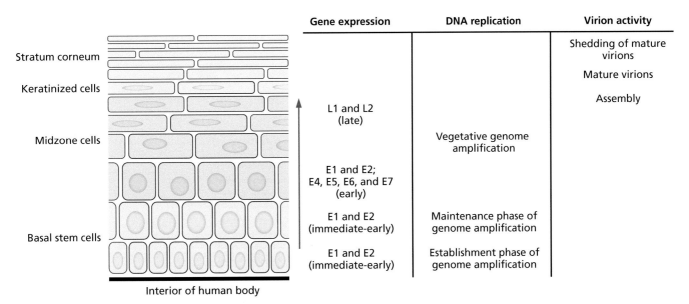

Figure 8.29 Papillomavirus gene expression and DNA replication in a stratified epithelium. HPV gene expression and genome replication are tied closely to the differentiation status of the host cells in a stratified epithelium. For example, the earliest genes are expressed in the lower layers of the epithelium (closest to the body cavity), whereas the later genes are expressed in the more distal layers. Vegetative DNA replication, resulting in hundreds of copies of the genome per cell, is finally triggered in the upper midzone cells. Viral late proteins are structural and accumulate after this point so that virion assembly occurs in the upper midzone cells. When the cells enter the stratum corneum, they are full of mature virions that are shed, thus spreading the infection to other susceptible hosts.

8.14 Human papillomaviruses encode about 13 proteins that are translated from polycistronic mRNA

As mentioned earlier, HPV proteins are grouped into immediate-early, early, and late classes. E1 and E2 are immediate-early proteins, and E5, E6, and E7 are examples of early proteins expressed after the immediate-early genes but before DNA replication has begun. The late proteins, such as the two capsomers, are expressed after DNA replication has amplified the number of genomes in the cell by a factor of 100 and are correspondingly first observed in the upper layers of the midzone, where that genome amplification occurs. Generally, the expression of immediate-early genes continues during early gene expression, and some early gene expression continues during late gene expression. This situation arises because all immediate-early proteins are encoded by polycistronic mRNAs that also encode one or more early proteins. Similarly, all early proteins are encoded by polycistronic mRNAs that encode an immediate-early protein or one or more late proteins (**Figure 8.30**). Typically, there are 12–25 different viral mRNA species, almost all of which are polycistronic. In most HPV isolates, the most abundant L1 capsomer (360 copies per virion) is the only one encoded by a monocistronic mRNA, which likely accounts for the protein's relative abundance.

Human papillomaviruses also encode variants of some of the proteins. For example, most encode E6, E6*I, E6*II, and E6*III proteins, where the * symbol indicates a truncation and the Roman numerals indicate that there are three different truncated E6* proteins (see Figure 8.30). Similarly, most encode E1^E4, where the ^ symbol indicates that there is a fusion between part of the E1 protein and part of the E4 protein. In most cases, the exact function of the variant proteins is not yet clear.

Figure 8.30 Map of HPV18 coding sequences and mRNAs. The map of the virus has been linearized in order to make it easier to see where the mRNA transcripts map onto the genome. Exons in mRNA are shown below the map.

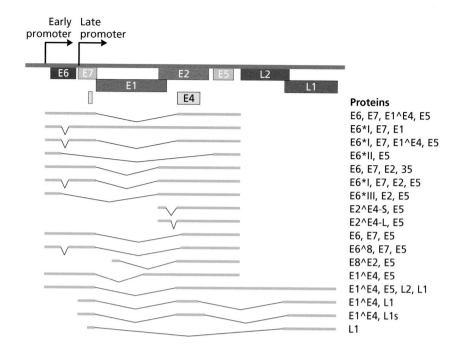

Proteins
E6, E7, E1^E4, E5
E6*I, E7, E1
E6*I, E7, E1^E4, E5
E6*II, E5
E6, E7, E2, 35
E6*I, E7, E2, E5
E6*III, E2, E5
E2^E4-S, E5
E2^E4-L, E5
E6, E7, E5
E6^8, E7, E5
E8^E2, E5
E1^E4, E5
E1^E4, E5, L2, L1
E1^E4, L1
E1^E4, L1s
L1

Molecular functions have been associated with many of the HPV proteins through biochemical work that defined their interacting partners or through genetic work in which mutations indicate the normal function of a gene by the phenotype of that mutant. The immediate-early **E6** and **E7** HPV proteins along with the early **E5** protein all manipulate either the cell cycle or apoptotic pathways or both in order to promote a cellular environment in which virus genome replication proceeds unimpeded. For example, the E7 protein binds to pRB, which causes derepression of promoters controlled by E2F (see Figure 8.20). The early **E1** protein is a helicase required to initiate and sustain viral DNA replication, reminiscent of the helicase activity provided by large T antigen in the polyomaviruses. The early **E2** protein has a DNA-binding domain and regulates viral gene expression. It is also important for distributing viral genomes equally into daughter cells as the cells enter upper layers of the epithelium. The **E1^E4** fusion protein is needed for the release phase of the virus life cycle. Finally, the late **L1** and **L2** proteins are components of the capsid with L1 outnumbering L2 5:1.

8.15 The long control region of HPV regulates papillomavirus transcription in which pre-mRNA is subjected to alternative splicing

Transcription of HPV genes is controlled by the **long control region**. The long control region contains one major early promoter and one major late promoter, both of which result in transcription moving in the same direction. The promoters are regulated by host factors that are specific to the differentiation status of the host cell; different types of cultured host cells exhibit different regulatory mechanisms. The viral E2 protein also regulates the early promoter through its interactions with host proteins. In some cultured cells, E2 represses the early promoter, whereas in others E2 activates the promoter, indicating that the host proteins determine the outcome of E2 activity. During natural infections, as opposed to ones in artificial tissue culture, the pattern of protein expression in stratified epithelia suggests that E2 represses the early promoter because high levels of E2 protein correlate with a switch to late proteins detectable in the host cells (see Figure 8.29).

All of the HPV transcripts are spliced; the DNA encodes at least five splice donor sites and six splice acceptor sites. Splice sites are composed of RNA, not DNA, and thus the genome is said to *encode* the donor and acceptor sites (see Sections 8.24 and 8.25). Splicing of HPV transcripts is controlled by host proteins that depend on the differentiation status of the host cell. The genome also encodes a single major polyadenylation site for immediate-early and early transcripts, as well as a cluster of polyadenylation sites for late transcripts. The particular polyadenylation site determines the 3' UTR, which may regulate the late transcripts' mRNA stability or translation. The result is many different mRNA transcripts (see Figure 8.30) with the possibility that more might be discovered, especially because different cultured host cells result in different species of mRNA.

8.16   Leaky scanning, internal ribosome entry sites, and translation reinitiation lead to the expression of papillomavirus proteins from polycistronic mRNA

All papillomavirus proteins are translated from polycistronic mRNAs that sequentially encode more than one protein; in many cases, the coding sequences for the different proteins overlap (see Figure 8.30). For example, the E1 coding sequence overlaps with that of E7 and E2, as does the E4 coding sequence. The clear pattern of protein expression correlating with host cell differentiation in a stratified epithelium (see Figure 8.29) indicates that translation from the polycistronic mRNA is likely regulated by host factors that change in correlation with the position of a cell in the epithelium. For example, the two immediate-early proteins are readily detectable in the basal stem cells, whereas early proteins are not, even though the immediate-early genes are encoded on polycistronic mRNAs that also include the coding sequence for one or more early proteins (see Figure 8.30).

There are three known mechanisms by which HPV translates more than one protein from its polycistronic mRNAs: leaky scanning (see Chapter 5), internal ribosome entry sites (see Chapter 5), and translation reinitiation. Briefly, leaky scanning occurs when the ribosome sometimes scans past the first AUG near the 5' end of the mRNA and instead initiates at a downstream AUG. Internal ribosome entry sites are sequences of RNA that have complex secondary structure and result in assembly of translation initiation factors at that internal site, which can be far downstream of the 5' end of the mRNA.

In HPV, translation reinitiation accounts for expression of E7 from mRNA that also encodes E6 or truncated E6 protein. The exact mechanism by which translation of E7 occurs in HPV is not known, but it is likely similar to an **RNA termination–reinitiation** mechanism found in other viruses. During RNA termination–reinitiation, translation of an upstream protein terminates in the normal way and that protein is released from the ribosome. Next, the large ribosomal subunit is also released. Normally, the small ribosomal subunit would then disassociate from the mRNA but during translation reinitiation, a reinitiating motif in the mRNA causes the small subunit to remain associated with the mRNA (**Figure 8.31**). Afterward, a large ribosomal subunit joins the complex and the ribosome reinitiates translation at an internal AUG, thereby expressing a downstream protein in a polycistronic message. Secondary structures found both upstream and downstream of the internal AUG are probably critical for reinitiation. The secondary structures may enable the mRNA to base pair to the 18S rRNA in the small subunit,

Figure 8.31 RNA termination–reinitiation. When the ribosome reaches the upstream stop codon, the nascent protein is released and the large ribosomal subunit dissociates. Because of reinitiation motifs in the RNA, which likely form secondary structures, the small ribosomal subunit remains tethered to the RNA in the vicinity of the downstream start codon. Subsequently, in a process that may involve eIF3, a large ribosomal subunit joins the complex, and translation of the downstream coding sequence follows.

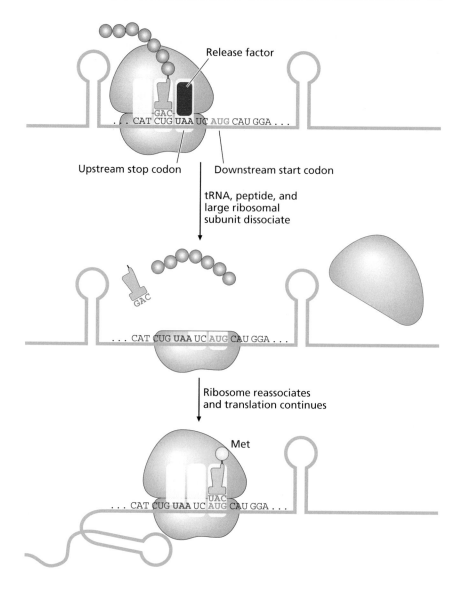

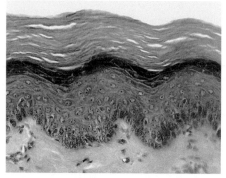

Figure 8.32 Layers of differentiated cells in a stratified epithelium. The distinctive layers of different cell types are evident in this section of tissue stained to show the cell bodies.

tethering it to the RNA, and may also enable host eIF3 to associate with the complex and enable a large ribosomal subunit to join the small subunit so that translation of the downstream coding sequence can ensue.

Although the detailed mechanisms by which different mRNA molecules are produced in different layers of the stratified epithelium remain unknown, there is substantial evidence to indicate that the expression of immediate-early, early, and late proteins occur in the lower, middle, and upper layers, respectively. For example, immunostaining (see **Chapter 3**) of tissue explants can be used to demonstrate that the late capsomer proteins are confined to the upper layer of an infected stratified epithelium (**Figure 8.32**).

8.17   DNA replication in papillomaviruses is linked to host cell differentiation status

Very little is known about the mechanism of DNA replication by papillomaviruses, owing in part to difficulty in finding an easily cultured host cell that can support the entire replication cycle and to the tight link between DNA

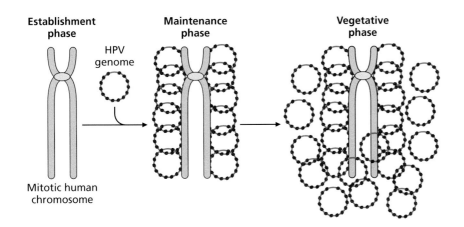

Figure 8.33 Three phases of papillomavirus genome replication. During the establishment phase, there are a few copies of the viral genome per cell. During the maintenance phase, the copies of the viral genome increase to many per cell and the viral DNA is physically associated with host chromosomal DNA so that it segregates with the host chromosomes during mitosis. During vegetative DNA replication, the nuclei fill with around 1,000 copies of the viral genome.

replication and host cell differentiation in a stratified epithelium during a real infection. Mimicking a stratified epithelium in cell culture is technically challenging and rarely achieved. It is known, however, that there are three phases of papillomavirus genome replication (**Figure 8.33**). The **establishment phase** occurs in the basal zone (see Figure 8.29) and results in a few copies of the genome per cell maintained as extrachromosomal episomes. In the basal zone cell nuclei, the circular episomal viral dsDNA is physically anchored to host chromosomes. In the second or **maintenance phase** for cells entering the midzone, the viral genome is replicated to 50–300 copies per cell and is partitioned along with the host chromosomes, thus passing the viral genomes on to daughter cells after mitosis. The viral E2 proteins are required for partitioning. Near the top of the midzone, however, the **vegetative DNA replication** (third) phase of HPV genome replication is triggered. As a result of vegetative DNA replication, the cells fill with about a thousand copies of the viral genome, and subsequently the assembly stage of the virus replication cycle follows.

Two viral proteins are known to be involved in the first amplifying type of genome replication that results in 50–100 copies of the viral genome per cell. These are the multifunctional immediate-early E1 and E2 proteins. The E2 protein binds to the origin of replication in the long control region and then recruits E1 to bind nearby. After that, there is a substantial conformational change in the E1 protein, which both ejects E2 from the DNA and results in formation of two E1 hexamers bound to the DNA in a head-to-head fashion. These hexamers are helicases, comparable to the polyomavirus large T antigen. It is possible that the papillomavirus DNA replication apparatus resembles that of the polyomaviruses in other ways as well, with host proteins supplying all other functions necessary for bidirectional (θ) DNA replication.

8.18 Papillomaviruses use early proteins to manipulate the host cell cycle and apoptosis

The cells in the two lowest layers of the stratified epithelium are actively reproducing, thus providing DNA replication proteins to the virus. The remainder of the epithelium consists of differentiated cells that are not regularly dividing as they are pushed toward the distal parts of the stratified epithelium (see Figure 8.28). In order to accomplish vegetative genome replication, then, the virus must force the quiescent host cells in the midzone to express host DNA replication proteins. Two virus proteins, E6 and E7, are crucial for this process (**Figure 8.34**). They interact with pathways controlled by p53 and pRB.

Figure 8.34 The HPV E6 and E7 proteins prevent apoptosis and bind to pRB, causing expression of S-phase genes. (A) The E6 protein covalently modifies the p53 protein, targeting it for degradation. The effect is to prevent apoptosis. (B) The E7 protein binds to pRB, which prevents it from interacting with E2F transcription factors. The effect is to increase expression of S-phase genes.

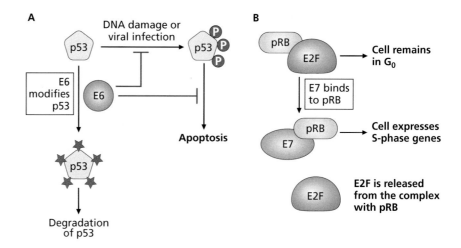

Normally, p53 promotes apoptosis in stressed cells such as those infected by a dsDNA virus. The E6 protein of papillomaviruses targets the p53 protein for proteolytic degradation, thereby preventing p53-triggered apoptosis. The E7 protein of papillomavirus, meanwhile, binds to pRB. When E7 has bound to pRB, pRB cannot bind to E2F transcription factors. Consequentially, the E2F transcription factors no longer repress the expression of DNA replication S-phase proteins. Although the mechanisms of inactivating p53 and pRB are different in polyomaviruses and in papillomaviruses, the end result is the same.

8.19 Comparing the small DNA viruses reveals similar economy in coding capacity but different mechanisms for gene expression, manipulating the host cell cycle, and DNA replication

Polyomaviruses and papillomaviruses are often considered together because of their similarities. For example, both have small circular dsDNA genomes that are complexed with nucleosomes. Both have genomes with a noncoding region that contains sequences for both transcription regulation and the origin of DNA replication. Both viruses encode early proteins that bind to pRB, thereby promoting expression of cellular S-phase proteins needed for the viruses to replicate their genomes. They also produce proteins that interfere with stress-induced apoptosis that would otherwise be caused by p53. Finally, both encode a protein that assembles into a hexameric complex during DNA replication where it functions as a helicase.

There are also important differences between the polyomaviruses and the papillomaviruses. The polyomavirus large T antigen is extremely multifunctional, exhibiting different molecular actions that interfere with the host cell cycle, regulate viral gene expression, and serve as an initiation factor and helicase during genome replication. In contrast, papillomaviruses use different proteins for these different functions, for example using E7 to manipulate the cell cycle and the E1 protein to serve as a helicase during genome replication. Furthermore, although papillomaviruses link viral gene expression and genome replication to the differentiation status of their host cells, there is no such linking in SV40 infections, which do not infect a stratified epithelium.

8.20 Adenoviruses are large dsDNA viruses with three waves of gene expression

Adenoviruses have linear dsDNA genomes that are four to eight times longer than the small DNA viruses already discussed in this chapter. Adenoviruses have been studied as models for expression and replication in viruses with linear dsDNA genomes, and they have also been studied intensively as potential gene therapy delivery agents (see **Chapter 13**). Adenoviruses are a major cause of the common cold; the model human mastadenovirus C (HAdV-C, genus *Mastadenovirus*) is a prototype for understanding the molecular and cellular biology of adenovirus infection. After entry into the host cell, the virus is progressively disassembled as it travels to the nucleus, where uncoating introduces the viral genomic DNA into the nucleus (**Figure 8.35**). The first wave of gene expression results in production of proteins that counteract host immunity in various ways and help create an environment for virus

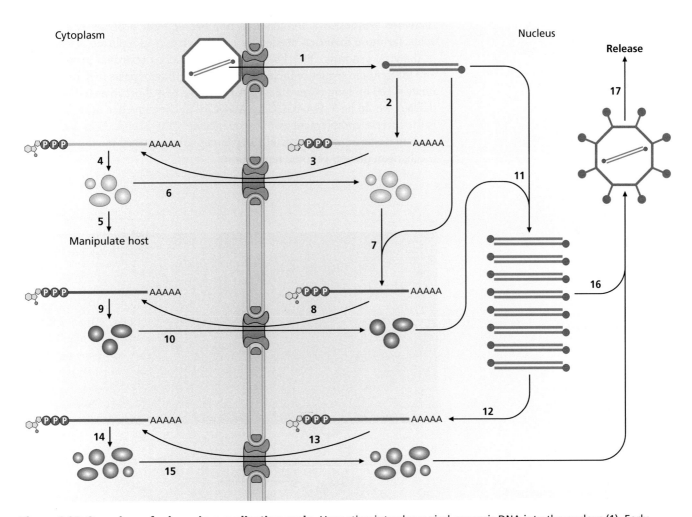

Figure 8.35 Overview of adenovirus replication cycle. Uncoating introduces viral genomic DNA into the nucleus (1). Early mRNAs are transcribed (2) and exported to the cytoplasm (3) and translated (4). Many early proteins manipulate the host (5), affecting immunity, the cell cycle, and apoptosis. Some early proteins enter the nucleus (6) and induce delayed-early transcription (7). Delayed-early mRNA is exported to the cytoplasm (8), where it is translated to produce delayed-early proteins (9). Some delayed-early proteins enter the nucleus (10), where they participate in genome replication (11). Intermediate and late mRNAs are transcribed from the new genomes (12) and exported to the cytoplasm (13), where they are translated to make late proteins (14). Late proteins enter the nucleus (15), where they assemble with new virions during maturation (16). Mature virions are released from the host cell by lysis (17).

replication. Intermediate gene expression induces replication of the viral genome. Late gene expression is focused on production of structural proteins that assemble into progeny virions in the nucleus and leave the cell through lysis. Remarkably, a single infected HeLa cell typically produces 100,000 progeny virions; this number of progeny is 10–100 times more than the burst size of most viruses that infect animal cells.

8.21 Adenoviruses have large naked spherical capsids with prominent spikes and large linear dsDNA genomes

Adenoviruses have naked icosahedral virions with prominent spikes at their vertices (**Figure 8.36**). The capsids are very complex, containing 11 different polypeptides, some at more than 500 copies per virion. The large linear genome is made up of 30–50 kbp of dsDNA; the diameter of the capsid is 90–100 nm, resulting in a particle volume that is six to eight times larger than that of the polyomaviruses. Host cells infected with adenovirus must therefore synthesize a tremendous number of viral proteins and nucleic acids, far more than that needed for smaller viruses to replicate. Each 5′ end of the viral genomic DNA has a covalently attached **terminal protein** (**TP**) and the genome sequence includes inverted terminal repeats that are approximately 100 bp long (**Figure 8.37**). An example of a short **inverted repeat** in dsDNA would be 5′ GAAGCCT…AGGCTTC; the sequence near the 5′ end is the reverse complement of the sequence near the 3′ end. As is usually the case, these terminal features of the genome are necessary for the particular mechanism of viral genome replication.

A

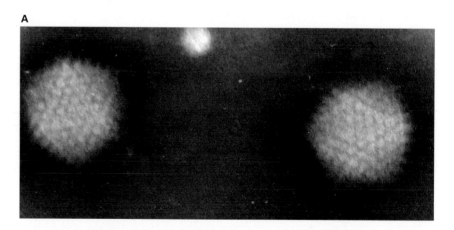

B

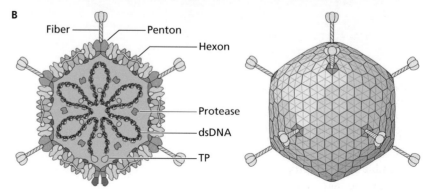

Figure 8.36 Human adenovirus C (HAdV-C). (A) Electron micrograph of adenovirus. Virions are about 90 nm in diameter. (B) HAdV-C virions. Adenoviruses are naked, have many different structural proteins, and have prominent spikes. (A, Courtesy of Graham Colm. Published under CC BY 3.0. B, Courtesy of Philippe Le Mercier, ViralZone, © SIB Swiss Institute of Bioinformatics.)

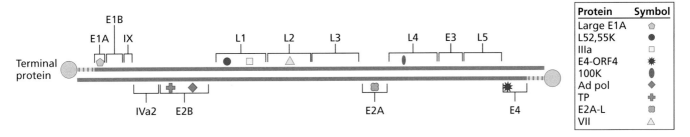

Figure 8.37 HAdV-C genome. The HAdV-C genome has two terminal proteins covalently attached to each 5′ end and inverted terminal repeats (yellow and blue). The brackets indicate transcriptional units, all of which result in various spliced mRNAs. Those above the genome are transcribed from left to right, whereas those below the genome are transcribed from right to left. The coding sequences for proteins discussed in detail in the text map to the area of the genome indicated by the symbols. At this scale, it is not possible to resolve all of the DNA encoding different introns and exons.

8.22   Adenoviruses encode early, delayed-early, and late proteins

HAdV-C encodes approximately 40 proteins with the possibility that more may be discovered. Although there is no systematic method for naming adenovirus proteins, in many cases the proteins are named for the mRNA transcript that encodes them. Rather than provide an exhaustive catalog of the known molecular functions of each, this section will consider adenovirus proteins in groups defined by the relative timing of expression. As was the case for other dsDNA viruses and bacteriophages, the earliest proteins produced are regulatory proteins or enzymes, whereas the late proteins produced after the onset of genome replication are needed for assembly or are structural. The first group are the **adenovirus early proteins**, which are small E1A and large E1A. E1A activates transcription of the **delayed-early genes**, encoding proteins such as the E1B, E2, E3, and E4 groups of proteins. Together, the early and delayed-early proteins are important for manipulating the host cell cycle and for preventing apoptosis; for example, E1A proteins can bind to pRB and cause misregulation of E2F transcription factor activity. There are three E2 proteins, all of which participate directly in DNA replication. A family of E3 proteins modulates the host immune response. The E4 proteins have various functions, from regulating alternative splicing and translation to affecting DNA replication and even preventing host cell apoptosis.

The **adenovirus intermediate and late proteins** are expressed at the same time as DNA replication, and they include all of the structural proteins. Both groups of proteins are transcribed from mRNA synthesized using new genomes as templates. Because of the exceptionally large burst size for adenovirus, it is important to consider the sheer number of structural proteins needed to form a single virion (see Figure 8.36). For example, in every virion there are 850 copies of the most abundant late protein, VII; it compacts the genome to fit inside the virion during packaging. It is also important for release of the genome into the nucleus during the last step of uncoating. Protein II, which forms the hexons in the capsid, is found in 720 copies per virion. Protein VI is found at 360 copies per virion, and protein III, which forms the penton base, is present in 300 copies per virion. Protein IX is found at 240 copies per virion and protein V is found at 157 copies per virion. Others found in fewer copies per virion include IIIa, IV, IVa2, VIII, and μ. The many structural proteins and the abundance of them required to assemble 100,000 progeny provide an extreme example of the capacity viruses have to take over a host cell's synthetic capacities.

Figure 8.38 Large E1A and mediator complex assembled with general transcription factors at a viral promoter. E1A stimulates Pol II transcription by interacting with host proteins and the mediator.

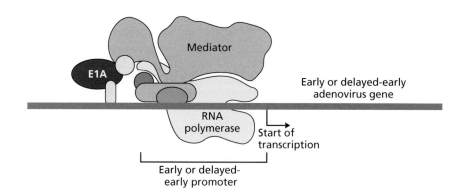

8.23   The large E1A protein is important for regulating the adenovirus cascade of gene expression

The cascade of adenovirus transcription is initiated by ubiquitous cellular transcription factors that bind to an enhancer upstream of the E1A gene resulting in production of **large E1A**. Large E1A is then responsible for activating its own promoter as well as those for the delayed-early genes. The mechanism by which the large E1A protein activates transcription has been studied in molecular detail. The protein regulates transcription through binding to a variety of other proteins, which themselves bind directly to Pol II, to DNA, or to the mediator. E1A does not bind directly to DNA but instead binds to one of several transcription factors that do bind to DNA, thus indirectly tethering the protein to DNA and, through the mediator, to Pol II. The activation by the E1A protein is known to be particularly strong, so these interactions result in substantial stimulation of transcription initiation at viral early promoters (**Figure 8.38**).

8.24   Splicing of pre-mRNA was first discovered through studying adenovirus gene expression

Adenovirus gene expression became famous in molecular biology because introns, exons, and splicing were first discovered by studying adenoviruses. No one expected that studying adenoviruses would fundamentally revolutionize our understanding of eukaryotic gene expression. The discovery of splicing joins the discovery of p53 in providing an example of how virology has had major impacts on our understanding of cellular life.

The 1970s research question that ultimately led to discovering splicing was: Where do the most abundant late adenovirus mRNA hybridize to the adenovirus genome? To address this question, the research team first purified some restriction enzymes from bacteria and then used those enzymes to digest the adenovirus genome. They prepared late mRNA from infected cells and then mixed the mRNA with digested genomes, reasoning that the mRNA would hybridize to the genomic DNA. Further, if they could break the DNA into known segments using restriction enzymes, they could see which genome segments hybridized to the late mRNA using electron microscopy and make careful measurements to determine how long the hybridizing DNA segment was. In the electron micrographs, however, they consistently found unexpected structures, where the mRNA hybridized to some of the DNA but intervening DNA segments were looped out and not base paired to any mRNA (**Figure 8.39**). They postulated that splicing certain parts of the mRNA together accounts for the looping patterns observed in

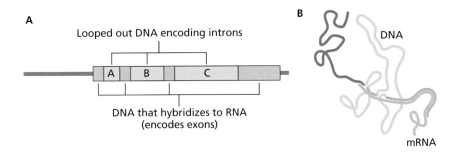

A

Looped out DNA encoding introns

DNA that hybridizes to RNA
(encodes exons)

B

DNA

mRNA

Figure 8.39 Discovery of spliced adenovirus mRNA by electron microscopy of an RNA–DNA hybrid. (A) The part of the adenovirus genome that encodes the hexon protein. The gene has four exons and three introns. (B) Single-stranded DNA from the adenovirus genome was mixed with mRNA encoding the hexon gene. The mixtures were examined by electron microscopy, which revealed that the mRNA hybridizes to only certain segments of the single-stranded DNA, which we now know correspond to exons. The DNA encoding introns does not hybridize to the mRNA, and so it forms large unpaired single-stranded loops. This line drawing represents the appearance of the one such hybrid molecule in an electron micrograph.

the micrographs. They further proposed that eukaryotic mRNA in general originates from pre-mRNA that must be spliced, which is now known to be the case, and this constituted a radical change in our understanding of gene expression.

8.25 Both host cells and adenovirus rely on alternative splicing to encode multiple proteins using the same DNA sequence

Most animal pre-mRNA is subjected to alternative splicing, a process in which an exon is sometimes included and other times is not recognized as an exon and so is spliced out (**Figure 8.40**). The result is that more than one mRNA, and therefore more than one protein, can be made from a single gene. Commonly, each exon encodes a specific functional domain or motif so that mixing and matching exons has the effect of mixing and matching domains that confer a unique combination of functions upon each protein product. Splicing of a particular mRNA in animals usually depends on the developmental timing or differentiation status of the cell in which the gene is being expressed. Alternative splicing is the reason that the human genome may encode 100,000 different proteins using only 25,000 genes.

Host cells are not the only entities that use alternative splicing to create multiple mRNAs from the same DNA transcript: adenoviruses are also masters of alternative splicing. There are 10 Pol II promoters in the adenovirus genome. Eight of these promoters drive expression of pre-mRNA that is extensively processed to give rise to multiple mature mRNA products through alternative splicing. As an example of this extensive splicing, this section examines two of the major late promoter's transcripts that give rise to one of two late proteins, IIIa or L52,55K.

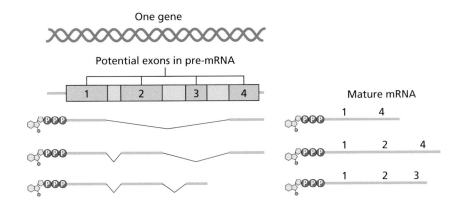

One gene

Potential exons in pre-mRNA

1 2 3 4

Mature mRNA

1   4

1   2   4

1   2   3

Figure 8.40 Overview of alternative splicing. A eukaryotic gene can have many potential exons, only some of which are ultimately part of the mature mRNA. In this example, three viral mRNAs are synthesized from one gene. One of the mRNAs contains exons 1 and 4, and the other two contain exons 1, 2, and 4, or 1, 2, and 3. The use of the same gene but different exons to synthesize different mRNAs from the same DNA sequence is termed alternative splicing. Alternative splicing is very common among eukaryotic viruses. Poly(A) tails have been omitted from the mRNA diagrams for clarity.

8.26   Regulated alternative splicing of a late adenovirus transcript relies on *cis*-acting regulatory sequences, on the E4-ORF4 viral protein, and on host splicing machinery

One of the most intensively studied alternative splicing events is the splicing that occurs in adenovirus in order to synthesize mRNA encoding either L52,55K or IIIa from the L1 mRNA transcript (**Figure 8.41**). Although both are late proteins, the L52,55K protein is produced earlier during infection than IIIa. The timing reflects the fact that L52,55K is needed for assembly while IIIa is a structural protein found in about 60 copies per virion. To make mRNA encoding the L52,55K protein, the virus uses the upstream 3′ splice acceptor site. Later during infection, to make instead the mRNA encoding the IIIa protein, an alternative downstream 3′ splice acceptor site is selected.

This downstream 3′ splice acceptor site is composed of two different *cis*-acting sequence elements that control splice choice, called the **IIIa repressor element** (**3RE**) and the **IIIa virus infection-dependent splicing enhancer** (**3VDE**). The 3RE serves as an assembly site for a hyperphosphorylated form of a host SR protein, which blocks the spliceosome from recognizing the downstream 3′ splice acceptor site for most of the adenovirus replication cycle (**Figure 8.42**). This repression of the downstream 3′ splice acceptor site is relieved very late during infection by the viral **E4-ORF4** protein, which causes dephosphorylation of SR proteins (**Figure 8.43**). The dephosphorylated SR proteins cannot bind to the 3RE and so use of the downstream 3′ splice acceptor site is derepressed. In this manner, accumulation of the E4-ORF4

Figure 8.41 The L1 pre-mRNA transcript with the two alternative mature mRNAs. The pre-mRNA has a leader sequence that is found in all mature late mRNA. The pre-mRNA can be spliced to produce mRNA that encodes L52,55K or IIIa. The L52,55K mRNA is produced using the upstream 3′ splice acceptor site. The IIIa mRNA is produced using the downstream 3′ splice acceptor site.

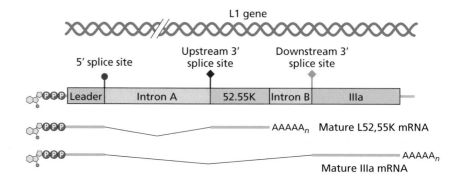

Figure 8.42 SR proteins interact with *cis*-acting sequences to determine splice site choice. The 3RE and 3VDE sequences in intron B are essential for regulating whether the mRNA will contain the 52,55K or IIIa sequences. Hyperphosphorylated SR proteins bind to the 3RE site, preventing the splicing factors from binding to the downstream 3VDE sequences in the 3′ splice acceptor site. Because the splicing factors cannot bind to 3VDE in intron B, the splicing machinery recognizes only the 3′ acceptor site at the end of intron A. As a result, the machinery selects the upstream 3′ splice acceptor site in intron A, ultimately resulting in mature 52,55K mRNA.

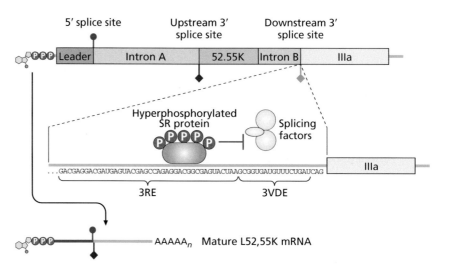

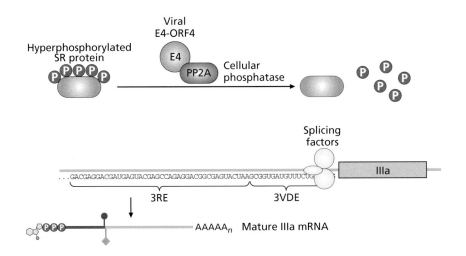

Figure 8.43 E4-ORF4 regulates the switch to IIIa production. The viral E4-ORF4 binds to the cellular PP2A phosphatase. The complex dephosphorylates the SR protein. The SR protein does not bind to 3RE unless it is phosphorylated. The result of dephosphorylating the SR protein is that splicing factors bind to the 3VDE sequences near the end of intron B. Because of the splicing factors, the splicing machinery skips over the upstream 3′ splice site and uses the downstream site instead, splicing out the entirety of intron A, the 52,55K coding sequence, and intron B. The result is mature IIIa mRNA.

delayed-early protein results in a switch from the production of L1 52,55K protein to production of IIIa.

The 3VDE *cis*-acting sequence also plays a role in splice site choice, and without it the downstream 3′ splice acceptor site needed to encode IIIa will not be selected (**Figure 8.44**). The 3VDE is composed of the IIIa branch site, a poor pyrimidine tract (which does not have as many pyrimidines as a constitutive splicing site), and the 3′ AG splice site. For other introns, the constitutive host splicing factor U2AF binds to the pyrimidine tract and subsequently stimulates splicing. In this case, the pyrimidine tract has so few pyrimidines that U2AF cannot bind to it. Instead, the viral protein **L4-33K** protein binds

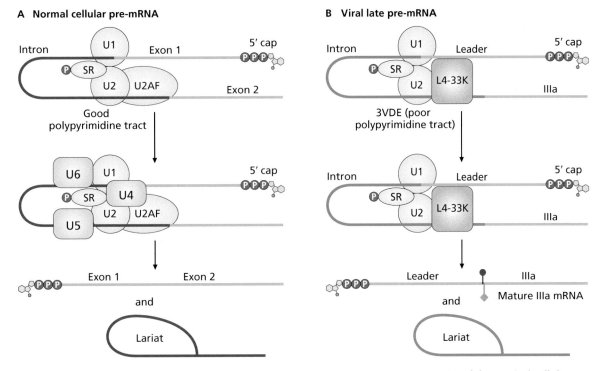

Figure 8.44 Role of 3VDE and L4-33K in alternative splicing of adenovirus late transcript. (A) In typical cellular mRNA, the cellular U2AF assembles on good polypyrimidine tracts to promote assembly of the spliceosome. (B) During viral infection, the U2AF cannot bind to the poor polypyrimidine tract in the 3VDE sequence. Instead the viral L4-33K protein binds to the 3VDE polypyrimidine tract, substitutes for U2AF, and promotes formation of the spliceosome. The result is production of IIIa mRNA. Poly(A) tails are omitted for simplicity.

to the weak pyrimidine tract in the viral downstream 3′ splice-acceptor site and thereby substitutes for U2AF. After this, the normal recruitment of U2 snRNP and all subsequent events in splicing can ensue. A combination of the relief of repression at 3RE and activation of splicing at the 3VDE is therefore required to correctly process mRNA encoding the late protein IIIa. This detailed example is probably similar to the splice-site choice reactions that take place throughout the adenovirus genome, in that the mechanism involves a combination of *cis*-acting sequence elements in the pre-mRNA, *trans*-acting proteins and RNA, and a combination of host proteins and virus proteins.

8.27 Adenovirus shuts off translation of host mRNA, while ensuring translation of its own late mRNAs through a ribosome-shunting mechanism

Once the adenovirus mRNAs are made, they must be exported from the nucleus and translated. Viral proteins translated in the cytoplasm then have to return to virus replication complexes in the nucleus, where the new viral

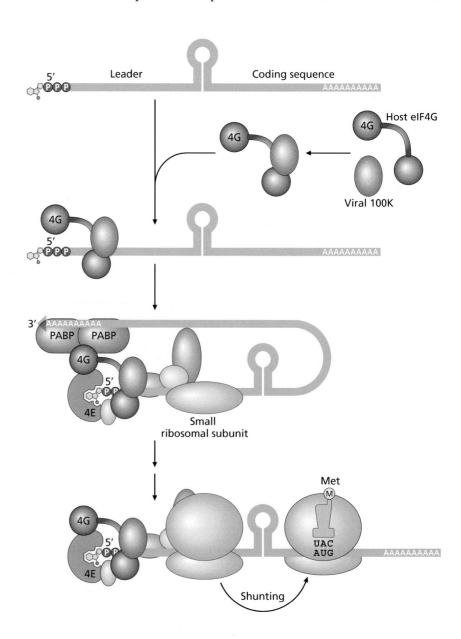

Figure 8.45 Ribosome shunting is used to translate late adenovirus mRNAs. The viral 100K protein binds to host eIF4G. It also binds to a leader mRNA sequence found in all late transcripts and thereby forms a complex with eIF4G and late viral mRNAs. Subsequently, the viral mRNA–100K–eIF4G complex recruits the small ribosomal subunit, other eukaryotic initiation factors, and the poly(A)-binding protein, which triggers assembly of a ribosome onto the viral mRNA. After assembly of the ribosome, secondary structure in the mRNA results in shunting in which the ribosome bypasses the 5′ end of the mRNA and instead initiates translation at an internal AUG found after the secondary structure.

genomes accumulate. Adenovirus shuts off translation of host mRNAs while forcing selective translation of its own messages, especially during late gene expression, when vast numbers of structural proteins are required. (To make 100,000 progeny requires >10^7 copies of several different capsomers.) To suppress the translation of host mRNA while simultaneously ensuring translation of its own late mRNA, the virus uses **ribosome shunting**. Ribosome shunting relies on the viral protein **100K** binding to host translation initiation factor eIF4G (**Figure 8.45**). When the 100K protein binds to translation initiation factor eIF4G, that factor cannot assemble to form a typical host 48S preinitiation complex (see **Chapter 5**). Because host mRNA requires the 48S preinitiation complex for translation, 100K prevents host mRNA translation. The 100K–eIF4G complex protein also binds to a leader mRNA sequence found in all late transcripts and thereby tethers eIF4G to late-viral mRNAs. Subsequently, the viral mRNA–100K–eIF4G complex recruits the small ribosomal subunit, additional eukaryotic initiation factors, and the poly(A)-binding protein, which triggers assembly of a ribosome onto the viral mRNA. Secondary structure in the mRNA results in shunting, a process in which the 5′ end of the mRNA is bypassed and the ribosome initiates translation at an internal AUG found after the extensive secondary structure. In this way, adenovirus selectively shuts off translation of host mRNAs while ensuring translation of its own late proteins, which are needed in tremendous abundance. It also shuts off translation of earlier viral mRNAs that lack the leader sequence, thus making the cell's amino acids available for the assembly of the extremely abundant structural proteins necessary for the assembly stage of the virus replication cycle.

8.28 DNA replication in adenovirus requires three viral proteins even though the genome is replicated in the host cell nucleus

Adenovirus late gene expression temporally coincides with viral genome replication, and late gene expression and DNA replication continue together until the host cell dies. DNA replication requires the expression of cellular S-phase genes, which is accomplished by viral manipulation of pRB and apoptosis, as occurs during infection by polyomavirus and papillomavirus. Unlike the smaller DNA viruses, however, the virus uses more viral proteins directly during DNA replication, and replication does not proceed using the normal cellular replication fork process.

Normal cellular replication forks take advantage of RNA primers synthesized by a cellular primase enzyme. These RNA primers must be removed from the DNA to complete chromosome replication, which causes the end-replication problem: the last RNA primer, once removed, leaves a single strand of DNA unreplicated (**Figure 8.46**). Cells solve this problem using telomeres, which are repeated sequences at the end of linear cellular chromosomes. Telomeres are synthesized by the enzyme telomerase rather than by DNA polymerase, so that telomerase maintains the length of chromosomes down through the generations. Telomerase is an enzyme made up of both protein and RNA components. The RNA provides a template and the protein component uses the RNA template in a reiterative process that adds many repeats of the telomere sequence to the ends of eukaryotic chromosomes. As a result, the coding sequence closest to the chromosome terminus is far away from the actual physical end of the chromosome and even if a small amount of DNA at the end of the chromosome does not get copied, daughter cells receive a complete copy of all coding sequences.

Figure 8.46 Telomerase solves the end-replication problem for eukaryotic cells. For clarity, only the template DNA (orange) and newly synthesized DNA (red) of the lagging strand are shown. To complete the replication of the lagging strand at the ends of a linear cellular chromosome, the template strand is first extended beyond the DNA that is to be copied (1). To achieve this, the enzyme telomerase adds more repeats to the telomere repeats at the 3′ end of the template strand (2). This allows cellular DNA synthesis to be initiated with the normal RNA primase-dependent process to synthesize new DNA (3). At the end of lagging strand synthesis, there is a short single-stranded region left at the 3′ end. Because telomerase adds many hundreds of copies of the telomere repeat sequence, the coding sequence for the nearest gene is far away from the end of the chromosome and is not disturbed by the single-stranded sequence at the 3′ end.

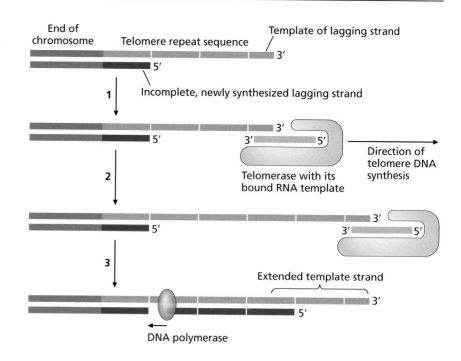

Adenoviruses have their own solution to the end-replication problem. Adenoviruses encode three viral proteins used directly during viral genome replication: a viral DNA polymerase (**Ad pol**), a single-strand binding protein (**E2AL**), and the **preterminal protein** (**pTP**), which is a precursor to the TP covalently attached to the 3′ ends of the genome. Use of pTP allows the virus to avoid the end-replication problem. Instead of relying upon RNA primers as host DNA polymerase does, Ad pol uses the pTP as a primer, which is similar to the use of VPg as a primer in picornaviruses (see **Chapter 5**).

The process of genome replication in adenovirus (**Figure 8.47**) begins with initiation, when pTP binds to one end of the genome and breaks hydrogen bonds to separate some of the base pairs. Next, the Ad pol binds to the pTP and forms a complex, and then catalyzes covalent addition of a C nucleotide to a certain serine side chain (OH group) in the protein. From there, the Ad pol–pTP complex displaces one of the strands and remains associated with the other strand, which serves as a template. Ad pol then extends the free 3′ OH attached to the pTP–C nucleotide. By copying the template, the other strand is completely displaced by these reactions and binds to a viral E2AL ssDNA-binding protein. At the end of copying the first strand, there is a new double-stranded molecule and one displaced nontemplate molecule bound to ssDNA-binding proteins. Use of pTP as a primer enables complete replication of a DNA genome with linear ends without the use of telomeres.

The inverted repeats at the end of the adenovirus genome cause the displaced adenovirus template molecule to circularize, creating a **panhandle** shape. This panhandle serves as an origin of replication so that another complex of the viral polymerase and pTP can bind and copy this strand as well. The net result is the creation of two dsDNA genomes with pTP covalently attached to the ends. The pTP is proteolytically processed during the assembly phase of the viral replication cycle, so that inside the virion it is the smaller TP that remains covalently attached to the 5′ ends of the genome.

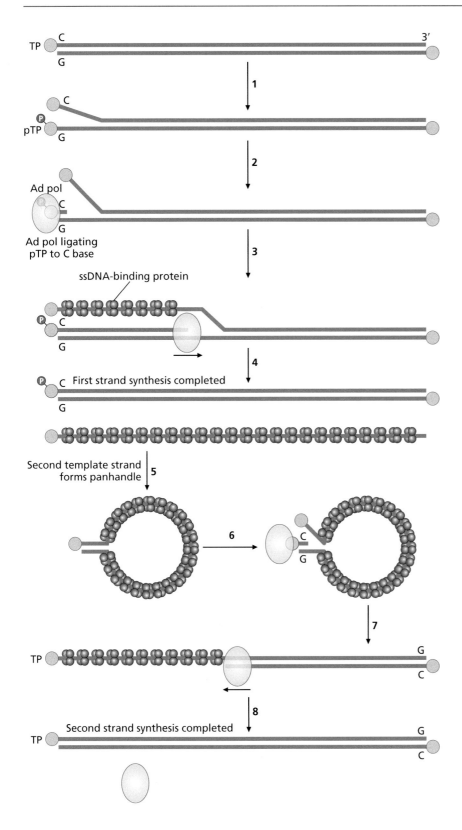

Figure 8.47 Ad pol and pTP are required for adenovirus genome replication. To initiate viral genome replication, pTP binds to one of the 3′ ends of the genome and breaks hydrogen bonds to separate some of the base pairs (1). The Ad pol binds to the pTP and forms a complex, and then catalyzes covalent addition of a C nucleotide to a certain serine side chain in the protein (2). Next, the Ad pol–pTP complex remains associated with the other strand, which serves as a template. Ad pol extends the free 3′ OH attached to the pTP–C nucleotide (3). At the end of copying the first strand, there is a new double-stranded molecule and one displaced nontemplate molecule bound to ssDNA-binding proteins (4). The displaced adenovirus template molecule then circularizes by virtue of inverted repeats at the end of the adenovirus genome, creating a panhandle shape (5). This panhandle serves as an origin of replication so that another complex of the viral polymerase and pTP can bind (6) and copy this strand as well. Because of the Ad pol activity, the template linearizes and the single-strand binding proteins are displaced as the enzyme copies the DNA (7). The net result is creation of two dsDNA genomes with pTP covalently attached to one strand (8).

8.29 Herpesviruses have very large enveloped virions and large linear dsDNA genomes

Herpesviruses such as **human herpesvirus 1** (subfamily *Alphaherpesvirinae*) have large spherical enveloped virions with several distinctive layers, including envelope spikes, an outer and inner **tegument**, and an inner icosahedral

A

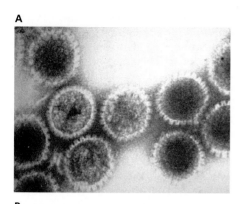

Figure 8.48 Herpes simplex virus 1. (A) Electron micrograph of herpes simplex virus 1. The virus is about 200 nm in diameter. (B) The herpes simplex virion. (B, Courtesy of Philippe Le Mercier, ViralZone, © SIB Swiss Institute of Bioinformatics.)

B

Envelope proteins — Outer tegument

Major capsid protein — Inner tegument

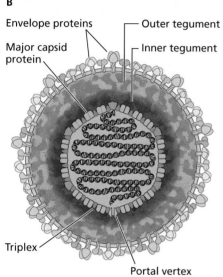

Triplex

Portal vertex

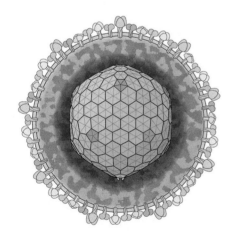

capsid (**Figure 8.48**). Until very recently, human herpesvirus 1 was known exclusively as **herpes simplex virus type 1** (HSV-1); because this is still the most common term used in publications, we will use the two names interchangeably. The large linear genome inside has 125–250 kbp of dsDNA (depending on the species of virus) with reiterated sequences such as indirect repeats and shorter **tandem repeats**, found at species-specific locations (**Figure 8.49**), with unique protein-coding sequences in between the repeats. The genome of the model HSV-1 is usually drawn with the **longer unique region** (**UL**) on the left and the **shorter unique region** (**US**) arm on the right. The UL region is flanked by one terminal repeat and one internal repeat. The US region is flanked by one internal repeat and one terminal repeat. The terminal repeats are called TR_L and TR_S, and the internal repeats are called IR_L and IR_S. Both the structure of the virion and the size of the genome account for the description of herpesviruses as large dsDNA viruses.

8.30 Lytic herpesvirus replication involves a cascade with several waves of gene expression

Herpesviruses have longer genomes than adenoviruses and even larger virions. Familiar herpesviruses include herpes simplex virus 1 (also known as human herpesvirus 1), varicella–zoster virus (causes chicken pox; also known as human herpesvirus 3), Epstein–Barr virus (causes mononucleosis, sometimes called kissing disease; also known as human herpesvirus 4), and Kaposi's sarcoma virus (also known as human herpesvirus 8). Although herpesviruses can cause latent infections (see **Chapter 13**), here we focus on lytic herpesvirus infections using HSV-1 as the model.

When the enveloped HSV-1 enters its host cell by fusion with the plasma membrane, the capsid is transported to a nuclear pore where viral DNA is released into the nucleus while the capsid remains in the cytoplasm. Cellular enzymes in the nucleus ligate the ends of the genome together, making circular DNA. As part of a host antiviral response, the host packages the DNA into heterochromatin. The virus has mechanisms to counteract this problem and change its genome into euchromatin during lytic infection.

HSV-1 transcription occurs in waves of virus gene expression. The first wave is called immediate-early, and herpes immediate-early genes promote the transcription of the next wave of gene expression, while also interfering with the host immune response. The next wave, called early gene expression, encodes proteins needed for replication of the viral genomes. DNA

Figure 8.49 Herpes simplex virus 1 genome. The herpes simplex virus 1 genome showing the long unique sequence (UL), the short unique sequence (US), the long and short terminal repeats (TR_L and TR_S, respectively), and the long and short internal repeats (IR_L and IR_S, respectively).

replication coincides with late gene expression; late genes encode the structural and assembly proteins that allow new virions to assemble and ultimately escape from the host cell by budding.

8.31   Groups of herpes simplex virus 1 proteins have functions relating to the timing of their expression

HSV-1 is known to encode at least 84 proteins and 16 regulatory miRNAs. The proteins are grouped into classes depending on the timing of their expression. The proteins produced first, from transcription of immediate-early genes, are called **α proteins**. The early proteins expressed next are classified as **β1** (**early-early**) and **β2** (**late-early**); these are ultimately followed by the late **γ1** (**leaky-late**) and **γ2** (**late-late**) proteins. There are various conventions for naming the proteins, and some HSV-1 proteins have more than one name; for simplicity, this section uses just one name for each protein discussed. As with adenovirus, groups of proteins are the primary focus.

The immediate-early α proteins are expressed first and are regulatory. Most have several functions and together they repress the host immune response and create a nuclear milieu favorable for viral gene expression and genome replication. The next two sets of proteins are the β1 and β2 groups. Most of these are proteins required for viral DNA replication, including enzymes to synthesize deoxribonucleotide triphosphates (dNTPs), a single-stranded DNA binding protein, and two DNA polymerases. The final two sets of proteins are the γ1 and γ2 proteins, which are mainly structural. An exception is the regulatory protein **VP16**, which is expressed late in order for it to be packaged into virions, where it plays no role. Instead, it plays a regulatory role in the next infection, when the infecting virion releases it into the cytoplasm. VP16 associates with host proteins and then is imported into the nucleus, where it stimulates early-early gene expression (**Figure 8.50**).

Another interesting protein carried by the virion that has immediate effects on the host cell when uncoating releases it into the cytoplasm is **VHS** (virion host shutoff). VHS is an RNase that selectively degrades host mRNA; the mechanism by which it distinguishes between host mRNA and viral mRNA is not well understood.

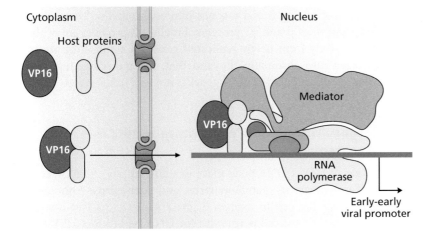

Figure 8.50 VP16 mechanism of activation. VP16 associates with cytoplasmic host proteins and then is imported into the nucleus, where it is responsible for activating early-early viral gene expression through interactions with host mediator.

8.32   Waves of gene expression in herpesviruses are controlled by transcription activation and chromatin remodeling

We will provide an overview of the regulation of HSV-1 gene expression. There is an ordered cascade of gene expression that includes waves before DNA replication and waves after the onset of DNA replication. As with other DNA viruses, early genes play regulatory roles, whereas late genes predominantly encode structural proteins. The α or immediate-early genes are expressed first as a result of the action of the VP16 activator protein that is carried along in the tegument part of the virion and enters the nucleus along with host transcription factors during uncoating. At first, the major α transactivating protein activates its own expression in a positive feedback loop, but soon it has the opposite effect and represses immediate-early gene expression. There are five major α proteins, including infected cell proteins **ICP0** and **ICP4**. Together the α proteins result in expression of the β1 and β2 waves of gene expression. ICP0 helps create virus replication compartments (VRCs) in the nucleus and also counteracts the host cell's attempts to silence viral genes by causing those host cell proteins to be degraded by a cellular machine called the proteasome (see **Chapter 14**). ICP4 is a major transcription regulator. Although it positively regulates some genes, it represses its own transcription. The β1 (early-early) and β2 (late-early) genes are turned on by the α proteins, and then the α and β genes together turn on the γ genes, which are also stimulated by DNA replication. Late gene expression can be reduced (leaky-late) or blocked entirely (late-late) by preventing genome replication. The γ proteins also turn off the expression of α and β genes.

Recent investigations suggest that the cascade of activation should be viewed instead in a more complex way, as a series of derepression and activation events that involve a central role for chromatin structure. When the naked HSV-1 DNA enters the nucleus, host histone proteins package the DNA in hypoacetylated nucleosomes that package the DNA into heterochromatin. This transcriptional silencing can be considered a host defense against herpesviruses. Both VP16 and ICP0 are important for counteracting this transcriptional silencing; VP16 activity affects the first waves of gene expression and ICP0 activity affects later waves of gene expression. For example, in the presence of ICP0, the histone deacetylases bound to the herpes genome nucleosomes are replaced by histone acetylases, which change the heterochromatin-silencing β and γ genes into euchromatin. These changes allow the expression of β and γ1 proteins. Finally, γ2 proteins are probably expressed exclusively from newly replicated genomes before they are completely packaged into chromatin. Beyond that, the molecular mechanism that accounts for this regulation of γ2 genes is not well understood.

8.33   Herpesvirus genome replication results in concatemers

The replication of HSV-1 DNA takes place in a different biochemical context from that of polyomavirus, papillomavirus, and adenovirus. Although the latter viruses force the cell to express S-phase genes as part of a strategy required to make the host cell permissive for DNA replication, HSV-1 does the opposite, even though its genome is also replicated in the nucleus. That is, the virus actively prevents the host cell from expressing S-phase genes, probably through the action of proteins encoded by some of the immediate-early genes. Therefore, HSV-1 encodes all the proteins needed for DNA replication because no analogous host proteins are available. These proteins include those that participate directly in replication as well as enzymes for dNTP synthesis. Although the virus does not force the cell to express S-phase genes, the virus does remodel the nucleus extensively, ultimately assembling intranuclear VRCs sometimes called factories.

This section focuses on the viral proteins needed to initiate genome replication and construct a functional replication fork. Keep in mind that prior to replication the herpesvirus genome was circularized by host proteins. The HSV-1 genome has an AT-rich *cis*-acting origin of replication that is absolutely required for genome replication. This origin binds to the viral **origin-binding protein** (**OBP**). Next, genome replication requires three viral proteins to form a complex that carries out the functions of both helicase and primase (which synthesizes RNA primers). Two different polypeptides encode the DNA polymerase, which is the same on both the leading and the lagging strands, and yet another protein encodes single-strand binding protein. Viral genome replication likely proceeds first through θ replication and secondarily through rolling-circle replication, ultimately filling infected cell nuclei with concatemers of viral DNA. Recently, concatemer formation has been associated with host DNA recombination proteins, and much work remains to be done to clarify the molecular mechanism of herpesvirus genome replication. In any case, the virus will ultimately process the concatemers into genome-length molecules during assembly.

8.34   Poxviruses are extremely large dsDNA viruses that replicate in the host cytoplasm

The poxviruses are the largest dsDNA viruses that infect human beings, whether evaluating the physical size of their virions or the length of their genomes. There are two well-known poxviruses that infect humans: **variola virus** and **vaccinia virus**. Variola virus caused the epidemic disease smallpox and was eradicated from natural circulation in 1979. Samples of the virus, guarded under lock and key, exist in the United States and Russia. Vaccinia virus is the name of the poxvirus used to immunize people to protect them against smallpox; it is so closely related to variola virus that it induces immunity against both variola virus and vaccinia virus, yet it does not cause disease in otherwise healthy humans. We study vaccinia virus as a model to understand how poxviruses complete their replication cycles.

Vaccinia virion structure is complex. Two different infectious vaccinia virions exist: the **intracellular mature virus** (**IMV**) and the **extracellular enveloped virus** (**EEV**) (**Figure 8.51**). The virions as examined with cryo-electron microscopy are approximately 250 nm × 270 nm × 336 nm, and in both forms they contain a **core**, a **core wall**, and **lateral bodies** surrounded by an **inner membrane**. The larger EEV form has yet another membranous **external envelope** with viral spikes. The large linear genome inside is composed of 194 kbp of dsDNA with inverted terminal repeats that form covalently closed hairpin termini (**Figure 8.52**).

After entering a host cell, the core of the vaccinia virion is released into the cytoplasm, where it will form a virus factory much like the VRCs discussed in **Chapters 5–7** on RNA viruses. Unlike all other groups of dsDNA viruses, vaccinia virus expresses its genome and replicates its DNA in the cytoplasm; this means that nuclear host proteins are not available for gene expression and genome replication. For example, host splicing machinery is confined to the nucleus and, because poxviruses do not encode splicing machinery, vaccinia mRNAs do not have introns. Poxvirus gene expression

Figure 8.51 Vaccinia virus. Transmission electron micrograph of virions, which are about 350 nm long.

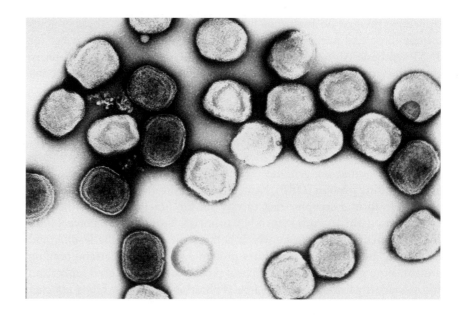

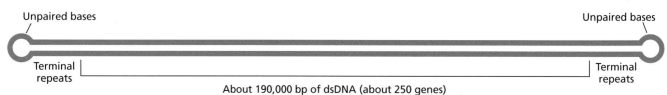

Unpaired bases

Unpaired bases

Terminal repeats

Terminal repeats

About 190,000 bp of dsDNA (about 250 genes)

Figure 8.52 Vaccinia virus genome. The genome is double-stranded DNA with covalently closed termini so that there are no free 5′ or 3′ ends. The genome thus resembles a collapsed circle.

occurs in waves, with early gene expression occurring until the core is completely disassembled and DNA replication has begun. This overlap blurs the distinction between the uncoating and gene expression stages of the virus replication cycle. Intermediate gene expression then ensues followed by late gene expression. Progeny virions assemble in dedicated cytoplasmic viral factories and are released by both lysis and budding.

8.35 Many vaccinia virus proteins are associated with the virion itself

Vaccinia virus encodes approximately 216 proteins that we will discuss in groups. They are expressed in early, intermediate, and late waves, where early transcription occurs before uncoating is complete and intermediate and late waves occur after uncoating and after the onset of DNA replication. One large group of proteins is those that have been detected in enveloped virions and that appear to be associated physically with the inner membrane or outer membrane; most of these are probably late proteins. The function of these 80 or so proteins is typically attachment, penetration, or assembly. Another 30 proteins are associated with the virion core. Many of these structural proteins are also enzymes. Poxviruses encode many enzymes because, thanks to their cytoplasmic replication cycle, they do not have access to nuclear host enzymes such as Pol II. Instead, poxviruses encode multiple subunits of an RNA polymerase. The virus also encodes all subunits needed for copying its genome, and some of these are also associated with the virus core. The nonenzymatic proteins associated with the virus core are most often required for viral assembly.

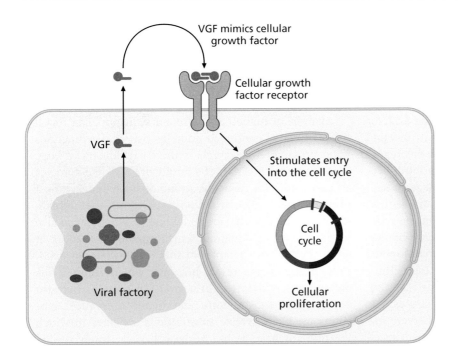

Figure 8.53 Vaccinia growth factor activity. VGF mimics cellular epidermal growth factor, so it binds to cellular epidermal growth factor receptors and triggers the cells to enter the cell cycle and proliferate.

Other vaccinia proteins include those that manipulate the host immune response for the virus's benefit or that prevent or subvert the apoptotic response to stress. A particularly well-understood example is **vaccinia growth factor** (**VGF**), which mimics cellular epidermal growth factor (**Figure 8.53**). The protein is encoded by an early gene and likely helps the virus counteract apoptotic signaling by sending a countervailing growth and proliferation signal to the host cells.

8.36   Vaccinia RNA polymerase transcribes genes in three waves using different transcription activators

The vaccinia RNA polymerase has eight subunits, each encoded by a different viral gene. Six of the viral RNA polymerase proteins are homologous to cellular proteins (**Technique Box 8.1**). The virus also encodes enzymes needed to synthesize the 5′ cap and to attach that cap to viral transcripts. Similarly, it encodes a polyadenylation enzyme that adds a normal poly(A) tail to viral transcripts. In this way, vaccinia produces mRNA that is very similar to host mRNA and so can be translated by host ribosomes.

TECHNIQUE BOX 8.1   BIOINFORMATICS

The field of **bioinformatics** has changed virology profoundly by making possible the analysis of protein and gene sequences, including entire genomes, using computational and mathematical methods that take sequences, apply algorithms or other mathematical manipulations to those sequences, and thereby suggest hypotheses regarding the function and evolutionary history of the sequences. Bioinformatics also increasingly includes the ability to compare three-dimensional models of protein structures. Proteins and genes are considered informational molecules because the order of subunits in their sequences provides information pertaining directly to their function and to their evolutionary history. As of 2023, more than 12 million virus genome segment sequences had been deposited in international databases, and the number of sequences is increasing exponentially over time.

One of the most fundamental goals of bioinformatics is to detect evolutionary relatedness and functional similarity through aligning multiple sequences. Most **alignment** algorithms seek to maximize the regions of similarity and minimize regions of difference. One biological assumption that supports this principle is that, in the absence of unusual mutagens, the chance that any given DNA or RNA nucleotide will be copied incorrectly during genome replication is very small and the chance of an insertion or deletion occurring would be even less likely. In **Figure 8.54**, for example, short sequences of Ebola viruses from 1999 and 2008 have been aligned, showing that most of this region is the same in the two viruses. The algorithm introduces insertions or deletions, called **indels**, wherever a gap is necessary to maximize the alignment. Note that an insertion in one sequence could instead be a deletion in the other sequence, so the term indel encompasses both possibilities.

Sequences that are similar are likely to be **homologous**, which means descended from a common ancestor. Thus, alignments can be particularly useful for determining how a newly isolated virus relates to known viruses. For example, during 2014 there was an outbreak of severe respiratory illness among children in the United States caused by a newly described virus. Its genome sequence revealed similarities to enteroviruses found in the *Picornaviridae* family; the virus is now known as enterovirus D68 (EV-D68). The respiratory epidemic coincided with a dramatic increase in acute flaccid myelitis among children, similar to the symptoms of polio, which is caused by a picornavirus. Because poliovirus is also in the *Picornaviridae* family, EV-D68 may cause acute flaccid myelitis in some children. The alternative to homology is convergent evolution, where two unrelated proteins or nucleic acids have similar sequences despite a lack of homology. Convergent evolution of informational molecules is extremely rare and almost never accounts for observed sequence similarities.

Bioinformatics can also be used to predict functions for nucleic acids and proteins. For example, given the sequence of all promoters used by bacteriophage T7 RNA polymerase, mathematical analysis can be used to determine the consensus binding site for that enzyme. Wet lab experiments can then be performed to determine how variations from this consensus affect the function of the promoter. Similarly, analysis of all of the RNA sequences that serve as an IRES for different picornaviruses can predict conserved stem–loop structures critical for ribosome binding.

Analysis of homologous protein sequences can reveal the presence of conserved domains that confer functions upon proteins. An example is the DnaJ domain of large T antigen, which is part of a family of DnaJ-like domains that all regulate a particular group of stress-induced host proteins. The particular amino acids within that domain that might be essential for its folding or function can also be predicted by finding out which ones are conserved in all of the known examples of that protein (**Figure 8.55**). Protein alignments reveal not only which amino acids are completely identical in a group of proteins but also conservative substitutions and semiconservative substitutions. Conservative substitutions are ones in which the proteins have amino acids with very closely related side chains, such as leucine and methionine, and semiconservative substitutions are those in which some aspects of the amino acids' side chains are similar, as is the case for serine and threonine. Identical and conserved amino acids are usually essential for a protein to fold properly or play a specific role in the function of that protein, for example by interacting with a substrate during catalysis. Knowledge of protein sequences that are conserved across many viral isolates can be used in combination with structural biology to find regions of viral proteins that are both conserved and likely to provoke a strong immune response because they are found on the surface of a virus. In turn, this knowledge can be applied to design of new vaccines.

Bioinformatic tools can also be used to trace the evolution of viruses. The same large T antigen alignment data as in **Figure 8.55** can be displayed as a tree in which the branch length separating two sequences is proportional to the amount of difference between the two sequences

```
Genome 1   GAACAACGCAU--UACGAGUCUUGAGAAUGGUCUAAAGCCAGUUUAUGAUAUGGCAAAAA
           ||||||||||  || || ||| |||||||||  ||||||||| | ||||| ||||| |||
Genome 2   GAACAACGCAUCAUA-GA-UCUAGAGAAUGGCUUAAAGCCAAUGUAUGACAUGGCUAAAG
```

Figure 8.54 RNA sequence alignment for a region of two Ebola virus genomes. Vertical lines indicate identical nucleotides and dashes indicate spaces that the computer algorithm introduced in an indel to maximize the aligned sequences.

```
Protein 1   MDKVLNREESMELMDLLGLDRSAWGNIPVMRKAYLKKCKELHPDKGGDEDKMKRMNFLYK
Protein 2   MDHLLTREESIRLMQLLELPMDEFGNFNAMRSQFHKQIKKMHPDKGGNPEQAKELISLYK
Protein 3   MDYTLTREESKLLMELLGLPMEQYGNFPLMRKAFLQKCKIMHPDKGGDEQTAKMLISLYK
Protein 4   MDHTLTREESKLLMELLGLPMEQYGNFPLMRKAFLQKCKIMHPDKGGDEQAAKMLISLYK
            **  *.****  **:** *   . :**: **. : :: * :******: :  *  : ***
```

Figure 8.55 Alignment of homologous protein sequences. Protein sequence alignment for the N-terminal region of large T antigen from four different viruses. Below the amino acid sequences, asterisks indicate conservation and (:) indicates the presence of conservative substitutions. The (.) symbol indicates a semiconservative substitution.

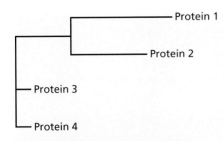

Figure 8.56 Alignment of large T antigen sequences can be used to calculate a tree. The tree shows the most likely evolutionary history of the sequences. The shorter the branches connecting any two sequences, the more they resemble one another, and the more likely that they share a more recent common ancestor with each other than with the other sequences displayed in the tree.

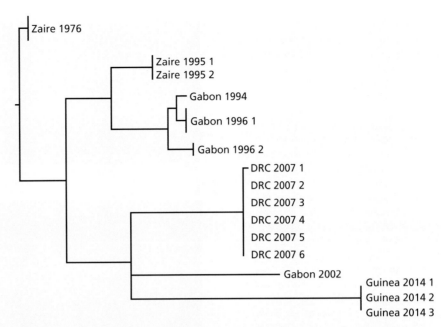

Figure 8.57 Most likely evolutionary history of Ebola virus strains isolated from the West African Ebola fever epidemic of 2014. Ebola viruses were isolated from patients in outbreaks in different years and preserved. Their genome sequences were determined and aligned. The resulting phylogenetic tree reflects their evolutionary history and indicates that strains from the 2014 West African outbreak are related to strains from Central Africa even though Central Africa is thousands of miles away. (DRC, Democratic Republic of the Congo.)

(**Figure 8.56**). In this example, proteins 3 and 4 are most similar, so the viruses encoding proteins 3 and 4 are probably descended from a recent common ancestor. Proteins 1 and 2 are more distantly related to 3 and 4. They are still homologous to proteins 3 and 4 but have a more distant common ancestor. A second example is the use of whole genome sequencing of Ebola viruses isolated from different patients during the 2014–2016 Ebola fever epidemic in West Africa (**Figure 8.57**). Sequence analysis showed that viruses from patients in Guinea in 2014 are most similar to viruses from previous epidemics in Gabon and the Democratic Republic of the Congo (DRC). The viruses from the 1976 outbreak in Zaire (now known as the DRC) are more distant relatives. This evolutionary history is interesting because Guinea and the DRC are 3,000 miles (5,500 km) apart, suggesting that the animal reservoirs for Ebola virus are geographically widespread.

The activity of the viral RNA polymerase is regulated by viral proteins in a cascade that results in early, immediate-early, and late genes. Each wave of gene expression is activated by its own dedicated transcription factors. An example is the **viral early transcription factor** (**VETF**) protein, which is made up of two viral polypeptides. This protein interacts with the viral RAP94 protein and RNA polymerase to direct transcription to initiate at early promoters that have a distinctive sequence compared with intermediate and late promoters. Similarly, there are **viral intermediate transcription factors** (**VITFs**) and **viral late transcription factors** (**VLTFs**) that target viral RNA polymerase to recognize distinctive intermediate and late promoters, respectively.

Another layer of regulation for intermediate and late promoters is suggested by the following observations. First, DNA replication is required for intermediate and late transcription; inhibiting viral DNA replication prevents intermediate and late transcription. Second, intact virus cores will not

Figure 8.58 Host translation factors localize to vaccinia virus replication factories. In both (A) and (B), uninfected cells (Un) are on the left, and the other two panels show infected cells (In). The cells were stained to detect a translation factor (green) or DNA (blue). The two middle and right-hand panels show an infected cell, stained to see DNA or both DNA and the translation factor, respectively. The nuclei have been labeled (N), as have the factories (F). Infected cells have cytoplasmic viral DNA in the virus factories, visible as large masses of blue in the cytoplasm. (A) Green fluorescence shows the translation factor eIF4G, which is cytoplasmic in an uninfected cell (on the left) and concentrated in a viral factory in an infected cell (on the right). (B) Green fluorescence shows a different translation factor, eIF4E, which also localizes to the viral factory in infected cells. (From Katsafanas GC & Moss B 2007. *Cell Host Microbe* 2:221–228. With permission from Elsevier.)

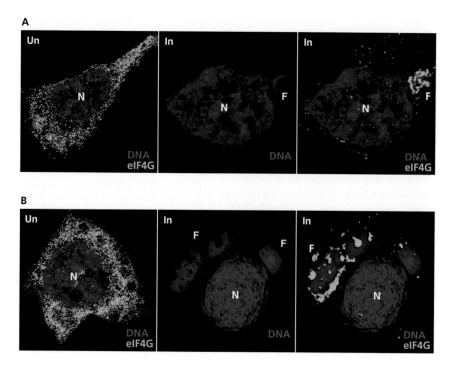

express intermediate or late genes. DNA extracted from virus particles can serve as a template for intermediate and late transcription, however. These observations suggest that, in the absence of DNA replication, some of the DNA is in a form that is not physically accessible to RNA polymerase and VITF and VLTF proteins. DNA replication likely makes the sites accessible by relaxing the DNA as a consequence of using helicase to expose the template nucleotides. It is possible that intermediate and late transcription can occur only on newly replicated templates that have not yet participated in any assembly processes. One unresolved challenge to this model is that intermediate and late genes are not clustered in the genome, raising the question of how they could be physically occluded without also affecting early genes.

As previously described, host translation initiation requires assembly of a complex of eIFs on the mRNA using both the polyadenylated tail and the 5′ methylated cap (see **Chapter 5**). For example, PABP binds to the poly(A) tail and to eIF4G, and these interactions are required for translation initiation. Vaccinia virus recruits most of the cellular eIF4G protein to its factories to such an extent that there is almost no eIF4G left in the cytoplasm for host translation to occur. This phenomenon can be demonstrated by fluorescence microscopy (**Figure 8.58**). It is likely that there are other examples of host translation factors recruited to poxvirus replication factories.

8.37 Vaccinia genome replication requires the unusual ends of the genome sequence

Vaccinia has a tremendous capacity for genome replication. A single infecting virion can give rise to 10,000 copies of the genome and more than 5,000 progeny virions; although this number of progeny may seem small compared with adenovirus, it is remarkable because of the much larger poxvirus genome and virion. Because poxviruses replicate their genomes in cytoplasmic factories and not in the nucleus, they must encode most proteins needed for DNA replication. Some host factors, such as ligase, do participate in pox-virus replication, however.

The leading model for the replication of vaccinia virus DNA is called self-priming with the repeated sequences at the end of the genome being key (**Figure 8.59**). The covalently closed termini each have a long-inverted repeat. Because the genome ends are covalently closed, there are no free 3′ OH groups until a nick occurs at one end of the genome inside one of the terminal inverted repeats. This free 3′ OH is opposite a template strand because

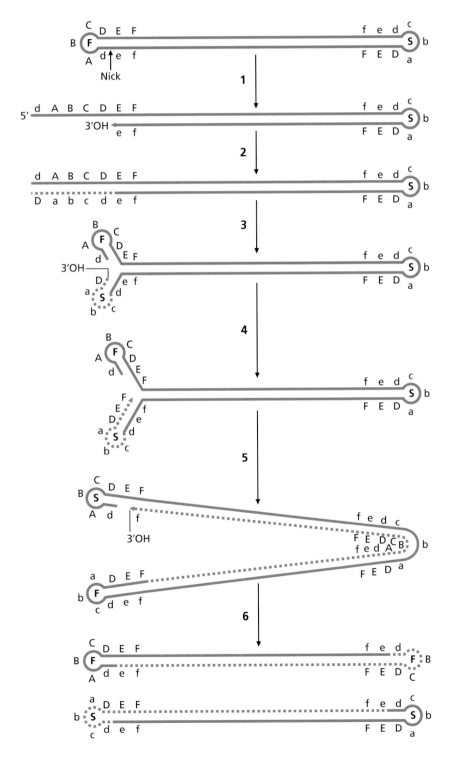

Figure 8.59 The self-priming model of vaccinia virus genome replication. The genome is a covalently closed hairpin. The termini have been exaggerated to show unpaired sequences in the termini (A, B, C) and the base-paired sequences in the inverted repeat (D/d, E/e, F/f, where D is complementary to d, and so on). The hairpins are designated F and S to indicate that replication occurs first at the fast hairpin (F) and later at the slow hairpin (S). A viral protein nicks the genome between the d and e sequences near the F hairpin, creating a free 3′ OH. When the hairpin F unfolds, the result is a 3′ OH that is opposite the D template sequence (1). The 3′ OH created by the nick is extended by viral DNA polymerase, copying template sequences D, C, B, A, and d (2). Because of this copying, the hairpin is destroyed and the DNA at the fast (F) end is double stranded. Next, the viral DNA spontaneously folds to form two hairpins, corresponding to the original F and to a new one, composed of newly synthesized DNA and designated S (3). The S hairpin has a free 3′ OH opposite a template strand. The next step is to extend that free 3′ OH by copying the e and f templates (4) and then the rest of the genome (5). The resulting concatemer is resolved into two individual genomes during a recombination event that occurs during assembly (6).

of the way that the virus genome folds on itself owing to its repeated terminal sequences. Replication starts by extending the 3′ OH through copying the opposite template strand and ultimately results in a concatemer that will be resolved during assembly.

8.38   The synthetic demands on the host cell make vaccinia a possible anticancer treatment

Because poxviruses are very efficient at forcing infected cells to produce thousands of large progeny virions that each contain 250,000 bp of DNA, they may be exploited to kill cancer cells. Vaccinia viruses are oncolytic, which means that they preferentially infect and kill cancer cells instead of normal cells. The molecular explanation for their preference for cancer cells as hosts is most likely related to the high demands poxviruses make on the cell's nucleic acid synthesis capacities. Unlike normal cells, tumor cells often have enhanced synthesis of dNTPs to support their more rapid proliferation compared with their noncancerous counterparts. To infect normal cells, the virus encodes a viral thymidine kinase enzyme that increases the pool of dNTPs in the cytoplasm. Genetic mutations that knock out the viral thymidine kinase gene make the virus more selective for tumors over normal cells. Only tumor cells, not normal cells, can compensate for the lack of this enzyme because tumor cells have overactive DNA replication and mitotic proteins. This trait of tumor cells permits viral replication in the absence of viral thymidine kinase. Vaccinia likely destroys tumors by direct infection and lysis of the cancer cells and the blood vessels that feed them, by induction of an immunological response against the tumor, or by some combination of these mechanisms. Certain picornaviruses and herpesviruses are also in clinical trials having their oncolytic capabilities tested.

Essential concepts

- The outcome of infection by a dsDNA virus might be a productive infection, host cell transformation, or latent infection.
- Most Baltimore Class I animal viruses express their genomes and replicate in the nucleus, relying on the host to provide the necessary enzymes.
- Class I viruses have to find ways to replicate their DNA even though eukaryotic DNA replication proteins are available only for a short time during S phase; to enable viral DNA replication using some host proteins, Class I viruses that replicate in the nucleus force their terminally differentiated host cells to re-enter the cell cycle.
- Polyomaviruses and papillomaviruses have small circular genomes with a high coding density of overlapping genes that encode a high number of proteins, many of which manipulate the host cell cycle and are necessary for regulated gene expression and genome replications.
- Class I viruses that express their genomes in the nucleus use alternative splicing to encode multiple proteins. In contrast, poxviruses do not use alternative splicing because they replicate in the cytoplasm, where host splicing factors are not available.
- Adenoviruses have much larger genomes than the polyomaviruses and papillomaviruses, but they still use splicing to encode a high density of proteins per nucleotide. They use alternative splicing to encode groups of proteins related by having some domains in common. Splicing of pre-mRNA in eukaryotes was first discovered through studies of adenovirus.

- Herpesviruses have even larger genomes than adenoviruses and thus overlapping coding sequences are less common. They are a model for eukaryotic viral gene expression that occurs in waves, similar to those found among the Class I bacteriophages, except that they use an ordered cascade of eukaryotic chromatin remodeling factors and transcription factors to cause coordinated waves of gene expression.
- Herpesvirus genome replication is similar to Class I bacteriophage genome replication because it uses circularized templates and results in concatemers that must be processed during assembly.
- Poxviruses are unusual in many respects, most especially because they have dsDNA genomes and they replicate in the cytoplasm instead of the nucleus. Replicating in the cytoplasm necessitates virally encoded RNA polymerase, transcription factors, and DNA replication proteins, likely contributing to the large poxvirus genomes. Despite the need to encode their own transcription factors, poxviruses are similar to herpesviruses in that gene expression occurs in waves.

Questions

1. Older classification schemes group the polyomaviruses and papillomaviruses together into a single family known as papovaviruses. Now, however, the two types of viruses are in separate families and are not even assigned to the same order. What are some plausible reasons that virologists once thought that polyomaviruses and papillomaviruses were close relatives? Similarly, what types of evidence most likely led to retiring the idea that papovaviruses are a single family?

2. In **Chapter 13**, we will examine more explicitly how some viruses cause cancer. Considering how some of the viruses in this chapter use host proteins during lytic infection, make some reasonable guesses about how polyomaviruses and papillomavirus infections could result in over-proliferation of host cells.

3. At least one poxvirus was recently discovered to manipulate the host cell cycle, forcing the otherwise terminally differentiated host cells into an S-phase-like state. Why was this discovery surprising, and what research questions would you pursue in order to find out how the poxvirus manipulates the cell cycle and what benefit this provides to the virus?

4. There are biotech companies that create new eukaryotic cell lines by expressing large T antigen in the cells. What effects could expression of large T antigen have on otherwise differentiated human cells, and what host proteins are likely involved in those effects? Construct a table that makes it easy to compare and contrast the polyomavirus, papillomavirus, and adenovirus proteins that manipulate the host cell cycle and apoptosis. Explain your reasoning.

5. Both herpes simplex virus 1 and human papillomavirus infect cells that are part of a stratified epithelium. Design an experiment to determine whether the waves of HSV-1 gene expression are coordinated with the differentiation status of human skin cells. What data would you collect and why?

6. If you were to genetically engineer host cells to express the adenovirus E4-ORF4 constitutively, predict how that modification would affect the adenovirus replication cycle. Address each stage and determine whether any stages would be disrupted. Explain the molecular basis for your prediction.

7. Viruses with larger dsDNA genomes encode proteins with more functions than smaller genomes do. Provide some examples of functions associated with the larger genomes that are missing from the smaller ones.

8. One of the first viral proteins expressed by RNA viruses is RNA-dependent RNA polymerase. Do you see any trends in the first viral proteins expressed by dsDNA viruses?

9. For any virus, if a protein is produced early during an infection and in relatively small amounts, what functions might that protein have? Provide one example from a eukaryotic virus with a single-stranded genome and one from a virus with a double-stranded genome.

10. For any virus, if a protein is produced late during an infection and in relatively abundant amounts, what functions might that protein have? Provide one example from a eukaryotic virus with an RNA genome and one from a virus with a DNA genome.

11. Animal chromosomes require telomerase for genome replication, yet none of the viruses in this chapter requires telomerase. Explain.

Interactive quiz questions

In addition to the questions provided above, this edition has a range of free interactive quiz questions for students to further test their understanding of the chapter material. To access these online questions, please visit the book's website: www.routledge.com/cw/lostroh.

Further reading

Polyomaviruses

Dynan W & Tijan R 1983. The promoter-specific transcription factor Sp1 binds to upstream sequences in the SV40 early promoter. *Cell* 35:79–87.

Gruss P, Dhar R & Khoury G 1981. Simian virus 40 tandem repeated sequences as an element of the early promoter. *Proc Natl Acad Sci USA* 78:943–947.

Keller JM & Alwine JC 1984. Activation of the SV40 late promoter: Direct effects of T antigen in the absence of viral DNA replication. *Cell* 36:381–389.

Waga S & Stillman B 1994. Anatomy of a DNA replication fork revealed by reconstitution of SV40 DNA replication *in vitro*. *Nature* 369:207–212.

White MK, Safak M & Khalili K 2009. Regulation of gene expression in primate polyomaviruses. *J Virol* 83:10846–10856.

Wiley SR, Kraus RJ & Zuo F 1993. SV40 early-to-late switch involves titration of cellular transcriptional repressors. *Genes Dev* 7:2206–2219.

Papillomaviruses

Doorbar J 2006. Molecular biology of human papillomavirus infection and cervical cancer. *Clin Sci* 110:525–541.

Spalholz BA, Yang YC & Howley PM 1985. Transactivation of a bovine papillomavirus transcriptional regulatory element by the E2 gene product. *Cell* 42:183–191.

Zheng ZM & Baker CC 2006. Papillomavirus genome structure, expression, and post-transcriptional regulation. *Front Biosci* 11:2286–2302.

Adenoviruseses

Ali H, LeRoy G, Bridge G & Flint SJ 2007. The adenoviral L4 33 kDa protein binds to intragenic sequences of the major late promoter required for late phase-specific stimulation of transcription. *J Virol* 81:1327–1338.

Berk AJ 2005. Recent lessons in gene expression, cell cycle control, and cell biology from adenovirus. *Oncogene* 24:7673–7685.

Jones N & Shenk T 1979. An adenovirus type 5 early gene function regulates expression of other early viral genes. *Proc Natl Acad Sci USA* 76:3665–3669.

Tribouley C, Lutz P, Staub A & Kedinger C 1994. The product of the adenovirus intermediate gene IVa is a transcriptional activator of the major late promoter. *J Virol* 68:4450–4457.

Whyte P, Buchkovich KJ & Horowitz JM 1988. Association between an oncogene and an anti-oncogene: The adenovirus E1A proteins bind to the retinoblastoma gene product. *Nature* 334:124–129.

Virologist Phillip Sharp's Nobel lecture on RNA splicing. http://www.nobelprize.org/nobel_ prizes/medicine/laureates/1993/sharp-lecture.html

Herpesviruses

Taylor TJ & Knipe DM 2004. Proteomics of herpes simplex virus replication compartments: Association of cellular DNA replication, repair, recombination and chromatin remodeling protein with ICP8. *J Virol* 78:5856–5886.

Wysocka J & Herr W 2003. The herpes simplex virus VP16-induced complex: The makings of a regulatory switch. *Trends Biochem Sci* 28:294–304.

Poxviruses

Broyles SS 2003. Vaccinia virus transcription. *J Gen Virol* 84:2293–2303.

Carrell JL 2004. *The Speckled Monster: A Historical Tale of Battling Smallpox*. Plume Books.

de Souza FG, Abrahão JS, & Rodrigues RA 2021. Comparative analysis of transcriptional regulation patterns: Understanding the gene expression profile in *Nucleocytoviricota*. *Pathogens* 10:935.

Fenn EA 2001. *Pox Americana: The Great Smallpox Epidemic of 1775–82*. Hill & Wang.

Henderson DA & Preston R 2009. *Smallpox: The Death of a Disease—The Inside Story of Eradicating a Worldwide Killer*. Prometheus Books.

Passurelli AL, Kovacs GR & Moss B 1996. Transcription of a vaccinia virus late promoter: Requirement for the product of the A2L intermediate stage gene. *J Virol* 70:4444–4450.

Rosales R, Sutter G & Moss B 1994. A cellular factor is required for transcription of vaccinia viral intermediate stage genes. *Proc Natl Acad Sci USA* 91:3794–3798.

Gene Expression and Genome Replication in the Single-Stranded DNA Viruses

9

Virus	Characteristics
Porcine circovirus 2	Model for circoviruses, which have circular ssDNA genomes. Encodes four proteins, despite having fewer base pairs than an average human gene.
Minute virus of mice	Model for parvoviruses, which have linear ssDNA genomes with hairpin termini. Uses alternative splicing to encode several proteins and a rolling hairpin mechanism for genome replication.

The viruses you will meet in this chapter and the concepts they illustrate

The Class II viruses of animals, such as circovirus and parvovirus, have single-stranded DNA genomes. So far, circoviruses and parvoviruses have only rarely been associated with human disease. Although circoviruses can sometimes be detected in humans, it is not clear that they cause disease symptoms, especially in otherwise healthy people. Circoviruses are important veterinary pathogens, however, infecting agriculturally important species such as pigs and chickens. One parvovirus (B19) has been associated with an infection called fifth disease, so named because it is the fifth common cause of infections that cause rashes in young children. But again parvoviruses are associated with much more serious veterinary disease, this time in the common companion animal, dogs.

Thus, the main motivation for studying circoviruses and parvoviruses is not to understand their pathogenesis but rather to learn about the molecular and cellular biology of their replication cycles. Both circoviruses and parvoviruses produce mRNA and replicate their genomes in the nucleus. Their single-stranded genomes must be used as templates to synthesize double-stranded DNA before the host transcription machinery can use them to synthesize viral mRNA because host RNA polymerase requires a double-stranded template. These viruses are infamous for being reproductively successful despite encoding a very small number of proteins. In this chapter, we focus on the Class II viruses that rely on **autonomous virus replication**, without help from another virus; there are other Class II **satellite viruses**, such as adeno-associated virus, that can complete their replication cycles exclusively in host cells that are co-infected with another virus that provides proteins essential for the replication of both viruses (see Chapter 15).

DOI: 10.1201/9781003463115-9

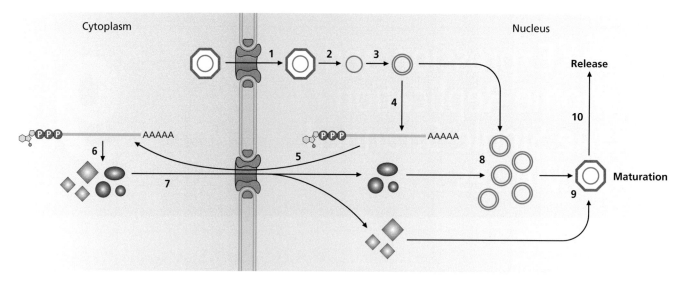

Figure 9.1 An ssDNA virus replication cycle. The replication cycles of circoviruses and parvoviruses are similar, although the circoviruses have circular genomes and the parvoviruses have linear genomes. The entire virion enters the nucleus (1), where uncoating releases the genome (2). Host enzymes synthesize complementary DNA so that the genome is now double stranded (3). Host enzymes synthesize mRNA (4), which is exported to the cytoplasm (5) and translated (6). The viral proteins enter the nucleus (7), where they are needed for genome replication (8) and assembly (9). Virions are released from the cell (10).

9.1 The ssDNA viruses express their genes and replicate their genomes in the nucleus

Gene expression and genome replication among the ssDNA viruses have many steps in common (**Figure 9.1**). The viruses enter through the plasma membrane but, during the penetration stage of the replication cycle, the virions have to be conveyed to the nucleus using the cytoskeleton. The genome is released into the nucleus as a result of uncoating. Cellular proteins convert the ssDNA into dsDNA, which is subsequently used by normal host machinery for transcription; viral proteins may also be part of this process. The genomes encode fewer than six proteins, which, although a bare minimum, are able to trigger all of the stages of the virus reproduction cycle. Viral mRNAs, which have typical eukaryotic 5′ caps and 3′ poly(A) tails because they are synthesized by host Pol II, are shuttled into the cytoplasm, where they are translated, and the viral proteins then enter the nucleus to participate in replication and other processes, such as assembly. After genome replication the dsDNA products can be used either for additional transcription or, late in infection, as a template to synthesize new ssDNA genomes that can be packaged into new virions.

9.2 Circoviruses are tiny ssDNA viruses with circular genomes

Circoviruses have the smallest genomes of any viruses infecting vertebrates and are therefore models for just how reduced a virus genome can be and yet still cause a productive infection. Circoviruses such as **porcine circovirus 2** (**PCV2**; genus *Circovirus*) have a small icosahedral capsid (**Figure 9.2**). Each capsid contains a single copy of the circular, single-stranded DNA genome, which is approximately 2 kb long (**Figure 9.3**). The genome is ambisense, meaning that different regions of the chromosome encode proteins on opposite strands (once the second strand of DNA has been synthesized in the

Figure 9.2 Electron micrograph of two different circoviruses. The larger virions (white arrows) are 26 nm in diameter and the smaller virions (black arrows) are 22 nm in diameter. (From Todd D 2000. *Avian Pathol* 29:373–394.)

host cell). The genome includes a short intergenic region with a conserved sequence that forms a stem–loop structure; the secondary structure serves as an origin for synthesizing dsDNA from the ssDNA genome. There are also transcription regulatory sequences once host enzymes have made the genome double stranded.

9.3 Although their genomes are shorter than an average human gene, circoviruses encode at least four proteins

The entire circovirus genome is smaller than the average size of a single human gene. Nevertheless, remarkably, the genome encodes at least four proteins. Three of the proteins are essential for completing the virus replication cycle in cultured cells and a fourth is involved in manipulating the host immune response. The three that are directly required for the replication cycle are the **Cap** protein, which forms the capsid and is therefore needed for attachment, penetration, uncoating, assembly, and release, and the **Rep** and **Rep′** proteins, which are needed for genome replication. The fourth, **ORF3**, is a nonstructural protein associated with manipulation of the host immune system and is dispensable in tissue culture (**Figure 9.4**).

Circoviruses produce one mRNA for each of their four known proteins as well as an additional seven more detectable RNA species, but so far evidence that these seven encode functional proteins is lacking. For example, proteins encoded by those mRNAs have not been identified in infected cells, and mutations that selectively destroy the coding capacity of these mRNAs do not affect replication in cultured cells.

How are the four known proteins encoded by this tiny genome? First, only double-stranded versions of the genome, which are also supercoiled (associated with nucleosomes) and are referred to as **closed supercoiled circles** (**cccs**), can serve as templates for transcription; these cccs can be considered the replicative form. The ccc has three promoters. Two of them flank the stem–loop structure and proceed in opposite directions around the genome away from the stem–loop so that the *cap* promoter points counterclockwise and the *rep* promoter points clockwise. The third is internal to the *rep* coding sequence and points counterclockwise (opposite the direction of the *rep* promoter). The two major promoters for *cap* and *rep* are understood in the most detail. Each has binding sites for common transcription factors found ubiquitously in host cells, so transcription proceeds as soon as the ccc is formed. The Rep′ protein is produced from an alternatively spliced mRNA, using the same start and polyadenylation signals as the Rep protein.

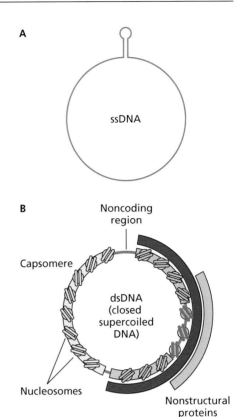

Figure 9.3 Porcine circovirus 2 genome. (A) The circovirus genome is single-stranded DNA with a hairpin. (B) The host converts the circovirus genome into double-stranded DNA complexed with nucleosomes.

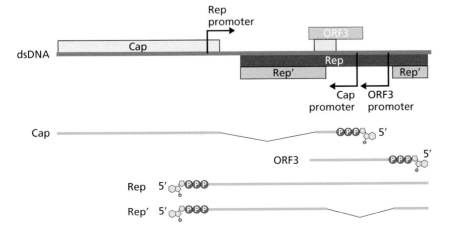

Figure 9.4 PCV2 promoters and coding sequences. Although the viral genome is circular, here the genome is depicted as linear to make the genes and promoters clearer. Rep and Rep′ are encoded from left to right, and Cap and ORF3 are encoded from right to left. Rep′ arises from the Rep gene through alternative splicing. The Cap pre-mRNA is also spliced. All of the mRNAs have a typical eukaryotic 5′ cap and poly(A) tail. Splicing is indicated by black lines connecting exons.

9.4   Both host and viral proteins are needed for circovirus genome replication

Circoviruses require that their host cells be in S phase in order to provide the virus with DNA polymerase. This requirement applies not only to genome replication but also to transcription because, for transcription to occur, the infecting genome must be converted into its transcriptionally active double-stranded ccc form. Although it was once thought that the virus passively waits for its host to enter S phase, new evidence suggests that nonstructural proteins may manipulate the host cell cycle and apoptotic pathways. Although it is clear that host DNA polymerase must produce the ccc, details of this process remain unknown. For example, host DNA polymerase requires a primer; the exact origin of this primer is still in doubt.

Once the ccc has formed, it can be used as a template for transcription or replication; here the focus is on replication (**Figure 9.5**). During replication, one strand of the ccc is designated the (+) strand and the other is designated (−) (these designations have nothing to do with transcription or mRNA). Replication proceeds through a rolling-circle mechanism (see **Chapter 4**). In the first step, a dimer of Rep and Rep′ binds to the ccc and nicks it, resulting in a free 3′ OH that can be extended by host DNA polymerase. The Rep and Rep′ proteins remain covalently bound to the original (+) strand while the host polymerase continues around the (−) template, creating a new (+) sense DNA identical to the genome found in the capsid. After the polymerase completes a full round of replication, the Rep and Rep′ proteins process the (+) strand that peeled away, creating a new genome that can be packaged into new virions or converted into a ccc by host proteins. Meanwhile, the Rep–Rep′ complex attacks the newly synthesized (+) strand and nicks it, so

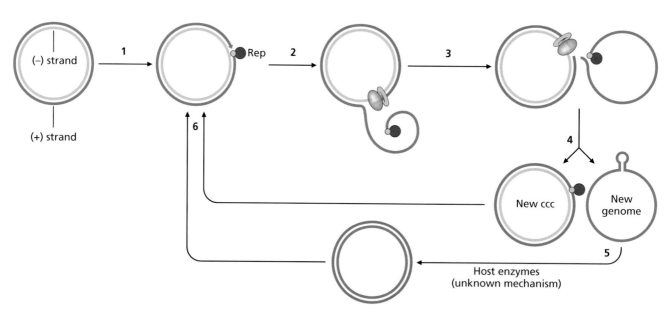

Figure 9.5 Circovirus genome replication. The ccc is a substrate for replication; it has one (+) strand and one (−) strand. Viral Rep and Rep′ proteins bind to an origin of replication and nick the DNA, releasing a free 3′ OH (1). They remain bound to the 5′ end of the nicked DNA. Host DNA polymerase assembles around the free 3′ OH and copies the (−) DNA, making a new (+) strand (2), while the original (+) strand, covalently attached to Rep and Rep′, peels away (3). After copying a complete new (+) strand, the Rep and Rep′ proteins ligate the old (+) strand to make a new genome with its characteristic hairpin structure in the origin, while also releasing a new ccc that is composed of one old strand and one new one (4). Probably during formation of the covalently closed new genome, the Rep and Rep′ proteins are transferred to the new ccc. The new (+) genome is converted into a second ccc by the same host enzymes active when the infection begins (5). The new double-stranded replicative forms are used for synthesis of more viral DNA (6).

that rolling-circle production of new (+) strands can continue. If the new genome is going to be converted into a ccc, host DNA polymerase requires a primer. It is thought that host polymerase somehow synthesizes this primer just before or during the reaction that involves Rep and Rep′ and releases the offspring (+) genomes, possibly taking advantage of repeated sequences within the origin of DNA replication. No information is available regarding the relative ratio of ccc to single-stranded genomes or the timing of production of these different molecules.

9.5 Parvoviruses are tiny ssDNA viruses with linear genomes having hairpins at both ends

Like circoviruses, **parvoviruses** have also been of interest because of their small genome size. Additionally they have unusual sequences at the 5′ and 3′ ends of their linear genomes. As is the case in every virus, these terminal features play a prominent role in the mechanism of genome replication. The **minute virus of mice** (**MVM**) has long been the model for parvovirus gene expression and genome replication. The adjective minute in the name MVM refers to the virus's tiny size and is pronounced "mi-**nyut**" or "mi-**noot**" (**Figure 9.6**). The genome inside the naked capsids is single-stranded DNA 4–6 kb long, with the ends arranged as **hairpins** (**Figure 9.7**); specifically, the MVM genome is 5,149 bases long. MVM has a genome with ends that differ from one another. The arbitrarily designated right end forms a simple stem–loop structure, and the left end is more complicated. Virions can contain either positive or negative strands of DNA, with the negative strand defined as complementary to viral mRNA. Synthesis of all viral mRNA uses just one of the strands as the template, after initial second-strand synthesis. As with circoviruses, it was long thought that the virus passively waits for its host to enter S phase, but recent experiments suggest that nonstructural viral proteins manipulate the host cell cycle and apoptotic pathways.

9.6 The model parvovirus MVM encodes six proteins using alternative splicing

The parvovirus MVM has a genome 5,149 bases long and yet encodes four different nonstructural proteins and two structural proteins. Structural proteins VP1 and VP2 are the capsomers, with VP2 found more abundantly in the capsids. The NS1 protein is active during genome replication and is also a transcription factor. The roles of the three NS2 proteins are less well understood. There are three versions of NS2 that vary at their C-termini. The NS2 proteins are known to be important for transcription of the genome, genome replication, accumulation of NS1 in the nucleus, assembly of the capsid, and assembly and release. Despite or perhaps because of all of these roles, further research is needed to understand how the NS2 proteins work on a molecular level.

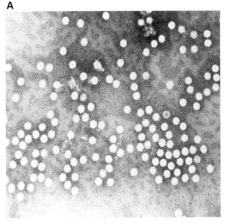

A

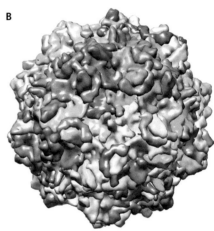

B

Figure 9.6 Parvovirus structure. (A) Electron micrograph of a parvovirus. The virions are about 20 nm in diameter. (B) Crystal structure of a parvovirus. (A, CDC. B, From PDB-101 [http://pdb101.rcsb.org]. Courtesy of Luuk T, Organtini LJ & Hafenstein SU. doi: 10.2210/pdb4qyk/pdb.)

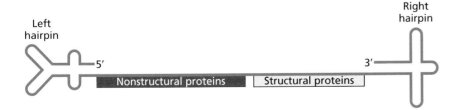

Left hairpin

Right hairpin

5′

3′

Nonstructural proteins   Structural proteins

Figure 9.7 Minute virus of mice genome. The MVM genome showing the terminal hairpin structures. The length of the hairpin structures is not to scale with the rest of the genome.

Figure 9.8 Gene expression in MVM, a parvovirus. The double-stranded replicative form of MVM has two promoters that are used by host proteins to drive synthesis of nine capped, polyadenylated mRNAs encoding six proteins.

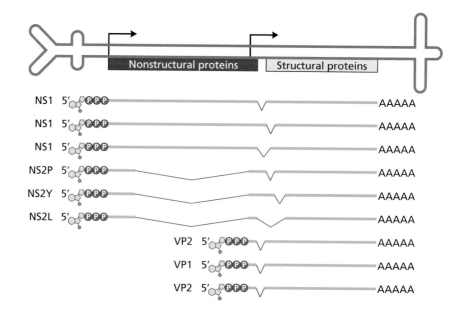

Gene expression in parvoviruses begins with formation of the double-stranded replicative form, which will be discussed further in **Section 9.7**. Once the replicative form has formed, it contains two Pol II-dependent promoters, each facing in the same direction (**Figure 9.8**). One drives expression of NS proteins, whereas the other drives expression of VP proteins. Three different splice variants encode NS1, and one splice variant each encodes the three different NS2 proteins. Similarly two splice variants encode VP2, and a single mRNA encodes VP1. Interestingly, the splice variants of encoding NS1 or VP2 have identical coding sequences but different 5′ or 3′ untranslated regions for which the particular molecular functions are not known. There is a single polyadenylation site near the 3′ of the original genome strand. The mRNAs are treated like normal cellular mRNA, from export to the cytoplasm to translation. Translated proteins are targeted back to the nucleus through normal nuclear localization signals, where they can participate in genome replication or in assembly and release.

9.7 The model parvovirus MVM uses a rolling-hairpin mechanism for genome replication

Genome replication in parvoviruses such as MVM is not as well understood as gene expression, although certain details have emerged (**Figure 9.9**). The unusual ends of the MVM genome are essential for its replication and interact with both viral and host proteins. To complete uncoating and allow gene expression to occur, host enzymes recognize the left end's 3′ OH opposite a single-stranded template, copy the template, and ligate the nick between the old DNA and the newly synthesized DNA. This replicative form is used for gene expression, resulting in production of viral NS1, which is needed for genome replication. Genome replication occurs through a **rolling-hairpin mechanism**, which is similar to a rolling-circle or σ mechanism but involves a linear genome with terminal hairpins instead of a circular template. To initiate genome replication, the viral NS1 protein binds near the right end and nicks the DNA, which results in DNA replication catalyzed by host DNA polymerase. The result is assembly of a unidirectional replication fork that moves toward the right end of the genome. Once the right end has been

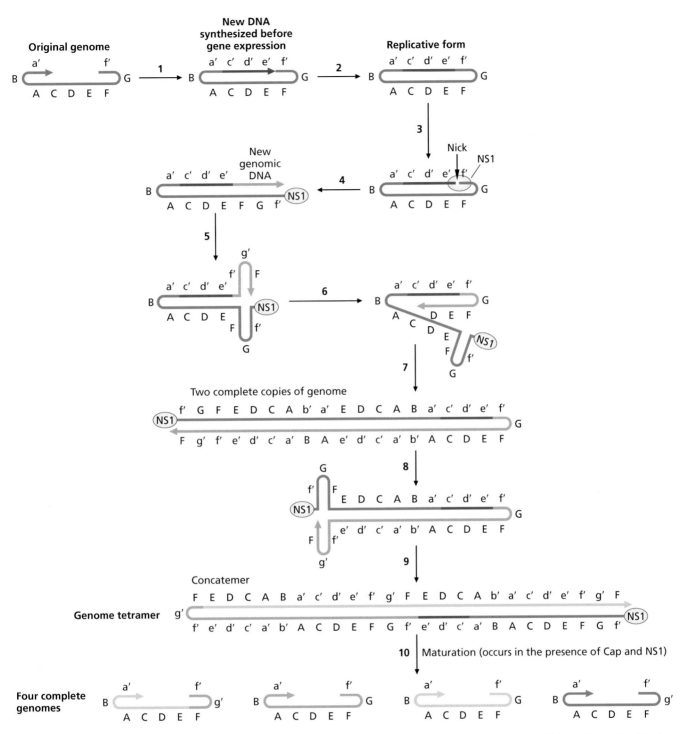

Figure 9.9 Genome replication in the parvovirus MVM. The hairpin structures at the left and right ends have been simplified to make the process of replication clearer. The genome has been labeled such that unique sequences are indicated by letters A, B, C, D, E, F, and G, and sequences complementary to A and F are labeled a' to f', respectively. The most stable form of the MVM genome provides a 3' OH opposite a template strand so that host enzymes copy the DNA and ligate the new DNA to the old DNA (1), creating a double-stranded replicative form (2). Viral NS1 nicks the template near the right-hand end, providing a 3' OH (not shown) opposite a template strand that can be recognized by host DNA polymerase (3). Host polymerase extends the DNA (4), which then spontaneously folds into a hairpin structure that again provides a 3' OH opposite a template strand (5). Host polymerase again extends the DNA (6), ultimately resulting in two complete copies of the genome (genome dimer) (7). Again, the terminus of the new DNA spontaneously folds into a hairpin structure that provides a free 3' OH opposite a template strand (8), ultimately resulting in formation of a genome tetramer (9). Viral proteins process the tetramer during assembly. Assembly is associated with an accumulation of Cap and NS1 (10).

copied, it is thought that the ends fold up in such a way that a free 3′ OH opposite a template strand exists and once again a host replication fork forms and copies the template to its end, resulting in a dimer. This folding-up reaction can occur only because of the repeats at the end of the genome. Again, the ends fold up in such a way that a new replication fork can copy the template, forming at this point a concatemer of four complete genomes. These tetramers will be resolved during assembly so that each virion gets a single complete genome with the correct hairpin ends. An interesting aspect of MVM replication is that the virion actually produces many more genomic strands than template strands, indicating that the overall process of rolling-hairpin replication is not completely understood because this one, the best model we have, indicates that both strands get produced in equal amounts.

Essential concepts

- Both circoviruses and parvoviruses are of interest as models of dense coding capacity in a small amount of nucleic acid and as models for the simplest independently replicating viruses of animals.
- Because of their tiny size, both circoviruses and parvoviruses depend heavily on host cell enzymes for genome replication and transcription. Nevertheless, both encode proteins required for viral genome replication, in part because host chromosomes are double stranded and so viral proteins are likely required in order to produce offspring ssDNA genomes.
- Although the DNA in each virion is single stranded, both circovirus and parvovirus have double-stranded replicative forms that are used as templates for transcription and genome replication.
- Neither circovirus nor parvovirus genome replication mechanisms are well understood. It appears that parvovirus replication occurs via a rolling-hairpin mechanism that requires the secondary structures at the ends of the genome and the viral NS1 protein, but this mechanism does not explain how the virus produces genome-sense strands more abundantly than complementary strands.

Questions

1. Which circovirus and parvovirus nonstructural proteins play similar roles during viral genome replication?
2. Compare and contrast the replicative forms in the circoviruses and the parvoviruses.
3. What is rolling-hairpin genome replication?
4. How is genome expression and replication in the circoviruses similar to the smallest Class I, Class IV, and Class V animal viruses?
5. How does genome expression and replication in the parvoviruses differ from the smallest Class I, Class IV, and Class V animal viruses?

Interactive quiz questions

In addition to the questions provided above, this edition has a range of free interactive quiz questions for students to further test their understanding of the chapter material. To access these online questions, please visit the book's website: www.routledge.com/cw/lostroh.

Further reading

Circoviruses

Cheung AK 2003. Transcriptional analysis of porcine circovirus type 2. *Virology* 305:168–180.

Cheung AK 2004. Palindrome regeneration by template strand-switching mechanism at the origin or DNA replication of porcine circovirus via the rolling-circle melting-pot replication model. *J Virol* 78:9016–9029.

Cheung AK 2015. Specific functions of the Rep and Rep′ proteins of porcine circovirus during copy-release and rolling-circle DNA replication. *Virol* 481:43–50.

Mankertz A, Caliskan R & Hattermann K 2004. Molecular biology of porcine circovirus: Analyses of gene expression and viral replication. *Vet Microbiol* 98:81–88.

Wen LB, Wang FZ, He KW, Li B, Wang XM et al. 2014. Transcriptional analysis of porcine circovirus-like virus P1. *BMC Vet Res* 10:1–8.

Parvoviruses

Adeyemi RO, Landry S & Davis ME 2010. Parvovirus minute virus of mice induces a DNA damage response that facilitates viral replication. *PLOS Pathog* 6(10):e101141.

Choi EY, Newman AE, Burger L & Pintel D 2005. Replication of minute virus of mice DNA is critically dependent on accumulated levels of NS2. *J Virol* 79:12375–12381.

Cotmore SF & Tattersall P 1998. High-mobility group 1/2 proteins are essential for initiating rolling-circle type DNA replication at a parvovirus hairpin origin. *J Virol* 72:8477–8484.

Fasina OO, Stupps S, Figueroa-Cuilan W & Pintel DJ 2017. Minute virus of canines NP1 protein governs the expression of a subset of essential nonstructural proteins via its role in RNA processing. *J Virol* 91:e00260-17.

Lorson C, Pearson J, Burger L & Pintel DJ 1998. An Sp1-binding site and TATA element are sufficient to support full transactivation by proximally bound NS1 protein of minute virus of mice. *Virology* 240:326–337.

Nüesch JP, Christensen J & Rommelaere J 2001. Initiation of minute virus of mice DNA replication is regulated at the level of origin unwinding by atypical protein kinase C phosphorylation of NS1. *J Virol* 75:5730–5739.

Zou W, Wang Z, Xiong M, Chen AY, Xu P et al. 2018. Human parvovirus B19 utilizes cellular DNA replication machinery for viral DNA replication. *J Virol* 92:e01881-17.

Gene Expression and Genome Replication in the Retroviruses and Hepadnaviruses

10

Virus	Characteristics
Human immunodeficiency virus 1 (HIV-1)	Model for retroviruses that have (+) ssRNA genomes but replicate via a reverse transcription mechanism. Viral cDNA must integrate into a human chromosome prior to gene expression. Cause of the AIDS pandemic.
Hepatitis B virus	Model for hepadnaviruses that have circular, partially dsDNA genomes containing a small amount of RNA and that replicate via a reverse transcription mechanism.

The viruses you will meet in this chapter and the concepts they illustrate

Retroviruses and hepadnaviruses share a dependence on a viral **reverse transcriptase** enzyme that can use an RNA template to synthesize complementary DNA (cDNA). Study of reverse-transcribing viruses was once considered rather obscure because there were no known reverse-transcribing viruses associated with important human diseases. Furthermore, the first reverse transcriptases described originated from Rauscher mouse leukemia virus and from Rous sarcoma viruses at a time when there were no viruses known to cause cancer in humans. This fact was even noted by the 1975 Nobel Prize in Physiology or Medicine committee when awarding the Nobel Prize for the discovery of reverse transcriptase: "Viruses causing tumours in man have not been demonstrated except in the case of wart virus. The type of tumours caused by this virus are of a benign nature." (The wart virus in question was human papillomavirus, now known to be a leading cause of cervical cancer; see **Chapter 13**.)

The 1970 discovery of reverse transcriptase earned a Nobel Prize not because of its medical significance but instead because it shook molecular biology to its core, by displacing DNA somewhat from its central throne in the discipline. That displacement made possible innovations in thinking, such as those that led to the idea of an RNA world as central to the story of cellular evolution (see **Chapter 15**). In addition, reverse transcriptases have many biotechnology applications that have enabled innovative experiments to better understand cellular life. An example is synthesis and cloning of cDNA starting with an mRNA template (**Technique Box 10.1**); the cDNA lacks introns and therefore can be used to synthesize eukaryotic proteins. A second example is reverse-transcription quantitative polymerase chain reaction (RT-qPCR), which is used to quantify mRNA very precisely and therefore is

DOI: 10.1201/9781003463115-10

useful for studies of gene expression. From our twenty-first-century vantage point, we know that reverse transcriptases have turned out to be not only common but also important for evolution, for basic cellular processes, and for human health. For example, the endogenous retroviruses that infect most vertebrates encode reverse transcriptase, as do other ubiquitous subviral entities such as retrotransposons (see **Chapters 1** and **15**), all of which have likely contributed to the diversification of eukaryotic life. Telomerase is a key cellular reverse transcriptase that is required for the basic process of replicating the ends of linear eukaryotic chromosomes (see **Chapter 8**).

As the quotation from the Nobel committee illustrates, the impact of reverse-transcribing viruses on human health in the twenty-first century was not anticipated when they were discovered in 1970. In our day and age, we focus on two model reverse-transcribing viruses, singled out because of their medical significance: **human immunodeficiency virus 1** (**HIV-1**), a retrovirus, and **hepatitis B virus** (**HBV**), a hepadnavirus (**Figure 10.1**). In 2015, approximately 37 million people were living with HIV infection and a staggering 2 billion have had HBV, leading to 350 million chronic HBV infections. HIV causes acquired immunodeficiency syndrome (AIDS) and HBV causes hepatitis, which in turn leads to liver cancer (see **Chapter 13**). Both AIDS and liver cancer are almost always fatal. Successful anti-HIV and anti-HBV pharmaceuticals target their reverse transcriptase enzymes selectively without affecting host telomerase. Another medical application of the study of reverse-transcribing viruses is gene therapy, in which a recombinant viral vector delivers a therapeutic gene to certain host cells. Because both HIV and HBV are adept at expressing genes from the nucleus, they may be useful gene therapy agents, although their compact genome size limits the amount of DNA that can be inserted into the vectors.

As always, the health effects of viruses are not the only reason to study them. Instead, we also study retroviruses and hepadnaviruses because of their fascinating molecular and cellular biology. For example, the mechanisms of reverse transcription in both cases are intricate, involving both *cis*-acting sequences and secondary structures in template mRNA, multiple active sites in the viral reverse transcriptase enzymes, and discontinuous synthesis, first introduced in **Chapter 5** (coronaviruses). Nevertheless, the two reverse transcriptases use different mechanisms, one of which relies on a tRNA for a primer, whereas the other relies upon a protein primer. Retroviruses manipulate nucleic acids even further than hepadnaviruses, encoding a viral integrase enzyme that inserts viral cDNA into host chromosomes. Gene expression in both groups of viruses is also interesting to investigate because both encode many proteins using compact genomes and exploit host transcription factors to stimulate viral mRNA production. Studies of HIV have increased interest in understanding not just transcription initiation but also transcription elongation as a cellular process that can be regulated. Studies of HIV have also stimulated interest in understanding how mRNA is trafficked out of the nucleus.

Studies of HBV are often focused on how the virus encodes eight proteins with a genome that is only 3.2 kbp long or on understanding a reverse transcriptase that is quite different from that of the retroviruses. Furthermore, genome amplification in HBV involves a novel pathway in which assembly of capsids is essential for the process. We will examine these topics in closer detail, beginning with the reverse transcriptase enzymes.

A

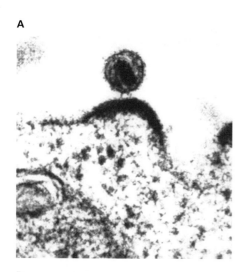

B

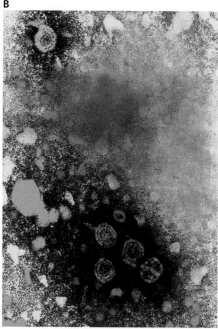

Figure 10.1 HIV-1 and HBV. (A) Digitally colorized electron micrograph of HIV-1. (B) Digitally colorized electron micrograph of HBV. (A, CDC/National Institute of Allergy and Infectious Diseases [NIAID]. B, CDC/Dr. Erskine Palmer.)

TECHNIQUE BOX 10.1   SYNTHESIS OF cDNA

It is often convenient to convert mRNA into a DNA copy, which is then termed **cDNA**. Because DNA is more stable than RNA, it is easier to analyze cDNA than RNA. Unlike mRNA, cDNA can be cloned into plasmids for various purposes. Furthermore, a common technique to quantify mRNA, namely RT-qPCR, relies upon first creating a cDNA copy of the mRNA in question. The RT-qPCR procedure is described in Technique Box 10.2. Common techniques such as microarrays and RNA-seq, which quantify all the different mRNAs in a complex sample, also rely upon first creating a cDNA copy of the mRNA; these techniques are discussed in Technique Box 10.3.

To make cDNA starting from an mRNA template, we use a reverse transcriptase enzyme (**Figure 10.2**). Recombinant enzymes cloned from retroviruses are available as are altered enzymes that have been changed so that they have better properties, such as favoring continuous synthesis of long nucleic acids greater than 10 kb in length. To synthesize cDNA, the reverse transcriptase is mixed with the template RNA, a primer, deoxyribonucleotide triphosphates (dNTPs), and a buffer that provides the necessary ions and pH conditions. The primer for synthesis of eukaryotic cDNAs is almost always a poly(T) sequence (**oligo(dT)**). Use of this primer allows copying of all eukaryotic mRNA because eukaryotic mRNA always has a poly(A) tail. Target sequence-specific primers can also be used to clone single mRNAs, even in the absence of a poly(A) tail. Such RNA includes viral sequences such as the genomes of (−) RNA viruses like rabies and influenza.

An oligo(dT) primer does not work to reverse transcribe bacterial or bacteriophage mRNA because bacterial mRNA is not polyadenylated. In that case, it is typical to use a target sequence-specific primer to reverse transcribe a specific bacterial mRNA. Or, if the goal is to examine all of the bacterial mRNA at once, we can use a random hexamer primer in which all six positions have been randomized and the whole population of primers is likely

to anneal to every mRNA. Sometimes it is more reliable, however, to collect all of the bacterial mRNA and then polyadenylate it *in vitro* using a poly(A) polymerase that adds a poly(A) tail to the 3′ end. After subjecting the mRNA to *in vitro* polyadenylation, an oligo(dT) primer can be used for reverse transcription.

Once cDNA has been produced there are several ways to detect a specific sequence or to detect many sequences in a population; there are also techniques for determining the quantity of a single cDNA or the amounts of thousands of different cDNAs in a complex sample. In addition, cDNA can be cloned into a plasmid, providing an archived copy of the coding version of the gene of interest. Cloning cDNA is particularly useful when studying eukaryotes in which the hnRNA transcribed from a gene contains introns that are not represented in the mature mRNA; that cDNA represents not the larger genomic DNA but instead the actual sequence that directly encodes the protein of interest. Cloning, however, requires synthesis of double-stranded DNA. The reverse transcriptase reaction produces single-stranded DNA that can be amplified by PCR, which produces enough DNA for cloning. It is easy to sequence a cDNA that has been cloned, and then that sequence can be subjected to bioinformatics. Plasmids can also be used to overexpress a protein so that it can be purified and studied *in vitro*. For example, it would be possible to clone three different cDNAs encoding the hepadnavirus large S, medium S, and small S proteins starting with mRNA isolated from an infected cell. If the cDNAs were cloned in an expression vector, the individual proteins could then be purified and subjected to biochemical analysis. It is not necessary to use cDNA to clone bacteriophage genes because bacterial genes do not have introns; instead, the genes of interest can be cloned starting with genomic DNA. Because it is already double-stranded DNA, it is easier to isolate and manipulate *in vitro* than mRNA.

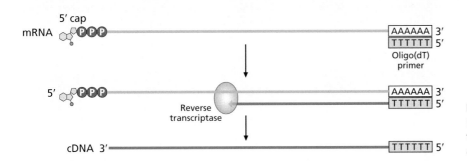

Figure 10.2 cDNA synthesis. Eukaryotic mRNA can be reverse transcribed using an oligo(dT) primer and reverse transcriptase *in vitro*.

10.1 Viral reverse transcriptases have polymerase and RNase H activity

Viral reverse transcriptase enzymes use an RNA template to synthesize DNA (reverse transcription) and use a DNA template to synthesize DNA (DNA polymerization). They also have an RNase H domain that removes RNA from RNA–DNA hybrids. In both HIV and HBV, the process of synthesizing the first cDNA strand is discontinuous, meaning that the reverse transcriptase begins copying one part of the (+) RNA template before jumping to a distal part of the template and continuing DNA synthesis, skipping over the intervening (+) RNA. The RNase H domain is essential for the jumping reactions. Both enzymes are also able to copy the (−) cDNA to synthesize (+) DNA, ultimately creating a double-stranded DNA copy of a (+) RNA template. Reverse transcriptases do not have 3′ to 5′ exonuclease editing activity (see **Chapter 5**, coronaviruses), so their mutation rate is about 10,000 times higher than that of the DNA polymerase used during host DNA replication. The result is that reverse-transcribing viruses give rise to offspring populations with tremendous genetic diversity. The reverse transcription reactions of HIV and HBV also have some unique variations that depend on the virus; we will consider the retrovirus model, HIV-1, first.

10.2 Retroviruses are enveloped and have RNA genomes yet express their proteins from dsDNA

Some steps of a **retrovirus** reproductive cycle occur in the cytoplasm and others occur in the nucleus (**Figure 10.3**). After attachment, there is partial uncoating of the virus genome followed by the long process of reverse transcription, in which the infecting (+) mRNA genome is converted into dsDNA. This viral dsDNA enters the nucleus, where a viral integrase enzyme inserts the dsDNA genome almost randomly into a host chromosome, thus setting the stage for gene expression and genome replication. Viral genes are transcribed by host Pol II, which produces typical eukaryotic capped and polyadenylated mRNA. Some viral mRNAs are spliced, whereas the full-length genomic RNA is not. The mRNA is exported to the cytoplasm, where translation occurs. Genome replication involves host Pol II transcribing the entire viral genome into a single (not spliced) mRNA, which is then encapsidated in the cytoplasm. Following assembly, the enveloped virus exits the host cell by budding; the last steps of maturation actually occur at the same time or slightly after budding.

Retroviruses (family *Retroviridae*) such as HIV-1 are spherical with distinctive ultrastructure that includes nucleocapsid and matrix layers inside an envelope with prominent spikes (**Figure 10.4**). Retroviruses have two copies of the same linear (+) strand RNA as their genomes; the (+) RNA has all the typical features of eukaryotic mRNA, namely a 5′ methylated cap and a 3′ poly(A) tail. Retrovirus genomes are about the same size as those of the picornaviruses or flaviviruses with between 7 and 12 kb of RNA and they have long 5′ and 3′ untranslated regions (UTRs) (**Figure 10.5**). The UTRs contain specific sequences needed for viral replication. In the 5′ UTR, these include one of the two copies of the direct repeat (R), a unique U5 sequence, and a primer binding site; the structure of the stem–loops in the 5′ UTR are also essential for replication. In the 3′ UTR, the essential *cis*-acting sequences include a polypyrimidine tract (PPT), a unique U3 sequence, and the second copy of R.

Whether considering a simple retrovirus or HIV-1, before any viral gene expression can occur, the genome must be reverse transcribed and then the cDNA inserted into a host chromosome. Therefore, these processes are addressed in **Section 10.3**, before we learn about how the retroviruses express their genes.

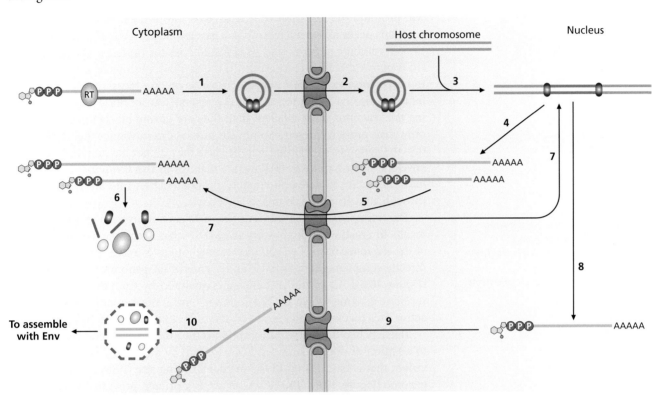

Figure 10.3 Overview of HIV-1 replication. After attachment and penetration, reverse transcription occurs (1) while the nucleic acids are conveyed to the nucleus (2). The viral dsDNA enters the nucleus along with a viral integrase enzyme that recombines the viral DNA with a host chromosome (3). Viral genes are transcribed by host Pol II, which produces typical eukaryotic mRNA with 5′ caps and poly(A) tails (4). The mRNA is exported to the cytoplasm (5), where translation occurs (6). Some viral proteins enter the nucleus, where they affect gene expression (7). For genome replication, host Pol II transcribes the entire viral genome into a single genome-length mRNA (8), which is then conveyed to the cytoplasm (9) where it is encapsidated (10). Other assembly and maturation steps are not depicted.

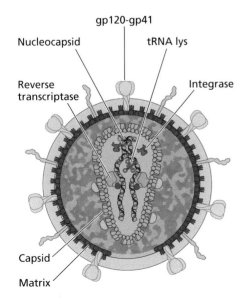

Figure 10.4 HIV-1. The key components of HIV-1.

Figure 10.5 Retrovirus genome. The virus genome showing the 5′ cap; the 5′ UTR with R, U5, and primer-binding site (PBS); the 3′ UTR with its polypyrimidine tract (PPT), U3, and R sequences; and the 3′ poly(A) tail. The sequences at the 5′ and 3′ ends have been exaggerated in length in order to depict them. The virion actually contains two copies of the genome. No scale bar is included because of the exaggerated length of the 5′ and 3′ sequences.

10.3 Reverse transcription occurs during transport of the retroviral nucleic acid to the nucleus, through a discontinuous mechanism

Following envelope fusion at the plasma membrane, a series of reactions converts the (+) ssRNA retrovirus into dsDNA and assembles with proteins that will promote the insertion of that dsDNA into a host chromosome. These reactions occur as the nucleocapsid is being transported to the nucleus using the microtubule network and its associated motor proteins. The nucleocapsid is permeable during transport to the nucleus, permitting cellular factors to have access to the HIV enzymes, such as reverse transcriptase. Host proteins are also co-opted to assist in the reverse transcriptase reactions, making the reactions more efficient than they are during biochemical tests with only virus proteins present. Some retroviruses can introduce their DNA only into the genomes of dividing cells because they do not have an intact nuclear envelope during mitosis. HIV, which belongs to the *Lentivirus* genus, does have the ability to deliver its double-stranded cDNA into an intact nucleus through a nuclear pore complex, however.

The retrovirus reverse transcriptase mechanism involves many steps and results in creation of double-stranded DNA that is complementary to most of the sequence of the (+) ssRNA genome but has 5′ and 3′ ends that are not directly complementary to the 5′ and 3′ ends of the genomic ssRNA template (**Figure 10.6**). All of the reactions are catalyzed by the virus reverse transcriptase enzyme, which is a heterodimer. One of the monomers has both the active sites that have polymerase or RNase H activity (**Figure 10.7**).

The first step in reverse transcription is the binding of reverse transcriptase to a region of double-stranded RNA formed by a tRNA, packaged with the virion, that is base paired to the **primer-binding site** (PBS) sequence of the genome (**Figure 10.8**). The tRNA serves as a primer, providing a free 3′ OH opposite an RNA template. Retroviral reverse transcriptase uses this primer to begin polymerizing deoxyribonucleotides from the host cell into cDNA. The enzyme copies the entirety of the 5′ end of the template RNA molecule, making (−) cDNA complementary to the 5′ end of the genome. The reverse transcriptase enzyme next degrades most of the 5′ end of the RNA template, except for the PBS. The next step is described as a jump, in which the tRNA–PBS hydrogen bonds are broken and the (−) cDNA complementary to the R

To see an animation showing discontinuous reverse transcription in HIV, please visit the book's website: www.routledge.com/cw/lostroh.

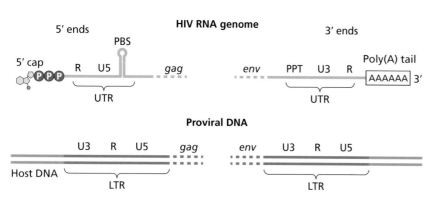

Figure 10.6 Comparison of the ends of the HIV RNA genome and the HIV proviral DNA. The 5′ end of the genome, defined as the 5′ end of the (+) sense DNA, differs in sequence from the 5′ long terminal repeat (LTR). The 3′ end of the genome is polyadenylated, whereas the 3′ end of the provirus is another copy of the LTR. The provirus is flanked by host DNA.

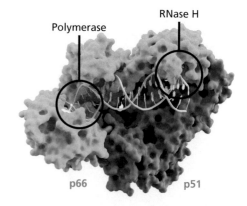

Figure 10.7 HIV reverse transcriptase. The crystal structure of the dimer has been modeled with double-stranded nucleic acid in its active sites. (Courtesy of Thomas Splettstoesser. Published under CC BY-SA 3.0.)

region base pairs with the (+) R RNA sequence adjacent to the 3′ poly(A) tail. The reverse transcriptase then uses the rest of the (+) RNA as a template to synthesize (−) cDNA complementary to all of the protein coding genes, while also degrading all of the remaining RNA template, except for the short PPT near the (−) cDNA complementary to the U3 sequence.

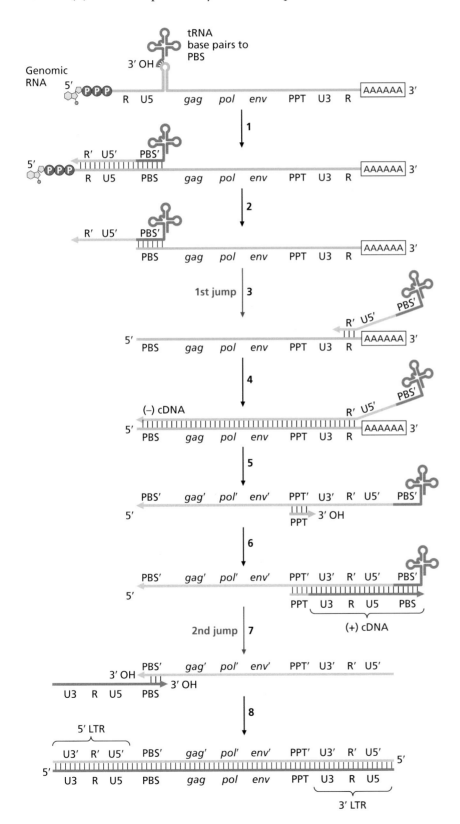

Figure 10.8 Reverse transcription in HIV. A tRNA packaged with the genome during assembly provides a primer for reverse transcriptase to synthesize the first (−) cDNA (1). The enzyme's RNase H activity degrades some of the RNA (2), permitting the first jump (3). The (−) cDNA complementary to R base pairs near the 3′ end of the RNA template, providing a primer that reverse transcriptase extends to synthesize more (−) cDNA (4). RNase H activity next leaves only the tRNA and a small amount of RNA complementary to the PPT′ (−) cDNA intact (5). As a result, the PPT RNA provides a primer for synthesis of (+) cDNA complementary to the 5′ end of the (−) cDNA (6). RNase H activity removes all remaining RNA and the template jumps again (7), this time providing two different primers with available 3′ OH groups opposite a template strand, which the viral polymerase extends (8). The result is double-stranded DNA with long terminal repeats (LTRs).

This small bit of RNA remaining provides a free 3′ OH opposite a (−) DNA template; the reverse transcriptase uses this (−) cDNA as a template to synthesize a second (+) DNA strand by copying all the way through the tRNA sequence that is covalently attached to the (−) cDNA, and degrading the tRNA (see Figure 10.8). At this point, another jumping reaction occurs. The new (+) cDNA sequence complementary to the tRNA jumps to the (−) cDNA of the PBS that was synthesized earlier in the reaction. This base pairing provides two free 3′ OHs opposite template single-stranded DNA, and the reverse transcriptase finishes copying both strands, resulting in dsDNA that has a directly repeated sequence, composed of U3, R, and U5, from 5′ to 3′, at both ends. This repeated sequence has a special name, the **long terminal repeat** or **LTR**.

10.4 Retroviral integrase inserts the viral cDNA into a chromosome, forming proviral DNA that can be transcribed by host Pol II

At the conclusion of the reverse transcriptase reactions, the HIV cDNA is in a complex with virus and host proteins called the **preintegration complex**. The complex includes at least some capsid proteins that enable the DNA and its associated proteins to dock with a nuclear pore complex, after which the preinitiation complex is released into the nucleoplasm. Viral gene expression cannot occur until the viral DNA becomes inserted into a host chromosome, a process that is catalyzed by the virus **integrase** enzyme that was packaged into the virion and traveled with the viral nucleic acids from the cell periphery to the nucleus.

The first step of integration, **3′-end processing**, is the removal of two nucleotides from each 3′ end of the dsDNA in the preintegration complex (**Figure 10.9**). These cleavages leave 3′ OH groups, which are used in the next step to attack a pair of adjacent phosphodiester bonds in the DNA of a host chromosome, resulting in a covalent bond between the host DNA and the retrovirus DNA. The sites attacked by the 3′ OH groups can be four to six nucleotides apart, depending on the virus. The 5′ overhangs from the viral DNA have to be removed, and nicks in the phosphodiester backbone are repaired by cellular ligase. Some of these later steps in the process are catalyzed by cellular enzymes instead of by the viral integrase. The resulting DNA of viral origin, once inserted into the host chromosome, is called the **provirus**. Once the provirus has formed, the stage is set for viral gene expression.

10.5 All retroviruses express eight essential proteins, whereas some such as HIV encode species-specific accessory proteins

Some retroviruses are simpler than others, so although we will otherwise use HIV-1 as the model retrovirus, we first learn about the proteins expressed by the simplest retroviruses because those components are necessary for replication of all retroviruses. All retroviruses express five nonenzymatic structural subunits: **SU**, the surface protein for receptor binding; **TM**, the transmembrane protein found in the viral envelope; **MA**, the matrix protein found between the envelope and the capsid; **CA**, the capsomer; and **NC**, the nucleoprotein found associated with the RNA genome inside the virion (see Figure 10.4). SU and TM together form the Env viral spike needed for attachment and penetration. All retroviruses encode three enzymes: reverse

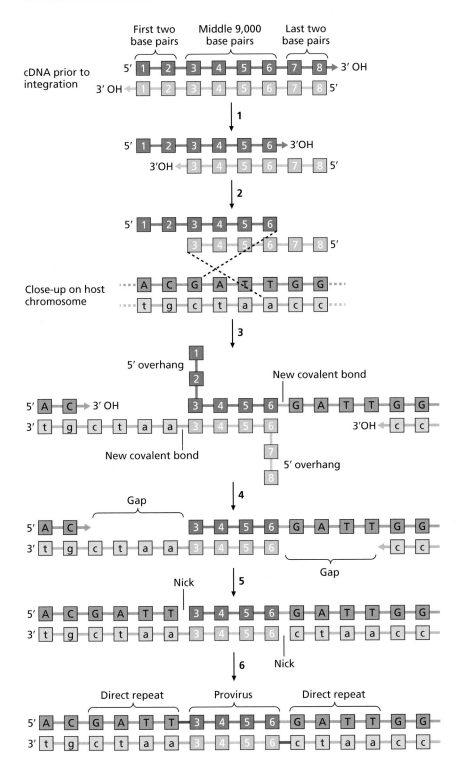

Figure 10.9 Integration of HIV cDNA catalyzed by viral integrase. DNA is represented as blocks connecting nucleotides to make the reaction clearer. The cDNA is much longer than shown, as is the host chromosome. In the first step, viral integrase processes the 3′ ends of the viral cDNA, leaving two unpaired bases at each 5′ end (1). Integrase then uses the two recessed 3′ OH groups in the preintegration complex to attack the DNA backbone of the recipient host chromosome (2), resulting in a covalent bond between the host DNA and the retrovirus cDNA (3). Host enzymes remove the 5′ overhangs from the viral DNA (4), extend the free 3′ OH groups to fill in the gaps (5), and repair the nicks in the DNA (6). The covalent bond synthesized by host ligase is in red. The result is integration of the proviral DNA lacking two base pairs at each end, flanked by direct repeats of host DNA that were created during the integration process.

transcriptase, integrase, and protease. Reverse transcriptase makes a cDNA copy of the viral genomic RNA, and integrase inserts this DNA into a host chromosome. Protease is necessary because HIV-1 encodes polyproteins similar to those employed by (+) RNA viruses such as picornaviruses (see **Chapter 5**). The retrovirus Gag polyprotein is processed into the MA, CA, and NC proteins, whereas the retrovirus Gag-Pol polyprotein can be processed into these three plus the three enzymes. The retrovirus Env polyprotein is processed into the SU and TM subunits.

HIV-1 is more complex in that it encodes accessory proteins not found in simple retroviruses. Retrovirus accessory proteins in general serve host-specific roles during infection; in the case of HIV, they provide nonstructural functions. For example, the HIV accessory protein **Tat** is a transcription regulator and **Rev** is important for export of viral mRNA from the nucleus. Other accessory proteins counteract specific host immune defenses or manipulate the cell cycle, enabling viral replication in host cells that are not actively dividing.

10.6 The retroviral LTR sequences interact with host proteins to regulate transcription

There are many sequences in the HIV-1 LTR that interact with host transcription factors (**Figure 10.10**). The promoter activity of the LTR can be measured in a variety of ways, most directly through detection and quantification of specific mRNA synthesized using the LTR as a promoter (**Technique Box 10.2**). The core HIV-1 promoter in the LTR is composed of a TATA box and many transcription-factor-binding sites. Some of the host transcription factors such as SP-1 are ubiquitous ones found in the nuclei of many different types of host cells. The enhancer region of the LTR contains binding sites for other transcription factors, including at least one that is active particularly in the specific differentiated immune cells that HIV-1 infects. Together these elements result in high levels of transcription initiation between the U3 and R sequences, and this single promoter drives expression of all HIV-1 mRNAs. Additionally, the identical 3′ LTR that follows the Env coding sequence regulates 3′-end processing of the pre-mRNA, which affects polyadenylation and transcription termination. The polyadenylation signal is part of the transcribed R sequence. This sequence is also found near the 5′ end of the pre-mRNA because its template is downstream of the promoter in the 5′ LTR. HIV suppresses the activity of the 5′ site so that coding mRNA can be produced. The different dedicated functions of the two LTRs illustrate how the context of a nucleic acid sequence influences its molecular function.

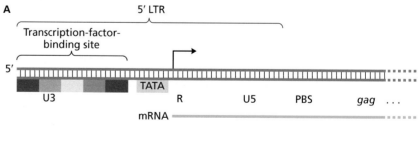

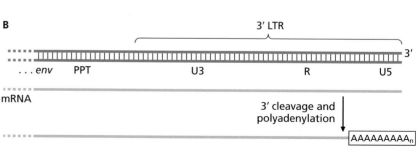

Figure 10.10 Sequences in the HIV LTRs that are essential for transcription. (A) Close-up of the 5′ LTR contains a promoter including a TATA box and transcription-factor-binding site and the transcription start site (bent arrow). (B) Close-up of the 3′ LTR, which is identical in sequence to the 5′ LTR but serves a different function. It encodes mRNA recognized as a 3′-end processing site for cleavage and polyadenylation.

10.7   The compact retroviral genome is used economically to encode many proteins through the use of polyproteins, alternative splicing, and translation of polycistronic mRNA

Retroviruses have small genomes, necessitating strategies to encode more than one protein using the same sequence (**Figure 10.11**). HIV-1, for example, encodes 16 proteins in about 9.7 kb. HIV uses three main strategies to encode so many proteins: polyproteins, alternative splicing to express proteins with overlapping coding sequences, and translation of polycistronic mRNA. It also uses just one promoter, which is economical because doing so enables most of the rest of the genome to encode proteins. Polyproteins are useful for expressing multiple proteins in a reduced amount of mRNA, compared with an average host protein, because the mRNA encoding a polyprotein has only one 5′ UTR, one start codon, one stop codon, and one 3′ UTR for the entire polyprotein, instead of one of each of these sequences for each of the individual proteins that will be released by proteolytic processing. In retroviruses, three genes, **gag**, **pol**, and **env**, encode polyproteins (**Figure 10.12**). There are three polyproteins: Gag, Gag-Pol, and Env. The viral protease enzyme processes Gag into MA, CA, NC, and p6 and processes Gag-Pol into MA, CA, NC, reverse transcriptase, integrase, and protease. Host proteases process Env to create SU and TM.

Alternative splicing is used to produce multiple HIV-1 pre-mRNA molecules starting from the same promoter, which is economical because it uses a single promoter and because some of the exons for one protein overlap with exons or introns from other genes (see Figure 10.12). Overlapping coding sequences are commonly used by viruses to encode less abundant proteins with regulatory or enzymatic functions. In the case of HIV-1, the accessory

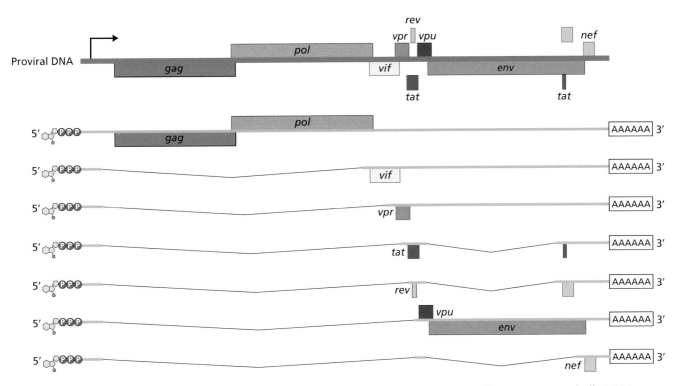

Figure 10.11 HIV-1 encodes three polyproteins and six accessory proteins. Using its small genome economically, HIV-1 uses one promoter to express its mRNA. It encodes the three polyproteins, Gag, Gag-Pol, and Env, found in all retroviruses, plus six accessory proteins. The coding sequences for the accessory proteins overlap with at least one other coding sequence.

Figure 10.12 The HIV-1 polyproteins.
(A) Viral protease processes Gag into MA, CA, NC, and p6. The crystal structure of CA is shown. (B) Viral protease processes Gag-Pol into MA, CA, NC, reverse transcriptase, integrase, and protease. The crystal structure of integrase (blue and purple) in a complex with viral DNA (red) and host DNA (orange) is shown, with the rest of the DNA represented with red or orange shading. (C) Host proteases process Env to create SU and TM, which together make up a trimeric spike. In the crystal structure of the spike, the SU protein is yellow with glycosylation indicated in orange. The TM protein is in red, with the transmembrane portion represented as rods crossing a membrane, because the crystal structure of that portion of TM is not known.

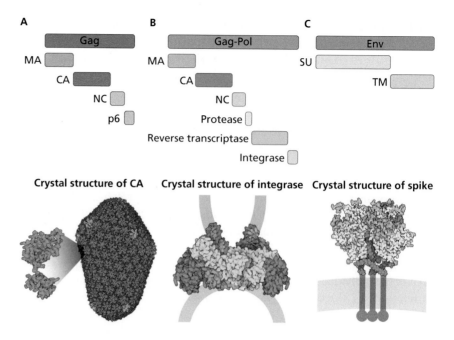

proteins have low abundance compared with the polyproteins. As is the case in many other viruses, the genes encoding the most abundant structural proteins (in this case, MA, CA, and NC), do not overlap with other coding sequences. In contrast, all of the accessory genes overlap with the *gag-pol* gene, with the *env* gene, or with other accessory genes.

There are also two cases in which a single HIV-1 mRNA encodes more than one protein by manipulating translation. In the first instance, the genome-length mRNA is used to translate the Gag polyprotein or the Gag-Pol polyprotein. Production of the less abundant Gag-Pol polyprotein requires a (−1) ribosomal frameshift (see **Chapter 5**). The second case is that of the accessory *vpu* gene, which overlaps the *env* gene. The Vpu protein is translated from the *vpu*/*env* mRNA through the use of an alternative start codon.

TECHNIQUE BOX 10.2   DETECTION OF INDIVIDUAL mRNA MOLECULES

Virologists might need to detect when a viral gene is transcriptionally active or to what extent a gene is being transcribed. For example, we might want to know if the HIV-1 promoter, found in the LTR, is equally active in different types of host cells such as T cells or macrophages.

The oldest technique for detecting a specific mRNA sequence is called a **northern blot** (**Figure 10.13**) in honor of the biologist Edward Southern, who in 1975 developed the Southern blotting method for detecting specific DNA sequences in a complex sample. Northern blotting, introduced in 1977, involves isolating all the RNA from a sample, size fractionating the sample using electrophoresis, and then blotting the RNA onto a supportive membrane (see Figure 10.13 and Technique Box 3.2). The membrane is then incubated with a DNA sequence called the probe. The probe is complementary to the target RNA sequence. The probe is also tagged in some way so that the position

of the target RNA sequence on the blot becomes visible; fluorescent probes are typically used today. A conceptually analogous technique called immunoblotting or western blotting was discussed in Technique Box 5.1 and allows detection of specific proteins in a complex sample.

Northern blotting, however, relies on manipulating RNA, which is not as stable as DNA. Therefore, we can run into problems when trying to analyze RNA directly. To solve this problem, there are alternatives to northern blotting. One technique used to quantify the activity of a promoter without having to measure RNA directly is to genetically engineer a **reporter gene** (**Figure 10.14**). A reporter gene encodes a protein that is easy to detect and that is found only when the promoter of interest is active. A typical reporter gene is *gfp*, which encodes the green fluorescent protein (GFP). A reporter gene can be used for many purposes, from microscopy to quantification of

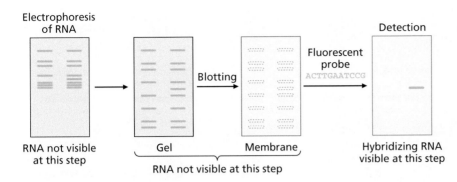

Figure 10.13 Northern blot. RNA is applied to a gel and separated by size using electrophoresis. The RNA is blotted to a membrane. The blot is immersed in a probe that binds selectively to a target sequence, revealing the position and abundance of the target RNA in the sample.

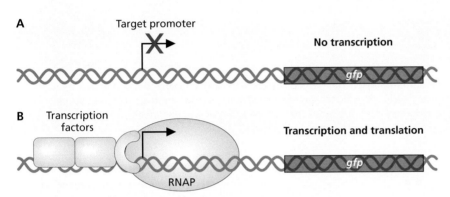

Figure 10.14 Reporter gene. A reporter gene is cloned under the control of a target promoter. (A) When the promoter is inactive, the reporter protein is not transcribed. (B) When the promoter is active, the cell transcribes the gene and translates its mRNA, and the reporter protein (GFP) can then be quantified. The amount of protein is proportional to the amount of promoter activity.

changes in gene expression in response to environmental conditions, developmental cues, or regulatory proteins.

A newer method for detecting and quantifying a specific mRNA species is **real-time quantitative reverse-transcription PCR (RT-qRT PCR)**. Unlike routine RT-PCR, in which the goal is to amplify as much of the target DNA as possible in order to obtain a large amount of cDNA, the goal of RT-qRT PCR is to quantify how much of a target sequence was present in a complex starting mixture (such as the contents of a cell infected by a virus). In order for such quantification to work, the amount of amplified target DNA must be proportional to the amount of target sequence in the original mixture. This requirement is met exclusively when the polymerase's substrates such as primers and dNTPs are in excess, so that the only limiting reagent in the reaction is the target sequence itself. This situation prevails during all polymerase chain reactions in the very early cycles, but the amount of DNA is so low during early cycles that it cannot be visualized on a typical agarose gel. Instead, an RT-qRT PCR instrument detects and quantifies fluorescence in addition to carrying out all the other functions of a thermocycler. Fluorescent signals can be detected reliably when the concentration of DNA in a sample is too low to be visualized using an agarose gel.

There are two ways to detect target DNA using fluorescence. The simplest method is to perform the RT-qRT PCR in the presence of a dye that fluoresces when it binds to double-stranded DNA (**Figure 10.15A**). In this case, as the amount of target sequence doubles with each cycle, the fluorescence doubles with each cycle. The more complicated method is to use a modified DNA probe in which one end of the probe has a covalently attached fluorescent dye and the other end has a covalently attached quencher molecule (**Figure 10.15B**). This probe is long enough to anneal specifically to the target sequence but short enough that the quencher is so close to the fluorescent dye that the signal from the dye is prevented (quenched). When DNA polymerase encounters the probe bound to its target, the polymerase's normal 5′ to 3′ exonuclease activity degrades the probe as it copies the template. It first degrades the end of the probe that is attached to the quencher; the quenching molecule diffuses away and the probe then emits a fluorescent signal until Taq degrades the entire probe. The quenching method is desirable when there are off-target sequences amplified during PCR or when detecting multiple target sequences in the same reaction using different colors of fluorescent dyes on a set of different sequence-specific probes.

In either case, the signal obtained by the qPCR machine is analyzed by making a semilogarithmic plot of the number of cycles (x-axis) compared to the amount of fluorescence expressed on log scale (y-axis). When the fluorescence in a sample first goes above a set amount, such as five times higher than the limit of detection, the cycle in which that occurs is called the quantification cycle or Cq (**Figure 10.16**). Cq is useful because the quantity of the target cDNA sequence doubles with every cycle and so

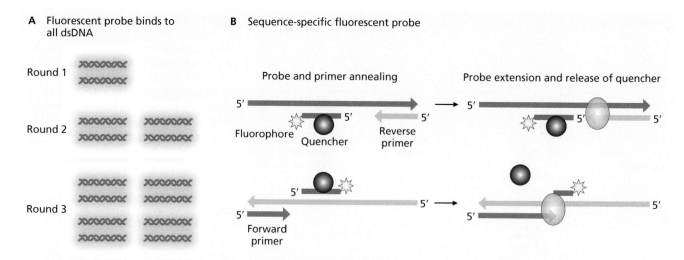

A   Fluorescent probe binds to all dsDNA

B   Sequence-specific fluorescent probe

Figure 10.15 RT-qPCR. RT-qPCR uses a modified thermocycler that detects fluorescence. (A) A fluorescent probe that binds to dsDNA sends a signal to the instrument after every cycle. (B) A fluorescent probe with a quencher binds to a certain sequence in the DNA. Taq polymerase degrades the probe, resulting in fluorescence that is quantified by the instrument.

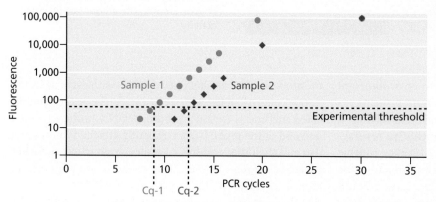

Figure 10.16 Cq. The quantification cycle (Cq) is the cycle in which the fluorescence reaches a predetermined threshold, such as five times above the limit of detection. Sample 1 reached Cq at cycle 10, whereas sample 2 reached Cq at cycle 13.

Cq can be used to compare the amount of target sequence across samples. For example, if the Cq of one sample occurs at cycle 10 and the Cq of a different sample occurs at cycle 13, it took three cycles for the target in the second sample to be as abundant as the target in the first sample. Because the target DNA doubles with every cycle, sample one had $2 \times 2 \times 2 (= 2^3) = 8$ times more target than sample two.

Although in principle the amplified DNA should double at every cycle, in later cycles the concentrations of reactants are no longer high enough for exact doubling to occur so that the amount of product at later cycles is not directly proportional to the amount of the target in the starting material. Use of Cq allows detection of the amount of target in earlier cycles, where the amount of the product is directly proportional to the amount of starting target in the sample. The use of fluorescence is necessary because it is sensitive enough to detect small amounts of DNA; DNA after only 10 or 15 cycles of PCR is often not abundant enough to be detected using standard electrophoresis.

10.8   The HIV-1 accessory protein Tat is essential for viral gene expression

Although we have often focused on transcription *initiation*, it is not sufficient to produce mRNA. Instead, transcription *elongation*, in which Pol II assumes a processive conformation able to copy thousands of nucleotides before dissociating from the template, must also occur. In the case of HIV-1, transcription elongation occurs very rarely in the absence of the viral Tat accessory protein. To promote transcription elongation, the Tat protein interacts with a stem–loop structure called the **Tar element**, which is found in all HIV-1 mRNA (**Figure 10.17**). This Tar element is composed of the first RNA bases polymerized. When it binds to the Tat protein, the whole complex then binds to two host proteins, one of which is a kinase. The result is that the kinase phosphorylates a certain region of Pol II, which promotes elongation. Because the Tat protein stimulates transcription of its own mRNA, Tat sets up a positive feedback loop (**Figure 10.18**) in which more and more HIV-1 mRNA is transcribed as levels of Tat increase. The production of mRNA encoding Tat in the first place (before there is Tat protein in the nucleus) occurs occasionally when the polymerase elongates the pre-mRNA despite the lack of Tat, which allows just enough Tat protein to be produced in order to trigger the positive feedback loop. The positive feedback loop triggered by Tat is necessary for the virus replication cycle to be completed.

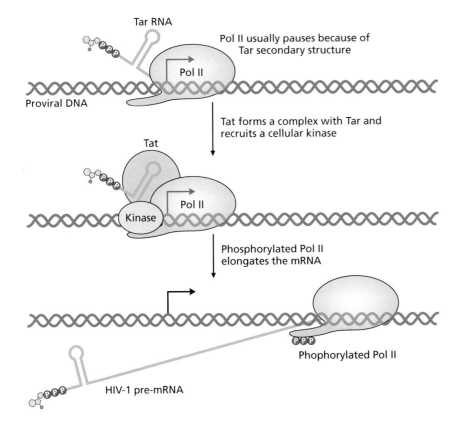

Figure 10.17 HIV-1 Tar RNA, Tat protein, and their effects on transcription elongation. The HIV-1 Tar RNA assumes a stem–loop structure. When the Tat protein binds to this stem–loop, the complex attracts host proteins including a kinase. The kinase phosphorylates Pol II, which promotes transcription elongation.

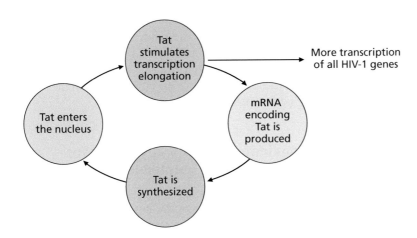

Figure 10.18 HIV-1 Tat positive feedback loop. Because Tat protein promotes elongation of mRNA encoding Tat, the effect of Tat is to trigger a positive feedback loop that results in production of more and more HIV-1 mRNA.

A

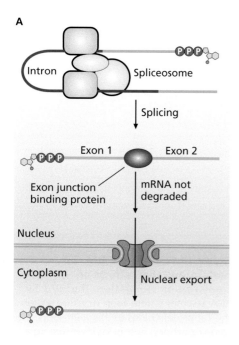

B

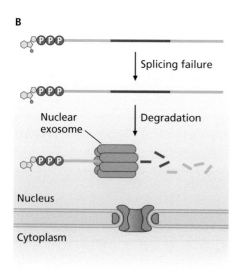

Figure 10.19 Quality control degrades pre-mRNA that is not spliced properly in the nucleus. (A) Normally, introns are removed from pre-mRNA during transcription. Protein complexes assembled at the exon–exon junction are essential for nuclear export of mature mRNA. (B) If a mistake occurs and a functional intron is left intact, the aberrant pre-mRNA is targeted for degradation, probably by a large protein complex called the exosome.

10.9   The HIV-1 accessory protein Rev is essential for exporting some viral mRNA from the nucleus

Although we have not discussed it previously, most human pre-mRNA is spliced at least once and that splicing is essential for the mRNA to be exported from the nucleus, where there is a mechanism for degrading pre-mRNA that has been improperly spliced (**Figure 10.19**). The purpose of this quality control is to prevent any improperly spliced mRNA from exiting the nucleus into the cytoplasm, where it could be translated into a potentially toxic protein while also consuming the cell's supply of amino acids without producing a functional protein. This quality control system poses a problem for HIV-1, which must avoid degradation of its longer mRNAs that must also be exported from the nucleus.

The HIV mRNAs are described as unspliced (genome-length mRNA), **incompletely spliced**, or completely spliced. The difference between incompletely and completely spliced transcripts is the removal of just one intron in the case of incompletely spliced transcripts compared with the removal of two introns from completely spliced transcripts. All unspliced and incompletely spliced longer mRNAs contain *cis*-acting sequences that are rich in AU dinucleotides that inhibit export of these mRNAs from the nucleus and make them vulnerable to degradation through the mRNA quality control process. HIV-1 encodes a protein called Rev that counteracts this problem. Rev is encoded by one of the completely spliced pre-mRNAs so that its own mRNA is readily exported to the cytoplasm and translated. Through a nuclear localization signal, Rev enters the nucleus, where it will allow export of incompletely spliced HIV-1 mRNAs. All incompletely spliced HIV-1 mRNA transcripts contain a sequence that folds into a complex stem–loop structure called the **Rev responsive element** or **RRE**, which occurs in the *env* coding sequence (**Figure 10.20**). The RRE serves as a scaffold for assembly of a complex of Rev proteins. The Rev proteins also subvert a host protein (CRM1) that normally has nothing to do with mRNA but is instead used for export of ribosomal RNA (not mRNA). The mRNP is then exported by relying upon the rRNA export pathway. The use of Rev also enables control over the amount of cytoplasmic HIV-1 mRNA that is incompletely spliced; high levels of incompletely spliced viral mRNA are needed late in the replication cycle to serve as new genomes.

10.10   Retrovirus genome replication is accomplished by host Pol II

Genome replication in retroviruses is simply transcription, catalyzed by host Pol II, to produce the longest unspliced RNA, which is sometimes referred to as genomic RNA (gRNA) when it is packaged into new virions instead of translated. The resulting genome sequence begins with R at the 5′ end and does not have LTRs (see Figures 10.6 and 10.8). Because it is synthesized by Pol II, the gRNA has a typical eukaryotic mRNA 5′ cap and poly(A) tail. The gRNA is exported from the nucleus as described earlier, using Rev, the RRE, and by co-opting the host CRM1 protein. The psi (ψ) sequence in the 5′ UTR is important for packaging the gRNA into new virions.

10.11   HIV-1 is a candidate gene therapy vector for diseases that involve the immune cells normally targeted by HIV

HIV-1 naturally infects differentiated cells of the immune system such as T cells and dendritic cells and therefore is able to express its genes in these cells

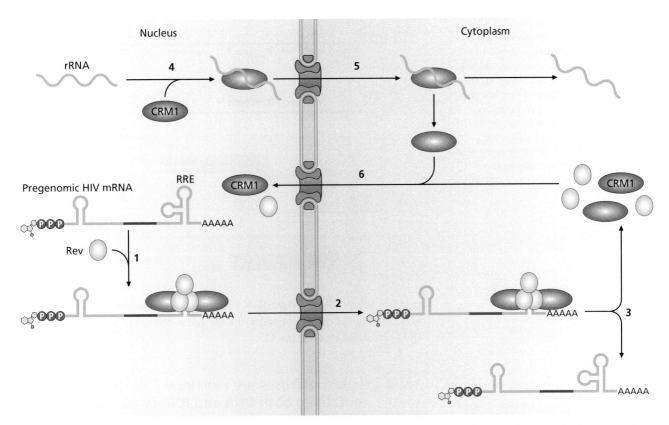

Figure 10.20 The HIV-1 Rev protein and the mechanism of HIV-1 mRNP export. Normally, the CRM1 protein is responsible for export of rRNA from the nucleus. HIV-1 co-opts this normal host pathway and uses it to export incompletely spliced mRNP containing Rev. Multimers of Rev assemble on the HIV mRNA (1) and then bind to CRM1, promoting the export of the HIV-1 mRNP (2). CRM1 and Rev dissociate from the mRNP (3) and are recycled to the nucleus for reuse (4). CRM1 transports rRNA out of the nucleus (5). CRM1 re-enters the nucleus and can be used repeatedly (6).

in particular. The 5′ LTR contains sequences that allow the viral promoter to be active in immune cells. The ability to express genes in these cells makes viral vectors that take advantage of this property ideal for gene therapies that manipulate immunity. For example, both T cells and dendritic cells are key for fighting cancer. Thus HIV-1 is an attractive candidate gene therapy vector, using the HIV-1 to genetically alter T cells or dendritic cells so that they can kill cancer cells more effectively. The idea here is that the HIV-1 sequence could be modified by replacing much of the genome with a **transgene** encoding a protein that enhances the body's immune reaction to the abnormal cancer cells. To create these modified HIV-1 virions containing an engineered genome, we can use a host cell line that expresses all proteins in the virion yet does not produce normal genomes. Instead, the cells produce only the engineered genomes (capable of expressing the transgene when they enter a new host) and they package these genomes into virions. The key genetic construct present in such a typical cell line is an HIV vector plasmid that includes LTRs needed for expression, the transgene, and the psi (ψ) signal needed to package RNA into virions. The remaining necessary helper plasmids express the Gag-Pol and Env polyproteins (**Figure 10.21**). The virions produced by such host cells carry the artificial genome because it is the only RNA with a ψ signal, and the virions are subsequently collected and used in gene therapy. When these artificial virions infect a host cell, they insert the transgene into its genome and then the transgene is expressed under the control of the LTR regulatory sequences. The major risk in using retroviral vectors is that the

Figure 10.21 HIV-1 compared with a gene therapy vector derived from it. Several helper plasmids are needed to provide the proteins necessary to amplify the recombinant genome containing the transgene and to package it into infectious virions. The ψ sequence is required for packaging. The transgene construct lacks the *tar* RNA sequence and so the Tat protein is unnecessary for gene expression. Plasmids are circular but have been depicted as linear to make their features clearer.

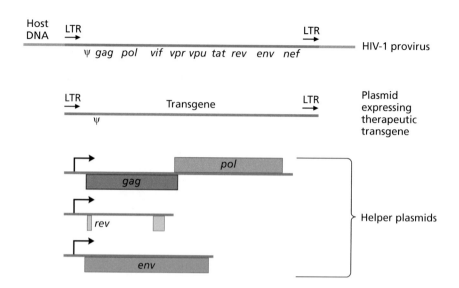

random integration of viral cDNA into the chromosome can occur in such a way that causes overexpression of host genes related to the cell cycle, resulting in cancer (see **Chapter 13**).

10.12   Hepadnaviruses are enveloped and have genomes containing both DNA and RNA in an unusual arrangement

Hepadnaviruses (family *Hepadnaviridae*), such as HBV, are spherical enveloped viruses with icosahedral capsids; the envelope has three different types of spike proteins (**Figure 10.22**). Hepadnavirus genomes are circular, double-stranded DNA, but with a gap where the DNA is single stranded (**Figure 10.23**). Additionally, the longer strand of DNA has a terminal protein covalently attached to the 5′ end, and the shorter DNA is actually a DNA–RNA chimera with a small number of RNA nucleotides at its 5′ end. Thus, hepadnaviruses are exceptions to the general rule that virions contain DNA or RNA, not both. There is also an area of triplex nucleic acids, consisting of two DNA strands and one RNA strand. In total, the DNA is only 3–3.3 kbp long, only slightly longer than circovirus genomes, yet these viruses encode up to eight proteins.

10.13   Hepadnaviruses use reverse transcription to amplify their genomes

A hepadnavirus replication cycle has many steps associated with gene expression and genome replication that occur in the nucleus (**Figure 10.24**). Gene expression and genome replication among the **hepadnaviruses** also involves reverse transcription, even though their genomes are predominantly DNA (see Figure 10.23). The enveloped virions attach to their host cells, and after entering host cells, the capsids with the genome inside are transported to the

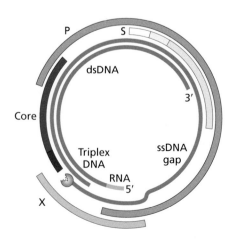

Figure 10.22 Hepatitis B virion. The key components of the HBV virion. (Courtesy of Philippe Le Mercier, ViralZone, © SIB Swiss Institute of Bioinformatics.)

Figure 10.23 Hepatitis B genome. Inside the virion, the 3.3-kbp genome is mostly made up of double-stranded circular DNA, although there is a gap with single-stranded DNA only, RNA at the 5′ end of the gapped strand, and a region of triplex DNA. The approximate locations of coding regions have been indicated using colored bars. There are three forms of the S protein, as indicated by the yellow shading. The HBeAg protein is an extracellular form of the core and is slightly larger, as indicated by the brick-red shading.

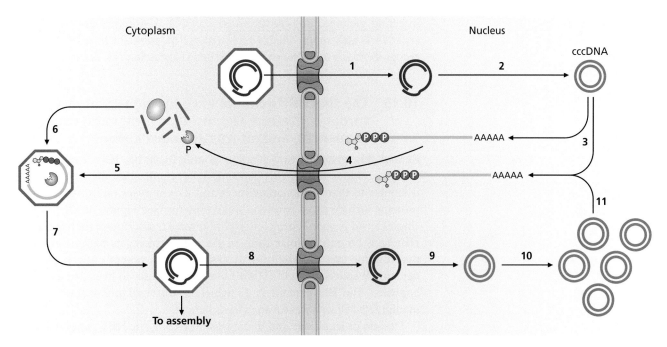

Figure 10.24 Overview of nucleic acid and protein synthesis during the hepadnavirus replication cycle. Uncoating releases the viral genome into the nucleus (1), where the host converts it into cccDNA (2). Transcription (3) results in mRNA exported to the cytoplasm, where it is translated (4). The long pregenomic RNA is exported but not translated (5) and then encapsidated (6). Inside the capsid it is reverse transcribed (7), resulting in a particle that can be trafficked to the nucleus (8), releasing the viral genome into the nucleus where it can be converted to cccDNA (9), which is subsequently amplified (10) and used to synthesize more mRNA (11). The particle can instead be trafficked to a membrane to continue the process of assembly.

nucleus along the microtubule cytoskeleton. After the relaxed circular DNA is released into the nucleus through a nuclear pore, host enzymes convert it into **covalently closed circular DNA (cccDNA)**. The cccDNA is maintained as an extrachromosomal episome throughout lytic infection. Gene expression then begins with host Pol II synthesizing viral mRNAs as well as viral **pregenomic RNA (pgRNA)**. Pregenomic RNA is encapsidated and then reverse transcribed inside the new capsid, ultimately forming new relaxed circular DNA. The new capsids can be trafficked back to the nucleus, where they release the new relaxed circular DNA into the nucleus thus amplifying the genome, or they can be targeted for assembly of new enveloped virions.

10.14 The cccDNA of HBV is not perfectly identical to the DNA in the infecting virion

After uncoating, host processes convert the infecting genome into cccDNA. The cccDNA is different from the infecting genome in that it lacks the 5′ terminal protein, does not have any duplicated (formerly overlapping) regions, and does not contain any RNA (**Figure 10.25**). The mechanism by which the

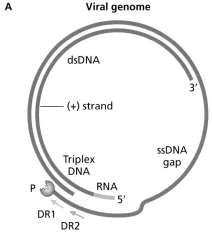

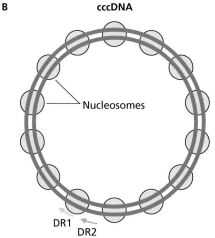

Figure 10.25 HBV infecting genome compared with cccDNA. (A) The infecting genome has P covalently attached to the 5′ end of the (−) sense DNA and RNA at the 5′ end of the (+) sense DNA. It also has a gapped (single-stranded) region and a triplex region. (B) The cccDNA, in contrast, is supercoiled dsDNA. It is not covalently attached to any proteins and does not contain any RNA. In both the infecting genome and the cccDNA, DNA complementary to the direct repeats that were critical for reverse transcription in the pregenomic RNA are marked DR1 and DR2. The host can only use cccDNA as a substrate for transcription.

infecting genomes are converted to cccDNA is not well understood. The cccDNA is also supercoiled, which makes it possible for host RNA Pol II to recognize it as a template so that viral gene expression can occur.

10.15   The tiny HBV genome encodes eight proteins through alternative splicing, overlapping coding sequences, and alternative start codons

Remarkably, given its tiny genome of only 3,200 bp, HBV expresses eight proteins (**Figure 10.26**). This particularly economical use of a small genome is similar to that in the circoviruses and parvoviruses (see **Chapter 9**). The proteins include three envelope spike proteins (see Figure 10.22). There is a single capsid protein (**core**), and a viral reverse transcriptase (polymerase; **P**). HBeAg is an extracellular form of the core protein; its name comes from the immune system's ability to recognize it as a potent antigen (Ag). The hepatitis B spliced protein (HBSP) likely contributes to liver damage during hepatitis. The last protein is X, which is not well understood but may manipulate cell survival for the virus's benefit.

Despite its small size and in contrast to HIV-1, the HBV genome has four different promoters (see Figure 10.26). It also has two enhancers and a single occurrence of DNA encoding a polyadenylation signal, so that all HBV transcripts have the same 3′ end because they use the same polyadenylation signal. The two enhancers stimulate transcription of the four promoters, using host transcription factors found commonly in differentiated hepatocytes (the host cell of choice). Expression of the three longest mRNAs is initiated at the first promoter, whereas transcription of the three shortest mRNAs is initiated

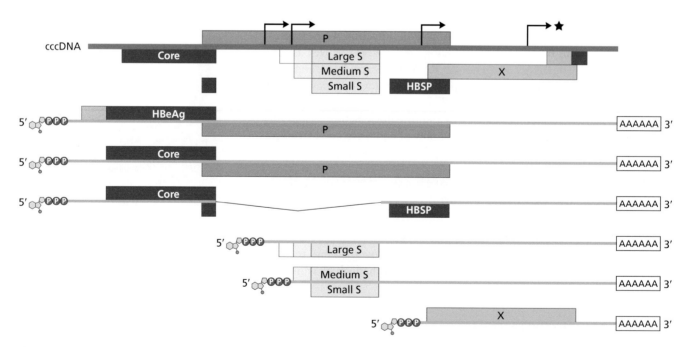

Figure 10.26 The HBV mRNAs with the proteins they encode. All of the mRNA is produced by host Pol II and therefore has typical eukaryotic 5′ caps and poly(A) tails. Although the cccDNA is circular, it is depicted here as linear in order to map the coding sequences to the genome more clearly. Note that the promoter marked with a star is the transcription start site for the mRNA encoding HBeAg and for the two mRNAs encoding core. Four of the mRNAs encode two proteins, whereas only one of the pre-mRNAs is subjected to splicing, as indicated by the angular black line. All of the transcripts are polyadenylated at the same site.

at a dedicated promoter in each case. Curiously, the two longer mRNA transcripts are longer than the genome itself, so that the polyadenylation signal is represented twice in those pre-mRNAs. Nevertheless, only the second signal near the 3′ end is recognized as a polyadenylation site. There is a correlation between protein abundance and the number of mRNAs that encode a protein, given that core is the only protein encoded by two mRNAs, and it is the most abundant protein per virion.

Only one of the smaller mRNAs, that encoding both core and HBSP, arises through splicing; none of the other HBV pre-mRNAs are spliced. Although the nuclear export of incompletely spliced HIV-1 mRNP has been studied in detail and involves a dedicated viral protein (Rev), it is not known how unspliced HBV mRNA is exported to the cytoplasm. Export may rely on a 500-bp sequence found in all HBV mRNAs, which is called the post-transcriptional regulatory element, but the viral or host proteins that interact with this sequence to enable viral mRNA export are unknown. It might be possible to examine host gene expression and how it changes during HBV lytic infection in order to identify candidate nuclear proteins that could be involved in nuclear export of mRNA containing the HBV post-transcriptional regulatory element (**Technique Box 10.3**).

The HBV genome is also economical in that every single coding sequence overlaps with the coding sequence for at least one other protein, enabling the production of twice as many proteins as are made by the similarly sized circoviruses (see **Chapter 9**). Most often, these overlaps also occur in the viral mRNA. Only two of the six mRNAs encode just a single protein, either the large S structural protein or X. The reason that X, a regulator needed in small amounts compared to structural proteins, has its own dedicated mRNA is not clear, though it is also expressed from a dedicated promoter that might be less active than the other three promoters. As was the case in many other viruses, the order of the coding sequences on the polycistronic mRNA molecules correlates with the abundance in which the proteins are produced. The more abundant proteins are encoded by the 5′ portion of the mRNAs, and the less abundant proteins are encoded by the 3′ portion of mRNAs. For example, the core protein is more abundant than the enzyme P, and the mRNA that encodes both core and P has the core coding sequence first, closest to the 5′ end of the mRNA. Leaky scanning, in which the ribosome occasionally skips the first AUG, is the mechanism used to translate the downstream proteins (see **Chapter 5**).

10.16   HBV genome replication relies upon an elaborate reverse transcriptase mechanism

Genome replication in HBV occurs through an RNA intermediate and involves the activity of the viral reverse transcriptase P. In some ways, HBV genome replication can be thought to be the opposite of reverse transcription in retroviruses. For retroviruses, reverse transcription is part of the uncoating process and results in dsDNA that enters the nucleus and inserts into the host DNA. For hepadnaviruses, reverse transcription is part of assembly and results in partially double-stranded DNA that is encapsidated and sent out of the cell to start new infections.

TECHNIQUE BOX 10.3   DETECTION OF MANY mRNA MOLECULES AT ONCE

There are many circumstances in which we might need to detect simultaneously all the mRNA that a particular virus or host cell can produce. For example, when studying a newly discovered bacteriophage with a long dsDNA genome, it is typical that 50% of the viral genes are unique and dissimilar to any previously discovered sequences in GenBank. In that case, classifying viral mRNAs according to the timing of their expression would provide clues as to the functions of the viral genes because of the well-known correlation between the timing of expression and the role of that bacteriophage protein, that is, whether the bacteriophage protein is regulatory, enzymatic, or structural. In addition, we might want to determine how a host cell attempts to block the progress of a viral infection. In that case, host cell mRNAs that increase during infection might encode proteins that interact with viral macromolecules and block their function.

In both these examples, it would be inadequate to monitor one or a few mRNAs at a time; instead, the research question demands use of a technique that can monitor many mRNA molecules at the same time. There are two techniques commonly used for this **gene expression profiling**: microarrays and RNA-seq. Both can analyze thousands of different sequences in a certain sample. Here we focus on the most common applications of these techniques.

A DNA microarray is a solid surface to which thousands of different DNA molecules have been attached in an ordered array of spots (**Figure 10.27**). Each spot has a different sequence and a computer keeps track of which sequences are represented in each spot. The DNA in each spot is referred to as the probe. Microarray experiments begin by collecting mRNA from two different sources and converting it into different populations of cDNA. For example, one sample might be from uninfected host cells, whereas the other sample might be from infected host cells. During the reverse transcription process, fluorescent dNTPs are used so that one sample of cDNA fluoresces red

and the other fluoresces green. In our example, imagine that the cDNA derived from the uninfected cells is red and the cDNA derived from the infected cells is green. After mixing the two populations of cDNA together, they are applied to the microarray, where the cDNA has the opportunity to bind to the DNA probes. A scanning device and computer then take a digital image of the red and green signals, which can be quantified over many orders of magnitude. The images can be superimposed (merged) so that spots with equal hybridization by both cDNAs appear yellow. Statistical analyses determine whether a given probe was bound by the same amount of cDNA from both samples or was instead bound preferentially by one of the two different cDNA populations. In our example, mRNA transcribed from host genes that are induced by viral infection will be more abundant in the cDNA that was labeled green, so those spots will be more green than red. From there, we can focus on discovering why viral infection induces those particular genes.

RNA-seq is a newer method for collecting mRNA from a complex sample and determining the sequence and abundance of all mRNAs in that sample. The method relies upon next-generation DNA sequencing methods that use clever engineering and software data analysis to sequence and quantify all the cDNA in a sample. There are many companies that compete to offer a variety of next-generation sequencing services, and because each relies on a different series of patented procedures, the details of how to accomplish the sequencing vary and will not be discussed in detail here. In every case, the protocol begins by collecting mRNA and converting it into cDNA. The cDNA is then manipulated using patented technologies that generate millions or even billions of short sequences, which are then assembled by software to reconstruct the original RNA sequences. The data set can be analyzed using statistics to determine the absolute abundance of cDNA in the sample or the relative abundance of cDNAs derived from different samples.

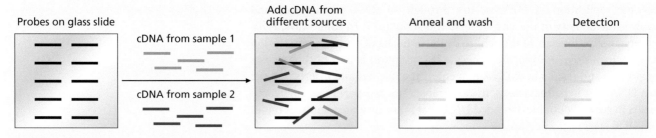

Figure 10.27 Microarrays used to study gene expression during viral infection. Single-stranded DNA probes with known sequences are arrayed on a solid surface such as glass. Fluorescent cDNA from different samples is applied to the array; when the two cDNAs are labeled either green or red, probes that hybridize to nucleic acids in both samples appear yellow. Subsequent quantification of the fluorescence signals (not shown) reflects different amounts of mRNA in the starting samples.

The second-longest HBV mRNA, encoding core and polymerase, is also known as pgRNA (pregenomic RNA) because it serves as the template for reverse transcriptase to synthesize the genome (**Figure 10.28**). This pgRNA has the interesting property of being both translated and used as the template for genomic DNA synthesis. As the RNA is translated, core protein accumulates in a ratio of approximately 300:1, compared with polymerase. These core proteins partially package the pgRNA and also serve an active, though currently undefined, role in reverse transcription (which cannot occur without core protein present). The features of pgRNA that are needed for genome replication include specific sequences and structures (see Figure 10.28). First, there are three direct repeats, with one near the 5′ end and the other two near the 3′ end. Two of them are perfectly identical in sequence and are designated DR1, whereas the third has a few different base pairs and is designated DR2. There are also two copies of a sequence called **epsilon (ε)**. The epsilon sequences fold into a stem–loop with a bulge (unpaired region) part way up the stem. These features of the pgRNA allow for discontinuous genome synthesis.

Within the maturing capsids, pgRNA and one molecule of P are co-localized. Unlike retroviral reverse transcriptase, the P protein has three major domains that play different roles during genome replication. One domain will play a role similar to that of the adenovirus terminal protein: it provides a covalent linkage to a single nucleotide that serves as the first primer for replication. The second will play a catalytic role, extending the protein primer while synthesizing DNA and extending other RNA and DNA molecules as they occur during the reverse transcription process, described shortly. This third domain has RNase H activity that degrades the RNA component of DNA–RNA hybrids.

To initiate genome synthesis, P becomes associated with the 5′ epsilon stem–loop structure (**Figure 10.29**). Next, the polymerase active site of P catalyzes a covalent bond between a tyrosine side chain in its terminal protein domain and a nucleotide complementary to the first template nucleotide in the ε bulge. Subsequently, the very same molecule of P uses the nucleotide covalently attached to the terminal protein domain as a primer and then copies three nucleotides of the bulge in the ε stem–loop, making a total of four nucleotides attached to P. This primer is also complementary to the nucleotides of DR1 in any (+) sense strand.

To continue with reverse transcription, the polymerase complex makes its first jump, which requires collaboration with the core protein (see Figure 10.29). The primer with its covalently attached terminal protein domain jumps almost to the 3′ end of the (+) orientation pgRNA template and binds to the very beginning of the DR1 repeat. Then P extends the primer, reverse transcribing the RNA through all nucleotides except the 5′ cap on the pgRNA. Because the same molecule of P that is covalently attached to the 5′ end of the primer catalyzes DNA synthesis, it is likely that the template RNA or new DNA loops out in some way so that the active site of the polymerase

Figure 10.28 Features of the pgRNA important for genome replication in HBV. There are three direct repeats, with one near the 5′ end and the other two near the 3′ end. There is also a sequence called epsilon (ε). The ε sequence folds into a stem–loop with an unpaired bulge part way up the stem. These features of the pgRNA allow for discontinuous genome synthesis. The bulge includes the sequence 5′ UUCA, which is also essential for genome replication.

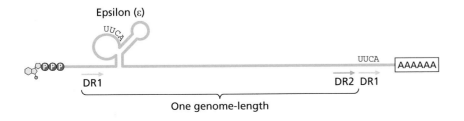

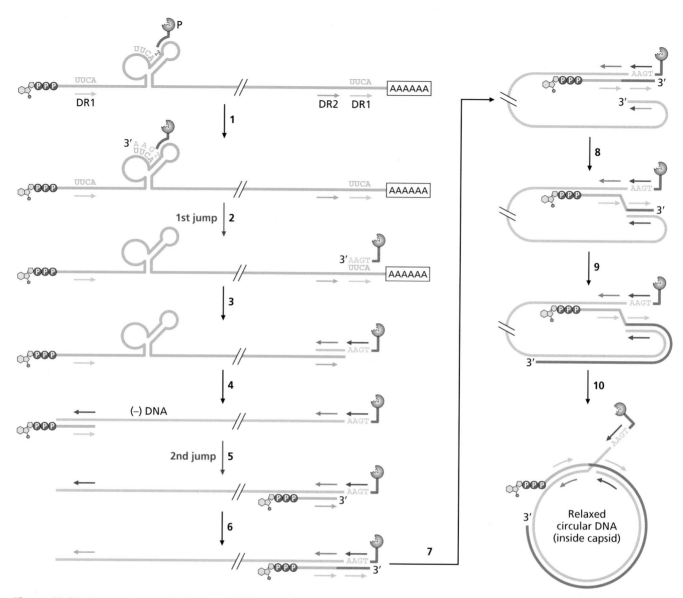

Figure 10.29 Reverse transcription of pgRNA to make a new HBV genome. From its position associated with the epsilon stem–loops, the P protein covalently modifies one of the amino acids in the terminal protein domain by attaching a T nucleotide. Then the P protein extends the 3′ OH on the T nucleotide, copying the bulge in the epsilon sequence (1). After reverse transcribing three more nucleotides, the P and its associated four nucleotides jump to the most 3′ copy of DR1, base pairing with it and providing a free 3′ OH opposite the RNA template (2). P extends this 3′ OH to synthesize more (–) DNA (3), while also degrading most of the RNA template (4). The RNA template jumps to base pair with the complement of DR2 near the 3′ end of the (–) DNA (5), and then the polymerase extends the 3′ OH on the RNA, synthesizing (+) DNA (6). The template circularizes (7), which enables the new (+) DNA to base pair with the 3′ end of the (–) DNA, so that the (+) DNA has a free 3′ OH opposite a (–) strand template (8). This template switch provides a long template strand. The P protein extends the (+) strand (9) for about a thousand nucleotides before stopping, resulting in relaxed circular DNA in the form found inside the virion (10). All the polymerization steps are catalyzed by the P protein that remains bound to the 5′ end of the (–) DNA throughout these reactions, so that looping of templates and new strands is probably necessary for the active site to be in the right position (not shown).

can move along the template, while the P domain remains covalently bound to the 5′ end of the new DNA. At the same time, the RNase H activity of the polymerase degrades most of the pgRNA but leaves some nucleotides at the extreme end intact. These include the complement of DR1 in the 5′ end of the pgRNA.

Next there is another jump where the 5′ end of the pgRNA moves and base pairs to the DR2 sequence in the new (–) strand of DNA. The P protein then synthesizes new (+) DNA using the 3′ end of the RNA as a primer and the new (−) DNA as a template. At about this same step, the template circularizes. Circularization allows template switching, so that the 3′ end of the (+) DNA can now provide a primer so that the P protein can copy around the circle of (−) DNA. The result is partially double-stranded DNA with one strand's 5′ end covalently attached to the P protein and the other strand's 5′ end composed of RNA with a 5′ cap.

The fate of this relaxed circular DNA (rcDNA) can take two paths. In a novel form of genome amplification, sometimes the capsids containing rcDNA go to the nucleus, where the rcDNA enters through a nuclear pore and is converted into cccDNA. As more and more such capsids go to the nucleus, the amount of cccDNA in the nucleus is amplified. Later during infection, the capsid with its rcDNA is instead further packaged and matures into enveloped virions that exit the host cell. The proportion of the time that one fate or the other occurs is undoubtedly regulated, but at this time there is no further information available about that process.

Essential concepts

- The retroviruses' genome must be reverse transcribed and the cDNA copy of the genome integrated into a host chromosome prior to the production of the first new viral mRNA.
- All retroviruses encode the three polyproteins, Gag, Gag-Pol, and Env, which are processed into components of the virion or the reverse transcriptase, integrase, and protease enzymes, which are essential for replication by all retroviruses.
- Despite their small genomes, more complicated retroviruses such as HIV-1 produce additional accessory proteins required for replication in certain hosts; examples in HIV-1 include Tat and Rev.
- Retrovirus genome replication occurs when the longest viral RNA is packaged instead of translated.
- Like retroviruses, hepadnaviruses encode a polymerase that has reverse transcriptase activity.
- The last step of uncoating in hepadnaviruses results in closed circular supercoiled DNA that host Pol II subsequently uses as a template for transcription.
- Hepadnaviruses use strategies such as splicing and leaky scanning to encode multiple proteins using a very compact genome (even smaller than that of the parvoviruses; see **Chapter 9**).
- Genome replication in HBV begins with a pgRNA template; the discontinuous mechanism of genome synthesis is catalyzed by the viral P polymerase.
- Both retroviruses and hepadnaviruses rely upon host Pol II for mRNA production and thus viral mRNAs have typical eukaryotic 5′ caps and poly(A) tails.
- Both retroviruses and hepadnaviruses rely on host proteins to activate the initiation of transcription.
- The mechanisms of reverse transcription in retroviruses and hepadnaviruses are distinct but both involve jumping reactions in which the polymerase copies regions of the templates that are not physically adjacent to one another in the template molecules.

Questions

1. Draw the genome of a simple retrovirus as it exists inside the virion; underneath, draw the genome as it exists once it has been integrated into a host chromosome. Compare and contrast the component parts of each nucleic acid.

2. What enzymes does HIV-1 encode and how is each employed?

3. What are the distinct roles for host transcription factors and the viral Tat protein in HIV gene expression?

4. If virologists made a mutant HIV virion containing a genome that cannot express the Rev protein but can express all other proteins normally, what would happen on a molecular and cellular level after infecting a host cell with such a virion? Exactly which stages of the reproduction cycle would be affected and which viral macromolecules would be present in the infected cell?

5. Describe the molecular mechanism that ensures that the Pol and Pro proteins are expressed at lower levels than the Gag proteins.

6. Define discontinuous synthesis of nucleic acids and list three viruses that use discontinuous synthesis.

7. Compare the use of a protein primer in hepadnaviruses with that in adenoviruses.

8. Compare and contrast the hepadnavirus reverse transcriptase protein P with host DNA polymerase, host RNA Pol II, and host primase.

9. Compare and contrast genome amplification in the hepadnaviruses and polyomaviruses.

10. For the sake of understanding how reverse transcription works, consider the case in which the sequences of the following elements in the genome are listed, from 5′ to 3′. (In this example, most of the sequences are much shorter than they would be in a real virus.) R = AAUg; U5 = CCgA; PBS = CAgU; PPT = UgAU; U3 = gAgC. What is the corresponding LTR sequence?

11. Which of the previous sequences, R, U5, PBS, PPT, or U3, is reverse transcribed into cDNA first?

12. Which of the previous sequences, R, U5, PBS, PPT, or U3, is reverse transcribed into cDNA last?

Interactive quiz questions

In addition to the questions provided above, this edition has a range of free interactive quiz questions for students to further test their understanding of the chapter material. To access these online questions, please visit the book's website: www.routledge.com/cw/lostroh.

Further reading

Retroviruses

Chavali SS, Mali SM, Jenkins JL, Fasan R & Wedekind JE 2020. Co-crystal structures of HIV TAR RNA bound to lab-evolved proteins show key roles for arginine relevant to the design of cyclic peptide TAR inhibitors. *J Biol Chem* 295:16470–16486.

Rafati H, Parra M, Hakre S, Moshkin Y, Verdin E & Mahmoudi T 2011. Repressive LTR nucleosome positioning by the BAF complex is required for HIV latency. *PLOS Biol* 9:e1001206.

Sobhian B, Laguette N, Yatim A, Nakamura M, Levy Y et al. 2010. HIV-1 Tat assembles a multifunctional transcription elongation complex and stably associates with the 7SK snRNP. *Mol Cell* 38:439–451.

Wang H, Liu Y, Huan C, Yang J, Li Z et al. W 2020. NF-κB-Interacting long noncoding RNA regulates HIV-1 replication and latency by repressing NF-κB signaling. *J Virol* 94:e01057-20.

Yamamoto T, de Crombrugge B & Pastan I 1980. Identification of a functional promoter in the long terminal repeat of Rous sarcoma virus. *Cell* 22:787–797.

The Nobel Prize in Physiology or Medicine 1975—Press Release. Nobelprize.org. Nobel Media AB 2014. 31 July 2016. http://www.nobelprize.org/nobel_prizes/medicine/laureates/1975/press.html

The Nobel Prize in Physiology or Medicine 2008—Press Release. Nobelprize.org. Nobel Media AB 2014. 31 July 2016. http://www.nobelprize.org/nobel_prizes/medicine/laureates/2008/press.html

Hepadnaviruses

Dörnbrack K, Beck J & Nassal M 2022. Relaxing the restricted structural dynamics in the human hepatitis B virus RNA encapsidation signal enables replication initiation *in vitro*. *PLOS Path* 18:e1010362.

Ko C, Shin YC, Park WJ, Kim S, Kim J & Ryu WS 2014. Residues Arg703, Asp777, and Arg781 of the RNase H domain of hepatitis B virus polymerase are critical for viral DNA synthesis. *J Virol* 88(1):154–163.

Qi Y, Gao Z, Xu G, Peng B, Liu C et al. 2016. DNA polymerase κ is a key cellular factor for the formation of covalently closed circular DNA of hepatitis B virus. *PLOS Path* 12:e1005893.

Sheraz M, Cheng J, Tang L, Chang J & Guo JT 2019. Cellular DNA topoisomerases are required for the synthesis of hepatitis B virus covalently closed circular DNA. *J Virol* 93:e02230-18.

Summers J & Mason WS 1982. Replication of the genome of a hepatitis B-like virus by reverse transcription of an RNA intermediate. *Cell* 29:403–415.

Assembly, Release, and Maturation

11

Virus	Characteristics
Tobacco mosaic virus (TMV)	Self-assembly of viruses first demonstrated; discovery of *pac* sites.
Bacteriophage T4x	Self-assembly demonstrated using lysates mixed *in vitro*; use of packaging motor to fill empty heads starting with genome concatemers.
Bacteriophage ΦX174	Example of sequential assembly and use of energy from DNA replication to fill procapsids.
(–) RNA viruses of animals	Examples of concerted assembly in which genomes associate closely with helical nucleocapsids during synthesis of genomic RNA.
Influenza virus	Concerted assembly; acquisition of envelope is temporally and spatially separated from assembly of capsomers and genomes.
Picornavirus	Naked virions that require proteolysis for maturation and use viroporins during exit by lysis.
Human immunodeficiency virus (HIV)	Enveloped virus that exhibits concerted assembly and envelope acquisition, requires viral protease for maturation, and leaves host cells through budding by subverting the cellular ESCRT pathway.
Bacteriophage λ	Packaging begins and ends at a *pac* site; lysis mediated by proteins dedicated to each of the three different layers in the host cell envelope (inner membrane, peptidoglycan, and outer membrane).
Coronavirus	Assembles in the endoplasmic reticulum (ER)–Golgi intermediate compartment.

The viruses you will meet in this chapter and the concepts they illustrate

In this chapter, we consider what happens at the conclusion of the virus replication cycle: assembly of capsids, genomes, and, when present, envelopes; maturation (required in some viruses but not in all); and release. At the conclusion of **Chapters 4–10**, we left each host cell full of virus components, including abundant structural proteins needed to manufacture the offspring virions and new genomes. On their own, the components are not infectious, which is why immune reactions often have the goal of killing an infected cell faster than the virus can complete its replication cycle. Instead, to be infectious the components must assemble in a very particular arrangement, creating infectious virions, and these must escape from the former host cell to find a new one.

DOI: 10.1201/9781003463115-11

11.1   The last stages of the virus replication cycle are assembly, release, and maturation

Viral assembly refers specifically to the assembly of capsids with genomes and so is universal to all viruses. Enveloped viruses must also acquire a membrane during assembly. Release is similar in all viruses in that the result is extracellular infectious particles, but there are two different methods of release, termed lysis and budding. Lysis is an option irrespective of the virus structure, whereas only enveloped viruses use budding. During **lysis**, the virus kills the host cell, which ultimately explodes because of osmotic pressure. Lysis releases all of the assembled virions in one large burst. During **budding**, in contrast, enveloped viruses exit the host cell by pinching off; the host cell may remain alive for some time until the energetic costs of viral replication take their toll and the cell dies despite not being lysed. Certain specific viruses have a **maturation** stage in which the assembled proteins undergo proteolysis, which is obligatory in order for the virion to be infectious. In other cases, during maturation the proteins must be rearranged in some way for the virion to be infectious. Either way, maturation events are irreversible and in eukaryotic viruses occur during or after release. In the literature about bacteriophages, readers may still encounter an older usage of the term "maturation," where it remains a synonym for assembly.

11.2   Unlike cells, viruses assemble from their constituent parts

A maxim for cell biology is "*Omnis cellula e cellula, id est,*" an idea first published by Rudolf Ludwig Carl Virchow in 1855, which means every living cell comes from a preexisting parental cell. More specifically, Virchow promoted the general principle that all cells arise from division of a parental cell. In contrast, although all viruses come from infected cells, those viruses arise by assembly, not by division of parental viruses. Study of one of the simplest known viruses, tobacco mosaic virus (TMV), first established that viruses assemble from their component parts. TMV has many copies of a single capsomer and one genome molecule. The X-ray crystallographer Rosalind Franklin, widely known for her crucial early images of DNA, was instrumental in determining that TMV virions are composed of 2,130 identical capsomers that are arranged in a helical pattern surrounding a single genomic RNA molecule (**Figure 11.1**).

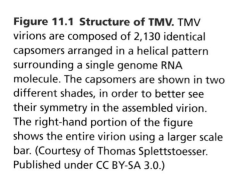
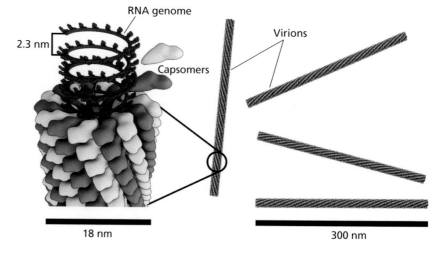

Figure 11.1 Structure of TMV. TMV virions are composed of 2,130 identical capsomers arranged in a helical pattern surrounding a single genome RNA molecule. The capsomers are shown in two different shades, in order to better see their symmetry in the assembled virion. The right-hand portion of the figure shows the entire virion using a larger scale bar. (Courtesy of Thomas Splettstoesser. Published under CC BY-SA 3.0.)

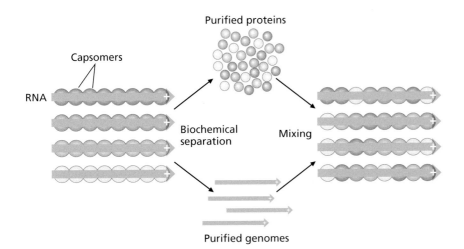

Figure 11.2 Spontaneous assembly of TMV *in vitro*. Purified TMV virions can be separated into protein and RNA components. The virions have been labeled with different colors in order to emphasize how new virions can be assembled from the separated components. Assembly is thermodynamically spontaneous and occurs readily at room temperature.

In the 1950s, Heinz Fraenkel-Conrat's group devised a method to separate the protein and RNA components of TMV *in vitro*. Doing this made it possible to demonstrate that the two component parts could be mixed back together, again *in vitro*, to reconstitute infectious virus particles (**Figure 11.2**). These experiments showed that TMV can be assembled from its component parts and so, in general, viruses do not reproduce by division but instead by assembly.

In vitro assembly of TMV is now understood in considerable molecular detail. The first step in formation of new virions is assembly of 34 capsomers to form a pair of discs (**Figure 11.3**). These discs bind to a special sequence in the RNA called the *pac* (for packaging) site. *Pac* sites are common in viral genomes and can be identified biochemically and genetically. For example, it can be demonstrated biochemically that mixing TMV RNA and proteins together at a ratio of approximately 1:50 leads to binding to the same 500 bp on every RNA molecule; that site is the *pac* site. Further examination indicates that the *pac* site contains multiple stem–loops, which are very likely important for binding to the discs. Genetic evidence also supports this definition of the *pac* site because mutations that disrupt the stem–loop structures

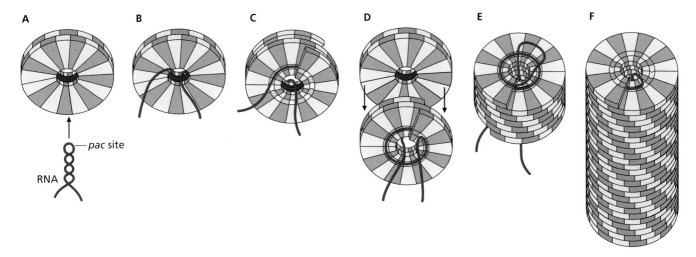

Figure 11.3 TMV assembly *in vitro* occurs in multiple steps. Double discs of 34 capsomers assemble (A). They bind to the *pac* site of the RNA (B), causing a rearrangement that makes the discs helical (C). Other discs join the top of the structure (D), drawing the RNA genome up the middle (E). Close-up on part of the assembled virion, which is a cylinder that is about 20 times longer than it is wide (F). (From Butler PJG 1999. *Philos Trans R Soc Lond B Biol Sci* 354:537–550. With permission from The Royal Society.)

interfere with virion assembly. Sometimes in other viruses *pac* sites are called ψ (psi), which is an abbreviation for packaging signal.

The association of the TMV nucleocapsid discs with the RNA causes a substantial conformational change in the discs, twisting them relative to one another so that they become narrower in diameter and form almost three full turns instead of two. At this point, additional 34-capsomer double discs can join on the top near the *pac* site and, as they join, they draw RNA up into the structure in a helical arrangement. They also assume the alternative three-turn shape as they join the growing virion. This process repeats itself until all the RNA has been packaged. Within the virion, the RNA is sandwiched between turns of each disc, leading to a very strong association between the capsid and genome.

11.3 Virions more structurally complex than TMV also reproduce by assembly, not by division

Replication through self-assembly of component parts makes viruses radically different from their host cells. The simplest known living cell, Synthia, has about 500 genes that were chemically synthesized from scratch and then placed into the cytoplasm of a living cell, where the synthetic genome replaced the resident natural genome. Even this simple cell cannot be separated into its component parts and then assembled from them. Despite its cute name, it is not a fully synthetic cell because it is not possible to synthesize all of a cell's component parts and then mix them together to create life. In addition, Synthia reproduces by division, as all cells do.

As the TMV example illustrated, viruses reproduce by assembly. Other experiments provide evidence that more complex viruses also reproduce by assembly. For example, the classic experiments of Hershey and Chase regarding the molecular basis of heredity helped establish that viruses do not reproduce by division. Hershey and Chase grew bacteriophages on bacteria living in media containing either radioactive dNTPs, to label the DNA, or radioactive amino acids, to label the proteins, and then they observed the result of infecting new host cells with phages that either had radioactive DNA (^{32}P) or radioactive proteins (^{35}S), which can be distinguished from one another. In particular, they sought to determine whether the DNA or the protein entered the host cell in order to direct synthesis of new bacteriophages. The DNA entered the host cells and directed synthesis of new bacteriophages, whereas the protein remained outside the host cell, far from the cytoplasmic site of bacteriophage production (**Figure 11.4**). This experiment showed that only part of the bacteriophage, namely the DNA, enters the host cell, strongly implying that this part of the bacteriophage acted independently of the parental phage proteins to direct production of offspring virions. If, instead, some form of division were responsible for virus replication, then the entire parental virus should have entered the host cell, and that was decidedly not

Figure 11.4 Only the DNA components of an infecting bacteriophage enter the host cell. Bacteriophages in which the DNA was labeled with ^{32}P and the protein with ^{35}S were prepared. They were used to infect host cells. Only the ^{32}P entered the host cells. Only the ^{32}P was ever passed on to offspring phages, and only rarely. Thus, viruses likely do not reproduce by division as their cellular hosts do.

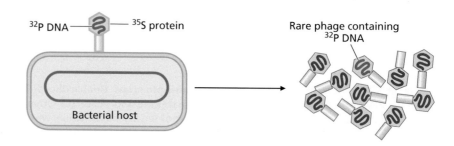

^{32}P DNA — — ^{35}S protein

Bacterial host

Rare phage containing ^{32}P DNA

the case. Furthermore, the ^{32}P and ^{35}S would both be distributed to offspring phages in about equal amounts, which was also not true.

Later, a combination of genetics and biochemistry showed that even the complex bacteriophage T4 with its icosahedral head, helical tail, and tail fibers could assemble spontaneously, despite its complexity (**Figure 11.5**). There were mutant T4 bacteriophages that could not complete their replication cycles because there was a block in one of the assembly steps. Host cells infected with one of these mutants would accumulate heads, or tails, or tail fibers without being able to make the rest of the phage. Lysates prepared from such cells could be mixed together *in vitro*, similar to the TMV experiments. In this way, investigators were able to demonstrate that a lysate with intact heads could complement a lysate with intact tail and tail fibers, so that mixing the lysates together created infectious virions, just as in the experiments with structurally simpler TMV.

11.4   Typical sites of assembly in eukaryotic viruses include the cytoplasm, plasma membrane, and nucleus

Different viruses assemble at various different subcellular locations (**Figure 11.6**). Bacteriophages, picornaviruses, and reoviruses assemble entirely in the cytoplasm. They also express and replicate their genomes in the cytoplasm. Polyomaviruses, papillomaviruses, and adenoviruses are naked viruses that assemble in the nucleus. Retroviruses and rhabdoviruses, which are enveloped, assemble at the plasma membrane, and coronaviruses and poxviruses, also enveloped, assemble on internal membranes. Herpesviruses and influenza virus are enveloped and assemble some components in the nucleus and other components at the plasma membrane or at an internal membrane site.

11.5   Eukaryotic virus assembly must take cellular protein localization into account

Because eukaryotic viruses assemble at different subcellular locations, protein localization is a very important aspect of viral biology. The synthesis and trafficking of cytoplasmic proteins, nuclear proteins, and transmembrane proteins are particularly important. Both cytoplasmic and nuclear proteins are synthesized on cytoplasmic polyribosomes. Nuclear proteins have a nuclear localization signal that interacts with importin, ultimately leading to import into the nucleus (see **Chapter 3**).

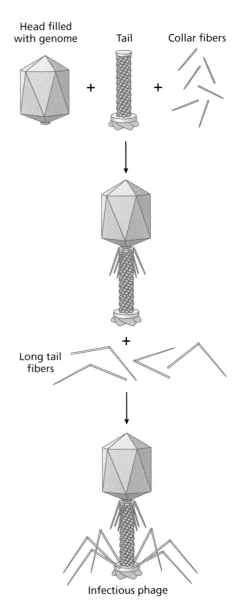

Figure 11.5 Assembly of T4 from lysates. Lysates prepared from host cells infected with different phage mutants contain intact heads, tails, collar fibers, or long tail fibers. Mixing the four lysates together reconstitutes infectious phages. (From Leiman PG 2003. *Cell Mol Life Sci* 60:2356–2370. With permission from Springer.)

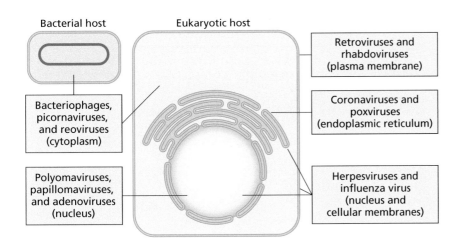

Figure 11.6 Subcellular locations of virus assembly.

In contrast, transmembrane proteins are synthesized on ribosomes associated with the rough endoplasmic reticulum (ER) so that insertion into the membrane is co-translational (see **Chapter 5**). Enveloped viral spike proteins are therefore synthesized on the ER. Because more than one ribosome binds to each mRNA, the viral transmembrane proteins end up in a patch of membrane that subsequently traffics through the cell, subverting the normal process for cellular vesicles moving through the endomembrane system (**Figure 11.7**). Normal cellular vesicles move proteins from the rough ER through the Golgi apparatus, where they acquire glycosylation, as do viral proteins passing through the Golgi apparatus. From the Golgi apparatus, glycosylated proteins in vesicles move to the plasma membrane, where the vesicles fuse with the plasma membrane so that the proteins in the vesicles become part of the membrane. The portions of the proteins that had been inside the lumen of the vesicle, where glycosylation enzymes modify the proteins, become the outside (distal) surface of the membrane and, in the case of viruses, will form the exterior surface of the viral envelope.

Keep in mind that the cytoplasm of a eukaryotic cell is very crowded and diffusion is much slower than in water. Diffusion is too slow to account for the ability of viral proteins to assemble and mature into a virion within the period of time needed for completion of the replication cycle. A eukaryotic cell is approximately 50–100 μm in diameter. In water, a poliovirus capsid could travel 10 μm in 3.85 sec. In the crowded cytoplasm, by contrast, it would take 1,800 sec or half an hour to travel the same distance, which is a significant barrier to encountering genomes and accomplishing assembly. Instead of depending upon diffusion for assembly, viruses rely on the cytoskeleton and motor proteins, especially on microtubules and kinesin. In fact, pharmaceuticals or other treatments that disrupt microtubules or microtubule-dependent motors interfere with viral assembly. Microtubules are arranged asymmetrically, with so-called minus ends at the centrosome near the nucleus and plus ends at the cell periphery. Different motor proteins carry cargo toward or away from the plus ends of microtubules. Kinesin carries cargo toward the plus ends, which are near the cell periphery, where enveloped viruses often assemble.

11.6 Capsids and nucleocapsids associate with genomes using one of two general strategies

Once capsomers and genomes arrive at the same subcellular location, they must still assemble. Virions have only a few basic arrangements of capsomers and nucleic acids, which determine the mechanism of assembly. For example,

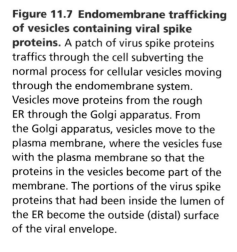

Figure 11.7 Endomembrane trafficking of vesicles containing viral spike proteins. A patch of virus spike proteins traffics through the cell subverting the normal process for cellular vesicles moving through the endomembrane system. Vesicles move proteins from the rough ER through the Golgi apparatus. From the Golgi apparatus, vesicles move to the plasma membrane, where the vesicles fuse with the plasma membrane so that the proteins in the vesicles become part of the membrane. The portions of the virus spike proteins that had been inside the lumen of the ER become the outside (distal) surface of the viral envelope.

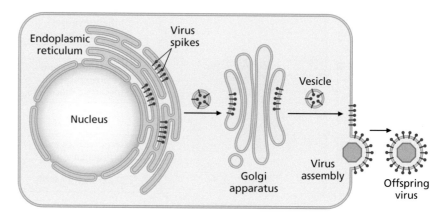

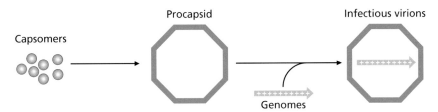

Figure 11.8 Sequential assembly. During sequential assembly, an empty procapsid assembles first and is subsequently filled with the genome.

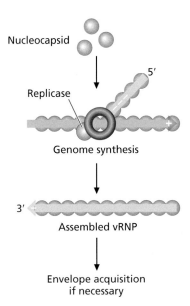

Figure 11.9 Concerted assembly. During concerted assembly, the capsid and the genome assemble together so that there are no empty procapsids. Instead, there is a nucleoprotein complex, such as viral ribonucleoprotein complex (vRNP) when the genome is RNA.

there might be a shell-like capsid with nucleic acid inside; this is the case for the icosahedral capsids. A second typical arrangement is a helical capsid in which the capsomers coat the genome, as is the case for TMV. The use of the term capsid to describe TMV is historical; in today's nomenclature such a tight physical association between the capsomers and genome would more likely be called a nucleocapsid. The third common arrangement is for a capsid or nucleocapsid to be surrounded by an envelope.

There are two general methods by which capsids and genomes associate with one another during assembly. In the first, the capsid forms separately from the genome and is subsequently filled in another step. During this **sequential assembly**, regular groups of capsomers such as pentamers or hexamers assemble first and then assemble together to make a hollow shell (**Figure 11.8**). These hollow **procapsids** are immature and usually contain proteins lacking in the mature virion. In the second strategy, namely **concerted assembly**, the capsomers and genome come together during genome synthesis assembly and there is no separate empty capsid to be filled (**Figure 11.9**). Concerted assembly is exemplified by the mononegaviruses, such as rabies virus. In that case, it is not possible to isolate assembled empty procapsids from infected cells because no such entities exist. Instead, the nucleocapsid protein encapsidates new genomes while they are being synthesized. Because the virus is enveloped, a later assembly step leads to acquisition of a matrix and envelope surrounding the helical nucleocapsid.

11.7 Assembly of some viruses depends on DNA replication to provide the energy to fill the icosahedral heads

Bacteriophages are models for assembly of icosahedral viruses with DNA genomes, including animal viruses such as polyomaviruses, papillomaviruses, adenoviruses, and herpesviruses. In general, icosahedral viruses tend to use sequential assembly by forming procapsids and then introducing DNA into the procapsid. The structurally simplest example understood in molecular detail is that of ΦX174, which relies upon rolling-circle (σ) DNA replication to provide the energy for assembly. The virion has spikes at the vertices and the interior has a single circular ssDNA genome and many copies of a basic protein that condenses the genome so that it can be packaged into the virion (**Figure 11.10**). In order to fill the procapsid with DNA, the template genome is closely associated with empty open procapsids (**Figure 11.11**). As replication occurs, the single strand of displaced template is pushed by the movement of the replication fork into the open procapsid. No other energy for packing the head full of nucleic acid is required. The DNA displaces virus proteins called scaffolds, discussed more in **Section 11.10**.

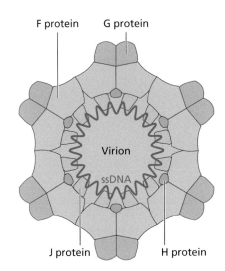

Figure 11.10 The ΦX174 virion. The ΦX174 virion has 60 copies each of proteins F, G, and J, and 12 copies of protein H. Protein G forms spikes. (Courtesy of Philippe Le Mercier, ViralZone, © SIB Swiss Institute of Bioinformatics.)

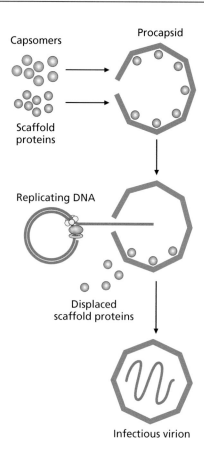

Capsomers

Procapsid

Scaffold proteins

Replicating DNA

Displaced scaffold proteins

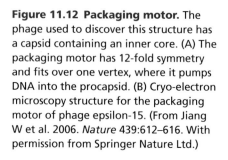

Infectious virion

Figure 11.11 Assembly of ΦX174 in a stepwise fashion. Procapsids containing scaffold proteins first assemble. DNA replication occurs in close association with the procapsids. As the movement of the replication fork provides energy to push DNA into the procapsid, the scaffold proteins are displaced, ultimately forming a mature infectious virion.

11.8 Assembly of some viruses depends on a packaging motor to fill the icosahedral heads

The structurally more complex dsDNA bacteriophages with tails also have an icosahedral head but fill them in a different way than ΦX174. In this case, the heads are filled after genome replication has concluded; it is not possible to use movement of the replication fork to provide force that could pack the heads full of DNA. Instead, these phages all have a special translocating vertex consisting of an opening and a **packaging motor** (Figure 11.12). The packaging motor uses hydrolysis of ATP to drive conformational changes that pump the DNA into the procapsid through the opening and is also responsible for cutting the DNA once a complete genome has been packaged. The remarkable packaging motor works against tremendous pressure; for example, the genome inside a full T4 head exerts 30 atm of pressure on the capsid.

All of the tailed dsDNA phages replicate their genomes in a way that results in long **concatemers** that must be processed during sequential assembly. There are three strategies used to ensure that each offspring virion gets a complete genome (Figure 11.13). The first is to have a unique *pac* site or sequence at each end of the genome, which is recognized by the packaging motor and results in site-specific cuts that package an entire genome in between two *pac* sites. Bacteriophage λ packages its DNA in this fashion; consequently, the ends of the genome inside each different λ phage is the same as that inside every other one. The second strategy to fill the heads is to begin packaging DNA at a certain *pac* site and then make sure that slightly more than one genome gets packaged after that, ignoring other *pac* sequences in the concatemer. Phage P22, for example, packages 104% of the genome. It begins at a single *pac* site but all later sites for cutting a concatemer are 104% away from the last cut, irrespective of nucleotide sequence. The consequence is that individual virions all have a complete copy of the genome but have different sequences at the termini of their linear genomes. The third strategy is that the packaging motor begins at a random site and uses a head-filling

Figure 11.12 Packaging motor. The phage used to discover this structure has a capsid containing an inner core. (A) The packaging motor has 12-fold symmetry and fits over one vertex, where it pumps DNA into the procapsid. (B) Cryo-electron microscopy structure for the packaging motor of phage epsilon-15. (From Jiang W et al. 2006. *Nature* 439:612–616. With permission from Springer Nature Ltd.)

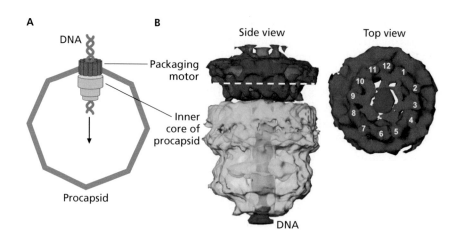

A

DNA

Packaging motor

Inner core of procapsid

Procapsid

B   Side view   Top view

DNA

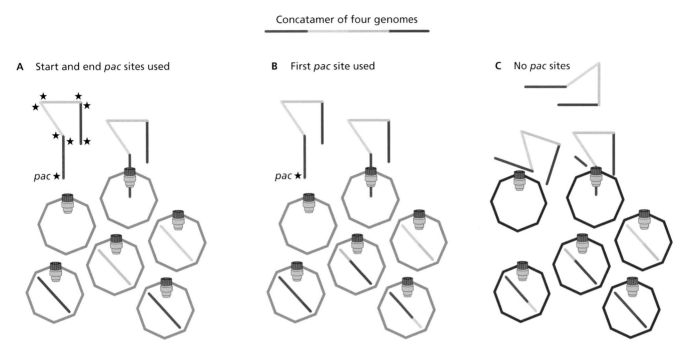

Figure 11.13 Three mechanisms of packing an icosahedral head starting with concatemers. (A) Packaging begins and ends at *pac* sequences. The genome inside each phage is the same. (B) Packaging begins with a *pac* sequence and then packages >100% of the length of the whole genome. The genome inside any individual phage has different 5′ and 3′ ends compared with those of other phages. (C) Packaging begins at a random site and packages >100% of the length of the whole genome. The genome inside any individual phage has different 5′ and 3′ ends compared with those of other phages.

mechanism to package more than 100% of the length of the genome, again producing a collection of phages in which the ends of the genomes inside any given head differ from those inside any other. In all cases, once the capsid has been filled, the packaging motor–DNA complex dissociates from it and then binds to a new procapsid to repeat the encapsidation process. There are specialized **head completion proteins** that bind where the translocating vertex had been, preventing exit of packaged DNA despite the enormous pressure exerted by the genome.

Herpesviruses are animal viruses with a similar packaging strategy to that of the tailed bacteriophages. In fact, the packaging motor proteins are homologous. During lytic replication, herpesvirus genome concatemers accumulate in the nucleus. After assembly of a procapsid, a packaging motor complex associates with one vertex, binds to a *pac* signal in the DNA, and fills the procapsid in a manner analogous to phage λ. Packaging of adenovirus genomes is also similar in overview. In contrast, naked icosahedral animal viruses with RNA genomes may not be assembled through a sequential pathway, though the evidence is not yet definitive.

11.9   Negative RNA viruses provide a model for concerted nucleocapsid assembly

Mononegaviruses such as rabies virus and Ebola virus have small (−) RNA genomes that are obligatorily associated with nucleocapsid protein N, forming viral ribonucleoprotein complexes (vRNPs) (**Figure 11.14A**; see **Chapter 6**). The nucleocapsid N protein is required during genome replication, and the new genomic RNA is encapsidated into RNP during RNA synthesis.

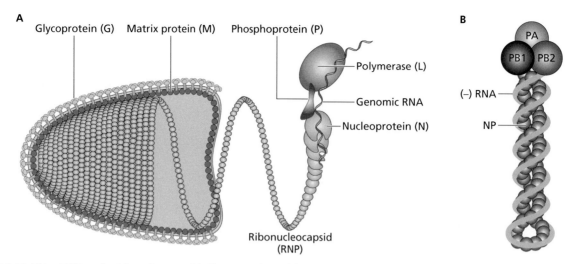

Figure 11.14 The vRNPs of rabies virus and influenza virus. (A) Rabies vRNP is a flexible helix composed of genomic RNA and N, P, and L proteins. (B) Influenza vRNP is a rigid helix composed of genomic RNA and NP, PA, PB1, and PB2 proteins. (Courtesy of Philippe Le Mercier, ViralZone, © SIB Swiss Institute of Bioinformatics.)

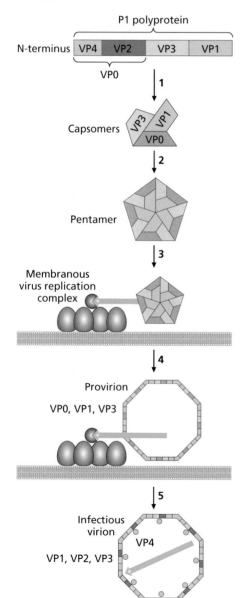

Influenza is also a (–) RNA virus, though it is not a mononegavirus. New helical influenza vRNP (**Figure 11.14B**) also assembles during RNA synthesis in this case.

There may also be concerted assembly of some icosahedral virions, especially those with RNA genomes. An example is poliovirus [(+) RNA; see **Chapter 5**]. It is not enveloped, but viral polymerase is enclosed by membranous viral replication complexes so that genome synthesis is associated with membranes. Although there is evidence that empty procapsids can be detected in lysates made from infected cells, increasingly these complexes are viewed as a dead end for viral assembly or perhaps as an artifact of methods used to lyse the cells, identify assembly intermediates, or both. Instead, it appears that capsid intermediates such as pentamers associate with the membranes used for viral RNA synthesis and that new genomes associate with the capsomers during synthesis (**Figure 11.15**). Picornaviruses use polyproteins for gene expression (see **Chapter 5**). Poliovirus, our example, has 60 copies each of the VP1 and VP3 capsomers, 59 copies each of the VP2 and VP4 capsomers, and 1 copy of the polyprotein VP0. The capsomers arise from the P1 polyprotein. First, the P1 is processed into VP1, VP3, and VP0. These associate in groups of five each to form pentamers. After the pentamers form they associate with newly replicated genomic RNA, forming procapsids that already have genomic RNA inside them. Then a maturation proteolysis occurs resulting in VP1, VP3, VP2, and VP4, which rearrange, forming a mature infectious virion in which the VP4 subunit is not displayed on the surface.

Figure 11.15 Concerted assembly of picornavirus. Picornaviruses such as poliovirus, the example here, use polyproteins for gene expression and exhibit concerted assembly. The P1 polyprotein is processed into the capsomers. First the P1 is processed into VP1, VP3, and VP0 (1). These associate in groups of five each to form pentamers (2). After the pentamers form, they associate with newly replicated genomic RNA (3), forming icosahedral procapsids that are not infectious (4). Then a maturation proteolysis occurs, resulting in VP1, VP3, VP2, and VP4, which then rearrange, forming a mature infectious virion in which the VP4 subunit is not displayed on the surface (5).

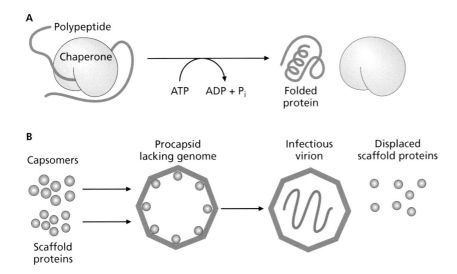

A

Polypeptide

Chaperone

ATP   ADP + P$_i$

Folded protein

B

Capsomers

Procapsid lacking genome

Infectious virion

Displaced scaffold proteins

Scaffold proteins

Figure 11.16 Chaperones and scaffolds are important for virus assembly. (A) Chaperones are proteins that make protein folding faster at the expense of ATP. (B) Scaffolds are proteins that are needed for parts of the virus to assemble but are not found in the mature virion.

11.10 To assemble, some viruses require assistance from proteins not found in the virion

Although many naked virions can be fractionated into their component parts and then reassembled simply by mixing the parts together, as in TMV, this experiment does not work for all viruses. One reason is that their assembly requires the presence of proteins that are not present in the final virion. An example of such proteins is a cellular or viral **chaperone** (**Figure 11.16A**). Chaperones are proteins that use NTP hydrolysis to catalyze protein folding faster than a protein can fold in the absence of catalysis, particularly in the crowded cellular milieu. Cellular chaperones have been found copurifying with many viral proteins including capsomers. Other viral proteins that assist in virion assembly are typically called **scaffold proteins** (**Figure 11.16B**) because they are required for early stages of assembly but are then displaced as assembly completes, much like a scaffold assists a house painter in finishing a job and is no longer present at the conclusion of the work.

11.11 Viruses acquire envelopes through one of two pathways

Enveloped viruses have more complex assembly than naked viruses for two reasons. First, they express both soluble cytoplasmic proteins and transmembrane proteins, which are translated at different subcellular sites yet must ultimately come together. Second, they acquire their membranous envelope at a site distinct from either the nucleus or the cytoplasm, where their offspring genomes accumulate. There are two different strategies by which some viruses acquire envelopes (**Figure 11.17**). In one case, acquisition of the envelope occurs sequentially after assembly of the capsid and genome has finished (see **Figure 11.17A**). The assembled capsid interacts with a cellular membrane that also contains virus spike proteins and pinches off; influenza virus uses this strategy. The other case is the formation of internal structures of the virion that occur in coordination with the envelope acquisition process; viruses such as HIV and SARS-CoV-2 mature in this manner (see **Figure 11.17B**).

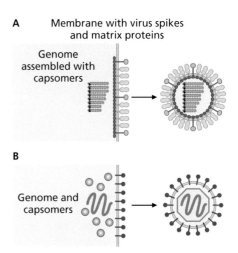

A   Membrane with virus spikes and matrix proteins

Genome assembled with capsomers

B

Genome and capsomers

Figure 11.17 Viral envelope acquisition. (A) Some viruses assemble a nucleocapsid or capsid prior to acquiring an envelope, as influenza does. (B) Some viruses acquire the envelope and assemble capsomers with the genome simultaneously, as HIV does.

11.12 The helical vRNPs of influenza virus assemble first, followed by envelope acquisition at the plasma membrane

A well-studied example of assembly is that of influenza virus, which has concerted genome–nucleocapsid assembly but sequential acquisition of its envelope. To understand assembly, it is important to review the structure of influenza virions (**Figure 11.18**). They have an envelope consisting of a lipid bilayer that includes four viral proteins: the HA spike, the NA spike, and the two matrix proteins M1 and M2. Internal to the lipid bilayer are eight different vRNPs, each consisting of a single different (–) sense RNA and the proteins PA, PB1, PB2, and NP. The NP coats the whole length of the RNA, whereas the other three form a trimeric complex found only once per vRNP (see **Chapter 6**).

The first stage of influenza assembly is for the PA, PB1, PB2, and NP proteins to form vRNP on new negative strands of RNA through concerted assembly (see Figure 11.9). For influenza virus, vRNPs arise from copy ribonucleoprotein (cRNP) templates and the genomic vRNPs are packaged into helical structures during genome synthesis (see **Chapter 6**). The virus must subsequently package vRNPs selectively with viral envelope rather than also packaging cRNPs (**Figure 11.19**). Remarkably, the ability to package vRNPs without packaging cRNPs appears to rely upon different interactions with PA, PB1, and PB2 between the two RNA molecules. This tiny difference

Figure 11.18 Influenza structure. Influenza viruses have an envelope consisting of a lipid bilayer including four viral proteins: the HA spike, the NA spike, and the two matrix proteins M1 and M2. Internal to the lipid bilayer are eight different vRNPs, each consisting of a single different (–) sense RNA bound to the proteins NP, PA, PB1, and PB2. (Courtesy of Philippe Le Mercier, ViralZone, © SIB Swiss Institute of Bioinformatics.)

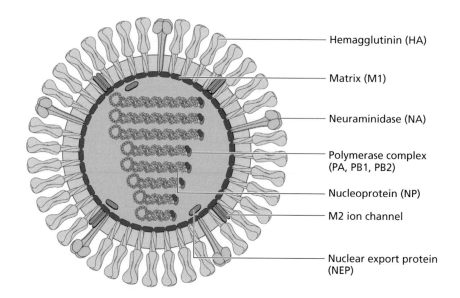

Hemagglutinin (HA)

Matrix (M1)

Neuraminidase (NA)

Polymerase complex (PA, PB1, PB2)

Nucleoprotein (NP)

M2 ion channel

Nuclear export protein (NEP)

Figure 11.19 Influenza vRNPs are targeted for export, whereas cRNPs are retained in the nucleus. Viral genomes are (–) sense, whereas the template to make more vRNP, called cRNP, contains (+) sense RNA. The (–) sense vRNPs have a slightly different interaction between the PA–PB1–PB2 complex and the RNA compared with that found in (+) sense cRNPs, which ultimately results in nuclear export of vRNPs. The cRNPs are selectively retained in the nucleus.

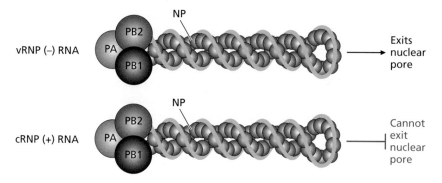

NP

vRNP (–) RNA   PB2 PA PB1   Exits nuclear pore

NP

cRNP (+) RNA   PB2 PA PB1   Cannot exit nuclear pore

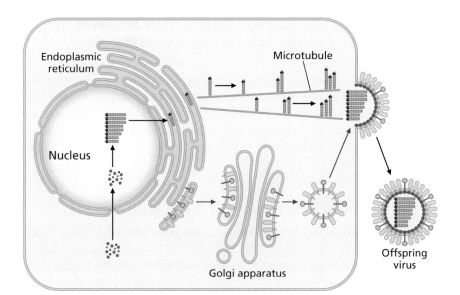

Figure 11.20 Envelope acquisition by influenza virions. Viral proteins are synthesized in the cytoplasm and on the endoplasmic reticulum. Viral proteins that will become part of new vRNPs must be trafficked to the nucleus (black arrows) while the viral envelope proteins traffic to the plasma membrane (red arrows). New vRNPs are synthesized in the nucleus and are then exported to the cytoplasm. They travel along the microtubule network at the periphery of the cell, bundling along the way. Eight different vRNPs gather under a patch of plasma membrane containing matrix and spike proteins. The new virion buds away from the surface of the host cell.

results in vRNP being competent for export out of the nucleus, whereas cRNPs are retained in the nucleus. The vRNPs are exported from the nucleus by co-opting the mechanism typically reserved for rRNA. Once exported, sequences near the 5′ and 3′ ends of the virus genome, which are conserved in all eight genome segments, target the vRNPs for packaging.

The virus does not package any combination of eight segments, however. Instead, it packages one of each of the eight different segments. Each of the different segments is needed for a successful subsequent infection because each encodes different essential genes. It appears that the eight RNA segments interact with each other to form a bundle through RNA–RNA interactions; this bundling has been demonstrated for two genome segments by creating mutations that destroy and then restore base pairing between the two, indicating that base pairing is important for bundling. The vRNPs travel through the cytoplasm using microtubules (**Figure 11.20**). They bundle along the way until they reach the plasma membrane where they coalesce under a patch of the plasma membrane that has been filled with the virion's four transmembrane proteins (two matrix proteins and two spikes). These were previously translated on the rough ER and then trafficked to the plasma membrane through the Golgi apparatus in the usual way. At this point the influenza virus is poised for release from the host cell.

11.13 Coronaviruses assemble in the ER–Golgi intermediate compartment

Coronavirus particles consist of a ribonucleoprotein center comprising its positive-sense RNA genome surrounded by an envelope with the viral S (spike), E (envelope), and M (membrane) proteins. SARS-CoV-2 and its close relatives assemble in the ER–Golgi intermediate compartment, sometimes known as "ERGIC." The ERGIC is a dynamic collection of tubules that deliver secretory cargo from the ER to the Golgi apparatus. In the case of coronaviruses, copies of the N protein assemble with genomic RNA and then the viral ribonucleoprotein interacts with the transmembrane structural proteins E and M. The spike protein may also contribute to virion formation, although this idea is still debated.

11.14   Some viruses require maturation reactions during release in order to form infectious virions

In general, maturation is defined as any irreversible process that occurs during or after viral release that takes the assembled virus from a noninfectious state to an infectious one. In two well-understood examples, picornaviruses and HIV, maturation is caused by a viral protease found inside the maturing procapsid. The maturation of picornaviruses was discussed as part of their concerted assembly (see Figure 11.15). The more complex situation in HIV is discussed in Section 11.15.

11.15   Assembly of HIV occurs at the plasma membrane

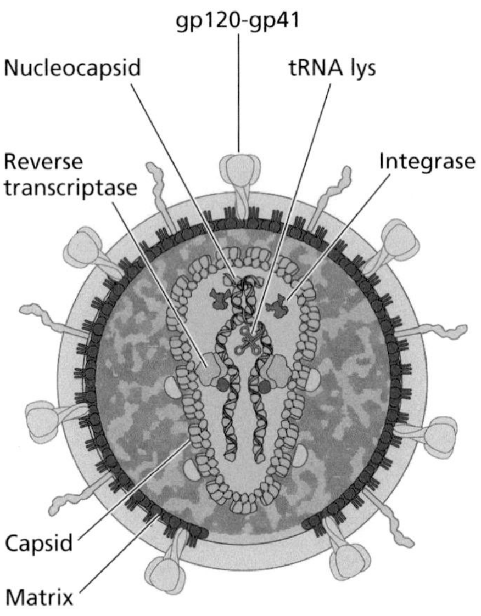

Every stage of the HIV-1 replication cycle has been studied in great detail because it was the cause of the AIDS pandemic and because molecular details about the stages make it possible to develop anti-HIV drugs that target each stage. Through this research, it has become clear that HIV-1 provides a classic example of concerted viral assembly. In order to understand how concerted assembly works, it is first important to review the components of an HIV virion and the relative arrangement of its constituent parts (**Figure 11.21**). Working from the periphery of the virion inward, the virion consists of the Env proteins SU and TM, followed by MA matrix protein. Inside the MA ring, there is a cone-shaped capsid made of the CA protein. The diploid RNA genome closely associates with the NC protein inside the CA. The virion contains the p6 protein along with the enzymes protease, reverse transcriptase, and integrase. HIV virions also contain a specific tRNA molecule annealed to the RNA genomes, and the accessory proteins Vif, Vpr, Vpx, and Nef are also found in the virions (see **Chapter 10**).

All the major structural proteins that form the internal parts of the HIV-1 virion are synthesized as polyproteins (**Figure 11.22A**). The Gag polyprotein can be processed into MA, CA, NC, and p6. The less abundant Gag-Pol polyprotein can be processed into the first three of these plus the enzymes protease, reverse transcriptase, and integrase. The Env protein is the precursor for the structural spikes SU and TM and goes through extensive glycosylation in the ER and Golgi apparatus before arriving at the plasma membrane.

Figure 11.21 HIV structure. The components of HIV-1 include an envelope, proteins, and RNA. (Courtesy of Philippe Le Mercier, ViralZone, © SIB Swiss Institute of Bioinformatics.)

Figure 11.22 Assembling HIV particle with polyproteins. (A) The Gag and Gag-Pol polyproteins give rise to proteins MA, CA, NC, p6, protease, reverses transcriptase, and integrase after proteolysis. (B) The Gag and Gag-Pol polyproteins assemble underneath patches of SU and TM spikes.

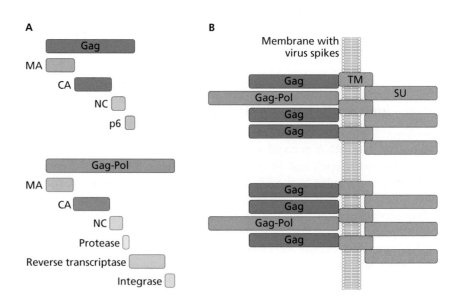

Because multiple ribosomes translate the Env RNA, SU and TM arrive at the plasma membrane in patches. It is underneath these patches of spike proteins that the Gag and Gag-Pol proteins assemble (**Figure 11.22B**). In a newly forming virion, the polyproteins are intact and form a sphere in which the MA protein associates with the lipid bilayer. The NC protein and future enzymes derived from Gag-Pol are found in the center of the sphere near the viral RNA. During or after budding (there is disagreement about when this takes place), maturation events initiated by the viral protease occur. The viral protease becomes active within the Gag-Pol polyprotein and cleaves the polyproteins into their final forms, which are MA, CA, NC, p6, protease, integrase, and reverse transcriptase (see Figure 11.22). These cleavages trigger a massive structural rearrangement, leading to the cone-shaped core that typifies HIV. Thus, this assembly process is considered to be concerted because no stable intermediate capsid structure lacking nucleic acid can be isolated. Furthermore, acquisition of the envelope also occurs at the same time that the nucleocapsid is assembling.

In addition to protein, the interior of the virion contains two types of RNA: two copies of the full-length genome and a specific tRNA. These occur in the virion because they interact with specific proteins. For example, the NC protein is responsible for packaging the diploid RNA genome. The RNA itself has a **dimer linkage sequence** responsible for dimerization of the RNA. There is a ψ sequence in the RNA as well, which folds into multiple stem–loops that are required for its interaction with NC; the NC packages the nucleic acid into the virion. A second key RNA packaging event is inclusion of host tRNA, which will be needed early during the next infection to prime reverse transcription (see **Chapter 10**). Virions each contain 50–100 copies of tRNA, although only two of the tRNAs are tightly associated with the RNA genome at the site that will prime DNA synthesis during the next infection.

11.16   Inhibition of HIV-1 maturation provides a classic example of structure–function research in pharmaceutical research

Maturation of HIV virions requires the action of HIV protease, which cleaves several protein targets and creates the final infectious form of the virion; without these cleavage events, the virions are not infectious. A major breakthrough in the treatment of HIV disease and in slowing the progression to AIDS was the development of **protease inhibitors**. Although most drugs are discovered by screening hundreds of thousands of randomly selected compounds, protease inhibitors were designed rationally, using the structure of the HIV-1 protease to intentionally design compounds likely to bind to the active site and to be uncleavable. More specifically, protease inhibitors were designed to mimic the transition state of the HIV proteins during cleavage. After an enzyme (E) binds to its substrate (S), it forms an ES complex. Subsequently, the ES complex enters a transition state where it becomes very likely that the substrate will react, in this case to be cleaved by protease. All HIV-1 protease inhibitors bind in the active site through an extensive hydrogen bond network (**Figure 11.23**). They also make extensive van der Waals interactions with the protease, binding to 30 different amino acids through such intermolecular forces.

Some protease inhibitors mimic peptides so closely that they can be cleaved by protease. Others have sufficiently different chemistry so that the enzyme's mechanism of action cannot break the bond. For example, the bond normally cleaved is a peptide bond consisting of —NH—CO—. The protease

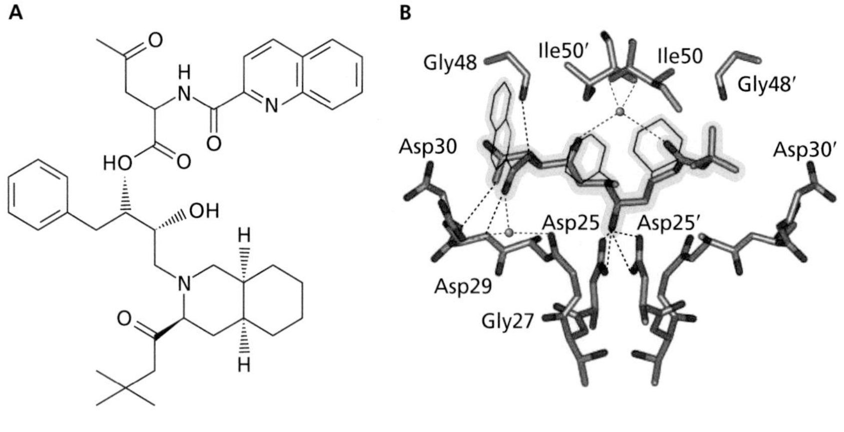

Figure 11.23 The protease inhibitor saquinavir hydrogen bonds extensively with HIV protease. (A) The structure of saquinavir. (B) In the chemical structures, red is oxygen and blue is nitrogen. Saquinavir is the molecule in the middle, highlighted in yellow and drawn showing the atoms of its ring structures in less detail. Protease is a dimer; the amino acids glycine 27 (Gly27), aspartic acid 29 (Asp29), Asp30, Gly48, and isoleucine 50 (Ile50) from one monomer are depicted, as are the same amino acids from the other monomer (Gly27′, Asp29′, Asp30′, Gly48′, and Ile50′). Hydrogen bonds are indicated by a dotted line. In some cases, the drug binds to the protein through hydrogen bonding with a water molecule (depicted as a red sphere with red dotted lines for hydrogen bonding). (A, Published under CC BY-SA 3.0. B, From Tie Y et al. 2007. *Proteins* 67:232–242. With permission from John Wiley & Sons.)

uses a water molecule and two aspartic acids in the active site to achieve proteolysis (**Figure 11.24**). In some protease inhibitors, these atoms are replaced by —CH$_2$—CH(OH)—, which cannot be cleaved by protease because the lack of the N and the carboxyl group prevents the enzyme's mechanism of action. In either case, the protease inhibitor competes directly for binding to the active site and thereby reduces the proportion of time that the active site is occupied by viral proteins.

Because the rational design of protease inhibitors was so commercially and medically successful, **rational drug design** in general has become much more common. The accelerated pace of solving the structures of proteins important in disease has also contributed to this switch. A more recent example of successful rational drug design in virology is that of drugs that target the protease of hepatitis C virus, which has revolutionized patient

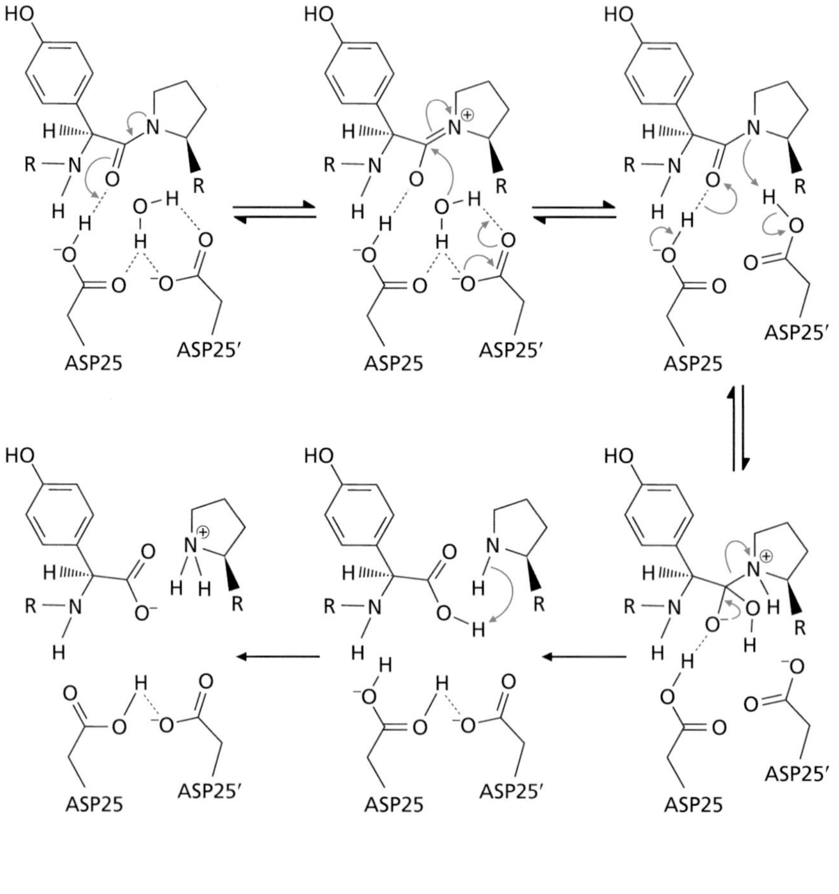

Figure 11.24 HIV-1 protease mechanism of action. The mechanism relies on water and two aspartic acids to hydrolyze the peptide bond. Asp25 is from one of the two subunits in the protease dimer and Asp25′ is from the other subunit. Movement of electrons is indicated by curved arrows using conventions common in organic chemistry. (From Brik A & Wong C-H 2003. *Org Biomol Chem* 1. doi: 10.1039/B208248A. With permission from Wiley-VCH Verlag GmbH & Co. KGaA.)

care for an infection that previously was impossible to treat in about 70% of patients. Untreated viral hepatitis led to fatal liver cancer in many instances. Rational drug design is also currently being used to find drugs to treat COVID-19. Rational design of additional antivirals is just around the corner, provided that there is a financial incentive to produce them.

11.17   Release from bacterial cells usually occurs by lysis

Most living things, including bacteria and plants, have rigid cell walls covering their plasma membranes. These cell walls squeeze back against osmotic pressure, protecting the plasma membrane from osmotic lysis. They also provide protection against viruses except in cases in which the viruses have evolved mechanisms to bypass the cell wall during penetration (see **Chapter 3**) and release of offspring virions.

To examine virus exit from bacterial host cells, this section focuses on **bacteriophage λ**. Bacteriophage λ infects bacteria that stain negative in the Gram reaction, which means that they have a particular type of cell wall external to the plasma membrane (**Figure 11.25**). An example of a common Gram-negative bacterium is *Escherichia coli*, which is the host for bacteriophage λ. The cell envelope consists of the plasma membrane, the **periplasm** filled with a loose network of a macromolecule called **peptidoglycan**, and then a second lipid bilayer, called the **outer membrane**. The peptidoglycan is a very large molecule that surrounds the cell in three dimensions like a net and protects the cell from osmotic lysis. In order to escape, offspring phages need to make it through the plasma membrane, the net-like peptidoglycan meshwork in the periplasm, and the outer membrane.

To lyse the host cell, bacteriophage λ encodes different proteins that each attack one of these layers. All three are expressed late in infection so that there is time for assembly of offspring phages before lysis occurs. The protein **holin** attacks the plasma membrane (**Figure 11.26**). It is a transmembrane protein that clumps together with other molecules of holin. At a high threshold amount the holin proteins rearrange themselves, forming huge pores in the plasma membrane. These can be on a micron scale, which is remarkable considering that the cell is approximately 2 μm long. By virtue of these pores, the second λ lysis protein, which targets the peptidoglycan, can access the periplasm. This protein is called **endolysin**, and it degrades the peptidoglycan (**Figure 11.27**), doing so to such an extent that the host cells can lose their

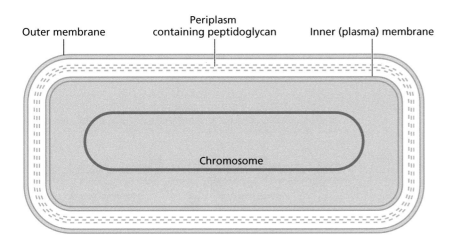

Outer membrane

Periplasm containing peptidoglycan

Inner (plasma) membrane

Chromosome

Figure 11.25 Diagram of Gram-negative cell envelope. The cell envelope of a Gram-negative bacterium such as *E. coli* consists of its plasma membrane, a surrounding periplasm filled with the macromolecule peptidoglycan, and an outer membrane.

Figure 11.26 Activity of holin.
Holin proteins aggregate in the plasma membrane. When they become very numerous, they rearrange, forming large aqueous pores through the plasma membrane.

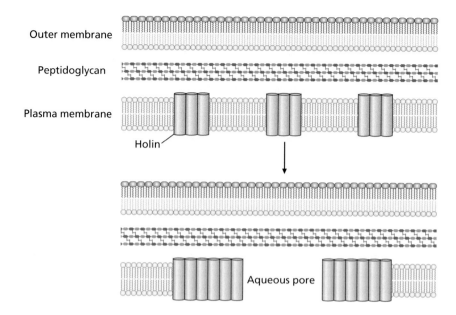

Outer membrane

Peptidoglycan

Plasma membrane

Holin

Aqueous pore

rod shape and become spherical. The result is that the mesh of the peptidoglycan becomes full of holes large enough to permit virions to escape.

Until recently it was thought that these two mechanisms alone could account for host cell lysis, with the idea that the compromised peptidoglycan would result in osmotic lysis by water rushing into the cell. Not all cells live in a hyperosmotic solution, however, and there is, indeed, a third protein needed for viruses to breach the outer membrane and escape. This protein is called **spanin**, because it has transmembrane segments embedded in the plasma membrane and in the outer membrane and thereby spans the periplasm (**Figure 11.28**). Spanin has the same function as viral fusion peptides: it causes membrane fusion (see **Chapter 3**). In this case, spanin fuses the inner membrane with the outer membrane, which effectively leads to the escape of offspring phage from the cytoplasm (see Figure 11.28). The discovery of spanin was surprising because it was thought that eukaryotes use membrane

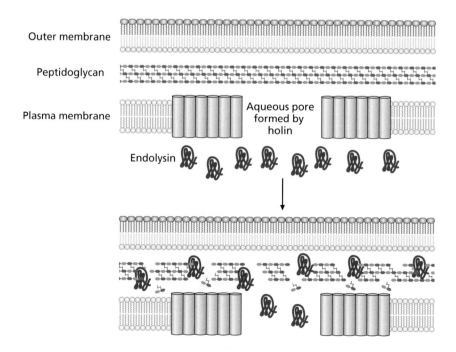

Outer membrane

Peptidoglycan

Plasma membrane

Aqueous pore formed by holin

Endolysin

Figure 11.27 Activity of endolysin.
Endolysin is an enzyme that can enter the periplasm only after holin creates large aqueous pores in the plasma membrane. After entering the periplasm, it binds to peptidoglycan and degrades the macromolecule.

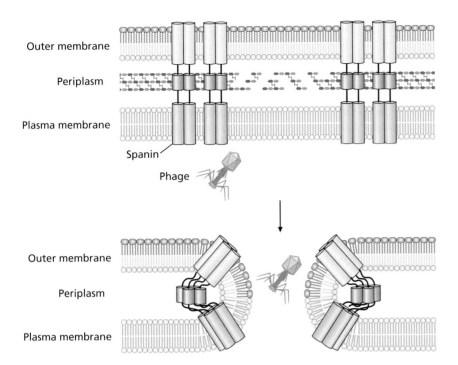

Figure 11.28 Activity of spanin. After holin and endolysin activity, spanin fuses the outer membrane to the plasma membrane, forming large channels through which phages escape.

fusion for many routine functions, whereas bacteria were not thought capable of membrane fusion. Examination of many bacterial genomes indicates that they can acquire genes from phages and that those genes can be adapted for a new purpose (rather than phage lysis in this case). Therefore, although the λ spanin is a phage protein, surely discovery of a bacterium with a spanin protein put to its own use is likely.

11.18   Release from animal cells can occur by lysis

Adenovirus is a classic example of a virus that exits an animal host cell through host cell lysis. Unlike bacteriophages and plant cells, there is no cell wall to obstruct virus egress. Instead, the cytoskeleton is key for maintaining cell integrity, and therefore viruses that exit through lysis must disrupt the cytoskeleton. An adenovirus late protein called **L3-23 K** is a protease that cleaves the **intermediate filament** cytokeratin so that the intermediate filament is no longer able to assemble properly. The effect is to deprive the cell of mechanical integrity and make it much more susceptible to lysis. There is a second protein, the **adenovirus death protein**, which kills cells when it accumulates during late stages of infection and causes lysis. The adenovirus death protein is a transmembrane protein that can be found in the nuclear membrane, ER, and Golgi apparatus. The mechanism by which it causes lysis is not understood. It is also not clear how lysis releases adenoviruses from the nucleus, where they can accumulate to more than 100,000 virions per infected cell. Other animal viruses that exit by lysis include picornaviruses, which encode viroporin proteins that enhance or cause lysis of the plasma membrane.

11.19   Release from animal cells can occur by budding

Chapter 3 demonstrated how a dynamic plasma membrane is essential for animal viruses to enter cells and to assemble; this section considers how the animal cell plasma membrane is also a critical determinant for how viruses exit an animal cell. Viral budding from the plasma membrane is the opposite of the process of a secretory vesicle fusing with the plasma membrane. In the case of a secretory vesicle, the fusion event results in the interior contents of the vesicle being released (**Figure 11.29A**). The components of transmembrane proteins that once faced the interior of the vesicle ultimately face the outside of the cell. In the case of a virus budding, the capsid, which is found in the cytoplasm, remains interior to the membranous structure that buds away (**Figure 11.29B**). Because of these differences, viruses cannot use the secretory machinery for budding; instead, they must use proteins designed for pinching off in such a way that the interior contents of a vesicle remain on the interior. Such a system can actually be found among the **endosomes** because the contents of endosomes become pinched off during normal endosome maturation to form **multivesicular bodies** or **MVBs** (**Figure 11.30**). The interior contents of the forming vesicular body, which is contiguous with the plasma membrane, remains interior to the vesicle after pinching off.

MVBs participate in the process of breaking down proteins marked for degradation in the lysosome. The MVBs are spherical vesicles that form in

Figure 11.29 Topology of secretory vesicle fusion versus viral budding. The contents of a cell have been simplified to show the cytoplasm and a secretory vesicle. (A) The interior contents of a secretory vesicle become externalized when the vesicle fuses with the plasma membrane. (B) In contrast, when a virus buds, the exterior surface of the plasma membrane remains exterior. The interior contents of the budding virus, which are contiguous with the cytoplasm, remain on the interior of the cell.

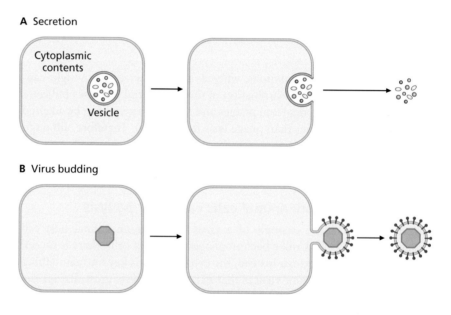

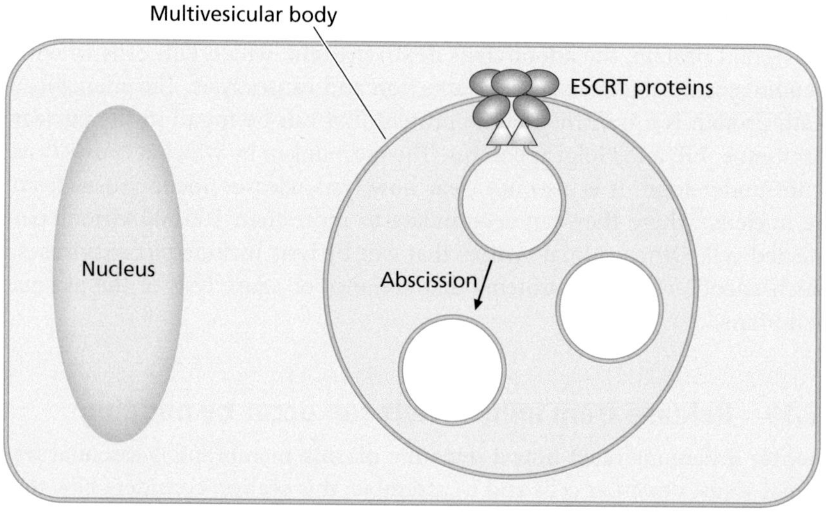

A Secretion

Cytoplasmic contents

Vesicle

B Virus budding

Multivesicular body

ESCRT proteins

Nucleus

Abscission

Figure 11.30 ESCRT proteins create multivesicular bodies. MVBs have interiors that are topologically identical to the cytoplasm. ESCRT proteins catalyze abscission to form MVBs.

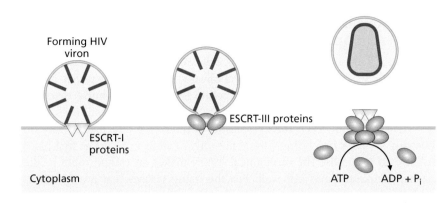

Figure 11.31 HIV budding from host cells using the ESCRT machinery. The Gag polyprotein (red lines) interacts with an ESCRT-I complex, initiating budding. The ESCRT-III machinery is recruited, and finally a cellular ATPase disassembles the ESCRT machinery, resulting in abscission and release of the enveloped virus. Proteolysis during maturation results in the characteristic trapezoidal shape of the HIV capsid.

the endosome and become more abundant in late endosomes, immediately prior to fusing with the lysosome, at which point the proteins are degraded and recycled. The MVB forms in the opposite way that a secretory vesicle joins the plasma membrane, and the membrane fusion event has a special name, **abscission**. Abscission also occurs when one cell undergoes mitosis and cytokinesis gives rise to two offspring cells. Abscission is catalyzed by protein complexes called the **ESCRT machinery** (see Figure 11.30), which is an acronym for endosomal sorting complexes required for transport.

HIV-1 illustrates the use of the ESCRT machinery to bud from and exit the cells (**Figure 11.31**). A portion of the Gag polyprotein has a short sequence called a **late domain**, which is essential for interacting with the ESCRT proteins. Many viruses that bud have these late domains, which are characterized by being four amino acids long and containing proline. In the case of HIV-1, the specific sequence is proline–threonine–alanine–proline, or PTAP using the universal one-letter amino acid codes. Interaction with the ESCRT-I complex begins the process of budding. The ESCRT-III machinery is recruited to form an ever-tightening spiral underneath the budding virion. Finally, a cellular ATPase disassembles the ESCRT machinery, which results in release of the enveloped virus.

Once virions have left the cell surface and successfully budded away through abscission, they can face a significant hurdle. The virus is escaping from the very cell its ancestor was primed to attach to and enter. What is to prevent the spike proteins from binding to the old host cell on the way out or binding in such a way that escape from the cell surface is impossible? Viruses have evolved mechanisms to prevent this scenario. In the case of HIV-1, two proteins called Vpu and Nef collaborate to remove the HIV receptor CD4 from the cell surface or to prevent it from being displayed there in the first place. In this way, HIV-1 prevents exiting virions from attaching to the very host cell that produced them.

Host cells infected with enveloped viruses produce thousands of virions. In most cases, budding in combination with the metabolic strain on the host cell causes its death, even though budding might at first seem gentler than lysis. HIV-1 is unusual in that the host cell can remain alive for many days despite viral replication and budding.

11.20   Release from animal cells can occur by exocytosis

New SARS-CoV-2 particles exit the ER–Golgi intermediate compartment by budding into a vesicle containing several virus particles. It is not clear whether this process uses the ESCRT system. Subsequently, SARS-CoV-2 particles most likely exit host cells through an exocytosis pathway that may involve lysosomes.

11.21   Release from plant cells often occurs through biting arthropods

Plant viruses have to surmount a thick cell wall when they exit the cell (see **Chapter 3**). Plant cell walls are metabolically inert layers external to the plasma membrane. They provide protection from osmotic shock, mechanical strength, and rigid support to the whole plant. The cell wall is so thick that viruses cannot reach the cytoplasm on their own and instead rely on mechanical injury or a biting animal not only to enter a cell in the first place but also to exit cells. The mechanical injury makes an escape route in the plasma membrane and cell wall. For the many viruses that are transmitted by arthropods, the animal sucks the viruses up into its digestive tract, where the virus may or may not replicate (see **Chapter 3**). Alternatively, plant viruses can spread from one cell to another through plasmodesmata (singular, **plasmodesma**), which are extensions of the ER of one cell that reach into a neighbor. Some plants can reproduce clonally so that, for example, a buried stem can form a new plant. For these plants, clonal reproduction can also result in clonal propagation of a virus.

Essential concepts

- Cells reproduce by division, whereas viruses replicate by assembly. Some viral assembly reactions are catalyzed by host chaperone proteins or viral scaffold proteins.
- In eukaryotes, the cytoskeleton is essential for bringing capsomers, genomes, and other virus components together because diffusion is too slow for all the component parts to encounter one another.
- During sequential assembly, virus capsids form separately and then are subsequently filled with genomic nucleic acids. During concerted assembly, virus capsomers and genomes come together during genome synthesis so that it is not possible to isolate empty procapsids.
- Enveloped viruses can acquire their envelopes either after the internal parts of the virion have assembled, as exemplified by influenza virus, or during the process of assembling the internal components of the virion, as exemplified by HIV.
- Protease inhibitors, developed using rational drug design, block the HIV protease from accessing its normal viral substrates. These drugs are critical for fighting the HIV/AIDS pandemic and their success is leading to more investment in rational drug design to treat other viral infections.
- The nature of the host cell surface determines the mechanism or mechanisms needed for new virions to exit their host cells. Viruses that must escape an animal cell face different biophysical challenges from those that must escape from a bacterial cell.
- The λ virus is a bacteriophage that serves as a model for exit from host cells with multiple exterior layers. Viral holin protein creates holes in the cytoplasmic membrane, viral endolysin degrades peptidoglycan in the periplasm, and spanin creates channels through the plasma membrane and outer membrane so that the virions can escape.
- Animal viruses such as adenovirus escape by weakening the cytoskeleton, which results in lysis. Other animal viruses escape by budding in which enveloped viruses co-opt normal host abscission pathways.
- Plant viruses rely on mechanical damage for spread, as well as on spreading from cell to cell throughout a plant using plasmodesmata or through clonal inheritance.

Questions

1. Compare and contrast self-assembly of TMV and picornaviruses.
2. What is the difference between a chaperone and a scaffold?
3. Which viral proteins in general are translated by ribosomes attached to the rough endoplasmic reticulum?
4. Why do enveloped viruses use the ESCRT system to bud from the plasma membrane, instead of using the cell's secretion mechanism?
5. Of all known enveloped viruses, only a fraction of them infect host cells that have cell walls. What specialized problems would an enveloped virus face by replicating in a host with a cell wall?
6. How is HIV assembly different from influenza assembly?
7. Which stages of the HIV life cycle are blocked by protease inhibitors?
8. Why does λ virus require three different proteins to exit host cells?
9. How could bioinformatics be used to examine a newly discovered enveloped virus in order to identify viral proteins that might be involved in budding?

 Interactive quiz questions

In addition to the questions provided above, this edition has a range of free interactive quiz questions for students to further test their understanding of the chapter material. To access these online questions, please visit the book's website: www.routledge.com/cw/lostroh.

Further reading

Nucleic acid–capsid interactions

Carlson CR, Asfaha JB, Ghent CM, Howard CJ, Hartooni N et al. 2020. Phosphoregulation of phase separation by the SARS-CoV-2 N protein suggests a biophysical basis for its dual functions. *Mol Cell* 80:1092–1103.e4.

Chou Y, Heaton N & Gao Q 2013. Colocalization of different influenza viral RNA segments in the cytoplasm before viral budding as shown by single-molecule sensitivity FISH analysis. *PLOS Pathog* 9:e1003358.

Liang Y, Huang T, Ly H & Parslow TG 2008. Mutational analysis of packaging signals of influenza virus PA, PB1, and PB2 genome RNA segments. *J Virol* 82:229–236.

Lu S, Ye Q, Singh D, Cao Y, Diedrich JK et al. 2021. The SARS-CoV-2 nucleocapsid phosphoprotein forms mutually exclusive condensates with RNA and the membrane-associated M protein. *Nat Commun* 12:502.

Muramoto Y, Takada A & Fujii K 2006. Hierarchy among viral RNA segments in their role in vRNA incorporation into influenza A virions. *J Virol* 80:2318–2325.

Reicin AS, Ohagen A & Yin L 1996. The role of Gag in human immunodeficiency virus type 1 virion morphogenesis and early steps of the viral life cycle. *J Virol* 70:8645–8652.

Xu J, Wang D, Gui M & Xiang Y 2019. Structural assembly of the tailed bacteriophage φ29. *Nat Commun* 10:1–6.

Acquisition of envelopes

Floderer C, Masson JB, Boilley E, Georgeault S, Merida P et al. 2018. Single molecule localisation microscopy reveals how HIV-1 Gag proteins sense membrane virus assembly sites in living host CD4 T cells. *Sci Rep* 8(1):1–5.

Exit from host cells

Adu-Gyamfi E, Smita SP & Jee CS 2014. A loop region in the N-terminal domain of Ebola virus VP40 is important in viral assembly, budding, and egress. *Viruses* 6:3837–3854.

Berry J, Rajaure M & Pang T 2012. The spanin complex is essential for lambda lysis. *J Bacteriol* 194:5567–5674.

Ghosh S, Dellibovi-Ragheb TA, Kerviel A, Pak E, Qiu Q et al. 2020. β-Coronaviruses use lysosomes for egress instead of the biosynthetic secretory pathway. *Cell* 183:1520–1535.e14.

He J, Melnik LI, Komin A, Wiedman G, Fuselier T et al. 2017. Ebola virus delta peptide is a viroporin. *J Virol* 9:e00438-17.

Hernandez-Morales AC, Lessor LL, Wood TL, Migl D, Mijalis EM et al. 2018. Genomic and biochemical characterization of *Acinetobacter* podophage petty reveals a novel lysis mechanism and tail-associated depolymerase activity. *J Virol* 92:e01064-17.

Madsen JJ, Grime JM, Rossman JS & Voth GA 2018. Entropic forces drive clustering and spatial localization of influenza A M2 during viral budding. *Proc Natl Acad Sci* 115:E8595-603.

Prescher J, Baumgartel V & Ivanchenko S 2015. Super-resolution imaging of ESCRT-proteins at HIV-1 assembly sites. *PLOS Pathog* 11(2):e1004677.

Rajaure M, Berry J, Kongari R, Cahill J & Young R 2015. Membrane fusion during phage lysis. *Proc Natl Acad Sci* 112:5497–5502.

Virus–Host Interactions during Lytic Growth

12

Virus	Property Related to Virus–Host Interactions
Phage T4	Uses HMC in mechanism to selectively degrade host DNA.
Rabies virus	Causes cytopathic effect known as Negri bodies.
Human herpesvirus 8 (HHV-8)	Blocks apoptosis using a v-FLIP.
Vaccinia virus	Blocks apoptosis using CrmA protease.
Human papillomavirus (HPV)	Triggers apoptosis by causing oligomerization of caspase-8.
Human immunodeficiency virus (HIV)	Triggers apoptosis by proteolytic degradation of procaspase-8. Subverts monoubiquitination for budding and gene regulation. Blocks endosome–autophagosome fusion using viral Tat protein.
Influenza	Subverts the unfolded protein response.
SARS-CoV-2	Triggers apoptosis and blocks autophagy using ORF3a.

The viruses you will meet in this chapter and the concepts they illustrate

In order to replicate, viruses interact with numerous host cell components and processes. Many viral effects on host cells contribute substantially to the efficiency of viral replication and tip the balance in favor of the virus. These effects include viral subversion of host transcription, translation, and replication-related machinery. For example, as described in **Chapter 6**, influenza virus simultaneously prevents translation of host mRNA and promotes production of its own proteins through the cap-snatching mechanism. Because all viruses include protein components, all viruses subvert cellular translation.

In animal hosts specifically, viruses cause **cytopathic effects** (**CPEs**), which are pathogenic changes in cellular structure associated with cell death or malfunction and observable through light microscopy. In many cases, CPEs cause the cells to have an abnormal shape and to detach somewhat from their culture dish or extracellular matrix (**Figure 12.1**). Most of these

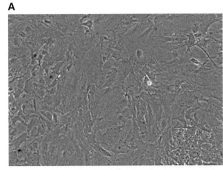

A

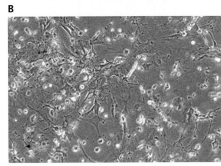

B

Figure 12.1 Cytopathic effects caused by a viral infection. (A) Uninfected cells. (B) Dead and dying host cells demonstrating the rounding-up cytopathic effects after 24 hours of infection. (From Zeng J et al. 2013. *Virol J* 10:157. Courtesy of BioMed Central Ltd. Published under CC BY 2.0.)

DOI: 10.1201/9781003463115-12

effects are harmful and are caused when the virus interferes with a normal host process, such as translation, apoptosis, protein degradation, or autophagy. Enveloped viruses also trigger a stress response called the unfolded protein response. Proliferation of intracellular membranes is another typical structural change in infected cells; most often it occurs because the virus uses host membranes to create virus replication compartments.

12.1   All viruses subvert translation

Viruses are protein coats enclosing nucleic acids, sometimes wrapped in a membranous envelope. Because of this composition, in order to cause the manufacture of thousands of offspring virions, all infecting virions must marshal the host cell's capacity to synthesize proteins. Once a virus infects a cell, newly synthesized host proteins are threatening to the virus in two ways. First, in general, they are detrimental because the virus needs all of the host's synthetic capacities to make thousands of offspring virions in the shortest period of time possible. Second, certain host proteins might also be detrimental because they have specific antiviral activities. Thus, blocking host translation is doubly important for virus replication. On the other hand, host cells must synthesize new proteins in order to remain alive; in the absence of synthesizing its own new proteins, a host cell will die. When a virus infects a host cell, there is a trade-off between the need for the host cell to remain alive long enough for the virus to exploit it and the need to take over all of the host cell's translation capacity. Once a virus has inhibited host protein synthesis, it is a race to complete viral replication before the inevitable death of the host cell.

12.2   Bacteriophages subvert translation indirectly

The ways that viruses bias the host toward production of viral proteins and away from production of host proteins depend mainly on whether the host is prokaryotic or eukaryotic. Bacteria will serve as our example of prokaryotes. Bacteriophages usually encode proteins that interfere with host translation indirectly, by blocking host transcription. This strategy makes sense because transcription and translation both occur in the cytoplasm. The most efficient mechanism to block all host transcription at once is to degrade the host chromosome because it is in the cytoplasm (**Figure 12.2**). Once the

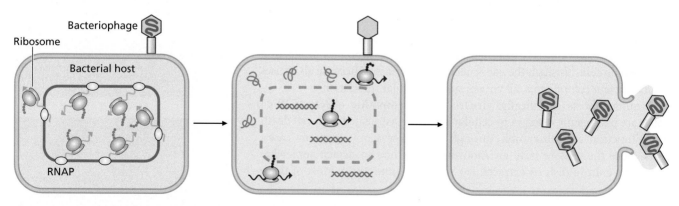

Figure 12.2 Bacteriophages block translation indirectly by degrading the host chromosome. In bacteria, transcription and translation both occur in the cytoplasm and are spatiotemporally linked. Degradation of the host chromosome prevents transcription of all host genes and the short half-life of bacterial mRNA means that there is little host translation within a few minutes of degrading the chromosome. Expression of the phage genome ultimately results in the release of offspring phages.

DNA template is gone, there can be no more transcription of host nucleic acids. Furthermore, on average, bacterial mRNA has a half-life of only 5 min. Because of its short half-life, once the template chromosome is gone, whatever host mRNA was present in the cytoplasm will be gone after only a few minutes without any further intervention from the phage. Degrading the host chromosome also makes deoxyribonucleotides available for viral genome replication.

One problem phages have in degrading the host DNA, however, is selectivity, namely how to distinguish host DNA from phage DNA. An example of just one solution to this problem is provided by bacteriophage T4, which encodes enzymes that convert the base cytosine into **hydroxymethylcytosine** (HMC) (**Figure 12.3**). This base behaves like a normal C for base pairing, but the hydroxymethyl group protects DNA from the phage's nuclease, which degrades only DNA with normal Cs, which is to say, it degrades the host chromosome.

Some phages also block host transcription in other ways (**Figure 12.4A**). In **Chapter 4**, for example, we saw how bacteriophage T7 phosphorylates host RNAP as part of the switch from immediate-early to delayed-early gene expression. The phosphorylated RNAP can no longer bind to host cell promoters; the phage encodes its own polymerase to transcribe phage genes. There are many other mechanisms by which different phages bias transcription toward their own genomes, for example through noncovalent modifications of RNAP in which phage proteins bind to and reprogram RNAP so that it recognizes only phage promoters; phage T4 uses this strategy (**Figure 12.4B**). The ubiquity of phage proteins that degrade host DNA and block host transcription strongly suggests that there has been evolutionary pressure that has favored phages that degrade host DNA and block host transcription. Selective degradation of the host chromosome, blocking host transcription, or both methods are effective strategies for preventing translation of bacterial host proteins while leaving ribosomes unaffected for use in translation of viral mRNA.

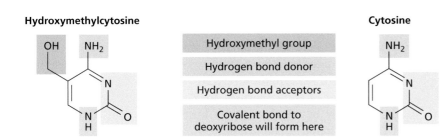

Figure 12.3 Bacteriophage T4 uses hydroxymethylcytosine in its DNA instead of cytosine. Hydroxymethylcytosine has an additional hydroxymethyl group compared with cytosine. It has the same hydrogen-bonding capabilities as cytosine so that it base pairs with guanine in DNA using the acceptors and donors indicated.

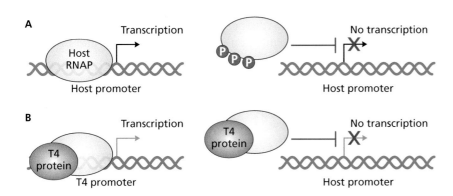

Figure 12.4 Bacteriophage strategies for blocking transcription. (A) Phage T7 phosphorylates RNAP so that it cannot recognize host promoters. (B) Phage T4 reprograms RNAP with phage proteins so that it can recognize only phage promoters, rather than host promoters.

12.3   Animal viruses have many strategies to block translation of host mRNA

Animal viruses have diverse strategies to prevent translation of host mRNA in general. In contrast to the ways that bacteriophages block synthesis of host proteins, animal viruses instead act at the level of translation itself to block host protein synthesis. Interfering at the level of translation instead of transcription probably results from differences in the average half-life of mRNA in bacteria compared with animal cells and with compartmentalization of transcription and translation in eukaryotes. The short half-life of bacterial mRNA means that blocking the production of new host mRNA blocks host translation almost immediately because any existing mRNA is soon degraded. In contrast, on average, animal mRNA has a half-life on the order of hours so that it could, in principle, be translated for most of the duration of the virus replication cycle. In eukaryotic cells, blocking host transcription would therefore take hours to have significant effects on global host protein translation. Furthermore, unlike transcription in bacteria, transcription in animal cells occurs in the nucleus, physically separated from the site of protein synthesis. Eukaryotic viral proteins that block transcription must use proteins that are actively localized to the nucleus. All proteins are translated in the cytoplasm, so viral proteins are passively co-localized to the same compartment in which translation occurs in eukaryotic cells.

Most eukaryotic host mRNAs are translated thanks to a multisubunit **48S translation preinitiation complex** that requires the poly(A) tail and the 5′ cap of the mRNA itself (**Figure 12.5**). It is common for viral proteases to target one of the proteins in this translation initiation complex. By proteolytically degrading one of the initiation proteins, that protein is no longer available to the complex, and all translation initiation by the 48S complex is impeded. A less common strategy is for a viral protein to interact with one of the subunits in the 48S complex and inactivate it by means other than proteolysis. For example, the viral protein might prevent one of the proteins from forming its normal protein–protein interactions with other host proteins in the 48S complex, sequestering it so that the 48S complex cannot form. Viruses that destroy translation initiation in these ways must have a way of bypassing the 48S complex for the purpose of translating their own mRNA. By far the most common solution to this problem is the use of an internal ribosome entry site, or IRES, in the viral mRNA transcripts (see **Chapter 5**).

A smaller number of animal viruses are known to shut off host mRNA translation indirectly by interfering with host transcription (**Figure 12.6**). RNA Pol II transcription is characterized by a large protein complex that mediates initiation as well as by specific phosphorylation events required for elongation to occur normally. One viral strategy to shut off transcription is to proteolytically degrade the TATA-binding protein (TBP), which is necessary for formation of all Pol II transcription-initiation complexes. Additionally, there are other mechanisms that disrupt transcription initiation, for example, by binding to one of the initiation-complex components and thereby

Figure 12.5 The translation preinitiation complex and viruses that inhibit its formation. Some viruses make proteases that cleave one of the subunits, marked with Xs. Examples include poliovirus and HIV, which both cleave PABP and eIF4G. Other viruses make proteins that bind to one of the subunits, sequestering it from participating in the 48S complex. An example is rubella virus (a togavirus), which sequesters PABP.

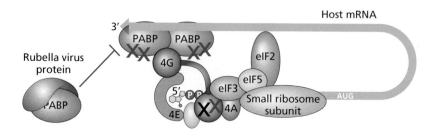

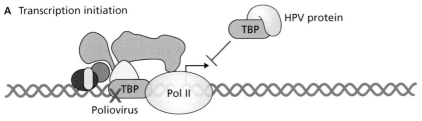

A Transcription initiation

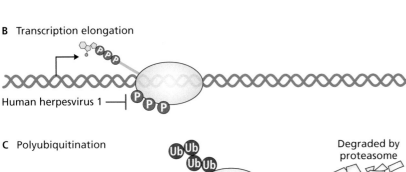

B Transcription elongation

C Polyubiquitination

Figure 12.6 Eukaryotic Pol II transcription and viruses that interfere with it. (A) Poliovirus proteolytically processes TBP, preventing its function. Human papillomavirus (HPV) sequesters TBP so it cannot join the initiation complex. (B) Human herpesvirus 1 blocks phosphorylation of Pol II so that elongation does not occur. (C) Influenza virus causes polyubiquitination of Pol II so that it is degraded by the proteasome.

sequestering it away from assembling into an initiation complex. Another strategy is to prevent phosphorylation of Pol II so that elongation cannot occur. Finally, there are viruses that indirectly cause the proteolytic degradation of Pol II, using the ubiquitin–proteasome system described in **Section 12.8.** Viruses that disrupt normal Pol II transcription must use other mechanisms to produce their own mRNA.

In addition, other viral host inhibition strategies target host mRNA directly rather than affecting transcription. These include mechanisms that act on mRNA that has already been synthesized, either by degrading that mRNA or by blocking its export from the nucleus. Degradation is typically catalyzed by viral endonucleases; whether there is any specificity for host mRNA is not clear. Another form of mRNA degradation is cap snatching, in which influenza A cleaves the first dozen or so nucleotides of host nuclear mRNAs to provide itself with a primer, while simultaneously blocking host translation by depriving host mRNA of its cap (see **Chapter 6**).

Nuclear export of mRNA is another target for viruses; normally it is mediated by proteins that interact with the nuclear pore complex, which is an enormous assemblage of many different proteins (**Figure 12.7**). Some viruses cause proteolysis of components of the nuclear pore complex, ultimately preventing the entire complex from functioning. An example is the poliovirus 2A protease, which degrades nuclear pore protein 98, preventing nuclear export of host mRNA without affecting tRNA export (which the virus needs for translation of its own proteins). Others make proteins that bind to specific subunits of the nuclear pore complex, again preventing them from assembling properly with the rest of the complex and thereby preventing the export of host mRNA. Yet another strategy is to interfere with pre-mRNA processing so that the host mRNA does not get spliced or polyadenylated properly. The multifunctional HIV-1 nonstructural protein Vpr provides an example; it binds to a ubiquitous host splicing factor, sequestering that factor from participating in normal splicing reactions and thereby blocking host splicing. Because splicing is a prerequisite for mRNA export, inhibition of splicing prevents nuclear export of unspliced host mRNA.

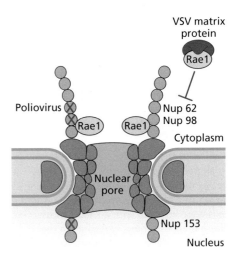

Figure 12.7 The nuclear pore complex and how viruses inhibit its function. The poliovirus 2A protease degrades nuclear pore (Nup) proteins 62, 98, and 153, preventing nuclear export of host mRNA. Vesicular stomatitis virus (VSV, a rhabdovirus) encodes a matrix protein that also binds to Rae1, preventing mRNA export.

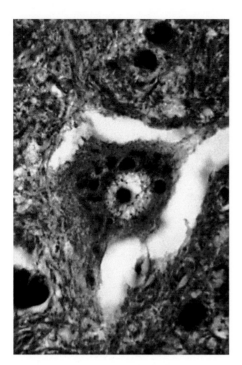

Figure 12.8 Negri bodies are an example of a CPE. Negri bodies can be detected using Negri's stain and light microscopy. The Negri bodies are the dark irregular shapes in the cytosol. (CDC/Dr. Daniel P. Perl.)

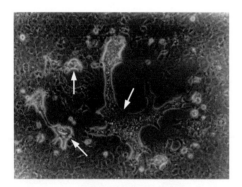

Figure 12.9 Syncytia. Normal cells are small (red arrow), but cells participating in a syncytium are enormous because they have fused together (white arrows). These cells were infected with a rhabdovirus. There is an enormous syncytium in the center of the image with about a dozen smaller ones visible in the image. (From Iwanowicz LR & Goodwin AE 2002. *Arch Virol* 147:899–915. doi: 10.1007/s00705-001-0793-z. With permission from Springer.)

12.4 Animal viruses cause structural changes in host cells referred to as cytopathic effects

Long before virologists had today's range of tools for molecular and cellular biology, they could use light microscopy to examine the effects of viral infection on host cells. They developed a panel of stains that could be used to make certain cellular or virological structures visible, again using light microscopy, and they could use electron microscopy as well. The negative effects of viruses on host cells, as documented by light or electron microscopy, were grouped together and collectively termed CPEs. In some cases, molecular and cell biological experiments have since been applied to better understand the underlying cause of the CPEs.

Inclusion bodies are very common CPEs and can be observed in the cytoplasm or nucleus. In general, an inclusion body is an aggregate of macromolecules that forms under unusual circumstances and that appears as densely staining foci in the microscope. In almost every case that has been investigated, viral inclusion bodies turn out to be dynamic virus replication complexes; one prominent example is that of the **Negri bodies** found in cells infected by rabies virus (**Figure 12.8**). In less frequent cases, the inclusion bodies are now known to be sites of viral assembly; a striking example is provided by adenovirus, which fills the nucleus with hundreds of thousands of virions arranged in a crystalline array.

Cells infected by viruses exhibit morphological changes in addition to inclusion bodies. These include shrinking of the nucleus, proliferation of cytoplasmic membranes, proliferation of the nuclear membrane, proliferation of cytoplasmic vacuoles, formation of **syncytia** (cell-to-cell fusion; **Figure 12.9**), and the rounding up and detachment of tissue culture cells. Not all of these effects are understood to the same degree. Proliferation of membranes typically occurs when those membranes are needed to form viral replication compartments. Syncytia typically occur when the fusion peptides of maturing enveloped viruses trigger fusion of neighboring cell membranes before the assembly process can be completed to produce budding virions, or they may occur to benefit the virus by enabling cell-to-cell spread from the cytoplasm of an infected cell into a neighboring cell following fusion. Rounding-up and detachment of tissue culture cells indicates cell mortality because healthy cells lie flat against the substratum and are attached to it. Rounding-up and detachment can be useful in the lab because they occur late during an infection, when there are many offspring virions. Thus, rounding-up and detachment can be a sign that it is time to harvest progeny virions.

12.5 Viruses affect host cell apoptosis

Apoptosis is a form of programmed cell death in which an individual cell dies in order to protect the entire organism of which it is a part or to accomplish tissue biogenesis during development; it can cause CPEs such as detachment from the plastic dish (*in vitro*) or extracellular matrix (*in vivo*). Although this book has mainly focused on the viruses in the host cell–virus interaction, keep in mind that the host cell is not helpless. One common response to viral infection is apoptosis, which does not save the infected cell but does in most cases minimize the number of infectious virions produced and thereby help protect the entire organism from the virus. Naturally, viruses have evolved anti-apoptotic proteins in response. Some have even evolved to subvert apoptosis rather than oppose it.

Several signal transduction cascades can trigger apoptosis in uninfected cells. All of them culminate in activation of the enzyme **caspase-3**, which is a protease that subsequently triggers cell death with the features characteristic of apoptosis: condensed chromatin, fragmentation of DNA, and cell shrinkage and blebbing to produce apoptotic bodies that are easily degraded by phagocytic immune cells (**Figure 12.10**).

Caspases including caspase-3 are produced as **proenzymes** that have the capacity to act as proteases, but only after they themselves have been proteolytically processed. Proteolytic processing of the proenzymes occurs in response to environmental signals. A common signal that triggers apoptosis occurs when an immune cell communication peptide called TNFα binds to receptors on the surface of a cell and then triggers a series of caspase enzymes that together result in caspase-3 activation (**Figure 12.11**). The first step is that the TNFα receptor causes proteolytic processing of procaspase-8, releasing **caspase-8**, which is an active protease. Caspase-8 is only effective when it oligomerizes with other caspase-8 molecules, at which point the caspase-8 complex becomes able to process procaspase-3, resulting in the release and

A Normal cells

B The same cells undergoing apoptosis

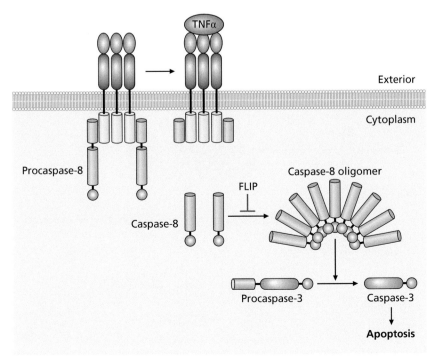

Figure 12.10 Apoptosis. (A) Normal cells that have formed a confluent monolayer. (B) Cells undergoing apoptosis that have become round and have visibly ragged edges because of membrane blebbing. The asterisk marks the position of a prominent bleb. (From Edelweiss E et al. 2008. *PLOS ONE* 3:e2434. doi: 10.1371/journal. pone.0002434. With permission from Public Library of Science.)

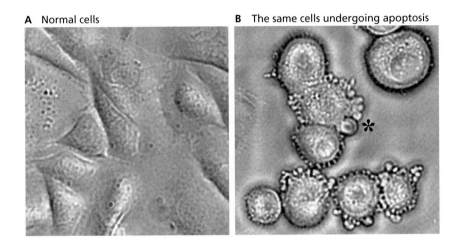

Figure 12.11 Apoptosis triggered by TNFα signaling. After TNFα binds to its receptor, the receptor causes proteolytic processing of procaspase-8, releasing caspase-8, which is an active protease. Caspase-8 is effective only when it oligomerizes with other caspase-8 molecules, at which point the caspase-8 complex becomes able to process procaspase-3, resulting in the release and activity of caspase-3. Pro-apoptotic TNFα signaling can be blocked by proteins called FLIPs that prevent caspase-8 from oligomerizing.

Figure 12.12 Apoptosis triggered by mitochondrial release of cytochrome c. When the mitochondria are stimulated to release cytochrome c into the cytoplasm, cytoplasmic cytochrome c acts through caspase-9 to cause caspase-3 activation. Cytochrome release from the mitochondria can be blocked by a protein called Bcl-2, yet another fail-safe mechanism.

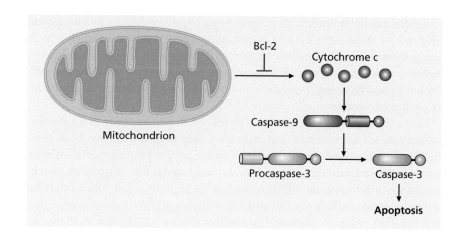

activity of caspase-3. Pro-apoptotic TNFα signaling can be blocked by proteins called **FLIPs**, which serve as a check on the system. FLIPs form a complex with caspase enzymes, ultimately preventing procaspase-3 from being converted into active caspase-3.

A second way that apoptosis can be triggered occurs when the mitochondria are stimulated to release **cytochrome c** into the cytoplasm; cytoplasmic cytochrome c acts through **caspase-9** to cause caspase-3 activation (**Figure 12.12**). Cytochrome release from the mitochondria can be blocked by a protein called **Bcl-2**, yet another fail-safe mechanism. Additionally, there are TNFα- and mitochondria-independent signal transduction pathways that can also activate caspase-3.

12.6 Some viruses delay apoptosis in order to complete their replication cycles before the host cell dies

Many viruses attempt to prevent or delay apoptosis in order to maximize the burst size. There are three mechanisms by which viruses inhibit apoptosis from occurring (**Figure 12.13**). In the first, a viral protein acts as a FLIP and prevents TNFα from sending a pro-apoptotic signal. An example is human herpesvirus 8, the cause of Kaposi's sarcoma (see **Chapter 13**), which encodes a viral FLIP (**v-FLIP**). Expression of v-FLIP slows apoptosis of infected host cells. In the second, a viral protein acts like Bcl-2 and prevents

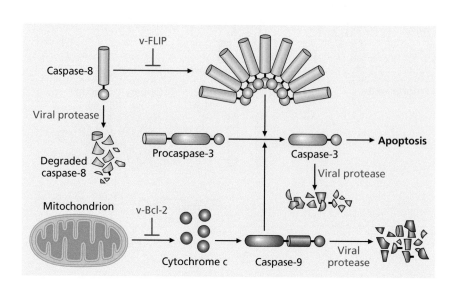

Figure 12.13 Three ways that viruses block apoptosis. Viruses can block apoptosis using virally encoded v-FLIPs or v-Bcl-2s. Other viruses can proteolytically degrade caspase-8 or caspase-9.

mitochondria from sending a pro-apoptotic signal. Another herpesvirus, for example, encodes **v-Bcl-2**, which delays apoptosis of its host cells. In the last case, viruses block the activity of the intermediary caspase-8 and caspase-9 enzymes. Some block the catalytic active site of these caspases, whereas others trigger the caspase enzymes to be degraded by the ubiquitin–proteasome system (see Section 12.8). An example of a virus that blocks the catalytic activity of caspases is vaccinia virus, which makes a protein called **CrmA**. CrmA delays apoptosis of infected cells by inhibiting host proteases including caspases. Still other viruses encode proteases that themselves degrade the caspase-8 or caspase-9 proteins. The effect in all of these cases is to prevent or delay apoptosis by preventing or reducing the presence of active caspase-3.

12.7   Some viruses subvert apoptosis in order to complete their replication cycles

Human papillomavirus (HPV) and HIV take the opposite tack. In an "If you can't beat 'em, join 'em" strategy, they subvert apoptotic proteins and use them for proviral purposes. Both of these viruses trigger apoptosis by activating caspase-8 (**Figure 12.14**). HIV does so through proteolytic degradation of procaspase-8, whereas HPV causes enhanced oligomerization of caspase-8. The benefit to the viruses of subverting apoptosis is an area of active research; there is disagreement regarding the benefits of promoting apoptosis. Apoptosis may allow the virions to be released after cell death. Alternatively, surrounding cells may take up the apoptotic bodies that are full of virions and thereby acquire the viral infection. Death by apoptosis avoids inflammation, which is characterized by an influx of immune cells, so HPV and HIV may also use apoptosis to avoid an immune response. There also may be a balance between apoptosis simultaneously enhancing and controlling the infection.

SARS-CoV-2 also induces apoptosis through the ORF3a protein inducing caspase-3 activation. The closely related SARS-CoV ORF3a protein induces apoptosis more strongly than SARS-CoV-2, which might contribute to a lower mortality rate for SARS-CoV-2. It is not yet clear whether induced apoptosis increases SARS-CoV-2 replication; the answer may depend in part on the particular infected host cell.

12.8   Viruses use the ubiquitin system to their advantage

One of the most common ways that a virus interferes with the action of a host protein is to cause that protein's degradation. We have so far focused on viral proteases that accomplish this degradation directly by binding to a target host protein and digesting it, but there is an indirect method that is

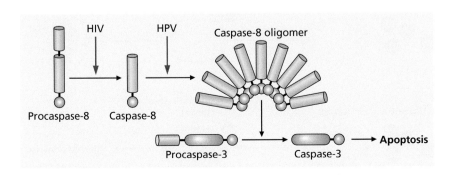

Figure 12.14 HIV and HPV trigger apoptosis using somewhat different mechanisms that ultimately activate caspase-8. HIV processes procaspase-8, releasing active caspase-8, whereas HPV promotes oligomerization of caspase-8, enhancing its activity.

nevertheless just as effective. In the indirect case, a viral protein adds a small abundant protein called **ubiquitin** to the R group (side chain) of the amino acid lysine found in a protein that is targeted for degradation (**Figure 12.15**). The enzymes that attach ubiquitin to targets are called **E3 ubiquitin ligases**. E3 ubiquitin ligases typically are specific for one or a few related substrates. Ubiquitin itself contains lysines, which themselves also become ubiquitinated. The result of this **polyubiquitination** process is a branched string of ubiquitin proteins attached to the protein of interest. Once the chain is four ubiquitins long, it is quickly recognized by a large cytoplasmic multiprotein machine called the **proteasome**. The proteasome is a hollow cylinder that uses adenosine triphosphate (ATP) to unfold and degrade polyubiquitinated proteins. The ubiquitin proteins themselves are recycled. The proteasome causes rapid degradation of tagged proteins, with degradation rates approximately matching the rate of amino acid incorporation by ribosomes.

There are four general ways that viruses use the ubiquitin–proteasome system to their advantage. First, many viruses use it to avoid detection by the immune system, for example by selectively degrading proteins that a host cell would normally use to alert its neighbors that trouble is afoot. The SARS-CoV-2 ORF7a protein is itself ubiquitinated and antagonizes the infected cell's innate immune response. In rotavirus, a multifunctional nonstructural protein called NSP1 causes the proteasome to degrade host proteins that would otherwise participate in an interferon immune response (interferon interferes with viral replication; see **Chapter 14**). The SARS-CoV-2 ORF8a protein similarly induces degradation of a host protein that would otherwise induce the interferon response. Second, retroviruses and possibly others subvert **monoubiquitination** (as opposed to polyubiquitination) by taking advantage of it during budding. The use of ubiquitin during budding was first discovered because experimental treatments that interfere with proteasome function also interfere with budding of HIV. Further study eventually revealed that monoubiquitination (not polyubiquitination) of the HIV Gag polyprotein is required for budding. Third, ubiquitination can be used to regulate viral transcription. Again, this process is found among the retroviruses, where monoubiquitination of retroviral transcription factors such as HIV-1's Tat or HTLV-1's (human T-lymphotropic virus 1's) Tax is also required for normal transcription regulation.

Fourth, some viruses subvert the ubiquitin–proteasome system in order to regulate major cellular processes such as apoptosis and the cell cycle (see **Chapter 8**). Virus interactions with apoptosis and the cell cycle involve not only ubiquitin but also the proteasome. In these cases, the virus encodes a protein that promotes the initial ubiquitination of a key cellular regulatory

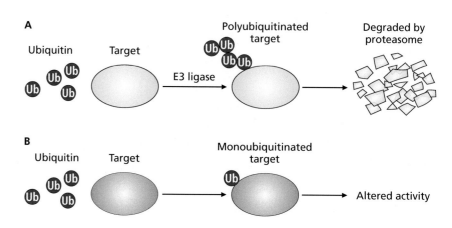

Figure 12.15 The ubiquitin–proteasome system. (A) Polyubiquitination leads to protein degradation. (B) Monoubiquitination is a regulatory event with many possible outcomes.

protein. That ubiquitination triggers further polyubiquitination as in canonical cellular degradation of proteins, and then the key cellular regulator, now polyubiquitinated, is degraded by the proteasome. If that regulator promotes apoptosis in its normal role, the virus has blocked apoptosis.

Similarly, if that regulator prevents progression through the cell cycle, its degradation typically promotes progression. An example is the cell cycle regulator p27, which helps maintain a host cell in G_0. For many viruses, the block between G_0 and G_1 is antiviral. A herpesvirus called myxoma virus infects host cells that are in G_0 and it encodes a protein that ubiquitinates p27. The result is that the cells degrade polyubiquitinated p27 and then progress into G_1, which favors viral replication.

12.9   Viruses can block or subvert the cellular autophagy system

Like the ubiquitin–proteasome system, **autophagy** is another complex pathway essential for host cell survival. Autophagy maintains cellular homeostasis by degrading old and damaged organelles. During autophagy, a damaged organelle becomes surrounded by a specialized vesicle with a double membrane and is ultimately degraded so that the cell can recycle the constituent parts. Autophagy results in degradation of structures too large or too complex for degradation by the proteasome. Formation of **autophagosomes** is tightly regulated and has been studied extensively so that we understand how some viruses avoid or subvert the system (**Figure 12.16**). Autophagy begins when the **Beclin-1 complex** triggers the **LC3-I** protein to become covalently attached to the lipid phosphotidylamine, resulting in formation of **LC3-II**. Consequently, pre-autophagosomal membranes form. When the membranes have completely surrounded their target, the compartment is known as the autophagosome. Fusion with the endosome causes a drop in the internal pH and the acidified compartment is now called an **amphisome**. The last step is that a lysosome fuses with the amphisome, providing the enzymes necessary for degradation of the engulfed cellular structure.

Autophagy can sometimes be part of an antiviral response in which the cell attempts to engulf and subsequently degrade virions or their component parts. In this case, the special term **xenophagy** is used to denote that it is not the cell's own normal components that are being engulfed and degraded but rather the virus or its component parts. Viral proteins that block many of the individual steps of xenophagy have been described. For example, human herpesvirus 1 (also known as herpes simplex virus 1) makes a protein that blocks Beclin-1 activation. HIV-1 has a nonstructural protein called Nef that binds to Beclin-1, preventing its activation. Another viral strategy to prevent xenophagy is to block the endosome from fusing with the autophagosome so that the autophagosome does not acidify properly. HIV prevents autophagosome acidification, though the molecular mechanism is not well understood. In some cells, the HIV-1 Tat protein appears to associate with proteins that host cells would otherwise use in order to accomplish endosome–autophagosome fusion. The Tat protein thereby blocks endosome–autophagosome fusion. The SARS-CoV-2 ORF3a protein also interferes with xenophagy through an unknown mechanism.

Conversely, an even larger group of viruses have been shown to need autophagic pathways for maximum virus replication, that is, they subvert rather than block autophagy. In some of these cases, the mechanism is not at all clear. In other cases, it seems that proliferation of internal membranes caused by autophagic pathways is beneficial during formation of virus replication

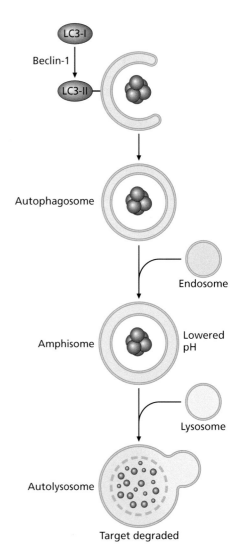

Figure 12.16 Autophagy. The Beclin-1 complex triggers the LC3-I protein to become covalently attached to the lipid phosphotidylamine, creating LC3-II. The result is the formation of pre-autophagosomal membranes that have not fully engulfed any other entities. At the next stage, the autophagosome has completely surrounded its target. Fusion with the endosome results in a lowered pH and the compartment is now called an amphisome. The amphisome fuses with a lysosome, and this autolysosome compartment finishes degradation of the engulfed cellular structure.

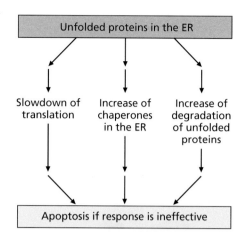

Figure 12.17 The three subpathways of the unfolded protein response. Detection of unfolded proteins triggers three pathways that can each trigger apoptosis if the unfolded protein stress continues for too long. The three pathways are a slowdown of translation, an increase of chaperones in the ER, and an increase in the rate of degradation of unfolded proteins.

compartments or during viral assembly and egress. It has even come to light that certain picornaviruses, traditionally thought to be naked, might exit cells surrounded by a double membrane of autophagic origin and this surrounding membrane might contribute to infectivity by enhancing spread to adjacent susceptible cells.

12.10   Viruses subvert or co-opt the misfolded protein response triggered in the endoplasmic reticulum

As we have seen in **Chapters 5–10**, all enveloped viruses synthesize transmembrane proteins using ribosomes docked to the rough endoplasmic reticulum (ER). The sheer amount of viral translation can lead to misfolding or delayed folding of proteins in the ER. Other viruses use ER membranes in their virus replication compartments, which also blocks normal ER function. The result in both cases is that the cell triggers an ER stress response called the **unfolded protein response** (**Figure 12.17**), which alerts the cell that the ER is overwhelmed. The stress response also attempts to eliminate the high burden of unfolded proteins. A normal unfolded protein response results in three subpathways that trigger (1) a slowdown of translation, which allows for more time for proteins to fold; (2) an increase of the ER folding capacity by increasing the number of chaperones in the ER lumen; and (3) increased degradation of unfolded and misfolded proteins. A prolonged, ineffective unfolded protein response can lead to apoptosis.

The ER unfolded protein response is tightly regulated by a complex signal transduction pathway. The many steps of the pathway give viruses many opportunities to interfere with the process, as with apoptosis and autophagy. Some viruses trigger the unfolded protein response unintentionally through gene expression. Others, such as SARS-CoV and SARS-CoV-2 coronaviruses, make nonstructural proteins that trigger one or more branches of the unfolded protein response. It is not known whether this nonstructural protein subverts the unfolded protein response to benefit the virus or whether instead cellular detection of this nonstructural protein triggers an antiviral unfolded protein response.

Some viruses block or subvert one of the three types of responses in the unfolded protein response. Viruses such as human herpesvirus 1 counteract the slowdown of translation probably because a reduction in translation efficiency would slow virus replication. Influenza A selectively triggers the unfolded protein response subpathway that results in expression of chaperones while simultaneously suppressing the subpathways that upregulate protein degradation or that can lead to apoptosis. The effect is subversion of the unfolded protein response to increase the abundance of correctly folded viral proteins, which were folded properly because of the expression of ER chaperones.

12.11   Viruses modify internal membranes in order to create virus replication compartments

The roles of virus replication compartments (VRCs) during viral gene expression and genome replication were discussed in **Chapters 5–8**, but VRC formation also affects the manipulation of host cell membranes, which we address here. The cytoplasm of animal cells is full of membranous structures, many of which viruses exploit for VRC formation. These structures include the ER, the ER–Golgi intermediate compartment, the Golgi apparatus (including *cis* and *trans* networks), clathrin-coated vesicles, early and late endosomes, lysosomes, and mitochondria. Cellular membranous organelles have specific functions to a large degree because specific transmembrane proteins and lumenal proteins are found in each different organelle.

The identity of the cellular membranes used for forming a VRC can be established not only by examining proximity to other membranes through electron microscopy but also through the use of fluoroscopy microscopy to look for specific proteins associated with the membranes. The idea is that if a VRC originates from the Golgi apparatus, it will still have Golgi-associated proteins in its lipid bilayer, and so on for each organelle of interest. Proteins can be detected using immunofluorescence and there is a standard cast of characters associated with most membranous organelles, making it possible to select the most useful antibodies. Another fluorescence microscopy strategy is to use fluorescent chimeras such as a viral polymerase–GFP fusion. In either case, the goal is to look for the membranous structures and also to co-localize them with viral replication proteins or newly replicated viral nucleic acids. An alternative approach that does not rely upon microscopy is proteomics, which is the identification of all of the proteins associated with a particular cell or subcellular compartment (**Technique Box 12.1**). It is possible to enrich a cellular lysate for VRCs and then use proteomics to identify all of the host proteins associated with the VRC. Proteomics can be used to inform microscopy experiments used to examine VRCs in intact host cells.

The formation of VRCs is not well understood, though formation likely comes about through interactions between viral nonstructural proteins and host lipids and proteins. There are many ideas of how a flat membrane could be induced to bend, which in turn is necessary for a membrane to form a pouch. One idea is that one or more cone-shaped transmembrane proteins could cause bending. Certain phospholipids also have more of a cone shape and can contribute to bending because of their shape. It is also possible for proteins to aggregate and form a scaffold that interacts with the membrane in such a way that it bends.

There are many possible reasons that replication in a membranous VRC could be advantageous. Among them are one or more of the following:

1. To increase the local concentration of proteins and nucleic acids needed for replication.
2. To provide a physical scaffold for the replication complex.
3. To keep genome and antigenome replication, especially RNA replication, confined to a certain location to avoid cellular antiviral responses.
4. To prevent the host cell from detecting viral dsRNA in particular, which would otherwise trigger an innate immune response.
5. To attach one end of viral genome to the membrane so that helicase activity to unwind a double-stranded template would be maximally effective.
6. To provide certain lipids that are needed for synthesis of particular viral genomes.

Any and all of these functions also disrupt the normal function of whatever organelle serves as the membrane of origin for the VRCs, which likely helps protect the virus from a negative host response. For example, where mitochondria trigger an antiviral apoptotic response, VRC formation using mitochondrial membranes may interfere with regulation of apoptosis.

TECHNIQUE BOX 12.1   PROTEOMICS

For some experiments, it can be crucial to identify all of the proteins expressed in a cell, known as that cell's proteome. **Proteomic** techniques are designed to determine the identity and sometimes also the relative abundance of proteins in a complex mixture. For example, we might use a proteomic approach to determine all of the viral proteins found in a poxvirus virion or all of the host proteins found in influenza virions. We also can use proteomic techniques to determine which of thousands of host proteins become more abundant during virus infection. This sort of identification is usually the first step in deciding what proteins to study in order to better understand the molecular and cellular biology of a particular virus.

Two techniques are fundamental in most proteomics: two-dimensional gel electrophoresis and mass spectrometry. In two-dimensional gel electrophoresis, proteins in a complex mixture are first solubilized and then separated according to their isoelectric point, also known as the pH at which the protein charge is perfectly neutral (**Figure 12.18**). Once separated in the pH gradient, the proteins are then separated using SDS-PAGE so that they separate according to size. Mass spectrometry is an analytical chemistry technique that allows proteins in a complex mixture to be identified. The proteins in a sample are ionized by bombarding them with electrons, which breaks them into peptides, each of which is likely unique to any given protein because of that protein's primary sequence. The peptides are then separated according to their mass-to-charge ratio, which can then be used to identify the proteins in the complex mixture. The technique can also quantify the proteins. Two-dimensional gels and mass spectrometry can be used in combination, for example by eluting a spot from a two-dimensional gel and then using mass spectrometry to identify proteins in the spot.

A less commonly employed proteomic technique is that of protein arrays (**Figure 12.19**). Proteins from a virus can be arrayed onto a solid substrate such as a glass slide and that slide can be used to monitor whether other proteins, nucleic acids, or drugs bind to those proteins. An example would be to array all of a virus's proteins on a slide and then use sera collected from recovering patients to determine which of the viral proteins provoked a strong antibody response. Alternatively, antibodies themselves can be arrayed so that a complex protein mixture can be applied and then proteins that bind to the antibodies can be detected and quantified. In general, protein microarrays are much more difficult to handle than nucleic acid microarrays, and they are employed much more rarely than mass spectrometry or two-dimensional gel analysis.

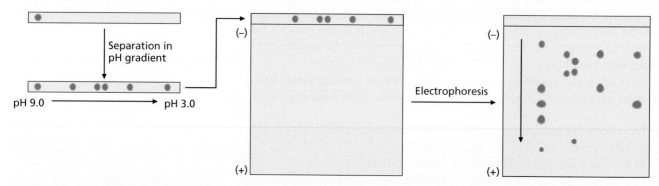

Figure 12.18 Two-dimensional gel electrophoresis. Proteins are first separated in one dimension according to their isoelectric points. Then the mixtures are separated by size using SDS-PAGE, resulting in two-dimensional separation (by isoelectric point and by size).

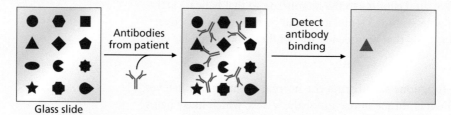

Figure 12.19 A protein array. In this example, 12 proteins from a virus have been arrayed on the slide. The slide is then incubated with antibodies taken from patients who were infected with this virus. The fourth protein (triangle) reacts most strongly to the antibodies, indicating that it provokes the strongest antibody response.

Essential concepts

- Viruses subvert translation, which interferes with production of host proteins and favors production of viral proteins.
- Bacteriophages typically block translation indirectly by blocking transcription through destruction of the host chromosome template, thereby ending coupled host transcription and translation.
- Animal viruses instead typically block translation itself, often interfering with one or more of the protein factors needed for translation initiation.
- Cytopathic effects are structural problems in infected animal cells that become visible using light microscopy. Inclusion bodies, syncytia, and rounded-up and detaching cells are examples.
- Some viruses attempt to block apoptosis, autophagy, or both because these processes would otherwise interfere with viral replication.
- Some viruses attempt to subvert apoptosis, the ubiquitin–proteasome system, or autophagy, meaning that instead of preventing these processes, they alter them and use them in order to improve viral replication.
- An unfolded protein response is triggered by many enveloped viral infections, which the viruses in turn can sometimes subvert for their own purposes.
- Viruses, even naked ones, usually rearrange and subvert intracellular membranes to create virus replication compartments or complexes (VRCs).

Questions

1. Eukaryotic viruses typically interfere with translation directly, whereas bacterial viruses interfere with translation indirectly. Why does this difference seem evolutionarily advantageous in the two different cases?
2. Provide an example of a mechanism by which an animal virus and a bacteriophage accomplish their own gene expression despite interfering with host translation.
3. What is the difference between blocking a complex pathway and subverting it?
4. Apoptosis can be triggered by a number of stimuli, yet ultimately all apoptotic pathways have common characteristics. What are some examples of similarities?
5. Provide two examples of mechanisms by which viruses block apoptosis.
6. What observations would you make to find out if a newly described retrovirus affects the ubiquitin–proteasome system in a way that is similar to the effects of HIV on that system?
7. How might a virus, in principle, use the ubiquitin–proteasome system to interfere with any cellular activity?
8. A newly described virus causes proliferation of membranes containing LC3-II. What would you predict about this virus given this information?
9. Why do viruses use microtubules and microfilaments for different purposes? Specifically, what is it about the different structures of microtubules and microfilaments that make them amenable for different purposes during virus replication?
10. Why is an uncontrolled unfolded protein response typical of many viral infections?
11. A newly described virus creates its VRCs using membranes derived from the Golgi apparatus. Explain two disadvantages this use causes the host cell.

Interactive quiz questions

In addition to the questions provided above, this edition has a range of free interactive quiz questions for students to further test their understanding of the chapter material. To access these online questions, please visit the book's website: www.routledge.com/cw/lostroh.

Further reading

Cytopathic effects

Lahaye X, Vidy A, Pomier C, Obiang L, Harper F et al. 2009. Functional characterization of Negri bodies (NBs) in rabies virus-infected cells: Evidence that NBs are sites of viral transcription and replication. *J Virol* 83(16):7948–7958.

Blocking translation of host mRNA

Banerjee AK, Blanco MR, Bruce EA, Honson DD, Chen LM et al. 2020. SARS-CoV-2 disrupts splicing, translation and protein trafficking to suppress host defenses. *Cell* 183(5): 1325–1339.e21.

Montero H, García-Román R & Mora SI 2015. eIF4E as a control target for viruses. *Viruses* 7(2):739–750.

Sharma A, Yilmaz A & Marsh K 2012. Thriving under stress: Selective translation of HIV-1 structural protein mRNA during Vpr-mediated impairment of eIF4E translation activity. *PLOS Pathog* 8(3):e1002612.

Tidu A, Janvier A, Schaeffer L, Sosnowski P, Kuhn L et al. 2020. The viral protein NSP1 acts as a ribosome gatekeeper for shutting down host translation and fostering SARS-CoV-2 translation. *RNA* 27:253–264.

Viruses and apoptosis

Madan V, Castelló A & Carrasco L 2008. Viroporins from RNA viruses induce caspase-dependent apoptosis. *Cell Microbiol* 10(2):437–451.

Ren Y, Shu T, Wu D, Mu J, Wang C et al. 2020. The ORF3a protein of SARS-CoV-2 induces apoptosis in cells. *Cell Molec Immunol* 17:881–883.

Yang Y, Wu Y, Meng X, Wang Z, Younis M et al. 2022. SARS-CoV-2 membrane protein causes the mitochondrial apoptosis and pulmonary edema via targeting BOK. *Cell Death Diff* 29:1395–1408.

Viruses and the ubiquitin–proteasome system

Faust TB, Li Y, Jang GM, Johnson JR, Yang S et al. 2017. PJA2 ubiquitinates the HIV-1 Tat protein with atypical chain linkages to activate viral transcription. *Sci Rep* 7(1):1–15.

White EA, Munger K & Howley P 2016. High-risk human papillomavirus E7 proteins target PTPN114 for degradation. *mBio* 7(5):e01530-16.

Viruses and autophagy

Campbell GR, Bruckman RS, Herns S, Joshi J, Durden DL et al. 2018. Induction of autophagy by PI3K/MTOR and PI3K/MTOR/BRD4 inhibitors suppresses HIV-1 replication. *J Biol Chem* 293(16):5808–5820.

Gassen NC, Papies J, Bajaj T, Emanuel J, Dethloff F et al. 2021. SARS-CoV-2-mediated dysregulation of metabolism and autophagy uncovers host-targeting antivirals. *Nat Comm* 12(1):1–15.

Halt AS, Olagnier D, Sancho-Shimizu V, Skipper KA, Helleberg M et al. 2020. Defects in LC3B2 and ATG4A underlie HSV2 meningitis and reveal a critical role for autophagy in antiviral defense in humans. *Sci Immunol* 5(54):EABC2691.

Richards AL, Soares-Martins JA, Riddell GT & Jackson WT 2014. Generation of unique poliovirus RNA replication organelles. *mBio* 5(2):e00833-13.

Viruses and microfilaments

Schudt G, Dolnik O, Kolesnikova A, Biedenkopf N, Herwig A et al. 2015. Transport of Ebolavirus nucleocapsids in dependent on actin polymerization: Live-cell image analysis of Ebolavirus-infected cells. *J Infect Dis* 323(suppl_2):S160–S166.

Viruses and endoplasmic reticulum stress

Johnston BP, Pringle ES & McCormick C 2019. KSHV activates unfolded protein response sensors but suppresses downstream transcriptional responses to support lytic replication. *PLOS Pathog* 15(12):E1008185.

Stahl S, Burkhart JM & Hinte F 2013. Cytomegalovirus downregulates IRE1 to repress the unfolded protein response. *PLOS Pathog* 9:e1003544.

Virus replication compartments

Fernandez de Castro I, Zamora PF, Ooms L, Fernandez JJ, Lai CMH et al. 2014. Reovirus forms neo-organelles for progeny7 particle assembly within reorganized cell membranes. *mBio* 5(1):e00931-13.

Laurent T, Kumar P, Liese S, Zare F, Jonasson M, Carlson A & Carlson L-A 2022. Architecture of the chikungunya virus replication organelle. *eLife* 11:e83042.

Wolff G, Limpens RW, Zevenhoven-Dobbe JC, Laugks U, Zheng S et al. 2020. A molecular pore spans the double membrane of the coronavirus replication organelle. *Science* 369(6509):1395–1398.

Viral proteomics

Gordon DE, Jang GM, Bouhaddou M, Xu J, Oberneir K et al. 2020. A SARS-CoV-2 protein interaction map reveals targets for drug repurposing. *Nature* 583(7816): 459–468.

Marion T, Elbahesh H, Thomas PG, DeVincnenzo JP, Webby R et al. 2016. Respiratory mucosal proteome quantification in human influenza infections. *PLOS One* 11(4):e0153674.

Soh TK, Darvies CTR, Muenzner J, Hunter L, Barrow HG et al. 2020. Temporal proteomic analysis of herpes simplex virus 1 infection reveals cell-surface remodeling via pUL56-mediated GOPC degradation. *Cell Reports* 33(1):108235.

Persistent Viral Infections

13

Virus	Characteristics
Phage λ	Model for lysogeny; complex regulatory cascades determine whether the phage replicates lytically or persists as a prophage.
Human immunodeficiency virus (HIV)	Causes a persistent infection, ultimately resulting in AIDS.
Human herpesvirus 1 (HHV-1)	Classic model for latent infections in human beings.
Human papillomavirus (HPV)	Oncogenic Baltimore Class I (dsDNA) virus; causes cancer via the E6 and E7 regulatory proteins.
HHV-4 (Epstein–Barr virus)	Oncogenic human herpesvirus [Baltimore Class I (dsDNA) genome]; causes cancer using viral structural mimics of host signal transduction proteins.
Hepatitis B virus (HBV)	Oncogenic Baltimore Class VII (reverse-transcribing) virus that causes chronic lytic infections that provoke chronic inflammation and oxidative damage, leading to cancer.
Hepatitis C virus (HCV)	Oncogenic Baltimore Class IV [(+) RNA] virus that causes chronic lytic infections that provoke chronic inflammation and oxidative damage, leading to cancer.

The viruses you will meet in this chapter and the concepts they illustrate

Most of this book has focused on lytic viral infections that proceed through all six stages of the virus replication cycle immediately upon entering a host cell. In contrast, this chapter focuses on **persistent infections**, also known as chronic infections. Persistent infections are those in which the virus, its genome, or parts of its genome persists in its host for the long term. There are two kinds of persistent infections that differ from lytic infections in their own ways. In the first type, lytic replication in a multicellular organism occurs at low levels and is ongoing for longer than the duration of the cell division cycle of most host cells. Such infections are peculiar to multicellular organisms such as humans in which the immune system takes a long time to eliminate the virus from the body. They differ from short-term lytic infections in that short-term lytic infections are eliminated from the human body on a time scale of days to weeks, whereas chronic infections persist for months, years, or even the duration of an infected person's lifetime. This chapter uses

DOI: 10.1201/9781003463115-13

HIV and hepatitis as examples of this first type of persistent lytic virus infection. The second type of persistent infections are **latent infections**. During a latent infection, a virus exits the lytic cycle and persists in host cells, usually in the form of a few nucleic acids and proteins, without causing production of new virions. Latent infections can occur in single-celled and multicellular organisms and they are reversible. Ironically, exit from latency and entry back into the lytic cycle, which will usually kill the host cell, typically occurs when the host cell encounters environmental stress (and might therefore die anyway). This chapter considers the lambda paradigm to discuss latent infections in bacteriophages and herpesviruses as models for latent infections in animals. An important consequence of some persistent infections in humans is oncogenesis, which can be caused by persistent lytic or latent viral infections. Two examples of DNA viruses that cause cancer include human papillomavirus and human herpesvirus 4. RNA viruses that cause cancer include hepatitis B virus, hepatitis C virus, and retroviruses.

13.1 Some bacteriophages are temperate and can persist as genomes integrated into their hosts' chromosomes

Bacteriophages are the most abundant evolving entities on Earth; it should come as no surprise, then, that in addition to the typical lytic reproductive cycle, some have an alternative strategy at their disposal. During **lysogeny**, a bacteriophage genome is inserted into a bacterium's chromosome and the host bacterium unwittingly replicates the entire phage genome during its normal reproduction. The bacteriophage genome inserted in the chromosome is called a **prophage** (**Figure 13.1**). A prophage can remain for an indefinite number of bacterial generations as it is passed down from one generation to the next. Bacteriophages that have the capacity for lysogeny are called **temperate** because they do not kill their host cells right away as lytic phages inevitably do. A bacterial host cell bearing a prophage is termed a **lysogen**.

13.2 Bacteriophage λ serves as a model for latency

Bacteriophage λ is a temperate phage; sometimes after uncoating, it undergoes lytic reproduction dominated by the Cro, N, and Q regulatory proteins (see **Chapter 4**). At other times, it undergoes conversion into prophage dominated by the regulatory protein CI (**Figure 13.2**). The decision to reproduce lytically or become a lysogen is binary: the phage must attempt one or the other. Nevertheless, the earliest stages of gene expression that will ultimately lead to lytic or lysogenic growth are the same. After the infecting phage

Figure 13.1 A lysogen contains a prophage. A lysogen is a bacterial cell that has a latent viral infection. A prophage is the viral genome after it has recombined with the host chromosome.

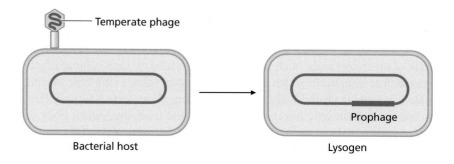

Temperate phage

Bacterial host

Prophage

Lysogen

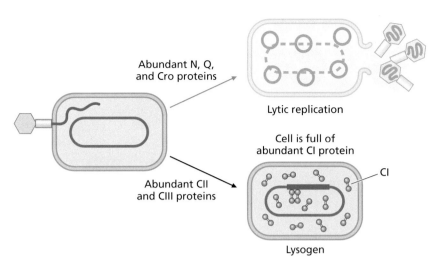

Figure 13.2 The λ phage either reproduces lytically or establishes a latent infection, known as lysogeny. If the infection produces abundant N, Q, and Cro proteins, lytic replication will occur. In contrast, if the infection produces abundant CII and CIII proteins, the virus will enter lysogeny. Lysogeny is maintained when the host cell contains abundant phage CI protein.

genome circularizes, immediate early gene expression occurs as a matter of course. The result is the production of low levels of Cro repressor protein and N antiterminator protein. Owing to N activity, a polycistronic mRNA encoding Cro, CII, O, P, and Q antiterminator is synthesized, bypassing terminator T_{R1} (**Figure 13.3**). The result is expression not only of the Cro and Q regulatory proteins but also of the CII regulatory protein. Shortly thereafter, N antitermination at the T_{L1} terminator results in a polycistronic message encoding the N and CIII regulatory proteins. The co-expression of Cro, N, and Q favors lytic replication, whereas the co-expression of CIII and CII favors lysogenic persistence. At this point, the phage has not yet committed to lytic or lysogenic replication and the host cell contains all of these proteins. Additional molecular events must occur in order to tip the balance toward abundant Cro, N, and Q or toward abundant CIII and CII. **Section 13.3** summarizes the events that occur so that the CIII and CII proteins predominate and will result in expression of the lysogeny-associated regulatory CI protein; **Chapter 4** explains the lytic cycle that occurs when Cro, N, and Q predominate.

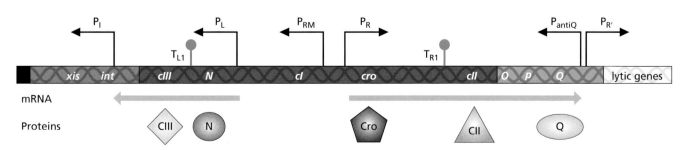

Figure 13.3 Immediate-early gene expression in λ. The λ genome is actually circular but it is represented here as linear to make it easier to see the position of promoters (black arrows) and terminators (blue lollipops), the genes, and the resulting mRNA and proteins. N antitermination results in polycistronic mRNA encoding the regulatory Cro, CII, and Q antiterminator proteins. At the same time, antitermination at the T_{L1} results in a polycistronic message encoding the N and CIII regulatory proteins.

Figure 13.4 Stability of CII determines the outcome of λ infection. (A) In rapidly growing cells, a host protease binds to CII and degrades it. The result is that lytic replication will occur. (B) In stressed host cells, the activity of the host protease is blocked by CIII. The result is that stable CII is highly active as a transcription activator.

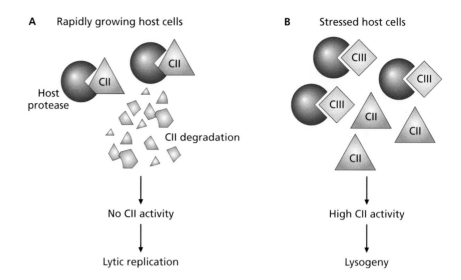

13.3 The amount of stable CII protein in the cell determines whether the phage genome becomes a prophage

CII is a DNA-binding protein that is intrinsically unstable; in the absence of **CIII** protein, it is rapidly degraded and has no effect on gene expression (**Figure 13.4**). This degradation occurs most often in host cells growing under environmental conditions that are favorable for host cell growth and reproduction. In that case, the CII protein is degraded and degradation of CII prevents lysogeny from occurring: lytic replication will occur. In host cells that are trying to grow under nutrient-limiting or other stressful conditions, the CII protein is more stable through the actions of the CIII protein. CII then accumulates and reaches concentrations at which it is sufficiently abundant to bind to regulatory DNA associated with three different promoters: P_{RE}, P_I, and P_{antiQ}. CII subsequently activates these promoters. The overall result is that lytic replication occurs in healthy host cells that were previously not experiencing stress, and lysogeny occurs in deprived host cells that might not have the necessary energy reserves to synthesize hundreds of offspring virions.

13.4 Activation of P_{RE}, P_I, and P_{antiQ} by CII results in lysogeny

The CII protein is a DNA-binding protein that can activate transcription from three λ promoters: P_{RE}, P_I, and P_{antiQ} (**Figure 13.5**). Together these promoters ultimately lead to the accumulation of the prolysogeny CI protein. CII activation of P_{RE} results in production of the CI protein, which does not occur during lytic growth. Abundant CI protein triggers a positive feedback loop that favors lysogeny. Beyond encoding protein, the CI mRNA has a second function, which is to reduce translation of Cro by binding to the complementary *cro* mRNA. Reduced Cro activity also results in reduced expression of the prolysis regulator Q. Expression from P_I results in production of integrase protein; the integrase enzyme takes the circularized phage genome in the cytoplasm and recombines it with the host chromosome forming a prophage. Meanwhile, expression from P_{antiQ} produces an RNA complementary to the mRNA encoding Q, which thereby further reduces levels of Q

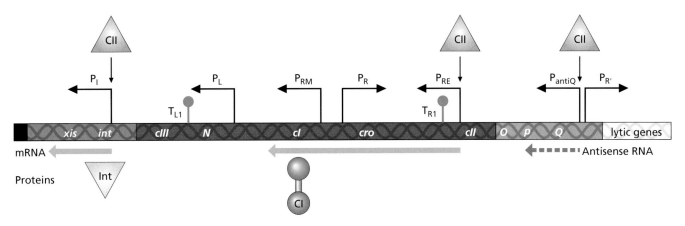

Figure 13.5 CII activates three promoters, which together tip the balance toward lysogeny. CII activation of P$_I$ leads to production of the integrase enzyme that recombines the λ genome with the host chromosome, creating a lysogen. CII activation of P$_{RE}$ leads to production of CI, which is a transcription activator that maintains lysogeny. CII activation of P$_{antiQ}$ results in production of an antisense RNA that binds to mRNA encoding Q and reduces the translation of Q antitermination protein.

protein; high levels of Q protein would otherwise bias the situation toward the lytic cycle.

The **CI protein**, also known as the **λ repressor**, accumulates as a result of CII activation of P$_{RE}$. Despite its name, mechanistically it is both a transcriptional repressor and activator, depending on where it binds relative to the start site of transcription. It binds to sites in the P$_L$ promoter region and represses P$_L$, resulting in a drop in N protein levels (**Figure 13.6**). A loss in N protein levels favors lysogeny by decreasing production of the Q antiterminator. The CI protein also binds to site in the P$_R$ promoter region and represses expression of O, P, and Q, again favoring lysogeny. Finally, CI binds to the P$_{RM}$ promoter and activates it. The mRNA produced by P$_{RM}$ activity encodes only the CI protein, thus initiating a positive feedback loop in which high levels of CI cause increasing production of CI. At very high levels of CI, however, expression of P$_{RM}$ is also repressed when CI binds to a repressing site in P$_{RM}$. This mechanism allows the positive feedback loop to function without filling the cell with a vast excess of CI protein.

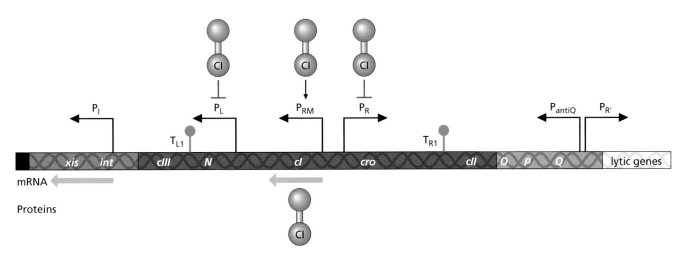

Figure 13.6 The activities of the CI protein favor lysogeny. CI binds to sites in the P$_L$ promoter region and represses P$_L$, resulting in a drop in N protein levels. The CI protein also binds to site in the P$_R$ promoter region and represses expression of O, P, and Q. Finally, CI binds to the P$_{RM}$ promoter and activates it, resulting in a positive feedback loop.

The molecular interactions between P_L, P_R, P_{RM}, and CI are understood in detail and provide a model for how a simple system consisting of one DNA-binding protein and a few binding sites with slightly different sequences can result in complex control of gene expression. In fact, understanding transcription regulation by CI was one of the foundational research programs in molecular biology. In the **P_L** promoter there are three CI-binding sites, also known as **operators**, named O_{L1}, O_{L2}, and O_{L3} (**Figure 13.7A**). Similarly, there are three CI-binding sites between the **P_R** and **P_{RM}** coding regions named O_{R1}, O_{R2}, and O_{R3}; these three operators control expression from both P_R and P_{RM}. The CI protein is a site-specific DNA-binding protein that binds to its sites as a dimer. It binds to certain sequences more tightly than to others. For example, CI binds better to the sequences of O_{L1} and O_{L2} than it does to O_{L3}. Moreover, the pair of dimers exhibits cooperatively so that a pair of dimers occupies O_{L1} and O_{L2} together, through dimer–dimer interactions, at lower concentrations of CI protein than would be predicted from each dimer binding to an artificially isolated O_{L1} or O_{L2} site by itself. As a consequence, at low levels of CI the protein occupies O_{L1} and O_{L2} but not O_{L3}; higher levels of CI are needed to occupy the binding sites, such as O_{L3} and O_{R3}, with a lower affinity for the protein. When CI binds to O_{L1} and O_{L2}, P_L is repressed.

The situation is similar at P_R and P_{RM}, where intermediate levels of CI lead to cooperative binding at O_{R1} and O_{R2} but leave O_{R3} unoccupied (see Figure 13.7A). In this configuration (low levels of CI), both P_L and P_R are repressed by CI binding, but the P_{RM} promoter, which is only a short distance from P_R, is activated by the CI dimer bound to O_{R2}. This situation leads to a positive feedback loop so that CI is the only viral protein found in a λ lysogen. At very high levels of CI, the CI protein binds to the O_{L3} and O_{R3} sites in both P_L and P_{RM} and forms a loop structure through protein–protein interactions among the six CI dimers (**Figure 13.7B**). The result is that all of the P_L, P_R, and P_{RM} are repressed and no further CI expression occurs. This configuration of DNA and proteins prevents the host cell from filling with excess CI during lysogeny; when CI levels drop, for example through cell division in which each new cell receives approximately half of the CI, expression from P_{RM} resumes, again raising CI levels so that lysogeny is maintained.

Figure 13.7 Interactions between CI and different DNA sequences result in complex regulation using just one regulatory protein. (A) Low levels of CI result in cooperative binding of dimers to O_{L1}, O_{L2}, O_{R1}, and O_{R2}, repressing P_L and activating P_{RM}. (B) High levels of CI result in cooperative binding of dimers to all operator sites, including to the low-affinity O_{L3} and O_{R3} sites, which causes formation of a DNA loop that represses transcription from P_L, P_{RM}, and P_R. Repression of P_{RM} at high concentrations prevents the cell from filling up with too much CI.

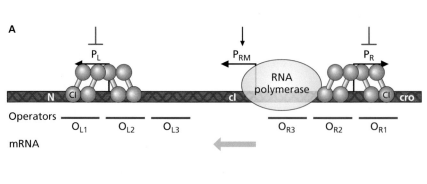

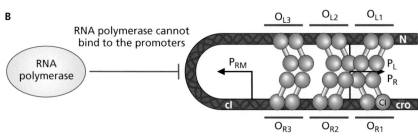

13.5   Stress triggers an exit from lysogeny

Although the positive feedback loop that maintains CI as the only λ protein in a lysogen is effective, it can be disrupted by environmental cues that indicate a stressful circumstance for the cell. If the cell might not survive the environmental assault, it is to the phage's advantage to excise from the chromosome and enter the lytic cycle, producing hundreds of offspring before the environmental assault kills the host cell. In response to stress, cells trigger a very well-known series of molecular events that begin with DNA damage and end with proteolytic degradation of the CI protein; collectively the chain of events is known as the **SOS response** (because the cells need help, as indicated by the international SOS Morse code signal; **Figure 13.8**). Environmental assaults that can cause DNA damage include ultraviolet radiation, ionizing radiation, desiccation, certain antibiotics, and chemical mutagens. The SOS response can repair damaged DNA so that the cell can continue growing and reproducing.

If the cell is a λ lysogen, however, the SOS response simultaneously leads to proteolysis of CI, which ultimately kills the host cell through inducing lytic reproduction of the phage (**Figure 13.9**). Proteolysis of CI destroys the capability of cooperative binding, so that the amount of CI remaining is no longer sufficient to bind to O_L and O_R sequences. As a result, the pro-lytic genes are expressed. A phage protein excises the prophage from the chromosome, forming a circular phage genome, and the entire lytic cycle proceeds, producing hundreds of offspring phages and killing the host cell. Ironically, the amount of DNA damage required to induce prophage excision is much lower than the amount of DNA damage it would take to kill a cell that is not a lysogen.

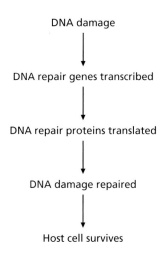

Figure 13.8 The SOS response in *E. coli* cells. DNA damage in an uninfected cell induces expression of DNA repair genes so that the cell may survive.

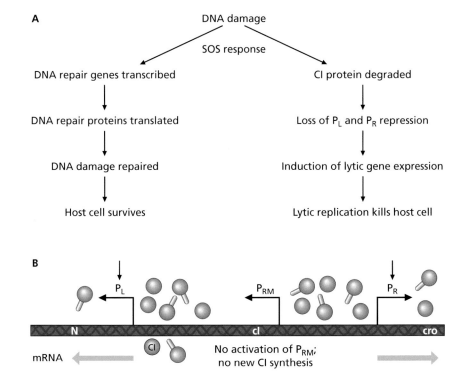

Figure 13.9 The SOS response in a lysogen triggers CI degradation. (A) DNA damage to a lysogen induces an exit from lysogeny so that lytic phage replication occurs and the host cell dies. (B) The molecular mechanism of exit from lysogeny is degradation of CI. CI degradation allows expression from P_L and P_R and prevents activation of P_{RM}. In the absence of activation, P_{RM} is not active, so no new CI protein is expressed. The result is expression of lytic genes.

Figure 13.10 Cholera toxin is encoded by a prophage. *Vibrio cholerae* must be a CTXφ lysogen to cause cholera. The genome of the prophage encodes cholera toxin, which is responsible for the signs and symptoms of cholera.

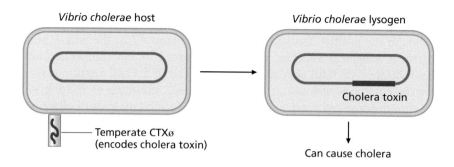

13.6   Some lysogens provide their bacterial hosts with virulence genes

The presence of the λ prophage does not seem to confer any advantages or new physiological capabilities upon its bacterial host. In contrast, other prophages are important for the unique physiological capabilities that enable some bacteria to cause disease, a property called pathogenesis. The species *Vibrio cholerae*, which causes the disease cholera, is perhaps the most well-known example. Pathogenic strains of *V. cholerae* encode cholera toxin, which is responsible for causing the life-threatening diarrhea associated with cholera. The cholera toxin is actually encoded by the genome of a temperate phage called CTXφ; *V. cholerae* strains that are themselves infected by this temperate phage are able to cause cholera (**Figure 13.10**). Strains that are not lysogens cannot cause cholera. **Virulence genes** such as those encoding cholera toxin contribute to the signs and symptoms of a particular disease or enable pathogenic bacteria to infect and survive in its animal host. It has become clear that prophages are a critical source of diverse virulence genes in many important pathogens such as *E. coli*, *Streptococcus pyogenes*, and *Staphylococcus aureus* (**Table 13.1**).

Table 13.1 Bacterial infections in which a virulence protein encoded by a prophage contributes to the signs and symptoms of the disease.

Disease	Toxic Protein	Bacteriophage	Bacterial Host
Cholera	Cholera toxin	CTXφ	*Vibrio cholerae*
Diphtheria	Diphtheria toxin	β-Phage	*Corynebacterium diphtheriae*
Strep throat	Superantigen	8232.1	*Streptococcus pyogenes*
Botulism	C1	Phage C1	*Clostridium botulinum*
Gastroenteritis	Enterotoxin P	φN313	*Staphylococcus aureus*
Gastroenteritis	SopE	SopEφ	*Salmonella typhimurium*
Dysentery	Shiga toxin	H-19B	*Escherichia coli*

13.7   Prophages affect the survival of their bacterial hosts

Prophages affect bacterial survival in other ways as well, as illustrated by antibiotics that cause prophage induction, biofilm formation, and sporulation. For example, in some cases exposure to antibiotics causes prophage induction and a simultaneous increase in the expression of prophage-encoded virulence genes. The induction of the prophage results in spread of that prophage and its virulence genes through the bacterial population. Prophages also affect **biofilm** formation. Biofilms are bacterial communities attached to a surface (**Figure 13.11**); they are more resistant to the immune system and to antibiotics than the same bacteria are when they are free-living outside of a biofilm. Biofilms consist of not only cells but also extracellular material that provides structure and mechanical strength to the biofilm. In the pathogen *Bacillus anthracis* (the cause of anthrax), the presence of prophages promotes biofilm formation to such an extent that strains lacking all prophages barely form biofilms. In the opportunistic pathogen *Pseudomonas aeruginosa*, the presence of prophages is associated not with biofilm formation but with biofilm dispersal, in which cells leave the biofilm and go on to colonize other surfaces and found the growth of new biofilms, thereby spreading the infection. Although the extracellular material in many biofilms is constructed from polysaccharides, in some biofilms extracellular DNA (eDNA) is a major component of the biofilm. In *Streptococcus pneumoniae*, which have biofilms composed substantially of eDNA, a prophage is essential for the presence of the eDNA and therefore for biofilm formation.

Some important human pathogens such as *B. anthracis* can **sporulate**, and sporulation is affected by prophages (**Figure 13.12**). **Spores** are tough survival forms of bacterial cells that can survive in a metabolically inert state for thousands or even millions of years under certain conditions; they are resistant to antibiotics and disinfectants, to environmental extremes of temperature, to ultraviolet light and ionizing radiation, and to desiccation. Pathogenic bacteria that can form spores include *B. anthracis*, *Clostridium botulinum*, *C. difficile*, *C. perfringens*, and *C. tetani*; because they form spores, they are very difficult to eliminate, even in hospitals with strict infectious disease prevention protocols in place. In many cases, the presence of prophages makes these bacteria more likely to form spores and more able to form spores abundantly and rapidly. There is also a curious situation in *Bacillus* species in which a critical sporulation gene called *sigK* is split into two parts by 48 kb of DNA; when the gene is in this form, the protein SigK cannot be produced (**Figure 13.13**). This intervening DNA can be excised precisely using site-specific recombination that takes advantage of a 5-bp repeat at each end. The recombination event restores the coding sequence of *sigK* so that the SigK protein can be synthesized and cause sporulation. The 48 kb of intervening DNA, called the skin (SigK intervening) element, do not encode an active prophage, but the presence of some phage genes and the precise recombination event itself strongly suggest that the intervening DNA evolved from a prophage. Such remnants that can no longer enter lytic replication are called **cryptic prophages**.

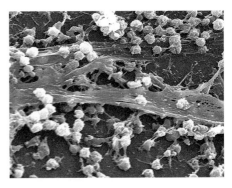

Figure 13.11 Biofilms. Biofilms are surface-associated communities of bacteria encased in an extracellular matrix that is synthesized by the bacteria themselves. The electron micrograph shows a biofilm of *Staphylococcus aureus* growing on a medical device. The spherical bacterial cells are surrounded by an extracellular matrix. (CDC/Rodney M. Donlan; Janice Haney Carr.)

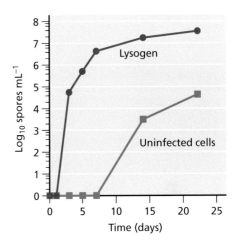

Figure 13.12 Sporulation and prophages. Lysogens of *Bacillus anthracis* that have a Wip2 prophage sporulate faster than uninfected counterparts. (From Schuch R & Fischetti VA 2009. *PLOS ONE* 4:e6532. doi: 10.1371/journal.pone.0006532. Published under CC BY 4.0.)

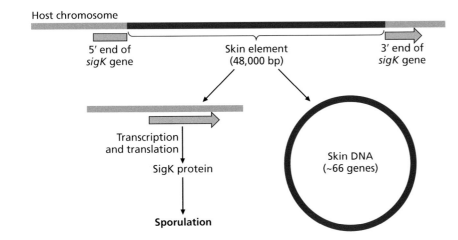

Figure 13.13 A cryptic prophage and the *sigK* gene in *Bacillus subtilis*. Without the SigK protein, sporulation cannot occur. The coding sequence for SigK protein is interrupted by a 48,000-bp cryptic prophage known as the skin element. Recombination similar to that needed for prophage induction can occur, removing the prophage and restoring the *sigK* coding sequence so that sporulation can occur.

Although the previous examples focus on pathogens, note that commensal and beneficial symbiotic bacteria are also strongly influenced by the presence of prophages. One study determined that 60%–70% of the bacteria in the human gut, of which the vast majority are beneficial symbionts or commensals that have no net effect on human health, are lysogens; the physiological activities (if any) affected by their prophages remain to be determined. Prophages are critical for the physiology and evolution of bacteria that live independently of animal hosts, too. One prominent example is that of the cyanophages that infect cyanobacteria. Cyanobacteria are the main photosynthesizers in the open ocean and as such are responsible for at least 25% of the primary production on earth. They are therefore key constituents of the carbon cycle. Cyanophages often encode proteins used during photosynthesis, so that bacteriophages are also fundamental to global carbon cycling.

13.8   Persistent infections in humans include those with ongoing lytic replication and latent infections

Long-term viral infections are common in humans. Sometimes, a healthy immune system eventually clears a chronic infection, eliminating the virus from the body, while other times, the chronic infection kills the person, as is the case in HIV. Latent infections such as those caused by herpesviruses, in contrast, resemble bacteriophage lysogeny; cold sores are common manifestations of latent herpesvirus infections that have sprung into lytic growth in epithelial cells. There are also a few groups of viruses in which persistent infections are oncogenic meaning that they cause cancer; in these cases, sometimes the infection is latent, whereas other times viral replication is persistent during oncogenesis.

13.9   Human immunodeficiency virus causes persistent infections

HIV emerged as a significant pathogen with epidemic potential in the 1980s; since that time the virus has been recognized as the etiological agent for the acquired immunodeficiency syndrome (AIDS) pandemic. Some of the early work on HIV showed that an initial acute infection is followed by years of a persistent infection before the CD4+ T cell hosts drop to such a low level that the person's immune system would collapse (**Figure 13.14**). The drastic reduction in viral load after the initial acute infection reflects the immune system's robust response, but despite that response, viral replication is ongoing. Because HIV is a retrovirus that relies on reverse transcription, which lacks any proofreading or editing functions, new genomes accumulate mutations very rapidly. The specific reverse transcriptase from HIV has an extremely high error rate that results in one or two substitutions in each new virion relative to its ancestor. During persistent lytic HIV, the detectable level of virions in the bloodstream may be less than 10,000 per mL but even these relatively low levels indicate that there are approximately 1–10 billion new virions produced every day. This high level of reproduction in combination with the high error rate means that HIV evolves during the persistent lytic phase of infection. The immune system puts tremendous selective pressure on the virus, ultimately selecting ones that escape the immune response. Antiviral medications also place huge selective pressure on the virus so that taking one medication at a time is almost never effective because the virus evolves resistance. HIV is most commonly treated with at least three antiviral drugs in combination in order to delay the evolution of drug resistance; the

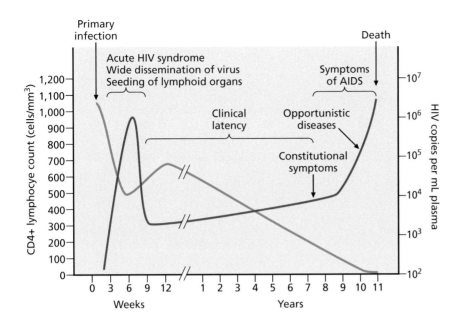

Figure 13.14 The course of viral replication, host cell numbers, and AIDS in someone with a chronic HIV infection. HIV RNA levels in blood plasma are plotted in red (on the right-hand axis), and CD4+ lymphocyte host cells are plotted in blue (on the left-hand axis). The phrase clinical latency refers to a long asymptomatic period that can last years.

chance of simultaneously evolving resistance to three different drugs at the same time is much smaller than the chance of evolving resistance to any one of them alone. Moreover, use of multiple drugs significantly slows HIV replication so that many fewer new viruses are produced each day and so there are many fewer opportunities for viral replication to produce a resistant virus.

Although HIV replicates in host cells such as CD4+ helper T lymphocytes, other host cells in the body become latently infected. These other cells have become the focus of intense scrutiny. Although actively replicating virus can be eliminated from the body with the right combination of drugs, latent virus that exists as a provirus is not actively being expressed and cannot be eliminated with drugs that target viral replication. There is a push to develop not just treatment for HIV but an actual cure, which would require eliminating not only actively replicating virions but also latently infected cells. So far, the leading strategy, sometimes called kick-and-kill, is attempting to eliminate HIV completely from the patient's body by inducing the latently infected cells to start producing virus, so that they, too, can be killed by the virus while the virus itself is inactivated by combination antiretroviral therapy.

13.10   Human herpesvirus 1 is a model for latent infections

The classic model of a latent infection in human beings is herpesvirus infection of sensory neurons; over 90% of adults globally have antibodies that recognize human herpesvirus (HHV) 1 or 2, indicating that most people have been exposed to at least one of these viruses. People with anti-herpesvirus antibodies but no detectable virions in their bodies have a latent herpesvirus infection. Latent herpesvirus infections begin with lytic infection of an epithelium, such as that occurring when HHV-1 or -2 is transmitted through sexual activity. The virus replicates in the epithelium and latency is initiated when some of the offspring virions infect the sensory neurons that innervate the epithelium (**Figure 13.15**). Virions that enter the terminally differentiated neurons traffic along the cytoskeleton to the cell nucleus, which is found in the spinal column (a very long distance to travel for a particle that is 200 nm in diameter). The nucleocapsid docks with a nuclear pore complex, which

Figure 13.15 Establishment of herpesvirus latency in a neuron. The virus replicates in the epithelium, which is represented here by a single layer of cells, and latency is initiated when some of the offspring virions infect the sensory neurons underlying the epithelium. Virions traffic along the cytoskeleton to the neuronal nucleus. The nucleocapsid docks with a nuclear pore complex and the genome is released into the nucleus. The viral genome circularizes and persists as an episome, establishing a latent infection.

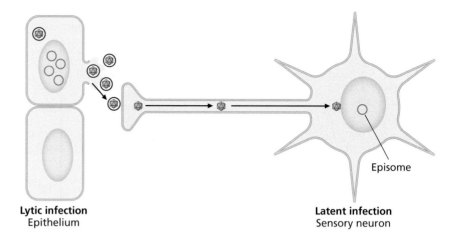

Episome

Lytic infection
Epithelium

Latent infection
Sensory neuron

induces a conformational change in the virion so that the genome is released into the nucleoplasm; the herpesvirus genome then circularizes and persists as an **episome** in the nuclei of latently infected cells.

In an epithelial cell, the viral VP16 protein activates the first wave of lytic gene expression, but this does not occur in neurons that will become latently infected (see **Chapter 8**). This activation requires collaboration with two nuclear transcription factors, Oct1 and Hcf. The Hcf protein is not found in neuronal nuclei, so VP16 cannot activate lytic gene expression in neurons. Moreover, conditions in the neuronal nuclei favor the formation of heterochromatin, which prevents the expression of lytic genes by condensing the DNA and preventing transcription factors from having access to the DNA. The one promoter required for latency, meanwhile, is not condensed into heterochromatin. It is instead activated by proteins uniquely found in neurons and results in transcription of an 8.3-kbp noncoding RNA. This RNA is differentially spliced, producing an unusually stable 2.0-kbp intron called **LAT** (for latency-associated transcript), which can itself be spliced further to make an additional 1.5-kbp LAT. The LAT RNAs are stable and accumulate to high levels in the host nucleus.

LAT RNAs affect many aspects of neuron physiology in ways that favor the maintenance of latency, though there is disagreement regarding many of the molecular mechanisms by which LAT causes these effects. For example, LAT interferes with host cell apoptosis, but the pathways affected by LAT are hotly disputed. LAT also affects chromatin by promoting formation of heterochromatin to block lytic gene expression, although again the molecular details are not known. The LAT transcript is complementary to mRNA encoding a major lytic transcription activator, and so it likely represses lytic reproduction in part through preventing translation of this transcription activator. Some of the controversies regarding exactly how LAT RNAs prevent expression of lytic genes and promote host cell survival likely arise because different animal and cell culture models are used to study herpesvirus infection.

Just as λ bacteriophage can exit lysogeny and re-enter lytic replication, so can herpesviruses. In the cases of HHV-1 and -2, **reactivation** is the term used to describe how the latent episome's transcriptional program shifts away from latency so that new virions exit the infected neuron and infect the epithelium innervated by that particular neuron (**Figure 13.16**). After that, lytic replication in the epithelium resumes. Reactivation is triggered by stressors, including emotional stress, exposure to ultraviolet light, and fever. It can even

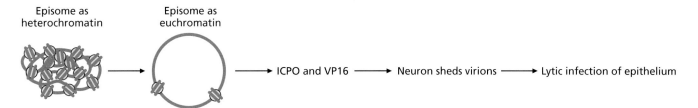

Figure 13.16 Herpesvirus reactivation. Reactivation causes chromatin remodeling of the episome so that the ICP0 transcription activator is expressed. ICP0 triggers the first wave of early gene expression. The VP16 transcription activator may also be required. The neuron sheds virions that will cause lytic infection of cells in the epithelium associated with the neuron terminus.

be triggered by the normal hormonal levels associated with menstruation in healthy women. Reactivation is also triggered by weakening of immunity brought about by aging or by infections such as HIV. In order to reactivate, the latent episome must express high levels of immediate early genes, which then set off a cascade of lytic gene expression; doing so must involve extensive chromatin remodeling in order to reverse the transcriptional silencing of lytic promoters. Chromatin remodeling then enables activation of lytic genes, probably beginning with the ICP0 transcription activator protein that triggers the wave of gene expression associated with lytic replication. Others have found that expression of VP16 is instead, or is additionally, needed for reactivation and that neuronal nuclei contain the crucial host Hcf transcription factor during reactivation. Virions that enter the epithelium replicate lytically, killing the epithelial hosts and spreading to other susceptible people. Disagreement regarding the roles of particular transcription activators during reactivation has most likely arisen because of the use of different animal or tissue culture models of reactivation.

Other human herpesviruses such as varicella–zoster virus (HHV-3, formerly VZV), Epstein–Barr virus (HHV-4, formerly EBV), cytomegalovirus (HHV-5, formerly CMV), and Kaposi's sarcoma-associated herpesvirus (HHV-8, formerly KSHV) also cause latent infections of certain host cells. The exact details of the molecular basis of latency and reactivation depend on the particular virus and its host cell. In general, these viruses replicate lytically when they first enter the human body and subsequently infect quiescent or terminally differentiated host cells where the viruses establish latency. Reactivation occurs upon weakening of the immune system or exposure to stress as in herpes simplex virus infections. A typical example is the extremely painful condition known as shingles that results from reactivation of HHV-3 (VZV), which causes chicken pox when a person is infected for the first time. Fortunately, vaccination is available to prevent chicken pox in the first place and to reduce the chance that an older adult with a latent VZV infection will develop shingles.

13.11 Oncogenic viruses cause cancer through persistent infections

Some persistent viral infections result in **cancer**, which is a collection of many different diseases that are all characterized by the abnormal proliferation of **malignant** cells that are invasive and ultimately kill the affected animal. Malignant cells evolve independently of the organism of which they are a part; they almost start to behave like a population of single-celled organisms instead of a population of differentiated cells that contribute to the physiology of a whole animal. For example, malignant cells exhibit several of the following properties. They are not as differentiated as the cells in the tissue from which they arise and so are sometimes called **dedifferentiated**. They are metabolically very active with an increased rate of nutrient transport, glycolysis, and the release of extracellular enzymes that degrade the supporting connective tissue proteins surrounding the cells. Cancer cells can more easily be cultured *in vitro*, where they require fewer growth factors than normal cells. Whereas normal cells spread out on the surface of growth flasks and stop proliferating once they become confluent (touching other cells on all sides), malignant cells may continue to proliferate despite confluence or may even be able to grow in suspension, detached from the culture dish. Malignant cells no longer respond to pro-apoptotic signals, whether from other cells or from within, and they become mobile and invade other tissues. They continue proliferating despite having unrepaired DNA damage, which tends to make them even more likely to obtain other favorable mutations that will further enhance their malignant properties. Cancer cells also exhibit genetic instability in which the normal controls on chromosome structure and number are missing and chromosomal loss and rearrangements become common. Another important property of malignant cells in a tumor is the ability to attract new blood vessel growth, called angiogenesis; angiogenesis not only provides the tumor with nutrients to support the cells' enhanced metabolic rate but also provides them with a way to escape to distant sites in the body carried along by the bloodstream. The process of becoming cancerous is known as oncogenesis or **transformation**; a population that has recently become transformed has fewer of the properties of malignant cells while a transformed population that has been evolving for a longer period of time exhibits more of the dangerous malignant properties that are ultimately fatal to the affected human being.

A small subset of viruses is known to be **oncogenic** (cancer-causing) in humans, including ones with dsDNA genomes (Baltimore Class I), one with a (+) ssRNA genome (Baltimore Class IV), one retrovirus (Baltimore Class VI), and one other reverse-transcribing virus (Baltimore Class VII). In some cases, transformation is caused by a chronic infection in which lytic replication is ongoing, whereas in most cases it is caused by a latent infection in which no offspring virions are produced. Cancers caused by latent infections are a dead end for the virus because no virions are produced that could infect other hosts. In contrast cancers caused by persistent lytic infections are not necessarily dead ends because the offspring virions have the chance of spreading to a new host.

There are two broad mechanisms by which viruses cause cancer (**Figure 13.17**). The first mechanism is that viral genes directly or indirectly (by affecting host gene expression) manipulate the cell cycle and apoptosis so that the virus immortalizes its host cell and causes host cells to proliferate even in the presence of mutations that ultimately lead to cancer (see also **Chapter 8**). The second mechanism is that the immune system's reaction to the presence of the virus, rather than the activities of any specific regulatory viral or host cell proteins themselves, actually causes the cancer. The role of chronic immune activation in many types of cancer is an increasingly active area of research. Although some viruses favor one mechanism over the other, in most cases both mechanisms likely contribute to causing malignant tumors. The rest of this chapter considers several oncogenic viruses about which we have the best molecular understanding, focusing on the first mechanism because, unlike a whole-body immune reaction to a foreign virus, it acts primarily at the molecular and cellular levels.

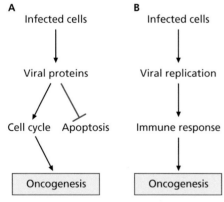

Figure 13.17 Two general mechanisms by which viruses cause cancer. (A) Viral proteins manipulate the cell cycle and apoptosis so that the host cell proliferates abnormally. (B) An immune reaction to the virus causes cancer.

13.12   DNA viruses transform cells with oncoproteins that affect the cell cycle and apoptosis

The DNA viruses known to cause cancer in humans include Merkel cell polyomavirus (MCPyV), human papillomavirus (HPV), Epstein–Barr virus (HHV-4 or EBV), and Kaposi's sarcoma-associated herpesvirus (HHV-8 or KSHV). This chapter will discuss in detail two of these viruses: HPV as a model for an oncogenic DNA virus with a small genome and EBV as a model for an oncogenic DNA virus with a large genome.

13.13   HPV oncoproteins E6 and E7 cause transformation

HPV causes cervical cancer; approximately 250,000 women die from HPV-associated cervical cancer in any given year, most of them in low-consumption countries. Although the average age of diagnosis is about 50, most women are exposed to the virus when they are younger, so that a persistent infection that endures for decades is responsible for cervical cancer. Most women infected by HPV do not go on to develop cervical cancer. Certain isolates of HPV, however, are more likely to cause cancer than others; among these dangerous types are HPV16 and HPV18. With circular dsDNA genomes that are approximately 8,000 bp long, these viruses encode approximately 12 proteins (new ones continue to be discovered; see also **Chapter 8**). The early E5, E6, and E7 proteins manipulate the cell cycle and apoptotic pathways in ways that promote viral replication (**Figure 13.18**). Malignant cervical cancer cells have continuous expression of E6 and E7 that counteracts the host proteins p53 and pRB, respectively (see **Chapter 8**). Because of their critical role in preventing many types of cancer, p53 and pRB are both known as **tumor suppressor proteins**. Because of their effects on these specific tumor suppressor proteins, the viral E7 protein causes increased cell cycling and proliferation and the viral E6 protein prevents apoptosis even in the presence of stresses such as unrepaired DNA damage. Because of their role in promoting cancer, E6 and E7 are examples of **oncoproteins**.

The key differences between dangerous and more benign oncogenic strains of HPV lie in the E6 and E7 proteins. In oncogenic strains, E6 and E7 have novel properties not found in the less dangerous isolates. For example, oncogenic E7 proteins are more abundant than E7 protein during a lytic infection (**Figure 13.19**). Furthermore, oncogenic E7 proteins bind more tightly to pRB than non-oncogenic E7 proteins, making them more effective at promoting

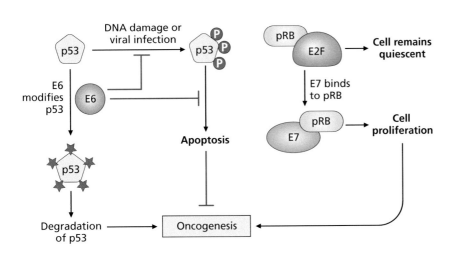

Figure 13.18 The HPV E6 and E7 proteins are oncogenic because they manipulate the cell cycle and block apoptosis. E6 interacts with p53, preventing apoptosis in response to DNA damage or to signals from surrounding tissue. E7 interacts with pRB, causing proliferation independent of signals from surrounding tissue. Together the two effects promote oncogenesis.

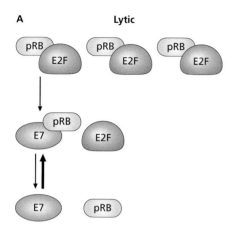

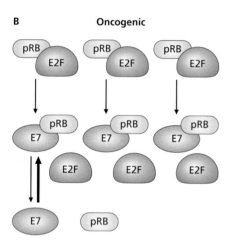

Figure 13.19 Comparison between the abundance and binding properties of lytic or oncogenic HPV E7 proteins. (A) During lytic infections, E7 is not as abundant and only binds to a small proportion of the cellular pRB. (B) During an oncogenic infection, E7 is more abundant and therefore binds to a larger proportion of the cellular pRB. Lytic E7 proteins also bind less tightly to pRB than oncogenic E7 proteins do.

cell cycling. Unlike non-oncogenic E7 proteins, oncogenic E7 proteins also target pRB for destruction by the ubiquitin–proteasome system, which further promotes cell proliferation (**Figure 13.20**). Similarly, novel properties associated with oncogenic E6 proteins include higher abundance and tighter binding to p53 as part of a complex with another host protein that targets p53 for destruction by the ubiquitin–proteasome system. Together, the effects of E7 and E6 on pRB and p53, respectively, strongly inhibit apoptosis and promote cellular proliferation. During lytic infections, the expression of E6 and E7 is controlled in part by the viral E2 protein, which normally represses E6 and E7 expression when it binds to the long control region in the viral genome; this binding is essential for the virus to switch from early-gene expression to late-gene expression (see **Chapter 8**). During oncogenesis, this function of E2 is lost so that the expression of E6 and E7 is constant; such constant expression is required for transformation.

13.14 HPV E6 and E7 overexpression occurs when the virus genome recombines with a host chromosome

One of the most common mechanisms for loss of E2 function is the integration of the viral genome into a host chromosome, which occurs in such a way that the E2 coding sequence is disrupted (**Figure 13.21**). During lytic infections, HPV does not recombine with host chromosomes, but integration of the HPV genome is found in the majority of cervical cancer cells. Integration promotes constant E6 and E7 expression. Recombination of the HPV genome and human chromosomes as well as the constant expression of oncogenic E6 and E7 proteins also cause genetic instability, further enabling the transformed cells to progress toward ever-greater malignancy. Recombination of the HPV genome with a host chromosome also seals the virus's fate because after that point the host cells no longer shed infectious virions and so the virus is no longer infectious. Fortunately, there is a vaccine that protects against the four most common types of oncogenic HPV; although its effects on persistent infections cannot yet be assessed because less than a decade has passed since the vaccine's introduction, it is already clear that the vaccine has reduced new HPV infections by more than 50%. It is also worth noting that the Pap smear diagnostic technique followed by removal of precancerous lesions is effective at preventing cervical cancer.

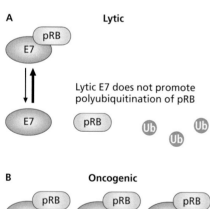

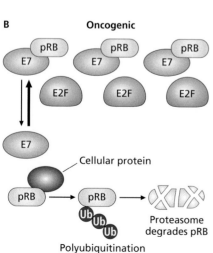

Figure 13.20 Comparison between the effects of lytic versus oncogenic HPV E7 proteins on pRB. (A) Lytic E7 proteins do not result in polyubiquitination of pRB. (B) Oncogenic E7 proteins form a complex with cellular proteins that targets pRB for destruction by the ubiquitin–proteasome system.

Figure 13.21 Recombination between HPV and the host chromosome occurs during oncogenesis. (A) During lytic infections, HPV has a circular genome that permits the production of E2 proteins and therefore new virions. (B) During oncogenic HPV infections, part of the HPV genome recombines with a host chromosome so that the E2 coding sequence is disrupted. As a result, E6 and E7 are overexpressed and the virus genome is no longer able to direct production of new virions.

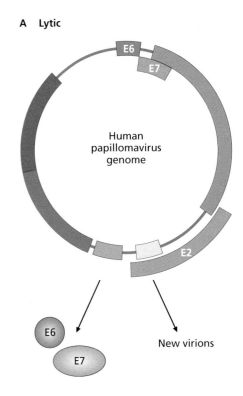

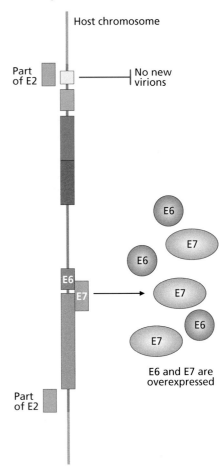

13.15   Merkel cell polyomavirus is also associated with human cancers

The other small dsDNA virus with a circular genome associated with human cancers is MCPyV, which is a polyomavirus. Polyomaviruses including MCPyV and SV40 have large T and small t antigens that regulate the cell cycle and prevent apoptosis during lytic viral replication (see **Chapter 8**). As with HPV oncogenesis, insertion of the MCPyV genome into a host chromosome is associated with transformation, as is the constant expression of the viral proteins that lead to proliferation and a lack of apoptosis. Also, similar to HPV, latent oncogenic MCPyV infections are a dead end for the virus because the tumor cells with integrated viral genomes do not shed infectious viruses; furthermore, only a small minority of MCPyV infections result in cancer.

13.16   Epstein–Barr virus is an oncogenic herpesvirus

The larger dsDNA herpesviruses encode about 10 times more proteins than the papilloma- and polyomaviruses (see also **Chapter 8**). For example, HHV-4, also known as Epstein–Barr virus (EBV), encodes approximately 80 proteins. About 90% of all people on Earth harbor a latent HHV-4 infection; only a tiny minority of these cases result in cancer. Nevertheless, HHV-4 is associated with several different malignancies. Here the focus is on just one of them, called **Hodgkin disease** (**HD**), also known as Hodgkin's lymphoma. Approximately 40% of all cases of HD are caused by HHV-4, and virus-associated HD is a classic model for how viral oncoproteins cause cancer through manipulation of signal transduction pathways that affect the cell cycle and apoptosis. Signal transduction pathways are the mechanism by which cells respond to extracellular signals and change their behavior (such as motility) or gene expression in response to that signal. An idealized signal transduction pathway begins with a ligand engaging a receptor, which causes the intracellular part of the receptor to become phosphorylated (**Figure 13.22**). A cytoplasmic signal transduction protein is then able to bind to the phosphorylated receptor, which triggers the cytoplasmic protein to become an active kinase itself. Like any kinase, it then phosphorylates its target proteins. Those target proteins go on to respond to the signal, for example by changing gene expression.

Infection by HHV-4 has a profound effect on signal transduction in latently infected cells as is evident by studying the malignant cells in HD, which are known as **Hodgkin and Reed–Sternberg cells** (**RS cells**). RS cells originate from B cells that were infected by HHV-4 and contain clonal viral genomes, meaning that the RS cells likely originate from a single ancestor that had been infected by a single virion; the viral genome persists for the lifetime of malignant cells. During latent infection of B cells, the viral genome exists as an episome and only a small subset of the viral proteins is expressed. This latency is in turn in associated with developing HD.

Figure 13.22 Idealized signal transduction pathway. A transmembrane receptor can be activated by an external ligand. Upon binding to the ligand, the receptor dimerizes and the intracellular component of the receptor becomes phosphorylated (1). Phosphorylation allows the receptor to associate with other proteins (2) that, when bound to the phosphorylated receptor, set off a cascade of protein–protein interactions and phosphorylations (3). The phosphorylated transcription factor dissociates from the receptor (4) and is targeted to the nucleus (5), which changes gene expression (6), thereby responding physiologically to the external signal.

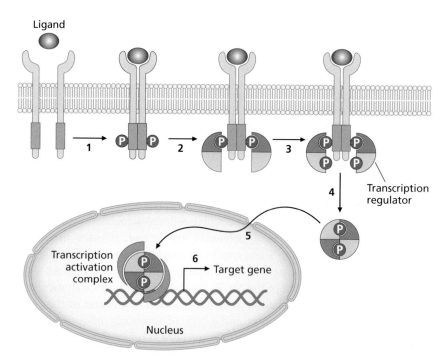

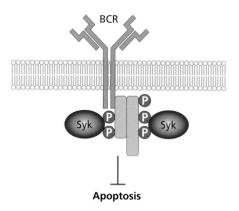

Figure 13.23 BCR signaling in normal B cells. In the absence of a ligand, the BCR sends a tonic anti-apoptotic signal through phosphorylation of certain tyrosines found in motifs on cytoplasmic signaling proteins. The Syk protein interacts with these when they are phosphorylated and ultimately prevents apoptosis.

Although they began as differentiated B cells, RS cells are dedifferentiated so that they lack proteins associated with functional B cells, such as the B-cell receptor (BCR). The latent viral infection causes dedifferentiation through manipulating BCR signaling. The BCR is critical for the B cell's role in immunity, but it is also essential simply for the survival of B cells. Independently of any ligand binding to the receptor, in a normal B cell the BCR sends a constant survival signal that counteracts apoptosis (**Figure 13.23**). If this signal is interrupted, the B cell undergoes apoptosis. It is therefore remarkable that RS cells, which lack the BCR, survive at all, let alone that they proliferate. The signal is propagated because the BCR associates with cytoplasmic signaling proteins that have specific motifs that can be phosphorylated on tyrosine side chains. After phosphorylation, these domains associate with other signal transduction proteins such as Syk, which ultimately prevent apoptosis through a series of other molecular events.

13.17 Latency-associated viral proteins are responsible for Epstein–Barr virus-induced oncogenesis

The survival and proliferation of RS cells depends on latency-associated viral proteins. These latently expressed proteins include LMP2A, LMP1, and EBNA1. LMP is an acronym for latent membrane protein and EBNA is an acronym for Epstein–Barr virus nuclear antigen. RS cells survive in the absence of constant BCR signaling largely owing to the presence of LMP2A. Although the BCR and LMP2A differ in structure, LMP2A mimics the normal function of BCR by sending an anti-apoptosis signal through the same cytoplasmic signaling cascade (**Figure 13.24**). It has motifs containing tyrosine that can be phosphorylated and then associated with the Syk protein, triggering it to prevent apoptosis. That is, LMP2A sends the signal that would otherwise be sent by the BCR; this so-called tonic signal is essential in RS cells that have dedifferentiated to such an extent that they lack the BCR. Importantly, LMP2A differs from the BCR in that it is not susceptible to any of the normal events that would otherwise cause cessation of the anti-apoptotic signal in uninfected B cells. Instead, constant LMP2A signaling keeps the malignant RS cells alive.

The viral LMP1 and LMP2A proteins are similar in that they both functionally mimic normal transmembrane B-cell signal receptors and that they send a constant on signal irrespective of extracellular signaling events. The LMP1 protein mimics the function, though not the structure, of a normal cellular transmembrane protein called CD40 (**Figure 13.25**). In normal B cells, CD40 is exclusively stimulated by helper T cells during an immune response that requires proliferation of that particular B cell so that the B cell will produce antibodies to attack a specific infectious agent (see **Chapter 14**). Through binding to CD40, the helper T cells stimulate several intracellular signal transduction cascades that ultimately drive proliferation; key for this stimulation is a PxPQxT motif in the intracellular domain of CD40, where P is proline, Q is glutamine, T is threonine, and x is any amino acid. In RS cells expressing LMP1, LMP1 stimulates the same intracellular signal transduction pathways in the absence of any T-cell signaling so that the infected cells proliferate abnormally in the absence of activation by helper T cells. The signal transduction occurs because LMP1 also has a PxPQxT motif on an intracellular domain.

Although the LMP proteins collaborate to cause survival and even proliferation, one additional protein is also expressed during all oncogenic latent HHV-4 infections. This protein, EBNA1, is a multifunctional protein that is required for oncogenesis (**Figure 13.26**). One of its functions is transcription regulation; it can bind to DNA and stimulate viral latent genes as well as cellular genes. It is also required for physically tethering viral episomes to host chromosomes and thereby for persistence of the episome when the host cell undergoes mitosis. Additionally, the cellular genes regulated by EBNA1 include those that manipulate the immune system and prevent the immune system from easily recognizing the infection, especially through the antigenic LMP1 and LMP2A proteins.

13.18   The Kaposi's sarcoma herpesvirus also causes persistent oncogenic infections

Although both the small oncogenic dsDNA viruses and the larger ones cause unregulated proliferation and cellular survival despite circumstances that would normally trigger apoptosis, the mechanisms by which they do

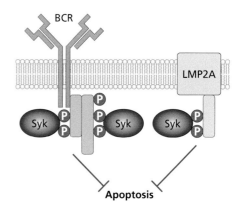

Figure 13.24 LMP2A signaling in an RS cell with a latent EBV infection. The LMP2A protein is phosphorylated on certain tyrosines found as part of motifs that mimic those important for tonic signaling by BCR. The Syk protein interacts with these and ultimately prevents apoptosis just as it does when regulated by the BCR.

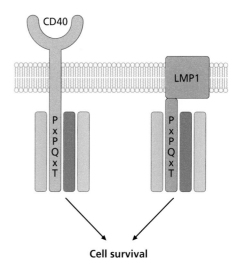

Figure 13.25 LMP1 signal transduction in RS cells with a latent EBV infection. Normal CD40 signaling occurs through interactions between cytoplasmic proteins and a PxPQxT motif on a cytoplasmic domain, resulting in associations with other signaling proteins that ultimately promote survival. LMP1 triggers the same signal transduction cascade through a PxPQxT motif in its cytoplasmic domain.

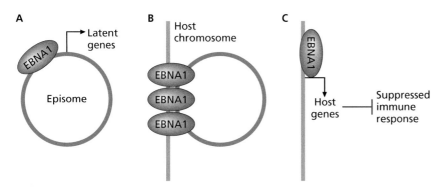

Figure 13.26 EBNA1 functions during latent EBV infection of RS cells. (A) The viral EBNA1 protein activates expression of latent viral genes. (B) It physically tethers viral episomes to a host chromosome. (C) It also activates expression of host genes, which has the effect of suppressing an immune response against EBV.

so are very different. The small viruses directly target p53 and pRB through direct viral–host, protein–protein interactions. In contrast, the larger HHV-4 (EBV) mimics normal signal transduction and indirectly interferes with normal cell cycling and apoptosis. HHV-8, formerly known as Kaposi's sarcoma-associated herpesvirus, is another large herpesvirus that can be oncogenic in its latent state. In that case, the virus makes a latency-associated protein called LANA-1 that directly binds to p53 and pRB, inactivating them. HHV-8 also encodes a viral cyclin that interacts with a cellular cyclin-dependent kinase and then phosphorylates pRB. Phosphorylation of pRB causes proliferation. Additionally, like EBNA1, LANA-1 tethers the viral episome to host chromosomes, ensuring mitotic inheritance of the viral episome, and LANA-1 is also a DNA-binding protein that regulates expression of viral and host genes. Kaposi's sarcoma therefore shares some properties with cancers caused by both the smaller oncogenic DNA viruses and with those caused by its closer relative HHV-4.

13.19   Hepatocellular carcinoma is caused by persistent lytic viral infections

The Class IV (+) RNA virus associated with human cancer is hepatitis C virus (HCV; **Figure 13.27A**), a flavivirus that causes a type of liver cancer called hepatocellular carcinoma (HCC). HCV is an enveloped Class IV virus with a medium-size genome that encodes three structural proteins and seven nonstructural proteins, including a protease, a viral RNA-dependent RNA polymerase (RdRp), and accessory proteins needed for it to function properly. A second virus, the reverse-transcribing Baltimore Class VII hepatitis B virus (HBV; **Figure 13.27B**) also causes HCC, so the two viruses will be discussed together in this section. Like HCV, the enveloped HBV virion also has a small coding capacity, encoding four structural proteins such as the core (capsomer); the small, medium, and large spike proteins; and four nonstructural proteins, including the viral polymerase P.

Both HCV and HBV cause the same type of cancer, HCC (**Figure 13.28**), using a primary mechanism that differs from those of the Class I DNA viruses. First, oncogenesis is associated with persistent ongoing viral lytic replication and not with latent infection. Second, HCV and HBV do not encode any mimics of signal transduction proteins. Instead, infection of the liver results in chronic inflammation and oxidative stress, which can progress

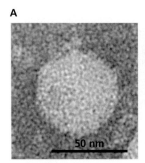
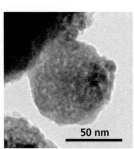

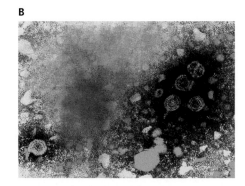

Figure 13.27 HCV and HBV cause hepatocellular carcinoma. (A) Electron micrograph of a hepatitis C virus (HCV), a spherical enveloped (+) RNA virus. (B) Digitally colorized electron micrograph of the reverse-transcribing, spherically enveloped hepatitis B virus (HBV), which is about 42 nm in diameter. (A, From Lindenbach BD & Rice CM 2013. *Nat Rev Microbiol* 11:688–700. doi: 10.1038/nrmicro3098. With permission from Springer Nature. B, CDC/Dr. Erskine Palmer.)

to a condition called cirrhosis. Chronic inflammation and oxidative stress cause cirrhosis irrespective of the cause of the inflammation and oxidative stress. Indeed, alcoholism also causes cirrhosis in the absence of any infectious agents; furthermore, cirrhosis can on its own be fatal even if it does not progress to cancer. During HBV and HCV infections, most people do not develop a robust immune response to the viruses and so the viruses persist without being cleared from the body. The immune reactions to HBV and HCV are areas of intensive research but will not be covered in detail in this book. Regardless of the cause, during cirrhosis the liver tissue becomes so damaged that it loses its ability to function; the liver tries to replace the damaged tissue through normal regeneration processes in which stem cells give rise to new hepatocytes. During a persistent lytic HCV or HBV infection, however, these processes are ultimately ineffective because of the chronic inflammation and oxidative damage, which prevent the regeneration from proceeding normally and from being effective. The ultimate result in some cases is HCC.

Although chronic inflammation and oxidative stress likely account for some of the transformation caused by HBV and HCV, upon recent closer inspection the viruses are also associated with increased cellular proliferation and survival, despite conditions such as damaged DNA that ought to trigger apoptosis. For example, both HCV and HBV infections promote genetic instability, which on its own can cause cells to progress toward malignancy. Although neither virus affects p53 or pRB as strongly as HPV does, both viruses do lead to reduced apoptosis and increased cell cycle progression. Both viruses are also associated with dedifferentiation of hepatocytes; this process likely contributes to transformation as well. The HCV proteins that most likely contribute to oncogenesis are NS3, NS5A, and core (capsomer), and the HBV proteins that most likely contribute to oncogenesis are HBx, pre-S, and S. Interestingly, core and S are structural proteins (capsomer and spike, respectively); it is unusual that a structural protein would contribute to oncogenesis. It is also increasingly accepted that sometimes parts of the HBV genome recombine with a host chromosome that is associated with stronger expression of HBx and more likely progression to HCC; as in the papillomaviruses and polyomaviruses, this recombination is a dead end for viral replication (resulting in a latent infection). The molecular pathways by which HBV and HCV proteins cause all these changes, however, are complex and not universally agreed upon, and so they are not covered in any detail here.

A better understanding of oncogenesis triggered by persistent lytic infection with HCV or HBV is important because approximately 170 million people are infected with HCV and an estimated 2 billion are infected with HBV, and some proportion of these people will go on to die of cirrhosis of the liver or HCC. There is an HBV vaccine available, although no vaccine for HCV is currently on the market. At least there are new anti-HCV drugs available to treat persistent lytic infections, and several promising anti-HBV drugs are in development. The first effective curative anti-HCV drug, sofosbuvir, targets the viral RdRp and was made commercially available in 2013. Its high cost has caused controversy.

13.20 Retroviruses have two mechanisms by which they can cause cancer

Retroviruses were the earliest viruses discovered to be oncogenic. The first one was Rous sarcoma virus (RSV), which causes cancer in chickens. The critical RSV oncoprotein is called v-Src and it is encoded in the viral genome.

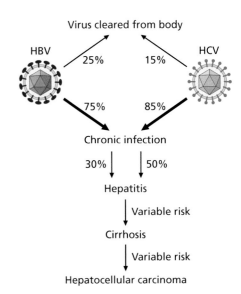

Figure 13.28 Hepatocellular carcinoma can be caused by HBV or HCV infection. The initial immune response may clear the infections but this is rare. More commonly both go on to cause a chronic infection with constant lytic replication. The result can be hepatitis followed by cirrhosis. An ongoing immune response causes inflammation and oxidative damage so that sometimes the cirrhotic liver develops hepatocellular carcinoma.

Figure 13.29 The c-Src and v-Src proteins affect signal transduction. Cellular c-Src is a cytoplasmic signaling protein that can be phosphorylated at different sites. It responds to extracellular growth factors by binding to an activated growth factor receptor, becoming phosphorylated at certain positions, and subsequently sending a proliferation signal. Cellular c-Src has a C-terminal regulatory region that can also be phosphorylated, but in that case phosphorylation is an inhibitory event that turns signaling off when the extracellular stimulus is withdrawn. The RSV v-Src protein, in contrast, lacks this C-terminal phosphorylation site, so that v-Src is constitutively active, sending a constant pro-proliferation signal irrespective of the presence or absence of growth factors.

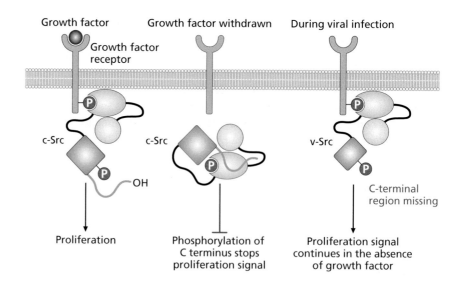

Study of oncogenes in retroviruses led to many fundamental discoveries in cell biology, especially when it was first demonstrated that DNA hybridizing to viral oncogenes is found in normal, uninfected cells. These normal genes were named **proto-oncogenes** to indicate both their similarities to and differences from viral oncogenes; the prefix v- refers to the viral oncogene and the prefix c- refers to its normal cellular counterpart. The name proto-oncogene also reflects the hypothesis that viral oncogenes originate in evolution from normal host genes acquired during viral replication. Retroviral oncogenes are almost always mutated versions of host genes that control the cell cycle or apoptosis, typically through signal transduction. For example, c-Src is a cytoplasmic signaling protein that participates in the stimulation of proliferation by extracellular growth factors; it has a C-terminal region that can be phosphorylated in order to turn signaling off when the extracellular stimulus is withdrawn (**Figure 13.29**). The RSV v-Src protein, in contrast, lacks this C-terminal phosphorylation site so that v-Src is constitutively active, sending a constant pro-proliferation signal irrespective of the presence or absence of growth factors. The study of oncogenes and proto-oncogenes has led to a detailed picture of proliferation- and apoptosis-related signaling in normal cells.

Study of oncogenic retroviruses led to classifying them into two groups: the **oncogene-transducing retroviruses** and the **oncogene-deficient retroviruses** (**Figure 13.30**). Infection with an oncogene-transducing

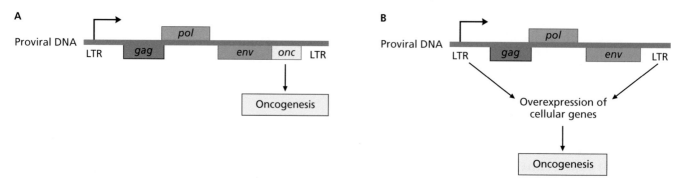

Figure 13.30 Oncogene-transducing and oncogene-deficient retroviruses. (A) Oncogene-transducing retroviruses carry an oncogene as part of their genome. (B) Oncogene-deficient retroviruses cause cancer when insertion of the long terminal repeat (LTR) causes overexpression of a cellular proto-oncogene.

retrovirus rapidly leads to cancer in nearly 100% of susceptible animals and leads to transformation of cultured cells in nearly 100% of cases. Infection with an oncogene-deficient retrovirus instead leads to cancer only some of the time, and does so slowly in infected animals; these viruses transform cultured cells in only a minority of cases. The difference between these two groups of viruses is as follows: the oncogene-transducing retroviruses carry a functional oncogene in their genomes (see **Figure 13.30A**) so that 100% of host cells carrying proviral DNA express the oncogene whenever the proviral promoter in the long terminal repeats (LTRs) is active (see also **Chapter 10**). The oncogene-deficient retroviruses, in contrast, do not encode an oncoprotein in their proviral DNA (see **Figure 13.30B**). Instead, oncogenesis is an accidental consequence of exactly where the proviral DNA is inserted in a host chromosome. In a minority of cases, the proviral DNA inserts near a proto-oncogene. The LTR then serves as an enhancer or even as a promoter that drives overexpression of the cellular proto-oncogene. This misexpression then causes overproliferation of the host cell, which sets it on a path toward becoming malignant. Because all retroviruses replicate through insertion of proviral DNA that includes an LTR, all retroviruses have the ability to cause cancer whether or not they encode an oncogene.

Although there are many known oncogenic retroviruses, most of them, like RSV, do not infect human beings. The most common oncogenic retrovirus in humans is human T-lymphotropic virus 1 (also known as human T-cell lymphotropic virus 1; HTLV-1). Typically, HTLV-1 in an infected person is found as a provirus inserted in a unique chromosomal site in around 10,000 different T cells. Of those infected, 5% will develop cancer following infection, and the cancer often takes decades to develop. HTLV-1 is therefore not a classic oncogene-transducing virus because it is associated with slow development of cancer in only a minority of infected people.

On the other hand, it is not simply a classic nontransducing oncogenic retrovirus either. Instead, the oncogenic properties of HTLV-1 are caused by two proteins, Tax and HBZ, encoded in the provirus (**Figure 13.31**). The mechanisms by which HTLV-1 causes cancer are not as clear-cut as they are for classic oncogene-transducing retroviruses, however. Tax is a transcription regulator, whereas HBZ is also regulatory in nature; neither one has a known homolog among normal cellular genes. In addition to activating viral genes, Tax also activates many host genes, including those that promote host cell proliferation and resistance to apoptosis. HBZ has similar effects on host cells and may use multiple mechanisms to cause them. It can interact with pRB, thereby promoting cell cycling. Tax may be essential for starting the process of oncogenesis, whereas HBZ may be responsible for maintaining

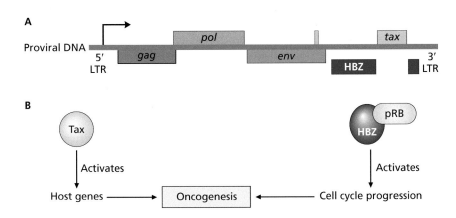

Figure 13.31 HTLV and oncogenesis. (A) The location of the Tax and HBZ coding sequences in an HTLV provirus. HBZ is encoded from right to left. (B) Tax is a transcription regulator that directly activates host genes, including those that promote proliferation and block apoptosis. HBZ binds to pRB, which promotes progression through the cell cycle.

it, because although the two proteins have similar effects on host cells, they appear to be active at different times over the course of persistent oncogenic infections.

13.21   Viral oncoproteins can be used to immortalize primary cell cultures

Study of viral oncogenesis has also transformed the disciplines of virology and cell biology by making it possible to immortalize cultured cells. When normal cells are explanted from an animal, they typically divide 60 or fewer times before they enter senescence and no longer divide. At this point, they cannot be used for any more experiments. Obtaining such **primary cell cultures** requires collaboration with a surgeon, and techniques to culture only a certain cell type from explanted tissue are notoriously difficult. Moreover, cells from one donor may differ from cells from another donor, complicating the interpretation of experimental results. In contrast, **immortalized cells** can continue to divide indefinitely and can be cryogenically preserved so they are preferred for many experiments. One way to obtain immortalized cells is to start with a tumor such as the cervical cancer that doctors took from Henrietta Lacks, creating the infamous HeLa cell line. These cells do not produce HPV virions but are cancerous because their ancestors were infected by HPV. The other way to obtain immortalized cells is to create them intentionally starting from a specific primary line.

Cell lines immortalized intentionally should be capable of proliferation beyond the capacity of primary cell lines and should be tolerant of cryogenic preservation, but they should ideally possess otherwise similar or even identical phenotypes and genotypes compared to their parental primary cells. Virology has been key in developing molecular tools to transform primary cultures in order to achieve this goal. Biotech companies make recombinant viruses that can infect primary cell lines and thereby introduce viral oncogenes such as SV40 large T antigen (see **Chapter 8**) or HPV's E6 and E7 proteins. The effect is that the cells become immortalized without having too many other characteristics modified. The use of viral oncogenes to immortalize cells is yet another contribution that virology has made to related disciplines.

13.22   The human virome is largely uncharacterized but likely has effects on human physiology

Through the use of next-generation sequencing technologies, we have only recently become aware that the human virome, the entire collection of viruses found in a human body at any given time, is composed of viruses that infect human cells and of viruses that infect all of the other microbes such as bacteria, archaea, protozoa, and fungi that also live in the human body. Many plant viruses are part of the virome because human beings eat plants. An inclusive definition of the human virome also includes DNA of viral origin found in human chromosomes.

Detecting the entire virome is difficult because there is no single gene similar to the cellular 16/18 s rDNA that can be found in all viruses; furthermore, strategies to detect DNA viruses are not as effective at detecting RNA viruses. It is clear, though, that ongoing exposure to the virome affects immunity in complex ways, some of which might even be beneficial. The ubiquitous presence of bacteriophages on human mucosal surfaces could be considered an aspect of immunity because those phages likely reduce the number of bacteria that reach the epithelium underlying the layer of mucin with its bacteriophages.

There could be other unexpected beneficial effects of the persistent viruses in the virome, which may be similar to those discovered in model organisms. For example, mice with certain latent herpesvirus infections are less susceptible to infections by the bacterial pathogens *Listeria monocytogenes* and *Yersinia pestis* than mice that are not infected by the herpesviruses. The story is even more convoluted when one considers that both *L. monocytogenes* and *Y. pestis* are lysogens in which virulence is enhanced by their own prophages.

Essential concepts

- There are two different kinds of persistent infections: those in which viral replication is ongoing for long periods of time in a multicellular organism such as an animal, and latent infections that can occur in unicellular or multicellular hosts. Latent infections are characterized by viral macromolecules persisting in a cell without producing virions.
- Bacteriophage λ can reproduce lytically or can cause a latent infection in which it persists as a prophage.
- Bacteriophage lysogeny is controlled by a complex cascade of events that ultimately results in expression of the λ repressor CI protein. Lysogeny can be reversed, especially if the host cells are stressed by DNA damaging agents; in that case the phage is excised from the chromosome and lytic replication ensues.
- After an acute infection characterized by high levels of replicating virus and the mounting of a strong immune response, HIV persists as a chronic infection in which there is both persistent lytic replication and some latently infected host cells.
- Because HIV continues to replicate during the lifetime of an infected person and because the reverse transcriptase enzyme causes one or two mutations in each offspring virion, HIV evolves to resist the immune system and antiviral drugs.
- Herpesviruses are the classic example of viruses that cause lifelong latent infections.
- Human herpesvirus 1 and 2, once known as herpes simplex viruses, replicate lytically in epithelial cells and then enter neurons, where the virus persists as an episome expressing the LAT transcripts. The LAT transcripts maintain latency and prevent host cell apoptosis.
- Oncogenic viruses cause host cells to progress toward malignancy, most often by expressing oncogenes that cause dysregulated growth and proliferation and that block apoptosis.
- Human papillomavirus causes cancer in a minority of infected people; the cancer is associated with oncogenic E6 and E7 proteins that substantially decrease levels of p53 and pRB in the cell.
- Human herpesvirus 4 (also known as Epstein–Barr virus) is an oncogenic herpesvirus infection in which a small proportion of the time a latent infection of B cells can lead to B-cell cancers such as Hodgkin disease.
- The LMP2A, LMP1, and EBNA-1 proteins are expressed during latent infection with HHV-4 and cause progression to malignancy.
- There are three main causes of hepatocellular carcinoma (HCC): alcoholism, persistent infection by hepatitis B virus, and persistent infection by hepatitis C virus.

- There are two types of oncogenic retroviruses: oncogene-transducing retroviruses and oncogene-deficient retroviruses.
- The human virome is huge and largely uncharacterized; it undoubtedly contributes to human physiology in ways that we cannot yet predict.

Questions

1. After the 2015 West African Ebola fever disease outbreak, some survivors complained of long-term symptoms such as problems with their vision. What would we look for to determine whether Ebola virus can cause a persistent or specifically latent infection?
2. What does the λ repressor CI have in common with latency-associated proteins in viruses that infect humans?
3. How might regulation of λ virus be altered if we swapped O_{R3} and O_{R1}? How could we test this prediction?
4. In a newly sequenced bacterial genome, how would we look for prophage sequences? For cryptic phage sequences?
5. If we could isolate a mutant HIV reverse transcriptase gene that had a mutation rate 1,000 times lower than that of the normal HIV, what effects on viral replication might occur if this mutant gene were to replace the wild-type gene? How could we test this prediction?
6. Why might the use of different animal or tissue culture models of herpesvirus latency give different results?
7. What are some of the most unusual features of herpesvirus LAT transcripts?
8. How is reactivation of herpes simplex virus similar to the reversal of lysogeny in λ virus?
9. Why are dysregulated cell proliferation and blocking of apoptosis common in most malignancies?
10. Design an experiment to show that the most important difference between oncogenic and non-oncogenic HPV isolates is the differences in their E6 and E7 proteins.
11. Classify oncogenic viral infections according to whether oncogenesis is a dead end for viral replication or not.
12. Why are both LMP2A and LMP1 important for survival of RS cells? Would a drug that blocks signaling by LMP1 be a good anticancer treatment?
13. If we engineered HHV-4 (also known as Epstein–Barr virus) in such a way that the cytoplasmic portions of LMP2A and LMP1 were swapped, what might happen regarding oncogenesis? How could we test this prediction?
14. What is dedifferentiation and what role does it play in oncogenesis?
15. Why might it be difficult to sort out to what extent HBV and HCV proteins might be oncogenic on their own, independent of the oncogenic effects of chronic cirrhosis of the liver?
16. What is the mechanism by which all retroviruses can cause cancer? Why is this a concern when using retroviral vectors in gene therapy?
17. What are some differences between viral oncoproteins and cellular proto-oncoproteins?
18. Compare and contrast a tumor suppressor gene and a proto-oncogene.
19. What factors make detecting the human virome using next-generation sequencing more challenging than it is to detect the bacterial component of the microbiome?

20. It is possible that SARS-CoV-2 can cause chronic infection in people who are immunocompromised by a genetic condition or another infection such as HIV. How could you go about investigating the frequency of SARS-CoV-2 chronic infection among long-COVID patients who have symptoms many months after an acute COVID-19 infection?

Interactive quiz questions

In addition to the questions provided above, this edition has a range of free interactive quiz questions for students to further test their understanding of the chapter material. To access these online questions, please visit the book's website: www.routledge.com/cw/lostroh.

Further reading

Bacteriophage λ

Brown S, Mitarai N & Sneppen K 2022. Protection of bacteriophage-sensitive *Escherichia coli* by lysogens. *Proc Natl Acad Sci* 119(14): e2106005119.

Lysogens

Al-Anany AM, Fatima R & Hynes AP 2021. Temperate phage-antibiotic synergy eradicates bacteria through depletion of lysogens. *Cell Rep* 35(8):109172.

Chen Y, Yang L, Yang D, Song J, Wang C et al. 2020. Specific integration of temperate phage decreases the pathogenicity of host bacteria. *Front Cell Infect Microbiol* 10:14.

Thompson LR, Zeng Q, Kelly L, Huang KH, Singer AU et al. 2011. Phage auxiliary metabolic genes and the redirection of cyanobacterial host carbon metabolism. *Proc Natl Acad Sci* 108:E757–E764.

Waldor MK & Mekalanos JJ 1996. Lysogenic conversion by a filamentous phage encoding cholera toxin. *Science* 272:1910–1014.

Persistent infection with HIV

Deng K, Pertea M, Rongvaux A, Wang L, Durand CM et al. 2015. Broad CTL response is required to clear latent HIV-1 due to dominance of escape mutations. *Nature* 517(7534):381–385.

Wedrychowski A, Martin HA, Li Y, Telwatter S, Kadiyala GN et al. 2022. Transcriptomic signatures of human immunodeficiency virus post-treatment control. *J Virol* 97:e0125422.

Herpesvirus latency

Campbell M, Yang WS, Yeh WW, Kao CY & Chang PC 2020. Epigenetic regulation of Kaposi's sarcoma-associated herpesvirus latency. *Front Microbiol* 11:850.

Ungerleider N, Concha M, Lin Z, Roberts C, Wang X et al. 2018. The Epstein–Barr virus circRNAome. *PLOS Pathogens* 14(8):e1007206.

Oncogenic infections

Chang EH, Furth ME & Scolnick EM 1982. Tumorigenic transformation of mammalian-cells induced by a normal human-gene homologous to the oncogene of Harvey murine sarcoma-virus. *Nature* 297:479–483.

Cherian MA, Baydoun HH, Al-Saleem J, Shkriabai N, Kvaratskhelia M et al. 2015. Akt pathway activation by human T-cell leukemia virus type 1 Tax oncoprotein. *J Biol Chem* 290(43):26270–26281.

Lu X, Lin Q & Lin M 2014. Multiple-integrations of HPV16 genome and altered transcription of viral oncogenes and cellular genes are associated with the development of cervical cancer. *PLOS One* 9:e97588.

Mahmudvand S, Shokri S, Taherkhani R & Farshadpour F 2019. Hepatitis C virus core protein modulates several signaling pathways involved in hepatocellular carcinoma. *World J Gastroneterol* 25(1):42.

Mancao C & Hammerschmidt W 2007. Epstein–Barr virus latent membrane protein 2A is a B-cell receptor mimic and essential for B-cell survival. *Blood* 110:3715–3721.

Morgan, EL, Scarth JA, Patterson MR, Wasson CW, Hemingway GC et al. 2021. E6-mediated activation of JNK drives EGFR signalling to promote proliferation and viral oncoprotein expression in cervical cancer. *Cell Death Diff* 28(5):1669–1687.

Sakaguchi AY, Naylor SL & Shows TB 1983. A sequence homologous to Rous-sarcoma virus v-SRC is on human chromosome-20. *Prog Nucleic Acid Res Mol Biol* 29:279–282.

Zhi W, Wei Y, Lazare C, Meng Y, Wu P et al. 2023. HPV-CCDC106 integration promotes cervical cancer progression by facilitating the high expression of CCDC106 after HPC E6 splicing. *J Med Virol* 95(1):e28009.

Use of T antigen in culture

Skloot R 2010. *The Immortal Life of Henrietta Lacks*. Crown Publishers.

Wang N, Zhang W & Cui J 2014. The piggyback transposon-mediated expression of SV40 T antigen efficiently immortalizes mouse embryonic fibroblasts (MEFs). *PLOS One* 9:e97316.

Human virome

Lim ES, Zhou Y, Zhao G, Bauer IK, Droit L et al. 2015. Early life dynamics of the human gut virome and bacterial microbiome in infants. *Nat Med* 21(10):1228–1234.

Pride DT, Salzman J, Haynes M, Rohwer F, Davis-Long C et al. 2012. Evidence of a robust resident bacteriophage population revealed through analysis of the human salivary virome. *ISME J* 6:915–926.

Tokuyama M, Kong Y, Song E, Jayewickreme T, Kang I et al. 2018. ERVmap analysis reveals genome-wide transcription of human endogenous retroviruses. *Proc Natl Acad Sci* 115(50):12565–12572.

Xu GJ, Kula T, Xu Q, Li MZ, Vernon SD et al. 2015. Comprehensive serological profiling of human populations using a synthetic human virome. *Science* 348(6239):aaa0698.

Viral Evasion of Innate Host Defenses

14

Immune response	Effect on viruses
Restriction enzyme	Degrades dsDNA phage genomes.
Pattern recognition receptors (PRRs)	Bind to pathogen-associated molecular patterns (PAMPs) and then trigger innate human immune responses.
Interferon	Cytokine that induces antiviral state in uninfected cells.
Neutrophils	Release antiviral neutrophil extracellular traps and cause inflammation.
Antigen presentation in major histocompatibility complex I (MHC-I)	Used by all cells to display endogenous epitopes; infected cells display viral epitopes.
Natural killer cells	Kill infected cells because they have reduced MHC-I display.
Complement	Forms a membrane attack complex (MAC) on viral envelopes and on cells with viral antigens in the plasma membrane.

The immune responses you will meet in this chapter and their effect on viruses

Virology shares historical roots with the discipline of immunology. We have seen how tools such as antibodies that derive from immunology are very important for investigating how viruses replicate. We have seen how virology was once inseparable from immunology, as when Pasteur studied rabies long before electron microscopy could reveal the structure of a rabies virus. The interconnected nature of immunology and virology continues to the present day so that cross-training in the methods and research questions of both fields is fundamental for anyone who wants to study virology. Immunology and virology are not connected merely by the human mind and tradition. They are physiologically connected because viruses must overcome immune responses in order to replicate. Furthermore, evolution fundamentally connects the two as host cells evolve to escape viruses and viruses evolve to replicate despite immune reactions.

Vertebrate molecular and cellular defenses can be categorized into two classes: innate and adaptive (also known as acquired). **Innate defenses** are omnipresent. Omnipresent means that they do not have to be triggered by exposure to a microbe; instead, they are present all of the time, so that if a microbe enters a bacterial cell or the human body, the innate defenses are immediately active. The innate defenses control replication by groups of viruses that share a property such as being enveloped or having a dsDNA

DOI: 10.1201/9781003463115-14

genome. For example, in people, the innate defenses that defend against (+) RNA viruses react the same way against poliovirus, rhinovirus, and hepatitis C virus because they all have similar replication strategies as a result of their shared genome characteristics. The innate defenses behave exactly the same way every time there is an infection; for example, the innate immune reaction against poliovirus is exactly the same the first, second, and subsequent times someone is exposed to poliovirus. The innate defenses are very powerful, which explains how there can be billions and billions of viruses yet only a short list of those that cause disease in otherwise healthy people. Typically, viruses that manage to reproduce in the human body must have ways of overcoming the innate defenses in order to have enough time to replicate. We will encounter many such viral evasion strategies in this chapter.

The **acquired** or **adaptive immune responses** are instead triggered by a specific infection or immunization and respond to specific antigens associated with an individual microbe. For example, an adaptive immune response to poliovirus does not develop until after someone has been exposed to poliovirus or to a vaccine that mimics poliovirus infection. An adaptive immune response against poliovirus has no effect on rhinovirus infection even though both are picornaviruses that share many features in common. The acquired immune responses also have memory. The adaptive response following the first exposure to a microbe can take as long as 2 weeks to fully develop and to be able to control the infection. In contrast, the response following a second exposure occurs within hours: the immune system remembers that a previous exposure was dangerous and responds quickly. The reaction after the second and subsequent exposures is also stronger than the reaction after the first exposure. An example relates to antibodies that are produced as part of an adaptive immune response. Antibodies that bind to antigens on a microbe are much more abundant after a second exposure to that same microbe. Immunological memory therefore refers to the faster and stronger reaction to a second exposure than to the first.

This chapter focuses on innate immunity, beginning with viruses that infect bacteria and ending with viruses that infect humans. **Chapter 15** covers adaptive immunity and **Chapter 16** discusses practical applications of molecular virology such as immunization.

14.1 Restriction enzymes are a component of innate immunity to bacteriophages

Bacteria have many innate defenses against bacteriophages. For example, bacteria actively modulate the availability and spacing of potential phage surface receptors in order to make phage attachment less likely. An important innate defense against dsDNA phages is **restriction modification**. This defense is so important that more than 90% of known bacterial genomes encode one or more restriction–modification systems. Such systems rely on epigenetic mechanisms to attack bacteriophages' genomes and degrade them selectively without degrading a bacterium's own chromosome. A restriction–modification system consists of two components: a restriction enzyme (**Figure 14.1**) and a modification enzyme. They work together to degrade phage DNA while leaving the bacterial chromosome untouched.

Restriction enzymes are familiar in molecular biology because they are site-specific endonucleases useful for a broad range of applications such as molecular cloning (inserting a gene into a plasmid). An example of a restriction enzyme and its target sequence is *Eco*KI, which binds to DNA with the following sequence 5′-AACN$_6$GTGC (N$_6$ indicates any six nucleotides).

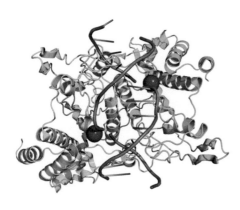

Figure 14.1 Restriction–modification system. Two subunits of a restriction enzyme (in teal and green) bound to DNA (backbone in orange). The restriction enzyme has hydrolyzed the DNA backbone. Ions necessary for the catalysis are in pink.

When *Eco*KI binds to this sequence in DNA, it catalyzes a double-strand break in the DNA. Restriction enzymes function as a critical antiviral defense. When dsDNA from a virus enters the cytoplasm of a cell that expresses a restriction enzyme, the restriction enzyme binds to its target sequence in that genome and degrades the viral DNA (**Figure 14.2**). The effect is to protect the cell from infection by degrading the phage's chromosome before phage gene expression can take over the cell's synthetic capacity. When an entering phage genome has no defense against a resident restriction enzyme, the phage infection is successful fewer than 1 in 10,000 times.

How does such a cell protect itself so that the restriction enzyme does not degrade its own chromosome? The answer is the second component, the **modification enzyme**. A modification enzyme **methylates** host cell DNA at the same target sequence as its cognate restriction site. Methylation occurs on sites that do not affect base pairing (**Figure 14.3**). For *Eco*KI, the methyltransferase modifies the first A in the sequence $5'$-AA*C(N_6)GTGC, where the * indicates methylation on the preceding nucleotide. This methylation prevents the *Eco*KI restriction enzyme from binding to that sequence of DNA and therefore protects the DNA from being hydrolyzed by the *Eco*KI restriction enzyme. In this way, the host's own methylated DNA is not degraded while incoming phage DNA, which is not methylated on the correct target site, is degraded (**Figure 14.4**). The name restriction enzyme comes from the observation that dsDNA phages are therefore restricted to multiplying in host cells in which the phage genome has been methylated to match the cellular chromosome's methylation pattern. For example, λ phages that replicate in *E. coli* K12 are called λ·12 and they are restricted to replicating in *E. coli* K12 host cells. Their genomes are methylated at all $5'$-AA*CN_6GTGC sites so that the *Eco*KI restriction enzyme does not degrade the phage genome. In contrast, *E. coli* B cells do not have the KI system but instead use a restriction–modification pair with a different target sequence (TGA*N_6TGCT). The λ·12 phages cannot replicate in *E. coli* B because of the *E. coli* B restriction enzyme. Similarly, λ·B phages are restricted to *E. coli* B host cells just as λ·12 phages are restricted to *E. coli* K12 host cells.

On the other hand, in 1 in 10,000 times, a phage infection can be successful even when the phage genome has the wrong methylation pattern. This rare event occurs when the host's cytoplasmic methyltransferase enzyme binds to the target sequences in the phage genome and methylates those sequences before the restriction enzyme binds to the same sequences and hydrolyzes the DNA (**Figure 14.5**). When the incoming phage genome becomes methylated, the methylation blocks the restriction enzyme from degrading the DNA, and so the phage genome is not degraded and the phage can mount a successful replication cycle. Offspring of successful phages take

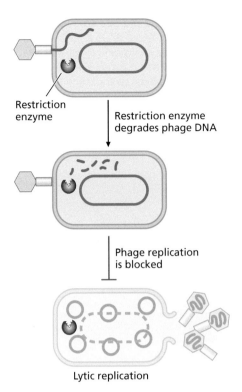

Restriction enzyme

Restriction enzyme degrades phage DNA

Phage replication is blocked

Lytic replication

Figure 14.2 A restriction enzyme degrades a phage genome during penetration. A restriction enzyme binds to the entering dsDNA genome of a phage and subsequently degrades the phage DNA, blocking phage replication.

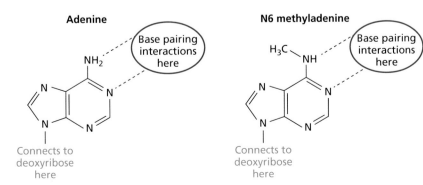

Adenine

Base pairing interactions here

NH_2

Connects to deoxyribose here

N6 methyladenine

Base pairing interactions here

H_3C NH

Connects to deoxyribose here

Figure 14.3 Methylation of nucleotides does not affect base pairing. An example of methylation is N6 methyladenine, which does not affect the portions of the adenine that participate in hydrogen bonding within a double helix.

Figure 14.4 λ·12 phages are restricted to λ·12 hosts. *E. coli* strain K methylation enables λ·12 to replicate in *E. coli* K12 strains (A) but not in *E. coli* B strains (B). Both the λ·12 phage genome and the K12 host have the same sequences methylated in their DNA, as indicated by the orange dots, whereas the methylated sequence in strain B is different, as indicated by the green dots.

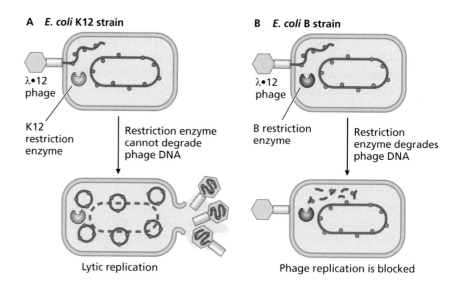

on the methylation pattern of their host cells because the cytoplasmic host methyltransferase enzyme modifies the new genomes, just as it would any newly replicated DNA. Thus, on the rare occasion that a phage λ·12 genome escapes restriction in an *E. coli* B cell, all of the offspring phage genomes take on the methylation pattern of this new host. Although their parent was λ·12, the offspring phages will be λ·B, restricted to *E. coli* B hosts.

This pattern of inheritance is unusual in that the actual DNA sequence of the λ·12 and λ·B phages is exactly the same: they have exactly the same genome sequence and yet have completely different properties. The inherited property of being restricted to one host or the other is caused by the inherited methylation, not by the sequence of the DNA. Inheritance of information in the form of modifications of the DNA rather than in the form of DNA sequences is called **epigenetic inheritance**. In this case, the offspring λ phages acquire the methylation pattern that matches their host cell and this trait is passed epigenetically to all offspring synthesized in that host cell. The methylation is an epigenetic modification that has profound effects on the offspring phage's ability to infect their next host cell.

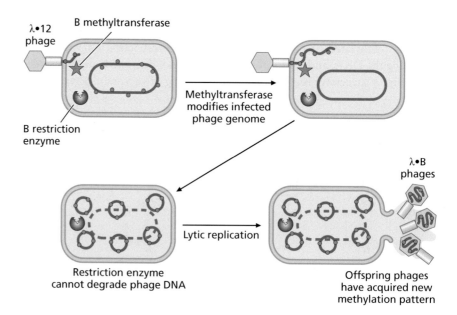

Figure 14.5 Rarely, a phage takes on a new methylation pattern and switches its restriction type. The resident methyltransferase enzyme modifies the phage genome as it enters the cell. The new modification prevents the action of the restriction enzyme. Offspring phages have gained the methylation pattern of their host cell. In this case, the offspring of λ·12 are methylated on the sequence determined by the B host strain.

14.2 Bacteriophages have counterdefenses against restriction–modification systems

Given the ubiquity of restriction–modification systems, it is unsurprising that some dsDNA phages have mechanisms to evade restriction. To do so, some phages encode their own methyltransferase enzymes, which interfere with restriction if the methylation overlaps a host's restriction site. The phage methyltransferase modifications usually occur frequently, such as every time a four-base sequence such as GATC appears in the genome. This sequence occurs at random every 1 in 256 bases [1 in $(4 \times 4 \times 4 \times 4)$]. The high frequency of the target sequence protects the phage from many restriction enzymes because the methylation is likely to overlap with a restriction site. Another viral evasion strategy is to encode a protein that binds to the active site of restriction enzymes, which blocks the enzymes' activity and thereby provides more time to allow the host's resident methyltransferase to modify the infecting virus genome so that restriction will not occur. The protein mimics the structure and chemistry of the phosphate backbone of DNA, which explains how it can bind to the active site of a restriction enzyme. Still other phages use alternative bases such as uracil or 5-hydroxymethyluracil to replace thymidine, which prevents restriction because restriction enzymes cannot bind to their target sites when the Ts are abnormal.

14.3 Human innate immune defenses operate on many levels

Vertebrate animals including humans have formidable defenses against infectious agents that operate on multiple levels. For example, at the whole organism level, skin covers the outer surface of the body with nonliving cells. Dead cells cannot serve as hosts for viral replication, providing innate immunity to viral infection. Immunity also operates at the molecular and cellular levels, sensing viral macromolecules within an infected cell and then triggering immune responses designed to kill the cell before the virus can complete its replication cycle or to prevent offspring viruses from infecting nearby cells. As in the rest of this book, here we focus on molecular and cellular reactions.

14.4 The human innate immune system is triggered by pattern recognition

Common structural elements found in groups of pathogens but not in uninfected human cells trigger molecular and cellular innate immune mechanisms. For example, infection by Class IV (+) ssRNA viruses (**Figure 14.6**) results in the presence of dsRNA with 5′ phosphate groups in the cytoplasm as a consequence of viral gene expression and genome replication. Uninfected human cells do not contain long dsRNA molecules with 5′ phosphate groups. Common structural features unique to microbes, compared with human beings, are called pathogen-associated molecular patterns, or PAMPs. Double-stranded RNA is a pattern in that all dsRNA has a helical shape with ribose in the backbone instead of deoxyribose irrespective of the sequence of that dsRNA. To trigger an innate immune response, PAMPs bind to **pattern recognition receptor (PRR) proteins** (**Figure 14.7**). PRRs are omnipresent, making cells immediately ready to mount an innate immune response. The cytoplasmic PRR that binds to this phosphorylated dsRNA is called RIG-I. RIG-I is also part of a family of RIG-I-like receptors that recognize different PAMPs. The Toll-like receptors are another family of structurally unrelated

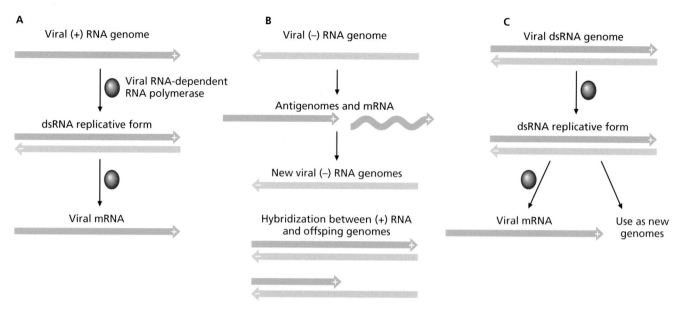

Figure 14.6 dsRNA during viral infection. (A) RNA viruses with (+) RNA genomes synthesize a dsRNA replicative form that is used as a template for mRNA synthesis. (B) RNA viruses with (–) RNA genomes synthesize a complete (+) strand during genome replication and also synthesize mRNA; the (–) and (+) strands can hybridize. (C) RNA viruses with dsRNA genomes use dsRNA replicative forms for mRNA synthesis and as new genomes.

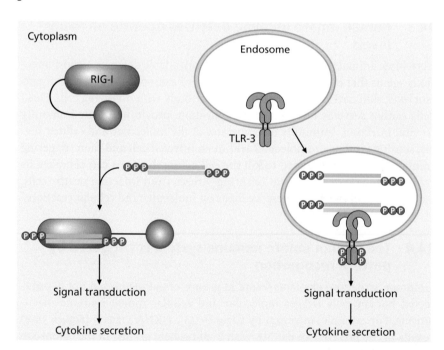

Figure 14.7 Pattern recognition receptor. Some pattern recognition receptors, such as RIG-I, are cytoplasmic. Others, such as the Toll-like receptors (for example TLR-3), are transmembrane proteins. Upon binding to a pattern associated with microbial infection, they trigger a signal transduction cascade that leads to an immune response, usually including cytokine secretion.

PRRs, some of which sense viral nucleic acids and glycoproteins. In general, when a PAMP binds to its PRR, the PRR undergoes a conformational change that triggers a signal transduction cascade, which in turn leads to an innate immune response. The immune response we will consider first is the secretion of a communication protein, known as a cytokine.

14.5   Viruses have counterdefenses against pattern recognition

Many viruses interfere with pattern recognition. An example is provided by SARS-CoV-2, which has an ssRNA (+) strand genome. Cells infected with a

Baltimore Class IV virus such as SARS-CoV-2 contain viral genomes, viral mRNAs, (–) RNA strands complementary to the genome, double-stranded replicative forms, and cytoplasmic RNA with unusual features such as 5′ phosphate groups or terminal poly(U) sequences (complementary to the poly(A) tails on viral mRNA). The SARS-CoV-2 nonstructural proteins create coronavirus replication compartments, also known as replication organelles, bounded by a double-membrane and connected to the rest of the cell through a proteinaceous pore. These replication compartments are the site of viral RNA synthesis and they sequester replication intermediates away from PRRs. Additionally, certain nonstructural SARS-CoV-2 proteins interfere with signal transduction pathways otherwise triggered by the presence of viral PAMPs. The SARS-CoV-2 virus also encodes a handful of nonstructural proteins that synthesize a 5′-cap like structure and subsequently attach it to the viral mRNAs, which protects that RNA from PRRs that would otherwise detect RNA with 5′ phosphate groups instead of a cap. The SARS-CoV-2 virus also encodes a nonstructural protein that degrades (–) sense strands with poly(U) sequences at their 5′ ends. All of these mechanisms allow SARS-CoV-2 to replicate early during an infection but ultimately the death of the infected cell releases their cellular contents including the contents of the replication compartments. The result is inflammation as the many viral PAMPs in the extracellular space trigger host PRRs. This inflammation contributes to the symptoms of COVID-19.

14.6   Innate immune responses include cytokine secretion

The innate immune response typically includes the release of **cytokines** (**Figure 14.8**), which are proteins that mediate communication about immunity and infection. One of the most important cytokines for controlling viral infections is **interferon** (**IFN**). These cytokines were named interferon because the first one discovered was able to interfere with viral infection. IFNα and IFNβ are cytokines that are produced by an infected cell in response to a virus and also trigger an antiviral response in neighboring cells that have not yet been infected. There are three main groups of IFNs: type I IFNs, such as **IFNα** and **IFNβ** (together known as IFNα/β), type II interferon (**IFNγ**) and type III interferon (IFNλ). We will focus on type I. Type I IFNs are produced by most nucleated cells as part of an innate immune response to viral infection. Type I IFN secretion is stimulated by the RIG-I PRR when it binds to dsRNA. IFNs are just one type of cytokine; others include the interleukins (ILs) such as IL-1.

14.7   Interferon causes the antiviral state

A cell in which RIG-I has been stimulated secretes type I IFNs, which diffuse to neighboring cells including those that have not yet been infected. Binding of IFNα/β to the IFN receptor on an uninfected cell causes a signal transduction cascade that results in the expression of hundreds of antiviral proteins, thus inducing an antiviral state in the affected cell (**Figure 14.9**). The antiviral state limits the viral infection from spreading through the body even if the virus manages to replicate in a few cells.

The genes induced by type I IFN signaling are called **interferon-stimulated genes** (**ISGs**). The **antiviral state** induced by the ISGs helps control viral infections in a variety of ways (**Figure 14.10**). There are proteins encoded by ISGs that control groups of viruses with a molecular property in

A   IFNα

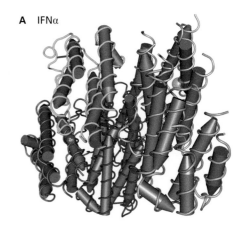

B   IL-1

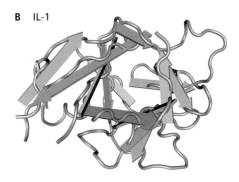

Figure 14.8 Cytokines. IFNα and IL-1 are examples of cytokines. (A) IFNα has many α helices, rendered as crayons. (B) IL-1 has many β-pleated sheets, rendered as arrows. (Courtesy of Nevit Dilmen. Published under CC BY-SA 3.0.)

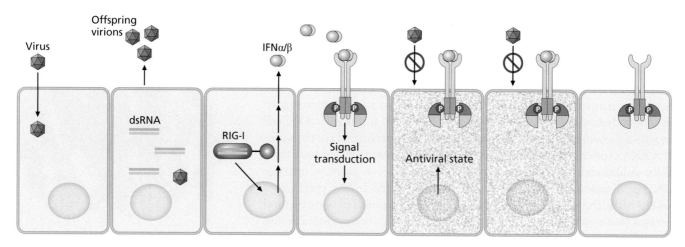

Figure 14.9 Interferon induces an antiviral state in neighboring uninfected cells. RIG-I detects dsRNA in a cell because of a viral infection. The cell responds by secreting IFNα/β. IFNα/β binds to its receptor on nearby cells, inducing in them an antiviral state. The antiviral state prevents offspring virions from replicating.

common. An example is **tetherin**, which interferes with budding by enveloped viruses that exit the cell through the ESCRT system (see **Chapter 11**). A second example is **Mx1**, a small protein with GTPase activity that blocks infection by viruses in which the nucleocapsid enters the cell by interfering with trafficking of nucleocapsids. A particularly well-known group of ISGs encodes three proteins (PKR, OAS, and RNase L) that block translation of mRNA synthesized by Baltimore Class III (ds), Class IV (+) and Class V (−) RNA viruses. Other ISGs encode proteins that protect the cells against most viruses irrespective of the type of genome or whether they are enveloped.

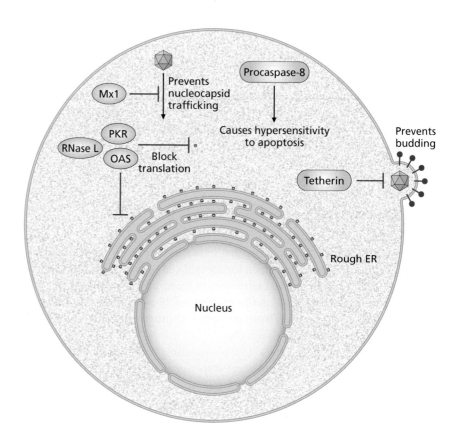

Figure 14.10 The antiviral state. The antiviral state includes interference with viral budding, intracellular nucleocapsid trafficking, and blocking translation if a cell becomes infected with a dsRNA, (+) RNA, or (−) RNA virus. Additionally, upregulation of pro-apoptotic proteins makes the cell more sensitive to control by the immune system. Proteins shown in green ovals are encoded by ISGs.

Examples include proteins that make the cell hypersensitive to pro-apoptotic signals, resulting in rapid death if the immune system detects that the cell has been infected by a virus. One such protein is the procaspase-8, which is a protease important for regulating apoptosis (see **Chapter 12**). Higher levels of procaspase-8 make cells more sensitive to apoptosis by increasing the chance that a pro-apoptotic signal will be propagated. There are many hundreds of ISGs, however, and their roles are still being worked out.

One of the best understood aspects of the antiviral state is the inhibition of translation caused by infection with a dsRNA, (+) RNA, or (−) RNA virus, which is caused primarily by three proteins encoded by three different ISGs. These are **double-stranded RNA-specific protein kinase R (PKR)**, **RNase L**, and **2′-5′-oligoadenylate synthetase (OAS)**. All these proteins are inactive unless the cell expressing them becomes infected by a virus. PKR's target is eukaryotic translation initiation factor 2α (eIF2α). When phosphorylated by PKR, eIF2α can no longer participate in cap-dependent translation initiation and so translation is blocked. PKR protein activity is itself regulated by the presence of the PAMP dsRNA; it is active only once it has bound to dsRNA (**Figure 14.11**). Similarly, RNase L is an enzyme that degrades all RNA and triggers both autophagy and apoptosis. RNase L must be activated by binding to the structurally unusual 2′-5′-oligoadenylate (**Figure 14.12**). This nucleic acid is not found in uninfected cells. Instead, it is synthesized from ATP by the OAS enzyme. The OAS enzyme must itself be stimulated by binding to the PAMP dsRNA (**Figure 14.13**), so that RNase L becomes active exclusively in cells that contain viral nucleic acids. RNase L blocks translation through degradation not only of mRNA but also of rRNA and tRNA. In this way, the antiviral state kills a cell exclusively if it becomes infected by a virus.

Although we have used (+) ssRNA viruses as the example, many viruses stimulate type I IFN secretion. These include viruses with (−) ssRNA genomes, such as influenza, and viruses with dsRNA genomes (such as rotavirus). Viruses such as herpesviruses and poxviruses with DNA genomes also stimulate IFN transcription, although through different PRRs that bind

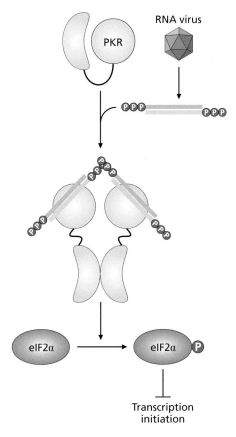

Figure 14.11 Regulation and activity of PKR. PKR is inactive until it binds to dsRNA, at which point there is dimerization and a conformational change. Normal, uninfected cells do not contain dsRNA, which is present when certain viruses replicate or express their genomes. The PKR then phosphorylates eIF2α, preventing cap-dependent translation.

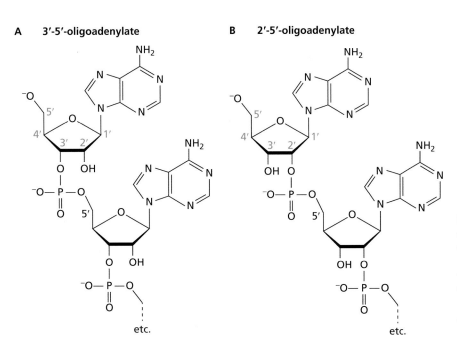

A 3′-5′-oligoadenylate

B 2′-5′-oligoadenylate

Figure 14.12 Structure of typical cellular 3′-5′-oligoadenylate RNA compared with 2′-5′-oligoadenylate RNA. (A) Cellular RNA, such as poly(A) tails, connects the nucleotides through a 5′ to 3′ linkage. (B) The 2′-5′-oligoadenylate RNA made in response to viral infection connects the nucleotides through a 5′ to 2′ linkage.

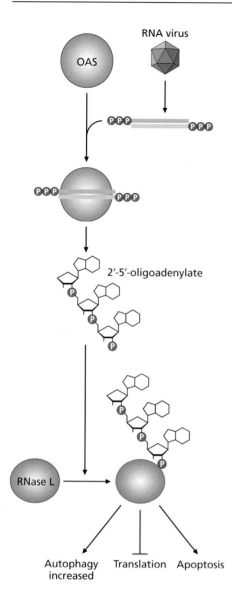

Figure 14.13 Regulation and activity of OAS and RNase L. The OAS enzyme is inactive until it binds to dsRNA. After binding to dsRNA it synthesizes 2′-5′-oligoadenylate. RNase L is inactive until it binds to 2′-5′-oligoadenylate. After binding to 2′-5′-oligoadenylate it degrades RNA (blocking translation) and promotes both autophagy and apoptosis.

to features associated with DNA viruses. Similarly, there are ISGs that specialize in limiting many different types of viral infections, not just infections by viruses with (+) ssRNA genomes.

14.8 Some viruses can evade the interferon response

Many viruses encode proteins that limit or counteract the IFN response. These viruses often block signaling through the IFN receptor, which would normally proceed as follows. IFNα/β binds to its receptor, which is associated with two signaling proteins called JAK1 and TYK2 (**Figure 14.14**). When stimulated by IFNα/β binding, this complex phosphorylates cytoplasmic signaling proteins STAT1 and STAT2. When phosphorylated, STAT1 and STAT2 dimerize. The phosphorylated heterodimer then associates with another cytoplasmic protein called IRF9, which is a transcription factor. Together, the complex comprised of phosphorylated STAT1, STAT2, and IRF9 enters the nucleus where the complex activates ISG expression.

The many players in the signal transduction cascade provide multiple cellular targets for viral evasion of the antiviral state (**Figure 14.15**). For example, a poxvirus makes an extracellular viral protein that binds to type I IFN before it can bind to its cellular receptor. SARS coronavirus and hepatitis C virus both induce ubiquitination and subsequent proteolysis of the IFN receptor, thus reducing a cell's ability to respond to IFN signaling. SARS-CoV-2 interferes with many aspects of IFN signaling, for example depleting

Figure 14.14 Type I IFN signal transduction cascade. When IFNα/β binds to its receptor, JAK1 and TYK2 are activated and form a complex with STAT1 and STAT2 (1). JAK and TYK2 then phosphorylate STAT1 and STAT1, which subsequently dimerize (2). When phosphorylated, the STATs dimerize (3). The phosphorylated heterodimer forms a complex with the IRF9 transcription factor (3). The complex enters the nucleus where it activates ISG transcription (4).

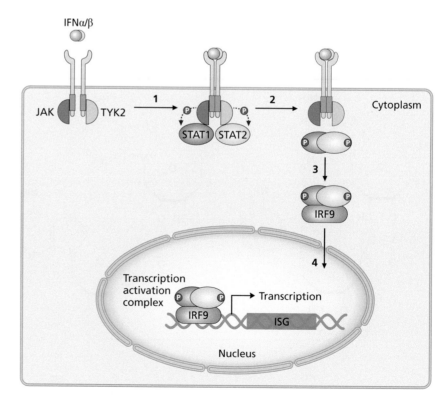

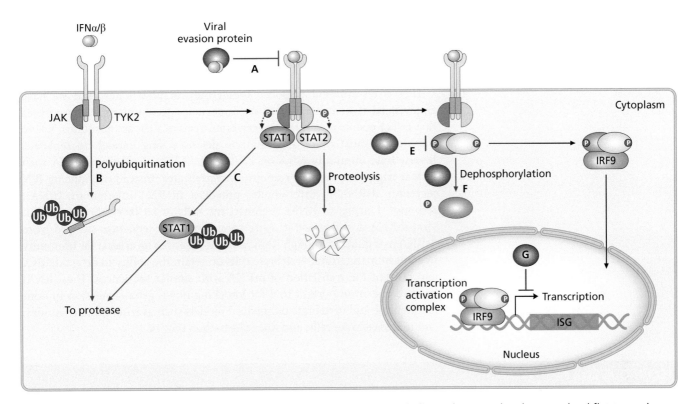

Figure 14.15 Viral evasion of type I IFN signaling. Black pointed arrows indicate the normal pathway and red flat-topped lines indicate viral evasion strategies. All viral proteins are represented as purple ovals, but they have been labeled with letters A to G to illustrate their different activities. (A) A poxvirus protein binds to type I IFN before the IFNα/β can bind to its cellular receptor. (B) Two viruses are known to induce ubiquitination and subsequent proteolysis of the IFN receptor, thus reducing a cell's ability to respond to IFN signaling. (C) Several viruses target STAT1 to the proteasome. (D) Some virally encoded proteases degrade STAT2. (E) Other viruses prevent JAK1 or TYK2 from phosphorylating the STATs. (F) Another strategy is for a viral phosphatase to dephosphorylate STAT1, blocking dimerization with STAT2. (G) Finally, there are viruses that block transcription of ISGs through interfering with IRF9 activity.

the cell of STAT proteins and preventing the phosphorylation of STAT1 and STAT2. The filovirus Marburg virus blocks the activity of JAK1, whereas human papillomavirus blocks signaling through TYK2. Several viruses target STAT1 to the proteasome, and vaccinia virus encodes a phosphatase that removes phosphate groups from STAT1 and thereby blocks its dimerization with STAT2. STAT2 is similarly a target for many viruses, typically through proteolytic degradation. Finally, human papillomavirus and reoviruses both block transcription of ISGs through interfering with IRF9 activity.

Viruses can also counteract the activities of specific proteins expressed in response to IFN signaling. For example, the PKR protein is a typical target for evading the antiviral state, and different viruses have different mechanisms for blocking activation of PKR or the consequences of activating PKR. Vaccinia virus and reoviruses encode proteins that bind to dsRNA and thereby reduce PKR activation by sequestering dsRNA away from it. Viruses such as hepatitis C and herpes simplex virus 1 (known as HSV-1 or human herpesvirus [HHV]-1) encode proteins that bind to PKR directly and prevent it from blocking translation. Herpes cytomegalovirus (known as CMV or HHV-5) interferes with the normal cytoplasmic localization of PKR by causing it to accumulate in the nucleus where it cannot block translation, and at least one virus induces proteasome-mediated degradation of PKR. The internal ribosome entry site (IRES), essential for translation of many

(+) RNA genomes, is also a way to evade PKR-mediated cessation of translation because normal host eIF2α is not needed to initiate translation using a viral IRES (see **Chapter 5**). Many other proteins induced by the antiviral state, including OAS and RNase L, are targets for yet other viral evasion strategies.

IFN is an antiviral response found only in vertebrates, and we have emphasized it here because of its importance for humans. Many viruses that infect people also infect invertebrate vectors such as mosquitos, where the viruses must evade the immune defenses long enough to replicate. Invertebrate immunity relies on PRRs and PAMPs, including dsRNA such as that triggering the IFN response in vertebrates. Instead of inducing IFN secretion, dsRNA in invertebrates provokes an RNA interference (RNAi) response. During an RNAi response, the cells of an invertebrate seek out viral mRNA and degrade it using complementary sequences derived from dsRNA as a guide. Although RNAi does not seem to be critical for immunity in adult human tissues, vertebrate cells do retain the ability to detect dsRNA and prevent the translation of mRNA with similar sequences. Thus, RNAi has become an important tool for knocking down gene expression in both invertebrate and vertebrate biomedical models such as fruit flies, nematodes, cultured vertebrate cells, and mice (**Technique Box 14.1**).

TECHNIQUE BOX 14.1   RNAi

RNA interference (RNAi) is the phenomenon in which cytoplasmic double-stranded RNA induces the degradation of mRNA complementary to the dsRNA or blocks translation of complementary mRNA. This **gene silencing** prevents the production of proteins despite mRNA synthesis. Induction of RNAi by exogenous viral nucleic acids is a critical component of immunity in plants and invertebrates, which do not have the IFN response. The role of RNAi in mammalian immunity is contested; it may be active in mammals during development (before birth) and in certain stem cells but is otherwise replaced by the IFN response in adult mammals. In the case of exogenous RNAi in invertebrates and plants, foreign dsRNA is processed into small interfering RNA (siRNA), which binds to a cytoplasmic complex called RNA-induced silencing complex (RISC) (**Figure 14.16**). RISC then uses the guide siRNA to bind to complementary mRNA and degrade it. Mammals, including humans, use endogenous miRNAs transcribed from their own genomes to regulate translation of cellular mRNAs (**Figure 14.17**). These RNAs are transcribed and processed extensively in the nucleus and then, after they are transported to the cytoplasm, they bind to RISC and either block translation of complementary mRNA or induce its degradation.

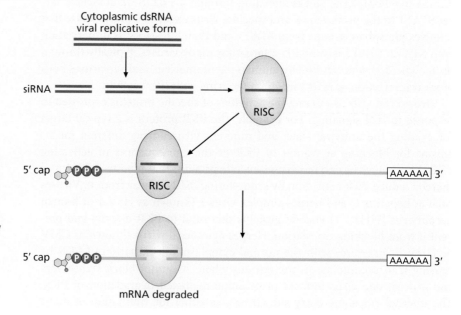

Figure 14.16 RNAi in invertebrates. Exogenous dsRNA caused by a viral infection is degraded into siRNA, which associates with RISC. RISC uses the siRNA as a guide to find complementary mRNA and degrade it, thus degrading the viral mRNA before it can be completely translated.

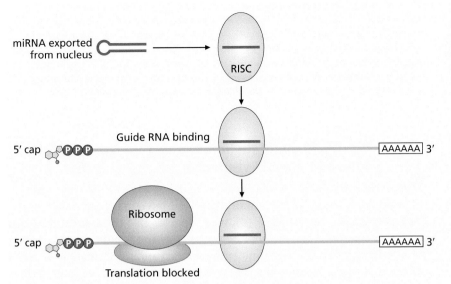

miRNA exported from nucleus

RISC

Guide RNA binding

5′ cap

PPP

AAAAAA 3′

Ribosome

5′ cap

PPP

AAAAAA 3′

Translation blocked

Figure 14.17 RNAi in mammals. Endogenously produced regulatory miRNAs associate with RISC once they are exported to the cytoplasm, where RISC uses the miRNA as a guide to find complementary mRNA and block its translation.

RNAi has become an important research tool because it can be exploited to knock down translation of specific mRNAs during an experimental test. Both endogenously produced and exogenously produced miRNAs and siRNAs end up in the cytoplasm bound to RISC, where they either degrade complementary mRNA or block its translation. We can use miRNAs or siRNAs to knock down protein expression in a cell or even in a whole animal or plant without needing to alter the chromosomal DNA of the organism. Plasmids that express artificial miRNAs or structurally similar short hairpin RNAs (shRNAs) can be introduced into a cell and the effect is to silence whatever gene we have selected. The effect of blocking translation of an mRNA using artificial miRNA or shRNA is called a knockdown. Use of RNAi is so convenient that it is possible to screen collections of thousands of artificially cloned miRNAs or shRNAs designed to target human mRNAs in order to find rare knockdowns that block the replication of a virus such as influenza A. Use of RNAi has been particularly helpful in cases in which engineering genetic knockouts is time consuming, expensive, or not possible because the knockout is lethal during development.

14.9 Neutrophils are active during an innate immune response against viruses

Neutrophils are short-lived phagocytic cells that are typically the first to respond to the site of an infection (**Figure 14.18**). They patrol the bloodstream for signs of PAMPs. Neutrophils have long been known to help control extracellular bacterial infections by phagocytosing and killing the bacteria. There are increasing reports of neutrophils playing a role in containing viral infections such as HIV, influenza, and Ebola virus primarily through the elaboration of neutrophil extracellular traps and the induction of inflammation.

In the presence of a viral infection, neutrophils produce **neutrophil extracellular traps** (**NETs**), which are a meshwork of extracellular fibers composed of chromatin and antimicrobial proteins (**Figure 14.19**). NETs are useful in controlling a viral infection in three ways. First, the extracellular NETs can bind to and thereby immobilize virions, physically preventing them from spreading through the body and entering new host cells. The histone components of the chromatin in the NETs are particularly important for this trapping because the histones are positively charged and most virions have negatively charged surfaces. Second, NETs include antimicrobial proteins, such as defensins, which inactivate viruses. Third, NETs stimulate other immune cells, including other neutrophils and cells of the adaptive immune system. These other immune cells are then triggered to increase their antiviral activities.

Multilobed nucleus

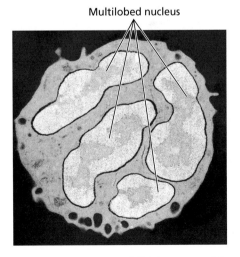

Figure 14.18 Neutrophil. The neutrophil in this false-colored transmission electron micrograph is 8–9 μm in diameter. Four lobes of the nucleus can be seen in this slice through the cell; the white and yellow nucleus is bounded by the purple nuclear membrane, which is in turn surrounded by teal cytoplasm. (Courtesy of University of Edinburgh.)

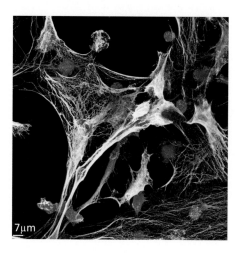

Figure 14.19 Neutrophil extracellular traps. The blue stains show the nuclei of neutrophils, some of which have burst open to create neutrophil extracellular traps. The intense white and yellow staining show regions where the neutrophil enzyme elastase and chromatin, no longer compacted in a nucleus, overlap. These are the fibrous neutrophil extracellular traps. Some of the nuclei in the image are still intact because neutrophil extracellular trap formation is asynchronous. (Courtesy of Volker Brinkmann, Max Planck Institute for Infection Biology.)

14.10 Viruses manipulate immune system communication to evade the NET response

Discoveries pertaining to viral evasion of NETs are cutting-edge research because the roles of NETs in controlling viral infections have only recently been recognized. Nevertheless, it already appears that there are at least two viral strategies to evade the NET response. The first is indirect and consists of interfering with the signaling pathways that induce formation of NETs in the first place. Neutrophils can be stimulated to make NETs in various ways, but they can also receive signals that prevent elaboration of NETs. A reduction in NETs is critical for stopping an immune response after it has served its purpose. One of the cytokines that suppresses NET formation is IL-10. Several herpesviruses encode secreted viral IL-10 homologs that may suppress NET formation. A second example of this strategy is provided by HIV, which stimulates immune cells to secrete authentic IL-10, effectively suppressing NET formation. A more speculative way that viruses may evade the NET response is through the use of DNases. It is well-known that bacterial DNase enzymes help bacterial pathogens evade NETs. Some large DNA viruses such as herpesviruses encode DNases that are involved in genome replication. It is possible that these enzymes, if released because of lysis of an infected cell, could degrade the DNA components of NETs, thus preventing the NETs from physically occluding virus spread through the body.

14.11 Inflammation is the hallmark of an innate immune response

Viral infections trigger **inflammation**, which is a physiological response to infection or tissue damage. Inflammation is an influx of fluids, signaling proteins such as cytokines, and immune cells, especially neutrophils. Inflammation is useful in that it helps confine an infectious agent to a small part of the body and eliminate that infectious agent from the body. Acute inflammation lasts only a few days and is capable of clearing most viral infections from the body. The four common symptoms of inflammation are redness, swelling, pain, and heat, which are exactly the common signs of a viral infection (redness, swelling, and pain in the part of the body infected by the virus). Neutrophils are particularly important for initiating and amplifying inflammation. Inflammation is triggered by PRRs that respond to PAMPs and by related receptors that respond instead to **damage-associated molecular patterns** (**DAMPs**). Examples of molecules perceived as DAMPs include extracellular DNA, extracellular RNA, extracellular histone proteins, and extracellular ATP. DAMPs are derived from the interior contents of cells that have died through **necrosis** rather than apoptosis (**Figure 14.20**). Cytolytic viruses typically kill cells through necrosis. Unlike apoptosis, necrosis is pro-inflammatory because of the release of DAMPs into the tissue. DAMPs are a sign of danger because they indicate lytic cell death, which does not occur during normal development. The influx of neutrophils, which produce NETs, also enhances inflammation because components of the NETs such as extracellular DNA and histones are DAMPs that further increase the inflammatory response. In contrast, during apoptosis cells break into smaller parts that remain membrane bound and the fragments are cleared from the body by **professional phagocytes** without inducing inflammation.

When an innate immune reaction is insufficient to clear a virus from the body, the adaptive immune response comes into play. Cells of the adaptive immune system are attracted to the site of inflammation and are also stimulated by cytokines and DAMPs that are present in inflamed tissues. Although acute inflammation is essential for controlling a viral infection, it can also damage healthy tissue. For example, antimicrobial chemicals and proteins used by immune cells can be found in extracellular fluids where some of them damage host and microbes alike. Sometimes this collateral damage can itself be fatal, as is the case during serious influenza infections. In order to limit tissue damage once the danger of an infection has passed, there are anti-inflammatory cytokines that return the tissue to normal once a virus has been cleared from the body. Sometimes the process of turning off inflammation malfunctions; uncontrolled chronic (long-term) inflammation such as that caused by persistent viral infections can also damage healthy tissue and can have pathological effects such as oncogenesis (see **Chapter 13**).

14.12   In order to be recognized as healthy, all cells present endogenous antigens in MHC-I molecules

All nucleated human cells have a cell surface molecule called the **major histocompatibility complex I** (**MHC-I**) (**Figure 14.21**). MHC-I is important for both innate and adaptive immune responses. The purpose of MHC-I is to display endogenous epitopes on the cell surface so that the immune system can test these epitopes in order to detect whether the cell is healthy, infected by an intracellular pathogen such as a virus, or cancerous. An epitope is the small part of an antigen recognized by the adaptive immune system; for example, antibodies bind to epitopes (see **Chapter 3**). Epitopes displayed by MHC-I are called endogenous because they were synthesized by the cell displaying them. In a healthy cell, MHC-I–epitope complexes are abundant on the cell surface; this normal level of abundance can be detected by the innate immune defenses. Furthermore, the displayed epitopes are simply the normal products

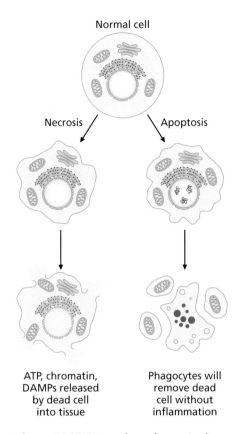

Figure 14.20 Necrosis and apoptosis are different forms of cell death. Necrosis promotes inflammation by releasing DAMPs into the surrounding tissue. Apoptosis does not promote inflammation because the dying cell breaks into small membrane-bound particles without releasing any DAMPs.

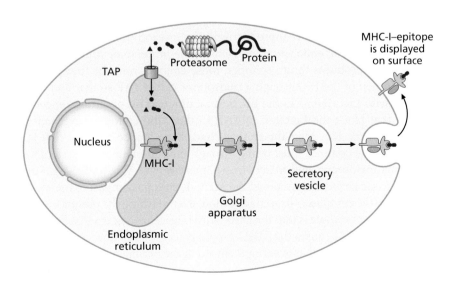

Figure 14.21 MHC-I loading. Endogenous proteins are degraded by the proteasome. Some of the degradation products enter the endoplasmic reticulum through the TAP protein. Some of the degradation products are loaded into the cleft of a newly synthesized MHC-I molecule in the endoplasmic reticulum. The MHC-I–epitope complex then traffics to the plasma membrane through the endomembrane system, ultimately displaying self-antigens on the surface of the cell.

of cellular recycling of old proteins by the ubiquitin–proteasome system. After degradation, some of the degradation products become epitopes that are loaded into the cleft in an MHC-I molecule in the endoplasmic reticulum (ER). After that, the MHC-I–epitope complex traffics to the plasma membrane through the endomembrane system, ultimately displaying self-antigens on the surface of the cell. Dedicated adaptive immune cells then detect these antigens and, if they are recognized as normal self-antigens, the immune cells leave the healthy cell alone to continue its normal functions.

14.13 Cells infected by viruses produce and display viral antigens in MHC-I

When a cell is infected by a virus, most of its protein-synthesizing machinery becomes dedicated to synthesizing viral proteins. This is part of the viral strategy to overwhelm the cell and replicate as quickly as possible before the individual cell or, in the case of an animal, the immune system, can stop viral replication. As a consequence of so much viral protein synthesis, some of the viral proteins will be processed just as normal endogenous antigens would otherwise be, and they will be loaded into MHC-I molecules. Thus, the infected cell displays viral antigens on its surface. These viral antigens can be detected as foreign, provoking an immune response against the infected cell. A typical immune response in reaction to foreign antigens displayed in MHC-I molecules is to kill the infected cell (see **Chapter 15**). Although the display of specific epitopes is critical for the adaptive immune system, the display of epitopes at all is important for innate immunity as explained in **Sections 14.14** and **14.15**.

14.14 Viruses have strategies to evade MHC-I presentation of viral antigens

Because presentation of viral antigens by MHC-I molecules is important for triggering antiviral immunity, many viruses have mechanisms for evading it. Multiple molecular events must occur for antigen display to work properly and there are viruses that evade each step. There are viral proteins that are difficult to process into epitopes so that they are never loaded onto the MHC-I for display. An example is provided by the EBNA1 protein from HHV-4 (also known as Epstein–Barr virus), which is expressed abundantly in cells harboring a latent HHV-4 infection. The protein is resistant to degradation by the proteasome, reducing viral epitope display in MHC-I (**Figure 14.22**). By occupying the proteasome unproductively, the evasion mechanism reduces MHC-I display of all endogenous antigens. If the proteasome functions normally, after it produces future epitopes, those epitopes must be transported into the lumen of the ER through a transporter called TAP so that they can associate with the MHC-I in the ER. Several different herpesviruses encode proteins that block the function of TAP so that viral epitopes never enter the lumen of the ER. After the epitopes have been loaded into the cleft in the MHC-I, the complex must traffic through the endomembrane system to the plasma membrane, and many viruses interfere with this trafficking. For example, some herpesviruses target the MHC-I–epitope complex to the lysosome where the complex is degraded without ever reaching the plasma membrane. Another example is that the herpesvirus varicella–zoster virus (known as VZV or HHV-3) causes the MHC-I–epitope complexes to accumulate in the Golgi apparatus so that fewer of them reach the plasma membrane.

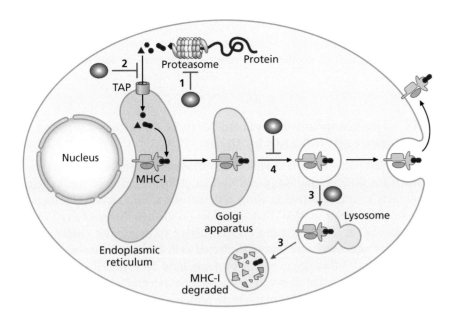

Figure 14.22 Viral evasion of MHC-I display of endogenous viral epitopes. Black arrows indicate the normal pathway, and red pointed arrows and red flat-head arrows indicate viral evasion strategies. One evasion strategy is for an abundant viral protein to be resistant to degradation by the proteasome (1). Another viral evasion strategy is to prevent transport of peptides from the proteasome into the ER through TAP (2). A third evasion strategy is to target vesicles containing MHC-I–epitope complexes to the lysosome so that the complexes get degraded and never reach the plasma membrane (3). A fourth evasion strategy is to prevent the MHC-I–epitope complexes from leaving the Golgi apparatus so that they never reach the plasma membrane (4).

14.15   Natural killer cells attack cells with reduced MHC-I display

Many viruses cause their infected cells to stop displaying peptides in MHC-I, thus leading to reduced display of viral epitopes. Malignant cells, including those that are malignant because of a viral infection, also typically have reduced amounts of MHC-I on their surfaces. Therefore, the innate immune system includes **natural killer (NK) cells** that are sensitive not to the specific epitopes but rather to the abundance of MHC-I displayed on another cell's surface. NK cells have a cell surface receptor that binds to MHC-I on a target cell. Binding to this NK receptor prevents the NK cell from killing the target. When a target cell has too few MHC-I molecules, this NK receptor is not engaged (**Figure 14.23**). When that happens, the NK cells release a collection of proteins that kill the abnormal cell. One of the proteins, perforin, forms an aqueous channel through which NK granzyme proteins enter the cell, causing apoptosis (**Figure 14.24**). Viruses evade the NK cells by manipulating signal transduction by NK cells themselves and by interfering with apoptosis in infected cells. An example of a virus manipulating NK signal transduction is an extracellular viral protein that is a molecular mimic of MHC-I, so that

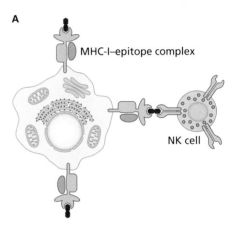

A

MHC-I–epitope complex

NK cell

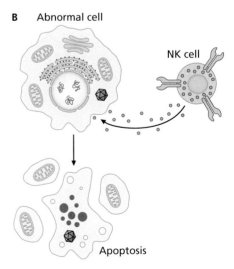

B   Abnormal cell

NK cell

Apoptosis

Figure 14.23 NK cell surveillance. (A) A normal cell has high levels of MHC-I–epitope complexes on its surface, so that the NK cell does not attack the normal cell. (B) An NK cell can detect an abnormal cell with lower levels of MHC-I–epitope displayed and will subsequently kill the abnormal cell.

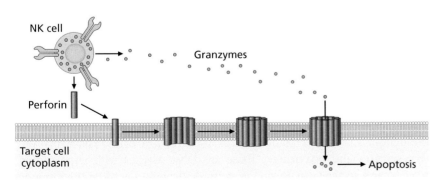

NK cell

Granzymes

Perforin

Target cell cytoplasm

Apoptosis

Figure 14.24 NK-mediated apoptosis. Perforin secreted by an NK cell oligomerizes and forms a pore. After that, granzymes secreted by the NK cell enter the cell through the perforin channels. The granzymes trigger apoptosis.

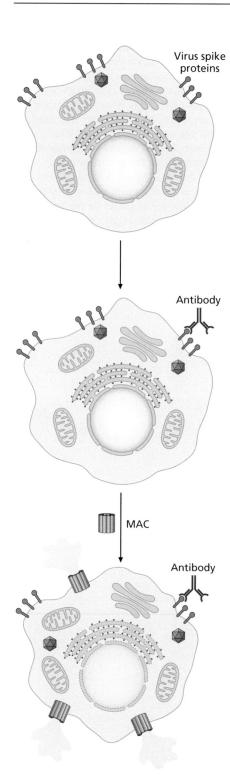

Figure 14.25 Complement stimulated by antibodies. A cell infected by an enveloped virus has virus spikes on the surface of its plasma membrane. Antibodies that bind to these spike proteins stimulate the complement system so that the MAC (membrane attack complex) forms on the infected cell. The infected cell dies before viral replication is complete.

the viral protein binds to the NK receptor and prevents the NK cell from detecting that the infected cell has abnormally low levels of MHC-I on its surface. Many viruses produce anti-apoptotic proteins, as was discussed in **Chapter 12**.

14.16 The complement system targets enveloped viruses and cells infected by them

The complement system is a collection of soluble proteins found in the bloodstream and in the lymphatic system. When activated, the system counteracts a viral infection in three different ways. First, it forms a membrane attack complex or **MAC** that assembles on the membrane of a cell with viral antigens on its surface and on the surface of enveloped virions. The MAC forms aqueous holes that kill infected cells and inactivate enveloped virions. Second, other complement proteins bind to extracellular virions, **opsonizing** them so that they will be engulfed and degraded by professional phagocytes. Third, still other complement proteins serve as chemoattractants that draw immune cells to the site of the infection, increasing inflammation in the affected tissue.

The complement is part of the innate immune response because it behaves exactly the same way every time it is stimulated, but it can be stimulated by antibodies that bind to the surface of virions or to cells with virion proteins (such as spikes) on their cell surface (**Figure 14.25**). Antibodies are produced during an adaptive immune response (see **Chapter 15**). Thus, the complement system is similar to the MHC-I–epitope display system in that they are both important for both innate and adaptive immune responses.

14.17 Some viruses can evade the complement system

Activation of the complement system requires a series of six or more molecular events that must occur in a specified order, much like a signal transduction cascade, even though the events are extracellular. Each event involves protein–protein interactions and often also involves proteolysis. Therefore, viruses have many different opportunities to interfere with the complement response by preventing any one of these molecular events from occurring properly. In general, interfering with early steps in the activation process prevents formation of the MAC. An example is provided by influenza; the M1 viral matrix protein prevents the first complement activation step from occurring. Dengue virus NS1 protein interferes with the second step in activation of the complement, whereas many viruses including herpes simplex virus I (also known as HHV-1) and smallpox virus make proteins that interfere with the third step. In all of these evasion strategies, MAC formation is prevented.

14.18 Viral evasion strategies depend on the coding capacity of the virus

Throughout this chapter, many of the examples of viral evasion have involved herpesviruses and poxviruses. These are DNA viruses with large coding capacity, and so they can encode proteins dedicated to evading the immune response. In contrast, viruses with smaller coding capacity, such as influenza, evade immune responses using multifunctional proteins that have additional more direct roles during replication. For example, the influenza matrix protein M2 is essential for both uncoating and assembly, but M2 also counteracts the antiviral state by binding to the host PKR protein. Coronaviruses such as SARS-CoV-2 have larger genomes than most groups of (+) RNA viruses and they encode many proteins with immune-evasion activities. Immune evasion permits high levels of virus replication, which subsequently can cause severe and even life-threatening symptoms.

14.19 In vertebrates, if an innate immune reaction does not clear an infection, adaptive immunity comes into play

Most viral infections are controlled by the innate immune system, which can be demonstrated by unfortunate circumstances in which someone is born lacking individual components of the innate defenses. People who lack normal levels of NK cells, for example, are hypersusceptible to viral infections, indicating that NK cells are important for clearing viral infections. Experimental animals that have been genetically engineered to lack individual components of the innate immune system are also hypersusceptible to viral infections. Despite the power of the innate defenses, there are some viruses that can be controlled only by the combined efforts of the innate and adaptive immune responses. **Chapter 15** therefore explains how an adaptive immune response is initiated with cooperation from the innate defenses.

Essential concepts

- Bacterial innate defenses include restriction–modification systems that selectively degrade phage genomic DNA without degrading host DNA. Phages can evade restriction modification by encoding proteins to methylate their DNA, which prevents restriction enzymes from binding to the DNA.

- In humans, the innate defenses are omnipresent, react to groups of microbes, and react the same way every time they are triggered. In contrast, the adaptive (or acquired) immune defenses must be induced by exposure to a specific microbe, react to the very specific microbe that induced the response in the first place, and have memory so that they react much faster and stronger upon a second or subsequent exposure to an infectious agent. The innate and adaptive immune responses are in constant communication and both are required for successful defense of a viral infection.

- Innate immune reactions at the cellular level are triggered by pathogen-associated molecular patterns, such as double-stranded RNA, which bind to pattern recognition receptors (PRRs) in cells infected by a virus.

- A common outcome of PRR signaling is the synthesis and secretion of cytokines such as type I interferon (IFNα/β). Interferon does not help the infected cell but it binds to neighboring cells and induces in them an antiviral state that makes it harder for a viral infection to spread through the body.

- One of the best understood aspects of the antiviral state is the block to cap-dependent translation that occurs through concerted action of the PKR, 2′-5′-oligoadenylate synthetase, and RNase L proteins.

- Viruses can evade the interferon response through interfering with signal transduction via the interferon receptor or through blocking the activity of specific interferon-induced proteins that contribute to the antiviral state.

- During a viral infection, acute inflammation is the result of tissue damage and of infected cells and neutrophils releasing cytokines that attract more immune cells to the site of an infection.
- Neutrophils respond to a viral infection by creating inflammation and elaborating NETs, both of which contribute to controlling viral infections. Viruses can evade the NET response by altering host cytokine-mediated communication.
- Infected cells display endogenous viral epitopes in MHC-I molecules, where foreign epitopes can be recognized by an adaptive immune response that kills infected cells.
- Some viruses evade MHC-I presentation of endogenous viral epitopes by reducing the number of MHC-I–epitope complexes that reach the cell surface. Natural killer cells detect cells with abnormally low levels of MHC-I–epitope complexes and kill them. Some viruses evade this response by producing MHC-I mimics that prevent the NK cells from recognizing reduced MHC-I–epitope display on infected cells.
- The complement system is a collection of soluble proteins found in the bloodstream and in the lymphatic system. When activated, the complement fights a viral infection in various ways such as lysing infected cells and attracting immune cells to the site of an infection. Some viruses evade the complement system by interfering with its activation.
- If the innate immune response does not clear a viral infection, the adaptive immune response comes into play.

Questions

1. Compare and contrast innate immunity in bacteria with innate immunity in humans.
2. List several important differences between PAMPs and epitopes.
3. What is the role of pattern recognition during the antiviral state inside cells stimulated by type I interferons?
4. How do professional antigen-presenting cells form a bridge between innate and adaptive immunity?
5. How does inflammation contribute to clearing a virus from the body?
6. Provide two examples of viral evasion of host immune responses that rely upon signal transduction.
7. Consider a newly emerging virus that was not previously recognized as causing disease in humans. An example could be MERS-coronavirus or a new enterovirus that causes acute flaccid paralysis. Would you expect that virus to have viral proteins that evade at least one innate immune response? Why or why not?

 Interactive quiz questions

In addition to the questions provided above, this edition has a range of free interactive quiz questions for students to further test their understanding of the chapter material. To access these online questions, please visit the book's website: www.routledge.com/cw/lostroh.

Further reading

Restriction modification

Isaev A, Drobiazko A, Sierro N, Gordeeva J, Yosef I et al. 2020. Phage T7 DNA mimic protein Ocr is a potent inhibitor of BREX defence. *Nucleic Acids Res* 48(10):5397–5406.

Viral PAMPs and pattern recognition

Fu YZ, Guo Y, Zou HM, Su S, Wang SY et al. 2019. Human cytomegalovirus protein UL42 antagonizes cGAS/MITA-mediated innate antiviral response. *PLOS Pathog* 15(5):e1007691.

Kouwaki T, Nishimura T, Wang G & Oshiumi H 2021. RIG-I-like receptor-mediated recognition of viral genomic RNA of severe acute respiratory syndrome coronavirus-2 and viral escape from the host innate immune responses. *Front Immunol* 12:700926.

Liu Y, Qin C, Rao Y, Ngo C, Feng JJ et al. 2021. SARS-CoV-2 Nsp5 demonstrates two distinct mechanisms targeting RIG-I and MAVS to evade the innate immune response. *mBio* 12(5):e02335-21.

Interferons

Lazear HM, Govero J, Smith AM, Platt DJ, Fernandez E et al. 2016. A mouse model of Zika virus pathogenesis. *Cell Host Microbe* 19(5):720–730.

Park ES, Byun YH, Park S, Jang YH, Han WR et al. 2019. Co-degradation of interferon signaling factor DDX3 by PB1-F2 as a basis for high virulence of 1918 pandemic influenza. *EMBO J* 38(10):e99475.

Sueng BP, Seronello S & Mayer W 2016. Hepatitis C virus frameshift/alternate reading frame protein suppresses interferon responses mediated by pattern recognition receptor retinoic-acid-inducible gene-I. *PLOS One* 11(7):e0158419.

RNAi

Kamath RS, Fraser AG & Dong Y 2003. Systematic functional analysis of the *Caenorhabditis elegans* genome using RNAi. *Nature* 421:231–237.

Karlas A, Machuy N & Shin Y 2010. Genome-wide RNAi screen identifies human host factors crucial for influenza virus replication. *Nature* 463:818–822.

Wooddell CI, Yuen MF, Chan HLY, Gish RG et al. 2017. RNAi-based treatment of chronically infected patients and chimpanzees reveals that integrated hepatitis B virus DNA is a source of HBsAg. *Sci Transl Med* 9(409):eaan0241.

Neutrophils

Funchal GA, Jaeger N & Czepielewski RS 2015. Respiratory syncytial virus fusion protein promotes TLR-4-dependent neutrophil extracellular trap formation by human neutrophils. *PLOS One* 10(4):e0124082.

Opasawatchai A, Amornsupawat P, Jiravejchakul N, Chan-In W, Spoerk NJ et al. 2019. Neutrophil activation and early features of NET formation are associated with dengue virus infection in human. *Front Immunol* 9:3007.

MHC-I endogenous antigen presentation

Luteijn RD, Hoelen H, Kruse E, Van Leeuwen WF, Grootens J et al. 2014. Cowpox virus protein CPXV012 eludes CTLs by blocking ATP binding to TAP. *J Immunol* 193(4):1578–1589.

Matschulla T, Berry R, Gerke C, Döring M, Busch J et al. 2017. A highly conserved sequence of the viral TAP inhibitor ICP47 is required for freezing of the peptide transport cycle. *Sci Rep* 7(1):1–13.

Complement

Conde JN, da Silva EM & Allonso D 2016. Inhibition of the membrane attack complex by dengue virus NS1 through interaction with vitronectin and terminal complement proteins. *J Virol* 90(21):9570–9581.

Fan B, Peng Q, Song S, Shi D, Zhang X et al. 2022. Nonstructural protein 1 of variant Pedv plays a key role in escaping replication restriction by complement C3. *J Virol* 96(18):e01024-22.

Viral Evasion of Adaptive Host Defenses

15

Immune response	Effect on viruses
CRISPR-Cas	Sequence-specific degradation of bacteriophage DNA.
Antigen presentation in MHC-II	Used by antigen-presenting cells to display exogenous epitopes; interacts with lymphocytes to initiate an adaptive response that can control a viral infection.
Professional antigen-presenting cell	Surveillance phagocyte that activates lymphocytes using MHC-II, leading especially to antibodies that can control a viral infection.
T helper lymphocyte	Activates lymphocytes and regulates immune responses that are needed to control viral infection.
B cell	Produces antibodies, which typically bind to structural virus proteins.
Antibodies	Neutralization and opsonization of virions.
Dendritic cell	Professional antigen-presenting cell that activates cytotoxic lymphocytes, which in turn kill cells infected with a virus.
Cytotoxic lymphocyte	Kills cells infected with a virus.

The immune responses you will meet in this chapter and their effect on viruses

The adaptive immune defenses are essential for clearing pathogenic viruses from the human body. The critical role of the adaptive responses is illustrated by acquired immunodeficiency syndrome (AIDS), caused by a virus that kills its host cells. These host cells are part of the adaptive immune system. When enough of them die because of human immunodeficiency virus (HIV) replication, an HIV+ person can no longer mount a typical adaptive immune response. Instead, an infected person becomes susceptible to many infections, ultimately leading to the symptoms of AIDS.

As introduced in **Chapter 14**, adaptive host defenses are those that develop only after a pathogen (or a vaccine that mimics it) has been encountered. Reactions against antigens, which are unique to an infectious agent, drive adaptive host defenses. In addition, adaptive immune defenses have memory, meaning that the immune reaction against the pathogen is both faster and stronger the second and subsequent times that a person is exposed to the same pathogen. Viruses also have mechanisms to evade adaptive responses,

DOI: 10.1201/9781003463115-15

however, underscoring the constant evolutionary interplay between viruses and their hosts (see **Chapter 17**).

This chapter begins with the unexpected discovery of adaptive defenses among the bacteria. Most of the chapter, however, emphasizes human adaptive defenses that combat viral infection. These include T helper cells, B cells and the antibodies made by them, and cytotoxic T cells.

15.1   CRISPR-Cas is an adaptive immune response found in bacteria

CRISPR-Cas was discovered 15 years before its function in immunity was appreciated. It was first noticed as a series of unusual repeats in the genomic DNA of *E. coli*. It is now known that bacteria and archaea use this system for adaptive immunity. Like adaptive immunity in people, the system specifically attacks particular infectious agents (such as a specific phage but not all related phages) and has memory so that future attacks by the same infectious agent are rebuffed. Its specificity is driven not by antigens but rather by nucleic acid hybridization.

CRISPR is an acronym for clustered regularly interspaced short palindromic repeats, and **Cas** is an acronym for CRISPR-associated. A typical CRISPR-Cas signature in the genome of a bacterium has a set of *cas* genes that encode helicases and nucleases followed by a series of palindromic repeats (**Figure 15.1**). Palindromic repeats have the same sequence from 5′ to 3′ on one strand as they do on the other; an example of a palindromic sequence in dsDNA is 5′-AATCGCGATT. In a real CRISPR-Cas locus, the palindromes are about 20–40 bp long and are not perfect. Between the palindromic repeats are unique **spacer** sequences of about the same length (20–60 bp). Before their use in adaptive immunity was appreciated, some of the spacer sequences in CRISPR-Cas systems were recognized to be the same as sequences found in mobile genetic elements including bacteriophage genomes. The association with *cas* genes that encode proteins with helicase and nuclease domains led to the proposal that CRISPR-Cas might act as an immune system in which the bacterium would be protected from infectious agents that have sequences that match the spacers. Soon after, microbiologists showed that bacteria are indeed protected from phage infections if the phage genome matches a spacer. They also demonstrated that bacteria can acquire new spacers in the rare event that a host cell processes the phage genome to make a new spacer faster than the phage could replicate and kill its host cell.

The mechanism by which CRISPR-Cas systems, specifically type I CRISPR-Cas systems, work to protect bacteria from infectious agents occurs as follows. The **Cas1** and **Cas2** proteins mount a defense against any incoming DNA by using that DNA to create a new spacer (**Figure 15.2**). Once a

Figure 15.1 A simplified diagram of a CRISPR-Cas locus. Genes encoding Cas proteins are followed by a promoter that drives expression of spacers separated by palindromic repeats. The palindromic repeats and spacers do not encode proteins.

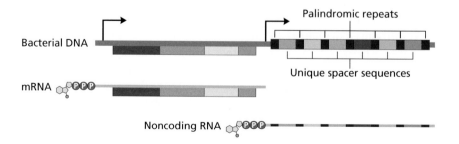

new spacer has been created, the DNA encoding the spacer can be used to synthesize RNA that will combat a viral infection. There is a leader sequence between the *cas* genes and the first palindromic repeat that serves as a promoter to drive synthesis of RNA using the palindromic sequences and spacers as the template. Cas proteins process the RNA to make short sequences called **crRNAs (Figure 15.3)**. The palindromic crRNAs form stem–loop structures that are further processed by other Cas proteins to make single-stranded RNA that is complementary to sequences in bacteriophage DNA represented by the spacers in the CRISPR-Cas site. The crRNA assembles with several Cas proteins to make **crRNP** complexes. The crRNP complexes scan along DNA looking for sequences that hybridize to the crRNA. If such a sequence is encountered, the Cas proteins in the crRNP complex degrade the hybridizing DNA; in some cases, the degradative enzyme joins the crRNP complex after it has bound to the hybridizing DNA. If this process occurs faster than the phage can replicate, the cell is protected from the phage. If another phage with the same genome sequence attempts to infect the cell, crRNP complexes resident in the cytoplasm remember the first attempted infection and use the guide RNA to recognize and degrade the phage quickly, before phage gene expression can ensue. Thus, CRISPR-Cas exhibits key aspects of adaptive immunity: specificity and memory. There are type II and type III CRISPR-Cas systems that differ mechanistically, but all CRISPR-Cas types use a guide RNA assembled into a complex with Cas proteins to bind to complementary DNA and then degrade the DNA.

The proteins that form the crRNP and bind to target DNA can be manipulated both *in vitro* and in other organisms besides bacteria, including model organisms such as *Drosophila*, zebrafish, and mice, as well as human cells. We have devised various means to use Cas proteins in combination with guide RNA to genetically modify organisms more simply, more affordably, and across more taxonomic groups than ever before. Research applications of CRISPR-Cas are discussed further in **Technique Box 15.1** on genome engineering, and medical applications are discussed in **Chapter 16**.

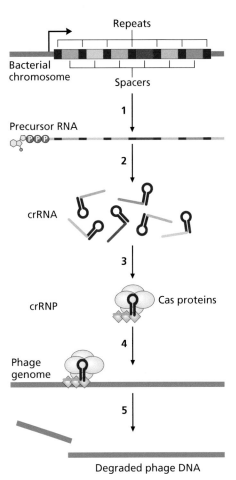

Figure 15.3 CRISPR-Cas processing and crRNP-based immunity. Transcription produces a precursor RNA that does not encode any proteins (1). In type I CRISPR systems, the RNA is processed into crRNA (2) consisting of 5′ guide RNA, corresponding to the unique spacers, with 3′ hairpins corresponding to the palindromic repeats. The crRNA assembles with many Cas proteins, forming crRNP (3). During an immune response, the guide RNA in the crRNP hybridizes to one strand of bacteriophage DNA (4) and one of the Cas proteins degrades the bacteriophage DNA (5).

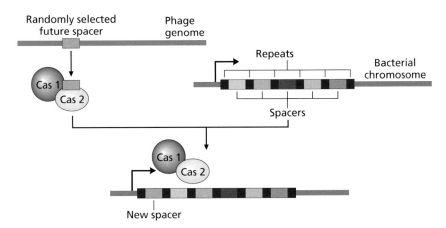

Figure 15.2 Creation of a new spacer by Cas1 and Cas2. After double-stranded DNA such as a bacteriophage genome enters the cytoplasm, the Cas1 and Cas2 proteins bind to a randomly selected fragment of the genome and recombine the segment with the host chromosome in between the promoter and the first repeat in the CRISPR array. The result is formation of a new spacer flanked by the palindromic repeat.

TECHNIQUE BOX 15.1   GENOME ENGINEERING USING CRISPR-Cas IN ANIMALS

Genetic engineering of model organisms such as mice has been very important for the discipline of virology. For example, mice engineered to lack the IRF7 gene were used to show that IRF7 is the most important transcription regulator that produces type I inteferon in response to pattern recognition receptor (PRR) signaling. Engineering knockouts in mice is time-intensive, expensive, and otherwise difficult, however. A major breakthrough in genetic engineering in 2013 revolutionized the production of engineered organisms, including mammals such as mice.

That revolutionary technique exploits the bacteriophage-related CRISPR-Cas system to use guide RNAs to make double-strand breaks in specific sequences of DNA. When the cell repairs the double-strand break, the repair can be manipulated to incorporate a new DNA sequence. Almost all cells are susceptible to CRISPR-Cas engineering so that we are not restricted to the traditional model organisms. So far, CRISPR-Cas has been used to alter genes in human cell culture, yeast, zebrafish, fruit flies, nematodes, plants, mice, monkeys, and human embryos. There have even been promising first clinical trials using CRISPR-Cas to treat human diseases such as transthyretin amyloidosis, sickle cell disease, and β-thalassemia. Additionally, agricultural scientists are eager to modify animals so that they are resistant to infections or produce more economically valuable meat, fur, eggs, or milk per animal, and to make higher-yield plant crops that are resistant to infections and environmental assaults. Such efforts have long been underway using traditional breeding methods, but CRISPR-Cas is making it possible to cause such changes in a more directed and rapid way than was previously imagined. There are increasing numbers of commercial agricultural products prepared from gene-edited plants, such as nutraceutical tomatoes that contain a high amount of γ-aminobutyric acid and mushrooms or apples that do not brown after slicing.

The most common CRISPR-Cas system used to engineer cells requires just one protein, **Cas9**. CRISPR-Cas9 uses both crRNA and a second molecule called tracrRNA. It is simple to make synthetic constructs in which the crRNA and tracrRNA are expressed together as a chimera on a single RNA molecule called guide RNA (**gRNA**). The gRNA binds to Cas9 (**Figure 15.4**) and the mature Cas9–RNA complex then binds to DNA complementary to the crRNA and makes a double-strand break in the DNA. The crRNA component of the gRNA can be altered to target any desired DNA sequence.

After the Cas9–gRNA complex binds to complementary DNA, the Cas9 protein induces a double-strand break in the DNA. Cells have two ways to repair double-strand breaks (**Figure 15.5**). One method is called nonhomologous end joining. The second method is

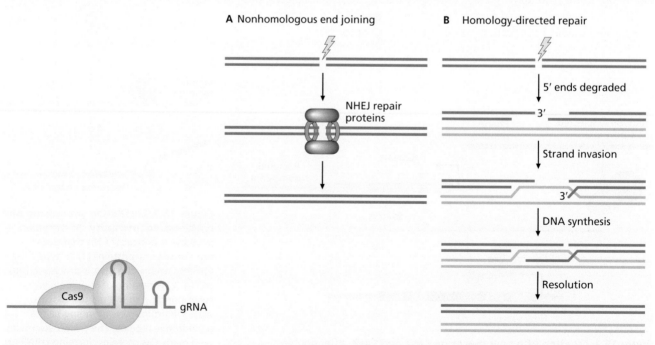

Figure 15.4 Cas9–gRNA complex. Cas9 protein binds to gRNA, which has some single-stranded regions and some components with stem–loops.

Figure 15.5 Cellular pathways to repair double-strand breaks in DNA. There are two ways to repair double-strand breaks. (A) Nonhomologous end joining (NHEJ) can repair some breaks. (B) Homology-directed repair takes advantage of a second copy of the broken DNA in a diploid cell and uses it to repair the double-strand break with DNA synthesis and ligation.

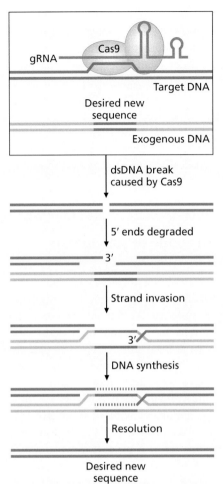

Figure 15.6 Use of exogenous DNA in genome editing with CRISPR-Cas9. After CRISPR-Cas9 induces a double-strand break, exogenous DNA provides a template for homology-directed repair, resulting in incorporation of new DNA sequences. The exogenous DNA on either side of the double-strand break must be homologous to the host chromosome.

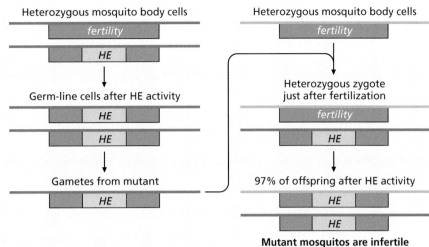

Figure 15.7 CRISPR-Cas used to create a heterozygous mosquito with a homing endonuclease (HE) gene replacing one allele of a gene needed for fertility. The heterozygous adult is fertile, but during development, the germ-line precursors to the gametes become homozygous because of the homing endonuclease. When these mutants mate with wild-type organisms, the homing endonuclease is active in the zygote. It replaces the other parent's chromosome with the homing endonuclease knockout allele, making a homozygous knockout in the fertility gene. The result is that 97% of the offspring are infertile.

homology-directed repair, which occurs when the cell uses homologous DNA flanking the double-strand break as a template to fill in the gap. This method can be used in combination with CRISPR-Cas to genetically engineer a cell. Providing abundant exogenous DNA can induce the cell to use the exogenous DNA as the homologous DNA during the repair process. The exogenous DNA is homologous to DNA on both sides of the double-strand break and contains whatever DNA is desired in between this complementary DNA. When the cell uses this artificially provided exogenous DNA to repair the double-strand break, the result is the introduction of whatever sequence is

desired into the site where there had been a double-strand break. Biotech companies already sell collections of tens of thousands of gRNAs targeting human and mouse genes, and the technique has the possibility of being so precise that some are already calling it **genome editing** (**Figure 15.6**).

Many medically important viruses such as yellow fever virus, dengue fever virus, and Zika virus are transmitted by mosquitos. Reducing contact between people and mosquitos reduces the prevalence of diseases caused by these viruses. It might be possible to use CRISPR-Cas9 to edit mosquito genomes in order to reduce populations of mosquito species that transmit infectious diseases. For example, genome editing can be used to make mosquitos capable of spreading sterility through a wild-type population using a gene drive. Release of sterilized male insects (treated with radiation to make them sterile) was already known to be useful to control the populations of other insects that transmit diseases. In this case, CRISPR-Cas was used to make heterozygote insects in which one allele of a gene conferring fertility was replaced with a gene encoding a selfish genetic element called a homing endonuclease (**Figure 15.7**). Homing endonucleases cause double-strand breaks that are repaired by homology-directed repair, the result of which is to copy the homing endonuclease gene into the other homologous chromosome (just like exogenous DNA causes introduction of new sequences after CRISPR-Cas treatment). This is called a gene drive because the homing endonuclease drives its own inheritance through the population.

When a fertile adult heterozygote forms gametes, the gametes are descended from germ-line progenitors that became homozygous because of the homing endonuclease. The adult can still mate, however. When such adults mate with a wild-type mosquito, almost all the offspring inherit a copy of the mutant allele, which then copies itself into the homologous chromosome from the other parent during development. The result is that 97% of the offspring of these matings are sterile. If these adults were released into a natural population, the sterility allele would likely spread quickly through the population, ultimately causing the mosquitos to go extinct in the local area. Concerns over unpredictable impacts on local ecosystems have prevented this technology from moving forward to help control infectious diseases for the time being.

The Nobel Prize in Chemistry in 2020 was awarded to two women, Emmanuelle Charpentier and Jennifer Doudna, for their work discovering CRISPR and its applications.

15.2 Some bacteriophages can evade or subvert the CRISPR-Cas system

Bacteria and bacteriophages have been evolving together for more than 4 billion years, so there are undoubtedly many bacteriophage strategies for evading or subverting CRISPR-Cas systems. We have been aware that CRISPR-Cas is a form of adaptive immunity for about 15 years and in that time only a few viral evasion proteins have been discovered. The first five **anti-CRISPR proteins** were discovered by studying phages that infect *Pseudomonas aeruginosa*. The five proteins have no sequence similarity to each other, yet, when overexpressed individually, each prevents CRISPR-mediated destruction of infecting phage genomes. None of these five anti-CRISPR proteins affects expression of Cas proteins or expression from the CRISPR-Cas promoter, so the proteins affect a process downstream of expression such as crRNP assembly or crRNP activity.

The mechanism by which three of these five proteins blocks immunity was recently elucidated. Two of the anti-CRISPR proteins (AcrF1 and AcrF2) block the ability of the crRNP complex to bind to DNA (**Figure 15.8**). The proteins differ in sequence because they interact with different protein components of the crRNP complex. AcrF2 binds to the surface of the complex that would otherwise bind to DNA, thus physically blocking the crRNP from assembling on DNA. Unlike AcrF2, AcrF1 binds elsewhere on the crRNP and changes its conformation so that the DNA binding surface is no longer active. The third, AcrF3, binds to the Cas3 nuclease and prevents it from assembling with the rest of the crRNP (**Figure 15.9**). Cas3 is an ATP-dependent

Figure 15.8 Anti-CRISPR proteins AcrF1 and AcrF2. (A) Typical degradation of phage genome in the presence of an active CRISPR system and complementary 5′ guide RNA. (B) AcrF1 is an anti-CRISPR protein. It binds to the crRNP and changes its conformation so that the guide RNA can no longer bind to DNA. AcrF2 binds to the surface of the complex that would otherwise bind to DNA. This particular type of CRISPR uses guide RNA that does not have a stem–loop structure in it.

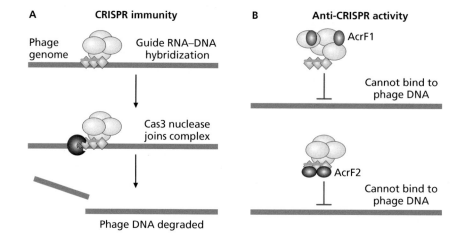

Figure 15.9 Anti-CRISPR protein AcrF3. (A) The nuclease Cas3 consumes ATP to provide energy for degrading DNA that hybridizes to guide RNA in crRNP. The ADP must be exchanged with ATP for the cycle to continue, ultimately degrading the DNA. (B) The anti-CRISPR AcrF3 viral evasion protein binds to Cas3 as a homodimer. Binding blocks Cas3 access to DNA. AcrF3 also prevents the ADP–ATP exchange necessary for Cas3 to become active after it has degraded some of the DNA. This particular type of CRISPR uses guide RNA that does not have a stem–loop structure in it.

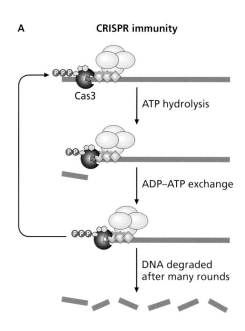

A   **CRISPR immunity**

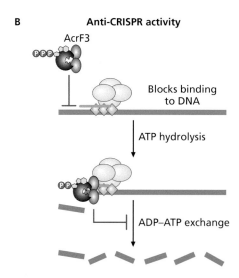

B   **Anti-CRISPR activity**

nuclease that would otherwise degrade the hybridizing DNA, thus destroying the infecting phage genome. Structural studies of AcrF3 in a complex with Cas3 revealed that a homodimer of AcrF3 binds to Cas3, which has two effects that account for its anti-CRISPR activity. First, AcrF3 locks Cas3 into an ADP-bound state. When AcrF3 prevents ADP from being exchanged for ATP, Cas3 activity is blocked. Second, AcrF3 blocks the Cas3 enzyme from binding to DNA by physically occluding the site that Cas3 would otherwise use to bind to the DNA.

Additional anti-CRISPR phage proteins have since been discovered studying other *Pseudomonas* phages. Again, they are not similar in sequence to each other or to other well-characterized proteins. Even more recently, new anti-CRISPR proteins were discovered in many bacteria across wide evolutionary distance, suggesting that not only phages but also their hosts employ anti-CRISPR proteins. Anti-CRISPR proteins might be important for hosts to acquire new DNA such as plasmids. Many other anti-CRISPR proteins will undoubtedly be discovered.

A bacteriophage that subverts CRISPR-Cas to enhance its own replication has also been discovered. In this case, the ICP1 phage infects *Vibrio cholerae*. The genome sequence of the lytic phage encodes its own CRISPR-Cas system. One of the phage spacers is identical in sequence to the host chromosome. When the phage CRISPR-Cas system is expressed, the phage-derived crRNP attacks the bacterial chromosome, degrading it.

15.3   The human adaptive immune response includes cell-mediated and humoral immunity

We now turn our attention from bacteria to the human adaptive immune system and its interactions with viruses. In general, the human acquired immune system includes both a cell-mediated response and a humoral response (**Figure 15.10**). Both the cell-mediated and humoral branches of the adaptive immune system are required to control viral infections. **Cell-mediated responses** are those involving T lymphocytes, whereas the **humoral response** refers to antibodies in bodily fluids such as the bloodstream. Antibodies are synthesized by B lymphocytes, but antibody synthesis itself requires B cells to interact with a subset of T cells known as T helper cells, also known as CD4+ T cells or T_H cells. Antibodies are needed to eliminate extracellular viruses from the body and to direct the complement to attack infected cells. T helper cells are needed to coordinate the immune attack against viral infections; for

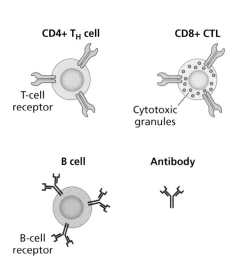

Figure 15.10 Cells and molecules of the adaptive immune response. Lymphocytes are the cells of the adaptive immune response and include CD4+ T_H cells, CD8+ cytotoxic T lymphocyte (CTL) cells, and B cells. B cells synthesize antibodies, which are responsible for humoral immunity.

example, they are needed to stimulate antibody production. There are also CD8+ cytotoxic T lymphocytes (CTLs). CTLs are needed to kill human cells that have already been infected by a virus.

Professional phagocytic cells such as dendritic cells and macrophages are sometimes considered part of the adaptive immune system. Their functions include surveillance, a process in which these cells engulf foreign particles such as viruses, degrade them, and alert the adaptive immune response of the presence of the foreign entities using antigen presentation. In this case, antigen presentation is similar to MHC-I–epitope display of endogenous epitopes, but these professional surveillance cells use a different **major histocompatibility complex II (MHC-II)** molecule that specializes in displaying **exogenous** epitopes instead. Exogenous epitopes are those that result from engulfed microbes that originated outside the cell that displays them. T helper cells and B cells have receptors that interact with MHC-II–epitope complexes and thereby initiate an immune response against viral epitopes. In contrast, CTLs react to endogenous epitopes displayed in MHC-I.

15.4 The human adaptive immune response has specificity because it responds to epitopes

A key difference between innate immunity and adaptive immunity is that innate immunity responds to pathogen-associated molecular patterns (PAMPs) associated with groups of pathogens that have shared molecular characteristics, whereas adaptive immunity responds instead to immunogenic components of antigens, known as **epitopes**. Most epitopes are chains of 9 to 25 amino acids. The specificity of an adaptive response arises from the uniqueness of epitopes derived from even closely related pathogens. As an example, consider poliovirus and rhinovirus. Both are closely related species in the *Enterovirus* genus of picornaviruses, which are naked icosahedral viruses with (+) RNA genomes (see **Chapter 4**). The virions are simple and consist of four capsid proteins known as VP1, VP2, VP3, and VP4. Before

```
Polio    6   LESMIDNTVRETVGAATSRDALPNTEASGPTHSKEIPALTAVETGATNPLVPSDTVQTRH   65
             +E+ ID  + E +        +PN + S  T S   P L A ETG T+ + P D ++TR+
Rhino    7   VENYIDEVLNEVL------VVPNIKESHHTTSNSAPLLDAAETGHTSNVQPEDAIETRY   59

Polio   66   VVQHRSRSESSIESFFARGACVTIMTVDNPASTTN-KDKLFAVWKITYKDTVQLRRKLEF  124
             V+ ++R E SIESF  R  CV I  +    N +D  F  WKIT ++   Q+RRK E
Rhino   60   VITSQTRDEMSIESFLGRSGCVHISRIKVDYTDYNGQDINFTKWKITLQEMAQIRRKFEL  119

Polio  125   FTYSRFDMELTFVVTANFTETNNGHALNQVYQIMYVPPGAPVPEKWDDYTWQTSSNPSIF  184
             FTY RFD E+T V      + GH    V Q MYVPPGAP+P K +D++WQ+ +N SIF
Rhino  120   FTYVRFDSEITLVPCIAGRGDDIGHI---VMQYMYVPPGAPIPSKRNDFSWQSGTNMSIF  176

Polio  185   YTYGTAPARISVPYVGISNAYSHFYDGFSKVPLKDQSAALGDSLYGAASLNDFGILAVRV  244
             + +G    R S+P++ I++AY  FYDG+     D ++   S YG+  ND G + R+
Rhino  177   WQHGQPFPRFSLPFLSIASAYYMFYDGYD----GDNTS----SKYGSVVTNDMGTICSRI  228

Polio  245   VNDHNPTKVTSKIRVYLKPKHIRVWCPRPPRAVAY  279
             V +      V     +Y K KH + WCPRPPRAV Y
Rhino  229   VTEKQKHSVVITTHIYHKAKHTKAWCPRPPRAVPY  263
```

Figure 15.11 An alignment of two enterovirus VP1 proteins shows the variation in potential epitopes derived from the viruses. The sequences of the poliovirus VP1 protein and the rhinovirus VP1 protein have been aligned using the single-letter amino acid code. In this type of alignment, when an amino acid is identical in two sequences it appears in the row between them. A plus sign indicates that the amino acids at that position have chemically similar side chains. A blank indicates that the amino acids are dissimilar at that position. The minus signs in the same rows as the amino acids show places where the software had to introduce a gap into the sequence to maximize the alignment between the two proteins.

attachment and penetration, only VP1, VP2, and VP3 are displayed on the virion's surface. A comparison between the amino acid sequences of the VP1 proteins for both poliovirus and rhinovirus illustrates how the same protein with the same function in closely related viruses can nevertheless give rise to very different epitopes (**Figure 15.11**). As you can see in the alignment, almost all stretches of nine or more amino acids have at least one, but more often more than three, different amino acids in the poliovirus sequence relative to the rhinovirus sequence. Even a difference of 1 in 25 amino acids in an epitope is enough to prevent an antibody from binding to an epitope well enough to control an infection. Thus, an adaptive immune reaction against poliovirus epitopes has no effect on rhinovirus and vice versa.

15.5   Professional antigen-presenting cells degrade exogenous antigens and display epitopes in MHC-II molecules

The professional phagocytes that surveil the body for foreign microbes and display exogenous epitopes in MHC-II molecules are known collectively as professional **antigen-presenting cells** (**APCs**). APCs include macrophages, dendritic cells, and B lymphocytes. Because of APCs that engulf virions, exogenous antigen presentation occurs during a viral infection (**Figure 15.12**). APCs are resident in most body tissues and are also found throughout the lymph system, where they sample the extracellular environment through constant phagocytosis. After a virion is ingested by an APC, the virion is degraded in a phagolysosome. After degradation, the virion's proteins have been broken down so that they are only about 9 to 25 amino acids long. Some of these peptides will become epitopes, which are subsequently loaded into the cleft of MHC-II molecules, and are then transported through the endomembrane system to the cell surface. Epitopes displayed in MHC-II are called exogenous because they originated from outside the APC (which engulfed an extracellular microbe in order to obtain the epitopes). Exogenous epitopes displayed in MHC-II can now initiate an adaptive immune response by interacting with lymphocytes. The lymphocytes are stimulated most strongly when damage-associated molecular patterns (DAMPs) or inflammation provide a second signal that the epitope is not only foreign but also

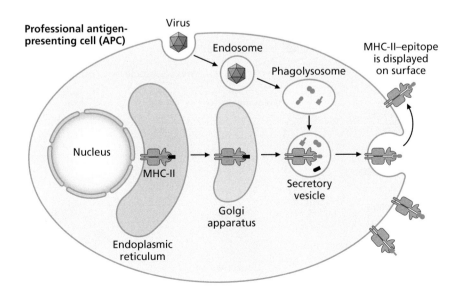

Figure 15.12 Exogenous antigen presentation by an APC. After a virion is ingested by an APC, it is degraded in a phagolysosome. After degradation, the virion's proteins have been broken down so that they are only about 9 to 25 amino acids long. Some of these peptides will become epitopes, which are subsequently loaded into the cleft of MHC-II molecules, and are then transported to the cell surface.

dangerous. The danger signal ensures that the lymphocytes do not overreact to the billions of harmless microbes that coat the gastrointestinal, upper respiratory, and genitourinary tracts. The harmless microbes do not cause inflammation and so do not produce DAMPs.

15.6 Some viruses evade MHC-II presentation

Viruses that infect APCs have evolved a variety of strategies for reducing the display of MHC-II–epitope complexes on the surface of the APCs (**Figure 15.13**). There is at least one herpesvirus that blocks transcription of MHC-II genes. Human cytomegalovirus, also a herpesvirus, encodes a viral protein that causes MHC-II to be degraded in the endoplasmic reticulum itself. HIV targets immature MHC-II molecules to the lysosome, leading to their degradation before they can be loaded with epitopes. Another example is that a vaccinia virus protein blocks the ability of processed epitopes to reach the vesicle where they could otherwise be loaded into the cleft of the MHC-II molecule.

A more general strategy for populations of viruses to evade the use of exogenous epitopes to trigger an effective adaptive immune response is **antigenic variation**, which is variation in the amino acids (primary sequence) of an antigenic protein. During viral replication, missense mutations occur that cause offspring virions to express slightly different proteins compared with the virus that first infected the person. Because in many cases one amino acid can substitute for another without altering the function of a protein, these variations usually do not affect viral replication directly. A small change in the primary sequence of an antigen can have profound effects on the immune response, however, because even a single amino acid change in one epitope compared with a second epitope can prevent antibodies from binding to the altered antigen. Thanks to antigenic variation, when the immune system mounts a response against epitopes derived from an original infecting population of virions, some of the mutant offspring virions escape detection because their epitopes are just different enough.

Antigenic variation is particularly important for two groups of viruses. The first group is composed of species that can cause new lytic infections in people, even if those people survived and controlled a previous infection by that same species of virus. Prominent examples include influenza, rhinoviruses, and

Figure 15.13 Viral evasion of MHC-II presentation. Viral evasion strategies (red) include reduction of transcription of MHC-II genes (1), degradation of MHC-II proteins in the endoplasmic reticulum (2), targeting the MHC-II from the Golgi body to lysosomes where the MHC-II will be degraded (3), and blocking processed epitopes from reaching MHC-II-containing secretory vesicles (4).

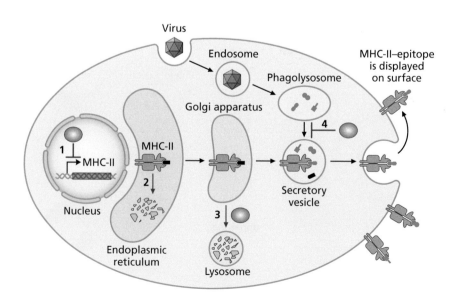

SARS-CoV-2. There are more than 100 antigenically distinct rhinoviruses, so that a person can catch the common cold, caused by an antigenically distinct rhinovirus each time, many times in a lifetime. In this case, the immune reaction that cured the person of rhinovirus number 77 will prevent future infections with rhinovirus number 77 but not from rhinovirus number 38, and so on. Similarly, antigenic variation is critical for the success of both influenza and SARS-CoV-2 as pathogens and has dramatic impacts on vaccination (see **Chapter 16**), epidemics, and pandemics (see **Chapter 17**).

The second group of viruses in which antigenic variation is particularly important is that of viruses that cause persistent lytic infections so that they must escape an infected person's immune response for many years. HIV provides a prominent example. During an HIV infection, viral replication is ongoing so that large populations of new virions are constantly being produced even as the immune system works to control the infection. In response to selection from the immune system, viral epitopes evolve and eventually the adaptive immune response is not able to control viral replication: the virus is said to escape from the immune response. Uncontrolled HIV replication ultimately causes HIV's preferred lymphocyte host cells to die and permits the onset of AIDS.

15.7   Lymphocytes that control viral infections have many properties in common

Lymphocytes (B cells, T helper cells, and CTLs) are all required for controlling viral infections and share certain traits. They all are stimulated by antigens presented to them in the form of epitopes associated with MHC molecules. They all are exquisitely specialized so that each individual lymphocyte reacts to just a single epitope. This degree of specialization is remarkable given the billions of different epitopes that can be recognized by the adaptive immune system. Any individual viral infection gives rise to hundreds if not thousands of different epitopes, some of which will be recognized by at least one lymphocyte. Another shared property is that all lymphocytes can be naive or activated. **Naive lymphocytes** have never been exposed to the epitope that could activate them and are in a resting state. **Activated lymphocytes** have been exposed to their stimulating epitope and as a consequence they have enhanced effector functions such as killing infected cells or secreting antibodies. Another property shared by all lymphocytes is that they can differentiate into memory lymphocytes at the end of an immune response. Memory lymphocytes persist in the body for many years and remain at rest, much like naive lymphocytes, unless they are activated by the antigen that triggered the first immune reaction. In that case, the population of memory cells quickly proliferates and some of the cells differentiate into activated effector cells that control the infection much faster than the lymphocytes that reacted to the first infection. Memory lymphocytes thereby account for the fact that an immune response is faster upon second and subsequent exposures to an antigen. The goal of preventative immunization is to create a population of memory lymphocytes.

15.8   CD4+ T helper lymphocytes interact with viral epitopes displayed in MHC-II molecules

Although lymphocytes share many properties in common, classes of lymphocytes have different functions during a viral infection. For example, there are two major classes of T cells that are critical for an adaptive antiviral response:

A   TCR on surface of CD4+ T$_H$ cell

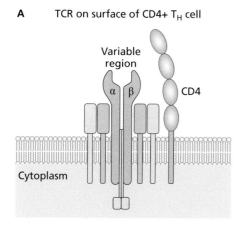

B   TCR on surface of CD8+ CTL

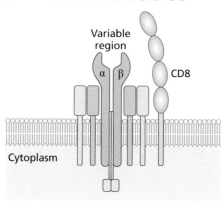

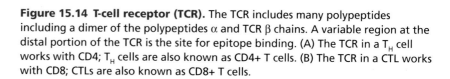

Figure 15.14 T-cell receptor (TCR). The TCR includes many polypeptides including a dimer of the polypeptides α and TCR β chains. A variable region at the distal portion of the TCR is the site for epitope binding. (A) The TCR in a T$_H$ cell works with CD4; T$_H$ cells are also known as CD4+ T cells. (B) The TCR in a CTL works with CD8; CTLs are also known as CD8+ T cells.

CD4+ T helper cells and CTLs. **CD4+ T helper (T$_H$) cells** play a role in activation of other lymphocytes and so they will be discussed first. T$_H$ cells use the **CD4** surface molecule and the **T-cell receptor (TCR)** to bind to MHC-II molecules displaying viral epitopes on the surface of professional APCs (**Figure 15.14**). Although there are two varieties of TCR, here we focus on the one that includes a dimer of the polypeptides TCR α and TCR β chains because this is the type needed for an acquired immune response against viruses. The TCR is a multisubunit complex with many other protein chains. There is a variable region at the distal portion of the TCR, which is the site for epitope binding; the variable region is different in one T cell compared with the next. The CD4 molecule binds to the outside surface of the MHC-II molecule, whereas the TCR binds to the center of the MHC-II (**Figure 15.15**). The CD4 molecule specifically binds to MHC-II but not to MHC-I, providing the structural basis that T$_H$ cells are restricted to activation by MHC-II–epitope complexes. If the viral epitope in the MHC-II binds to the variable portion of the TCR then the interaction stimulates the T cell, which will next go on to help control the viral infection.

Even when the epitope is a match for the TCR, however, an additional verification signal provided by the APC is required to activate the T$_H$ cell; this signal can be mediated by a variety of cell surface-associated proteins and helps ensure that the immune response is specific to dangerous situations. One of the most common co-stimulatory proteins on the surface of APCs is the B7 protein (**Figure 15.16**). The B7 protein is more abundant

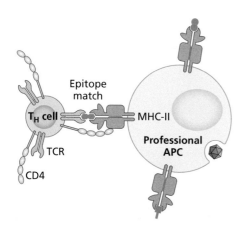

Figure 15.15 A T$_H$ cell interacts with an APC through interactions between the TCR, CD4, and the MHC-II–epitope complex. The APC has presented viral epitopes in its MHC-II molecules. The TCR recognizes one of the epitopes in the MHC-II groove, whereas the CD4 molecule binds to the outside surface of the MHC-II molecule.

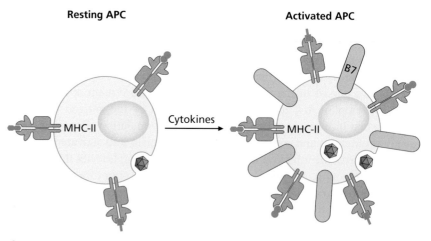

Figure 15.16 The B7 receptor is upregulated by APCs during an innate immune response. During an immune response, cytokines trigger APCs to upregulate the B7 transmembrane protein so that it becomes more abundant on the cell surface.

on the surface of APCs that have themselves been stimulated by an innate immune response against a virus. Abundant B7 protein on the surface of the APC is therefore a sign of danger. The use of a co-stimulatory danger signal ensures that the adaptive immune system does not overreact to the billions of harmless microbes that coat most surfaces of the human body.

When stimulation by both an epitope and co-stimulation occur, a naive CD4+ cell is activated and secretes cytokines that cause its own proliferation. The resulting population differentiates into one of several classes of T_H cells, each of which is specialized for different functions such as regulating B cells so that antibodies can be produced (**Figure 15.17**). We focus on this class of T_H cells because antibodies are critical for controlling viral infections.

The goal of activating B cells during a viral infection is to make antibodies that bind selectively and tightly to viral antigens. The first step is for an activated T_H cell to stimulate the naive B cells that have the capacity to make these antibodies. B cells use the **B-cell receptor** (**BCR**) to recognize epitopes; the BCR is a transmembrane form of the antibodies secreted by an activated B cell (**Figure 15.18**). When an activated T_H cell interacts with a naive B cell, the first molecular reactions determine whether the BCR recognizes the epitope that stimulated the T_H cell in the first place. If the BCR recognizes that epitope, the naive B cell will proliferate and its offspring will differentiate into antibody-producing **plasma cells** and memory cells (**Figure 15.19**). The role of antibodies during a viral infection and their production are discussed in **Section 15.9**.

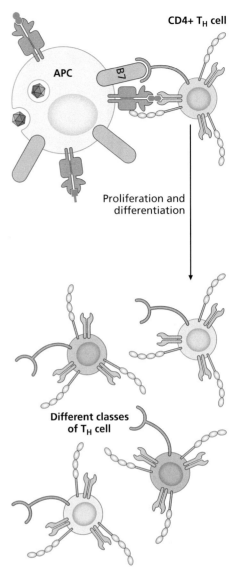

Figure 15.17 Consequences of activating a naive CD4+ cell. Activated CD4+ cells proliferate. The resulting population differentiates into different classes of T_H cells, such as those that regulate B cells.

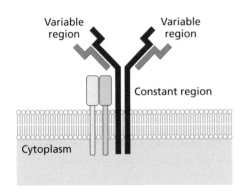

Figure 15.18 B-cell receptor. The BCR includes several polypeptides, including four with transmembrane segments. Two variable regions at the distal portion of the BCR are sites for epitope binding; the two variable regions in any single BCR are identical to each other. The extracellular portion of the BCR more proximal to the membrane is known as the constant region.

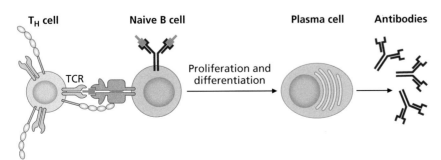

Figure 15.19 An activated T_H cell can activate a naive B cell. If the TCR and the BCR recognize the same epitope as that displayed in the B cell's MHC-II, the T_H cell activates the B cell and it proliferates. The population differentiates into plasma cells that secrete antibodies.

A

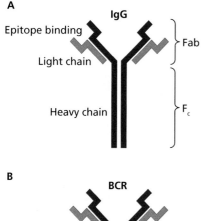

B

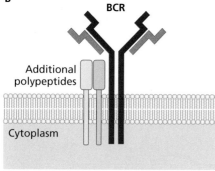

Figure 15.20 Antibody compared to BCR. (A) An IgG antibody consists of one F$_c$ constant region and one Fab variable region. The Fab has two epitope-binding sites. The antibody is composed of two polypeptides: the heavy chain and the light chain. (B) The BCR shares many components with an antibody but has additional features such as a transmembrane segment, cytoplasmic tails, and additional transmembrane polypeptide chains that are important for signal transduction.

15.9 Antibodies are soluble B-cell receptors that bind to extracellular antigens such as virions

Antibodies are complex proteins made of many polypeptides, some of which are shared with the BCR (**Figure 15.20**). An abundant class of antibody, immunoglobulin G (IgG), consists of two heavy-chain polypeptides and two light-chain polypeptides. One end of the antibody has two variable regions, each composed of part of the light chain and part of the heavy chain, that are identical to each other in any single antibody and bind to epitopes. This region of the antibody is known as the Fab. The constant region of the antibody, which is the same in all IgG molecules irrespective of the epitope they bind, is known as the F$_c$. All IgG antibodies have identical constant regions that are the same as the cell-proximal parts of the BCR. IgG antibodies have different variable regions, which are the same as the distal portion of the BCR and enable different antibodies to bind to different epitopes.

Antibodies are useful in controlling a viral infection (**Figure 15.21**). They flood the lymphatic and circulatory systems and, when they encounter virions, they bind to epitopes on the virions. Many antibodies can bind to the surface of the virion at the same time, which usually has the effect of **neutralizing** the virion (see **Figure 15.21A**). A neutralized virion can no longer bind to its normal receptor on a host cell because the large antibodies surrounding it physically occlude the viral spike proteins and prevent them from reaching their receptors. Furthermore, the constant region of the neutralizing antibodies can be bound by phagocytes such as macrophages, which subsequently engulf the coated virions and degrade them in a process known as **opsonization** (see **Figure 15.21B**).

Antibodies also help control a viral infection because they bind to viral proteins on the surface of infected cells. Examples of abundant viral proteins on the surface of infected cells include envelope spikes that are found in patches that mark the future site of viral budding. Antibodies binding to the viral spikes on the surface of a host cell target that cell for destruction by the complement system, resulting in formation of the membrane attack complex (MAC) on the infected cell (see Figure 14.25). The result is that the host cell dies through necrosis and, when it bursts open, it releases many PAMPs (from the replicating virus) and DAMPs into the surrounding tissue. The increased inflammation that results contributes to controlling the viral infection.

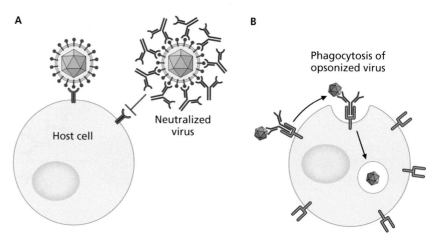

Figure 15.21 Antibodies help control a viral infection. (A) When a neutralizing antibody binds to a virion, the antibodies physically prevent the virion from approaching its receptor. (B) Opsonization. Professional APCs use a cellular receptor that binds to the constant region of antibodies to phagocytose and destroy a virus.

15.10 During an antiviral response, B cells differentiate to produce higher-affinity antibodies

Antibodies detected during a viral infection often bind more tightly to their target viral epitopes than the TCR that initiated the response binds to those epitopes. This situation arises in an interesting manner that relies upon interactions among B cells, antigens, dendritic cells, and T_H cells. First, it is important to know that B cells are professional APCs that display epitopes in MHC-II. They acquire their exogenous epitopes in a special manner instead of by phagocytosis (**Figure 15.22**). To acquire exogenous epitopes, each B cell makes a single type of BCR that has a unique variable region and is therefore specialized in binding to a single epitope. When the BCR binds to a virus with such an epitope on its surface, the B cell internalizes the BCR with its attached virion. The virion is then degraded and ultimately the same epitope that allowed the BCR to bind to the virion is displayed in MHC-II molecules on the surface of the B cell.

In order for antibodies to be produced abundantly during an immune response to a virus, B cells, T_H cells, and APCs congregate and create **germinal centers** (**Figure 15.23**). Naive B cells expressing a single type of BCR enter the dark zone of a germinal center where they are stimulated to proliferate if activated by a T_H cell. During proliferation, the genes encoding the variable region of the BCR undergo a process known as somatic hypermutation so that the offspring of a single naive B cell is a population of cells that each make a slightly different BCR. The new BCRs are all variations of the ancestral BCR. Because somatic hypermutation makes random changes in the variable region of the BCR, some of these BCRs bind to the epitope with higher affinity than the original, whereas others bind to their epitope with lower affinity. After somatic hypermutation, the B cells enter the light zone of the germinal center where they encounter APCs known as follicular dendritic cells and also T_H cells. B cells with BCRs that have decreased affinity for the epitope die through apoptosis. In contrast, B cells with BCRs that have increased affinity for the epitope are stimulated by the follicular dendritic cells and T_H cells. These useful B cells proliferate into a clonal population in which every cell has the same BCR. Some of these cells differentiate into plasma cells that secrete antibodies and others differentiate into memory cells. The plasma cells secrete abundant antibodies, which have the same variable region as the BCR and therefore bind to the

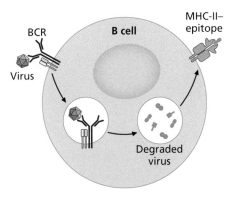

Figure 15.22 Antigen presentation by a B cell. When the BCR binds to an epitope associated with an antigen on the surface of a virus, the B cell internalizes the virus. The B cell degrades the virus and ultimately the same epitope that allowed the BCR to bind to the virus in the first place is displayed in MHC-II molecules on the surface of the B cell.

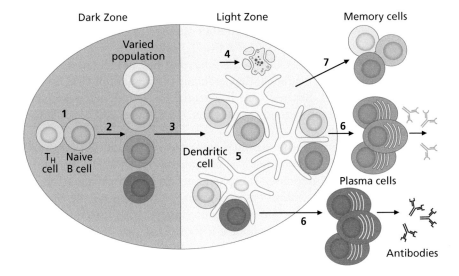

Figure 15.23 B-cell differentiation in a germinal center. Germinal centers are created during an adaptive immune response and consist of a dark zone and a light zone. Different activities occur in each zone. Naive B cells expressing a single type of BCR enter the dark zone; if they are stimulated by a T_H cell (1) they proliferate (2) because the BCR recognizes an epitope important for the immune response. Somatic hypermutation occurs during B cell proliferation in the dark zone. The B cells with altered BCRs enter the light zone (3), where those with BCRs that have lower affinity for the epitope undergo apoptosis (4). B cells with BCRs that have higher affinity for the epitope interact with follicular dendritic cells and T_H cells (5) and are stimulated to proliferate. Some of the population differentiates into plasma cells (6), which secrete large numbers of antibodies that bind to the epitope with high affinity. Others become memory cells (7).

viral epitope with high affinity. Such antibodies can opsonize and neutralize viruses, and they can direct the MAC to attack infected cells. Memory B cells can be stimulated rapidly following a second exposure to the microbe that first stimulated the immune reaction, quickly proliferating and differentiating into plasma cells to control the infection, often before an infected person develops symptoms.

15.11 Viruses have strategies to evade or subvert the antibody response

Viruses have mechanisms to avoid being neutralized or opsonized by antibodies. For example, both hepatitis C virus (HCV) and human herpesvirus (HHV) 1 and 2 infect new cells by spreading directly from one cell into a new host cell. Antibodies are extracellular, so cell-to-cell spread is a mechanism to evade the antibody response (**Figure 15.24**). In the case of HCV, the process is called cell–cell contact-mediated transfer. It depends on viral spike proteins expressed on the surface of the infected cell binding to cellular receptors on the target uninfected cell. HHV-1 and -2 also use cell-to-cell spread in an epithelium by passing through the tight junctions that connect epithelial cells to one another. In the case of retroviruses, the viruses are described as spreading through a virological synapse where retroviral spike proteins in an infected cell interact with receptors on an uninfected cell, but in this case, it is thought that virions transfer from one cell to the next through the intracellular space. The small size of the space between the cells and the short duration of the time that the virions are extracellular contribute to evading neutralizing antibodies. SARS-CoV-2 can also spread in the body through cell–cell contact-mediated transfer.

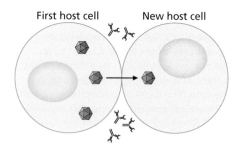

Figure 15.24 Cell-to-cell spread evades the neutralizing antibody response. Viruses that spread from one cell into another without spending much time in the extracellular space cannot be neutralized by antibodies.

Viruses that do not use cell-to-cell spread often exhibit variation in surface proteins over the course of infecting a whole population of people. This variation is naturally selected because it allows them to evade neutralization by antibodies that were induced by a different epitope. Another immune evasion mechanism is glycosylation. Viral spike proteins can be glycosylated in such a way that epitopes on the spike are masked and antibodies cannot bind to them; this is called a glycan shield (**Figure 15.25**). The shield interferes with antibody binding because it prevents the antibody's variable region from approaching the epitope; furthermore, the shield is often negatively charged, which also interferes with the intermolecular forces that would otherwise allow antibody binding. Some viruses also evade MAC formation stimulated by antibodies by interfering with activation of the complement (see **Chapter 14**). Another strategy to evade an antibody response is to prevent activation of naive T_H cells in the first place. For example, when HIV infects CD4+ cells, the viral Nef protein causes CD4 to be endocytosed and degraded, thus reducing the amount of CD4 on the surface of the T_H cell. Because CD4 is essential for T_H cell activation and T_H cells are essential for antibody production, the ultimate effect is to block antibody production.

Several viruses, most prominently dengue fever virus and HIV, exhibit antibody-dependent enhancement, a phenomenon in which the virus subverts the antibody response. This phenomenon is better understood in dengue fever. During assembly of offspring dengue fever virions, the viral M protein is exposed on the surface of immature virions, which are sometimes released from host cells, as for example if those host cells are lysed by an immune response. Some antibodies that bind to the M protein can enable those immature virions to replicate in the following manner. The antibodies

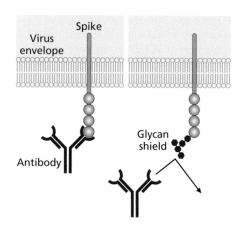

Figure 15.25 Glycosylation can mask epitopes on viral spike proteins. The antibody would normally bind to the epitope on the viral spike, but the glycan shield interferes with antibody binding.

bind to the M protein, and then when phagocytic surveillance cells bind to the constant regions of the antibodies and ingest the immature virions, the virions replicate inside the phagocytes. Thus, the antibodies to the M protein make the infection worse—the virus has subverted the antibody response.

15.12   CD8+ cytotoxic T lymphocytes are crucial for controlling viral infections

The third class of lymphocytes that control viral infections are the **CD8+ cytotoxic T lymphocytes** (**CTLs**). Activated CTLs use the CD8 surface molecule and the TCR to bind to MHC-I molecules displaying epitopes on their surface (**Figure 15.26**). The MHC-I molecules display endogenous epitopes (see **Chapter 14**). Endogenous epitopes derived from normal proteins indicate that the cell can be ignored by the CTL. Endogenous epitopes derived from viral proteins trigger the CTL to cause the death of the infected cell.

Activation of CTLs requires certain APCs known as dendritic cells, which have the special property of displaying exogenous viral epitopes in both their MHC-II and their MHC-I molecules, even when they have not themselves been infected by the virus (**Figure 15.27**). This special situation makes it possible for a T$_H$ cell and a CTL to be stimulated at the same time by the same antigen displayed by a dendritic cell. The T$_H$ cell is stimulated by the epitope displayed in MHC-II and the CTL is stimulated by the same epitope displayed in MHC-I. CTLs also require a co-stimulatory signal to be activated, just as T$_H$ cells did. The B7 protein on a dendritic cell can provide this verification signal. During the trinary interaction between a dendritic cell, a T$_H$ cell, and a CTL, both the T$_H$ and CTL cells release a cytokine that triggers proliferation of the T cells. After activation and proliferation, the CTLs patrol the body, identify cells displaying the viral epitope in MHC-I molecules, and induce those cells to undergo apoptosis.

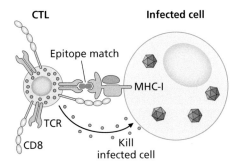

Figure 15.26 Activated CTLs use the CD8 surface molecule and the TCR to bind to MHC-I molecules displaying epitopes on their surface. If the endogenous epitope is recognized as foreign, the CTL kills the infected cell.

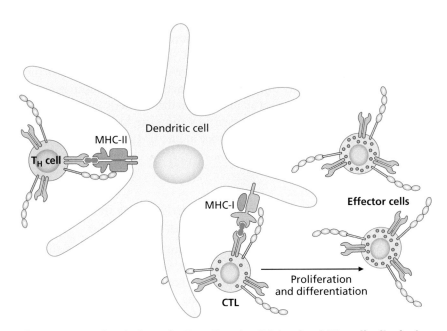

Figure 15.27 Stimulation of a T$_H$ cell and a CTL by dendritic cells displaying viral epitopes in both MHC-II and MHC-I. Dendritic cells display the same viral epitope to T$_H$ cells in MHC-II and CTLs in MHC-I, allowing the T$_H$ cell and dendritic cell to activate the CTL.

15.13 Some viruses can evade the CTL response

Large viruses with dsDNA genomes are particularly known for evading the CTL response. Prominent examples are found among the herpesviruses, which are very successful pathogens and typically cause persistent infections, during which time they must resist an effective immune response. In fact, most people have persistent infections with at least one herpesvirus such as human herpesvirus 1 (herpes simplex virus 1; HHV-1 or HSV1), human herpesvirus 2 (herpes simplex virus 2; HHV-2 or HSV2), human herpesvirus 3 (varicella–zoster virus; HHV-3 or VZV), human herpesvirus 4 (Epstein–Barr virus; HHV-4 or EBV), human herpesvirus 5 (cytomegalovirus; HHV-5 or CMV), or human herpesvirus 8 (Kaposi's sarcoma-associated herpesvirus; HHV-8 or KSHV). Herpesviruses evade the CTL response in several ways. Some herpesviruses express very few or even no proteins during latent infection, severely reducing the availability of viral epitopes for MHC-I display to CTLs. Another group of herpesvirus evasion mechanisms reduces MHC-I display on the surface of infected cells (see **Chapter 14**), thereby reducing the ability of a CTL to detect viral epitopes in an infected cell. At least one herpesvirus blocks apoptosis of infected cells so that when the CTL attempts to induce apoptosis after recognizing a viral epitope, the CTL response is ineffective. Other herpesviruses evade CTLs indirectly by interfering with their activation, which requires functional T_H cells and dendritic cells expressing epitopes in both MHC-I and MHC-II. Evasion of MHC-I presentation was discussed in **Chapter 14** and evasion of MHC-II presentation was discussed in **Section 15.6**. There are several herpesviruses that affect dendritic cell abundance or function. Patients infected with EBV (HHV-4), which can infect the precursors of dendritic cells before they differentiate, have reduced levels of dendritic cells because the virus inhibits their precursors from maturing properly during blood cell development. Patients infected with KSHV (HHV-8), which can infect both dendritic cells and their precursors, not only have reduced levels of dendritic cells but also have abnormal mature dendritic cells. Mature dendritic cells infected with KSHV do not respond as strongly to cytokines from T_H cells, which prevents normal activation of CTLs. EBV (HHV-4) can evade T cell activation in general by encoding a soluble protein that prevents the TCR and MHC-II from interacting correctly. Finally, herpesviruses interfere with activation of CTLs through manipulating the cytokine interleukin (IL-10). One of the functions of IL-10 is to suppress immune cell activity at the end of an infection so that formerly infected tissues can go back to normal. Some herpesviruses stimulate host cells to secrete IL-10, and others encode viral proteins that mimic IL-10. The effect in either case is to suppress the activation of CTLs.

15.14 Viruses that cause persistent infections evade immune clearance for a long period of time

Persistent viruses that cause long-term infections must evade being eliminated from the body for weeks, months, years, or even decades, often for the entirety of an infected person's life (see **Chapter 13**). Examples include situations in which viral replication is ongoing over months or years as well as latent infections in which viral nucleic acids persist in the body but viruses are rarely produced. Persistent infections with ongoing lytic replication include some forms of hepatitis and infection with HIV. Hepatitis B virus (HBV) and HCV, which infect hepatocytes and cause liver damage, evade clearance by virus-specific lymphocytes for years even though infected people make such cells in abundance. The molecular and cellular biology of how HBV and HCV resist immune clearance remains poorly understood. It is clear that damaged liver tissue at the site of virus replication contains many natural killer (NK) cells and neutrophils, yet those cells are not effective in eliminating the virus. Indeed, the inflammation that they both provoke and reflect can be detrimental and lead to cancer (see **Chapter 13**). HIV directly infects T_H cells and APCs where the virus directs lytic replication. By infecting and killing immune cells, HIV interferes with most aspects of adaptive immunity. Furthermore, HIV has tremendous antigenic variation, resulting in evasion of adaptive immune responses including humoral immunity.

HHV-1 (or HSV1), HHV-2 (or HSV2), HHV-3 (or VZV), and HHV-4 (or EBV) cause lifelong latent infections in which lytic viral replication is rare and episodic and the viruses persist in tissues primarily as genomes expressing only a small number of viral RNAs, proteins, or both (see **Chapter 13**). Most adults are infected with more than one different herpesvirus, demonstrating the success of latency as a strategy for evading the immune system. HHV-1, HHV-2, and HHV-3 use lytic replication in epithelial cells to access the underlying neurons, which they subsequently infect. They are then maintained as latent infections in a neuron and can persist there for the duration of the infected person's lifetime. During latency, they make none or very few viral proteins so that there are few if any endogenous viral antigens available for loading onto MHC-I molecules. Moreover, the nervous system is a privileged site in the body where conditions are intrinsically anti-inflammatory because inflammation that damages the spinal cord or brain would be very harmful to the infected person.

HHV-4 (EBV) establishes a latent infection not in neurons but instead in naive B cells and then causes those B cells to develop further in the absence of normal stimulation by T_H cells, ultimately becoming long-lived memory B cells in which the virus persists for the lifetime of an infected person. The two viral proteins expressed in latently infected B cells are EBNA1 and LMP2A (see also **Chapter 13**). EBNA1 has many functions during infection; one of them is to interfere with the proteasome so that MHC-I presentation of antigens is impaired (**Figure 15.28**; see **Chapter 14**). EBNA1 protein is about 640 amino acids long and has a 230-amino-acid region where the primary sequence is composed exclusively of glycine and alanine. This region of the protein binds to the proteasome so that the proteasome cannot pull the protein through its chamber and degrade the protein. The effect is to inhibit endogenous epitope presentation in general because any proteasomes that attempt to degrade EBNA1 become nonfunctional. The transmembrane LMP2A protein evades the adaptive immune response through a different mechanism (see Figure 15.28). The LMP2A strategy relies on the fact that, even though many antigenic peptides are produced during the process of loading MHC molecules in general, only a subset of these peptides end up presented on the cell surface as epitopes. The selection process is not completely understood, but the immune response mounted by any given person is skewed toward the immunodominant epitopes that are selected for display. The response to immunodominant peptides interferes with developing responses against other epitopes, which are therefore termed subdominant. Through an unclear molecular mechanism, the LMP2A peptides are subdominant to normal endogenous peptides, leading to immune evasion despite the expression of LMP2A by infected cells.

Figure 15.28 During latent infection, HHV-4 (EBV) evades immune detection of EBNA1 and LMP2A. The proteasome is unable to degrade EBNA1, so that EBNA1 epitopes are never loaded into MHC-I. LMP2A can be degraded by the proteasome, but its epitopes are subdominant to cellular peptides and are rarely selected for display because they are subdominant. (ER, endoplasmic reticulum.)

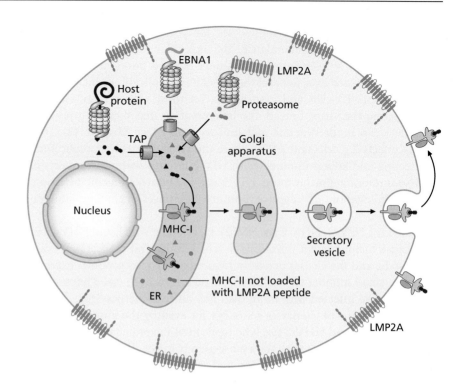

15.15 The immune response to influenza serves as a comprehensive model for antiviral immune responses in general

The immune response to infection with lytic influenza A virus (IAV) provides a comprehensive example that includes viral evasion strategies (**Figure 15.29**). These evasion strategies are very effective, as illustrated by the success of IAV as a pathogen. One measure of evolutionary success is the number of people infected by a virus. Every year, influenza infects between 350 million and 1.75 billion people. Such a high number of infected people indicates that the virus evades the immune response long enough for high levels of replication and spread to new hosts. In more than 99% of cases, the immune system prevails and the infected person survives the infection. IAV thus provides a well-understood example of a successful immune response in the face of temporarily effective viral evasion strategies.

The immune system reacts against IAV proteins and nucleic acids (see Figure 15.29). IAV is an enveloped Class V (−) ssRNA virus that has a segmented genome and replicates in the nucleus (see **Chapter 6**). The genome segments are always associated with proteins, forming viral ribonucleoprotein (vRNP). IAV has small coding capacity, encoding just 12 proteins. Structural proteins include matrix proteins M1 and M2, the nucleoprotein NP, and the two glycoprotein spikes hemagglutinin (HA) and neuraminidase (NA). Three more IAV proteins are found in the vRNPs and are essential for synthesis of mRNA, antigenomes, and genomes: PB1, PB2, and PA. The genome also encodes four nonstructural proteins: NS1, NS2, PA-X, and PB-F2.

IAV has mechanisms for evading both the innate immune response and the adaptive immune response. Immune system control of IAV relies upon several aspects of the innate immune response including pattern recognition receptors (PRRs) binding to PAMPs, phagocytes, and NK cells (see Figure 15.29). Influenza A replicates in respiratory epithelial cells, where viral

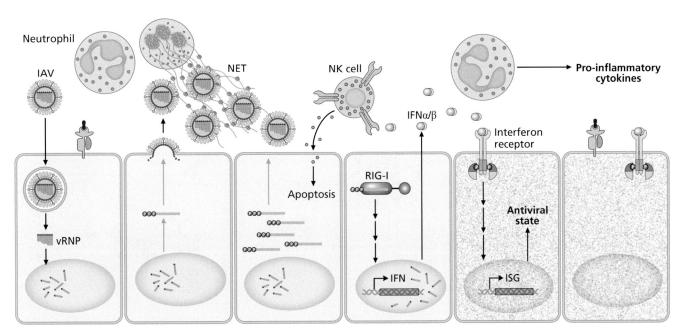

Figure 15.29 Innate immune responses to IAV. During infection of a respiratory epithelium there are many innate immune responses, as illustrated by IAV infection. Neutrophils are the first immune cells to respond, and they produce neutrophil extracellular traps (NETs) and cytokines. Loss of MHC-I on the cell surface triggers infected cells to undergo apoptosis. Cytoplasmic viral RNA that occurs as a result of viral genome replication binds to RIG-I, inducing a signal transduction cascade that causes synthesis and secretion of IFNα/β. Extracellular IFNα/β triggers nearby uninfected cells to enter an antiviral state. Interferon attracts neutrophils to the site of infection, which in turn cause more inflammation, attracting more neutrophils and other cells such as NK cells to the infected tissue. Blue arrows indicate viral replication. (ISG, IFN-stimulated gene.)

nucleic acids such as its genome trigger signal transduction by PRRs. The dominant PAMP detected by PRRs during influenza infection is cytoplasmic viral RNA that has a triphosphate group at its 5′ ends. This PAMP is detected by RIG-I family receptors, causing signal transduction that induces the infected cells to synthesize and secrete interferon (IFN)α/β (see **Chapter 14**). Subsequently, the secreted IFNα/β triggers nearby uninfected cells to enter an antiviral state that will limit the spread of influenza through the epithelium. IFN is pro-inflammatory, so IFN signaling also induces inflammation that attracts neutrophils to the site of infection, which in turn cause more inflammation in a positive feedback loop (see **Chapter 14**). The neutrophils release neutrophil extracellular traps that interfere with viral spread. NK cells are also important for controlling influenza, as can be demonstrated by comparing infection in normal versus genetically modified animal models in which the mutants lack normal NK cell functions. Animals lacking NK cells quickly succumb to influenza infection.

IAV also provokes a strong adaptive immune response (**Figure 15.30**). The lungs where IAV replicates have resident APCs known as alveolar macrophages, which trigger a strong T_H response. The T_H cells activate B cells; antibodies directed against the HA and NA spikes are critical for clearing the infection. Antibodies that bind to HA neutralize and opsonize the virions, which are subsequently ingested and destroyed by APCs such as alveolar macrophages. Additionally, antibodies that bind to NA inhibit its enzymatic activity, which is to degrade the sialic acid sugar components of glycoproteins. When NA is inhibited, the virus cannot escape from the surface of infected cells. Antibodies that bind to either spike protein on the surface of infected cells direct the complement to form a MAC on those infected cells,

Figure 15.30 Adaptive immune responses to IAV. Antibodies that bind to HA neutralize (1) and opsonize (2) the virions, increasing phagocytosis and destruction of the virions. Antibodies that bind to NA inhibit its ability to degrade the sialic acid component of mucus, trapping the viruses in the mucus (3). Antibodies that bind to spike proteins on the surface of infected cells direct the complement to form a MAC on the infected cell, which lyses it (4). CTLs detect viral antigens in MHC-I and trigger infected cells to undergo apoptosis before the virus completes its replication cycle (5).

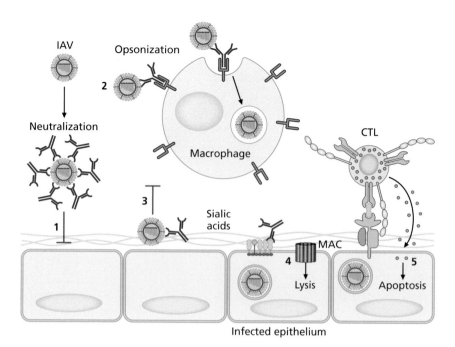

which kills the infected cells. IAV also causes a strong CTL response and the CTLs trigger infected cells to undergo apoptosis before the virus completes its replication cycle. Alveolar macrophages subsequently ingest the nonviable fragments of the apoptotic host cells and degrade immature virus particles (see Figure 15.30).

Although in most cases the immune reactions control the infection, sometimes a strong immune reaction contributes to mortality during fatal influenza infections. When influenza replication damages the respiratory epithelium directly, the inflammatory response and alveolar macrophages also cause collateral damage to the epithelium. If too much of the epithelium is damaged, the damage impairs the function of the lungs, which is to obtain O_2 from the air and release the waste product CO_2 into the air. It is thought that the 1918–1920 pandemic influenza strain had an unusually high mortality rate among otherwise healthy adults because the immune reaction it provoked caused too much tissue damage to the lungs, which interfered with breathing.

15.16 Influenza provides a model for how a lytic virus evades both innate and adaptive immunity long enough to replicate

Most people infected with influenza survive when the innate and adaptive immune defenses collaborate to clear the virus from their bodies before these defenses, the immune reaction, or both do irreparable harm to the lungs. Nevertheless, the virus has many immune evasion strategies that delay the effectiveness of these defenses and that contribute to cases in which IAV infection is fatal. Both the innate and adaptive responses are critical for controlling an IAV infection and, in order to be comprehensive, we will discuss all of them here. The fastest host responses are the innate ones, and IAV must overcome these defenses in order to persist in the body long enough to replicate. IAV strategies to evade the innate responses include inhibition of PRR signaling that would otherwise result in IFNα/β secretion by infected cells, evasion of the antiviral state in newly infected cells that had previously been stimulated by IFN, interference with IFN receptor signal transduction, and evasion of the NK cell response.

The use of multifunctional proteins to evade immune responses is common for viruses with small coding capacity and is exemplified by IAV evasion of innate immune responses (**Figure 15.31**). The multifunctional viral NS1 protein blocks several different steps in PRR signaling, and the PB2 protein plays dual roles as a nucleic acid synthesis protein and as a protein that blocks PRR signaling. The viral NP protein reduces the amount of PRR signaling by encapsidating newly synthesized viral RNA when it is still in the nucleus, reducing the binding of cytoplasmic RNA to RIG-I. Because most PRRs are cytoplasmic, nuclear replication may itself be viewed as an anti-immune system strategy. The viral matrix protein M2 plays direct roles in viral uncoating and in blocking protein kinase R (PKR) activation in cells that are attempting to achieve the antiviral state.

Influenza strategies to evade the adaptive responses include evasion of both the antibody and CTL responses. For example, the multifunctional NS1 protein interferes with dendritic cell maturation and function, which thereby disrupts lymphocyte activation. The major mechanism of evading the adaptive response is antigenic variation, namely changes in the primary sequence of viral proteins. Antigenic variation arises because the viral polymerase makes approximately one error for each complete genome replicated. Every infected cell releases 10,000 viruses, some of which have missense mutations that affect the coding sequence of the proteins. Of these, some change the amino acids found in the epitopes that had been used earlier in the infection to activate T_H cells, B cells, and CTLs. Some of the major epitopes that activate a CTL response are found in the NP. There are more frequent amino acid substitutions in the antigenic epitopes of the NP than there are in the parts of the protein that are not typically displayed as epitopes, indicating that the immune system apparently selects for these variants because they escape the CTL response.

As is evident from the influenza A case study, even a lytic virus with small coding capacity can have profound effects on the immune response. Study of viruses will undoubtedly continue to make fundamental contributions

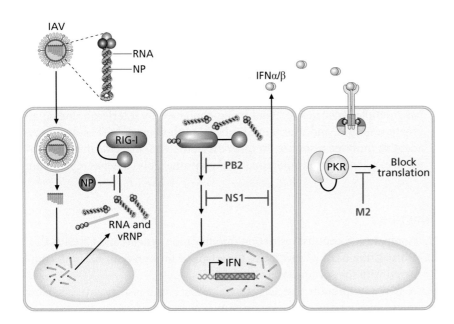

Figure 15.31 IAV evasion of innate immune responses. Viral NP reduces the amount of PRR signaling by encapsidating newly synthesized viral RNA, which interferes with RIG-I binding to RNA. Viral NS1 and PB2 both block PRR signaling in infected cells. Viral protein M2 blocks PKR activation in cells that are attempting to achieve the antiviral state.

not only to virology itself but also to the closely related discipline of immunology and to the intentional manipulation of the human immune system for the public good. This topic, specifically immunization, will be covered in **Chapter 16**, as will other medical applications of virology.

Essential concepts

- Adaptive immune responses develop after exposure to a virus, react to epitopes encoded by that virus rather than to PAMPs, and have memory.
- CRISPR-Cas is a form of adaptive antiviral immunity found among bacteria. Some bacteriophages can evade the system by encoding proteins that disrupt the function of the Cas proteins.
- In humans, viral infections trigger multiple components of the acquired immune system, provoking both a cell-mediated response (T_H lymphocytes and CTLs) and a humoral response (antibodies). T_H cells regulate immune responses against viruses, CTLs kill cells that have been infected by viruses, and antibodies neutralize viruses, allow viruses to be opsonized, and direct the complement to kill infected cells.
- Professional APCs engulf virions and present their component parts as exogenous epitopes in MHC-II molecules, where they are used to activate T_H lymphocytes. Because T_H lymphocytes regulate all other lymphocytes, MHC-II presentation is critical for all adaptive immune responses. Viral antigenic variation is an evolutionary response to MHC-II antigen presentation and antibodies.
- Lymphocytes can be naive or activated. Naive cells are already specialized in that they express many copies of a single TCR or BCR, but they have never been exposed to an antigen that could activate them. Activated lymphocytes have been exposed to an epitope that matches their TCR or BCR and have enhanced activities such as killing infected cells or secreting antibodies.
- Memory lymphocytes develop at the end of an immune reaction and persist in the body for many years; when they encounter the antigen that first stimulated their ancestors, they respond faster and more strongly to control the infection, usually before the person experiences symptoms. Memory lymphocytes are the cellular basis of immunization.
- Viruses that cause persistent infections must manipulate immune responses in order to persist; viral latency can be viewed as an extreme form of immune evasion.
- Influenza A virus is very successful in that it infects millions of people every year and replicates to high levels in most of them, thus indicating success at evading both the innate and adaptive immune responses. IAV has multiple strategies to avoid being contained by interferon during an innate response, and it also evades both antibodies and CTLs during an adaptive immune response. In most cases, the viral proteins that contribute to evasion are multifunctional and play other roles in virus replication.

Questions

1. Compare and contrast restriction–modification systems and CRISPR-Cas.
2. What is the basis of specificity for adaptive immunity in bacteria compared with the specificity in vertebrates?
3. A new Cas protein that has RNase activity was just discovered. How might such a protein provide immunity against viruses?

4. A new Cas protein that has a predicted domain structure similar to reverse transcriptases has been discovered. What role might such a protein play in protection against bacteriophages?

5. List several important differences between PAMPs and antigens.

6. How are PAMPs and antigens similar?

7. Virions are intracellular parasites, yet APCs and T_H cells are essential for controlling a viral infection. Explain.

8. How do professional antigen-presenting cells form a bridge between innate and adaptive immunity?

9. How are antibodies useful in clearing a viral infection?

10. Patients with X-linked agammaglobulinemia have abnormal B-cell development and produce fewer antibodies than normal in response to an infection. They are more susceptible than other people to specific viruses, such as poliovirus and rotavirus, which infect people through the gastrointestinal tract. What do these observations suggest about the adaptive response to poliovirus and rotavirus in people who do not have this disorder?

11. Compare and contrast the TCR and BCR.

12. Provide an example of a multifunctional viral protein that plays a role in gene expression or genome replication and in evading the immune response.

Interactive quiz questions

In addition to the questions provided above, this edition has a range of free interactive quiz questions for students to further test their understanding of the chapter material. To access these online questions, please visit the book's website: www.routledge.com/cw/lostroh.

Further reading

Bacterial CRISPR-Cas

Barrangou R, Fremaux C & Deveau H 2007. CRISPR provides acquired resistance against viruses in prokaryotes. *Science* 315:1709–1712.

Gao L, Altae-Tran H, Böhning F, Makarova KS, Segel M et al. 2020. Diverse enzymatic activities mediate antiviral immunity in prokaryotes. *Science* 369(6507):1077–1084.

Jinek M, Chylinski K, Fonfara I, Hauer M, Doudna JA & Charpentier E 2012. A programmable dual-RNA–guided DNA endonuclease in adaptive bacterial immunity. *Science* 337(6096):816–821.

Rauch BJ, Silvis MR, Hultquist JF, Waters CS, McGregor MJ et al. 2017. Inhibition of CRISPR-Cas9 with bacteriophage proteins. *Cell* 168(1-2):150–158.

Genome engineering with CRISPR-Cas

Hu JH, Miller SM, Geurts MH, Tang W, Chen L et al. 2018. Evolved Cas9 variants with broad PAM compatibility and high DNA specificity. *Nature* 556(7699):57–63.

Wang H, Yang H & Shivalila CS 2013. One-step generation of mice carrying mutations in multiple genes by CRISPR/Cas-mediated genome engineering. *Cell* 153:910–918.

Viral evasion of cell-mediated immunity

Duette G, Hiener B, Morgan H, Mazur FG, Mathivanan V et al. 2022. The HIV-1 proviral landscape reveals that Nef contributes to HIV-1 persistence in effector memory CD4+ T cells. *J Clin Inv* 132(7):e154422.

Murer A, Rühl J, Zbinden A, Capaul R, Hammerschmidt W et al. 2019. MicroRNAs of Epstein–Barr virus attenuate T-cell-mediated immune control *in vivo*. *mBio* 10(1):e01941-18.

Viral evasion of humoral immunity

Caniels TG, Bontjer I, van der Straten K, Poniman M, Burger JA et al. 2021. Emerging SARS-CoV-2 variants of concern evade humoral immune responses from infection and vaccination. *Science Adv* 7(36):eabj5365.

Zhou T, Doria-Rose NA, Cheng C, Stewart-Jones GB, Chuang GY et al. 2017. Quantification of the impact of the HIV-1-glycan shield on antibody elicitation. *Cell Rep* 19(4): 719–732.

Latency as an extreme evasion of the immune response

Juillard F, de Miranda MP, Li S, Franco A, Seixas AF et al. 2020. KSHV LANA acetylation-selective acidic domain reader sequence mediates virus persistence. *Proc Natl Acad Sci* 117(36):22443–22451.

Medical Applications of Molecular and Cellular Virology

Vaccine type	Description
Attenuated vaccine	Produced by passaging a virus *in vitro* under environmental conditions not found at the site of replication in the human body or by genetic engineering; provokes a strong immune response.
Inactivated vaccine	Produced by treating a virus with physical or chemical agents that render it unable to replicate; virus provokes an immune response but requires use of an adjuvant.
Subunit vaccine	Produced by manufacturing virus proteins *in vitro*; provokes an immune response but requires use of an adjuvant.

Virus	Characteristics important for this chapter
Influenza A	Exhibits very high antigenic variation; universal vaccine and better antiviral drugs needed.
Human immunodeficiency virus (HIV)	Cause of the AIDS pandemic; exhibits very high antigenic variation; better vaccines needed; combination drug therapy targeting many viral proteins available.
SARS-CoV-2	Cause of current COVID-19 pandemic.
Adenovirus	Commonly used as a gene therapy vector with very high packaging capacity; dsDNA genome does not recombine with host chromosome.
Retrovirus	Commonly used as a gene therapy vector with medium packaging capacity; RNA genome converted into dsDNA provirus before recombining with a host chromosome.
Parvovirus	Commonly used as a gene therapy vector with low packaging capacity; ssDNA genome converted into dsDNA; persists as nuclear episome without recombining with a host chromosome.
Cytomegalovirus	Herpesvirus source of promoter commonly incorporated into gene therapy vectors.

Antiviral drug	Effect on viral enzymes
Competitive inhibitor	Reversible enzyme inhibitor that binds to active site; many antiviral drugs such as protease inhibitors are competitive inhibitors.
Noncompetitive inhibitor	Reversible enzyme inhibitor that binds to enzyme and deforms the active site; some antiviral drugs are noncompetitive inhibitors.

DOI: 10.1201/9781003463115-16

Figure 16.1 Edward Jenner vaccinating a boy. Oil painting by E.-E. Hillemacher, 1884. Jenner is injecting the boy with vaccinia virus, which provides lifelong immunity against the related variola virus (smallpox). This scene was painted 61 years after Jenner's death. (Courtesy of Wellcome Trust. Published under CC BY 4.0.)

Figure 16.2 AIDS protesters urging faster development of anti-HIV drugs. On May 21, 1990, AIDS activist group ACT UP stormed the US National Institutes of Health to protest the slow pace of research, especially to develop treatments that could slow or prevent death from AIDS. (Courtesy of Donna Binder.)

Medical concerns intersect with molecular and cellular virology because better understanding of viruses results in better clinical strategies for preventing, curing, or lessening the negative effects of viral infections. Basic research in virology has led to medical interventions such as vaccination, gene therapy, and antiviral drugs, as well as the possibility of therapeutic uses for CRISPR-Cas.

Historically, the first technology developed to prevent death caused by a virus was vaccination, named for the poxvirus (vaccinia) used in the immunization procedure (**Figure 16.1**). Better knowledge of molecular and cellular virology is necessary to invent vaccines that we still lack, such as one that could prevent human immunodeficiency virus (HIV) infection or one that could protect against all strains of influenza.

Antiviral medications are another important arm of twenty-first century medicine. Antiviral drugs took much longer to develop than the first antibiotics that could cure bacterial infections, perhaps in part because of the success of immunization to prevent the most common serious viral infections. In the 1980s and 1990s, AIDS activists pushed for development of new antiretrovirals (**Figure 16.2**). Research on antivirals for AIDS led to greater interest in developing antivirals to treat other viruses. Antiviral medications work by binding to specific viral proteins and impairing their function. Discovery of new antivirals relies heavily on molecular virology through careful study of viral proteins to select those that would be the best targets for new drugs.

Understanding the molecular and cellular biology of viruses has also contributed to the emergence of gene therapy as a way of treating diseases caused by mutations that negatively affect a person's health. Recombinant viruses used in gene therapy are genetically altered to confer useful properties upon them, such as the ability to deliver a therapeutic gene to a stem cell or to lyse cancer cells.

The last topic addressed in this chapter is the potential use of genome editing to improve human health. Genome editing relies on CRISPR-Cas, a system derived from bacterial adaptive immunity (see **Chapter 15**). The goal of genome editing is to make very precise changes in a genome, and it can be used to modify both cellular and viral genomes. Genome editing could be used to create better gene therapy treatments. So far, CRISPR-Cas treatments for sickle cell disease, β-thalassemia, and certain cancers are in clinical trials and showing promise.

16.1 Vaccines are critical components of an effective public health system

The World Health Organization estimates that vaccines prevent 2.5 million deaths each year. Many of these deaths would otherwise be caused by the viral diseases measles, mumps, rubella, influenza, rotavirus infection, rabies, and yellow fever. A **vaccine** is a medical treatment that provokes an immune response. **Prophylactic** or **preventative vaccines** prevent disease, whereas **therapeutic vaccines** treat an existing infection. Vaccines against human papillomavirus (HPV) and hepatitis B virus (HBV) prevent cervical cancer and hepatocellular carcinoma, respectively. Vaccines against human herpesvirus 3 (HHV-3), formerly known as varicella–zoster virus, prevent not only chicken pox but also the very painful condition called shingles, which can occur when HHV-3 is reactivated from latency. Effective vaccination campaigns have led to the eradication of two dangerous viruses: rinderpest virus, which caused plagues among cattle (and therefore famine among people), and smallpox (variola) virus, which caused plagues among humans.

Poliovirus, which can cause lifelong musculoskeletal problems and is sometimes fatal, may soon be eradicated through the use of effective vaccines and public health measures.

There still remains an urgent need for new or better vaccines to prevent viral infections. There are no vaccines to prevent HIV infection, which untreated leads to fatal AIDS nearly 100% of the time. We need more effective SARS-CoV-2 vaccinations that can prevent disease despite the extreme variability of the viral surface proteins. We lack vaccines to prevent Middle East respiratory syndrome and severe acute respiratory syndrome caused by SARS-CoV-1; otherwise these infections are fatal 10%–50% of the time. We do not have vaccines to prevent cancers caused by hepatitis C virus (HCV), Kaposi's sarcoma herpesvirus (HHV-8), or Epstein–Barr herpesvirus (HHV-4). We need a vaccine against respiratory syncytial virus (RSV, a paramyxovirus) because although the infection is mild in most people, it is very serious in newborn and premature babies; vaccination of people in general would reduce the amount of RSV circulating in the population and thus reduce risk to babies. A vaccine to prevent herpes simplex virus 2 (HHV-2) infection would reduce rates of genital herpes, which in turn might reduce rates for other sexually transmitted infections because genital herpes increases susceptibility to other sexually transmitted infections. Vaccines that protect against viruses that cause common but mild upper respiratory infections could reduce the significant economic and educational losses that occur when people miss work or school. Such viruses include rhinoviruses, adenoviruses, certain coronaviruses, and parainfluenza viruses. **Sections 16.2–16.4** discuss the types of vaccines currently in use before turning to efforts to develop better influenza and HIV vaccines as examples of vaccine development.

16.2   Attenuated vaccines are highly immunogenic because they can still replicate

The three different types of vaccines are attenuated, inactivated, and subunit. They are listed from most to least immunogenic. **Attenuated** vaccines are made from microbes that can still replicate to some extent in the human body but cannot cause a symptomatic infection in healthy people. Attenuation is accomplished by propagating a virus in a nonhuman host or host cell under different environmental conditions than those in the human body. In order to replicate, the virus must acquire mutations that enable replication in the new host under these new conditions. After collecting offspring viruses and repeating the infection many times, called **passaging**, the virus becomes attenuated because it is now adapted to favor infection of nonhuman cells under the new environmental conditions (**Figure 16.3**). Viruses

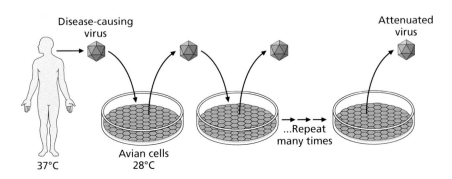

Figure 16.3 Attenuation of a virus by passaging. Viruses are collected from a patient and then propagated in avian tissue culture cells at a temperature cooler than that in the human body. Shed viruses are collected and this procedure is repeated many times. As the virus adapts to the *in vitro* passaging, it becomes attenuated for replication in the human body.

can also be attenuated through genetic engineering that deletes genes needed to cause infection in humans but that are dispensable for growth in culture. Recombinant COVID-19 vaccines comprising genetically engineered non-replicating adenoviruses expressing SARS-CoV-2 antigens are a prominent example of genetically engineered attenuated vaccines. The viral vector delivers a gene encoding part of the viral spike protein into the nucleus of susceptible cells and subsequent expression provokes an immune reaction against the spike. Attenuated vaccines to protect against measles, mumps, rubella, seasonal influenza, chicken pox, polio, rotavirus infection, yellow fever, and rabies are also currently in use.

Most attenuated vaccines replicate slowly in the human body, giving the adaptive immune response time to develop. At the same time, viral replication causes enough destruction that damage-associated molecular patterns (DAMPs) are produced (see **Chapter 14**). These trigger the innate immune system to stimulate the adaptive branch of immunity. Because attenuated vaccines replicate in the body, they provoke long-lasting immunity and require few booster shots to repopulate the pool of memory lymphocytes. For attenuated viruses that can replicate in the body, when administered to millions of people, an attenuated virus like this can cause a small number of vaccinated people to become sick. This can happen because the person has an underlying immune deficiency or had another infection that altered the response to the vaccine, because the attenuated vaccine acquired a new mutation that somewhat restored its ability to replicate in humans, or because the vaccine strain recombined with a natural virus that had co-infected the affected person. Because of these dangers, inactivated vaccines are sometimes favored over attenuated ones.

16.3 Inactivated vaccines are composed of nonreplicating virions

Inactivated vaccines are safer than attenuated ones because the virus in the vaccine has been treated with physical assaults (heat, radiation) or chemicals that render it unable to replicate (**Figure 16.4**). An example is treating poliovirus with formaldehyde, a chemical that forges covalent bonds between capsomeres, preventing the virus from uncoating. An inactivated vaccine provides the immune system with foreign antigens that can provoke a specific response against the virus. Because the inactivated virus cannot replicate, it does not induce cell death and so, although it may be recognized as foreign, it is not recognized as dangerous. Thus, it does not result in a strong innate immune response, which in turn is required for a long-lasting adaptive immune response because of the mechanism by which naive T helper cells are activated (see **Chapter 15**). To overcome this problem, inactivated vaccines contain **adjuvants**, which are chemicals that enhance the immune response by simulating danger. For many years the molecular basis of adjuvants was not understood, but it is increasingly clear that the most commonly used adjuvants result

Figure 16.4 Inactivated vaccines. Inactivated vaccines are produced by treating a disease-causing virus with physical or chemical agents that render it unable to replicate in the human body.

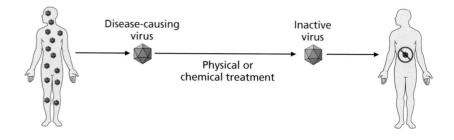

in the release of DAMPs (see **Chapter 14**) in the area of the injection. Although the adjuvant does improve the immune response, inactivated vaccines typically require more boosters over a person's lifetime in order to repopulate the pool of memory lymphocytes. Inactivated vaccines currently in use include one of the two types of polio vaccine and certain COVID-19 vaccines in use in China and Russia. Occasionally, however, there have been instances in which something went wrong during the inactivation process, and people were injected with infectious virions. This danger makes subunit vaccines more attractive because they are even safer, though less immunogenic.

16.4   Subunit vaccines are composed of selected antigenic proteins

Subunit vaccines have some of the component parts of the infectious microbe (**Figure 16.5**). An example in current use protects against HBV. To make the vaccine, the HBV S gene was cloned in yeast, thereby overexpressing the large S spike protein. The S protein was then purified and used in combination with an adjuvant in a vaccine. Subunit vaccines are very safe because they do not originate as infectious microbes, but they do not provoke a very strong DAMP response on their own, so boosters and strong adjuvants are necessary. Another example of a common subunit vaccine is one to prevent COVID-19. This vaccine comprises domains of the viral spike protein expressed in the yeast species *Pichia*. It was developed to be both effective and especially low cost to manufacture.

A variation of subunit vaccination is to immunize someone with mRNA or DNA encoding viral antigens. These are called **mRNA** or **DNA vaccines**, respectively. When RNA is the basis of the vaccine, that RNA is chemically modified (modRNA) to make it more stable in the cytoplasm. The modification delays cellular degradation of the foreign modRNA in the vaccine. The modRNA COVID-19 vaccines encode immunogenic variations of the viral spike protein. The strategy for DNA vaccinations is similar; introducing DNA into someone's body allows some of the DNA to enter the nucleus of some cells, leading to expression of that DNA. Ultimately, both modRNA and DNA vaccines result in the display of foreign endogenous antigens in major histocompatability complex I (MHC-I) molecules, mimicking a viral infection (see **Chapters 14** and **15**). There is one DNA vaccine approved for use in people in India; this vaccine protects against COVID-19. There is also a veterinary DNA vaccine that protects horses from West Nile virus (WNV, a flavivirus; see **Chapter 5**). The DNA encodes the WNV matrix and envelope proteins. Another veterinary DNA vaccine encodes an enzyme that is more common in cancer cells and is used to prolong the life of dogs with melanoma.

Another variation of subunit vaccination is the use of **viruslike particles** (**VLPs**), which are assembled from one or more viral capsomeres but do not contain any genetic material (**Figure 16.6**). VLPs are more immunogenic than

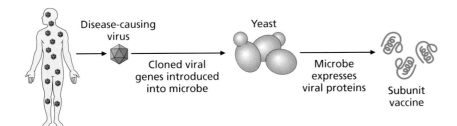

Figure 16.5 Subunit vaccines. Viral proteins can be used as a subunit vaccine. Subunits vaccines are typically produced by overexpressing viral proteins in a microbe such as yeast or bacteria.

Figure 16.6 Viruslike particles. VLPs are assembled from viral capsomeres but do not contain any genetic material.

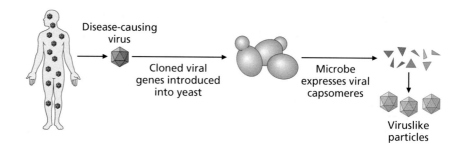

Disease-causing virus

Cloned viral genes introduced into yeast

Microbe expresses viral capsomeres

Viruslike particles

the same capsomeres if they are not first assembled into a VLP. The current prophylactic HPV vaccines are composed of VLPs that contain capsomeres L1 and L2 (see **Chapter 8**).

16.5 Although seasonal influenza vaccines are useful, a universal flu vaccine is highly sought after

Influenza A virus (IAV) is a major pathogen that causes millions of serious infections and 250,000–500,000 deaths every year. As an RNA virus, IAV has an RNA-dependent RNA polymerase (RdRp) with a high intrinsic misincorporation rate, so IAV exhibits high rates of antigenic variation (see **Chapters 5, 15**, and **17**). Furthermore, IAV has a segmented genome in which major antigenic proteins are encoded by different segments (**Figure 16.7**). This genome configuration makes it possible for two different influenza viruses to co-infect the same cell and have recombinant offspring with a new combination of genome segments, which is the origin of most pandemic influenza strains (see **Chapter 17**). Both of these forms of genetic variation are problematic for vaccine development. Moreover, every year manufacturers produce about 500 million doses of vaccine, so that the sheer volume of vaccine required means that vaccine manufacturers must begin many months before flu season. Vaccine manufacturers make an educated guess about which forms of influenza will be circulating during the following flu season. Sometimes their guess is on target, but other times there is a mismatch between the vaccine and the most prevalent form of influenza in any given year. This situation can even cause the alarming effect of discouraging the public from getting vaccinated at all.

It is also possible for influenza to jump from animals to humans. When this occurs, the **zoonotic** influenza strain is at first not adapted to humans.

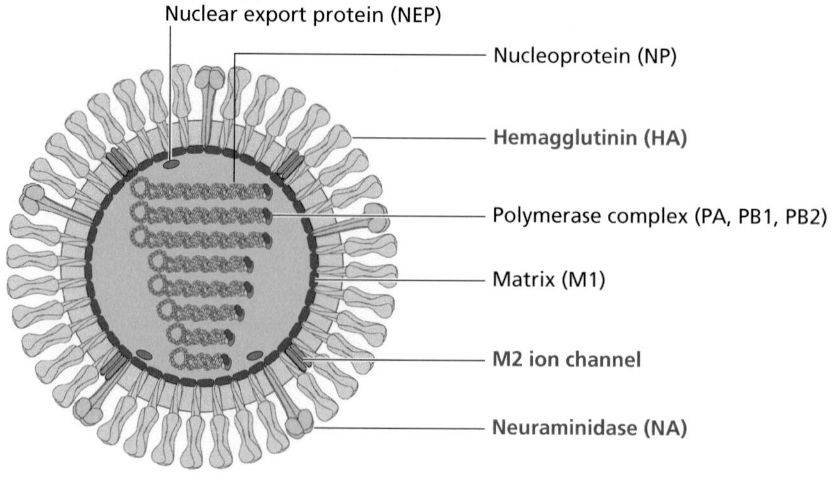

Figure 16.7 Antigenic proteins of influenza A virus. The two spikes are hemagglutinin (HA) and neuraminidase (NA). In addition to the eight genome viral ribonucleoprotein complexes, the virion contains proteins NP, M1, M2, NEP, PA, PB1, and PB2. Influenza vaccines often provoke a strong antibody response against HA and NA; vaccines in development may also employ the M2 protein. Available anti-influenza drugs target the NA and M2 proteins.

Nuclear export protein (NEP)

Nucleoprotein (NP)

Hemagglutinin (HA)

Polymerase complex (PA, PB1, PB2)

Matrix (M1)

M2 ion channel

Neuraminidase (NA)

Although at first it may seem like such a virus would be unable to replicate in humans, sometimes the opposite is true and instead the virus replicates so ferociously that the death rate is much higher than it is for human-adapted influenza strains. H5N1 avian influenza is an example. It has a 60% mortality rate in people but can only be contracted directly from birds; it cannot be transmitted from human to human. In addition to the high mortality rate, another danger with zoonotic influenza is the possibility that with just a few mutations in the right places, the virus might gain the ability to be passed from human to human while still maintaining a higher mortality rate than typical seasonal influenza (which is <0.1%). Reducing the number of people infected by avian influenza is important for limiting the virus's opportunities to evolve the ability to be transmitted from person to person. Zoonotic IAVs that have gained the ability to be transmitted from one human to another are very dangerous because they cause pandemics with high fatality rates. **Gain-of-function experiments** that start with avian viruses and then select those with improved transmission among lab mammals are considered controversial because of the danger that a laboratory-selected strain could escape the lab. The point of these experiments is to determine the genetic changes that might be necessary for an avian strain to become a pandemic so as to predict the likelihood that a pandemic will occur and also to understand the molecular biology of transmission. There is disagreement regarding whether the risks of doing gain-of-function transmission experiments outweigh the public health benefits of the knowledge gained by doing such research.

Thus, a universal influenza vaccine that would protect people and agricultural animals against most forms of IAV including both seasonal and zoonotic strains is desperately needed. One strategy is **rational vaccine design**, which involves studying the immune response to a virus in order to determine which viral antigens are capable of inducing a strong sterilizing adaptive immune response. In this context, sterilizing means that a vaccinated person would be completely protected from acquiring the viral disease. Then these viral antigens are classified as those that exhibit high and those that exhibit low levels of antigenic variation. The best candidates for an effective universal vaccine are the proteins that are capable of evoking a strong adaptive immune response and that also exhibit low antigenic variation. In principle, these are antigens that should provoke an immune response that would protect against all IAVs, enabling development of a universal IAV vaccine.

Candidate universal influenza antigens include the stalk of the influenza hemagglutinin (HA) spike protein (**Figure 16.8**). Unfortunately, it turns out that epitopes in the head are usually immunodominant to the epitopes in the

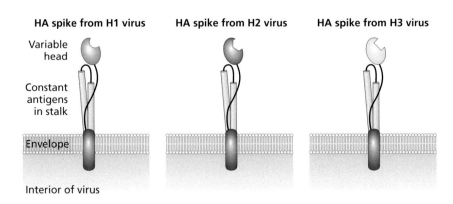

Figure 16.8 Candidate universal antigens are found in the influenza HA stalk. The head of HA exhibits extreme antigenic variation, but the stalk is more constant from one virus to the next. There are nearly universal epitopes in the stalk that might be targeted by a universal influenza vaccine.

stalk, reducing the number of neutralizing antibodies that bind to the stalk (see **Chapter 15**). Yet, the rare individual who can generate antibodies against the stalk is able to neutralize the virion. Such antibodies can also prevent membrane fusion and thereby prevent internalized virions from escaping the endosome. Now that these antibodies have been detected and characterized, a goal is to develop a vaccine that provokes them in everyone by somehow bypassing the immunodominance of the head epitopes in most people. One strategy in development is to use a viral spike protein that normally does not provoke a strong immune response and create a chimera between its globular head and the influenza stalk, thereby avoiding the problem that the HA head epitopes are normally immunodominant. Another idea is to create a recombinant adenovirus that expresses the IAV stalk antigens in such a way that they are processed into potent epitopes.

Recombinant vaccines such as this IAV vaccine are attenuated vaccines in which an attenuated virus has been engineered to express proteins from another species; we have already discussed a COVID-19 vaccine of this type. Adenoviruses are dsDNA viruses that have the capacity to incorporate large amounts of DNA from other sources and so they have long been favored as recombinant vaccine vectors and gene therapy agents (see **Sections 16.16** and **16.17**).

Another candidate universal influenza antigen is the M2 matrix protein. The virus uses this protein during penetration when M2 forms an ion channel that allows acidification to trigger membrane fusion. During viral replication, the M2 protein can be found on the surface of infected cells at the sites of future virion assembly. Antibodies that react to a constant surface-exposed region of the M2 protein bind to the surface of infected cells and induce killing of infected cells by the complement system (see **Chapters 14** and **15**). These same antibodies interfere with viral budding. Anti-M2 antibodies are not neutralizing, however, because M2 is not accessible to antibodies when it is incorporated into infectious virions (see Figure 16.7). Although anti-M2 antibodies cannot prevent influenza infection, they can inhibit replication, which might be useful in combination with other antibodies such as neutralizing ones that bind to the HA stalk. A recombinant adenovirus has a large enough coding capacity in principle to express both HA stalk and M2 antigens at the same time, suggesting a way forward for developing such a vaccine.

16.6   Preventative HIV vaccines are in development

Influenza is not the only globally worrisome virus for which we lack a universal preventative vaccine. For example, approximately 35 million people are currently living with HIV, most of them in sub-Saharan Africa where access to medical care can be limited (**Figure 16.9**). The lifetime risk of acquiring HIV is 4 individuals out of 5 for a black 15-year-old South African girl living in the state of KwaZulu-Natal. In North America, AIDS is the leading cause of death among African American women aged 25–34, and it is the second leading cause of death, behind suicide, for gay and bisexual men. Currently, the US Centers for Disease Control and Prevention predicts that 1 in 51 American men will be diagnosed with HIV in their lifetimes, with higher rates for African American men (1 in 16), Native Hawaiian and Other Pacific Islander men (1 in 33) and Latino men (1 in 36) and extremely high rates for gay African Americans (1 in 2), gay Latinos (1 in 4), and gay white men (1 in 11). A vaccine that could prevent transmission of HIV would be an extremely valuable tool in bringing the AIDS epidemic under better control.

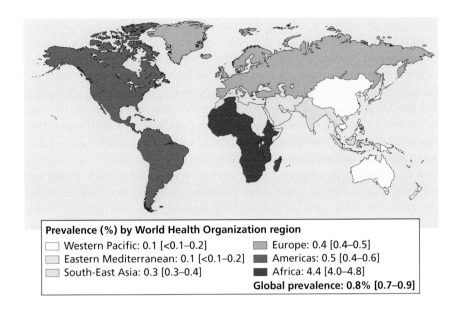

Prevalence (%) by World Health Organization region

☐ Western Pacific: 0.1 [<0.1–0.2]   ▨ Europe: 0.4 [0.4–0.5]
▨ Eastern Mediterranean: 0.1 [<0.1–0.2]   ▨ Americas: 0.5 [0.4–0.6]
▨ South-East Asia: 0.3 [0.3–0.4]   ■ Africa: 4.4 [4.0–4.8]
Global prevalence: 0.8% [0.7–0.9]

Figure 16.9 Global prevalence of HIV infection, 2015. Percent of adults aged 15–49 years with HIV. HIV is 10 times more prevalent in sub-Saharan African than in the region with the next-highest prevalence, the Americas. But the map does not show the significant variability in prevalence rates within different populations. (From Adult HIV prevalence [15–49], 2015 By WHO Region [Map]. World Health Organization, 2016.)

When HIV was isolated and shown to be the cause of AIDS in the early 1980s, prospects for an effective vaccine seemed promising. At that time, it had been less than a decade since vaccination had led to the eradication of natural smallpox infections, and there were many safe, effective attenuated vaccines to prevent infections that used to kill people during childhood. No one could have foreseen that experiment after experiment based on past vaccine development strategies would fail. It wasn't until a trial that lasted from 2003 to 2006 and was published in 2009 that an HIV vaccine candidate showed any efficacy at all. This RV144 HIV vaccine trial resulted in an HIV infection rate that was 31% lower in people who received the vaccine than in people who received a placebo. By comparison, most commercially available vaccines result in a rate of infection that is at least 50% lower in vaccinated people than in unvaccinated people, leaving room for considerable improvement.

Preventative HIV vaccines must use one or more of the 15 viral proteins to provoke a protective response; the RV144 vaccine relies on several of these proteins. As a reminder, HIV-1 (**Figure 16.10**) is an enveloped virus that encodes two polyproteins: Gag and Gag-Pro-Pol (see **Chapter 10**). The Gag polyprotein is the precursor for the matrix, capsid, nucleocapsid, and p6 proteins, and the longer Gag-Pro-Pol polyprotein is the precursor for all of the Gag proteins and also for the protease, integrase, and reverse transcriptase enzymes. The virus spike proteins are encoded by the *env* gene and are known as gp41 and gp120.

In the RV144 trial, Thai volunteers were immunized with six treatments over 6 months. They were first immunized with four injections of a recombinant canarypox ALVAC-HIV vector that was engineered to express HIV Gag, protease, and Env proteins. Canarypox is a type of poxvirus that functions like an attenuated vaccine in human beings (see **Chapter 8**). It can enter mammalian cells where some viral proteins are expressed, but no infectious virions are produced because replication cannot proceed past an early phase of the replication cycle. Canarypox vectors can be produced in cultured avian cells. The details of the ALVAC-HIV vector's exact genome are owned by a pharmaceutical company and so are not available, so we do not know exactly how the genetic engineering was accomplished. In the RV144 trial, volunteers then received two more injections, each containing the recombinant

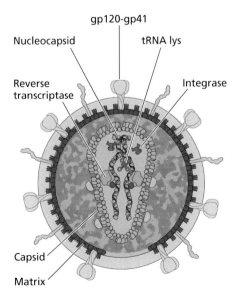

Figure 16.10 Major proteins of HIV-1.

canarypox as well as a subunit vaccine consisting of purified gp120 spike protein with an adjuvant. The theory behind this regime is that the subunit vaccine would provoke a strong antibody response, whereas the recombinant canarypox would provoke a strong cytotoxic T lymphocyte (CTL) response. Vaccines that protect against other viruses cause both neutralizing antibodies that prevent infection and a strong CTL response that clears infection caused by the few viruses that evade the neutralizing antibodies. Variations on this vaccination strategy are being developed and tested, for example, using a different adjuvant and vaccination schedule.

A challenge in developing a preventative vaccine against HIV is that the immune system almost never cures a natural HIV infection. For example, although infected people have many HIV antibodies in their systems, those antibodies do not eliminate HIV from the body. In contrast, for polio, measles, mumps, rubella, chicken pox, and all other viral infections for which we have effective vaccination, surviving a natural infection leads to development of an immune response that removes the virus from the body. Furthermore, in these cases, natural infection protects that person from a second infection for most of the rest of their life. But, there are a very small number of people who have been exposed to HIV over and over again and yet remain HIV negative (HIV−); studying them may lead to better anti-HIV vaccination strategies. In some cases, these people are resistant to being infected with HIV because they have a naturally occurring variation in a cellular co-receptor for attachment and penetration, but in other cases, they have a very rare and effective sterilizing immune response to HIV even though almost no one else does. Comparing these unusual individuals to the RV144 trial's volunteers has led to certain insights about why it has been so difficult to develop a preventative HIV vaccine, as covered in **Section 16.7**.

16.7   Extreme antigenic variation is a problem for developing an HIV vaccine

From studying animal models, volunteers who participated in vaccine trials, and rare highly exposed people who remain HIV−, we have learned that HIV vaccination is challenging because of the extreme antigenic variability of the virus (see **Chapter 13**). An untreated chronic HIV infection results in 10^9–10^{10} new virions every day, and each of them has one or two nucleotide substitutions relative to its parent. Thus, any population of HIV viruses is a collection of viruses with antigenic variation. Because of such extensive antigenic variation, antibodies that react to epitopes in a vaccine are not always able to neutralize natural isolates of HIV, especially during natural infections where someone likely gets infected by a population of tens of thousands viruses rather than by a single one. Nevertheless, the best hope for an effective vaccine may be to trigger in vaccinated people the production of **broadly neutralizing antibodies**, which are defined as antibodies able to neutralize a virion despite antigenic variation. Broadly neutralizing antibodies are provoked by all known effective antiviral vaccines.

About 20% of people infected with HIV develop broadly neutralizing antibodies, but they only do so after several years of infection, when it is too late for the antibodies to eliminate the virus from the body because of latently infected cells. The VRC01 antibody is an example of a broadly neutralizing anti-HIV antibody. It binds to a relatively invariant part of the gp120 spike protein, partially mimicking the shape of the HIV receptor CD4. Clinical trials to test whether infusion of this antibody can prevent or treat HIV are underway. The time it takes for neutralizing antibodies to develop during a natural infection indicates that the B cell developmental program needed to produce these antibodies is slow and may be more complex than typical B cell development (see **Chapter 15**). Could a vaccine trigger the development of these B cells in more people and more quickly, so that the vaccine would produce broadly neutralizing antibodies fast enough to clear the body of HIV and prevent latent infection? Recent work has also discovered more HIV epitopes that can trigger formation of broadly neutralizing antibodies, at least in some people (**Figure 16.11**). Discovery of these epitopes may enable vaccines to target them specifically or to target several different epitopes at the same time.

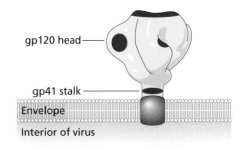

Figure 16.11 Epitopes on the HIV gp41–gp120 spike. Epitopes (red circles) that bind to broadly neutralizing antibodies are found on both the gp120 head and the gp41 stalk.

16.8 An effective HIV vaccine may require stimulating a strong CTL response

It is probably most effective for a vaccine to elicit both antibody and CTL responses, as occurs in natural viral infections that are eliminated (cleared) from the human body. Development of recombinant HIV vaccines focused on eliciting a CTL response was slowed by trials in 2007 that ended early because the experimental vaccines showed no efficacy. Retrospective analysis showed that the vaccines actually increased a volunteer's risk of developing HIV, possibly by activating the immune system so that there were more host cells available to HIV when it entered the volunteer's body. Stimulating a CTL response might still be an important component of a preventative vaccine, however, because highly exposed but uninfected people have a strong CTL response against HIV antigens. Therefore, a variety of recombinant vectors and vaccination schedules are in development to elicit CTL responses against HIV. The idea is to use vectors derived from viruses known to provoke a CTL response such as herpesviruses and poxviruses. Recent results using a recombinant herpesvirus engineered to express many HIV proteins did not prevent infection of experimental animals in the first place but enabled about 50% of them to clear the virus after several months. These animals had a strong CTL response against the retroviral antigens. Vaccines derived from an attenuated vaccinia virus may also cause a stronger or more enduring CTL response than the existing canarypox vaccine and so may be better than the recombinant vector used in the RV144 trial.

It may be that cautious optimism about HIV vaccines is finally warranted almost 40 years into the AIDS pandemic. This optimism occurs at the same time that highly effective drug regimens allow HIV+ people to live a nearly normal life span. Furthermore, some of these same drugs can be used to treat someone after acute exposure in order to prevent infection or as a daily preventative medication that protects against infection. These innovations in the use of pharmaceuticals came about in order to find preventative measures despite the absence of an effective vaccine. Antiviral drugs including these are the topic of **Sections 16.9–16.14**.

16.9 Antiviral drugs target proteins unique to viruses and essential for their replication cycle

The study of the molecular biology of viruses has been essential for drug development because antiviral drugs bind to and inhibit viral proteins. In the ideal situation, an antiviral drug does not bind to any normal human proteins, though in practice this goal is rarely achieved. Instead, a good antiviral drug binds to its target virus protein more tightly than it binds to human proteins so that, at levels of the drug achieved in the body, most of the drug binds to

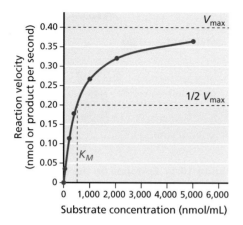

Figure 16.12 Michaelis–Menten enzyme kinetics. The Michaelis–Menten equation describes the line formed by measuring v (the rate of product formation) when the amount of enzyme is held constant and the amount of substrate is varied. V_{max} is extracted from the graph by finding the line approached by the asymptotic curve. K_M is extracted from the graph by finding the substrate concentration when the velocity is equal to $\frac{1}{2}V_{max}$.

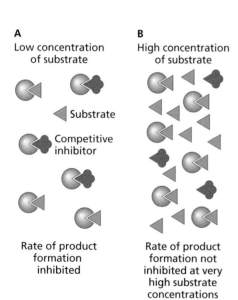

Figure 16.13 A competitive inhibitor binds reversibly to the active site of an enzyme. (A) The enzyme's normal substrate binds to the active site. A competitive inhibitor also binds to the active site. (B) At very high concentrations of substrate, the competitive inhibitor has no effect.

viral proteins. Viral proteins that play an essential role at any point during replication are particularly good candidate targets because inhibiting them slows viral replication. Examples include fusion proteins, proteases, and polymerases. Effective antiviral drugs also have good pharmacological properties. The pharmacological properties of a drug include whether the drug has side effects, whether it can be taken orally, how long it persists in the human body, and how the human body breaks down or otherwise modifies the drug. These properties are not typically predictable using molecular biology and so they will not be discussed further. Instead, this chapter focuses on the molecular biology of drugs that target viral polymerases, drugs to treat influenza, drugs to treat hepatitis viruses, and drugs to treat or prevent HIV infection.

Most antiviral drugs bind to viral enzymes. Antiviral drugs slow an enzyme's kinetics, also known as the rate of product formation. Enzyme activity is described by the following reaction, where E is the enzyme, S is the substrate, ES is the enzyme–substrate complex prior to the formation of product, and P is the product:

$$E + S \underset{k_r}{\overset{k_f}{\rightleftharpoons}} ES \xrightarrow{k_{cat}} E + P$$

The constants k_f, k_r, and k_{cat} are unique to specific enzymes, substrates, and products, and are determined experimentally. They can be affected by a drug that binds to the enzyme.

Antiviral drugs that target enzymes interfere with the rate of product formation, which can be studied using the Michaelis–Menten equation. The Michaelis–Menten equation describes how the rate of product formation is related to the concentration of substrate, k_f, k_r, and k_{cat}. The equation states that the velocity of the reaction (v), also known as the rate of product formation ($d[P]/dt$), is related to the maximum velocity of product formation (V_{max}), the substrate concentration ([S]), and the substrate concentration at which the enzyme reaches half its maximum velocity (K_M). K_M is a mathematical combination of the rate constants but is easier to measure than the constants themselves.

$$v = \frac{d[P]}{dt} = \frac{V_{max}[S]}{K_M + [S]}$$

The V_{max} and K_M values are obtained experimentally by holding the amount of [E] constant, varying [S], and measuring reaction velocity. The data are then plotted to see velocity as a function of substrate concentration, and the values V_{max} and K_M are extracted from the graph (**Figure 16.12**). V_{max} is the velocity attained at the highest concentration of substrate. It is proportional to the amount of enzyme in the reaction, because more enzyme molecules make the reaction faster. K_M is determined by finding the point at which the velocity is $\frac{1}{2}V_{max}$. The substrate concentration corresponding to $\frac{1}{2}V_{max}$ is K_M.

There are three general classes of reversible enzyme inhibitors, classified according to whether they affect K_M, V_{max}, or both, which in turn reflect the physical mechanism with which the inhibitor interacts with E or ES. **Reversible inhibitors** bind to enzymes but do not covalently modify them. Many antiviral drugs belong to the first class of reversible enzyme inhibitors, known as competitive inhibitors. **Competitive inhibitors** bind to the active site of an enzyme, where they compete with the normal substrate for binding (**Figure 16.13**). They result in an increased K_M, which can be observed by holding the concentration of enzyme and drug constant, varying [S], and measuring velocity (**Figure 16.14**). Higher amounts of substrate overcome the effect of the drug, allowing the enzyme to achieve V_{max}, as normal.

Other antiviral drugs are **noncompetitive inhibitors**. Noncompetitive inhibitors do not compete with substrate for binding to the active site. Instead, they bind elsewhere on the target enzyme and cause a conformational change that alters the active site so that the normal substrate cannot bind (**Figure 16.15**). The effect is to take some of the enzyme out of commission, essentially lowering the concentration of active enzyme. Noncompetitive inhibitors lower the V_{max} of the enzyme because the rate of product formation is proportional to the concentration of E; higher amounts of enzyme increase the rate of product formation (**Figure 16.16**). The active sites that are available, however, are completely normal and so the reaction has the same K_M as the uninhibited enzyme (**Figure 16.17**). In contrast to a competitive inhibitor, at a constant concentration of enzyme and drug, raising the substrate concentration [S] cannot overcome the effects of a noncompetitive inhibitor (see Figure 16.15). The third class of enzyme inhibitors are **uncompetitive**, meaning that the inhibitor binds to the ES complex and not to E. Uncompetitive inhibitors lower both V_{max} and K_M. Few if any antiviral drugs are uncompetitive inhibitors.

Antiviral drugs that are reversible enzyme inhibitors slow the rate of product formation. Because the virus requires the product to complete its replication cycle, a lack of product is obviously detrimental to viral replication. For example, consider an antiviral drug that inhibits viral RNA synthesis. With slower RNA synthesis, viral gene expression is delayed, slowing the entire replication cycle. Reversible inhibitors also lead to an accumulation of substrate, which can itself be useful to the human body, especially when the viral enzyme's substrate is a viral protein that can trigger an immune response. For example, imagine that a drug's target is a protease that should cleave a viral polyprotein. An antiviral that competes for the active site of the protease delays processing of the polyprotein. Because of the delay, the polyprotein substrate is available to be targeted to the cell's proteasome for a longer period of time, ultimately resulting in more viral epitopes displayed in MHC-I on the cell's surface (see **Chapter 14**). Any delay in the replication

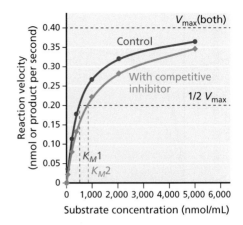

Figure 16.14 A competitive inhibitor increases K_M. In this experiment, the amount of enzyme and inhibitor are held constant while the amount of substrate is varied. A competitive inhibitor increases the substrate concentration necessary to achieve a velocity of ½V_{max} so that the control K_M 1 increases to K_M 2 in the presence of the competitive inhibitor.

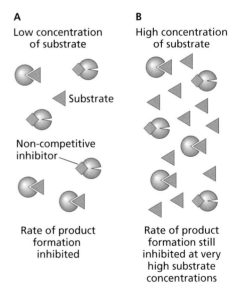

Figure 16.15 A noncompetitive inhibitor binds reversibly to an enzyme. (A) The enzyme's normal substrate binds to the active site. A noncompetitive inhibitor binds elsewhere to the protein, deforming the active site. (B) At very high concentrations of substrate, the competitive inhibitor still reduces the effective concentration of the enzyme, which reduces the velocity of product formation.

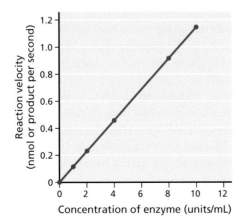

Figure 16.16 The velocity of an enzyme-catalyzed reaction depends on the concentration of enzyme. In this experiment, the amount of substrate is held constant and the amount of enzyme is varied. The concentration of substrate is very high (at least 10 times K_M). The velocity of product formation is proportional to the amount of enzyme in the reaction.

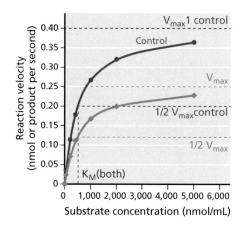

Figure 16.17 A noncompetitive inhibitor decreases V_{max}. In this experiment, the amount of enzyme and inhibitor are held constant while the amount of substrate is varied. A noncompetitive inhibitor (blue) decreases V_{max} without affecting K_M.

cycle also gives the immune system a longer period of time to recognize and kill an infected cell, and to produce neutralizing antibodies.

The different ways that antiviral drugs inhibit viral polymerases provide good examples of competitive and noncompetitive inhibition and are addressed in Sections 16.10, 16.12, and 16.13.

16.10   Many antiviral drugs are nucleoside or nucleotide structural analogs that target the active site of viral polymerases

Many antiviral drugs on the market are competitive inhibitors of viral polymerases. They are **structural mimics** of nucleosides, nucleotides, or nucleotide triphosphates. An example is **azidothymidine**, also known as **AZT**, which is a mimic of thymidine (**Figure 16.18**). It has a similar overall shape and chemistry to thymidine, yet is different enough that it cannot substitute for normal thymidine during DNA synthesis. AZT was the first antiviral drug commercialized in order to delay HIV replication and was the primary drug used to prevent mother-to-child transmission of HIV in sub-Saharan Africa for most of the 1990s. Drugs that are nucleosides and nucleotides must be modified by cells to mimic the nucleotide triphosphate polymerase substrates before they can inhibit their targets; cells convert AZT into its active form AZT-triphosphate. Because they are or become structural mimics of the natural nucleotide triphosphate substrates, these drugs bind to the active site of viral polymerases and are competitive inhibitors. AZT inhibits HIV replication by binding to the reverse transcriptase polymerase.

In addition to competing for the active site of a polymerase, most effective nucleoside, nucleotide, and nucleotide triphosphate mimics are incorporated into the growing chain of nucleic acid. In the case of normal nucleotide polymerization, the reaction synthesizes a covalent bond between the 3′ carbon on the deoxyribose or ribose sugar and one of the oxygen groups associated with the phosphates on the incoming nucleotide triphosphate. The reaction mechanism requires a hydroxyl (OH) group on the 3′ carbon (**Figure 16.19**). Drugs that mimic nucleotide triphosphates in their active form often lack a hydroxyl group in the position necessary for the enzyme to add the next nucleotide to the growing chain. When those drugs are incorporated into a growing chain, they act as **chain terminators** because extension of the new nucleic acid cannot continue in the absence of a reactive 3′ hydroxyl. Therefore, in addition to competing for substrate binding, these drugs prevent synthesis of complete viral nucleic acids, effectively delaying viral replication even more. AZT is a chain terminator.

Occasionally an effective antiviral nucleotide triphosphate mimic is not a chain terminator, but instead has altered base-pairing properties. This is the case for the first antiviral produced commercially, **idoxuridine** (**Figure 16.20**). Idoxuridine has a 3′ OH and so it can be extended by DNA polymerase. The

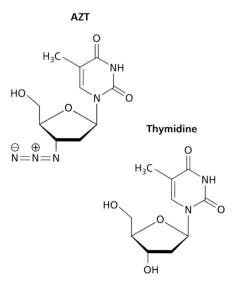

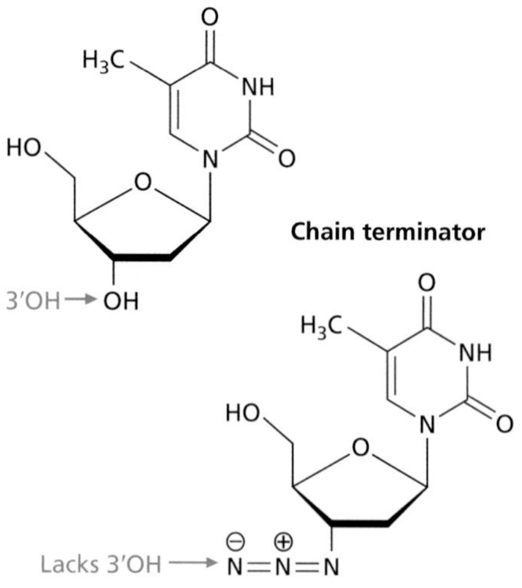

Figure 16.18 Azidothymidine. AZT is a pyrimidine structural mimic.

Figure 16.19 The 3′ OH is critical for nucleic acid polymerization. A free OH on the 3′ carbon is required to react with the incoming nucleotide. If this OH is missing, polymerization cannot continue. Structural mimics that get incorporated into nucleic acid and lack a 3′ OH are chain terminators, preventing the addition of any new nucleotide subunits.

iodine in the cyclic base does not base pair properly, however, so the altered DNA can't be used as a faithful template for producing mRNA or DNA. The effect is to inhibit viral replication by interfering with both gene expression and genome replication. Idoxuridine is useful for treating herpesvirus infections but can only be applied topically (to the skin) because it has harmful pharmacological properties when ingested orally.

Nucleic acid synthesis is not the only target for antiviral drugs, however. In fact, drugs that treat influenza target other aspects of viral replication, as discussed in the next section.

16.11   Drugs to treat influenza target the uncoating and release stages of viral replication

Influenza is one of the most important viral infections in that it causes millions of dangerous infections every year and as many as 500,000 deaths even during nonpandemic years when vaccination is 50% effective at preventing infection. Thus antiviral drugs to treat influenza are highly desirable. Existing drugs that treat influenza infections target M2 and neuraminidase (NA), which are envelope-associated proteins unique to the virus (see **Chapter 6**). The influenza M2 matrix protein is essential for penetration. After the influenza virus enters the cell through endocytosis, the host cell acidifies the endosome as a matter of course. The M2 matrix protein is an ion channel that allows protons to cross the viral envelope, which triggers rearrangements of the HA viral spike. This activates the viral fusion peptide so that the viral envelope fuses with the host endosome membrane and the genome is released into the cytoplasm (see **Chapter 3**). NA is one of two viral spikes; it is an enzyme needed during release from host cells (see **Chapter 11**). During budding, the HA spikes on newly formed virions get stuck to neuraminic acid on the surface of their former host cell unless the NA enzyme cleaves the surrounding sialic acid sugars to release the virus into the extracellular space.

There are five anti-influenza drugs on the market. Two of them (amantadine and rimantidine) inhibit the M2 protein. Four M2 polypeptides assemble into an aqueous pore that permits the flow of protons. The drug amantadine blocks the pore by binding noncovalently through interactions with particular valine, alanine, serine, and glycine amino acids that line the aqueous channel (**Figure 16.21**). It binds to the center of the pore formed by M2 and prevents the movement of protons. Rimantidine probably also binds to M2, though the mechanism is not clear. The other three anti-influenza drugs (oseltamivir, zanamivir, and peramivir) are competitive inhibitors of the NA enzyme, so they block budding.

A newer anti-influenza drug targets the viral endonuclease (PA) of the influenza virus polymerase complex. During infection, PA binds to host mRNA in the nucleus and removes its 5′ cap. This cap-snatching activity reduces the expression of host antiviral proteins and also provides the primer for replication of new influenza virus RNA. The active form of this medicine is baloxavir acid.

New anti-influenza drugs are needed because some of the existing ones are not very efficacious. For example, most seasonal IAVs are resistant to amantadine and rimantidine. Use of the drugs has selected for variant M2 proteins that do not bind to the drugs. There is contradictory evidence as to whether oseltamivir is useful for preventing influenza-related hospitalizations and deaths. Similarly, zanamivir may shorten the duration of influenza symptoms by about 1 day if administered early during infection, but may not reduce the number of hospitalizations or deaths. Peramivir is a very new drug that

Idoxuridine

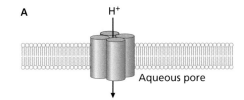

Figure 16.20 Idoxuridine. The first antiviral, idoxuridine, is a thymidine analog that can be incorporated into DNA. It blocks base pairing.

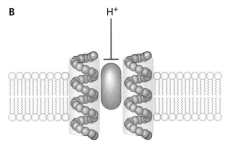

Figure 16.21 The action of amantadine. (A) Four M2 polypeptides assemble into an aqueous pore that permits the flow of protons, which is essential for influenza uncoating. (B) The drug amantadine blocks the pore by binding noncovalently, especially through interactions with particular valine, alanine, serine, and glycine amino acids (orange) that line the aqueous channel.

also shortens the duration of symptoms by 1 day, but it must be administered intravenously in a hospital, and whether it reduces risk of serious complications or the number of deaths is not known. Thus, new anti-influenza drugs targeting viral enzymes such as the RNA-dependent RNA polymerase are in development.

16.12 Drugs to treat COVID-19 target the viral polymerase or one of the viral proteases

Medicines that interfere with the replication of SARS-CoV-2 include molnupiravir and nirmatrelvir. The human body converts molnupiravir into an active form where it serves as a ribonucleoside analogue. It is a competitive inhibitor of the viral RdRp that can be incorporated into the growing viral RNA chain. The result is aberrant base pairing, ultimately causing error catastrophe by introducing so many mutations in the new viral nucleic acids that the mRNA and genomes do not encode functional proteins. An additional aspect of this nucleoside analog is that the viral 3′ to 5′ exonuclease editing activity cannot remove the analog from the growing chain of RNA. In contrast, nirmatrelvir mimics the shape and chemistry of the viral proteins that are targets for the viral protease M^{pro} (also known as 3CL^{pro}). Competitive inhibition of M^{pro} results in severely reduced viral replication.

16.13 Drugs to treat hepatitis C virus target the viral polymerase

Hepatitis C virus (HCV) is a serious public health problem. The World Health Organization estimates that there are 170 million people who have chronic HCV infections, of which 5%–7% will die because of the infection. Co-infection with HIV and HCV is particularly dangerous and is very common among IV drug users. The earliest treatments for HCV were interferon or a combination of ribavirin and interferon. Interferon has terrible side effects, as might be predicted given its actions as a pro-inflammatory cytokine. Ribavirin is metabolized into a nucleotide triphosphate mimic that can be incorporated into RNA, where it does not base pair properly and thus scrambles the genetic code. These treatments are effective only 50%–75% of the time, require 24–48 weeks of treatment, and have many harmful side effects. Newer, safer, more effective anti-HCV drugs are protease inhibitors. HCV is an enveloped Class IV (+) RNA virus that has a polyprotein that must be processed to release active viral proteins; two different viral proteases are required for these processing events (see **Chapter 5**). The protease inhibitors are either competitive or noncompetitive inhibitors of the HCV proteases. The newest anti-HCV drugs inhibit the viral RdRp (NS5B) that synthesizes viral mRNA, genomes, and antigenomes, or they bind to NS5A, an essential multifunctional nonstructural viral protein. The drugs that target the RdRp include competitive and noncompetitive inhibitors of the polymerase.

There is a risk that HCV will evolve resistance to these new, more efficacious drugs. Because HCV is an RNA virus, its polymerase does not have editing functions. It is somewhat less sloppy than the HIV-1 reverse transcriptase and makes approximately one or two substitutions per 10 offspring virions. Even so, this is a very high mutation rate and the collection of all the virions in someone's body is referred to as a quasi-species because of the tremendous antigenic variation. This level of mutation means that beneficial mutations that inactivate a drug, for example by altering the amino acids to which the drug binds, are common. Therefore, the state-of-the-art treatment for HCV is to use a combination of two or three of the following: an NS5A inhibitor, a protease inhibitor, and a polymerase inhibitor. This combination is 91%–100% effective at curing HCV infection.

Combination therapy works against the evolution of resistance because a single virus would need at least one mutation in each of the different genes that encode the protein target for each drug in order to be resistant to the treatment. For a triple cocktail, viruses that evolve resistance to one of the drugs are still eliminated by the other two. Where possible, combination therapy for any virus with a high intrinsic mutation rate is preferable to monotherapy because it will delay the evolution of resistant mutants. For example, combination therapy that would prevent serious complications and deaths from influenza is urgently needed.

Although combination therapy to cure HCV is a medical breakthrough, it is not necessarily available to everyone who needs it. For example, in the United States combination therapy lasts 24 weeks and costs approximately $200,000 (four times the US median household income). Indian drug manufacturers plan to sell the same cure for about $2,000 (USD).

16.14   Drugs to treat HIV target many stages of the virus replication cycle

The HIV/AIDS pandemic spurred antiviral drug development as nothing else ever has. As a result of intense study of the virus at a molecular level, there are drugs on the market or in development that target every stage of the virus replication cycle, described briefly here. HIV is an enveloped virus that uses a spike to bind to its receptor CD4 (see **Chapter 3**). Binding to CD4 causes a conformational change in the spike so that it subsequently binds to one of several co-receptors. Binding to the co-receptors leads to another conformational change that activates the viral fusion peptide, and the nucleocapsid is released into the host cell (see **Chapter 3**). Next, the viral reverse transcriptase enzyme uses the viral RNA genome to synthesize a DNA version of the genome (see **Chapter 10**). After entry into the nucleus, the viral integrase enzyme recombines the viral DNA with a chromosome, creating proviral DNA (see **Chapter 10**). In the nucleus, the viral Tat protein is essential for transcription from the proviral DNA. The virus synthesizes two different polyproteins that must be processed by viral protease in order for new viruses to assemble at the plasma membrane, where they escape through budding (see **Chapters 10** and **11**).

Drugs to treat HIV include attachment inhibitors, co-receptor antagonists, fusion inhibitors, integrase inhibitors, nucleoside-analog reverse transcriptase inhibitors (NaRTIs), nonnucleoside reverse transcriptase inhibitors (NNRTIs), and protease inhibitors. Attachment inhibitors that bind either to CD4 or to gp120 are in development; they must interfere with HIV binding without inhibiting the normal function of CD4 (which is essential for T helper lymphocyte function; see **Chapter 15**). A co-receptor antagonist called maraviroc binds to the CCR5 co-receptor and forces it into a conformation that cannot be bound by HIV. The fusion inhibitor enfuvirtide was designed to block specific interactions among α-helices in the viral spike that are essential for fusion to occur. The drug is a synthetic peptide that forms an α-helix that disrupts the natural interactions that would otherwise catalyze fusion between the viral envelope and the plasma membrane.

Reverse transcriptase acts during the long uncoating process while the viral nucleic acids are conveyed to the nucleus. NaRTIs are competitive inhibitors of reverse transcriptase, whereas NNRTIs are noncompetitive

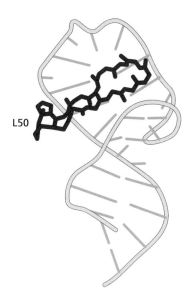

L50

Figure 16.22 The L50 candidate antiviral drug binds to TAR. TAR is an HIV stem–loop RNA structure needed for viral gene expression. The drug L50 is depicted in red; it binds to the surface of TAR that would otherwise bind to the HIV protein Tat. (From Lalonde MS et al. 2011. *PLOS Pathog* 7:e1002038. doi: 10.1371/journal.ppat.1002038. Published under CC BY 4.0.)

inhibitors of reverse transcriptase. Most of the competitive inhibitors are also chain terminators. The last viral enzyme active during uncoating is integrase. Commercially available integrase inhibitors bind to the cations found in the integrase active site, where they function as competitive inhibitors.

After formation of the provirus, transcription ensues. There are no commercially available drugs that target transcription of HIV, but drugs that interfere with the function of the viral Tat protein are in development. HIV mRNA elongation does not occur in the absence of the Tat protein. The Tat protein interacts with a stem–loop structure in the HIV mRNA known as TAR (see **Chapter 10**). After Tat binds to TAR, the whole complex binds to a host enzyme that phosphorylates host RNA Pol II in such a way that elongation can occur (see **Chapter 10**). The anti-Tat drug in development mimics the structure of a particular β-sheet in Tat. This β-sheet is the surface that normally interacts with TAR, so the drug competes with binding to TAR (**Figure 16.22**).

During maturation, the viral protease cleaves the two viral polyproteins. There are competitive and noncompetitive protease inhibitors available. There are also drugs blocking other steps of maturation and release in development, but none is available commercially at this time. One example is a drug that binds to the Gag polyprotein and prevents protease from binding to it (rather than binding to the protease itself as protease inhibitors do).

The development of protease inhibitors was one of the first blockbuster uses of **rational drug design**, in which the discovery of a new drug begins with a crystal structure for a target protein. Computer modeling is then used to test binding of small molecules to the target. Protease inhibitors were designed to mimic the transition state of the enzyme. The transition state of any enzyme is the unstable intermediate that forms during the process of converting substrate into product. The success of rational drug design in developing protease inhibitors inspired the pharmaceutical industry to make the use of crystal structures, chemical structures, and computer modeling a fundamental strategy for drug discovery.

Use of combination therapy for treatment of HIV is critical because the virus exhibits extreme antigenic variation. This variation allows for very rapid emergence of drug resistance in a person taking just one antiviral. It can also be dangerous because it can lead to the circulation of drug-resistant viruses, which then reduces treatment options for newly infected people. Combination therapy in HIV was originally called **highly active antiretroviral therapy** or **HAART**, which has a new name: **cART** for **combination antiretroviral therapy**. Combination ART is beneficial for an individual patient in two ways. First, it slows viral replication so much that the host cells for viral replication are not killed as rapidly, significantly delaying the onset of AIDS. Second, by slowing viral replication, the amount of antigenic variation per day is significantly reduced, which slows both the emergence of drug resistance and the emergence of viruses that escape the immune response. Combination ART is also beneficial on a population level in two ways. First, it reduces the proportion of drug-resistant viruses in circulation. Second, because people in treatment release billions of times fewer viruses, it reduces the chance that the patient will spread HIV to another person. This idea has been called treatment as prevention.

Drugs that target HIV can also be used in regimens termed postexposure prophylaxis and preexposure prophylaxis. The purpose of **postexposure prophylaxis** (**PEP**) is to prevent someone who was exposed to HIV from becoming HIV+, as might happen if a healthcare provider gets a needle stick while treating an HIV+ patient or after a sexual assault in which the assailant was HIV+. PEP is a combination of two or three drugs with different targets and must be taken within 72 hours of exposure (and taking it sooner is more effective). PEP has been especially useful for preventing mother-to-child transmission of HIV and for healthcare workers, although anyone who has been exposed can use PEP to avoid becoming HIV+. The purpose of **pre-exposure prophylaxis (PrEP)** is to have drugs in the system that prevent HIV replication if HIV should happen to enter the body. People using PrEP take the drug every day, analogous to taking a birth control pill; research is under-way to find out if it might be effective to take the medicine less frequently, for example as late as 24 hours before exposure. PrEP has been especially effective for gay men including HIV– men whose partners are HIV+. New drugs and new delivery systems, such as vaginal rings or infrequent injection of long-acting drugs, are being tested to find PrEP regimens that would work better for other populations, such as South African girls and women.

16.15   Viral evolution occurs in response to selective pressure from antiviral drugs

As was mentioned in the discussions of drugs to treat influenza, HCV, or HIV, antiviral drugs exert strong selective pressure on viruses and these pres-sures result in the emergence of viruses that are resistant to the drug that was used to treat them. Typically, these resistant viruses have genetic variations that alter the amino acids that would otherwise participate in noncovalent intermolecular forces between the target protein and the drug (**Figure 16.23**). The amino acids that bind to a drug are termed **surface-exposed amino acids** or **solvent-exposed amino acids**, meaning that a drug dissolved in water can contact those amino acids. Alternatively, alteration of an amino acid near the binding site can change the shape complementarity so that the drug cannot bind in exactly the same way and thus is not able to form hydro-gen bonds or other intermolecular forces with the protein. An example is provided by the structure of the influenza NA spike bound to the sialic acid-mimic drugs zanamivir and oseltamivir (**Figure 16.24**). The drugs bind in a cleft where noncovalent interactions between them and the solvent-exposed amino acids glutamic acid at position 276 and histidine at position 274 hold the drug in the pocket. Mutations that change the histidine at position 274 to a tyrosine result in a loss of drug binding and efficacy; viruses with this alter-ation in NA are resistant to the drugs zanamivir and oseltamivir. The drugs can still bind to the altered target, but only at much higher concentrations so that under clinical conditions the drug is ineffective.

There are promising strategies to use molecular biology during develop-ment of a drug to try to reduce the chance that resistance can evolve. For some viruses such as HIV, there is so much natural antigenic variation that it is possible to examine large data sets of existing protein sequences to find amino acids that never or almost never vary. When such amino acids are also found in the active site of an enzyme or in other surface-exposed regions of the protein, small molecules that bind to those amino acids can be discovered.

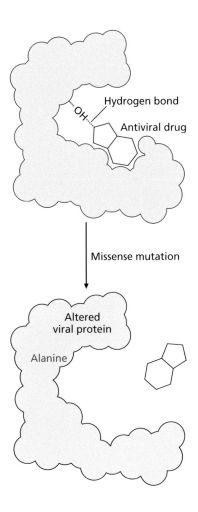

Figure 16.23 Noncovalent interactions are responsible for drug–target binding. Noncovalent intermolecular forces such as hydrogen bonds are very important for drug–target binding. A mutation that alters an amino acid can block drug binding by eliminating one or more noncovalent intermolecular forces between the drug and the target.

Figure 16.24 Zanamivir and oseltamivir binding to influenza neuraminidase. (A) The drugs zanamivir and oseltamivir are sialic acid analogs that bind in the active site of influenza neuraminidase (NA) spike. (B) Superposition of zanamivir (yellow), oseltamivir (green), and sialic acid (gray) further reveals their similar structures. The small molecules bind to the active site of NA through intermolecular forces such as hydrogen bonds with glutamic acid 276 (Glu276). (C) A missense mutation substituting the normal histidine at position 274 with tyrosine causes oseltamivir resistance. The figure shows an extreme close-up of the binding cleft with the protein as a three-dimensional surface with red to indicate negative solvent-exposed surfaces and blue to indicate positive ones. Oseltamivir (green) bound to the normal protein cannot bind to the antigenic variant in exactly the same way and is instead slightly shifted (cyan), which significantly reduces the noncovalent intermolecular forces between the protein and the drug, resulting in drug resistance. (From Das K et al. 2010. *Nat Struct Mol Biol* 17:530–538. doi: 10.1038/nsmb.1779. With permission from Springer Nature.)

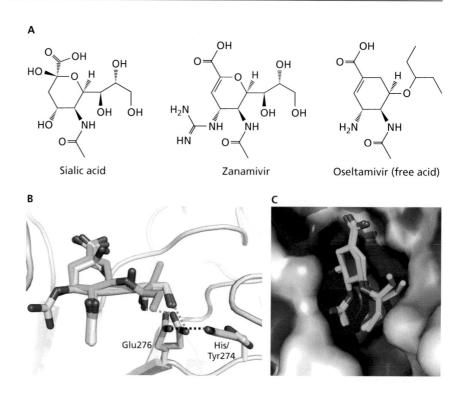

Some of these may inhibit the viral protein and thus be developed into drugs to which the virus is unlikely to evolve resistance. There are examples of HIV protease inhibitors that were developed in this manner. A second strategy is to propagate a virus *in vitro* and select for antiviral-resistant mutants. The variants encoded by these mutants also reveal which surface-exposed amino acids must remain constant for the viral protein to function, providing targets for the development of new antiviral drugs.

16.16 It might be possible to develop bacteriophage therapy to treat people with antibiotic-resistant bacterial infections

Viruses are not the only microbes that can evolve resistance to drugs used to treat them. Indeed, antibiotic resistance among pathogenic bacteria such as *Mycobacterium tuberculosis*, *Staphylococcus aureus*, and *Neisseria gonorrhoeae* (among many others) is an increasing problem. For almost 100 years, microbiologists have been interested in developing bacteriophages as pharmaceuticals that could treat infection. The Soviet Union was particularly innovative in developing phages as a treatment for bacterial infections, and phages are still in common use in the food industry to reduce bacterial loads during food processing. An advantage to using phages to treat bacterial infections is that, unlike antibiotics, phages are extremely specific and so kill the pathogen without affecting the commensal and beneficial microbiota. On the other hand, precisely because phages are so specific, a disadvantage is the difficulty of creating a collection of phages that could reasonably be expected to kill a particular pathogen during a natural infection.

The most recent innovation in using phages to treat bacterial infections is the development of phage lysins and tailocins as drugs. **Lysins** are enzymes that degrade peptidoglycan, which is critical for the release of offspring phages (see **Chapter 11**). These enzymes are useful in killing Gram-positive bacteria such as *S. aureus*, including multidrug-resistant *S. aureus*. Such enzymes are less specific than phages and can reliably treat an entire genus. **Tailocins** are headless phages. When a tailocin binds to its target cell, it creates a hole in the cell membrane, which dissipates the membrane potential; collapse of membrane potential kills the bacterial cell by preventing many transport and metabolic reactions including ATP synthesis. Tailocins can be genetically modified to improve their capacity to kill particular hosts.

16.17   Engineered viruses could in principle be used for gene therapy to treat cancer and other conditions

Some diseases are caused by the inheritance of faulty alleles that encode nonfunctional or dysfunctional proteins or mRNAs. These include such disorders as ornithine transcarbamylase (OTC) deficiency, adenine deaminase deficiency (which causes severe combined immunodeficiency), and familial lipoprotein lipase deficiency. In principle, it should be possible to treat such diseases through **gene therapy**, by providing a functional wild-type allele to the cells that need it. Because viruses are expert at binding to and introducing genes into host cells, they are ideal **vectors** for the selective delivery of **therapeutic transgenes**. If the therapeutic transgenes could alter the genetic code of stem cells, then a single treatment could in principle cure an inherited disorder by providing a functional gene expressed by the correct cells. Gene therapy can also be used to treat cancer, for example by transforming cancer cells with a gene to make them more sensitive to apoptosis or sensitive to a pharmaceutical.

The first gene therapy trials in humans occurred in the last quarter of the twentieth century, and there have been about 2,000 such trials since that time. In the early days, knowledge about the molecular biology of the viral vectors was less complete and a series of setbacks seriously affected the field. In addition to financial problems arising from a lack of **clinical efficacy** (benefit to patients), there was a particularly notorious death caused by a clinical trial; this death understandably brought a much greater sense of caution to the whole enterprise. At the age of 18, Jesse Gelsinger participated in a Phase I (safety) clinical trial that involved injecting his liver with billions of modified adenoviruses. Instead of correcting the symptoms of his mild OTC deficiency, the virus triggered a massive systemic immune response that resulted in his death. After Gelsinger's death, an investigation determined that there were several ethical problems with the trial; for example, the university and the head researcher had financial stakes in a positive outcome of the trial. Unfortunately, it is still not clear exactly what about the adenovirus vector caused the fatal reaction; although the virus had had the E1 and E4 genes deleted, the other early genes were still intact (**Figure 16.25**).

Despite its early failures, research into gene therapy remains popular. The most common gene therapy vectors in development today are modified adenoviruses (dsDNA Class I; see Figure 16.25), parvoviruses (ssDNA Class II), or retroviruses (Class IV) (**Figure 16.26**). Although adenoviruses and parvoviruses can infect both dividing and nondividing cells, the most common retroviral vectors in use infect only dividing cells. The lentivirus group of retroviruses, to which HIV belongs, can infect nondividing cells, and lentiviral vectors are therefore increasingly popular. Although all three deliver packaged DNA into the nucleus during the final stages of uncoating, neither adenovirus nor parvovirus genomes recombine with host chromosomes, leading to loss of viral DNA when the cells proliferate. The integrated proviral DNA from a retrovirus is instead maintained for the lifetime of

Figure 16.25 First-generation adenovirus gene therapy genome. (A) Typical adenovirus genome with terminal proteins (purple), inverted terminal repeats (highlighted yellow), and a packaging signal (Ψ). (B) In first-generation adenovirus gene therapy vectors such as the one used in the Phase I trial that killed Jesse Gelsinger, the therapeutic transgene replaced the E1 region of the viral genome.

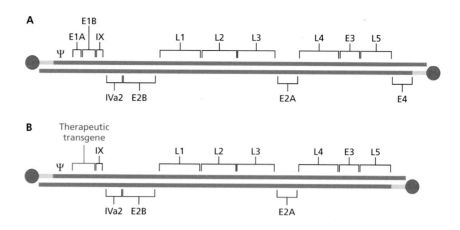

Figure 16.26 Genomes of a parvovirus and retrovirus. Adenoviruses (see Figure 16.25), parvoviruses, and retroviruses are the most common vectors in development as gene therapy agents. (A) Adeno-associated virus-2 is an example of a parvovirus. Its single-stranded genome is about 4.6 kbp long and encodes four nonstructural proteins and three capsomeres. (B) A lentivirus is an example of a retrovirus. Its RNA genome is converted into proviral DNA, which is about 9.2 kbp long. Gag and Env encode structural proteins such as the capsomeres and transmembrane spikes. Pol encodes enzymes such as the reverse transcriptase, protease, and integrase. Lentiviruses also encode accessory proteins such as Vif.

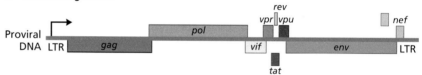

the host cell and so provides the longest lasting expression of the therapeutic gene. Insertion of proviral DNA can also cause cancer, however, and so carries risks (see **Chapter 13**).

Viral genomes have a size limit based on the capsid as well as the mechanism of packaging the genome during assembly and maturation. The **packaging capacity** of a gene therapy vector refers to the length of nonviral genes that can be inserted into the vector. For parvoviruses, the packaging capacity is only 4.5 kb, whereas for retroviruses the packaging capacity is about 8 kbp. An average human cDNA is about 2,300 bp long, so parvovirus and retrovirus vectors can usually carry one therapeutic gene along with sequences to regulate its expression. The capacity of adenovirus vectors is larger and depends on how much the vector has been modified. Adenovirus genomes naturally range from 26 to 45 kb and normally encode E1, E2, E3, and E4 families of early proteins (see **Chapter 8**). They have inverted terminal repeats important for replication and a packaging signal (Ψ) used for filling new capsids with the genome. The oldest first-generation adenovirus vectors replace the E1 sequences with a therapeutic transgene and a strong promoter to drive its expression. Third-generation adenovirus vectors are called gutless because all of the viral genes are missing; instead, up to 36 kb of therapeutic DNA is inserted between the inverted terminal repeats, also leaving Ψ intact (**Figure 16.27**).

To manufacture viral vectors containing a genome that is missing viral genes otherwise essential for viral replication, it is necessary to use engineered host cells that provide viral proteins and the heavily modified genome so that only the therapeutic genome is packaged into infectious virions. In the case of gutless adenovirus vectors, the host cells must be co-infected with

A   Adenovirus genome

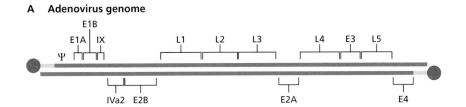

B   Gutless adenovirus genome

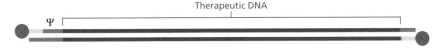

Figure 16.27 A gutless adenovirus vector genome. (A) Adenovirus genome. (B) In a gutless adenovirus, the adenovirus genes are missing so that the genome is composed of the inverted repeats and packaging signal with therapeutic DNA.

a helper adenovirus that provides the viral proteins. In that case, the cells and helper virus are engineered to delete Ψ from the helper virus genomes so that only the gutless vector is packaged. This feat is accomplished using the bacteriophage P1-derived Cre-*lox* system (see **Chapter 4**), where the helper virus genome is engineered to have *loxP* sites flanking Ψ (**Figure 16.28**). The host cell expresses the recombinase Cre, which deletes any DNA in between two *loxP* sites and therefore deletes Ψ from the helper genomes. Without Ψ, the helper genomes cannot be packaged. Deletion of Ψ from the helper

A   Helper virus genome

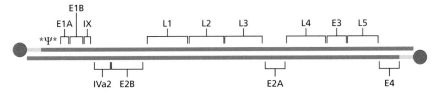

B   Gutless genome cloned in plasmid

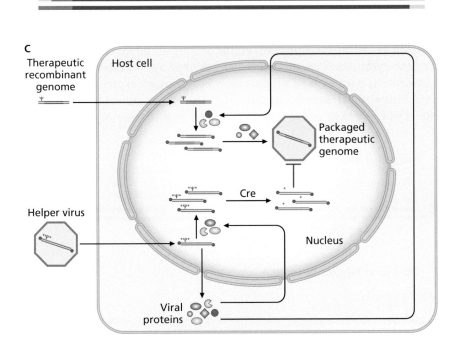

Figure 16.28 Gutless adenovirus vector production. (A) Genome of a helper virus. The helper virus genome encodes all the proteins necessary to replicate adenovirus genomes and package them into infectious virions. The helper virus genome is engineered so that *loxP* sites (red asterisks) flank Ψ. (B) The gutless vector has inverted terminal repeats and Ψ but has therapeutic DNA replacing all of the other adenovirus sequences. (C) The host cells are engineered to express the bacteriophage Cre enzyme. They are transfected with a plasmid that provides the gutless genome. At the same time, they are infected with the helper adenovirus. The helper adenovirus proteins amplify the adenovirus genomes and cause production of therapeutic virions, which are therapeutic genomes packaged into adenovirus capsids. Expression of Cre deletes the Ψ from 99% of the helper virus genomes, so that 99% of the therapeutic virions contain just the therapeutic recombinant genome and not the genome of the helper virus.

genomes is successful about 99% of the time. The therapeutic virions must be purified away from the contaminating 1% of helper viruses before they can be used in patients.

16.18   Gene therapy and oncolytic virus treatments currently in use

There are eight gene therapy or oncolytic virus treatments approved for use in the United States, Europe, China, the Philippines, and Russia. The treatments in use are derived from adenoviruses, retroviruses, adeno-associated viruses (a type of parvovirus), and herpesviruses. Because they are commercialized, the molecular details of the vector features, including the genome structure, are not always publicly available.

The **rAD-p53** treatment is an example of gene therapy derived from adenovirus. Adenoviruses are dsDNA viruses that express and replicate their genomes in the nucleus and do not insert viral DNA into any host chromosomes (see **Chapter 8**). The rAD-p53 pharmaceutical is a recombinant adenovirus that expresses the p53 gene (**Figure 16.29**). The normal version of the p53 gene promotes apoptosis; when the p53 cell detects unrepaired DNA damage, cells that have mutations do not survive to become oncogenic (see **Chapter 8**). More than 50% of all cancers have nonfunctional or misfunctional p53, so the theory behind this therapy is to restore p53 function in cancer cells. In China, rAD-p53 is used to treat head and neck cancers, although it has not been approved for use by European or US regulatory bodies.

A second anticancer treatment based on adenovirus, **H101**, is also used to treat head and neck cancers in China. Instead of providing a gene that prevents cancer cells from reproducing, H101 preferentially kills cancer cells. H101 is **oncolytic** because the adenovirus has a deletion that removes the part of the genome encoding a regulatory early protein called E1b-55K (**Figure 16.30**). The normal function of E1b-55K during adenovirus infection is to inhibit the host protein p53. If p53 is not inhibited, the virus cannot complete lytic replication. Because of the deletion, H101 cannot replicate in normal host cells that have functioning p53. It can, however, replicate in host cells that are already deficient in p53, as many cancer cells are, thus restricting viral replication to cancer cells. Replication of adenovirus is lytic, killing the host cell.

Rexin-G (retroviral expression vector bearing an inhibitory construct of the gene-cyclin G) is an anticancer gene therapy agent based on a murine retrovirus; its purpose is to kill the cancer cells through expression of a toxic gene (**Figure 16.31**). Retroviruses are RNA viruses that convert their genomes into cDNA and then insert that cDNA into a host chromosome at a nearly

A   rAD-p53 genome

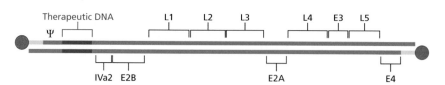

Figure 16.29 The rAD-p53 gene therapy agent. (A) The adenovirus vector has the E1a, E1b, pIX, and E3 genes deleted. (B) The therapeutic DNA includes a strong promoter from cytomegalovirus (CMV), the human *p53* cDNA, and a TPL sequence that, once transcribed, enhances translation of the mRNA.

B   Close-up of the therapeutic DNA

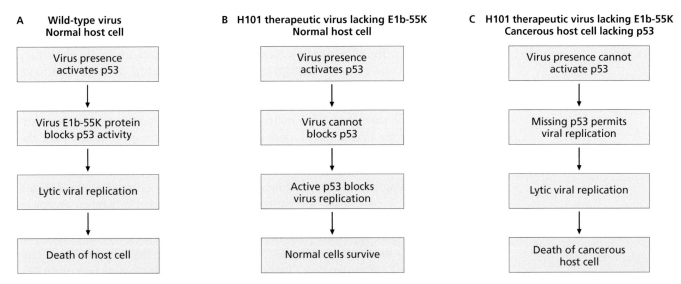

A   Wild-type virus
Normal host cell

Virus presence
activates p53

↓

Virus E1b-55K protein
blocks p53 activity

↓

Lytic viral replication

↓

Death of host cell

B   H101 therapeutic virus lacking E1b-55K
Normal host cell

Virus presence
activates p53

↓

Virus cannot
blocks p53

↓

Active p53 blocks
virus replication

↓

Normal cells survive

C   H101 therapeutic virus lacking E1b-55K
Cancerous host cell lacking p53

Virus presence cannot
activate p53

↓

Missing p53 permits
viral replication

↓

Lytic viral replication

↓

Death of cancerous
host cell

Figure 16.30 The principle behind the H101 gene therapy agent. H101 is an adenovirus with the E1b-55K coding sequence deleted. (A) The normal function of E1b-55K is to permit viral replication in normal host cells by counteracting the p53 protein. (B) H101 cannot replicate in normal host cells because the p53 host protein blocks its replication. Thus, H101 does not kill normal host cells. (C) H101 can replicate in cancer cells that lack normal p53 function. H101 therefore selectively lyses its cancerous host cells.

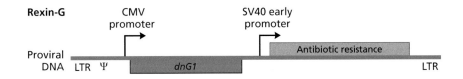

Figure 16.31 The Rexin-G genome.
The recombinant provirus includes long terminal repeats (LTRs), DNA encoding a packaging signal (Ψ), and a strong cytomegalovirus (CMV) promoter driving the expression the dnG1 gene encoding a toxic variant of the human cyclin G protein. It also encodes an antibiotic resistance gene useful for selecting cells that host the provirus during production of the gene therapy agent. The antibiotic resistance gene is driven by the early SV40 (polyomavirus; see **Chapter 8**) promoter. The genome does not encode any retroviral proteins, so proteins needed for attachment, penetration, and uncoating must be supplied during manufacture by a helper virus or an engineered host cell.

random position. The particular Rexin-G retrovirus can only infect actively dividing cells because its preintegration complex of viral cDNA cannot cross the nuclear envelope (see **Chapter 10**). This property confers some specificity upon the vector because most cancer cells divide more frequently than most normal cells. The virus has been modified extensively to confer upon it properties that make it effective. First, it has been pseudotyped to express a modified viral spike protein (see **Chapter 3**) that binds preferentially to collagen, which is overrepresented in tumors because of the way that cancerous cells modify their extracellular environment. Thus, when the vector is infused into a target site, the vector is more likely to infect cells in a tumor than to infect cells in nearby normal tissue. Second, the genome expresses a toxic version of a human cyclin protein. Cyclins regulate the cell cycle. In this specific case, the altered cyclin prevents cell division and ultimately kills the infected cell. The recombinant vector does not encode any retroviral proteins so during production these must be provided by engineered host cells or by co-infection with a helper virus. Philippine regulatory authorities approved Rexin-G for use in treating all solid tumors. In the United States, Phase II trials using Rexin-G to treat metastatic pancreatic cancer tripled patient survival from 3 months to 9 months and use of Rexin-G to treat pancreatic cancer, metastatic breast cancer, and sarcoma is in US clinical trials now.

Other approved gene therapies based on retroviral vectors include treatments for severe combined immunodeficiency disease (SCID) and graft-versus-host disease (GVHD). Patients with SCID do not have lymphocytes and succumb to opportunistic infections. One type of SCID is caused by an enzyme deficiency, which can be corrected with a retroviral vector. The vector is engineered to express adenine deaminase, the protein lacking in this type of SCID. In this case, in a process termed *ex vivo*, the patient's hematopoietic stem cells are collected and then transduced with the virus in a laboratory. The altered stem cells are returned to the patient's body, where they persist and give rise to offspring cells for years, leading in some cases to an apparent cure.

GVHD is a serious and often fatal complication that can occur after a bone marrow transplant; bone marrow transplants are typically used to treat cancers of the blood. The purpose of a bone marrow transplant is to replace the patient's hematopoietic stem cells with those from a donor, thus eliminating the source of a blood cancer, while also providing the patient with blood cells including all cells of the immune system. GVHD occurs when the donated CD8+ T cells attack the patient's other cells, which can happen except in the very rare cases in which the donor and recipient are homozygous twins. There is an approved gene therapy treatment that uses a retroviral vector to modify the donated cells prior to their infusion into the recipient. The vector causes T cells to express a herpesvirus thymidine kinase gene (**Figure 16.32**). The gene is therapeutic because the herpesvirus thymidine kinase enzyme turns the pharmaceutical acyclovir into toxic metabolites that kill the cells expressing the herpesvirus thymidine kinase. If a patient transplanted with transformed cells shows signs of GVHD, ganciclovir can be used to kill the problematic T cells.

An example of an approved gene therapy based on adeno-associated virus (AAV), a type of parvovirus, is **alipogene tiparvovec**. Adeno-associated viruses have ssDNA genomes and are called satellite viruses because they cannot replicate in the absence of co-infection with another virus such as adenovirus or herpesvirus (see **Chapter 17**). Instead, in the absence of co-infection, the host converts their small (<5 kb) genomes into dsDNA that enters the nucleus and persists as an episome, where genes can be expressed but the genome cannot direct lytic reproduction without a helper virus. Gene therapy vectors derived from AAV are advantageous compared with retroviruses in that the DNA never inserts into a host chromosome, minimizing the risk of the therapy itself causing cancer.

Alipogene tiparvovec treats familial lipoprotein lipase deficiency, a metabolic disorder that disrupts normal processing of fats and causes severe inflammation of the pancreas, leading to repeated hospitalizations and excruciating pain. Patients are injected with alipogene tiparvovec intramuscularly, where the therapeutic lipoprotein lipase is expressed. In this case, the therapeutic transgene encodes a rare naturally occurring variant of the enzyme that is associated with lower rates of cardiovascular disease because it is even more efficient at processing lipids than the more common one. The recombinant AAV genome is altered so that expression of the transgene is driven by a strong immediate-early promoter originally derived from cytomegalovirus (HHV-5)

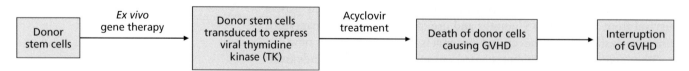

Figure 16.32 Cells transformed with the vector express herpesvirus thymidine kinase. Patients who receive bone marrow from a related but not identical donor can develop graft-versus-host disease (GVHD), caused by CD8+ T cells derived from the donated stem cells. To prevent GVHD, the donor cells are transformed *ex vivo* so that CD8+ T cells from the donor stem cells express herpesvirus thymidine kinase. After transplantation, the recipient is treated with acyclovir to kill CD8+ T cells from the donor, which would otherwise cause GVHD.

and enhanced by a post-transcriptional regulatory element derived from a woodchuck hepadnavirus (**Figure 16.33**). This sequence enhances cytoplasmic accumulation of the therapeutic transgene. The treatment is approved for use in Europe and the US and at the time of its release was the most expensive pharmaceutical at 1 million euros per treatment.

Talimogene laherparepvec is a viral treatment for melanoma that is based on a herpesvirus. The recombinant vector is injected directly into tumors, where it is oncolytic. Its replication is restricted to tumor cells because the ICP34.5 gene is deleted. The normal function of ICP34.5 during herpes simplex virus (HHV-1 or -2) infection is to counteract the interferon-stimulated antiviral state, which is often nonfunctional in cancer cells. Thus, the virus is restricted to replicating in abnormal cells. The ICP47 gene is also deleted from the vector because its normal function is to inhibit the MHC-I display of viral epitopes, which is critical because displaying viral epitopes is essential for stimulating an immune response against the cancer cells. In furtherance of this goal, the virus has also been engineered to express a cytokine known as GM-CSF, which stimulates antigen-presenting cells and, in this case, is intended to cause CD8+ T cells to attack the cancer cells. Talimogene laherparepvec is approved to treat advanced melanoma in Europe and the US and is in clinical trials, sometimes in combination with other drugs, to treat other cancers.

Another gene therapy agent derived from a herpesvirus is **pCMV-VEGF**, which has no viral capsid but is instead a plasmid in which a strong immediate-early promoter originally derived from cytomegalovirus (HHV-5) drives expression of vascular endothelial growth factor (VEGF). VEGF is useful to stimulate blood vessel growth in disorders in which additional blood flow would be clinically beneficial. Examples include bone grafts, where increased blood flow has the potential to improve the survival of grafted bone. It is approved for use in Russia to treat lower limb ischemia, a condition in which blocked arteries lead to reduced blood flow to the lower legs and feet. The condition can require amputation and is particularly severe in elderly patients. During the treatment, the plasmid is injected into the site where blood vessel growth is needed. The tissue expresses VEGF, stimulating blood vessel growth.

Although the commercialization of a handful of gene therapy and oncolytic viruses is a positive sign for the use of viruses as therapeutic agents, we may not be on the brink of an explosion of such treatments. In thousands of cases, gene therapy clinical trials have shown either a lack of safety or, especially, a lack of efficacy. Research in molecular and cellular virology has improved the safety of viral vectors by advancing understanding of immune reactions to viruses and the locations where viral DNA integrates into human chromosomes. Vector safety has been improved by developing gutless vectors and the cell lines needed to package viral genomes expressing only a small number of viral genes. Now it is necessary to move beyond merely safe treatments toward efficacious ones. Therapeutic genes might be more efficacious if they could be expressed in a higher proportion of cells, for a longer period of time, or in the optimum clinical amount. Expression of a therapeutic gene in long-lived stem cells is often a goal, but obtaining those cells and targeting a virus to infect them remain challenging. When a medical treatment is approved in one country but not in others, there is usually a dispute over efficacy if one regulatory body has determined that the treatment is effective but another has made the opposite determination. Other challenges facing the gene therapy research community are economic, because treatments that help a tiny number of people are unlikely to be as profitable as those that can benefit large numbers of patients.

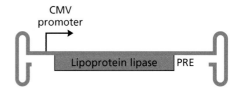

Figure 16.33 The genome of alipogene tiparvovec. The genome is a modified parvovirus. Expression of the therapeutic enzyme is driven by a CMV promoter and enhanced by DNA encoding a post-transcriptional regulatory element (PRE) from a hepadnavirus.

16.19   Therapeutic applications of CRISPR-Cas technology

Genome editing technologies, including those derived from bacterial CRISPR-Cas adaptive immunity (see **Chapter 15**), can be used to modify the genomes of human cells. Genome editing is particularly straightforward for deleting a gene. Diseases that could in theory be treated by deleting a cellular gene include HIV. Some people are naturally resistant to HIV infection because they do not express a normal version of the CCR5 co-receptor, which is required for viral attachment and penetration (see **Chapter 3**). Lack of CCR5 causes no apparent detrimental effects in people who naturally lack the protein and are resistant to HIV. Thus, an idea to treat HIV is to remove hematopoietic stem cells, edit them to delete CCR5, and then return the modified cells to the patient's body. Ideally, the stem cells will give rise to HIV-resistant T cells and thereby significantly reduce HIV replication and delay the onset of AIDS.

Another idea is to treat sickle cell anemia and β-thalassemia with genome editing. These inherited diseases are called β-hemoglobinopathies because they are caused by pathogenic β-hemoglobin proteins. Hemoglobin is the most abundant protein in red blood cells, where it shuttles oxygen to tissues and carbon dioxide to the lungs. Humans have the capacity to express many different hemoglobins. Typical adults have hemoglobin composed of two α-hemoglobin and two β-hemoglobin polypeptides (**Figure 16.34**). During fetal development, however, hemoglobin is instead composed of two α-hemoglobin and two γ-hemoglobin polypeptides. This fetal hemoglobin has a slightly higher capacity for carrying oxygen than adult hemoglobin, facilitating transfer of oxygen from the mother's circulation to the fetus. After birth, the γ-hemoglobin is no longer expressed and so adults have α/β hemoglobin rather than α/γ. But, about 1 in 1,000 people have a benign condition called hereditary persistence of fetal hemoglobin (HPFH), in which they continue to express γ-hemoglobin during adulthood. A small number of people who have a β-hemoglobinopathy also have HPFH, which alleviates their symptoms by replacing some of their faulty α/β hemoglobin with α/γ. Thus, a proposed treatment for β-hemoglobinopathies is to edit a patient's hematopoietic stem cells to provide a promoter mutation that leads to lifelong expression of γ-hemoglobin as occurs naturally for people with HPFH. In this case, gene editing introduces a point mutation rather than introducing a deletion. The plan is to collect some of a patient's bone marrow, edit the hematopoietic stem cells *ex vivo*, and then put the altered cells back into the patient.

Bone marrow cells are not the only stem cells that can be edited. It is possible to edit pluripotent stem cells (PSCs), for example. PSCs have the capacity to proliferate and, for their descendent populations, to differentiate into any cell type. PSCs can be created from an adult's cells, in which case they are called induced PSCs (iPSCs). It may be possible to collect cells from a patient, use them to create iPSCs, and then edit the cells *ex vivo* to correct a problematic genetic variation or to introduce a therapeutic genetic change. The cells could then be induced to proliferate and differentiate before being introduced back into the patient's body. Because the cells are derived from the patient, there is a low risk of immune-mediated complications.

Human embryonic stem cells have also been subjected to genome editing. Editing human embryos is controversial because it could in theory introduce a genetic change into every cell in the human body, including gametes that could be passed on to the next generation. The first human embryos that were edited were nonviable triploid embryos that could not survive even if

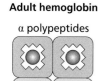
Adult hemoglobin

α polypeptides

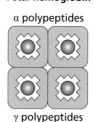

β polypeptides

Fetal hemoglobin

α polypeptides

γ polypeptides

Figure 16.34 Adult and fetal hemoglobin. Hemoglobin is a tetramer of polypeptides. Adult hemoglobin is composed of two α and two β subunits, whereas fetal hemoglobin is composed of two α and two γ subunits. Both forms of hemoglobin can transport O_2 and CO_2, as represented by the red dots in each subunit.

a woman had agreed to have them transplanted into her uterus. There has also been editing of potentially viable diploid human embryos, which can be propagated in a laboratory for about 2 weeks, ultimately forming a blastocyst. After that the embryos are not viable and are discarded. So far, the stated goals for editing human embryos are to study the first 2 weeks of development *in vitro*, and to learn more about reliably editing human embryonic stem cells. It remains to be seen when or even if the technology will be used to create embryos for reproductive purposes.

16.20   Antibodies to treat viral infections

The early twenty-first century has seen rapid development of antibodies produced in a laboratory to treat viral diseases. Prominent examples include antibody infusions to treat COVID-19 in children and adults or respiratory syncytial virus disease in infants. These therapeutic antibodies are produced through genetic engineering. The antibodies can neutralize viruses and accelerate their destruction by the immune system. But there is also strong selective pressure for viruses to give rise to escape mutants that no longer bind to such antibodies. This phenomenon has unfortunately negated the therapeutic effect of many commercialized anti-SARS-CoV-2 antibody infusions.

Essential concepts

- Effective prophylactic vaccines produce long-lived memory lymphocytes that prevent infection if someone is exposed to the natural infectious agent.
- Attenuated vaccines cause microbes to replicate very slowly in the immunized person and provoke a strong immune response, but they do not usually cause disease.
- Inactivated vaccines inactivate the microbe by physical or chemical means and prevent it from replicating; these vaccines typically require a stronger adjuvant or more boosters than attenuated vaccines.
- Subunit vaccines are very safe and are composed of antigenic components of microbes. They require stronger adjuvants and more boosters than inactivated and attenuated vaccines.
- Antigenic variation is a serious problem for developing protective influenza, HIV, and SARS-CoV-2 vaccines. A universal influenza vaccine would protect against all strains of influenza A by provoking broadly neutralizing antibodies. Vaccines to prevent HIV are in development and likely must provoke both a neutralizing antibody and a CTL response. Vaccines that prevent COVID-19 and hospitalization due to COVID-19 are available.
- Most antiviral drugs are competitive or noncompetitive inhibitors of viral enzymes that are essential for virus replication and that are not similar to host cell proteins.
- A large class of antiviral drugs is nucleotide structural mimics. They are incorporated into viral nucleic acids, where they cause chain termination or scramble the genetic code so that offspring virions cannot synthesize proteins.
- Drugs to treat influenza target either the neuraminidase spike or the M2 matrix protein and ion channel. Drugs to treat hepatitis C target the viral polymerase, proteases, or other nonstructural proteins. Drugs to treat HIV target most stages of the virus life cycle and can be used for treatment, postexposure prophylaxis, and preexposure prophylaxis.

- Antiviral drugs exert powerful selective pressure on viruses, which can evolve very quickly. RNA viruses evolve particularly rapidly because of the high error rates of reverse transcriptases and RdRps. RNA virus infections are best treated with multiple drugs that target different proteins in order to slow the evolution of resistant variants.
- Bacteriophages and proteins derived from them may be developed into effective antibacterial reagents to treat infections in people.
- Viruses can be genetically altered to deliver therapeutic genes either *in vivo* or *ex vivo*, providing treatments for inherited conditions or cancer. Other viral vectors are oncolytic because they have been genetically altered to infect or kill cancer cells selectively.
- CRISPR-Cas is being used therapeutically to perform *ex vivo* editing and then to introduce the edited cells into a patient's body. Genome editing may be particularly useful when used to alter stem cells including induced pluripotent stem cells. Several clinical trials evaluating the safety and efficacy of genome-editing treatments are underway.
- Antibodies manufactured in a laboratory are available to treat COVID-19 or respiratory syncytial virus. Viruses can evolve to escape binding to these antibodies.

Questions

1. What is the molecular explanation for why attenuated vaccines require less adjuvant and fewer boosters than subunit vaccines?
2. What is intended to be universal about a universal flu vaccine?
3. What are two mechanistically different ways that drugs mimicking nucleoside triphosphate interfere with viral replication?
4. Why is it important to treat viruses such as hepatitis C and HIV with more than one drug at the same time?
5. What are some disadvantages of using phages and proteins derived from phages as antibacterial treatments?
6. When testing a candidate antiviral drug biochemically, you find that the drug lowers the K_M of a viral protease. What kind of inhibitor is it and where does it bind to the enzyme? Explain.
7. Compare and contrast a genetically engineered virus used for oncolytic therapy versus gene therapy.
8. Compare and contrast a virus genetically engineered for vaccination with one genetically engineered for gene therapy.
9. In order for the Rexin-G proviral DNA to integrate into a host cell during gene therapy, what viral proteins must have been provided by the previous host cell during vector manufacture?
10. Provide two examples of safety concerns that led to alterations in an existing medical technology.

Interactive quiz questions

In addition to the questions provided above, this edition has a range of free interactive quiz questions for students to further test their understanding of the chapter material. To access these online questions, please visit the book's website: www.routledge.com/cw/lostroh.

Further reading

Vaccine development

Mezhenskaya D, Isakova-Sivak I, Matyushenko V, Donina S, Rekstin A et al. 2021. Universal live-attenuated influenza vaccine candidates expressing multiple M2e epitopes protect ferrets against a high-dose heterologous virus challenge. *Viruses* 13(7):1280.

Tomalka JA, Pelletier AN, Fourati S, Latif MB, Sharma A et al. F2021. The transcription factor CREB1 is a mechanistic driver of immunogenicity and reduced HIV-1 acquisition following ALVAC vaccination. *Nature Immunol* 22(10):1294–1305.

Antiviral drugs

Kumar G, Cuypers M, Webby RR, Webb TR & White SW 2021. Structural insights into the substrate specificity of the endonuclease activity of the influenza virus cap-snatching mechanism. *Nuc Acids Res* 49(3):1609–1618.

Owen DR, Allerton CM, Anderson AS, Aschenbrenner L, Avery M et al. 2021. An oral SARS-CoV-2 Mpro inhibitor clinical candidate for the treatment of COVID-19. *Science* 374(6575):1586–1593.

Todd B, Tchesnokov EP & Götte M 2021. The active form of the influenza cap-snatching endonuclease inhibitor baloxavir marboxil is a tight binding inhibitor. *J Biol Chem* 296:100486.

Gene therapy

Bennett J, Wellman J, Marshall KA, McCague S, Ashtari M et al. 2016. Safety and durability of effect of contralateral-eye administration of AAV2 gene therapy in patients with childhood-onset blindness caused by RPE65 mutations: a follow-on phase 1 trial. *Lancet* 388(10045):661–672.

Lewis R 2012. *The Forever Fix*. St. Martin's Press.

Pasi KJ, Rangarajan S, Mitchell N, Lester W, Symington E et al. 2020. Multiyear follow-up of AAV5-hFVIII-SQ gene therapy for hemophilia A. *New Eng J Med* 382(1):29–40.

Stolberg SG 1999. The Biotech Death of Jesse Gelsinger. *New York Times Magazine* 28:136–140, 149–150; see also http://www.nytimes.com/1999/11/28/magazine/the-biotech-death-of-jesse-gelsinger.html, accessed December 24, 2016.

Therapeutic genome engineering with CRISPR-Cas

CRISPR: The good, the bad, and the unknown. Archive at http://www.nature.com/news/crispr-1.17547, accessed December 27, 2015.

Gaudelli NM, Komor AC, Rees HA, Packer MS, Badran AH et al. 2017. Programmable base editing of A•T to G•C in genomic DNA without DNA cleavage. *Nature* 551(7681):464–471.

Traxler EA, Yao Y & Wang Y 2016. A genome-editing strategy to treat β-hemoglobinopathies that recapitulates a mutation associated with a benign genetic condition. *Nat Med* 22:987–990.

Yang L, Güell M & Niu D 2015. Genome-wide inactivation of porcine endogenous retroviruses (PERVs). *Science* 350:1101–1104.

Viral Diversity, Origins, and Evolution

17

Virus	Characteristics
Picornaviruses	Order of viruses substantially updated after a metagenomic study of invertebrate RNA viruses.
Mimivirus	Nucleocytoplasmic large DNA virus once thought to be a cell because of its enormous size; encodes thousands of proteins including some used during translation.
Adeno-associated virus (AAV)	Satellite virus that depends on co-infection with a helper virus for replication.
Influenza A	Example of zoonotic virus that periodically causes pandemics and provides examples of the roles of mutation, recombination, genetic drift, and selection during viral evolution.
Simian immunodeficiency virus (SIV)	Ancestor of HIV.
Human immunodeficiency virus (HIV)	Example of zoonotic virus recently acquired from nonhuman animals; provides examples of the roles of genetic drift, selection, and post-transfer adaptation during viral evolution.
SARS-CoV-2	Zoonotic coronavirus possibly acquired from bats.
MERS-CoV	Zoonotic coronavirus acquired from camels.
Ebola virus	Zoonotic filovirus acquired from bats.
Endogenous retroviruses	Defunct retroviruses inherited vertically through the germ line; ubiquitous in the human genome. Some encode proteins adapted for new functions important for the evolution of pregnancy.

The viruses you will meet in this chapter and the concepts they illustrate

Most of this book has focused on case studies in which the molecular biology of viral replication is well understood. In this chapter, we turn instead to viral diversity, origins, and evolution, which will introduce some new characters and examine familiar viruses in a new light. Viral diversity is breathtaking. Viruses infect every known species of cellular life. Although there are self-replicating subviral nucleic acids smaller than most cellular genes, there are also viruses so enormous that they were once thought to be cells. Viral origins are ancient, and there is no theory that provides a unified explanation of how every known group of viruses originated. Some may have precellular origins, whereas others may be derived from cells; some may have provided the selective pressure that drove the emergence of cells themselves.

DOI: 10.1201/9781003463115-17

Like cells, viruses and subviral entities can evolve because they have genomes consisting of nucleic acids, which are therefore subject to heritable change. Viral evolution operates on the same general principles as cellular evolution. For example, it requires genetic variation, which can arise through mutation or recombination. **Mutation** refers to changes in the nucleotides that comprise a genome, whereas **recombination** refers to larger-scale changes in a viral genome that bring new alleles of different genes together. An allele is a variation of a gene that has a different sequence as compared to another allele of the same gene. Viruses have very large population sizes, which also lead to high levels of genetic variation. The frequency of alleles in a virus population can change because of random factors, in which case the change is known as **genetic drift**, or because of **natural selection**. Genotypes with the highest **fitness** result in more offspring being transmitted to new hosts and thus are favored by natural selection.

Although the molecular biology pertaining to the origins of viruses is not clear, the evolution of influenza virus has been examined almost since the first discovery of viruses as subcellular particles. The emergence of influenza pandemics provides a good example of how the molecular biology of viral replication affects viral evolution. Pandemic influenza, HIV and SARS-CoV-2 are medically dangerous viruses—contracted originally from nonhuman animals. The molecular biology of the changes in HIV from its ancestral virus that infects other great apes to a virus that is human-specific provides a comprehensive example of zoonotic viral adaptation to a new host. We have also observed SARS-CoV-2 evolution very closely as it has spread to become an important global pathogen. The final sections of this chapter discuss some molecular and cellular examples of the tremendous effects that viruses and subviral entities have had on the evolution of cellular life.

To appreciate viral evolution, we turn first to a comprehensive catalog of viral diversity.

17.1 The viral world is extremely diverse

Viruses are diverse by any measure. For example, the size, composition, and shape of virions varies widely. They range in size from 40 to 1,000 nm in diameter. Some have envelopes, but others do not. They can have simple helical or icosahedral shapes, or they can have more complicated structures, such as the tailed bacteriophages that resemble a lunar lander or bricklike poxviruses. Their most stunning diversity is found at the level of molecular genetics. They use every possible strategy for encoding proteins, and many employ unique gene expression and genome replication strategies not otherwise found among their cellular hosts.

Although most newly discovered viruses are similar to known viruses, we may yet discover completely new ones. An example is the mimivirus, which was first detected in 1992 but was not recognized as a virus until 2003. For more than a decade after its discovery, mimivirus was mistakenly thought to be a cell because of its complex structure and size (**Figure 17.1**). It is an enveloped virion that is 0.7 μm in diameter and it has a dsDNA genome of 1.2 million base pairs; both of these size measurements put it well within the cellular range. Many other megaviruses have since been discovered, and now it appears that they are not unusual at all but are instead ubiquitous in aquatic environments. The largest megaviruses are 1.5 μm in diameter (larger than *Escherichia coli*) and encode more than 2,500 genes in more than 2.5 million base pairs of dsDNA. Megaviruses even have some translation-related genes, thus blurring the distinctions between cells and viruses even further.

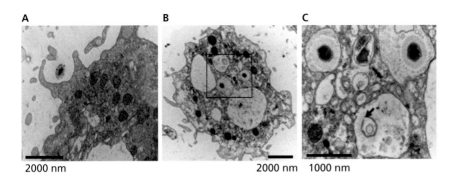

Figure 17.1 Mimivirus. Mimivirus is an example of a gigantic nucleocytoplasmic large DNA virus. (A) Mimivirus enters its ameba host by phagocytosis. (B) Several viral particles within cellular vacuoles. (C) Higher magnification of the boxed area in (B) that shows an empty particle after uncoating (black arrow). (From Suzan-Monti M et al. 2007. *PLOS ONE* 2:e328. doi: 10.1371/journal.pone.0000328. Published under CC BY 4.0.)

Megaviruses have already changed the way that we see cellular evolution. For example, megavirus replication factories resemble cell nuclei in some regard, leading to speculation that a megavirus could be the origin for the eukaryotic nucleus. Most megavirus genes are not similar to any known cellular genes, leading to the hypothesis that megaviruses may be descended not from any extant cells or even from their **last universal common ancestor (LUCA)** but rather from a cellular domain that went extinct before LUCA. Megaviruses can even be infected by their own viruses, termed virophages. Although they have many unique attributes, analysis of their coding sequences revealed that megaviruses and poxviruses are related and belong to an order-level taxonomic group provisionally called **nucleocytoplasmic large DNA viruses**. They are so named because some express their genes in the cytoplasm, whereas others express their genes in the nucleus.

Although some members of the virus world are unexpectedly large, there are several classes of subviral entities that are simpler than viruses yet behave in some ways like viruses. For example, they reproduce inside hosts and spread from one host to another or from one host chromosome to another. Examples include satellite viruses, viroids, transposons, and introns. Satellite viruses depend on co-infection with another virus to replicate. Satellite viruses and viroids spread from one host to another, whereas transposons move within a single host genome. Through the sequencing of more and more genomes, the interrelatedness of viruses and subviral entities is increasingly recognized as fundamental for understanding viral diversity, origins, and evolution, and so Sections 17.2–17.4 define some of these entities.

17.2   Satellite viruses and nucleic acids require co-infection with a virus to spread

Satellite viruses require co-infection with another virus to complete their replication cycle and spread to a new host. An example is the dependovirus adeno-associated virus (AAV), which requires co-infection with adenovirus or herpesvirus to complete its replication cycle. AAV virions are naked and contain either (−) DNA or (+) DNA, but not both. The ends of the linear DNA are complementary to each other and self-complementary so that they can form a hairpin. The single-stranded AAV genome is very short at 4.7 kb long, yet it encodes four nonstructural Rep proteins and three structural VP proteins (**Figure 17.2**). These seven proteins are encoded by six mRNAs that are generated from three different promoters that are expressed at early, middle, and late time points during infection. All of the proteins are encoded on the same strand, and the mRNAs are generated by alternative splicing. Two of the structural proteins are encoded by the same mRNA through a leaky scanning mechanism.

Figure 17.2 Adeno-associated virus-2 genome and the proteins it encodes. AAV2 has a ssDNA genome with hairpin ends. Once the host converts it into dsDNA, the genome encodes seven proteins using six mRNA molecules that are transcribed from three promoters.

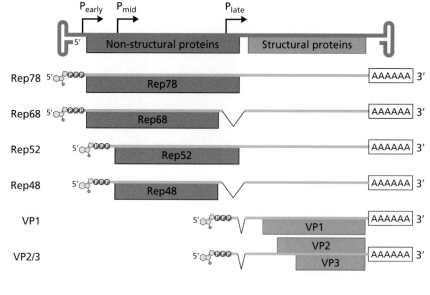

In the absence of co-infection by a helper virus, the AAV genome enters the nucleus where it is converted into dsDNA and persists as an episome; this property makes it a popular gene therapy vector (see **Chapter 16**). In the presence of helper viruses, replication occurs in nuclear viral replication compartments.

AAV genome replication is thought to proceed by a rolling-circle mechanism even though the genome is not circular (**Figure 17.3**). Upon uncoating, the host extends one of the hairpins, resulting in a double-stranded structure. During genome replication, an AAV rolling-circle Rep protein nicks this double-stranded template and then a strand-displacement mechanism allows copying of the hairpin sequences at one end. A similar process occurs at the other end, resulting in a linear double-stranded template with no secondary structure. This structure spontaneously rearranges so that hairpins form at one end or the other and then these hairpins are used to initiate synthesis of more DNA because they have a free 3′ OH opposite a template strand. The AAV, helper virus, and host proteins that participate directly in genome replication are not well characterized.

Satellite nucleic acids are similar to satellite viruses in that they depend on a helper virus for replication, but they differ in that they do not encode a capsomer. Most satellite nucleic acids are even smaller than the genomes of satellite viruses. There is one known satellite DNA, and there are three classes of satellite RNAs. The classes differ from each other in genome length and in the arrangement of their genomes. Some have linear genomes, and others have circular genomes. Sometimes they do not encode any proteins at all. In that case, they probably serve regulatory functions as noncoding RNAs.

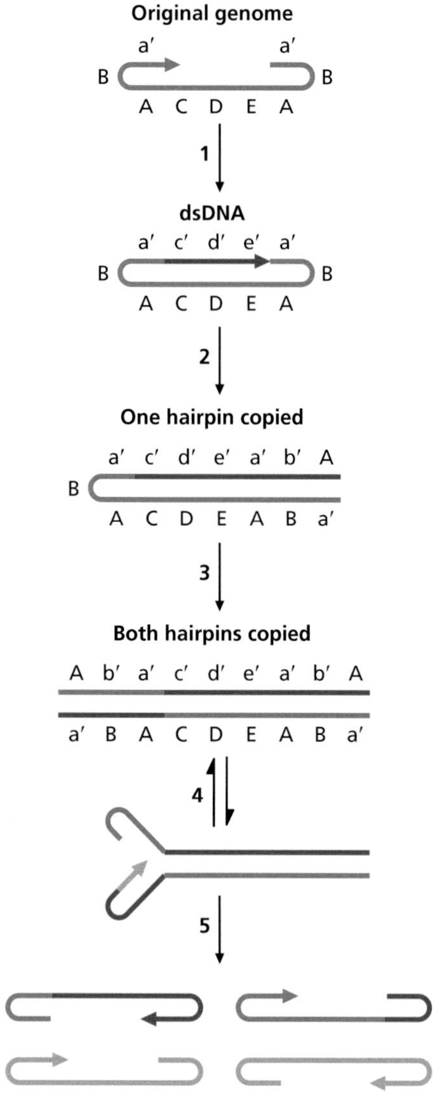

Original genome

a′   a′

B   B

A   C   D   E   A

1

dsDNA

a′   c′   d′   e′   a′

B   B

A   C   D   E   A

2

One hairpin copied

a′   c′   d′   e′   a′   b′   A

B

A   C   D   E   A   B   a′

3

Both hairpins copied

A   b′   a′   c′   d′   e′   a′   b′   A

a′   B   A   C   D   E   A   B   a′

4

5

Figure 17.3 Genome replication in AAV. Because the protein components that catalyze these reactions are not well characterized, this diagram emphasizes only the nucleic acids. Upon uncoating, the host converts the ssDNA with hairpin ends into dsDNA with sealed ends (1). During genome replication, one of the AAV Rep proteins nicks the dsDNA and DNA complementary to one hairpin is synthesized (2) using a strand-displacement mechanism. The other hairpin is similarly copied, resulting in a linear double-stranded template with terminal repeats but no hairpin structures (3). This structure spontaneously rearranges and hairpins form at one end; these hairpins are extended to synthesize more DNA (4). Replicated molecules are processed into full-length genomes (5) before packaging. The processing steps are not well understood.

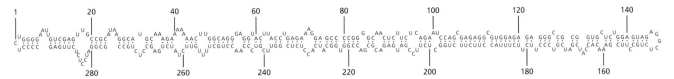

Figure 17.4 Structure of a viroid. The *hop stunt viroid*, which infects grapes, is an example of a viroid. Viroids are RNA with a secondary structure that confers a rod-shaped structure. (From Navarro B et al. 2009. *PLOS ONE* 4:e7686. doi: 10.1371/journal. pone.0007686. Published under CC BY 4.0.)

17.3 Viroids are infectious RNA molecules found in plants

Viroids are small RNA molecules that have extensive secondary structure, encode no proteins, and yet infect plants. They are closed circles in which each half of the circle is complementary to most of the other half so that they form rodlike structures of RNA (**Figure 17.4**). Viroids infect plants through mechanical damage and do not require encapsidation for transmission. Viroid interaction with plant proteins and processing into RNAi (see **Chapter 14**) results in disease symptoms. Some viroids replicate in the cytoplasm, whereas others replicate in the nucleus. Their replication is autonomous in that it does not rely on proteins provided by any viruses; instead, host RNA Pol II, which is normally restricted to DNA templates, copies the viroids through an unknown mechanism. During viroid replication, concatemers form. In some cases, host RNases process the concatemers into mature offspring viroids. In other cases, the viroid itself has ribozyme activity and cleaves and ligates itself to make offspring from concatemers.

Hepatitis D virus (**HDV**) is the only known viroidlike element that infects humans. It has properties similar to both viroids and satellite viruses. HDV has a minuscule 1.7-kb genome that is folded up on itself like a viroid. Also like a viroid, host RNA polymerases copy the genome. One end of the genome is particularly similar to viroids and encodes a ribozyme that processes concatemers into mature new genomes. Unlike viroids, HDV forms enveloped particles. It also encodes two nearly identical proteins called hepatitis δ antigen large and small; they differ in that the long antigen is 20 amino acids longer than the short one. These antigens are found in HDV particles. Furthermore, HDV can replicate only in the presence of a helper hepatitis B, which provides the envelope spike proteins and other proteins necessary for HDV's replication cycle. For unknown reasons, co-infection with HDV and hepatitis B virus makes the symptoms of hepatitis more severe. HDV has never been detected in the absence of hepatitis B co-infection.

17.4 Transposons and introns are subviral entities

Transposons (also known as **transposable elements**) are DNA sequences that can move from one place to another in a DNA genome. They are similar to viruses in that they replicate at the expense of their host. Furthermore, some of the proteins that they encode are homologous to viral proteins. All transposable elements are mutagens because they can insert into functional DNA such as promoters or coding sequences and thereby disrupt the function of that DNA. In fact, mutations caused by transposon insertion are responsible for some forms of hemophilia and severe combined immunodeficiency, among other human diseases. Despite their potential to cause deleterious effects, transposons are found in almost every known cellular genome. Class I transposons, also known as **retrotransposons**, move via a reverse transcription-based **copy-and-paste mechanism** in which the retrotransposon is first copied into RNA, then reverse transcribed, and then inserted into a new location, leaving the template retrotransposon in its original location (**Figure 17.5**). The reverse transcriptase is usually encoded by the retrotransposon itself. In vertebrates including humans, genomes have many copies of retrotransposons. Class II **DNA-based transposons** move instead through recombination of DNA. Most DNA-based transposons encode a transposase enzyme that uses a **cut-and-paste mechanism** (to excise the transposon from the DNA and insert it somewhere new). Some of these cut-and-paste transposase enzymes are homologous to viral enzymes that recombine viral genomes with a host chromosome. Other DNA-based transposons called helitrons use a copy-and-paste mechanism in which an original copy of the transposon remains at its site while another copy is inserted elsewhere in the genome (**Figure 17.6**). The copy-and-paste DNA transposition is mediated by a protein homologous to viral rolling-circle replication (Rep) proteins (see **Chapter 9**).

Until very recently, it was thought that no transposon could encode a capsomer. The discovery of a new class of very large transposons has brought this idea into question. These large DNA-based transposons, called polintons, encode 10 proteins including a DNA polymerase and an integrase homologous to retrovirus integrases. Polintons encode two capsomers. Thus, polintons may form infectious particles under some as yet to be discovered conditions. In support of this hypothesis, a new group of polintonlike viruses that do encode capsomeres was discovered in 2015. Although they encode capsomers, they do not encode an integrase. The molecular biology of polinton and polintonlike virus replication cycles is unknown.

There are four types of **introns**, some of which are mobile like transposons. **Nuclear spliceosomal introns** are the kind that interrupt mRNA in eukaryotes and in many eukaryotic viruses such as adenovirus (see **Chapters 8–10**). Although they are not mobile in the present, they have shared ancestry with

Figure 17.5 Retrotransposon. Retrotransposons move through a copy-and-paste mechanism that relies on reverse transcriptase.

Figure 17.6 DNA-based transposon. (A) Most DNA-based transposons move using a cut-and-paste mechanism catalyzed by transposase. (B) Rarer helitrons move using a copy-and-paste mechanism that uses a DNA intermediate.

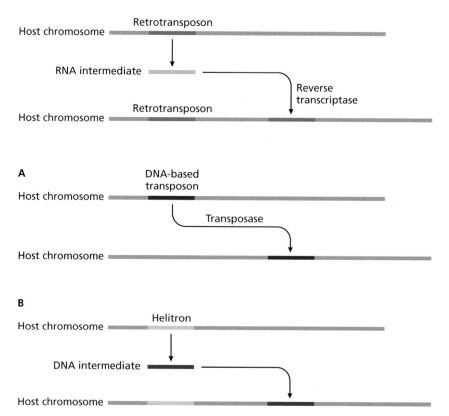

Group II introns. Group II introns are autocatalytic **ribozymes** (enzymatic RNA molecules) that splice themselves out of rRNA, tRNA, and mRNA in a wide range of organisms, but they probably originated in bacteria. Some of them encode a reverse transcriptase that is homologous to all other known reverse transcriptases.

17.5 Viruses have ancient origins

Viruses and related subviral entities are the most abundant members of the biosphere and encode the most genes and the most diverse genes. They also use the most diverse range of strategies for encoding genetic information. These observations all suggest ancient origins for viruses and subviral entities because only replication over vast periods of time could have generated such diversity. There are three simplified views of how viruses and subviral entities could have originated (**Figure 17.7**). The first is that they were once cells; these cells degenerated and lost most of their coding capacity and became viruses. The second is that viruses and subviral entities are escaped genes that first came from cells, originating from cellular chromosomes, plasmids, or mRNA. The nucleic acids acquired capsids and sometimes envelopes and became viruses. The third is that viruses and subviral entities are remnants of precellular life where nucleic acids and proteins existed before there were cells. When such nucleic acids acquired capsids and sometimes envelopes, they became viruses. An emerging consensus is that viral and cellular origins are intimately intertwined and that the three simplified versions of viral origins have likely all occurred.

Sequence comparisons that reveal underlying homology provide the main evidence used to support theories of viral evolutionary origins. Structural similarities in addition to primary sequence similarities are used to examine viral history because viruses are so ancient and reproduce so quickly. Because the history of viral evolution is so ancient, proteins may still have a shared structure that reflects homology even though the primary sequences have diverged so much that we no longer detect statistically significant similarity. A problem with this approach is that proteins can also have similar structures not because of homology but rather because of **convergent evolution**—they lack shared ancestry but have similar structure for other reasons, such as natural selection, for the same functional properties. The search for homology using primary sequences is a less controversial approach and is the focus here; as we learn thousands and even millions more virus sequences, discovering homologous proteins in distantly related viruses becomes more likely.

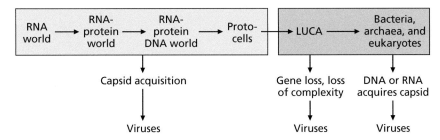

Figure 17.7 Three theories of viral origins. Some viruses are remnants of precellular life and may have originated at any of the periods between the RNA world and LUCA. Some viruses are descended from ancient cells such as LUCA or from bacteria, archaea, or eukaryotes through loss of genes and simplification of structure. Some viruses are escaped genes that originated as cellular DNA or mRNA but then acquired the ability to encode a capsid.

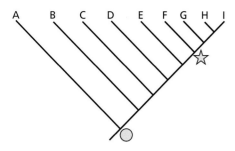

Figure 17.8 Monophyletic relatedness. Monophyletic sequences share a single common ancestor. Proteins F, G, H, and I are monophyletic with their common ancestor marked by a star. All of the proteins are monophyletic with respect to the ancestor marked by a circle.

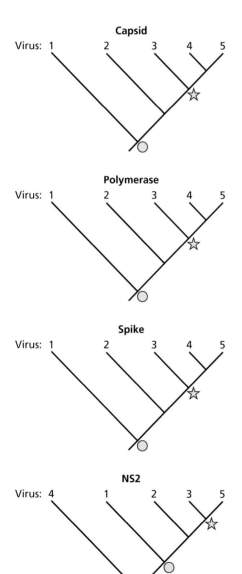

17.6   Viral hallmark proteins can be used to trace evolutionary history

Because all cells share homologous 16S/18S rRNA, it has been possible to construct a universal tree of cellular life, showing that, despite extensive horizontal gene transfer, all living cells are descended from cells that used similar ribosomes. Because there are no genes shared by all viruses, such an analysis of the relatedness of all viruses is not possible. The lack of such a gene provides evidence that all viruses do not share a common ancestor. It has been possible to identify **viral hallmark genes**, however, that are useful for tracing the evolutionary relatedness of viruses. Although viral hallmark genes are not conserved universally among all viruses, they do share common properties. For example, all viral hallmark genes encode proteins and are shared by many diverse groups of viruses and have no homologs or only very distant homologs in cellular organisms. Any given viral hallmark gene is not only homologous but also **monophyletic**, meaning that all the viral sequences have a single shared last common ancestor (**Figure 17.8**). Viral hallmark proteins play direct roles in viral genome replication, packaging, or assembly and so are essential for the virus life cycle, and from one virus to the next the same viral hallmark protein plays a similar role (much like 16S/18S rRNA playing a similar role in all cells). Finally, viral hallmark proteins are the most ubiquitous proteins in the viral world.

The major types of viral hallmark proteins are jelly-roll capsid proteins, α-helical major capsid proteins, superfamily 3 helicases, RNA-dependent RNA polymerases (RdRps) and reverse transcriptases (which are a single monophyletic group), rolling-circle DNA replication initiation proteins, viral DNA primases, UL9-like superfamily 2 helicases, genome packaging ATPases, and the ATPase subunit of viral terminases (**Table 17.1**). These proteins are found broadly distributed among viruses, which means that the groups of viruses that encode them either are descended from a common ancestor or have shared genes extensively through horizontal gene transfer. Software that aligns sequences and displays their relatedness as a tree can detect horizontal gene transfer by comparing the trees for several different genes shared by two viruses. If one gene has a tree that is very different from the trees generated by comparing the other genes, the gene with the unusual tree was probably distributed among the viruses by horizontal gene transfer (**Figure 17.9**). There are examples of both ancient common ancestry and horizontal gene transfer accounting for the distribution of viral hallmark genes. It is also possible that because viral hallmark proteins are not homologous to cellular proteins, all viral hallmark genes originated before LUCA.

Figure 17.9 Horizontal gene transfer can be detected by comparing trees representing the similarity of multiple proteins. In this hypothetical scenario, the evolutionary trees were created to investigate the ancestry of viruses 1–5. Using sequences based on the viral capsid, polymerase, and spike, the viruses share a certain pattern of ancestry. In contrast, using the NS2 sequences, the pattern of ancestry is different as can be seen by observing the position of NS2 in the tree relative to the ancestors marked by the star and the circle. The most likely explanation is that the NS2 gene moved into virus 4 through horizontal gene transfer (recombination).

Table 17.1 **Viral hallmark proteins.**

Protein	Function	Viruses	Examples
Jelly-roll capsomere	Major capsid protein	Picornaviruses, herpesviruses, nucleocytoplasmic large DNA viruses (NCLDVs), papovaviruses, parvoviruses, icosahedral DNA phages, some archaeal viruses	Poliovirus VP1, VP2, and VP3; vaccinia D13
α-Helicase capsid protein	Major capsid	Tristromaviruses	Thermoproteus tenax virus 1 major capsid protein
HK97-fold capsid protein	Major capsid	Certain phages, adenoviruses	PRD-1 capsid; adenovirus hexon
Superfamily 3 helicase	Genome replication	Picornaviruses, NCLDVs, eukaryotic ssDNA viruses, some dsDNA phages, and some plasmids	Poliovirus 2C^{ATPase}; SV40 large T antigen
UL9-like superfamily 2 helicase	Genome replication	Herpesviruses, NCLDVs, some dsDNA phages	HSV-1 UL9; phage T5 D2
RdRp/reverse transcriptase	Replication of RNA genomes	(+) RNA viruses, dsRNA viruses, reverse-transcribing viruses, (−) RNA viruses, retrotransposons	Poliovirus 3D^{pol}, HIV reverse transcriptase
Rolling-circle initiator	DNA genome replication	Parvoviruses, other small ssDNA viruses, some ssDNA phages, plasmids, and certain DNA-based transposons	Minute virus of mice NS1
Archaeo-eukaryotic viral DNA primase	Genome replication	Herpesviruses, NCLDVs, some dsDNA phages	HSV-1 UL52
Packaging ATPase	Assembly	Adenoviruses, NCLDVs, some ssDNA and some dsDNA phages	Adenovirus IVa2; phage Phi29 packaging motor
Terminase ATPase subunit	Assembly	Herpesviruses, tailed dsDNA phages	HSV-1 UL15; phage T4 gp17

Although we use the Baltimore Classification System to organize viruses, that system was not developed to reflect viral evolutionary history. It is based instead on convenience because viruses with similar genome arrangements have similar ways of expressing and replicating their genomes. The Class I dsDNA poxviruses, herpesviruses, adenoviruses, papillomaviruses, and polyomaviruses, for example, do not appear to share a single common ancestor, even though it was convenient to consider their gene expression strategies together in a single chapter because they all rely on host RNA Pol II for mRNA synthesis. There is an emerging consensus that there are six "realms" of viruses that are united by specific viral hallmark genes and could reflect at least six independent origins for different groups of viruses (Table 17.2).

Understanding of viral evolutionary history is likely to change dramatically in the immediate future because of the discovery of thousands of new viral genome sequences made possible by metagenomic technologies. **Metagenomics** is the study of all genetic information sampled from a habitat such as a liter of sea water, a gram of soil, or a small invertebrate. DNA or RNA can be isolated directly from the sample without first culturing the organisms or viruses in the sample, in which case the collection of all DNA or RNA in the sample is referred to as the **metagenome** for that sample. Innovations in bioinformatics will also radically transform how we interpret sequences gathered using metagenomics. The metagenome revolution is addressed in the next section.

Table 17.2 **Six viral realms.**

Realm	Viral Hallmark Protein	Example
Adnaviria	α-Helical capsid protein	Thermoproteus tenax virus 1
Duplodnaviria	HK97-fold major capsid protein	Enterobacteria phage T7, herpes simplex virus
Monodnaviria	HUH superfamily endonuclease	ΦX174, human papillomavirus
Riboviria	RdRp and reverse transcriptase	Bacteriophage Qβ, tobacco mosaic virus, poliovirus, SARS-CoV-2, rabies virus, influenza virus, HIV
Ribozyviria	Deltavirus ribozyme	HDV
Varidnaviria	Jelly-roll major capsid protein	Bacteriophage PRD1, adenovirus

17.7   Metagenomics is revolutionizing evolutionary understanding of viruses

Metagenomics results in discovery of thousands of new viral genomes every year and the pace of discovery is accelerating. The genomes can be transcribed and translated by software and then the proteins they encode can be subjected to bioinformatics analysis, all without ever culturing a single virus. Sampling diverse environments leads to discovery of many new virus genomes. The human gut is a rich source of plant virus genomes because of our omnivorous diets. Deep-sea hydrothermal vents have diverse virus communities with especially large populations of dsDNA bacteriophages and archaeal phages.

One instance of the revolutionary potential of viral metagenomics is provided by a 2016 study that determined the viral RNA metagenomes associated with 220 different invertebrate species. The sampled invertebrates include crickets, mosquitos, spiders, crabs, flies, shrimps, clams, octopuses, nematodes, and worms (among others). The study discovered 1,445 distinct virus genomes that encode an RdRp. The 1,445 new genome sequences did not always group with previously known genomes so that it was necessary to create new groups of RNA viruses that must substantially change the existing orders recognized by the International Committee on Taxonomy of Viruses (ICTV). For example, the ICTV had recognized three orders of (+) ssRNA viruses including the *Picornavirales*, whereas many other families of viruses such as the *Caliciviridae* could not be assigned to an order. Because of the new sequence information from this 2016 metagenomics study, it is now clear that the caliciviruses belong in the same order as the picornaviruses, which as of 2023 is called **Picornavirales**. On the other hand, the new genomes also provide strong evidence that some previously recognized orders remain monophyletic and should not be changed by the new genomes; an example is the order *Mononegavirales* ([−] ssRNA viruses with nonsegmented genomes).

The 2016 study of invertebrate RNA virus metagenomes also discovered that the evolutionary history of the RdRp proteins often differs from the evolutionary history of capsomeres or glycoprotein spikes, indicating extensive recombination among viruses (**Figure 17.10**). The study observed recombination across wide evolutionary differences including horizontal gene

RdRp phylogenies

Capsid phylogenies

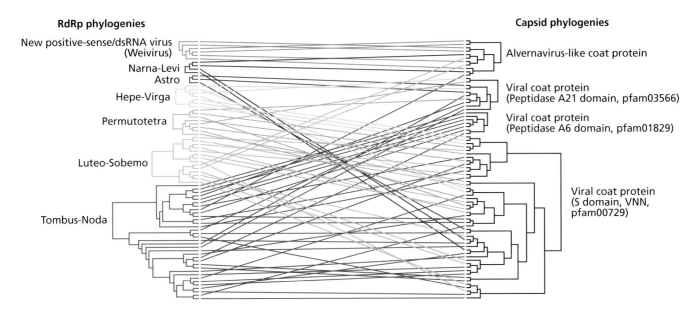

Figure 17.10 The ancestry of viruses according to RdRp sequences disagrees with that reconstructed from capsomers. Phylogeny based on the RdRp is on the left and phylogeny based on the capsomers is on the right. Colored lines connecting different viruses in the phylogenies highlight disagreement. The most likely explanation is horizontal gene transfer (recombination). (From Mang S et al. 2016. *Nature* 540:539–543. doi: 10.1038/nature20167. Courtesy of Springer Nature.)

transfer of glycoprotein spike genes between viruses from different orders of (−) ssRNA viruses, between viruses from different orders of (−) ssRNA and (+) ssRNA viruses, and even between (−) ssRNA and DNA viruses. In order for this mixing of genes to occur, two very different viruses would have to infect the same cell in the same animal and then genes from both parents must recombine. Recombination seems very unlikely for (−) ssRNA genomes and (+) ssRNA genomes, let alone for (−) ssRNA genomes and DNA viruses. Nevertheless, the extremely large population size of viruses, their rapid reproduction cycles, and the long period of time over which they have been evolving makes even improbable events possible. The molecular mechanisms that account for horizontal gene transfer among viruses with such different genetic systems are unknown.

The discovery of viral nucleic acid and protein sequences is increasing exponentially, but we still know about only a tiny fraction of viruses. The 2016 invertebrate RNA virus metagenome study provides a single example of how a mere 1,500 new genomes can substantially change our view of virus evolution. Thus, when we consider the evolutionary history of viruses and especially their origins, it is important to keep in mind that discovery of new viral genomes will undoubtedly change our understanding of viral origins.

17.8 Viral genetic diversity arises through mutation and recombination

There are two major sources of change in nucleotide sequences over time: mutation and recombination. Mutation occurs when a mistake during replication or a physical or chemical assault changes a nucleotide to something different from that in the ancestral genome. Cellular DNA polymerases have editing functions that keep the intrinsic rate of nucleotide substitutions very low, but many viral polymerases have no editing functions. Editing refers to the process of excising a misincorporated nucleotide and replacing it with the correct nucleotide during replication. Because most viral polymerases

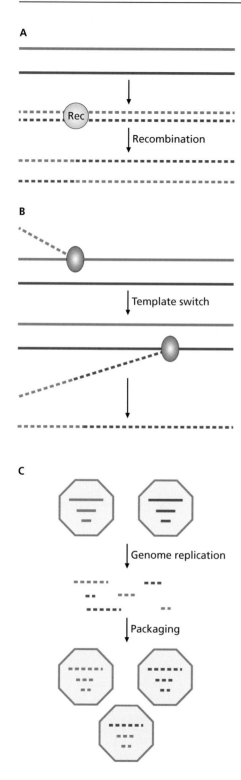

Figure 17.11 Viral genetic recombination. In each example, the cells have been infected with two viruses. The genome of one virus is blue and the other is red. Infecting genomes are solid and offspring genomes are dashed. (A) Recombination of DNA viruses using homologous recombination. The genomes are double stranded, and cellular or viral recombination proteins catalyze homologous recombination. (B) Recombination by template switching. (C) Recombination of segmented RNA virus.

lack editing functions, viruses that rely on viral polymerases accumulate mutations at a much higher rate per generation than cells do. Viral RdRps are particularly mutagenic; they misincorporate the wrong base approximately 1 in 10,000 times, with a range of 1 in 1,000 to 1 in 100,000, depending on the RdRp. These values are between 10 and 10,000 times higher than the misincorporation rate of cellular DNA replication. An example of viral mutation caused by misincorporation is the diversity of HIV in an infected person's body (see **Chapter 16**). Misincorporation is responsible for much more genetic change than physical and chemical assaults, but UV light is still an important source of mutation. High doses of UV light can be used as a virucidal treatment because the radiation prevents the genome from functioning properly after uncoating. UV light causes formation of covalent bonds between atoms of adjacent pyrimidines. These atoms would otherwise contribute to hydrogen bonding during synthesis of new nucleic acids. The disruptions to hydrogen bonding cause alternations to the viral mRNA, new genomes, or both, so that functional viral proteins cannot be synthesized. Small insertions and deletions that occur during copying mistakes are also common and are referred to as **indels**.

Recombination occurs when two viruses co-infect the same cell and an offspring virus contains a combination of genes that originate with both parents instead of just one parent. These are larger-scale genetic changes than those typically classified as mutations. Horizontal gene transfer is an example of recombination. Homologous recombination (**Figure 17.11A**), initiated by proteins that recognize regions of identity between two DNA genomes, is one mechanism of recombination among the DNA viruses. Recombination also may occur when a viral polymerase switches templates during synthesis of new nucleic acids; this form of recombination accounts for recombination in the coronaviruses such as SARS-CoV-2 (**Figure 17.11B**). Finally, viruses with segmented genomes can undergo recombination when two different viruses infect the same host cell and segments of different origins assort together into new virus particles (**Figure 17.11C**). An example of viral recombination can occur when two different influenza A viruses co-infect the same cell. Influenza viruses have segmented genomes, and recombinant offspring of two different influenza A viruses have genomes composed of a new combination of genome segments, some from one of the parental viruses and one or more from the other (see **Chapter 16** and **Section 17.12**).

The evolutionary history of influenza A has been studied intensively because of its effects on human health. Influenza A provides examples of extensive diversity, mutation, and recombination and will therefore be considered in **Sections 17.9–17.13**.

17.9 Genetic diversity among influenza A viruses arises through mutation and recombination

The genetic diversity of influenza A virus has been studied for many years because of the global impact of influenza. In typical years, influenza A virus causes millions of serious infections and 250,000–500,000 deaths. During influenza pandemics, morbidity and mortality are even higher. There have been seven influenza pandemics in the 120 years between 1889 and 2009 (the most recent influenza pandemic). Pandemic influenza A strains infect a high proportion of people and cause high mortality rates greater than 0.1%. The worst flu pandemic of the twentieth century, the 1918–1920 pandemic, infected 25% of the world's population and had a mortality rate of 2%–20%. A similar attack rate (25%) and mortality rate (2%) would result in 370 million deaths today. The 2009 pandemic influenza A virus also infected about 25% of the world's population but its mortality rate (0.1%–5%) was lower, at least in wealthy countries with good healthcare systems. The influenza A mortality rate in communities that lack access to good hospital-based care is unknown.

The evolution and origins of pandemic influenza strains are therefore of great interest. The molecular and cellular biology of influenza virus control its evolution. Influenza A viruses have (−) ssRNA genomes consisting of eight different segments; each segment encodes between 1 and 4 of the 17 influenza A proteins (**Figure 17.12**; see also **Chapter 6**). Because of the misincorporation rate for its viral RdRp, influenza A virus replication results in about one mutation per offspring genome. Every infected cell gives rise to 10,000 mutants. Such high genetic diversity provides plenty of variation for both natural selection and genetic drift to cause changes in allele frequencies very rapidly with respect to human time scales such as a few years. For influenza in particular, variation in protein sequences caused by mutation is known as **antigenic drift** (**Figure 17.13A**). Mutation causes the amino acid sequences of the immunogenic proteins to change slowly enough that adaptive immune response provoked by the evolving virus will usually protect against infection by one of its descendants for years.

Influenza A viruses are particularly prone to recombination because they have segmented genomes. When two influenza A viruses co-infect the same cell, some of the 10,000 offspring virions produced by the cell will have genome segments copied from both parental viruses (see Figure 17.11C). For example, a recombinant virus might have segment 4, encoding the hemagglutinin (HA) spike protein, from parent 1 and segment 6, encoding the neuraminidase (NA) spike protein, from parent 2. For influenza in particular, genetic variation caused by the recombination of one or more genome segments with segments from a different virus is known as an **antigenic shift** (**Figure 17.13B**). When an antigenic shift that changes the combination of

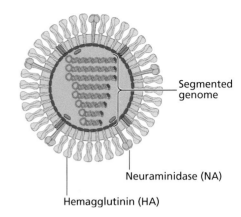

Figure 17.12 Influenza A virus. The virus has a segmented genome and prominent HA and NA spikes. (Courtesy of Philippe Le Mercier, ViralZone, © SIB Swiss Institute of Bioinformatics.)

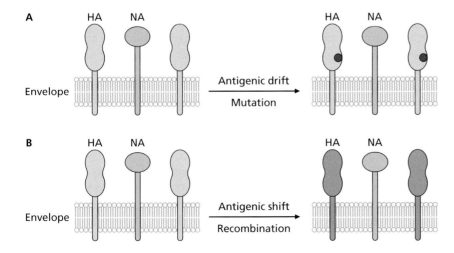

Figure 17.13 Influenza A virus antigenic drift and antigenic shift. (A) Antigenic drift occurs when mutations cause influenza A antigens such as the HA spike protein to change a small amount over several years. (B) Antigenic shift occurs when recombination causes influenza A antigens such as the combination of HA and NA spike proteins to change dramatically in a short period of time (such as a single transmission event).

HA and NA genes occurs, the result can be a pandemic because an adaptive immune response against one of the parent viruses is not effective against the recombinant offspring.

17.10   Influenza A spike proteins are particularly diverse

The diversity and evolution of the HA and NA spikes are affected by both antigenic drift and antigenic shift. The molecular functions of HA and NA are critical for their evolution. **HA** is the virus attachment protein that binds to the sugar neuraminic acid on host cells (see **Chapter 3**). **NA** is required for budding from host cells (see **Chapter 11**). In some ways, the HA and NA spikes have antagonistic functions because HA binds to neuraminic acid in order for an infecting virus to enter cells, and NA degrades neuraminic acid so that offspring virions can escape from their former host. Both HA and NA provoke a strong adaptive immune response that includes production of polyclonal antibodies that bind to HA and NA (see **Chapter 15**).

Humoral responses to HA and NA have long been used to classify influenza A viruses according to their reaction to libraries of human serum. The serum contains polyclonal antibodies. This practice originated before its molecular basis was elucidated. Type 1 HA viruses bind to the polyclonal antibodies found in the serum of someone who has survived a type 1 infection. Similarly, type 2 HA viruses bind to the polyclonal antibodies in the serum of someone who has survived a type 2 infection, and so on. Different types of HA proteins consist of such different epitopes that polyclonal antibodies that bind to one HA type do not bind well to other HA types. The same is true for the NA types (see **Chapter 16**). There are 18 known HA and 11 known NA **immunological types**, resulting in $18 \times 11 = 198$ possible different combinations; in practice, there are fewer than 10 combinations in circulation among humans. Examples of influenza strains include the H1N1 and H3N2 viruses used to manufacture 2013–2014 seasonal influenza vaccines for the northern hemisphere.

As is illustrated by the few immunological types in common circulation among humans, the major reservoir for influenza A genetic diversity is not humans. Instead, migratory aquatic birds are the main hosts for influenza A virus and they spread influenza A viruses on a global scale. Wild avian influenza strains can spread to other animals such as agricultural fowl and swine. In all known cases, human influenza pandemics have been caused by nonhuman animal strains that have evolved the ability to infect people and to be transmitted from person to person. New human pandemic strains arise (**emerge**) because of mutation or recombination followed by natural selection that favors viruses that are more fit for replicating in human populations than they are in nonhuman animals. Viral transmission from a nonhuman animal to a human is called **zoonotic transmission**. The molecular biology of the HA protein is critical for zoonotic transmission of influenza A.

17.11   Variations among influenza A viruses reflect genetic drift and natural selection

The only influenza pandemic to occur during the era of genomics began in 2009, so we have the most information about influenza diversity and evolution through studying that pandemic H1N1 (pH1N1) strain. When pH1N1 was first detected over a period of 20 days in the spring of 2009 in the United States, viruses were isolated from all 642 of the original patients and the viral genomes were sequenced. All were >99% identical, indicating that a single

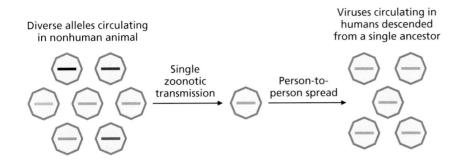

Diverse alleles circulating in nonhuman animal

Single zoonotic transmission

Person-to-person spread

Viruses circulating in humans descended from a single ancestor

Figure 17.14 Evolutionary bottleneck caused by zoonotic transmission. When a pandemic virus originates from a single zoonotic transmission, the genetic diversity in the viral population experiences a severe bottleneck.

zoonotic transmission event, probably from a pig to a person, was responsible for the origin of the new virus. The H1N1 virus population therefore experienced an evolutionary bottleneck when it jumped from nonhuman animals to people. A **bottleneck** is any sharp reduction in genetic diversity (**Figure 17.14**). In this case, the diversity of the large H1N1 population circulating among swine was reduced to the sequence of just a single virus that successfully gained the ability to infect humans and to be transmitted efficiently from person to person. Bottlenecks cause genetic drift because whatever genetic sequences were in the tiny founder population are those found in the immediate offspring.

After that, however, the population of pH1N1 viruses in humans began to diversify thanks to mutation (antigenic drift). Within the first year of its existence, it spread to 60 million people in the United States alone. Someone with an active influenza A infection exhales approximately 10 viruses per minute, so a conservative estimate of the number of viruses shed by each person is about 60 million × 10 viruses/minute × 60 minutes/hour × 24 hours/day × 5 days = 4.3×10^{12} viruses. Each of those viruses has on average one genetic change compared with its immediate ancestor.

Genetic drift and natural selection then acted on this diversity within the pH1N1 populations, which can be observed by focusing on the HA protein. The HA gene is subjected to natural selection because the protein it encodes must bind to the host receptor (neuraminic acid) and because antibodies that bind to HA reduce the number of offspring virions, either by blocking or slowing infection inside a single person or on a population level by preventing a person from being re-infected by the same strain of influenza A. Thus, HA evolves to continue binding to the receptor and to evade the immune response both within a single infected individual and on a human population level. Different HA genes from pH1N1 isolates can be cloned, sequenced, and expressed to study whether variations in the primary sequence affect the function of HA or its susceptibility to binding to H1 antisera. Changes that have no effect on the function or antibody binding of HA are attributed to genetic drift, whereas changes that improve its attachment or fusion functions or reduce its binding to antibodies are attributed to natural selection.

17.12 Pandemic influenza A strains have arisen through recombination

The origin of the influenza A strains that have caused pandemics since 1918 has been studied using contemporary approaches that rely on genome sequencing and bioinformatics. Viral genomes have been recovered from the tissues of influenza victims from 1918–1920 and from samples preserved from all subsequent pandemics. Increasing samples are available from swine, domesticated birds, and migratory birds. Wherever the evolutionary history can be reconstructed, every pandemic was caused by antigenic shift in which avian influenza recombined with other strains of influenza (**Figure 17.15**). The 1918 H1N1 virus originated in birds, but it is not known if it was caused by antigenic shift. It is known that after the human pandemic began, humans transmitted it to swine, which is important for the evolution of later pandemic strains. The 1957 H2N2 virus is a recombinant: five of its genome segments descend from the 1918 H1N1 virus and three segments, including the two encoding HA and NA, originated from a different avian influenza strain. The 1968 H3N2 pandemic was also caused by a recombinant virus; in this case, an H2N2 virus descended from the 1957 pH2N2 when it acquired two genome segments from an avian virus, including one encoding HA type 3. Because of the ancestry of the 1957 pH2N2, the 1968 pH3N2 strain retained five genome segments directly descended from the 1918 pH1N1 strain.

The 2009 pH1N1 pandemic strain is a multiple recombinant. Three of its gene segments including HA originate from swine H1N1, which was itself contracted from humans in the 1918 pandemic. So this virus, too, has genome segments directly descended from the 1918 pH1N1 strain. Two of its genome segments including the one encoding NA originate from a different swine virus that has recent ancestors in birds. Two of the remaining genome segments are from another avian influenza virus, and the last genome segment is descended from H3N2 human viruses (which themselves originate with the 1968 pandemic).

Recombination between different influenza A viruses occurs often in swine (**Figure 17.16**). Pigs are more susceptible to avian influenza than humans are, and they are also susceptible to human influenza. When avian influenza replicates in pigs, there is selection for offspring viruses that are more likely to infect humans. This ability arises in part from the molecular biology of influenza attachment (see **Chapter 3**). HA mediates attachment to its receptor neuraminic acid, also known as sialic acid. There are different types of sialic acids, and different HA variants preferentially bind to certain sialic acids. The terminal (distal) sialic acid on human respiratory cells, for example, is connected to the adjacent galactose through an α-2,6 linkage. This linkage has a distinctive shape compared with other types of linkages, such as α-2,3. The HA protein of human influenza A viruses binds preferentially to α-2,6-linked sialic acid that is most common on nonciliated human respiratory cells. In contrast, avian influenza HA binds better to α-2,3-linked sialic acids. The swine respiratory tract has primarily α-2,6-linked sialic acid so that when an avian virus infects pigs, there is selection for attachment to humanlike respiratory epithelia. An example is provided by the 1968 pH3N2 virus where the HA protein evolved by mutation in pigs from an avian HA protein. An example involving recombination is provided by the

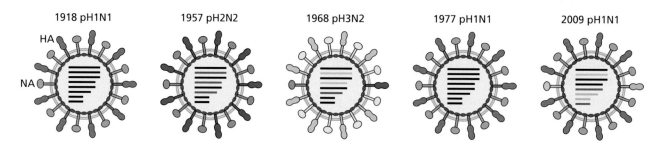

Figure 17.15 Genetic recombination and antigenic shift among pandemic influenza strains. The viral genome segments from different years are represented in different colors, as are the corresponding HA and NA spike proteins. The 1957, 1968, and 2009 viruses are all recombinants. The 1977 strain is descended directly from pH1N1 1918.

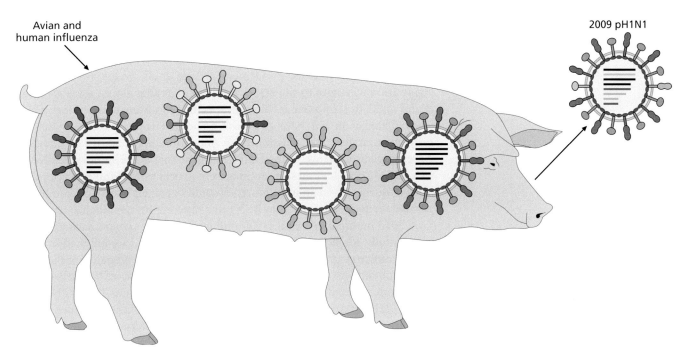

Figure 17.16 Influenza recombination in pigs. Pigs can be co-infected with avian, swine, and human-adapted strains of influenza. They can recombine, leading to the emergence of strains that can cause human pandemic influenza.

2009 pH1N1 strain in which the gene encoding the HA protein capable of infecting humans and swine was already present in swine. This genome segment recombined with two other avian virus genomes and a human-adapted H3N2 genome to produce the 2009 pH1N1 strain.

The pandemic influenza strain from 1977 is a possible exception to the rule that pandemic strains originate from antigenic shift. It is a direct descendent of the 1918 pH1N1 strain and, in fact, 1977 H1N1 strains are so similar to H1N1 viruses archived in the 1950s that the most logical conclusion is that the 1977 epidemic was caused by release of a frozen laboratory stock of old H1N1 viruses. People born after 1957 rarely developed antibodies to H1N1 and so the 1977 virus caused a pandemic that was mainly restricted to people younger than 25. This human-made pandemic killed approximately 700,000 people.

17.13   New pandemic influenza A strains may be able to arise through mutation

Recombination resulting in antigenic shift may not be necessary for a dangerous strain of influenza to leap from birds or swine to humans. Instead, antigenic drift may enable avian viruses to gain the ability to infect human cells and be transmitted from person to person. Of particular concern are H5N1, H7N7, and H7N3 strains of avian flu, which have caused recent human infections with a high mortality rate (up to 60%). For example, between 2003 and 2016, 850 people (mostly in Asia) contracted H5N1 directly from birds, with 53% mortality. None of these viruses has been transmitted from person to person, however, which raises the question of whether H5N1 viruses have the capacity to evolve person-to-person transmission. One aspect of transmission is attachment to host cells. More than 15 years of **gain-of-function experiments** using cultured human cells, cultured animal cells, and animal models have found that mutation can alter the H5N1 HA protein so that the virus can attach to human cells more efficiently. Gain-of-function experiments are risky because they intentionally create viruses with pandemic potential, so they must be performed (if done at all) under strict biosafety supervision. Changes to the H5N1 HA gene that alter amino acids K189 to R and Q222 to L improve its binding specificity to human respiratory sialic acids. Changes in HA have also been associated with other traits such as droplet respiratory transmission from one laboratory mammal to another. An example is a change in the H5N1 HA gene that alters amino acids N220 to K and Q222 to L. If these mutations were to occur in natural avian populations or during a human infection, they could in principle lead to a pandemic without an antigenic shift from H5N1 to a different immunological type.

A 2013 outbreak of avian H5N1 influenza among Cambodians provides a case study for how research in molecular virology in combination with global influenza genome sequencing is being used to prepare for the next pandemic influenza A strain. During the outbreak, some of the viruses isolated from infected patients had the potentially dangerous K189R, N220K, and Q222L variations of the H5 HA protein. When these sequence variations were detected in the human H5N1 isolates, international teams went to Cambodia to study the situation. They determined that the potentially dangerous sequence variations were absent in H5N1 viruses isolated from migratory and agricultural birds and thus arose through mutation of the viruses while they replicated in infected people. They also determined that all the human cases were acquired directly from birds, not from another person, so that these variants were apparently unable to spread from person to person. These studies also enabled development of a vaccine candidate that could be manufactured quickly in the event that an H5N1 virus evolves human-to-human transmission.

Selection of gain-of-function mutants that would allow H5N1 to become transmissible from human to human is more likely when H5N1 viruses have more generations replicating in humans. Humans usually contract avian influenza from domestic poultry where influenza A from wild migratory birds causes **epizootics** (pandemics in nonhuman animals). The 2022–23 winter saw a widespread epizootic in birds and wild mammals caused by H5N1. The more epizootics in poultry, the more the chance that H5N1 will be transmitted to humans. The more human infections there are, the more likely that antigenic drift will make it possible for a new virus to cause a pandemic. Similarly, widespread prevalence of H5N1 among domesticated birds during an epizootic increases the chance that such viruses will infect swine, where recombination with other viruses could lead to antigenic shift and a new pandemic strain. One way to stop the evolution of more dangerous descendants of H5N1 avian flu is therefore to cull infected poultry when there is an epizootic.

Study of the evolution of influenza A has contributed substantially to understanding the roles of mutation and recombination during viral evolution. Study of influenza A evolution has also contributed to understanding the selective factors that affect the evolution of viruses more generally.

17.14   Selective pressures and constraints influence viral evolution

As exemplified by the influenza A example, the frequency of mutation and recombination combined with very large population sizes means that viral genetic diversity is very high. Nevertheless, viruses do fall into easily distinguishable groups including species such as influenza A. Therefore, there must be some **constraints** that prevent viruses from diversifying so much that such groups would not be discernible. Natural selection imposes many of these constraints. Selective pressures that affect the evolution of all viruses include pressures to maintain the functionality of their proteins and nucleic acids. Other selective pressures include the need to infect and manipulate hosts and transmission to new hosts. Viruses with icosahedral capsids are also constrained by capsid size.

Selection against nonfunctional viral nucleic acids and proteins provides severe constraints on viral evolution, especially for RNA viruses with intrinsically high mutation rates. Throughout the biosphere, the chance that a random mutation will be neutral or even detrimental is always higher than the chance that a random mutation will be beneficial. Consequently, most mutations are neutral or detrimental. In the case of viruses in which a very high proportion of the genome either encodes proteins or serves a regulatory or functional role, it is thought that random mutations are even more likely to be detrimental than neutral. The impact of selection to maintain genome functionality is seen in the intrinsic mutation frequency of different viruses. The mutation frequency correlates very strongly with genome size (**Figure 17.17**). Larger genomes have lower mutation frequency, whereas smaller genomes have higher mutation rates. This pattern occurs because viruses with larger genomes can tolerate only a low number of mutations in order to avoid creating multiple mutations in each new genome. In contrast, viruses with a smaller genome can have a higher mutation frequency that will still result in a low number of offspring with multiple mutations because the genome is smaller. A large virus with a genome of 50,000 bases would need a mutation frequency of about 1 per 50,000 bases replicated in order to have an average of one mutation per offspring genome. In contrast, a virus with a small genome of 5,000 bases could tolerate a mutation frequency of 1 per 5,000 bases replicated and still have an average of one mutation per offspring genome. The mutation frequency of the smaller virus is 10 times higher than that of the larger virus, but the effect is to keep the proportion of mutant offspring with about one mutation per genome about the same.

All viruses must infect host cells to replicate, which exerts selective pressures in various ways. Some of these are contradictory, as for example, when viruses experience selection in favor of binding to host receptors. Such binding limits variations in viral attachment protein spikes because host cell receptors evolve slowly relative to viral replication. On the other hand, host defenses select for greater variation in viral attachment proteins, either to adapt to changing expression of host surface proteins (for bacteriophages)

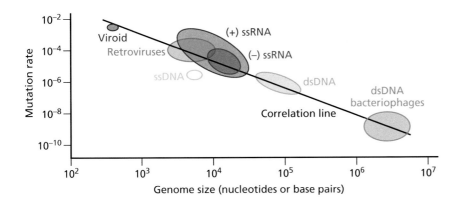

Figure 17.17 Genome size versus mutation rate. There is a strong correlation between viral genome size and mutation rate. Groups of viruses with different genomes are indicated by ovals. Mutation rate is defined as number of substitutions per nucleotide per generation. Genome size is defined as nucleotides for single-stranded genomes and as base pairs for double-stranded genomes.

or to evade the adaptive immune response (for animal viruses). Selection also affects the overall rate of virus replication. Viruses that kill their host cells too quickly do not leave as many offspring as viruses that kill their hosts more slowly, but very slow replication can allow host immune defenses to control the infection entirely. Viral fitness is thus a balance between these two pressures. Many molecular interactions between viruses and their hosts affect how fast a virus expresses its genes, copies its genome, and assembles offspring virions. Interactions between viral macromolecules and host macromolecules constrain the evolution of viral macromolecules because they must retain their ability to bind to host macromolecules. For example, when a viral protein binds to a host protein in order to co-opt it for viral genome replication, variations of the viral protein that do not bind well to the host protein are selected against. Viruses with small coding capacity (<20 proteins) often face extreme constraints because they use multifunctional proteins that play direct roles in several aspects of virus replication, immune evasion, and co-optation or subversion of host processes. Multifunctional viral proteins interact with various different host macromolecules, severely constraining their evolution.

Viral transmission to new hosts imposes important constraints too, as was exemplified by the influenza example. Transmission is related to viral abundance, the abundance of susceptible hosts, and factors such as viral attachment and the ability of a virus to replicate in a certain host cell once penetration and uncoating have occurred. The benefits of viral abundance selects in favor of high viral replication but, in the case of viruses that infect people, rapid viral replication can cause such severe symptoms that sick people do not come into contact with potential new hosts. This pressure selects for slower viral replication and also for transmission before a person develops symptoms.

The evolution of viruses with icosahedral capsids is subjected to an additional constraint, namely capsid size. Although helical capsids can in principle package nucleic acids of any length, icosahedral capsids have a defined internal volume. For example, the capsomers for the polyomaviruses assemble into icosahedrons so tiny that they can enclose approximately 5,000 base pairs of double-stranded DNA. It is unlikely that polyomavirus capsids could ever evolve to be much larger than they are now because the three different capsomers must interact extensively with each other to assemble into an icosahedron, and mutations that would allow one or all of them to become larger yet still maintain their ability to interact are very improbable. Thus, polyomavirus capsids select for small genomes. Almost every nucleotide in a polyomavirus genome is critical for transcription, critical for genome replication, or encodes proteins. In fact, many of the sequences encode more than one protein using a variety of overlapping gene strategies (see **Chapter 8**). Overlapping genes constrain evolution in much the same way that multifunctionality constrains the evolution of viral proteins. In contrast, the much larger icosahedral capsids of adenoviruses, herpesviruses, and some bacteriophages do not select for such small genomes, which allows for evolutionary events such as gene duplication and horizontal gene transfer to occur much more often among icosahedral viruses with larger capsids.

17.15   Some viruses and hosts coevolve

Coevolution occurs when two or more species have reciprocal effects on each other's evolution over the long term by exerting selective pressure on each other. An example would be a host's evolution of a new defense mechanism such as a restriction enzyme, which in turn selects for evolution of a new restriction enzyme evasion mechanism in a virus that infects that host. Coevolution can be observed under experimental conditions in the lab. For example, exposure to a bacteriophage selects for rare mutant hosts that have an alteration to the phage receptor to which the phage no longer binds (**Figure 17.18**). These mutants escape from the phage and proliferate, leaving many offspring. The new population of hosts in turn selects for bacteriophages with altered attachment proteins that enable the phage to infect the new population. This cycle can continue for many generations and occurs in nature.

When considering animals instead of bacteria, the term coevolution is usually used to describe situations in which the two interacting populations cospeciate. The coevolution of certain lentiviruses and African monkeys provides an example. Lentiviruses are complex retroviruses that include two species of HIV and more than 40 simian immunodeficiency virus (SIV) species. SIVs are as ancient as the origins of primates and they are specialists, infecting just one primate host species. Host switching, in which a virus jumps from one species to another, is very rare, even though it accounts for the origin of HIV. Coevolution of SIVs and their hosts has occurred, which

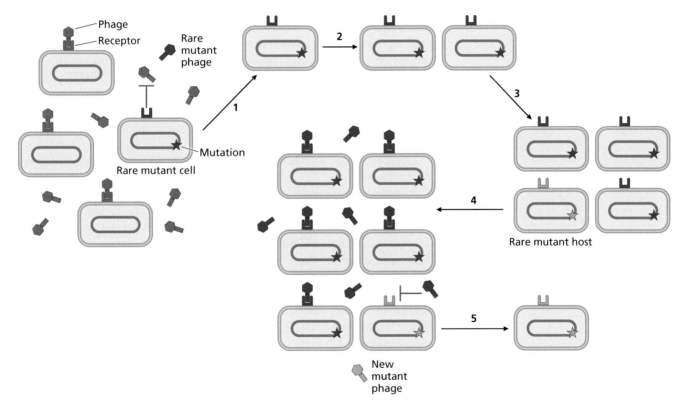

Figure 17.18 Coevolution of host receptors and bacteriophage attachment proteins through cycles of proliferation and selection. A bacteriophage kills all hosts except rare hosts. These unique hosts have a preexisting mutation (star) that alters the phage receptor-binding site, rendering the mutant host cell resistant to the phage. Because the population of phages is very large, however, there will also be rare mutant phages capable of recognizing the new receptor (1). The surviving host cells proliferate (2). Occasionally this gives rise to host cells that have further small changes to the receptor protein, but because the host cells are rare, they have not yet encountered a phage (3). When the host cell and phage populations become abundant, phages infect and kill all host cells, except new mutants that are resistant because of alterations to the receptor protein (4). The altered host cells are not susceptible to the previous generation of phages and so they survive (5). These alternating cycles of selection continue, with the phage selecting for mutant hosts, and the hosts selecting for phages with new receptor-binding properties in alternating generations.

Figure 17.19 Coevolution of the SIV capsid and the host restriction factor TRIM5 in monkeys. Species-specific SIV strains infect monkeys. The TRIM5 restriction factor in each host species blocks replication of most SIV. Only SIV strains that have adapted to overcome their host's TRIM5 protein can replicate. For example, SIV^mac replicates in macaques but not in baboons, whereas SIV^bab replicates in baboons but not in macaques because of this mutual adaptation.

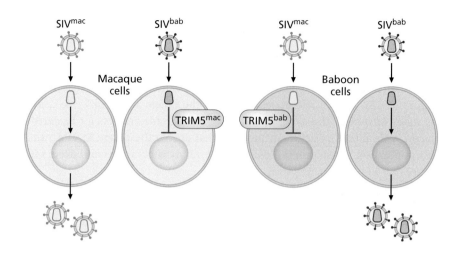

can be observed at the molecular level by examining viral proteins and the **host restriction factors** they interact with. Host restriction factors are innate animal defense proteins that limit or prevent infection by particular groups of viruses.

A key example is provided by the host restriction factor TRIM5 and the viral protein it targets, namely the capsid. After a lentiviral capsid enters the cytoplasm, TRIM5 binds to the capsid and blocks subsequent steps of viral replication such as reverse transcription or transport of the cDNA to the nucleus, thus preventing infection. Alterations to the capsid proteins that prevent TRIM5 from binding can overcome this defense. TRIM5 proteins contribute to species barriers that prevent lentiviruses from infecting novel hosts; for example, HIV-1 cannot infect rhesus macaque monkeys because the macaque TRIM5 binds to the HIV-1 capsid after penetration releases the capsid into the cytoplasm (**Figure 17.19**).

Compensatory alterations in TRIM5 and SIV capsid proteins have been observed for SIV species that infect different species of African Cercopithecine monkeys such as macaques and baboons. Each monkey species has a unique TRIM5 variant that binds to the capsid of one SIV species but not others. SIV capsid proteins in return exhibit variations that selectively overcome the TRIM5 defense in their particular host monkey. The best explanation for these observations is that the genes encoding TRIM5 and SIV capsids have coevolved (see Figure 17.19). Phylogenetic trees that describe the evolutionary history of Cercopithecine monkeys and the SIV species that infect them have the same branching patterns so that each speciation (branch point) in the history of the hosts is accompanied by speciation of the viruses. Such congruent evolutionary histories are a typical sign of coevolution.

Although some specialist viruses with narrow host ranges and their hosts have coevolved, host switching can also occur. HIV evolved from chimpanzee-adapted SIV strains that switched to human hosts. Many emerging viruses such as HIV originate in nonhuman animals and these origins are the subject of the next section.

17.16 Medically dangerous emerging viruses are zoonotic

Ecologists who study the emergence of HIV and other viruses often point out that the tremendous human population on the Earth, now more than 7 billion, means that humans are inevitably entering new landscapes and disturbing them with new housing and economic opportunities. This situation puts humans in contact with many animal viruses, some of which will be able to jump the species barrier. Zoonotic viruses such as severe acute respiratory syndrome coronavirus (SARS-CoV) and Middle East respiratory syndrome coronavirus (MERS-CoV) are recent examples along with HIV and pandemic influenza A. The 2013–2016 Ebola fever epidemic in West Africa is another example; that outbreak was started by a single zoonotic transmission from a bat to a toddler. All of the other 28,000 human infections during the epidemic originated from the progenitor virus that infected the toddler.

Most emerging viruses originate in mammals, and the molecular and cellular biology of such viruses help to explain this trend. Viruses that infect mammals such as bats, camels, chimpanzees, and swine are already adapted to many of the conditions that would make it possible for them to replicate in humans. For example, they can replicate at humanlike body temperatures, meaning that their proteins have the right balance of flexibility and rigidity at human body temperature. Viruses that infect mammals have also evolved to respond to mammalian defenses. Finally, some animal viruses such as influenza A are generalists with broad host ranges, whereas others are much more specific. Most emerging zoonotic viruses originate from populations of generalists because they have molecular features that make replication possible in a wide range of host cells. Viral generalists, for example, bind to receptors that are more conserved across species than are the receptors favored by viral specialists. Animal viruses with broad host range are thus more likely to be able to use a human receptor.

In fact, mutations that alter virus–receptor interactions are important for most known zoonotic viruses. Influenza was one example. Another example is SARS-CoV. SARS-CoV jumped from its normal animal reservoir (bats) to civets and from there to humans, much the same way that influenza A jumps from its normal animal reservoir (migratory aquatic birds) to domesticated animals and from there to people. Coronaviruses are enveloped and use spikes for attachment. SARS-CoV attaches to a transmembrane protein called ACE. The rare forms of SARS-CoV isolated from civets and humans have changes in the spike gene relative to the much more abundant and diverse viruses circulating among bats. These changes result in variant spike proteins with four altered amino acids; these alterations change the binding properties of the spike so that it binds more than 1,000 times better to human ACE than the bat coronavirus spike proteins do.

Although changes to virus attachment proteins may be necessary for zoonotic transmission, they are not necessarily sufficient for person-to-person transmission. Avian influenza viruses that are transmitted from birds to humans are a well-characterized example (see **Section 17.12**). So far, none of these viruses has been transmitted from person to person even though the viruses have HA spike variations that have enabled them to switch from infecting bird cells to infecting human cells. The molecular and cellular factors that affect transmission thus cannot be reduced to attachment–receptor interactions. Instead, the molecular and cellular factors that affect transmission are poorly understood for any virus, even influenza A. Influenza A viruses infect the gastrointestinal tract of birds and are spread by the oral–fecal route, whereas influenza A viruses replicate in the human respiratory tract and are spread by droplets in the air. Despite these dramatic differences in transmission and many decades of study, the molecular and cellular factors that contribute to these differences are unknown.

In order for a virus to jump from one species to another, it most likely has to adapt to optimize its fitness in the new host. This phenomenon is called post-transfer adaptation. Because evolution requires diversity, and viral diversity increases with increased viral replication, there is tremendous interest in controlling zoonotic outbreaks before post-transmission adaptation can occur. Examples of control measures include culling poultry to reduce outbreaks of avian influenza and quarantine to contain SARS in the early twenty-first century and the west African Ebola fever epidemic of 2013–2016. Other non-pharmaceutical interventions include physical distancing, isolation after exposure, and medical-grade face masks to reduce the spread of SARS-CoV-2. In all of these cases, comparisons between viruses isolated from sick patients and those isolated from animal reservoirs show molecular changes consistent with post-transfer adaptation. By definition, these changes occur during replication in humans, after whatever initial changes were necessary for the original zoonotic transmission event. Thus, a feature of post-transfer adaptation is that isolates from patients infected later during an epidemic have more of these adaptations. In the case of enveloped Ebola virus, the glycoprotein spikes from human isolates have specific amino acid variations that result in increased tropism for human cells and decreased tropism for bat cells. Some of these variations are more common in viruses isolated later in the epidemic, indicating that they evolved after transfer.

We have especially clear examples of post-transfer adaptations in the case of SARS-CoV-2. The COVID-19 pandemic has been monitored by molecular virologists since its beginning in 2019. There are millions of complete genome sequences derived from patient samples from all over the world, shared through a genomics database called GISAID. The original strain of the virus that infected people in China quickly gained post-transfer adaptations that improved its transmission from person to person. One example of such a change is a missense alteration D614G that changed aspartic acid 614 in the spike protein to glycine. Viruses with a genome encoding this change took over the global population of SARS-CoV-2 in March 2020 (**Figure 17.20**).

The D614G substitution has functional consequences that make sense in light of this global takeover. The spike protein is a trimer and each subunit of the trimer as two conformations (**Figure 17.21**). In one conformation, termed "closed," the receptor-binding domain is folded down, making it unlikely that

Figure 17.20 Global frequency of the SARS-CoV-2 D614G variant in 2020. Among 52,292 independent genome sequences of SARS-CoV-2 collected during the first 6 months of the COVID-19 pandemic, mutants encoding the D614G amino acid substitution took over the global viral population in March. (Adapted from Yurkovetskiy L et al. 2020. *Cell* 183:739–751. doi: 10.1016/j. cell.2020.09.032. With permission from Elsevier.)

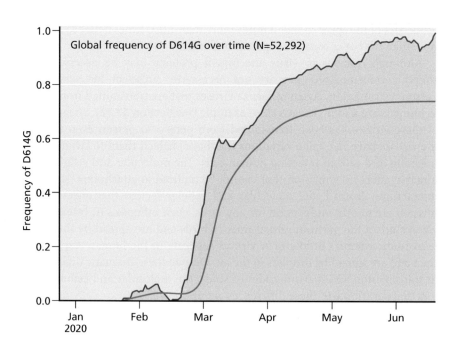

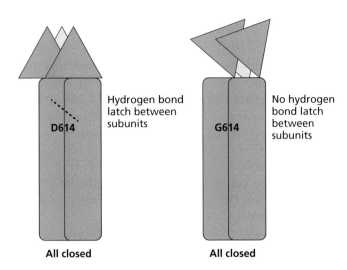

Figure 17.21 Structural consequences of the D614G substitution in the SARS-CoV-2 spike protein. The spike protein of SARS-CoV-2 is a trimer, depicted here with each subunit in a different color. In the original spike protein, a hydrogen bond (dotted line) connects D614 in one subunit to the adjacent subunit. When this bond is present, the spike is said to be in the closed position. The D614G variant cannot form this hydrogen bond and so each spike subunit is more likely to be in the "open" conformation, which favors binding to the cellular ACE2 receptor.

the receptor-binding domain can interact with the cellular ACE2 receptor. In the other conformation, termed "open," the receptor-binding domain is flipped up, making the receptor-binding domain available for binding to the receptor. The closed conformation is stabilized by a hydrogen bond between one aspartic acid 614 to the adjacent subunit of the trimer. When there is a D614G substitution, this hydrogen bond cannot form, making the open conformation more likely. Spike proteins with this substitution are more able to mediate entry into host cells expressing ACE2. Experiments with hamsters demonstrated that a virus with the D614G change spreads more readily through the air from an infected hamster to a hamster living in an adjacent cage. Thus, our emerging understanding of the impact of the D614G variation is that it facilitates the spread of COVID-19 and so it is a key post-transfer adaptation to spreading among people.

Post-transfer adaptation can result in a virus becoming so adapted to humans that it circulates among us and is no longer contracted directly from animals. Many ubiquitous human viruses have zoonotic origins even though they are no longer acquired from nonhuman animals in the present day. In these cases, adaptation to humans is associated with long periods of close contact between humans and the nonhuman source of the virus. Examples include measles virus, mumps virus, rotavirus, and variola virus, which were all originally contracted from domesticated animals. Post-transfer changes that allowed variola to adapt to humans and cause smallpox likely included changes in viral proteins that counteract innate immune defenses, whereas the post-transfer adaptations evolved by the other examples are not as well understood.

It is likely that new viral diseases will continue to emerge as humanity's population continues to rise and as our impacts on surrounding ecosystems grow ever more profound. Because HIV-1 emerged during the era of molecular and cellular biology and causes a pandemic with high mortality, immunological samples and virus samples from almost the entire epidemic are available, as are viruses and viral sequences from related monkey viruses (SIV). Thus, the molecular and cellular factors that affected HIV emergence and evolution are understood in some detail and serve as a model for how other viruses became epidemic and eventually endemic over the course of human history. **Sections 17.17** and **17.18** therefore address HIV diversity, HIV origins, and the molecular factors that are important for its adaptation to humans.

17.17 HIV exhibits high levels of genetic diversity and transferred from apes to humans on four occasions

The human immunodeficiency viruses that cause acquired immunodeficiency syndrome (AIDS) originated in chimpanzees or gorillas within the last 150 years. There are two species of HIV: HIV-1 and HIV-2. HIV-2 is less widely distributed and less pathogenic, whereas HIV-1 drives the global HIV/AIDS pandemic. Lentiviruses such as HIV are retroviruses that exhibit high genetic variation caused by an intrinsically high rate of misincorporation by the viral reverse transcriptase. High viral population levels also contribute to diversity; for example, a person with an untreated HIV infection produces 10^9 new viruses every day. HIV/SIV infections are lifelong, providing ample time for tremendous viral diversification.

All retroviruses are also prone to recombination because of their molecular biology. First, retroviruses package two copies of the genome into each virion and, because of mutation, the two copies often differ slightly from one another. During co-infection with two or more strains of HIV, copies descended from two different parental templates can be co-packaged into virions. Second, the retrovirus reverse transcriptase uses a discontinuous method of DNA synthesis (see **Chapter 10**). During jumps from one region of a single RNA template to another region of the same molecule, the reverse transcriptase sometimes switches to a different template entirely so that the cDNA is derived from a combination of two or more different genome segments. This, too, can happen during co-infection with two or more parental virions, resulting in recombination.

We are aware of about 40,000 complete HIV genome sequences and more than 3,000 complete SIV genome sequences. Comparison of these sequences indicates that HIV-1 crossed the species barrier four separate times, with zoonotic transmission from chimpanzees to humans in the early twentieth century in West Africa being responsible for HIV-1 subtype M (for the main form of HIV in circulation). Two other zoonotic transmission events account for less prevalent subtypes of HIV-1 such as type O (outlier) and type N (non-M, non-O) and for HIV-2. The virus probably crossed species through hunting, butchering, and eating a chimpanzee infected with SIV. An evolutionary tree shows how HIV-1 isolates are related to SIV strains adapted to other primates (**Figure 17.22**).

Figure 17.22 Genetic relatedness of HIV-1 and SIV strains. HIV-1 and SIV were isolated from people, one of two subspecies of chimp (*Pan troglodytes* or *schweinfurthii*), or gorillas, and their genetic relatedness determined by sequencing complete genomes. HIV-1 types M and N are most similar to SIV^cpz, indicating that they were zoonotically acquired from the chimpanzee subspecies *troglodytes*. HIV-1 type O is similar to SIV^gor, and so it was zoonotically acquired from a gorilla. SIV^gor originated recently by crossing the species barrier from chimpanzees into gorillas.

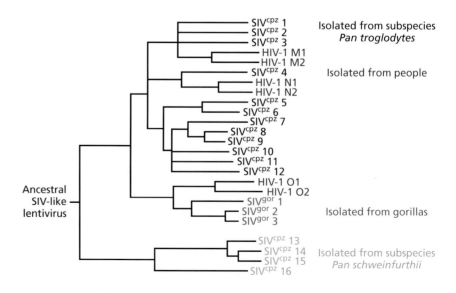

17.18   HIV-1 has molecular features that reflect adaptation to humans

All species of HIV and SIV infect CD4+ cells of the primate immune system. Different strains of HIV and SIV have different tropisms, however, and do not replicate equally well in all CD4+ cells. Study of this phenomenon led to the discovery of host restriction factors that prevent certain retroviruses from successfully replicating in their hosts and to the discovery of retroviral proteins that overcome these restrictions in particular host cells. The TRIM5 host restriction factor was already discussed to explain how SIV and certain African primates have coevolved (see Figure 17.19). The host proteins SAMHD1, APOBEC3, and tetherin are additional restriction factors that limit the replication of retroviruses and contribute to host adaptation. Lentivirus accessory proteins counteract these restriction factors. For example, some species of SIV have an accessory protein called **Vpx** that counteracts the effects of **SAMHD1**. Vpx is missing from the chimpanzee-specific SIV (SIV^{cpz}) that is the immediate ancestor of HIV-1 and, in fact, HIV-1 tissue tropism is therefore restricted to cells that intrinsically express low levels of in SAMHD1.

Both species of HIV and all species of SIV encode a **Vif** accessory protein that counteracts the **APOBEC3** host restriction factor. In the well-understood case of HIV-1, there are different human alleles encoding somewhat different host APOBEC3 proteins. There are also many different HIV-1 alleles encoding variants of the Vif protein. Interactions between the various APOBEC3 and Vif proteins determine how well a specific strain of HIV-1 replicates in a particular person's cells. Furthermore, evolution of the Vif protein was key for zoonotic transmission of SIV^{cpz} to humans as follows. The APOBEC3 protein of hominids such as chimpanzees and humans differs from that in other primates such as monkeys. The Vif proteins in SIV^{cpz} and HIV-1 have a small unique region at the C-terminal end that is not found in monkey SIV. The SIV^{cpz} and HIV-1 Vif proteins specifically counteract hominid APOBEC3 but not monkey APOBEC3. Thus, the infection of SIV^{cpz} in chimpanzees in particular made it more likely that SIV would be able to cross from other primates to humans specifically by selecting for Vif proteins that could counteract hominid APOBEC3.

HIV-1 also exhibits post-transfer adaptations to replication in humans. For example, a single amino acid change in the HIV matrix protein has evolved in multiple HIV-1 lineages, where it significantly increases viral replication in human cells. The accessory proteins have continued to evolve in response to human-specific variations in host restriction factors. Other adaptations have altered the way that the host cells display HIV-1 antigens in MHC-I (see **Chapter 15**). There are thousands of slightly different alleles encoding MHC-I molecules in the human population, although individual people express just three types of MHC-I. These variations have dramatic effects on the selection and display of endogenous antigens, including viral antigens, and therefore on the cytotoxic T lymphocyte (CTL) response to viral infections (see **Chapter 15**). Through their effects on the CTL response, variations in MHC-I have strong effects on the course of HIV-1 infection. When HIV-1 first crossed the species barrier into humans, it was not adapted to human MHC-I display. Since that time, the CTL response has selected for HIV-1 viruses that have peptides that are less likely to be displayed than those of the ancestral SIV^{cpz} responsible for the zoonotic transmission. These viruses evade the CTL response more effectively and thus replicate to higher levels in the human body.

Just as African monkeys have themselves evolved in response to SIV infection, humans may yet evolve in response to HIV-1. Humans and our ancestors have certainly evolved in response to other viruses and subviral entities, which is the concluding topic in this chapter.

17.19   Viruses and subviral entities are common in the human genome

Most DNA in the human genome is derived from repeated sequences, many of which are subviral entities or remnants of viral genomes recombined with the DNA. These repeated sequences include Class I retrotransposons and Class II DNA-based transposons (see **Section 17.4**). DNA transposons are ubiquitous in cellular organisms and they make up approximately 3% of the human genome. Nevertheless, human DNA-based transposons have been studied much less intensively than human retrotransposons.

A much larger percentage of the human genome is composed of four types of retroelements: endogenous retroviruses and three types of retrotransposons (**Figure 17.23**). **Endogenous retroviruses** are proviral DNA inserted into a chromosome (see **Chapter 10**); they are inherited endogenously, meaning that they are found in the germ line and inherited from parent to child through sexual reproduction. Almost always they also have inactivating mutations that prevent them from replicating as viruses. They are abundant, making up between 5% and 8% of the human genome.

The longest type of retrotransposon is known as an **LTR retrotransposon**; the sequences of these transposons closely resemble retrovirus proviral DNA. Retrovirus proviral DNA is always flanked by direct repeats because of the mechanism of integration (see **Chapter 10**). Retrovirus proviral DNA always has LTRs because of the mechanism of reverse transcription, and these LTRs flank coding sequences for the *gag*, *pol*, and *env* genes that are found in all retroviruses. Like proviral DNA, LTR retrotransposons are flanked by direct repeats and include two LTRs. They also have *gag* and *pol* genes in between; however, they lack *env* sequences.

The shorter **long interspersed nuclear elements** (**LINEs**) are the most common retrotransposons in the human genome, making up approximately 20% of the DNA. They are similar to proviral DNA in that they are flanked by direct repeats, but they differ in that they do not have LTRs or *gag*, *pol*, or *env*

Figure 17.23 Retroelements in the human genome. Endogenous retroviruses are flanked by direct repeats (DRs), have long terminal repeats (LTRs), and have *gag*, *pol*, and *env* genes; at least one of these genes is typically nonfunctional so that the DNA does not encode an active virus. LTR retrotransposons are flanked by direct repeats, have LTRs, and have *gag* and *pol* sequences. LINEs are flanked by direct repeats and encode proteins with endonuclease (EN) and reverse transcription (RT) domains; they also have a poly(A) region at one end. SINEs are flanked by direct repeats and have a poly(A) region at one end. They have sequences derived from RNA Pol III transcripts such as rRNA.

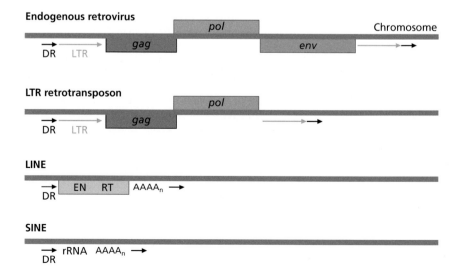

sequences. Instead, they encode an endonuclease and a reverse transcriptase followed by a poly(A) tract. Of the 850,000 copies of LINEs in the human genome, fewer than 100 are actively able to transpose, which requires both the endonuclease and the reverse transcriptase they encode. About 13% of the human genome is composed of the last group of retrotransposons, known as **short interspersed nuclear elements** (**SINEs**). SINEs are flanked by direct repeats of host DNA. There are in the order of a million SINEs in the human genome. SINEs depend on endonucleases and reverse transcriptases encoded by LINEs for their replication and transposition. Their DNA sequence is not viral but is descended instead from the sequences of ubiquitous cytoplasmic RNAs that are transcribed by host Pol III.

17.20 Viruses and subviral entities have strongly affected the evolution of organisms including humans

Although endogenous retroviruses and transposons were once considered junk DNA, closer scrutiny has revealed examples in which they have obviously played a crucial role in the evolution of animals including humans because of their profound effects on physiology. For example, pregnancy depends on endogenous retroviruses and transposons in several different ways. Humans are placental mammals, meaning that we bear live young rather than reproducing through eggs as other mammals such as platypuses do. The placenta is an organ that forms specifically during pregnancy, when it allows a developing embryo to connect to the uterus and exchange nutrients, gases, and waste with the pregnant woman. Thousands of genes must be expressed and regulated properly in order for the placenta to form and for the uterus to sustain a pregnancy. Many of these are regulated by transcription factors that bind to transposons: transposons actually coordinate the thousands of genes that must work together for pregnancy to occur. Furthermore, two endogenous retroviruses are essential because they encode proteins required for pregnancy. One of them encodes a protein, **syncytin-1**, needed for development of the placenta. Formation of a placenta requires fusion of cellular plasma membranes, creating multinucleated cells known as **syncytia**. Syncytia are one of the common cytopathic effects observed when enveloped viruses infect cultured tissue culture cells (see **Chapter 12**). Syncytin-1 catalyzes membrane fusion during placenta formation and is homologous to retroviral Env proteins, which catalyze envelope–host plasma membrane fusion (see **Chapter 3**). Meanwhile, a different endogenous retrovirus encodes another Env homolog that has also been adapted for a new function. It is essential for preventing the woman's immune system from attacking the foreign developing embryo.

A DNA transposon is implicated in the evolution of the adaptive branch of the vertebrate immune system. In this case, a particular transposon is the ancestor of the *RAG1* and *RAG2* genes. RAG proteins catalyze the recombination events that lead to the production of mature T-cell receptors, B-cell receptors, and antibodies. Thus, a DNA-based subviral entity is responsible for the evolution of adaptive immunity, which itself evolved to help vertebrate hosts escape from pathogens such as viruses.

Introns, which are related to retrotransposons, have played critical roles in the evolution of all eukaryotes. Spliceosomal eukaryotic introns are one of four known types of intron and are implicated in eukaryogenesis. When the ancestral proto-eukaryote acquired mitochondria, group II introns from the endosymbiotic bacteria invaded the genome of the proto-eukaryote and became the first spliceosomal introns. The selective pressure to survive the invasion of these introns may have caused the formation of the nucleus itself to segregate slow splicing away from rapid translation. The nucleus prevents translation of unspliced mRNA that still contains introns, which would lead to a nonfunctional protein product. Mobile introns may have also driven the evolution of multidomain proteins in which each domain is encoded by a different exon; multidomain proteins are much more common in eukaryotes than in bacteria or archaea. There is even the possibility that introns allowed the formation of linear chromosomes by enabling their faithful maintenance from one generation to the next, which requires a reverse-transcribing enzyme known as telomerase. Telomerase is a reverse transcriptase and all reverse transcriptases are homologous. About half of all group II introns encode a reverse transcriptase, so that both spliceosomal introns and telomerase may have originated with one or more group II introns that entered the first proto-eukaryote when the mitochondrial endosymbiosis began.

17.21   Virology unites the biosphere

This book began with the premise that a major motivation for studying viruses is to better understand cells. Yet viruses tell us so much about molecular and cellular biology precisely because all cells evolved awash in viruses and subviral entities. As we have seen, viruses are actually the central components of the biosphere. The discipline of virology reveals the fundamental connectedness of all nucleic acid-based life.

Essential concepts

- The virosphere is huge and diverse; we have direct knowledge of the gene sequences for only a minuscule fraction, which could impede understanding the evolutionary origins and interrelatedness of viruses.
- Subviral entities that cannot replicate without helper viruses or that are composed of nucleic acids include satellite viruses, satellite nucleic acids, viroids, transposons, and introns.
- Analysis of viral hallmark genes suggests that viruses have ancient polyphyletic origins that are deeply intertwined with every major evolutionary transition in the history of cellular life.
- Genetic variation among viruses arises through mutation and recombination.
- Influenza A diversity and evolution have been subjected to intense scrutiny because of the desire to understand influenza pandemics. Key molecular factors affecting influenza evolution are antigenic drift, antigenic shift, and HA–host receptor interactions.
- Major selective pressures that constrain viral evolution include the need to maintain nucleic acid and protein functions, the need to manipulate hosts through intermolecular interactions, and selection in favor of transmission to new hosts. The genome size of viruses with icosahedral capsids is constrained by the size of the capsid.
- Viruses and their hosts sometimes coevolve, which involves molecular changes in which a change in the host selects for a compensatory change in the virus and so on over long periods of time.
- Medically dangerous emerging viruses are typically zoonotic.
- SARS-CoV-2 is a zoonotic virus that emerged in 2019 and is undergoing adaption to human hosts. Molecular features of the SARS-CoV-2 adaptation to humans include changes to the spike protein.

- HIV-1 is a zoonotic virus that has emerged from nonhuman hominids on several occasions and is now adapted to human hosts. Molecular features of the HIV-1 adaptation to humans include changes to the virus that enable it to overcome host restriction factors and the CTL response.
- The human genome is made up of many copies of endogenous retroviruses, transposons, and introns, all of which originated as viruses or subviral entities. Viral sequences have influenced human evolution in various way, including the emergence of placental mammals, the development of vertebrate immunity, and even eukaryogenesis itself.
- There has never been cellular life in the absence of viruses and subviral entities; viruses are the most abundant components of the biosphere.

Questions

1. How does a lack of knowledge of the majority of viral genomes influence attempts to understand viral evolution?
2. Compare and contrast viral hallmark genes and the genes used to trace cellular evolution.
3. Provide an example of viral evolution through mutation.
4. Provide an example of viral evolution through recombination.
5. Provide evidence that humans and retroviruses have an intimate shared evolutionary history.
6. How are LINEs similar to endogenous retroviruses?
7. Compare and contrast viroids and satellite viruses.

 Interactive quiz questions

In addition to the questions provided above, this edition has a range of free interactive quiz questions for students to further test their understanding of the chapter material. To access these online questions, please visit the book's website: www.routledge.com/cw/lostroh.

Further reading

Viral diversity and origins

Krupovic M & Koonin EV 2017. Multiple origins of viral capsid proteins from cellular ancestors. *Proc Natl Acad Sci* 114:E2401–E2410.

Shi M, Lin XD & Tian JH 2016. Redefining the invertebrate RNA virosphere. *Nature* 540:539–543.

Shi M, Lin D, Chen X, Tian JH, Chen LJ et al. 2018. The evolutionary history of vertebrate RNA viruses. *Nature* 556:197–202.

Zayed AA, Wainaina JM, Dominguez-Huerta G, Pelletier E, Guo J et al. 2022. Cryptic and abundant marine viruses at the evolutionary origins of Earth's RNA virome. *Science* 376:156–162.

Subviral entities

Shimizu A, Nakatani Y, Nakamura T, Jinno-Oue A, Ishikawa O et al. 2014. Characterisation of cytoplasmic DNA complementary to non-retroviral RNA viruses in human cells. *Sci Rep* 4:1–9.

Wu J, Zhou C, Li J, Li C, Tao X et al. 2020. Functional analysis reveals G/U pairs critical for replication and trafficking of an infectious non-coding viroid RNA. *Nucleic Acids Res* 48:3134–3155.

Evolution of influenza A

Liu M, Huang LZ, Smits AA, Büll C, Narimatsu Y et al. 2022. Human-type sialic acid receptors contribute to avian influenza A virus binding and entry by hetero-multivalent interactions. *Nat Comm* 13:4054.

Yasuhara A, Yamayoshi S, Kiso M, Sakai-Tagawa Y, Koga M et al. 2019. Antigenic drift originating from changes to the lateral surface of the neuraminidase head of influenza A virus. *Nat Microbiol* 4:1024–1034.

Evolution of HIV

Fari NR, Rambaut A, Suchard MA, Baele G, Bedford T et al. 2014. The early spread and epidemic ignition of HIV-1 in human populations. *Science* 346:56–61.

OhAinle M, Helms, Vermeire, J, Roesch F, Humes D et al. 2018. A virus-packageable CRISPR screen identifies host factors mediating interferon inhibition of HIV. *elife* 7:e39823.

Evolution of SARS-CoV-2

Hou YJ, Chiba S, Halfmann P, Ehre C, Kuroda M et al. 2020. SARS-CoV-2 D614G variant exhibits efficient replication *ex vivo* and transmission *in vivo*. *Science* 370:1464–1468.

Rochman ND, Wolf YI, Faure G, Mutz P, Zhang F & Koonin EV 2021. Ongoing global and regional adaptive evolution of SARS-CoV-2. *Proc Natl Acad Sci* 118:e2104241118.

Yurkovetskiy L, Wang X, Pascal KE, Tomkins-Tinch C, Nyalile TP et al. 2020. Structural and functional analysis of the D614G SARS-CoV-2 spike protein variant. *Cell* 183:739–751.

Human evolution and mobile genetic elements

Britten RJ 2010. Transposable element insertions have strongly affected human evolution. *Proc Natl Acad Sci* 107:19945–19948.

Dunlap KA, Palmarini M & Varela M 2006. Endogenous retroviruses regulate periimplantation placental growth and differentiation. *Proc Natl Acad Sci* 39:14390–14395.

Faulkner GJ, Kimura Y & Daub CO 2009. The regulated retrotransposon transcriptome of mammalian cells. *Nat Genet* 41:563–571.

Lynch VJ, Leclerc RD & May G 2011. Transposon-mediated rewiring of gene regulatory networks contributed to the evolution of pregnancy in mammals. *Nat Genet* 43:1154–1158.

Mananey M, Renard M & Schlecht-Lour G 2007. Placental syncytins: Genetic disjunction between the fusogenic and immunosuppressive activity of retroviral envelope proteins. *Proc Natl Acad Sci* 104:20534–20539.

Nakajima R, Sato T, Ogawa T, Okano H & Noce T 2017. A noncoding RNA containing a SINE-B1 motif associates with meiotic metaphase chromatin and has an indispensable function during spermatogenesis. *PLOS One* 12:e0179585.

Viruses and Public Health

This chapter can be found online at www.routledge.com/cw/lostroh.

The discipline of virology intersects with the health of populations. Two especially important perspectives to consider are global health and One Health. These perspectives help us understand epidemics, pandemics, and epizootics. COVID-19 and AIDS are examples of zoonotic epidemic viral diseases. While vaccination and antiviral drugs are crucial measures against viral pandemic disease, nonpharmaceutical interventions such as clean air can also ameliorate outbreaks. The COVID-19 pandemic illustrates how global health and One Health approaches provide insight into epidemic viral disease.

DOI: 10.1201/9781003463115-18

Glossary

100K Adenovirus protein that both prevents formation of the translation 48S preinitiation complex on host mRNA and stimulates translation of viral mRNA through binding to a leader sequence found in all late viral mRNA.

2′-5′-oligoadenylate synthetase (OAS) Innate immune protein that is encoded by an interferon-stimulated gene. Enzyme that synthesizes 2′-5′-oligoadentylate from ATP when stimulated by dsRNA indicative of a viral infection.

3′-end processing A step during the retrovirus cDNA integration process.

^{35}S-methionine Radioactive derivative of the amino acid methionine commonly used to label proteins during a pulse-chase analysis.

3CD^{pro} Picornavirus protease enzyme necessary to process the viral polyprotein.

3D^{pol} Picornavirus protein with RdRp activity needed for synthesizing replicative forms and new (+) RNA.

48S translation preinitiation complex Eukaryotic multiprotein complex essential for translation initiation; often targeted by viruses to block host translation.

5′ to 3′ exonuclease activity Hydrolysis of nucleotides from the 5′ end of a nucleic acid.

(+) strand A single-stranded viral genome that cannot base pair to viral mRNA.

(−) strand A single-stranded viral genome that can base pair to viral mRNA.

−1 direction During a ribosome frameshift, the movement of the ribosome backward toward the 5′ end of the mRNA template.

α protein Herpesvirus proteins encoded by immediate-early genes; mRNA synthesized during the first wave of viral gene expression.

Abscission Membrane fusion event catalyzed by ESCRT proteins; during viral exit, abscission allows enveloped viruses to exit the host cell.

Accessory proteins Coronavirus proteins unique to each species of coronavirus, probably needed for host-specific functions such as defeating the host immune response.

Acquired immune response *See* **Adaptive immune response**.

Activated cell In vertebrate immunity, an immune cell that has been activated by cytokines and thus has enhanced activities.

Ad pol Adenovirus enzyme that catalyzes DNA replication using the preterminal protein as a primer.

Adaptive immune response Component of the vertebrate immune response that is induced by exposure to a specific microbe and includes T cells, B cells, and antibodies.

Adenovirus death protein Transmembrane viral protein needed for adenovirus lysis.

Adenovirus early proteins The E1A large and E1A small proteins, which are transcription regulators that activate delayed-early adenovirus gene expression.

Adenovirus intermediate and late proteins Group of adenovirus proteins that includes the very abundant structural proteins required to make a large capsid with more than 2,500 polypeptide components.

Adenoviruses Large naked dsDNA (Baltimore Class I) virus with prominent spikes that is infamous for extensive alternative splicing and uses a protein-priming mechanism for replication of its linear DNA. Common vector for gene therapy.

Adjuvants A chemical that enhances the immune response against the microbial components of a vaccine.

Affinity chromatography Chromatography technique in which specific molecular bait is attached to inert column material and then macromolecules that have high affinity for that bait adhere selectively to the bait.

Alignment Output of a statistical algorithm that finds sequence similarities between proteins or nucleic acids.

Alipogene tiparvovec Gene therapy based on recombinant adeno-associated virus; treats familial lipoprotein lipase deficiency.

Alternative splicing Production of different mature mRNA molecules from the same immature transcript using one or more different exons in each case.

Alternative start codon Start codon with an unusual sequence that specifies an amino acid other than methionine; causes a reduction in translation initiation compared with use of the typical AUG start codon.

Ambisense Viral genome in which one part is complementary to viral mRNAs but another part has the same sequence as other viral mRNAs.

Amphisome Intermediate compartment that forms during autophagy; has an acidic interior as a consequence of fusion with an endosome.

Anti-CRISPR proteins Bacteriophage proteins that evade or subvert a host adaptive immune response.

Antibody Large protein made up of several polypeptides; binds to epitopes as part of a vertebrate adaptive immune response.

Antigen-presenting cells (APCs) In vertebrate immunity, cells that are specialized for displaying exogenous epitopes to lymphocytes.

Antigenic drift Slow changes in the influenza virus genome caused by misincorporations that occur during viral genome replication.

Antigenic shift Major changes in the influenza virus genome caused by mixing of genome segments that originate from different parental virions.

Antigenic variation Phenomenon in which the same viral protein, such as a capsomer, has different epitopes because escape from vertebrate adaptive immunity has selected for variation.

Antigenome Full-length complement of a single-stranded viral genome.

Antiviral medicine Pharmaceutical that inhibits the replication of a virus and can be used safely in humans or other animals.

Antiviral state Cellular condition induced by exposure to type λ interferons; in this condition cells are primed to resist viral replication.

APOBEC3 Primate host restriction factor that restricts replication of some retroviruses.

Apoptosis Eukaryotic programmed cell death used especially during animal development and immune responses so that damaged cells can be removed from the body without damaging healthy cells. Often occurs during viral infection and can cause cytopathic effects in tissue culture.

Arenavirus Group of viruses with single-stranded ambisense genomes.

Assembly Association of viral nucleic acids and proteins, sometimes with a membrane, which results in formation of an offspring virion.

Atomic force microscopy Form of scanned probe microscopy with very high resolution, useful for imaging topographic features of small biological samples such as viruses.

Attachment First stage of virus replication cycle in which a virion physically associates with a potential host cell through noncovalent molecular forces such as hydrogen bonds and ionic bonds.

Attenuated vaccine A vaccine comprising a microbe that can replicate slowly and provoke a strong immune response yet does not normally make healthy people sick.

Attenuated virus Virus that has been propagated outside its normal host or environment, resulting in reduced replication rates and reduced pathogenicity; sometimes useful in vaccination procedures.

Attenuation Process of selecting increasingly less virulent microbe by passaging it through a host cell, host animal, or environmental condition that it does not normally encounter in nature; the goal of attenuation is to produce a microbe that can still replicate enough to provoke a strong immune response during its use as a vaccine but that is unable to cause disease in an otherwise healthy person.

Autolysosome Final compartment that forms at the end of autophagy; site of degradation for an engulfed organelle or virus.

Autonomous virus replication Viral replication cycle that can occur when a single type of virus has infected a cell.

Autophagosome Membraneous organelle that surrounds old organelles targeted for degradation and recycling in the eukaryotic cytoplasm.

Autophagy Eukaryotic cellular process in which old organelles are engulfed by membranes that target the organelles for degradation.

Avidity The cumulative strength of many noncovalent intermolecular forces that hold a macromolecular complex, such as a virus spike and cellular receptor, together.

Axenic A microbial culture in which only a single species is present; also known as a pure culture.

Azidothymidine (AZT) Structural mimic of thymidine; inhibits the reverse transcriptase enzyme of HIV; first drug approved to treat HIV disease.

β1 (early-early) Herpesvirus proteins encoded by early-early genes; mRNA synthesized during the second wave of viral gene expression.

β2 (late-early) Herpesvirus proteins encoded by late-early genes; mRNA synthesized during the third wave of viral gene expression.

B-cell receptor (BCR) Receptor found on the surface of B lymphocytes; recognizes epitopes; transmembrane form of an antibody.

Background signal Low level of a signal, such as fluorescence, that occurs in the absence of manipulating a sample.

Bacteriophage Virus that infects bacteria or archaea; synonym of phage.

Bacteriophage λ Virus that infects *Escherichia coli* bacteria and serves as a model for gene expression, and for the mechanisms necessary for bacteriophages to exit a Gram-negative cell that has two cell wall layers external to the plasma membrane.

Baltimore Classification System A systematic way of classifying viruses according to the type of genome found in the virion and the mechanism of mRNA synthesis.

Bcl-2 Eukaryotic protein that can block apoptosis despite the presence of cytoplasmic cytochrome c.

Beclin-1 complex Eukaryotic protein complex that triggers autophagy.

Beet necrotic yellow vein virus Baltimore Class IV (+) RNA plant virus that is a model for viruses with multipartite genomes.

Biofilm Bacterial community that is attached to a surface through an extracellular matrix.

Bioinformatics Analysis of biological sequence data using computation.

Biosafety Level 4 Highest level of biosafety used to handle pathogens that can be acquired through an aerosol route and for which there is no treatment and no preventative vaccine. An example of a virus handled under Biosafety Level-4 conditions is Ebola virus.

Blot Paper-like membrane used after gel electrophoresis in procedures such as Southern, northern, and western blotting. Macromolecules separated by electrophoresis are transferred to a blot because it is less fragile than a gel and so can be further manipulated in order to detect the macromolecule of interest.

Bottleneck In evolutionary biology, an event that restricts the genetic variation in a population.

Broadly neutralizing antibodies Antibodies that neutralize many different isolates of a virus or other microbe.

Budding Exit of enveloped viruses from a host cell, such as an animal cell, that lacks a cell wall or cuticle.

Burst Period of time during which offspring virions exit the host cell.

CA Retrovirus capsomere.

Cancer Disease caused by overproliferation of abnormal cells in an animal.

Cap Circovirus capsomere.

Cap snatching Mechanism by which influenza virus acquires a 5′ cap.

Capsid Proteinaceous covering that protects a virion's genome; composed of capsomeres.

Capsomere Protein subunits that together make up a viral capsid. Synonym for capsomer.

Cargo Macromolecule shuttled into the nucleus by association with a complex containing importin.

Cas Protein components of a bacterial antiviral adaptive immune system; acronym for CRISPR-associated proteins.

Cas1 Bacterial CRISPR-associated protein needed to create new spacer sequences as part of an antiviral adaptive immune response.

Cas2 Bacterial CRISPR-associated protein needed, along with Cas1, to create new spacer sequences as part of an adaptive immune response.

Cas9 Bacterial antiviral defense protein used in genome editing because it makes site-specific cuts in DNA.

Caspase-3 Eukaryotic protease essential for triggering apoptosis.

Caspase-8 Eukaryotic protease essential for triggering apoptosis; active when oligomerized.

Caspase-9 Eukaryotic protein that senses cytoplasmic cytochrome c and triggers apoptosis in response.

Catenated Connected like links in a chain; offspring bacterial chromosomes are usually catenated at the end of θ replication.

Caveolin Eukaryotic protein needed for formation of caveosomes, which internalize particles.

Caveosome Eukaryotic vesicle coated with the protein caveolin.

CD4 Receptor found on the surface of T helper lymphocytes.

CD4+ T helper (T$_H$) cells T lymphocytes that coordinate an adaptive immune response in vertebrates.

CD8+ cytotoxic T lymphocytes (CTLs) T lymphocytes that specialize in killing cells that have been infected by a virus and cancer cells.

cDNA library Collection of DNA molecules derived by reverse transcribing mRNA collected from a specific tissue or population of cells.

Cell-mediated responses In vertebrate immunity, component of an adaptive immune response mediated by T lymphocytes.

Cellular transformation Cellular gain of traits associated with cancer cells, such as heightened proliferation and less responsiveness to proapoptotic signals.

Chain terminators Structural mimics of nucleotide triphosphates that prevent a polymerase from extending the new chain of nucleic acids once they have been incorporated into the macromolecule.

Chaperone Protein that catalyzes folding of other polypeptides by preventing inappropriate protein–protein interactions during maturation of a protein.

Chase Interval of time during a pulse-chase experiment when the cells are provided with a nonradioactive counterpart to the labeled reagent used during the pulse, which thereby confines the label to a short period of time (the pulse).

Checkpoint Times in the eukaryotic cell cycle when there are regulatory opportunities to continue the cell cycle, attempt to fix errors in DNA replication, exit the cell cycle, or undergo apoptosis.

Chimera In molecular biology, one protein that originates by artificially fusing the coding sequences of two proteins together in the absence of an intervening stop codon so that the proteins are translated as one; common examples include fusions to green fluorescent protein and epitope-tagged proteins.

Chronic infection *See* **Persistent infection**.

CI Bacteriophage λ protein that is critical for establishing and maintaining lysogeny; also known as the λ repressor.

CII Bacteriophage λ protein that is critical for establishing lysogeny.

CIII Bacteriophage λ protein that is critical for establishing lysogeny.

Circovirus Tiny naked animal virus with a circular ssDNA genome.

Circulative persistent transmission Transmission of a plant virus that occurs when the virus is ingested by an insect vector, passes through the insect's digestive tract, and is transported to the salivary glands, where it will subsequently be passed to another plant. The virus does not replicate in the insect.

***cis*-responsive RNA element (CRE)** RNA secondary structural element in a picornavirus genome that is necessary for genome replication.

Clamp loader Protein that helps assemble a DNA replication machine by connecting the sliding clamp to the machine.

Class I virus A virus with a double-stranded DNA genome in the virion.

Class II virus A virus with a single-stranded DNA genome in the virion.

Class III virus A virus with a double-stranded RNA genome in the virion.

Class IV virus A virus with a positive-sense single-stranded RNA genome in the virion.

Class V virus A virus with a negative-sense single-stranded RNA genome in the virion.

Class VI virus A virus with a diploid positive-sense, single-stranded RNA genome in the virion; genome must be reverse transcribed prior to viral gene expression. Synonym for retrovirus.

Class VII virus A virus with a gapped circular double-stranded DNA genome in the virion; the viral genome is synthesized starting with an RNA template.

Clathrin Eukaryotic protein needed for endosome formation.

Clinical efficacy Benefit to patients treated with a drug or gene therapy agent.

Closed supercoiled circle (ccc) Replicative form of the circovirus genome needed as a template for transcription and genome replication.

Cloverleaf RNA secondary structural element in which there are three stem–loops, resembling the leaf of a clover plant.

Co-receptor Host macromolecule that is required, along with another different macromolecule, for irreversible attachment.

Coevolution In virology, the process in which a virus adapts to its host, which then evolves to respond to the virus, thus triggering new changes in the virus so that the cycle of infection can continue.

Combination therapy Treating a microbe, especially a virus, with more than one drug that targets more than one protein, thus delaying replication and selection for drug resistance.

Combination antiretroviral therapy (cART) Treatment for HIV infection in which three or more drugs that target different viral proteins are taken at the same time, to severely reduce viral replication and thus delay viral replication and viral evolution.

Competitive inhibitors Reversible inhibitor that binds to the active site of an enzyme and increases K_M.

Complementation When two or more virions, each with a different inactivating mutation, co-infect the same host cell and the wild-type genes in one viral genome compensate for the mutant genes in the other, allowing a productive infection.

Concatemer Many viral genomes encoded one after another on the same contiguous nucleic acids.

Concatemers Long nucleic acid that contains two or more sequential copies of a viral genome linked end to end; must be processed into one-unit lengths during assembly.

Concerted assembly Viral assembly process in which genomes and capsomeres assemble at the same time so that there are no empty procapsids.

Confluent cells In eukaryotic cell culture, cells that have grown and multiplied so that they are physically touching other cells in the same culture, resulting in a nearly continuous lawn of cells.

Conformational epitope An epitope formed by amino acids that are near one another in the tertiary or quaternary structure of a protein but not in the primary sequence of the polypeptide.

Constant region Segment of an antibody that is invariant for antibodies made by a particular species.

Constitutive splicing machinery Eukaryotic protein–RNA complexes that routinely splice introns, resulting in mature mRNA.

Constraints In evolutionary biology, features that confine populations to certain changes and prevent others.

Convergent evolution Process by which unrelated viruses or organisms evolve similar characteristics because of natural selection.

Copy-and-paste mechanism Mechanism of helitron movement in which the DNA transposon copies itself into a new locus without using reverse transcriptase.

Core wall Ultrastructural feature of large poxvirus virions.

Core Ultrastructural feature of large poxvirus virions. Or hepadnavirus capsomere. Or capsomere of hepatitis C virus.

Coronaviruses Group of Baltimore Class IV (+) RNA viruses with especially large genomes; they express their genomes via multiple (+) viral RNAs and appear to encode a replication machine that has proofreading activity.

***cos* sites** Single-stranded ends of the otherwise dsDNA linear genome of λ bacteriophage.

Cosmid Large hybrid plasmid consisting of cloned DNA flanked by λ cohesive end (*cos*) sites.

Covalently closed circular DNA (cccDNA) In hepadnaviruses, the dsDNA form of the genome that is used as a template during gene expression.

CRISPR Genomic component of a bacterial antiviral adaptive immune system; acronym for clustered regularly interspaced short palindromic repeats.

CrmA Vaccinia virus protein that inhibits the activity of proapoptotic host caspases, thus blocking or delaying apoptosis.

cRNP In any Baltimore Class V (−) RNA virus, the (+) sense antigenome used as a template for genome synthesis; must be in a complex with viral nucleocapsid proteins.

Cro Bacteriophage λ protein that controls gene expression through binding to DNA and repressing transcription initiation.

crRNA Short RNA sequences with palindromic and unique regions that are used by CRISPR-Cas systems during an adaptive immune response against bacteriophage infection.

crRNP RNA–protein complex used by CRISPR-Cas systems during an adaptive immune response against bacteriophage infection; degrades bacteriophage DNA that **hybridizes** to the unique sequence in the RNA component of the complex.

Cryo-electron microscopy A method to image biological samples that have been preserved using low temperatures to preserve their shape prior to imaging. Can provide superior 3D renderings of biological specimens that are too large to image with X-ray crystallography, and images of enveloped viruses.

Cryo-electron tomography Technique in which two-dimensional images obtained using cryo-electron microscopy are subjected to computational analysis that uses them to create a three-dimensional image of the biological sample (such as a virus).

Cryo-EM *See* **Cryo-electron microscopy**.

Cryptic prophages DNA sequences descended from prophages that have mutations that prevent them from entering into a productive lytic cycle.

Cut-and-paste mechanism In transposons, the process by which some DNA-based transposons are cut out from a chromosome and moved to a new locus without creating a new copy of the transposon.

Cuticle In plants, a waxy layer that prevents water loss and is thick enough to prevent viruses from attaching to cells underneath an intact cuticle.

Cyclin D Eukaryotic regulatory protein that causes the transition from G_1 to S phase.

Cyclin-dependent kinase (Cdk) Eukaryotic regulatory protein that, when bound to its cognate cyclin, phosphorylates other proteins and is necessary for progression through the cell cycle.

Cytochrome c Mitochondrial protein; cytoplasmic release triggers apoptosis.

Cytokines Secreted proteins used for intercellular communication, especially for communication during an immune response.

Cytopathic effects (CPEs) Pathogenic changes in cellular structure associated with cell death or malfunction and observable through light microscopy.

Damage-associated molecular patterns (DAMPs) Molecules that should be intracellular in the absence of necrosis; when found in the extracellular space, they enhance inflammation. Examples include DNA, RNA, histones, and ATP.

de novo Synthesis of nucleic acids without a primer.

Decatenation Process by which catenated offspring chromosomes are physically separated after genome replication.

Dedifferentiation Process of reversing differentiation, accompanied by a loss of specialized functions and often a hallmark of oncogenesis.

Delayed-early gene Temporally, the second wave of gene expression in viruses, such as adenovirus, following immediate-early gene expression and before intermediate and late gene expression.

Developmental switch Molecular events that cause one state to switch to another; an example is a switch from lysogeny to lytic replication.

Diffraction The spread of electromagnetic waves, such as X-rays, when they encounter an obstacle, such as atoms, in a crystal.

Dimer linkage sequence In retroviruses, the RNA sequence used to dimerize the genome so that two copies are packaged into each virion.

Discontinuous mechanism Mechanism of nucleic acid synthesis in which the replication complex jumps from one region of a template to another without using the intervening nucleotides as a template.

Dissociation constant In biochemistry, an equilibrium constant that reflects the propensity of a macromolecular complex held together by noncovalent intermolecular forces to dissociate. Lower dissociation constants reflect a tighter association between the two macromolecules.

DnaA Bacterial protein that controls the switch from θ replication to σ replication in bacteriophage λ.

DNA-based transposons Transposons that do not use reverse transcription during movement.

DNA ligase Enzyme that catalyzes covalent bond formation between adjacent nucleotides in a nucleic acid that contains a nick.

DNA polymerase I Bacterial enzyme that catalyzes DNA synthesis, filling in the gaps left by removal of RNA primers.

DNA polymerase III Bacterial enzyme that catalyzes DNA synthesis during chromosome replication.

DNA vaccine A vaccine composed of DNA encoding protein subunits of a microbe.

Double-layered particle (DLP) Rotavirus particle inside a host cell consisting of two concentric capsids; synthesizes RNA.

Double-stranded RNA-specific protein kinase R (PKR) Eukaryotic protein that is an essential component of the innate defenses against viral infections.

Downstream hairpin loop (DLP) RNA structure unique to the sgRNA of togaviruses; required for cap-independent translation of that RNA.

Dynamin Eukaryotic ATPase protein that is required for endocytosis.

E Coronavirus envelope transmembrane protein.

E (end) sequence Regulatory sequence in mononegavirus genome that affects the termination of mRNA synthesis; found in between each protein-coding sequence along with I and S regulatory RNA.

E1 human papillomavirus protein Early HPV protein that functions as a helicase for DNA replication.

E1 protein A flavivirus envelope protein.

E1^E4 human papillomavirus protein HPV protein encoded by part of the E1 and part of the E4 coding sequence; necessary for mature virions to be released from the stratified epithelium.

E2 human papillomavirus protein Early HPV protein that regulates viral gene expression.

E2 protein A flavivirus envelope protein.

E2F transcription factor Eukaryotic transcription factors that are responsive to pRB and regulate gene expression necessary for progression through the cell cycle.

E3 ubiquitin ligases Enzyme that attaches the ubiquitin protein to target proteins, thus modifying their localization or activity.

E4-ORF4 Adenovirus protein that causes dephosphorylation of host SR proteins, ultimately resulting in production of the L4-33K transcript by alternative splicing late during infection.

E5 human papillomavirus protein HPV protein that manipulates the host cell cycle and apoptosis to make it possible for the virus to replicate.

E6 human papillomavirus protein HPV protein that manipulates the host cell cycle and apoptosis to make it possible for the virus to replicate.

E7 human papillomavirus protein HPV protein that manipulates the host cell cycle and apoptosis to make it possible for the virus to replicate.

Early endosome Endosome that formed recently and has not acidified.

Early protein Viral protein expressed early during infection, typically prior to the onset of viral genome replication. Most early proteins are enzymes or regulators.

Ebola virus Example of a filovirus; causes hemorrhagic fever in humans.

Eclipse period Period of time during an infection when no virions can be detected but the viral-replication cycle is ongoing.

Electron microscopy Form of microscopy used to examine viruses using an electron beam instead of light.

Elongation factor Nonribosomal protein required for translation elongation.

Emerge To arise or first occur, as when a new infectious agent, such as HIV, causes disease in appreciable numbers of humans for the first time.

Emergent infectious disease An infectious disease that occurs more frequently or among more people than it did historically, or than it was recognized to do so historically. Examples include acquired immunodeficiency syndrome (AIDS) and Middle East respiratory syndrome (MERS).

Endogenous retrovirus DNA that is complementary to the genome of a retrovirus and that is inherited through the generations of an animal because a chromosome in one of the gametes contains a DNA copy of the genome of that retrovirus.

Endogenous retroviruses Proviral cDNA now inherited vertically through the germ line; descended from a retrovirus, yet no longer produces infectious particles.

Endolysin In bacteriophages, a phage protein that degrades host peptidoglycan; needed for viral escape from the host cell.

Endoplasmic reticulum Eukaryotic organelle that serves as the site for synthesis of secreted and transmembrane proteins.

Endosome Eukaryotic intracellular compartment that is part of the endomembrane system; it can be formed from the plasma membrane after internalization of viruses or signaling molecules. Endosomes mature into multivesicular bodies using the ESCRT system.

Energy of activation Energy required for a favorable reaction to overcome the activation energy.

env Retrovirus gene that encodes the Env polyprotein.

Env The spike of a retrovirus, which has two component parts: SU and TM, also known as gp120 and gp41.

Epigenetic inheritance Inheritance of patterns of covalent modification to a chromosome but not to the nucleotide sequence of the DNA.

Episome In herpesviruses, the form of the circular genome found in latently infected cells.

Epitope-binding site Distal portion of an antibody that binds to an epitope on another macromolecule (most often a protein).

Epitope Component of antigens that provoke an adaptive immune response.

Epizootic A pandemic among nonhuman animals.

Epsilon (ε) Secondary structure in hepadnavirus pgRNA that is required for reverse transcription.

ESCRT machinery Protein complex that catalyzes abscission.

Establishment phase Earliest of three phases of human papillomavirus genome replication in which viral genomes are maintained at a few copies per cell in the lower layers of a stratified epithelium.

Etiological agent Agent responsible for a disease; for example, poliovirus is the etiological agent of poliomyelitis.

Eukaryotic translation initiation factor Eukaryotic protein necessary for the initiation of translation.

ex vivo Treating stem cells *in vitro*, for example with a gene therapy agent, before introducing them into a patient.

Exogenous Originating from outside a cell or organism; in the case of vertebrate immunity, an extracellular microbe such as a virion in the blood or other extracellular fluids.

ExoN Coronavirus nonstructural protein required for proofreading during viral RNA synthesis; also known as nsp14.

Exonuclease Enzyme that degrades nucleic acids by removing subunits from one end.

External envelope Outermost poxvirus membrane found in extracellular enveloped viruses; spikes protrude from the external envelope.

External enveloped virion In poxviruses, the infectious form with a complex structure that includes two membranes.

Extracellular enveloped virus (EEV) Infectious form of poxviruses found extracellularly; has an extra layer of envelope compared with intracellular mature viruses.

F One of the two spike proteins of the paramyxoviruses, along with HN.

Fenestrated Having perforations; in the case of dsRNA viruses, the double-layered particles are fenestrated so that NTPs can enter and mRNA can exit.

Fidelity How well a polymerase copies its template. Misincorporation of nucleotides that are not complementary to the template reflects low fidelity. Conversely, polymerases with high fidelity have very low error (misincorporation) rates.

Filoviruses Baltimore Class V (−) RNA mononegaviruses characterized by two matrix proteins, both a structural spike protein and a soluble form of that spike, and a transcription regulator. The class includes Ebola fever virus, which causes a hemorrhagic fever with high mortality rates.

Fitness In evolutionary virology, the potential of a virus to produce many offspring.

Fixation Chemical process that crosslinks macromolecules prior to examining a biological sample, such as a virus with certain forms of microscopy, including electron microscopy.

Flaviviruses Group of Baltimore Class IV (+) RNA viruses that encodes transmembrane proteins so that its genome must be translated in association with the eukaryotic host endoplasmic reticulum; hepatitis C virus is a prominent example that infects humans.

FLIPs Eukaryotic proteins that can block apoptosis by preventing formation of active caspase-3.

Fluorescence The property of absorbing light at a particular wavelength and subsequently emitting light at a longer wavelength. Useful in many biological applications such as fluorescence microscopy.

Fluorescence microscopy Type of microscopy in which light is used to excite and detect fluorescent molecules such as GFP.

Fourier transformation Mathematical process that allows use of a diffraction pattern to determine the three-dimensional structure of a crystalized molecule such as a protein.

Fusion Coming together of two biological membranes so that their lipid and protein components mix and become a single membrane.

Fusion pore Aqueous connection between two compartments after membrane fusion.

γ1 (leaky-late) Herpesvirus proteins encoded by leaky-late genes; mRNA synthesized during the fourth wave of viral gene expression.

γ2 (late-late) Herpesvirus proteins encoded by late-late genes; mRNA synthesized during the fifth (and final) wave of viral gene expression.

G Specifically in the mononegaviruses, the glycoprotein spike.

G_0 A period of the cell cycle in which cells are quiescent and have exited G_1. G_0 typically occurs when cells have differentiated.

gag Retrovirus gene that encodes the Gag polypeptide and part of the Gag-Pol polyprotein.

Gain-of-function experiments In virology, the study of mutant viruses that have gained a new property, especially the ability to infect a different species.

Gene expression profiling Molecular technique to measure the quantity of many different mRNA species all at once.

Gene silencing Natural or artificial process that prevents the production of a protein product even though the DNA encoding that protein is intact.

Gene therapy Clinical treatment that manipulates a patient's cells in order to express a therapeutic gene in the patient's body.

Genetic drift Slow changes in a genome caused by misincorporations and other small mutations that occur during replication of a genome.

Genome A complete set of genes for a certain virus or organism; for viruses the genome can be composed of either RNA or DNA.

Genome editing Use of CRISPR-Cas and other technologies to attempt to change a genome with high precision in order to introduce a specific mutation.

Genome packaging Process of incorporating viral nucleic acids into a forming capsid.

Germinal centers Specialized sites in lymph nodes and spleen where antigen-presenting cells and lymphocytes interact producing an adaptive immune response.

GFP *See* **Green fluorescent protein.**

Gibson cloning Recombinational cloning method in which a bacteriophage protein is used to recombine DNA molecules without using restriction enzymes or ligase.

gp2.5 Bacteriophage T7 single-strand binding protein needed for phage genome replication.

gp4 Bacteriophage T7 protein with helicase and primase activity needed for phage genome replication.

gp5 Bacteriophage T7 DNA polymerase needed for phage genome replication.

gp41 Component of the retrovirus spike that is approximately 41 kilodaltons in mass, that has a segment that passes through the lipid bilayer of the viral envelope, and that includes the fusion peptide. Synonym for TM.

gp120 Component of the retrovirus spike that is approximately 120 kilodaltons in mass and that is responsible for attachment to host cells. Synonym for SU.

GP Filovirus glycoprotein spike.

Green fluorescent protein (GFP) Fluorescent protein that has been adapted for many uses in molecular biology, especially in combination with fluorescence microscopy and reporter gene technologies.

gRNA A guide molecule made of RNA; used in genome editing along with CRISPR-Cas technologies.

Growth factor Signaling molecule, typically a hormone or protein, that stimulates a cell to exit G_0 and begin proliferation.

H101 Oncolytic gene therapy agent used to treat head and neck cancers in China.

Hairpin In parvovirus genomes, a dsDNA structure at the end of an otherwise ssDNA linear genome. More generally, any structure in DNA in which a single strand folds back on itself, forming a base-paired structure that resembles a hairpin.

Head completion protein Viral protein that fills in the space used by the packaging motor at the conclusion of sequential assembly, thus completing the assembly stage.

HeLa cells A type of cultured human cervical cancer cells that are easy to handle under laboratory conditions and that were derived from a patient sample obtained from Ms. Henrietta Lacks without her consent in the 1950s, before laws required consent for tissue donation for medical research.

Helicase Enzyme that disrupts hydrogen bonds that otherwise hold double-stranded nucleic acids together.

Hemagglutinin (HA) Influenza spike protein needed for attachment, penetration, and uncoating; important for determining the immunological type of influenza viruses.

Hepadnavirus Baltimore Class VII enveloped virus with a genome that is part DNA and part RNA and that replicates via a reverse transcription mechanism; viral DNA does not recombine with the host chromosome.

Hepatitis B virus (HBV) Model Baltimore Class VII retrovirus that replicates through a reverse transcription mechanism.

Hepatitis C virus (HCV) An enveloped Baltimore Class IV (+) RNA virus with a genome that is translated in association with the eukaryotic host endoplasmic reticulum because it encodes many transmembrane proteins; it causes hepatitis and can cause liver cancer.

Hepatitis D virus (HDV) Infectious RNA similar to a viroid but that infects human cells.

Herpes simplex virus type 1 Synonym for human herpesvirus 1; herpes simplex virus 1 and herpes simplex virus type 1 are older terms that the International Committee on Taxonomy of Viruses has retired but that are still in common use.

Highly active antiretroviral therapy (HAART) First combination therapy developed to treat HIV.

Histone acetyl transferase Eukaryotic enzyme that acetylates nuclear histone proteins, resulting in formation of euchromatin, which in turn is associated with transcription.

Histone deacetylation enzyme Eukaryotic enzyme that removes acetyl groups from nuclear histone proteins, resulting in formation of heterochromatin, which in turn is associated with silenced transcription.

Hit In bioinformatics, a database sequence that is similar to a query sequence.

H One of the two spike proteins of the paramyxoviruses, along with F.

Hodgkin and Reed Sternberg cells (RS cells) Malignant cells found in Hodgkin disease; descendants of B cells that are infected with human herpesvirus 4.

Hodgkin disease (HD) Form of B-cell cancer caused by human herpesvirus 4 (also known as Epstein–Barr virus).

Holin In bacteriophages, a phage protein that creates pores in the inner membrane; needed for viral escape from the host cell.

Holoenzyme Bacterial RNA polymerase complex containing four subunits: two α, one β, and one β′. The holoenzyme cannot initiate or terminate transcription on its own. For example, σ factors are required for transcription initiation.

Homologous protein Two proteins that have a common ancestor.

Homologous Descended from a common ancestor.

Homology Property of sharing common ancestry. In molecular virology, homology is usually indicated by sequence similarity.

Horizontal gene transfer Transfer of genetic material between organisms independent of sexual reproduction.

Host range The diversity of host organisms that can be infected by a virus.

Host restriction factor An antiviral innate immune protein that prevents a virus from replicating in a host cell, especially after the attachment stage of viral replication.

Human herpesvirus 1 Large enveloped dsDNA virus that causes the disease herpes and is a model for DNA viruses that cause latent infections. Synonym for the more common term herpes simplex virus 1.

Human immunodeficiency virus 1 (HIV-1) Model Baltimore Class VI retrovirus that replicates through a reverse transcription mechanism.

Human milk oligosaccharide Collection of diverse short-chain oligosaccharides found in human breast milk that serve as a probiotic to support development of a healthy gut microbiome; also plays a role in preventing transmission of viruses through breast milk.

Human papillomavirus (HPV) Model papillomavirus that is the cause of genital warts and most forms of cervical cancer; certain types are associated with high risk of cervical cancer but a preventative vaccine is available.

Humoral response In vertebrate immunity, component of an adaptive immune response mediated by antibodies.

Hydroxymethylcytosine Covalently modified form of cytosine used by some dsDNA bacteriophages in order to selectively degrade unmodified host DNA while leaving phage DNA intact.

I (intergenic) sequence Regulatory sequence in mononegavirus genome that affects the termination and initiation of mRNA synthesis; found in between each protein-coding sequence in between E and S regulatory RNA.

ICP0 Herpesvirus α protein that, together with ICP4, causes early-early and late-early gene expression.

ICP4 Herpesvirus α protein that, together with ICP0, causes early-early and late-early gene expression.

Idoxuridine Nucleotide mimic to treat herpesvirus infections; causes altered base pairing so that DNA containing idoxuridine cannot be used as a template to produce functional mRNA or genomes.

IFNα Type I interferon produced by most nucleated cells as part of an innate immune response to viral infection.

IFNβ Type I interferon produced by most nucleated cells as part of an innate immune response to viral infection.

IFNγ Type II interferon important for regulating innate and adaptive immune responses.

IIIa repressor element (3RE) Regulatory sequence in adenovirus RNA that regulates alternative splicing to encode either the 52,55K protein or the IIIa protein. Binding site for hyperphosphorylated host SR proteins.

IIIa virus-infection-dependent splicing enhancer (3VDE) Regulatory sequence in adenovirus RNA that regulates alternative splicing to encode either the 52,55K protein or the IIIa protein. Binding site for viral L4-33K protein.

Immortalized cells Animal cells that will proliferate indefinitely in culture because they have some cancerlike properties.

Immune system System of molecules, cells, and tissues that protects a multicellular organism from pathogenic microbes such as viruses.

Immunization Clinical treatment that stimulates immunity against an infectious microbe such as a virus or bacterium.

Immunoblotting Technique in which antibodies are used to selectively stain a specific protein. Typically, proteins are separated by size using electrophoresis before blotting. Synonym for western blotting.

Immunohistochemistry Technique in which tissue sections are stained with antibodies and then visualized with microscopy.

Immunological types In virology, subspecies of influenza that are categorized based on the immune response they provoke.

Immunostaining Technique in which antibodies are used to selectively stain a specific protein in a biological specimen. Often used in combination with microscopy or flow cytometry.

Importin Eukaryotic protein needed for transport across a nuclear pore.

in silico By computer, or using software, specifically in contrast to doing an experiment *in vivo* (in a living organism) or *in vitro* (in a test tube or flask).

Inactivated vaccine A vaccine composed of chemically or physically treated microbes that cannot replicate.

Inclusion bodies An aggregate of macromolecules that appears as densely staining foci in a light microscope; often the sites of viral replication in infected cells.

Incompletely spliced mRNA HIV-1 mRNA in which just one intron has been removed.

Indel When comparing two nucleic acid sequences in an alignment, an insertion or deletion in one sequence compared with the other.

Infected cell proteins Proteins found uniquely in host cells infected by herpesvirus; most are encoded by the herpesvirus genome.

Infectious dose The amount of a virus that must enter a multicellular organism in order to cause an infection.

Inflammation Innate immune response to infection or tissue damage resulting in an influx of fluids, cytokines, and immune cells, especially neutrophils.

Influenza virus Baltimore Class V (−) RNA virus+A296:C336 with a segmented genome; it can cause epidemics and pandemics.

Initiation factor Nonribosomal protein required for translation initiation.

Initiator-binding proteins Collection of host cell proteins that bind to the noncoding control region of polyomaviruses and regulate the switch from early to late transcription of viral genes.

Innate defenses Omnipresent components of the vertebrate immune system that include neutrophils and interferon.

Inner capsid In reoviruses, the most interior capsid closest to the genome.

Inner membrane Membrane in poxvirus virions that surrounds the core, core wall, and lateral bodies.

Inoculation Process of adding an infectious agent, such as a virus, to a culture of host cells.

Integrase A retroviral enzyme that recombines host chromosomal DNA with viral cDNA.

Interferon (IFN) Cytokine that is essential for an innate immune response against viral infection.

Interferon-stimulated genes (ISGs) Genes that are transcribed in response to signaling from type I interferons.

Intermediate capsid In reoviruses that have three capsids, the middle capsid.

Intermediate filament Eukaryotic cytoskeletal filaments that are needed especially for structural stability; lytic viruses often disrupt intermediate filaments in order to exit the host cell.

Internal mature virion In poxviruses, an intermediate form of the virus that occurs during assembly and can be observed inside infected cells.

Internal ribosome entry site (IRES) Secondary structure in (+) RNA that allows that RNA to be translated even when cap-dependent eukaryotic translation initiation is not possible.

Intracellular mature virus (IMV) Infectious form of poxviruses found inside infected cells.

Introns Sequences that interrupt RNA and must be removed in order for the RNA to be functional.

Inverted repeat Sequence in nucleic acid that has two parts. The upstream part is the reverse complement of the downstream part. Inverted repeats are often components of regulatory regions, such as those necessary for controlling viral transcription or genome replication. Inverted repeats can assemble into hairpin structures because of the complementarity of bases on the same chain.

Irreversible binding Viral attachment to a host cell receptor through many weak intermolecular forces that together add up to a strong interaction, resulting in the initiation of the penetration and uncoating stage of replication.

Kozak sequence Sequence surrounding the AUG start codon in eukaryotic mRNA; influences the amount of cap-dependent translation initiation.

λ repressor Synonym for λ CI protein; critical for establishing and maintaining lysogeny.

L protein A component of the rhabdovirus RdRp complex that has the active site for nucleic acid synthesis; forms a complex with P protein.

L1 human papillomavirus protein Late HPV protein that is the more abundant of the two capsomeres.

L2 human papillomavirus protein Late HPV protein that is the less abundant of the two capsomeres.

L3-23K A late adenovirus protein that cleaves host intermediate filaments, which are part of the cytoskeleton. Necessary for host cell lysis.

L4-33K Regulatory adenovirus protein that binds to the 3VDE sequence in mRNA, biasing the splicing reaction toward production of the viral IIIa maturation protein.

Large E1A Early adenovirus protein that is critical for initiating the adenovirus cascade of gene expression.

Large T antigen Multifunctional polyomavirus regulatory protein that regulates the host cell cycle and apoptosis, and also participates directly in regulation of gene expression and in viral genome replication.

Last universal common ancestor (LUCA) The population of cells that gave rise to all extant lineages of cells.

LAT In herpesviruses, an RNA transcript that is expressed during latency and is not translated.

Late domain Amino acid signature found in viral structural proteins; binds to the ESCRT machinery so that abscission results in viral budding.

Late protein Viral protein expressed after the onset of viral genome replication. Many late proteins are structural proteins.

Latent infection Viral infections in which a host cell contains the viral genome and a few other viral macromolecules but does not produce virions for long periods of time.

Lateral bodies Ultrastructural feature of large poxvirus virions.

Lawn A layer of host cells covering a culture dish or nutritive surface, such as agar in a petri dish.

LC3-I Form of the eukaryotic LC3 protein that is important for the first step of autophagy.

LC3-II Form of the eukaryotic LC3 protein that is associated with phosphotidylamine and is critical for the autophagy pathway to continue.

Leader (le) sequence Regulatory sequence near the 3′ end of a mononegavirus genome; regulates nucleic acid synthesis and is partially complementary to the trailer sequence.

Leader sequence Coronavirus sequence found near the 5′ end of the genome; required for discontinuous RNA synthesis.

Leaky scanning Viral regulatory mechanism in which the ribosome sometimes begins translation at the most 5′ start codon but other times begins at a start codon closer to the 3′ end because that stop codon has a better Kozak sequence.

Light microscopy Technique in which visible light and lenses are used to increase the magnification of tiny objects such as cells.

Linear epitope An epitope formed by amino acids that are near one another in the primary sequence of a polypeptide.

Long control region   Noncoding region of DNA in papillomavirus genomes that regulates transcription in response to the differentiation status of host cells in a stratified epithelium.

Long interspersed nuclear elements (LINEs)   Most common retrotransposons in the human genome; they are flanked by direct repeats and encode an endonuclease and a reverse transcriptase.

Long terminal repeat   In retroviruses, a sequence repeated in viral cDNA at both ends of the genome, adjacent to host chromosomal DNA. Includes sequences necessary for regulation of viral gene expression.

Longer unique region (UL)   Longest of two regions of the herpesvirus genome that encodes proteins.

LTR   *See* **Long terminal repeat**.

LTR retrotransposons   Retrotransposons that are descended from retroviral proviral DNA. They include long terminal repeats, and they include retroviral *gag* and *pol* sequences but lack *env* sequences.

Lymphocytes   B and T cells crucial for an adaptive immune response in vertebrates such as human beings.

Lysins   Bacteriophage enzymes that degrade the peptidoglycan macromolecule found in bacterial cell walls; can be used to treat bacterial infections.

Lysogen   Bacterial host that contains one or more prophages.

Lysogeny   State in which a host cell contains viral macromolecules but is not actively synthesizing offspring virions.

Lytic cycle   Temperate phage cycle in which offspring phages are produced and the host cell lyses.

Lytic infection   Viral infection that results in death of the host cell, most typically within the time span of a normal cell division cycle.

M   Coronavirus matrix protein.

M1   Influenza virus matrix protein.

M2   Influenza virus ion channel.

M13   Bacteriophage with a circular ssDNA genome. Its replicative form is the basis for many common cloning vectors.

MA   Retrovirus matrix protein.

MAC   Membrane attack complex; immune defense triggered by the complement system; forms pores in the membranes of infected cells, killing them before a virus can complete its replication cycle.

Maintenance phase   Middle of three phases of human papillomavirus genome replication in which viral genomes are maintained at 50–300 copies per cell in the midzone layers of a stratified epithelium.

Major histocompatibility complex I (MHC-I)   Surface molecule abundant on vertebrate cells; displays endogenous peptides to immune cells. Viral infection results in display of foreign endogenous peptides, ultimately causing an immune response against infected cells.

Major histocompatibility complex II (MHC-II)   In vertebrates, a protein complex used by professional phagocytic cells to display epitopes that originate from degradation of exogenous antigens (extracellular pathogens). Presents epitopes to B lymphocytes and T_H lymphocytes.

Malignant   Life-threatening, especially when referring to cancer.

Matrix protein (M)   A mononegavirus structural protein.

Maturation   Any molecular events that occur after assembly and are needed for the virion to be infectious; an example is proteolysis of HIV proteins inside the virion during or immediately after release.

Maturation cleavage   Proteolytic processing of viral polyproteins that occurs after a polyprotein has been released from the ribosome, in contrast to nascent cleavage, which occurs during synthesis of a polyprotein.

Maturation protein   Phage protein found in one copy per virion in the (+) RNA phages; required for attachment and penetration.

Measles virus   Baltimore Class V (−) RNA paramyxovirus responsible for measles, one of the most infectious diseases known to infect humans.

Mechanical inoculation   In plant viruses, infection of a plant by physically disrupting a plant tissue and allowing a virus access to the cells.

Membrane   In molecular biology, a solid substrate made of nitrocellulose or a similar polymer that can be used in blotting to detect specific proteins or nucleic acids.

Messenger ribonucleoprotein complexes (mRNPs)   mRNA in complex with proteins that control its localization, translation, and degradation.

Metagenome   Collection of genetic information characteristic of a specific environment.

Metagenomics   Analysis of all the DNA isolated from an environment, such as a gram of soil or a milliliter of sea water.

Metastable   Biochemical state in which it is favorable for a macromolecule or macromolecular complex to undergo a chemical or structural transformation, but a high activation energy makes that change to a lower energy state unlikely even though the reaction is thermodynamically favorable.

Methylation   Covalent modification of a macromolecule by addition of a methyl group.

Minute virus of mice (MVM) Model parvovirus.

miRNA Regulatory microRNA that is usually about 20 nucleotides long, forms a stem–loop structure, and regulates eukaryotic gene expression.

Misincorporation Covalent addition of a nucleotide that is not complementary to its template.

Missense mutation Mutation that alters a nucleotide in an mRNA molecule so that the resulting protein has a single amino acid difference, compared with the protein encoded by the wild-type counterpart gene.

Model A particular virus, gene, protein, or organism that is studied intensively in order to make discoveries that are broadly applicable to other entities that have not been studied as intensively.

Modification enzyme Enzyme that binds to a specific nucleotide sequence in DNA and covalently modifies adenine bases by methylating them on positions that do not interfere with base pairing.

Molecular and cellular virology Research discipline emphasizing the proteins, genes, and cellular features that are important for virus replication and pathogenesis.

Molecular mimicry Situation in which a macromolecule mimics the three-dimensional shape of a completely different molecule, using a different and unrelated primary sequence.

Mononegavirales Order of Baltimore Class V (−) RNA viruses that share conserved proteins, gene order, and gene expression and genome replication strategies. Includes rabies virus.

Monopartite Virus in which the genome is composed of a single molecule of nucleic acid.

Monophyletic Sharing a common ancestor.

Monoubiquitination Covalent attachment of a single ubiquitin protein to a target; does not result in degradation of that target and is involved in retrovirus budding.

mRNA vaccine A vaccine comprised of chemically-stabilized mRNA that encodes one or more antigens.

MS2 Bacteriophage with a (+) RNA genome in which secondary structure controls the relative amounts of proteins translated from the genome.

Multipartite Virus in which the genome is found in different segments, each carried by a physically distinct capsid.

Multiplicity of infection (MOI) Ratio of infectious virions to host cells during an experiment.

Multivesicular bodies (MVBs) Late endosomes that contain internal membrane-bound vesicles that originate by abscission of the vesicular membrane through the action of the ESCRT system.

Mutation Change in the nucleotide sequence of a genome.

Mx1 Antiviral protein encoded by an interferon-stimulated gene; Mx1 interferes with intracellular nucleocapsid trafficking.

N protein Bacteriophage λ protein that controls gene expression through binding to RNA and preventing transcription termination.

Naive 1. Animal that has never been exposed to an infectious agent such as a virus. 2. Lymphocyte that has never been exposed to its specific cognate antigen.

Naked virion A virion that does not have an envelope.

Nascent Just recently synthesized; nascent polypeptides are typically still associated with the ribosome, whereas nascent nucleic acids are typically still associated with a polymerase complex.

Natural killer (NK) cell Immune cell that kills cells displaying too few MHC-I molecules as part of an antiviral or anticancer response.

Natural selection Evolutionary force in which viruses or organisms that are more fit than their relatives leave more offspring in the next generation.

NC Retrovirus nucleocapsid protein.

Necrosis Form of cellular death in which the internal contents of a cell are released into the surrounding tissue, typically provoking inflammation.

Negri bodies Inclusion bodies that form during rabies virus infection; the sites of rabies virus gene expression and genome replication.

Nested set Set of nucleic acids of different lengths in which the sequence of the smallest one is found in all the others, and the sequence of the next smallest is found in all the larger ones, and so on.

Neuraminidase (NA) Influenza virus spike protein responsible for escape from the host cell surface after budding; important for determining the immunological type of influenza virus.

Neutralize During an adaptive immune response against a viral infection, the binding of antibodies to the surface of a virion so that it cannot bind to its cellular receptor.

Neutrophil extracellular trap (NET) Meshwork of extracellular fibers composed of chromatin and antimicrobial proteins; produced by neutrophils during an immune response.

Neutrophils Innate immune cells that defend against viruses by producing neutrophil extracellular traps and by enhancing inflammation.

Nick Missing covalent bond between the phosphate groups of adjacent nucleotides in a nucleic acid.

Noncoding control region (NCCR) Polyomavirus DNA sequence that occupies 8% of the virus's tiny genome, is necessary for transcription regulation, and serves as the origin for DNA replication.

Noncompetitive inhibitor Reversible inhibitor that increases the V_{max} of an enzyme.

Nonpermissive cell A cell that does not permit a virus to complete its replication cycle.

Nonpersistent transmission Transmission of a virus that occurs when the virus adheres to the mouth parts of an animal vector and is subsequently injected into the next plant that the animal bites.

Nonstructural protein Viral protein that is not found in the virion; typically plays a regulatory or enzymatic role.

Northern blot Procedure in which labeled nucleic acids are used to detect complementary RNA that had first been size-fractionated using electrophoresis.

NS2 The second nonstructural protein encoded by a virus; in hepatitis C virus, NS2 is a site-specific protease.

NS3 The third nonstructural protein encoded by a virus; in hepatitis C virus, NS3 is a site-specific protease.

NS5B One of the hepatitis C virus nonstructural proteins; it has the active site for RNA-dependent RNA replicase activity.

nsp7 Coronavirus nonstructural protein that is a component of the RNA-dependent RNA synthesis and proofreading machine; works with nsp8 to form a sliding clamp that improves the processivity of the replicase.

nsp8 Coronavirus nonstructural protein that is a component of the RNA-dependent RNA synthesis and proofreading machine; works with nsp7 to form a sliding clamp that improves the processivity of the replicase.

nsp9 Part of the coronavirus RNA replication machine that stimulates replication.

nsp10 Part of the coronavirus RNA replication machine that stimulates the exonuclease activity of the proofreading subunit ExoN (nsp14).

nsp12/14/10 complex Complex of coronavirus nonstructural proteins that has RNA polymerization and proofreading activities.

nsp14 Coronavirus nonstructural protein required for proofreading during viral RNA synthesis; also known as ExoN.

NSP2 (rotavirus) One of the two rotavirus proteins, along with NSP5, needed for the formation of viroplasm.

NSP5 (rotavirus) One of the two rotavirus proteins, along with NSP2, needed for the formation of viroplasm.

Nuclear export protein (NEP) Influenza virus protein needed during viral gene expression and genome replication.

Nuclear localization signal Short amino acid sequence that binds to importin, targeting a protein with the signal to the nucleoplasm.

Nuclear spliceosomal introns Eukaryotic introns that must be removed from pre-mRNA before the mRNA can be translated properly.

Nucleocapsid A viral capsid that contains the viral genome.

Nucleocapsid (N) protein The nucleocapsid protein in the rhabdoviruses such as rabies and other mononegaviruses.

Nucleocapsid protein (NP) For viruses in which the genome occurs as an obligatory part of a ribonucleoprotein complex, the most abundant protein that packages the RNA into the ribonucleoprotein complex.

Nucleocytoplasmic large DNA viruses Order of viruses with DNA genomes that includes some members that replicate in the cytoplasm and others that replicate in the nucleus.

Nucleoplasm Aqueous interior of the eukaryotic nucleus.

Obligate intracellular parasite A virus or organism that must live inside another host cell in order to replicate (viruses) or survive and reproduce (organism).

Oligo(dT) A primer that anneals to poly(A) tails and so can be used to reverse transcribe eukaryotic mRNA.

Oncogene-deficient retroviruses Retrovirus that can cause cancer by virtue of misexpressing host genes because of proximity to the proviral long terminal repeat. These retroviruses do not encode oncogenes.

Oncogene-transducing retroviruses Retrovirus that encodes an oncogene as part of its genome; none that infect humans are known.

Oncogenic The ability to cause cancer or to initiate a series of cellular events that may result in cancer.

Oncolytic Virus or other microbe that preferentially infects cancer cells and lyses them.

Oncoproteins Viral proteins that cause cancer by dysregulating the cell cycle, apoptosis, or both.

One-step growth experiment Classic experiment using bacteriophages to discover the six stages of a typical viral-replication cycle.

Operators In bacterial transcription regulation, a DNA sequence that binds to a regulatory protein, especially a repressor.

Opsonization During an adaptive immune response, binding of many antibodies to the surface of a microbe, which increases the likelihood that it will be engulfed by professional phagocytes such as macrophages.

Opsonize To coat a viral particle with immune system proteins, such as complement or antibodies, targeting them for destruction by immune cells such as phagocytes.

Optical sectioning Process in which lasers and computers are used to image sections of a thick sample without having to physically section the sample first. Commonly used in combination with fluorescence microscopy.

ORF3 Circovirus protein that manipulates host immunity.

Origin binding protein (OBP) Herpesvirus protein that binds to the origin of viral DNA replication.

Origin of DNA replication (*ori*) Any site on a DNA genome that is necessary and sufficient for the assembly of proteins that initiate replication.

Origin of replication (*ori*) Sequence in a genome that is both required for the initiation of genome replication and the site where genome replication begins.

Orphan virus Virus that can be isolated from a plant or animal but is not associated with any disease symptoms.

Orthomyxoviruses Viruses that have enveloped virions containing segmented negative-sense ssRNA genomes. Influenza is a model orthomyxovirus.

Osmotic lysis Lysis of host cells that occurs when water external to the cell diffuses in, creating pressure that bursts the host cell. Lysis occurs when bacterial cell walls are compromised because bacteria live in dilute aqueous solution, necessitating an intact cell wall to push back against osmotic pressure.

Outer capsid In reoviruses, the most exterior capsid furthest from the genome.

Outer membrane In a Gram-negative bacterium such as *Escherichia coli*, the outer membrane is the most distal layer of the cell envelope and is composed of proteins and lipids, with lipopolysaccharide in the outer leaflet.

ΦX174 Bacteriophage with a circular ssDNA genome. It was the first nucleic acid completely sequenced.

φ Abbreviation for the packaging signal in a viral genome; the sequence that initiates genome packaging.

P1 polyprotein Picornavirus polyprotein encoded closest to the 5′ end of the genome; will be processed into the VP1, VP2, VP3, and VP4 proteins.

P2 polyprotein Picornavirus polyprotein encoded between the sequences specifying P1 and P2; will be processed into three nonstructural proteins: 2Apro, 2B, and 2C.

P3 polyprotein Picornavirus polyprotein encoded closest to the 3′ end of the genomic RNA; will be processed into six proteins: 3AB, 3CDpro, 3A, 3B, 3C, and 3D^{pol}. 3B

is also known as VPg, which is found in the virion. All other P3 proteins are nonstructural.

p53 Eukaryotic regulatory protein that controls apoptosis.

P (hepadnavirus) In hepadnaviruses, P is the polymerase, which has reverse transcriptase activity.

P protein A component of the rhabdovirus RdRp complex that can be phosphorylated; forms a complex with the L protein.

P/V/C Collective term for three different paramyxovirus proteins that are encoded by overlapping RNA through leaky scanning (C) or RNA editing (V).

PA Influenza virus protein that is one component of the RNA-dependent RNA polymerase; has the active site necessary for cleaving host mRNA during cap snatching.

Packaging capacity Amount of new genetic material that can be incorporated into the genome of a gene therapy vector.

Packaging motor Viral nanomachine that uses ATP hydrolysis to pump a genome into a procapsid during assembly.

Panhandle Structure in adenovirus genomic DNA that serves as an origin of replication for the second strand of new adenovirus DNA.

Papillomavirus Small naked dsDNA (Baltimore Class I) virus that infects stratified epithelia and is a model for how viruses can encode many regulatory functions despite having tiny genomes.

Paramyxoviruses Family of Baltimore Class V (−) RNA mononegaviruses characterized by two glycoprotein spikes (F and HN) instead of G and regulatory proteins V and C; examples include measles virus.

Parvovirus Tiny naked animal virus with a linear ssDNA genome with hairpin ends.

Passaging Propagating a virus in tissue culture for many generations.

Pattern recognition receptor (PRR) proteins Eukaryotic proteins that bind to molecules such as dsRNA that are commonly associated with pathogens and not with animal cells.

PB1 Influenza virus protein that is one component of the RNA-dependent RNA polymerase; has the active site for catalyzing RNA synthesis.

PB2 Influenza virus protein that is one component of the RNA-dependent RNA polymerase; has the active site necessary for binding to the cap on host mRNA during cap snatching.

PCBP Eukaryotic protein found in most host mRNP.

pCMV-VEGF Therapeutic plasmid based on cytomegalovirus and used to stimulate blood vessel growth.

Penetration Entry of the virion or some of its components into a host cell.

Peptidoglycan Bacterial macromolecule resembling a chain-link fence and composed of carbohydrates and amino acids; responsible for cell shape and for preventing osmotic lysis when the cell is in a hypotonic solution (relative to the cytoplasm).

Periplasm In a Gram-negative bacterium such as *Escherichia coli*, the compartment between the plasma membrane and the outer membrane. It contains peptidoglycan and many proteins.

Permissive host cell A host cell that permits replication of a virus.

Persistent infection Long-term infection that lasts a long time relative to the typical life span of the host cell. For single-celled organisms, persistent infections are those that last longer than a single host-division cycle. For multicellular organisms, persistent infections can last for weeks, months, or even years.

Persistent transmission Transmission of a plant virus that occurs when a virus persists in many tissues of an insect for many days or weeks but does not replicate in the insect.

Phagocytosis Eukaryotic process in which the cytoskeleton deforms the plasma membrane, forming projections that reach out above the plane of the rest of the plasma membrane and engulf particles or microbes.

Phosphorimager Instrument that can quantify the amount of radioactivity in a band on a gel or blot.

Picornavirales Order of viruses discovered in 2016 through metagenomics.

Picornaviruses Large group of Baltimore Class IV (+) RNA viruses that infect animals and plants; poliovirus is a prominent example that infects humans.

P$_L$ Bacteriophage λ promoter that drives expression of the N protein during lytic replication and is repressed during lysogeny.

Placenta Mammalian organ that mammals with live birth develop exclusively during pregnancy in order to provide for gas, nutrient, and waste exchange between the embryo or fetus and the mother.

Plaque Visible hole in a lawn of host cells, indicating the presence of lytic viral replication.

Plaque assay Quantitative measure of the number of infectious viruses in a suspension.

Plaque-forming unit Number of virions necessary to form a plaque; for some viruses the plaque-forming unit is 1 whereas for others the number is greater than 1 because all virions in a suspension may not be competent for replication.

Plasma cells Differentiated B cells specialized in manufacturing and releasing large quantities of antibodies.

Plasmodesma Channel of cytoplasm that passes from one plant cell into a neighboring cell through the cell walls; singular form of the word *plasmodesmata*.

pol Retrovirus gene that encodes part of the Gag-Pol polyprotein.

Poliovirus A picornavirus that can infect human beings and cause poliomyelitis; model for gene expression and genome replication in Baltimore Class IV (+) RNA viruses.

Polycistronic Bacterial and archaeal mRNA that encodes more than one protein sequentially (without overlapping); classic example is encoded by the *lac* operon of *E. coli*.

Polyomavirus Small dsDNA (Baltimore Class I) virus that infects animal cells and is a model for the smallest possible viruses that infect human cells.

Polyprotein Viral protein that will be proteolytically digested in a specific way in order to release individual protein components that have unique activities.

Polyubiquitination Process of covalently attaching many ubiquitin proteins to a single target protein; usually involves covalently attaching ubiquitins to both the target and to other ubiquitins that are themselves already attached to the target protein.

Porcine circovirus 2 (PCV2) Model circovirus.

Postexposure prophylaxis (PEP) Drugs taken soon after exposure to HIV in order to prevent the virus from establishing itself in the body.

pp1a Shorter coronavirus polyprotein.

pp1ab Longer coronavirus polyprotein.

PP2A Eukaryotic regulatory protein that is a phosphatase enzyme. It removes phosphate groups from other regulatory proteins, such as pRB, contributing to regulation of cellular proliferation.

P$_R$ Bacteriophage λ promoter that drives expression of Cro during lytic replication and is repressed during lysogeny.

pRB Eukaryotic tumor suppressor protein first discovered because the gene encoding it is often nonfunctional in retinoblastoma patients. It regulates progression through the cell cycle.

Preexposure prophylaxis A medical treatment that must be done prior to exposure to an infectious agent and that prevents infection by that agent.

Preexposure prophylaxis (PrEP) Anti-HIV drug taken before exposure to HIV in order to prevent the virus from establishing itself in the body in the event that someone is exposed to HIV.

Pregenomic RNA (pgRNA) In hepadnaviruses, the RNA that is used as a template for synthesizing new genomes.

Preintegration complex In retroviruses, a complex between viral cDNA and the viral integrase protein, prior to that cDNA recombining with a host chromosome.

Preterminal protein (pTP) Adenovirus protein used to prime viral genome replication; proteolytically processed during maturation to result in the terminal protein (TP) covalently attached to the 5′ ends of the linear adenovirus dsDNA genome.

Preventative vaccine A vaccine that prevents disease.

Primary antibody During immunoblotting and related procedures, the antibody that is incubated with the blot first and which binds to the epitopes on the antigen of interest.

Primary cell cultures *In vitro* cell cultures recently derived from a living animal; normally proliferate only a few times before they die.

Primary sequence Order of monomers in DNA, RNA, or proteins.

Primary tissue culture cells Cultured animal cells that were recently derived from a clinical sample and can divide for only a few generations.

Primary transcription Production of mRNA using the infecting vRNP as a template specifically in the mononegaviruses.

Primary viremia First wave of viral replication in a multicellular organism.

Primase Enzyme that synthesizes an RNA primer during replication of DNA templates.

Primer-binding site In retroviruses, a specific genome sequence required for the primer to hybridize and initiate reverse transcription.

P<small>RM</small> Bacteriophage λ promoter that drives expression of CI, also known as the λ repressor, during lysogeny.

Probe Labeled macromolecule, such as DNA, that binds specifically to a molecule of interest in order to detect that molecule on a blot or in a micrograph.

Procapsid Immature viral particle that does not yet contain genomes during a sequential assembly process.

Processivity The number of nucleotides a polymerase can polymerize before it dissociates from its template.

Proenzymes Protein that must be covalently modified before it becomes catalytically active; a common modification is proteolytic cleavage.

Professional phagocytic cells In vertebrate immunity, cells that are specialized for phagocytosing microbes and displaying epitopes derived from them to lymphocytes.

Programmed −1 ribosomal frameshift Viral regulatory mechanism in which features of the RNA cause the ribosome to sometimes shift position, resulting in translation of a different protein had the ribosome not shifted.

Proofreading Activity found in some nucleic acid synthesis machines in which the complex can recognize and remove when the polymerase has added a noncomplementary nucleotide to the 3′ end.

Propagative persistent transmission Transmission of a plant virus that occurs when the virus is ingested by an insect vector, where it actively replicates in the animal so that its numbers increase substantially, thus enhancing transmission to other plants bitten by the insect vector.

Prophage Bacteriophage genome integrated into a host chromosome during lysogeny.

Prophylactic Preventative. Prophylactic vaccines or medications are administered before a person becomes infected with a virus, to prevent that viral infection.

Prophylactic immunization Immunization administered prior to exposure to an infectious agent in order to prevent infection.

Prophylactic vaccine A vaccine that prevents disease.

Protease inhibitors Antiretroviral drugs that bind to the viral protease and inhibit its activity; used as a treatment for HIV infection.

Proteasome Large nanomachine that degrades old and misfolded proteins; in eukaryotes, the proteasome degrades ubiquitinated proteins.

Proteolytically processed Subjected to degradation of one or more specific peptide bonds. The consequence is typically to release proteins with new activities or to activate a protein.

Proteome All of the proteins expressed by a cell, tissue, or community.

Proto-oncogenes Normal human genes that, when mutated or misexpressed, can cause cancer.

Provirus Retrovirus cDNA that has inserted into a host chromosome.

Pseudoknot RNA structural element having at least two stem–loop structures in which bases from one of the stems hydrogen bond with bases that are otherwise parts of the other stem–loop.

Pseudotyping Procedure to create an enveloped virion that expresses the attachment protein from a different viral species, which changes the attachment properties of the modified virus.

Pulse-chase analysis Method for analyzing changes to macromolecules that occur over time by labeling during a brief pulse and watching what occurs afterward during the chase.

Pulse labeling Procedure in which radioactive precursors, such as amino acids, are provided to cells in order to label the macromolecules that are actively being synthesized during the pulse.

Pulsed Provided with a label for a pulse-chase analysis; the label is most often a radioactive amino acid or nucleotide.

Q protein Bacteriophage λ protein that controls gene expression through binding to DNA and preventing transcription termination.

Qβ Bacteriophage with a (+) RNA genome used as a model to understand viral RNA-dependent RNA polymerase structure and function.

Query In bioinformatics, the sequence used to find other similar sequences in a database.

Rabies virus Baltimore Class V (−) RNA virus with a broad host range; it can serve as a model for the mononegaviruses.

rAD-p53 Recombinant adenovirus used to treat head and neck cancers in China; provides a therapeutic copy of the p53 gene to cancer cells.

Rational drug design Use of crystal structures and modeling to design small molecules that might inhibit a protein of pharmacological interest.

Rational vaccine design The use of knowledge of immunology to design a vaccine more likely to provoke a desired immune response.

Reactivation In herpesviruses, process in which a latent infection is triggered to become a lytic infection.

Real-time quantitative reverse-transcription PCR Molecular technique for measuring the quantity of a specific mRNA.

Receptor Macromolecule on the surface of a host cell that serves as the site for a virus to attach; for animal cells, receptors are often glycoproteins.

Recombinant vaccines Vaccines produced using recombinant DNA technology (molecular cloning).

Recombination Large-scale changes to a viral genome so that alleles once associated with different parental genomes are found combined in a recombinant offspring virion.

Reductionism Research strategy that seeks to understand something complex by breaking it down into its simplest components; molecular virology seeks to understand viruses by investigating the macromolecules that are necessary for their replication.

Reiterative transcription Mechanism in which a viral transcriptase copies a series of Us in a template many times, resulting in a poly(A) tail on viral mRNA.

Release Exit of virions from the host cell.

Release factors Cellular proteins that recognize stop codons and cause translation termination.

Reoviruses Family of dsRNA viruses that contains the best-studied genera.

Rep Circovirus protein needed for genome replication.

Rep′ Circovirus protein needed for genome replication.

Replicase Viral enzyme that synthesizes antigenomes or genomes from a viral nucleic acid template.

Replication protein A Eukaryotic DNA replication protein that binds to single-stranded DNA in the origin of replication.

Replicative form Double-stranded form of a viral genome that is single stranded in the virion.

Reporter gene A gene that is easy to detect and so can be used to measure the activity of a promoter; common examples include *gfp* and *lacZ*.

Resolution In microscopy, the minimum distance necessary to distinguish two adjacent objects.

Restriction modification Bacterial innate defense against bacteriophages; uses epigenetic covalent modification of DNA to prevent host DNA from being degraded by a restriction enzyme while unmodified phage DNA is degraded by that same enzyme.

Retrotransposon Transposon that moves via a mechanism that requires reverse transcriptase.

Retrovirus Baltimore Class VI enveloped virus that replicates via a reverse transcription mechanism; viral cDNA must recombine with a host chromosome prior to gene expression.

Rev HIV-1 accessory protein needed for nuclear export of some viral mRNA.

Rev responsive element (RRE) Sequence in HIV-1 mRNA that has secondary structure and enables Rev protein to assemble on the mRNA, thus targeting it for export from the nucleus.

Reverse transcriptase Polymerase that can use DNA as a template to synthesize RNA.

Reversible binding Viral attachment to a host cell surface through a few weak intermolecular forces that are easily disrupted by Brownian motion.

Reversible inhibitors Compounds that interact with an enzyme temporarily, reducing the activity of that enzyme.

Rexin-G Anticancer gene therapy treatment based on using a retrovirus to express a toxic version of a human cyclin protein.

Rhabdoviruses Group of Baltimore Class V (−) RNA viruses that includes rabies virus.

Rho (ρ) protein Protein that binds to mRNA and interacts with holoenzyme, causing transcription termination in bacteria.

Ribosome-binding site Synonym for Shine–Dalgarno sequence, which is a sequence in bacterial mRNA that binds to a ribosome in order to initiate translation.

Ribosome shunting Viral translation mechanism that enables adenovirus to simultaneously block translation of host messages while ensuring translation of viral mRNA. Requires the viral 100K protein and *cis*-acting sequences in the viral mRNA.

Ribozymes Catalytic RNA.

RNA-dependent RNA polymerase (RdRp) Viral enzyme that uses an RNA template to synthesize complementary RNA.

RNA editing Mechanism in which a viral transcriptase inserts one or more specific nucleotides into a specific site in an mRNA in the absence of a template that specifies that insertion so that the mRNA is longer than its corresponding template; found in the paramyxoviruses (among others).

RNA interference (RNAi) Process in which an RNA molecule selectively silences expression of a complementary mRNA.

RNA secondary structure Hydrogen bonds among the nucleotides of a single strand of RNA cause the RNA to fold up in two dimensions.

RNA termination–reinitiation Viral mechanism for encoding different proteins on a polycistronic mRNA in eukaryotic cells, which do not normally have polycistronic mRNA. Used to express HPV E7 protein.

RNA tertiary structure Hydrogen bonds among the nucleotides of a single strand of RNA cause the RNA to engage in long-range interactions between nucleotides that are not near one another in the primary sequence.

RNase L Innate immune protein that is encoded by an interferon-stimulated gene. When bound to its ligand 2′-5′-oligoadenylate synthetase, degrades RNA and thereby prevents infection by viruses with RNA genomes.

Rolling-hairpin mechanism Hypothesis (model) for the mechanism of parvovirus replication.

Rolling circle Form of DNA replication in which a single strand of template DNA peels off a double-stranded DNA template; also called σ replication because of its resemblance to the Greek letter.

Rotavirus Species of reovirus that is pathogenic in humans, especially infants and young children.

Round up Pathogenic change in tissue culture cells that have detached from the substrate, indicating that those cells are dying.

Route of infection Physical route a virus takes to infect a multicellular organism; an example in humans is the respiratory route of infection in which a virus is contracted by inhalation.

RT-qRT PCR *See* **Real-time quantitative reverse-transcription PCR.**

S (start) sequence Regulatory sequence in mononegavirus genome that affects the termination of mRNA synthesis; found in between each protein-coding sequence following E and I regulatory RNA.

SAMHD1 Primate host restriction factor that restricts replication of some retroviruses.

SARS-CoV-2 *See* **SARS coronavirus 2**.

SARS coronavirus 2 (SARS-CoV-2) Coronavirus that causes severe acute respiratory syndrome.

Satellite virus A virus that can only complete its replication cycle when the cell is co-infected with another virus.

Scaffold protein Viral protein that is an obligatory component of the procapsid during assembly but is missing from the mature virion.

Scanned probe microscopy Technique in which a physical probe is dragged over the surface of a sample and then computational techniques render a three-dimensional image of that sample; can have atomic or even subatomic resolution.

Scanning electron microscopy Form of electron microscopy that generates an image of the surface properties of a sample such as a virus.

Screen Genetic procedure to sort through a collection of mutants in order to find ones with a desired phenotype.

Secondary antibody During immunoblotting and related procedures, an antibody that binds to the constant region of a primary antibody and is tagged so that the primary antibody (and the antigen to which it is bound) can be detected. Use of tagged secondary antibodies amplifies the signal so that small amounts of an antigen can be detected.

Secondary transcription Production of mRNA using newly synthesized vRNP as a template specifically in the mononegaviruses.

Secondary viremia Second wave of viral replication in a multicellular organism; often occurs in a different tissue than the site of primary viremia.

Segmented Viral genome in which there are many independent nucleic acids that encode different proteins; for example, the influenza A virus has eight segments.

Semipersistent transmission Transmission of a plant virus that occurs when a virus persists in the foregut of an insect for many days but does not replicate in the insect.

Sequential assembly Viral assembly process in which empty procapsids assemble and are later filled with genomes.

Sequestration Process in which a viral protein binds to a host target protein, thus preventing that host target protein from interacting with other host proteins. The effect is to block the activity of the host protein.

sGP Longer soluble form of the filovirus glycoprotein spike, which may be responsible for some of the symptoms and lethality of Ebola hemorrhagic fever disease.

Shine–Dalgarno sequence Synonym for ribosome-binding site, which is a sequence in bacterial mRNA that binds to a ribosome in order to initiate translation.

Short interspersed nuclear elements (SINEs) Shortest retrotransposons in the human genome; they do not encode proteins but have sequences that originate from Pol III transcripts such as rRNA.

Shorter unique region (US) Shorter of two regions of the herpesvirus genome that encodes proteins.

Sigma (σ) factor Protein that binds to holoenzyme, enabling RNA polymerase to bind to promoter DNA.

Signal peptidase Eukaryotic protease found in the endoplasmic reticulum that removes signal sequences from proteins translated on the rough endoplasmic reticulum.

Signal recognition particle (SRP) Component of the eukaryotic system required for synthesizing proteins on ribosomes docked to the endoplasmic reticulum.

Signal sequence Short amino acid sequence that targets a protein to a specific subcellular location; a prominent example in eukaryotes is the sequence in transmembrane proteins that targets the ribosome to synthesize the protein on the endoplasmic reticulum.

Simian vacuolating virus 40 (SV40) Model polyomavirus that has been studied intensively in order to understand polyomavirus gene expression and genome replication.

Sindbis virus Togavirus that is a model for a Baltimore Class IV (+) RNA virus that synthesizes full-length and subgenomic (+) RNA.

Single-strand binding proteins Proteins that bind to single-stranded nucleic acids during genome replication.

Site-specific protease Enzyme that degrades a peptide bond found in the context of a specific amino acid sequence.

Sliding clamp Portion of host DNA replication machinery that keeps the DNA polymerase closely associated with its template.

Slippery sequence Sequence in viral mRNA that is necessary for a programmed −1 ribosomal frameshift; it is considered slippery because it is the site where the ribosome sometimes shifts.

Small t antigen Polyomavirus regulatory protein that regulates the host cell cycle.

SNARE Animal protein found especially in the nervous system where it is essential for fusion of vesicles at nerve termini.

snRNP Small nuclear ribonucleoprotein; many snRNPs are complexes required for splicing.

Solvent-exposed region Region of a protein that is exposed to water in a cell or virus.

SOS response In bacteria, the networks of proteins that respond to DNA damage.

Southern blot Procedure in which specific DNA sequences can be detected in a complex sample. Named for its inventor, Dr. Edwin Southern (b. 1938).

Spacer Unique 20- to 60-bp sequences flanked by palindromic repeats in the CRISPR-Cas bacterial adaptive immune system; these sequences provide a record of viral infection and protection against viruses with similar nucleic acid sequences.

Spanin In bacteriophages, a phage protein that fuses the outer membrane and inner membrane, enabling escape of phages.

Spike protein Protein that protrudes away from the surface of a virion and is typically used for attachment.

Spliceosome Eukaryotic complex of many protein and RNA molecules that catalyzes splicing (intron removal).

Spores Metabolically inactive, differentiated bacterial cells that can survive environmental assaults such as dehydration and irradiation that would kill an actively dividing cell.

Sporulate Process in which certain bacteria create and release survival structures known as spores.

SR protein Nuclear protein containing many serine and arginine residues; phosphorylation of SR proteins controls splicing.

ssGP Shorter soluble form of the filovirus glycoprotein spike, which may be responsible for some of the symptoms and lethality of Ebola hemorrhagic fever disease.

Stain Chemical treatment that enhances contrast or detectability so that a biological sample can be visualized more clearly.

Start–stop model Currently accepted mechanism to explain how rhabdovirus mRNA is generated in a pattern that results in decreasing abundance for each mRNA as the transcriptase moves toward the 5′ end of the template RNA.

Stem–loop structure Form of RNA secondary structure in which part of the RNA base pairs because it is complementary. The stem is the base-paired region, whereas the loop is composed of unpaired nucleotides.

Structural mimics Drug that is similar in structure to a natural enzyme substrate but does not participate in the same chemical reaction.

Structural protein Viral protein that serves a structural, nonenzymatic, and nonregulatory role in the virion; an example is a capsomere.

SU One of two main parts of a retrovirus spike. SU is an abbreviation for surface, because the spike is on the surface of the virions.

Subgenomic (sg) Viral (+) sense RNA that is shorter than the genome.

Subunit vaccine A vaccine manufactured from purified component parts of a microbe, such as viral capsomers.

Subversion Situation in which a virus uses a host protein for a novel function that assists with phage replication.

Suppression of translation termination Viral regulatory mechanism in which a tRNA binds to a stop codon, resulting in continued translation instead of translation termination; togaviruses use this mechanism.

Surface (SU) Component of the retrovirus spike that is approximately 120 kilodaltons in mass and that is responsible for attachment to host cells. Synonym for gp120.

Surface-exposed region Region of a protein that is exposed to a solvent, most commonly water.

Syncytia Plural of syncytium. *See* **Syncytium**.

Syncytin-1 Human protein encoded by an endogenous retrovirus; essential for placenta formation.

Syncytium A large multinucleate eukaryotic cell that forms as a result of plasma membrane fusion between cells; enveloped viruses often cause the formation of syncytia that are detectable as a cytopathic effect.

T4 ligase Bacteriophage T4 enzyme that can catalyze a covalent bond to repair a nick in dsDNA.

T4 polynucleotide kinase Bacteriophage T4 enzyme that catalyzes addition of a phosphate group to the 5′ end of DNA.

T7 bacteriophage Baltimore Class I DNA bacteriophage that infects *Escherichia coli* and is a model for lytic replication.

T7 RNA polymerase (RNAP) Large protein encoded by bacteriophage T7, which catalyzes transcription initiated at Class II and Class III T7 promoters.

T-cell receptor (TCR) Receptor found on the surface of T lymphocytes; recognizes epitopes.

Tailocins Bacteriophage tail lacking a head; can attach to bacteria and kill them.

Talimogene laherparepvec Oncolytic gene therapy agent based on herpesvirus; used to treat melanoma in Europe and the US.

Tandem repeats In nucleic acids, two or more repeated sequences that are adjacent to each other. The repeated sequences may be exactly identical or sometimes have a few unique bases.

Tar element Stem–loop structure in HIV-1 mRNA that is essential for gene expression.

Tat HIV-1 accessory protein needed for gene expression.

Tegument Layer of herpesvirus proteins in the virion that occurs between the envelope and genome.

Temperate Bacteriophage that can replicate lytically or enter into lysogeny.

Temperate phage Bacteriophage that has the genetic capacity to cause a latent infection.

Ter Sequence in a genome that both is required for the termination of genome replication and is the last sequence copied during genome replication.

Terminal differentiation Cellular state in which an animal cell has completed a developmental program in order to specialize in a certain function; such cells have usually exited the cell cycle and do not express DNA replication proteins unless a virus manipulates them to do so.

Terminal protein (TP) Adenovirus protein that is covalently attached to the 5′ strands of the double-stranded linear DNA genome.

Terminal redundancy Feature of viral genomes in which the ends of the dsDNA are repeated.

Termination factor Nonribosomal protein required for translation termination.

Tertiary interactions Interactions between different stem–loop structures in RNA.

Tetherin Antiviral protein encoded by an interferon-stimulated gene; tetherin interferes with viral budding and reduces the number of virus particles that escape from the cell surface.

Therapeutic immunization Immunization to treat an existing infection.

Therapeutic transgenes In gene therapy, a gene that, once expressed in a target cell, provides clinical benefit.

Therapeutic vaccine A vaccine that treats a disease.

Theta (θ) replication Replication of circular dsDNA templates in which replication forks proceed bidirectionally away from *ori*.

Tissue tropism The tissues in a multicellular organism that contain host cells that can be infected by a virus.

Tobacco mosaic virus (TMV) Baltimore Class IV (+) RNA virus with a naked helical capsid and a tRNA-like sequence in its genome that mimics the structure of host tRNA and might regulate translation or replication of the genome.

Togaviruses Group of Baltimore Class IV (+) RNA viruses that replicate via production of two different (+) sense RNAs.

Topoisomerase Enzyme that introduces or removes supercoils in DNA.

Trailer (tr) sequence Regulatory sequence near the 5′ end of a mononegavirus genome; regulates nucleic acid synthesis and is partially complementary to the leader sequence.

Transcriptase Viral enzyme that synthesizes mRNA from a viral nucleic acid template.

Transcription regulating sequence (TRS) Coronavirus sequence that regulates production of sgRNAs.

Transduction Horizontal gene transfer mechanism in which a virus transfers genetic material from one organism to another.

Transfection Experimental introduction of nucleic acids into a eukaryotic host cell, usually for purposes of genetic engineering or RNA interference.

Transformation The process of becoming cancerous.

Transgene In gene therapy, a therapeutic gene that replaces some or all of the coding sequences in a viral vector's genome.

Translational repression Repression of translation caused by binding of a protein to the RNA, thus physically preventing the ribosome from binding to the RNA in order to initiate translation.

Translational scientist Researcher who takes insights from basic science research and develops them into practical technologies, such as vaccines, gene therapy, or antiviral medicine.

Translocon Eukaryotic protein complex found in the endoplasmic reticulum and necessary for synthesis of secreted and transmembrane proteins that are synthesized on the rough endoplasmic reticulum.

Transmembrane (TM) Component of the retrovirus spike that is approximately 41 kilodaltons in mass, that has a segment that passes through the lipid bilayer of the viral envelope, and that includes the fusion peptide. Synonym for gp41.

Transmission electron microscopy (TEM) Form of electron microscopy that generates an image of the interior of thin sections of a subject such as a cell infected by a virus.

Transposable elements *See* **Transposons**.

Transposons Also known as transposable elements; these subviral entities consist of DNA that can catalyze movement from one locus to another in a host genome.

tRNA-like sequence (TLS) Viral RNA structure that mimics the structure of host tRNA and might regulate translation or replication of the genome; found prominently among the potyviruses including tobacco mosaic virus.

TRS-B Coronavirus sequence found adjacent to coding sequences in (+) RNAs; forms secondary structures and is necessary for discontinuous RNA synthesis.

TRS-L Coronavirus sequence found adjacent to the leader in (+) RNA; forms secondary structure, and is necessary for discontinuous RNA synthesis.

Tumor suppressor proteins Proteins that prevent oncogenesis and regulate the cell cycle and apoptosis in normal animal cells.

Ubiquitin A small very abundant eukaryotic protein necessary for the routine recycling of old or unfolded proteins; when covalently attached to target proteins, ubiquitin alters their localization or activity.

Uncoating All molecular events that occur between initial penetration and expression of the first viral gene; in eukaryotes often includes transport of the genome to a specific subcellular site.

Uncompetitive Enzyme inhibitor that binds to the complex formed by the enzyme and its substrate; the effect is to alter both K_M and V_{max}.

Unfolded protein response Eukaryotic stress response in which an overabundance of misfolded proteins in the endoplasmic reticulum causes the cell to initiate three mechanisms for reducing the amount of misfolded proteins: to slow translation, to increase the number of chaperones in the endoplasmic reticulum, and to increase the degradation of misfolded proteins.

Uridylylating Covalent addition of a molecule of uridine, especially to an amino acid.

v-Bcl-2 Viral protein that mimics a host Bcl protein and thereby delays apoptosis.

v-FLIP Viral protein that has the same activity as a host FLIP protein: it blocks apoptosis.

Vaccine Medical treatment to provoke a strong adaptive immune response against a pathogen without making the subject sick.

Vaccinia growth factor (VGF) Regulatory poxvirus protein that manipulates host apoptotic pathways.

Vaccinia virus Poxvirus (dsDNA, Baltimore Class I) used in preventative immunization to eradicate smallpox. Poxviruses have very large genomes and virions, and replicate in the cytoplasm, even though their genomes are composed of dsDNA.

Variola virus Poxvirus (dsDNA, Baltimore Class I) that causes smallpox; only known remaining samples are stored in research facilities in the United States and Russia.

Vector 1. An animal that carries an infectious agent such as a virus from one host to another. 2. Virus that has been genetically engineered to deliver a therapeutic or experimental gene into a cell.

Vegetative DNA replication Last of three phases of human papillomavirus genome replication in which viral genomes increase to thousands of copies per cell in the upper midzone cells of a stratified epithelium.

Very late transcript Polyomavirus mRNA encoding agnoprotein; produced during the maturation stage of the viral replication cycle.

VHS Herpesvirus regulatory protein packaged into virions and released during uncoating; results in selective degradation of host mRNA by an unknown mechanism.

Vif HIV and SIV accessory protein that counteracts a host restriction factor known as APOBEC3.

Viral early transcription factor (VETF) Regulatory poxvirus protein that stimulates early transcription.

Viral envelope Lipid bilayer that covers the capsid of some virions.

Viral hallmark genes Genes useful for tracing viral evolution because they are found in many diverse groups of viruses that have no close homologs in cellular organisms.

Viral intermediate transcription factor (VITF) Regulatory poxvirus protein that stimulates intermediate transcription.

Viral late transcription factor (VLTF) Regulatory poxvirus protein that stimulates late transcription.

Viral ribonucleoprotein (vRNP) Viral proteins and nucleic acids tightly associated with each other through noncovalent intermolecular forces such as hydrogen bonds; usually has a highly ordered structure as found in influenza A virus.

Viral ribonucleoprotein (vRNP) complex Any complex in which at least one RNA and one protein are held together by noncovalent intermolecular forces such as hydrogen bonds.

Virion Infectious form of a virus containing at a minimum the capsid and genome; some also have an envelope.

Viroids RNA molecules that infect plants without involving a capsid.

Viroplasm Collection of viral and host proteins surrounding rotavirus double-layered particles; subcellular sites of rotavirus gene expression and genome replication.

Virulence genes Genes that enhance the ability of a pathogen to infect a host or cause symptoms of disease.

Viruslike particles (VLPs) Viral capsids assembled *in vitro* in the absence of any nucleic acids so that they cannot replicate; can be used as vaccines.

Virus replication compartment (VRC) Membrane-bound compartment that serves as a site for viral nucleic acid synthesis.

Virus replication complex (VRC) Assemblage of proteins, viral nucleic acids, and host membranes that is the site of viral gene expression and genome replication in a eukaryotic cell. Most typically used to describe sites of replication for RNA viruses.

VP1/VP3 flower complex Rotavirus protein complex responsible for synthesis of capped mRNA.

VP16 Regulatory herpesvirus protein expressed late and packaged into nascent virions so that it can activate early-early gene expression.

VP24 One of the two filovirus matrix proteins, along with VP40.

VP30 Filovirus protein that regulates mRNA synthesis.

VP35 Filovirus protein that is a component of the replicase and transcriptase.

VP40 One of the two filovirus matrix proteins, along with VP24.

VPg protein Picornavirus protein that is covalently linked to the 5′ end of the (+) RNA genome.

Vpx SIV accessory protein that counteracts a host restriction factor known as SAMHD1.

Wave A group of viral genes that are all expressed at the same time.

Western blotting *See* **Immunoblotting**.

X-ray crystallography Technique in which crystallized biological samples, such as naked viruses, are bombarded with X-rays to collect the diffraction pattern that occurs when the X-rays interact with the sample. Computational techniques use the diffraction pattern to construct the three-dimensional shape of the sample.

Xenophagy Use of autophagic processes to engulf and destroy an infecting virus or its products.

Zoonotic An infection that typically reproduces in a nonhuman animal, especially a vertebrate, but that can also be transmitted to humans.

Zoonotic transmission Transmission of an infectious microbe, such as a virus, from a nonhuman animal population to a human.

Answers

Chapter 1

1. There are many possible correct answers. Similarities: Both have DNA genomes; both the virion and the cell are made up of proteins and nucleic acids. Both the virus and the cell can evolve (the cell could become cancerous, for example). Differences: The virion lacks lipids and RNA. The virus is an obligate intracellular parasite, but the cell is not. The virus reproduces by assembly, whereas the cell reproduces by growth and mitosis. The shape and dimensions of the virus and the cell are very different. The cell has many highly structured internal features, but the virion has few structural features. The virion can probably survive outside of a host cell long enough to infect another cell, whereas the epithelial cell would quickly die outside of the animal unless it were kept alive by artificial means such as cell culture.

2. There are many possible correct answers. Similarities: Both have capsomeres and have genomes made of nucleic acids. Both can evolve, are obligate intracellular parasites, and reproduce by assembly. Both have virions composed of nucleic acids and proteins, but the Class VII virus also contains lipids. The Class V virus has a (−) RNA genome, whereas the Class VII virus has a double-stranded DNA genome with gaps in the strands. The Class V virus synthesizes new genomes from (+) antigenomes, whereas the Class VII virus synthesizes new genomes from an mRNA template. The Class V virus must encode an RNA-dependent RNA polymerase, whereas the Class VII virus must encode a reverse transcriptase (RNA-dependent DNA polymerase).

3. Viruses are not considered to be alive because virions are metabolically inert in that they do not have any energized membranes. Because viruses are composed of informational molecules (nucleic acids and proteins), which are expressed in living host cells, those molecules can change over time as a result of mutation in the nucleic acids encoding the proteins. Changes in hereditary information over time are by definition evolutionary changes.

4. There are many possible correct answers. Pragmatically speaking, definitions of life are useful in proscribing the objects of study within the science of biology. Human beings tend to treat living organisms differently from the treatment of nonliving organisms; for example, in some religious faiths there is an ethical imperative to harm as few living organisms as possible. The answer to the question of whether the last remaining infectious smallpox virions should be destroyed would be different depending on whether we think the virions are alive. The question of whether viruses are living might also be important to evolutionary biologists trying to determine when life began; if there were viruslike entities before there were cells, were those viruslike entities alive? Finally, plans to attempt to detect life elsewhere in the solar system or in the rest of the universe might be different if those attempts include looking for viruslike aliens.

5. The 10^7 dilution is obviously unusable. The 10^9 dilution has a larger sampling error than the 10^8 dilution, so use the 10^8 dilution to calculate that (242 plaques/0.100 mL) $\times 10^8 = 2.4 \times 10^{11}$ plaque-forming units per mL. In general, always use the single best dilution, with best defined as the most countable plaques, because this dilution has the lowest calculated error rate. The error in these measurements is typically calculated as the square root of all of the plaques counted and then converted into a percentage of the total number of plaques counted. The square root of 243 is 15.6, which is a 6% error. The square root of 31 is 5.6, which is an 18% error. In practice, it is common to use three independent measurements to determine the PFU/mL, although the more times the population is sampled, the closer one gets to the actual value.

6. Although there are some enormous viruses with diameters that approach the 220 nm resolution of light microscopy, most are four or five times smaller than that in diameter.

7. In order for an alignment to be useful in this regard, many generations of reproduction must have occurred in order to make it possible for changes to occur and be selected against. In other words, comparing proteins that are too closely related can be less informative than comparing those that are less closely related, depending on the purpose of the alignment.

Chapter 2

1. About 35 minutes, the period of time when extracellular virions are below the limit of detection.

2. At the time of inoculation, there was one virion. After the replication cycle, there are 200 viruses. Therefore, the burst size is 200.

3. Measuring the viruses by microscopy would be difficult because the viruses would have to be purified from the broth, concentrated, applied to EM grids, and stained for imaging. Although this could be done in principle, it would be unnecessarily arduous. Instead, a plaque assay would be a better way to measure PFU/mL at every sampling period. In that case, the amount of virus could easily be measured using a dilution series to sample, making it possible to detect even tiny numbers of extracellular virions.

4. Shut off of host transcription, host protein synthesis, or both; production of viral mRNA; production of viral proteins; production of viral genomes; assembly of new virions; and degradation of host cell wall (for bacteriophages).

5. Acidification of endosome; fusion between viral envelope and endosome membrane, causing release of vRNPs; transport of vRNPs to nucleus; entry of vRNP into nucleus through nuclear pores.

6. Both are long-term infections. Chronic infections occur exclusively in multicellular hosts because a chronic infection is one in which lytic viral growth is constant, but the host can respond to the infection for a while by replacing the dead cells killed by the virus. Persistent infections can occur in multicellular or single-celled hosts, because the infection does not result immediately in lytic production of viruses. During chronic infections, a virus is replicating and so it can evolve to resist the immune system or antiviral medications, whereas in latent infections of animals and plants, the virus is not reproducing. For latent infections of bacterial cells, the virus evolves at the same pace as its host cell because the prophage is passed from generation to generation through normal chromosomal DNA replication.

7.

Stage	Organelles
Attachment	Plasma membrane
Penetration and uncoating	Endomembrane system; Cytoskeleton; Nucleus[a]
Synthesis of viral mRNA	Nucleus[a]
Synthesis of viral proteins	Ribosomes; Rough endoplasmic reticulum[b]; Golgi body[c]
Synthesis of viral genomes	Nucleus[a]
Assembly	Cytoskeleton; Endomembrane system; Plasma membrane; Nucleus[a]
Release	Plasma membrane; Cytoskeleton

[a] For viruses with genomes localized to the nucleus.
[b] For viruses that encode transmembrane proteins.
[c] For viruses that encode glycoproteins.

Chapter 3

1. Its target host cells may be absent or found very infrequently at the wrong site, or the physical conditions, chemical conditions, or both in the wrong entry route might destabilize or otherwise inactivate the virion, making it not infectious.

2. There are many correct ways to compare and contrast the genetic, biochemical, and immunological approaches. All three approaches require the ability to culture permissive host cells. The genetic approach is the only one that also requires nonpermissive host cells that can support virus replication if the genome is introduced into the cell. The biochemical and genetic approaches involve fractionating permissive cells. In the biochemical case, the fraction collected is that of the plasma membrane proteins, whereas in the genetic case, the fraction collected is that of the mRNA in order to make cDNA from the mRNA in the permissive cells. The genetic approach uses reverse transcriptase, the biochemical approach uses a column packed with some kind of inert material attached to the virion or its spike, and the immunological approach uses antibodies. The immunological approach probably takes the longest because of the time it takes to raise antibodies and then create immortalized plasma cell hybridomas in order to collect monoclonal antibodies that recognize just one epitope. Any conclusions drawn using one of the three approaches is strengthened by a similar conclusion arrived upon using either of the other two approaches.

3. SNAREs, the proteins that facilitate fusion in neurons, form a four α-helix bundle that is very stable and have a transmembrane component so that some of the

proteins are embedded in the vesicle and others in the plasma membrane. Through these structural elements, the SNAREs catalyze membrane fusion using a hemifusion intermediate. The HIV fusion peptide is similar in that once it has been activated, it also forms a four α-helix bundle that is very stable, and it has transmembrane segments so that the protein is embedded in the virion and in the target host membrane, which in the case of HIV is the plasma membrane.

4. Unlike SNAREs, the influenza fusion peptide is activated exclusively by low pH conditions encountered in late endosomes during the uncoating process. Unlike SNAREs, the fusion peptide begins with only one transmembrane segment, in this case embedded in the virion envelope, but then a massive rearrangement resulting from the low pH conditions causes the fusion peptide to become active, which means that it both forms a transmembrane helix that passes through the target host vesicle membrane and forms the helical bundle typical of fusion peptides and SNAREs.

5. The fusion peptide is probably activated by conditions found in a phagolysosome. Those could include proteolysis by enzymes activated by the low pH conditions in the phagolysosome or the presence of oxidizing agents also typical of phagolysosomes.

6. All dsDNA viruses except poxviruses (Baltimore Class I), all ssDNA viruses (Baltimore Class II), orthomyxoviruses (influenza; Baltimore Class V), retroviruses (Baltimore Class VI), and reversiviruses (Baltimore Class VII) have to deliver their genomes into the nucleus. To test which are dependent on importin, we would attempt to interfere selectively with importin activity and then measure whether there is a loss in virus replication after selectively interfering with importin compared with appropriate controls. The most common way to interfere with a specific protein today is probably RNAi, which involves transfecting cells with dsRNA encoding importin, and results in a dramatic loss of translation of importin and thereby depletes importin from the host cells.

7. In their plant hosts, they do not use any receptor molecules. In their animal hosts, where they also replicate, they use receptors for attachment, just as any other animal viruses would. Plant viruses transmitted in this way can be thought of as both plant and animal viruses.

8. All of them have a very high mol percent C because they are all organic. Lipids have the highest mol percent H. Carbohydrates and proteins both have a high mol percent O. Proteins and nucleic acids have a high mol percent N. Nucleic acids have a high mol percent P, whereas proteins are the only macromolecules with a high presence of S. These are only general trends; any given particular protein, carbohydrate, or lipid (and so on) might be an exception.

9. Molecule being separated and examined: Southern = DNA; northern = RNA; western = protein. Type of probe: Southern and northern = DNA; western = antibody. All three involve an initial electrophoresis step (with different types of gels for each), followed by blotting the separated molecules onto a sturdier membrane that makes probing possible. Southern blot is the only one that is preceded by digesting the target molecule (DNA), whereas northern and western blottings are typically intended to detect the full-length target molecule.

Chapter 4

1. Host RNA polymerase (all dsDNA and ssDNA phages in **Chapter 4**); host ribosomes and translation factors (all phage in **Chapter 4**); DnaA in λ replication; EF-Tu and EF-T in Qβ replication; host DNA polymerases and ligase (all DNA phages in **Chapter 4**); trx in T7 replication.

2. Similarities: Both use phage-encoded proteins to control transcription; both result in waves of gene expression; both use a circular dsDNA template. Differences: T7 encodes its own RNAP, which has a totally different binding site from host RNAP, whereas λ encodes proteins that control the activity of host RNAP. T7 controls transcription initiation, whereas λ uses control of termination at least some of the time. There may be other legitimate answers.

3. During θ replication, there are two replication forks moving around a circular template in opposite directions. During rolling-circle replication, there is only one replication fork moving in a single direction around a circular DNA template. Bidirectional replication produces two double-stranded DNA molecules with one new strand and one old strand each. Unidirectional replication produces single-stranded DNA that must be converted into dsDNA by other enzymes. Bidirectional replication stops after one round of replication, but unidirectional replication can continue for many rounds, producing concetamers.

4. Although the B gene overlaps the A gene in phage ΦX174, the B gene is actually expressed from mRNA synthesized using a promoter upstream of B. Another example is that of the K gene, which overlaps with B; in this case, the mRNA for K has its own ribosome-binding site, which is recognized at a low frequency. The D and E genes also overlap and use different ribosome-binding sites in the mRNA. In MS2, an RNA phage, the lysis protein overlaps the coat protein. The coat protein has to be translated many times, whereas the lysis protein is needed only rarely. It is expressed from its own ribosome-binding site and only from newly synthesized (nascent) genomes that haven't yet folded.

5. No; although they form extensive stem–loop structures, they are not completely double stranded.

6. λ is linear in the phage heads but circular in a host. Design primers that face each other around the circular form, but point outward, away from each other, when the phage genome is linear.

7. ααββ′ make up the core enzyme with enzymatic activity to polymerize RNA using a DNA template. The sigma factor binds to these and enables the whole complex to bind to promoters with a certain sequence and to melt those promoters.

8. It is a single polypeptide chain, whereas host RNAP has five; also, it has different promoter sequence specificity.

9. The GTPase protein EF-Tu carries the aminoacyl-tRNA into the free site of the ribosome. If the codon/anticodon match is correct, EF-Tu catalyzes the hydrolysis of GTP, releasing Pi but retaining GDP in its catalytic site and releasing the aminoacyl-tRNA so that the peptide bond can form. After the EF-Tu leaves the ribosome a second elongation factor, EF-T, stimulates the release of GDP from the EF-Tu so that the EF-Tu can bind to a new molecule of GTP and continue to participate in translation. Finally, the EF-G elongation protein is also needed to catalyze ribosomal translocation at the expense of more GTP.

10. T7: Trx used as processivity factor (host protein); λ phage: DNA Pol I, Pol III, ligase, single-strand binding protein, helicase, primase; M13: same as λ; MS2: EF-T and EF-Tu. The only trend seems to be that at least one host protein is involved in every case; for two-thirds of the DNA viruses in this list, many host proteins are involved, but λ is much larger than M13 so there is no size correlation. Meanwhile, λ is similar in size to T7 yet T7 encodes its own replication machinery.

Chapter 5

1. The virus needs the RdRp in order to synthesize nucleic acids that can serve as templates for more protein synthesis or genome synthesis.

2. An IRES is a sequence and structure in the 5′ UTR of picornaviruses. Its function is to allow the assembly of a functional translation initiation complex on the genomic RNA, despite the absence of a normal 5′ cap (which is required for translation initiation of host mRNA).

3. The amino acid sequence of the first protein in the region of interest is Pro-Phe-Leu-Asp-Glu-Leu. The amino acid sequence of the second protein in the region of interest is Pro-Phe-Phe-Arg-Arg-Ala.

4. All have short (+) RNA genomes and use polyproteins and proteolysis to encode multiple proteins in a genome that has only one translation start site. They all encode an RdRp that is one of the first proteins translated and released from the polyprotein. They replicate their genomes inside membranous VRCs through a double-stranded replicative form.

5. A 3′ poly(A) tail is normally required for translation initiation because of its role in forming mRNP. So, the 3′ UTR must enable translation in the absence of host PABP. The 3′ UTR in HCV might also be needed for genome replication because secondary structures in the genomes of other Class IV viruses are important for genome replication.

6. There are many correct answers. Similarities between picornaviruses and coronaviruses include having a (+) RNA genome that is translated after uncoating introduces the viral genome into the host cytoplasm. They each have at least one polyprotein that must be proteolytically digested in order to form mature proteins. They both encode an RdRp responsible for synthesis of both (−) and (+) orientation copies of the genome. Both have features near the 5′ and 3′ ends of their genomes that are essential for translation, replication, or both. Picornaviruses differ from coronaviruses because the picornavirus genome is small and encodes a single polyprotein, whereas coronaviruses encode two different polyproteins and multiple single proteins. Picornaviruses do not have any proofreading or editing during synthesis of new nucleic acids, whereas coronaviruses do. Cells infected with picornaviruses contain full-length genomes and antigenomes, whereas cells infected with coronaviruses contain not only full-length genomes and antigenomes but also many sgRNAs in both the (+) and (−) orientations. Picornaviruses have just one replicative form, whereas coronaviruses use full-length and shorter replicative forms. Unlike picornaviruses, coronaviruses use an unusual discontinuous synthesis method to create the (−) strand templates for subgenomic length replicative forms that encode the structural proteins.

7. There are many correct answers. The IRES near the 5′ end of the genome of picornaviruses enables cap-independent translation of viral genomes, which makes it possible for the virus to poison host translation by blocking cap-dependent translation. The Vpg protein covalently bound to the 5′ end of the picornavirus genome is used as a primer for synthesis of nucleic acids. The picornavirus CRE is needed for genome replication. HCV has an unusual tertiary structure in its 3′ UTR rather than a poly(A) tail, but a function for this structure was not provided in the text. Togavirus DLPs are needed for cap-independent translation of togavirus sgRNA. The repeated TRS-B and TRS-L sequences in coronavirus genomes are essential for (−) strand synthesis because they facilitate discontinuous RNA synthesis. The tRNA-like sequences

(TLSs) in plant viruses are needed for translation, RNA synthesis, or both, depending on the virus.

8. There are many possible correct answers. Leaky scanning occurs at the initiation of translation, whereas suppression of translation termination occurs at the termination of translation. In suppression of translation termination, an aminoacylated tRNA sometimes binds to a stop codon instead of the release factor binding, but in leaky scanning the ribosome sometimes assembles on a downstream start codon instead of on the start codon closest to the 5′ cap because the most 5′ Kozak sequence is not as good as the most 3′ Kozak sequence.

9. There are many possible correct answers. In all cases, a major justification for the choice of protein is that it is expressed and active in infected cells and not in uninfected cells, so that a drug that targets it should selectively affect infected cells. A second major justification is that the targeted protein is essential for the virus to complete its replication cycle, so that blocking it prevents the cycle from being completed. (1) The RdRp; it would be beneficial if the RdRps of multiple viruses were similar enough that a single drug could target many different viruses. (2) The 3CD^{pro} protease of picornaviruses. (3) The uridylylation active site of picornavirus 3D^{Pol}. (4) The 2C^{ATPase} of picornaviruses. (5) Any of the proteases of flaviviruses, togaviruses, or coronaviruses. (6) EndoN of coronaviruses. (7) Any nonstructural proteins that synthesize 5′ caps for viral nucleic acids.

10. The SRP system is needed to synthesize transmembrane proteins, and all enveloped viruses have transmembrane proteins.

11. To determine how a protein synthesized during the pulse changes during the chase, as, for example, by becoming shorter or by changing its localization. This chapter focused on experiments that used pulse-chase to examine proteolysis rather than localization.

12. The purpose is to detect where the primary antibody has bound to the membrane. The secondary antibody binds to the constant region of the primary antibody. The secondary antibody is tagged in some way that makes it possible to visualize where it bound to the primary antibody; the advantage to tagging the secondary antibody instead of just tagging a primary antibody and not using a secondary antibody at all is that the secondary antibody amplifies the signal so that it is possible to detect very small amounts of the antigen in question.

13. The advantages are that we learn not only whether the protein is present or absent but also what size the protein is. Although antibodies are very specific, sometimes they cross-react with other epitopes that are part of an irrelevant antigen. Use of SDS-PAGE allows focus on antigens that are of the correct size, while ignoring signals from antigens that are not really part of the protein of interest. Sometimes such cross-reacting antigens can even be experimentally useful by demonstrating that approximately the same amount of protein was loaded in each lane. In that case, the absence of a signal in the size range of the protein of interest can be confidently attributed to a lack of the protein of interest instead of to experimental error, which may have led to not loading the same amount of protein in each lane of the gel.

Chapter 6

1. There are many correct answers, such as the rhabdovirus RdRp, the Ebola virus RdRp or transcription factor, or the influenza cap-snatching subunit of the RdRp. In general, one might choose targets for diseases that are nearly always fatal, or for diseases that affect many people, or some combination. A profit-driven pharmaceutical company would likely select targets for which there will be a large market, such as influenza, over smaller or poorer markets, such as those for rhabdovirus or Ebola virus.

2. N: encapsidate nucleic acid; P: work with L to form functional RdRp; M: matrix protein that helps provide structure to the virion; G: glycoprotein spike for attachment; L: major RdRp functions such as polymerization of RNA.

3. Rhabdovirus N = filovirus NP; rhabdovirus P = filovirus VP35; rhabdovirus M = filovirus VP40 and VP24; rhabdovirus G = filovirus GP; rhabdovirus L = filovirus L.

4. Similarities: (−) RNA genomes; enveloped; RdRp that uses encapsidated RNA as a template; each makes monocistronic mRNAs encoding a single protein; both have encapsidated antigenome templates; both use a reiterative transcription mechanism to add poly(A) tail; both make mRNA with normal caps and tails. Differences: Rhabdovirus replicates in the cytoplasm, influenza in the nucleus; rhabdovirus L makes the cap and tails, whereas influenza has the cap-snatching mechanism to make caps for its mRNAs; rhabdovirus has one genome segment, whereas influenza has many. There are probably additional correct answers.

5. The RNA in cRNP is an exact and complete copy of the entire genome, whereas the mRNA is lacking certain sequences corresponding to the 5′ and 3′ ends of each genome. But the cRNP and mRNA both have (+) sense RNA. The cRNP is bound to the NP over most of its length and to PA, PB1, and PB2. The mRNA is instead bound to the poly(A) binding protein and to cap-binding proteins, at least in the cytoplasm. The cRNP's function occurs in the nucleus, whereas the mRNA's function occurs in the cytoplasm.

6. The arenavirus genome segments cannot be translated directly; instead, an RdRp carried in the virion has to make a subgenomic mRNA before the rest of the replication cycle can occur.

7. Reiterative transcription means using a template of several Us to synthesize As, and using the Us over and over again. The poly(A) tail on host eukaryotic mRNA is synthesized without using a template and using an enzyme separate from RNA polymerase.

8. Probably very similar, though activation of transcription is likely different because of the transcription factor in filoviruses.

9. No; the viruses make mRNA with normal caps and poly(A) tails so the host mRNA translation initiation program must be intact for the mononegaviruses to express their genomes. It is possible that they might use a mechanism similar to the DLP of togaviruses, however, in which some aspect of the mRNA folds into a structure that can substitute for a eukaryotic initiation factor.

10. PB2 binds to mRNA in the nucleus. PA cuts the cap and some nucleotides off, providing a primer for influenza mRNA synthesis. PB1 is also part of the enzyme complex that synthesizes influenza RNA.

11. The rhabdovirus RdRp is comprised of one enormous polypeptide (L) and a co-factor (P); L does the capping and tailing as well. The influenza RdRp's catalytic active site for RNA synthesis maps to the PB1 subunit but, in fact, three polypeptides (PA, PB1, PB2) collaborate to form the RdRp. Influenza virus RdRp does not synthesize a cap but instead snatches one from host mRNA, though both influenza and rhabdovirus use a reiterative transcription mechanism to make poly(A) tails.

12. Activation domain fused to protein #1; DNA binding domain fused to protein #2.

13. VP35 and VP30.

14. Both replicate in the cytoplasm; in both cases the viral genome encodes an RdRp; rhabdovirus genomes are similar in size to those of the smallest (+) RNA viruses; both have ssRNA genomes.

Chapter 7

1. Both have segmented RNA genomes but the influenza genome is (−) sense and single stranded, whereas the rotavirus genome is double stranded. Influenza mRNA production occurs in the nucleus, whereas the mRNA production of rotaviruses occurs in the cytoplasm. Influenza mRNA has both a cap and a poly(A) tail; rotavirus mRNA has a cap, but it lacks a poly(A) tail. There are other possible correct answers.

2. Primary transcription occurs in the double-layer particle formed by the infecting virion. In contrast, secondary transcription occurs within newly formed double-layer particles once enough viral proteins and nucleic acids have accumulated.

3. They are constituent parts of the virion.

4. They have enzymatic activity, such as synthesis of a 5′ cap and synthesis of RNA.

5. The viral protein NSP3 substitutes for PABP by binding to a conserved sequence in all of the mRNAs.

Chapter 8

1. Before genome sequencing was common, traits other than DNA sequences were used for taxonomic classification. In the case of polyomaviruses and papillomaviruses, they both have circular dsDNA genomes of similar sizes, they both have naked capsids of similar sizes, and they both infect animal cells. Greater reliance on direct genome comparisons leads to the conclusion that, because they are not genetically similar, the viruses should not be assigned to the same order.

2. In order to promote lytic infection, polyomaviruses use the large T antigen and the small t antigen to manipulate the host cell cycle. In normal, uninfected animal cells, the cell cycle is tightly regulated not only to control proliferation but also to ensure that every time a cell reproduces, it is genetically normal. Mutations are important contributors to cancer (oncogenesis). Manipulating the cell cycle, therefore, might lead the cells to proliferate abnormally, giving more opportunities for mutations and causing an abnormally high number of those cells in the body.

   In order to promote lytic infection, papillomaviruses use the E5, E6, and E7 proteins to manipulate the host cell cycle, which could cause similar effects to those caused by large T and small t antigens in polyomaviruses.

3. This discovery was surprising because poxviruses replicate in the cytoplasm and do not use host nuclear DNA replication polymerases for their replication. One way to find out how the poxvirus manipulates the cell cycle is to compare the abundance of cell cycle proteins found in uninfected and infected cells, looking for those that have abnormal abundance.

   To find out what benefit this provides to the virus, you could try to block expression of the host protein or proteins targeted by the virus and then determine whether that treatment reduces poxvirus replication. If replication is reduced, analyze the viral mRNA and proteins in infected cells and compare them to normal infected cells, to see if there is an obvious defect in which viral mRNA, proteins, or both are not produced in the correct amount. An alternative is to use microscopy to attempt to see at what stage of viral replication the treatment blocks viral infection or to determine the timing of that block. Yet another alternative is to try to find out which host proteins are found in the cytoplasmic virus replication complexes and then to determine whether any of them are normally regulated by the cell cycle. This would enable you to find out if the virus manipulates the cell cycle, making that protein available during viral replication.

4.

Protein	Virus	Host Cell Target	Effect
Large T antigen	Polyomavirus	p53 and pRB	Prevents host apoptosis via effects on p53; promotes expression of host S-phase proteins via pRB.
Small t antigen	Polyomavirus	PP2A	Through altered phosphorylation of pRB, causes expression of host S-phase proteins.
E6	Papillomavirus	p53	By degrading p53, prevents host apoptosis.
E7	Papillomavirus	pRB	By preventing pRB from binding to E2F transcription factors, causes expression of host S-phase proteins.
E1A	Adenovirus	pRB	Misregulation of host E2F transcription factors (cell cycle manipulation).

Explanation of the table: The interaction of the large T antigen with pRB leads to an appropriate expression of S-phase genes, which could then lead to DNA replication and to dysregulation of the cell cycle in general, thus contributing to proliferation in the absence of growth factor stimulation. Interactions of the large T antigen with p53 lead the cell to be less sensitive to pro-apoptotic signals that likely occur because of abnormal cell cycle regulation, proliferation, or both, again contributing to abnormal proliferation. The small t antigen interacts with p53, reducing the chance that the cell will enter apoptosis despite the viral infection. E6 and E7 proteins together affect pRB and p53 in a similar way, with similar outcomes. Adenovirus E1A targets pRB, again affecting the cell cycle. There are also adenovirus E4 proteins that prevent apoptosis by targeting p53, but they were not discussed in this chapter.

5. A first step would be to use immunostaining to look for HSV-1 proteins associated with different waves in an infected epithelium, similar to the HPV experiment shown in **Figure 8.29**. The experiment could be enhanced by co-immunostaining not only for viral proteins characteristic of certain waves of gene expression but also for host proteins that are known to be markers of differentiation within the epithelium. These data would show if HSV-1 proteins expressed during the α, β1, β2, γ1, and γ12 waves are preferentially expressed in certain layers of the stratified epithelium. An alternative to immunostaining could be to infect a stratified epithelium with HSV-1 mutants engineered to express a fluorescent protein under the control of promoters that are active exclusively during the different waves, and then to examine the fluorescence of different cells in the epithelial layers using fluorescence microscopy. Although the first suggestion might be done with clinical explants from patients, the second experiment would probably require an animal model or an *in vitro* tissue culture system for making and infecting a stratified epithelium in the lab.

6. The E4-ORF4 protein is normally expressed late during infection, when it dephosphorylates host SR proteins and thereby biases the splicing of late transcripts so that there is a switch from producing L52,55K protein to making instead the IIIa protein. If E4-ORF4 were instead expressed all the time, it would accumulate to high enough levels to cause this switch earlier during gene expression and genome replication of viral infection. As a consequence, the cells would produce less L52,55K protein and more IIIa protein than they would during a normal adenovirus replication cycle. Because L52,55K is needed for maturation, it is likely that maturation would be adversely affected. The structural IIIa protein is found at 60 copies per virion, so it is possible that having too much IIIa could disturb assembly.

7. More DNA replication proteins, such as single-stranded binding protein and polymerases, are found in the larger viruses. There are also more transcriptional regulators in viruses with larger genomes.

8. For polyomaviruses, papillomaviruses, and adenoviruses, the early proteins typically manipulate the cell cycle. For all the viruses in this chapter, at least one early protein regulates expression of later proteins.

9. Proteins produced early during infection in small amounts are nonstructural proteins that can be divided into enzymes and regulatory proteins. An early enzyme for all ssRNA viruses is the RdRp, and an early enzyme from the dsDNA bacteriophage T7 is an RNA polymerase. Other early enzymes for single-stranded viruses would be proteases. Other early dsDNA bacteriophage enzymes degrade host DNA and the early enzymes in poxviruses include RNA polymerase and capping enzymes. There are many other possible correct

answers. Early regulatory protein among the ssRNA viruses include the accessory proteins of coronaviruses, and regulatory proteins in the dsDNA viruses include the polyomavirus T antigens, many papillomavirus E2 proteins, adenovirus E1A, herpesvirus α proteins, and poxvirus VETF, VITF, VLTF, and VGF proteins. There are many other possible correct answers.

10. It is very likely that the protein is either a structural protein or a scaffolding protein needed to build the virion. An example from a virus with an RNA genome is the capsid proteins of rotavirus. An example from a DNA virus is the capsid proteins of polyomavirus. There are many other possible correct answers.

11. The dsDNA viruses in this chapter are polyomavirus, papillomavirus, adenovirus, herpesvirus, and poxvirus. Although they all have dsDNA genomes, they avoid the need for telomerase in three ways. Polyomavirus, papillomavirus, and herpesvirus genomes are all circular prior to the initiation of DNA replication. Thus, their genomes have no linear ends, and no telomerase is required. The dsDNA poxvirus genome is linear dsDNA but, instead of having free 5′ or 3′ ends on each strand, the ends of the genomes form closed hairpins, again avoiding the need for a telomerase-based mechanism to prevent chromosome shortening from one generation to the next. Finally, adenovirus uses an unusual priming method in which a protein that is covalently attached to a nucleotide provides the viral DNA polymerase with its initiation template, and this combination can begin replication directly at the end of the template, again avoiding the end-replication problem solved by telomerase.

Chapter 9

1. The circovirus Rep complex plays a similar role in genome replication as that played by the parvovirus NS1 protein.

2. In both cases, the replicative form is a double-stranded version of the genome that is synthesized by host enzymes. It is required for transcription because host RNA polymerases and transcription factors interact exclusively with double-stranded, not single-stranded, DNA. It is required for replication because the template for ssDNA synthesis is the complementary strand in the replicative form. The replicative form of circoviruses is circular and is supercoiled, while the replicative form of parvovirus is linear with covalently closed ends.

3. The hairpin in rolling-hairpin genome replication is the unusual secondary structure at the ends of the otherwise single-stranded parvovirus genomes. The folding resembles the shape of an old-fashioned hairpin. Rolling-hairpin genome replication is the process

by which viral NS1 and host proteins cooperate to use one of the hairpins to initiate genome replication and ultimately produce concetamers that will be processed into single genomes during assembly.

4. The smallest Class I viruses are the polyomaviruses, which are similar to circoviruses in that host transcription factors stimulate host Pol II to transcribe the viral genomes. Both use alternative splicing and both produce hostlike mRNA with normal caps and poly(A) tails. Both express and replicate their genomes in the nucleus. Both use a double-stranded template for genome replication. The smallest Class IV viruses are the picornaviruses, which are similar to circoviruses in that secondary structure in the genome is essential for copying the genome and their mRNA equivalents have normal poly(A) tails. The smallest Class V viruses are the rhabdoviruses, which are very different from the circoviruses. One similarity is that the mRNA produced by each virus is polyadenylated. There are many other possible correct answers.

5. The smallest Class I viruses are the polyomaviruses, which differ from parvoviruses in many ways. For example, the polyomaviruses force their hosts into S phase, whereas parvoviruses do not. The smallest Class IV viruses are the picornaviruses, which differ from parvoviruses in that they express and replicate their genomes in the cytoplasm using a viral RdRp instead of host enzymes. The smallest Class V viruses are the rhabdoviruses, which also differ from parvoviruses in that they express and replicate their genomes in the cytoplasm using a viral RdRp instead of host enzymes. There are many other possible correct answers.

Chapter 10

1. The virion contains two identical copies of RNA with all of the features of normal host mRNA. Starting with the 5′ end, the features of the genome are: 5′ cap, R, U5, primer-binding site (PBS), gag, pol, env, PPT, U3, R, and poly(A) tail. The virus as it exists integrated into a host chromosome instead has the following sequences: U3, R, U5, PBS, gag, pol, env, PPT, U3, R, U5. The genome in the virion is RNA, whereas the genome in the host chromosome is DNA. The RNA genome does not have long terminal repeats and instead has only one copy of each of the U5 and U3 sequences; in contrast, the integrated DNA genome has long terminal repeats composed of U3, R, and U5 sequences. The RNA genome and the integrated DNA genome both have exactly two copies of the R sequence and exactly one copy of the PBS, PPT, gag, pol, and env sequences.

2. HIV-1 encodes three enzymes: reverse transcriptase, integrase, and protease. Reverse transcriptase and

integrase are used during uncoating while the virion is disassembling in the cytoplasm and the nucleic acids are being trafficked to the nucleus. Protease is used during gene expression and during maturation.

3. Host transcription factors stimulate the promoter that drives expression of HIV mRNA. Tat interacts with that mRNA in such a way that the mRNA elongation can occur. Without Tat, the mRNA produced would be truncated because elongation past a certain point does not occur.

4. Attachment, penetration, and uncoating would occur as normal, culminating in integration of a DNA version of the genome into a host chromosome. Transcription initiation would occur as normal and the small completely spliced mRNA encoding Tat would leave the nucleus and result in production of Tat protein. The Tat protein would enter the nucleus and allow production of the longer incompletely spliced mRNAs. Because the longer mRNAs are incompletely spliced and require Rev to leave the nucleus, they would accumulate in the nucleus. Without translation of these longer messages, the polyproteins could not be produced, and without transport of the longest message, there would not be any new genomes found in the cytoplasm.

5. The mRNA that carries the gag, pol, and pro sequences is usually translated to release the Gag polyprotein. A −1 ribosome frameshift, which occurs rarely, is required to synthesize the longer gag–pol–pro polyprotein.

6. Discontinuous synthesis of nucleic acids describes the process in which copying a template occurs at one particular site and then stops, so that the nascent nucleic acid jumps to base pair somewhere else on the template genome and then synthesis continues from this new site. Discontinuous synthesis occurs during coronavirus synthesis of sgRNA, during replication of influenza genomes, during reverse transcription in retroviruses, and during genome replication in hepadnaviruses.

7. Among the hepadnaviruses, the terminal protein primer is the same as the polymerase, whereas among the adenoviruses, the terminal protein primer is a dedicated protein called terminal protein. They both enable copying a linear template by allowing synthesis of nucleic acids to begin with the very first template nucleotide, although in fact the first nucleotide copied in the hepadnaviruses is not at the end of the template but is instead part of a bulge on an internal stem–loop structure.

8. Protein P can use either RNA or DNA as its template whereas all three of the other proteins (host DNA polymerase, host RNA Pol II, and host primase) are restricted to using DNA templates. Protein P uses a protein primer, while host primase and RNA Pol II require no primer and host DNA polymerase can use either RNA or DNA as a primer. Protein P is the only one of the four that catalyzes discontinuous synthesis of nucleic acids.

9. Hepadnavirus genome amplification occurs through an mRNA intermediate (pgRNA), occurs partly in the cytoplasm and partly in the nucleus, requires a capsid protein, and requires that an assembled capsid carries DNA to the nucleus. In contrast, polyomavirus genome amplification occurs through copying a circular DNA template and does not involve any RNA. Polyomavirus genome amplification occurs through θ replication. Hepadnaviruses use viral P protein to make a cDNA copy of the pgRNA during genome amplification, although nuclear host enzymes are also required to convert the rcDNA into cccDNA. In contrast, polyomaviruses use the viral large T antigen, which is not a polymerase, during DNA replication but rely on host proteins to provide the polymerization functions.

10. The order of sequences in the LTRs is U3 R U5, and the genome order for the proviral DNA is LTR–PBS–gag–pol–env–PPT–U3 R U5. Neither the PBS nor the PPT is part of the LTR.

   If the bottom strand runs from 5′ to 3′ and matches the sequence of the viral genome except that Us substitute for Ts, then the sequence of the bottom strand would be as follows, where the LTRs are in boldface.

   Written with dashes separating each sequence:

   5′-**gAgC**-**AATg**-**CCgA**-CagT-gag-pol-env-TgAT-g**AgC**-**AATg**-**CCgA**

   Written contiguously:

   5′-**GAGCAATGCCGA**CAGT…….
   TGAT**GAGCAATGCCGA**

   The sequence of just the LTRs: 5′-**GAGCAATGCCGA**.

11. After the tRNA binds to the PBS, the adjacent U5 sequence is reverse transcribed first.

12. PBS. The PBS sequence is not reverse transcribed into cDNA until after the reverse transcriptase jumps.

Chapter 11

1. In picornaviruses, the P1 polyprotein is processed into different capsomeres, which assemble first into protomers and then pentamers and then come together in the absence of the RNA genome to form procapsids. The procapsids are then filled with genomic RNA. In contrast, in TMV, only 34 capsomeres assemble to form double-layered discs. These discs cannot assemble to form empty capsids but instead interact with the

RNA genome to form the helical shape of the virion. Furthermore, TMV assembly does not require the action of any proteases.

2. Chaperones are typically encoded by host genes, whereas scaffolds are typically encoded by viral genes. Another difference is that proteins that serve as interior support inside a forming virion are almost always scaffolds, not chaperones.

3. Transmembrane proteins are translated by ribosomes docked to the rough endoplasmic reticulum; viral spike proteins typically have transmembrane segments, as do some viral matrix proteins.

4. The ESCRT system produces extracellular vesicles in which the interior contents are the same topologically as the contents of the cytoplasm. The cell's secretion system instead releases the interior contents of the vesicle by fusion with the plasma membrane. The secretion system cannot cause membrane fusion in such a way as to release an entire membrane-enclosed vesicle.

5. Organisms with cell walls have their plasma membranes interior to the cell wall. This organization would pose a problem for attachment and membrane fusion, as the enveloped virus would need to somehow overcome the physical cell wall barrier in order to attach to and fuse with the plasma membrane. This organization would also interfere with viral exit from the cell because the enveloped virus would have to leave the surface of the cell and also somehow get past the physical barrier of the cell wall after being released from the cell surface.

6. HIV assembly occurs exclusively in the cytoplasm and it is entirely concerted; the Gag and Gag-Pro-Pol polyproteins assembly underneath Env in the cytoplasm and around the viral genome associated with NC protein. During or after budding, the viral protease cleaves its substrates causing a tremendous reorganization that results in mature infectious virions. In contrast, influenza assembly occurs both in the nucleus and near the plasma membrane, and some stages of assembly are independent of the other stages. HIV packages two identical RNA genomes, whereas influenza packages eight different RNA genome segments. Although HIV RNA is associated with NC protein and with some enzymes, the influenza vRNPs have much tighter RNA–protein interactions. There are probably other correct answers.

7. Protease is active during maturation and budding, so these are blocked by protease inhibitors. For immature virions in which protease inhibitors have prevented maturation, entry into the next host cell is also prevented.

8. Its host has a three-layer cell envelope and each of the different proteins is dedicated to disrupting each of the three layers (plasma membrane, peptidoglycan in the periplasm, and the outer membrane).

9. A virologist could look for proteins with proline-rich regions similar to late domains found in other viruses.

Chapter 12

1. Many eukaryotic viruses express and replicate their genomes in the cytoplasm, where the translation machinery exists. Moreover, translation is highly regulated in the eukarya already, which gives viruses many opportunities to intervene. It is advantageous to target translation directly because of the relative location of virus expression and replication and because of this complexity of regulation. For bacteria, however, translation initiation is simpler and also occurs almost simultaneously with transcription, so that blocking translation is achieved efficiently by affecting transcription rather than translation; interfering with host transcription also makes more nucleotide triphosphate precursors available for viral gene expression or replication.

2. The most common mechanism in animal viruses is to have an IRES in the mRNA, which permits cap-independent translation initiation. Among the bacteriophages, there are many options; phage T4, for example, has covalently modified genomic DNA so that a phage enzyme can selectively degrade the host chromosome. Phage T7 encodes its own RNAP and also a protein that phosphorylates host RNAP, thereby blocking host RNAP's activity.

3. Blocking a pathway prevents it from occurring; subversion of a pathway involves using proteins or other manifestations of that pathway to advance the virus replication cycle in some way.

4. The outcome of all of them is cell death; all of them converge on making active caspase-3 enzyme and all of them involve protein–protein interaction. There are likely other correct answers.

5. Some viruses encode a FLIP analog that prevents TNFα signaling. Some viruses encode a Bcl-2 analog that prevents mitochondria from triggering apoptosis. Others block the active sites of caspases so that the caspases cannot transduce a pro-apoptotic signal. Viruses can also cause caspases to be degraded by the proteasome system, or they can encode proteases that themselves directly degrade caspases, again so that the caspases cannot transduce a pro-apoptotic signal.

6. The HIV Gag polyprotein has to be monoubiquitinated for budding to occur. The HIV-1 regulator Tat is also monoubiquitinated. One possible method would be to infect host cells with the novel retrovirus and use a technique such as immunoprecipitation to isolate the

viral Gag protein and any transcription regulators that appear to be encoded by the viral genome (according to bioinformatics analysis). Then some kind of chemical technique could be used to detect to what extent the Gag and activator proteins are monoubiquitinated. This might be possible with SDS-PAGE and immunoblotting, if the monoubiquitinated proteins migrate at a different enough size compared with their unmodified counterparts. In that case, primary antibodies that bind to a Gag, to the transcription factor, or to both could be used to visualize and quantify the total amount of those specific proteins, and primary antibodies directed against ubiquitin can be used to visualize and quantify the number of those proteins that are monoubiquitinated. There might even be primary antibodies that can distinguish between mono- and polyubiquitination.

7. The virus could encode an E3 ligase analog that ubiquitinates a protein that is needed for the cellular activity; the result would be blocking that cellular activity. In principle, a virus could affect almost any cellular process by doing this, including translation and transcription in addition to the text's examples of cell cycle regulation and apoptosis.

8. The virus subverts autophagy and encodes at least one protein that somehow stimulates autophagic pathways.

9. Microtubules and microfilaments are both directional fibers in that they have one end that is chemically distinctive from the other end. In differentiated cells infected by most animal viruses, the directionality of microtubules is used to create an intracellular transport network that allows the cell to move its contents toward or away from the nucleus. The directionality is of obvious utility to viruses that must transport virions or their component parts toward or away from the nucleus during different stages of the replication cycle. Although microfilaments also have this intrinsic directionality, host cells do not use actin in the same way; instead, actin filaments are bundled into different larger structures depending on the cell type and their subcellular localization. In most cells, microfilaments are important for the shape and movement of the cell surface, so that viruses interact with actin during the stages of replication in which they interact with the cell surface, namely during attachment and penetration, and during exit.

10. All enveloped viruses produce at least one protein that must fold co-translationally in the ER; virus gene expression is typically very abundant and very fast so that the normal chaperones in the ER are usually overwhelmed by viral translation.

11. One substantial disadvantage is the sequestration of viral nucleic acids and proteins in these compartments, where they cannot be detected by innate cellular immune surveillance processes that might trigger apoptosis if they were activated. A second notable disadvantage is that the function of the Golgi apparatus is probably compromised, leading the cell to have a loss in normal protein glycosylation and transport.

Chapter 13

1. If the infection is chronic without being latent, there should be ongoing viral replication, which might be detected by isolating infectious virions from affected tissues. If the infection is latent, there should be some part of the viral genome in cells from affected tissues in the absence of being able to detect virions. Experiments to detect these two possibilities in human volunteers would be very challenging.

2. The protein regulates gene expression, as do latency-associated proteins such as LMP2A and LMP1.

3. Binding to O_{R1} occurs at lower levels of CI than binding to O_{R3}. Swapping the two would therefore result in repression of CI because the original location of the O_{R3} sequence overlaps the RNAP binding site at P_{RM} so that when CI binds there, P_{RM} is repressed. This would set up a situation in which CI represses its own transcription at lower levels of CI, which would likely interfere with the formation of lysogeny. The prediction about the activity of the control and altered P_{RM} promoters could be tested using a technique such as RT-qPCR or a reporter gene, and the prediction about lysogeny could be tested using infection of host cells and a plaque assay to see whether fewer lysogens form when the sites are swapped. (Lysogens appear as tiny colonies inside a plaque and can be counted.)

4. Prophages should encode proteins such as capsomers and enzymes to manipulate DNA (for example for integration or excision). Cryptic phages can be detected the same way but often there will be nonsense mutations or other mutations that destroy the coding capacity of one or more of the phage proteins.

5. The virus would accumulate mutations at a much lower rate, so that it would likely evolve resistance to drugs or the immune system more slowly. This could be tested using an animal model (such as SIV infection of chimpanzees or HIV infection of humanized mice.) It might also be tested in cell culture by determining the frequency of drug-resistant mutants in the supernatant of infected cells, comparing the mutant virus to the wild-type virus in both cases.

6. The genes expressed in the same cells of different animals might be quite different. Similarly, different kinds of cultured cells might also have significant differences in their gene expression.

7. Although they are introns, instead of being degraded rapidly, they are very stable and build up to high levels in latently infected cells.

8. They are both triggered by stress. They both require large shifts in viral gene expression.

9. The more often a cell proliferates, the more frequently there is an opportunity for the cell to obtain a mutation that sets it on a path toward malignancy. DNA damage is typically surveyed at a cell cycle checkpoint and, if there is unprepared damage, a normal cell undergoes apoptosis. Preventing apoptosis thereby allows cells with mutations to continue replicating, which further biases them toward greater genetic instability and toward some of their offspring having beneficial mutations that contribute toward their evolving as malignant cells free of the normal controls exerted by the rest of the body.

10. First, determine the amino acid sequence of E6 and E7 proteins from HPV isolates that are either associated with cancer or not associated with cancer. Perhaps there would be some changes that were uniquely found in those that are associated with cancer. It would then be possible to take a non-oncogenic E6 or E7 gene and engineer it to have the specific sequence variations that appear to be associated exclusively with oncogenic E6 and E7 genes. The oncogenicity of the different E6 and E7 variants then could be determined, for example by studying their ability to transform a certain kind of cultured cell line. There are many other experiments that could be done—for example to compare the stability of p53 or pRB in cells infected with engineered HPVs that are different only because of their E6 or E7 genes.

11. Oncogenic infections that are not a dead end are those in which the virus continues to replicate despite oncogenesis. These include oncogenic retrovirus infections and infections by HCV and HBV. For HPV and MCPyV, transformation of host cells is a dead end.

12. The role of LMP1 is to trigger ongoing cell proliferation in RS1 cells; LMP2A is more responsible for survival than for proliferation. Nevertheless, a drug that blocks LMP1 signaling might slow cell proliferation and cause the cancer to progress more slowly or possibly even allow the immune system to catch up to it, so such a drug might be a good anticancer treatment.

13. Provided that there are approximately equal amounts of LMP2A and LMP1 in typical RS cells, swapping the cytoplasmic portions should have no effect as they would be present in basically the same way and amount as they were originally.

14. Dedifferentiation describes the situation in which differentiated cells lose some of the characteristics, including protein and gene expression, that they had when they were differentiated. Cancer cells are often dedifferentiated; examples include the RS cells in HD and the hepatocytes in HCC.

15. A model that allows isolation of their oncogenic effects in the absence of cirrhosis is needed; this could likely be achieved in tissue culture but then it is difficult to go back to an animal model and prove that what happened in tissue culture is equally important in the animal. Perhaps it could be done by engineering lab animals that express HBV or HCV oncoproteins to see whether they trigger proliferation of liver cells even in the absence of a chronic immune reaction against HBV or HCV.

16. All retroviruses can cause cancer by inserting proviral DNA in the vicinity of a proto-oncogene, resulting in higher and misregulated expression of that proto-oncogene.

17. Viral oncoproteins differ from their cellular proto-oncoprotein counterparts by having activity that is unregulated. For example, a viral oncoprotein might send a constant pro-proliferation signal, whereas the cellular proto-oncoprotein version would send the pro-proliferation signal only when triggered to do so by a growth factor.

18. A tumor suppressor gene is one that prevents progress through the cell cycle unless the cell is healthy and not mutated, or one that causes apoptosis when something goes wrong. A proto-oncogene is one that encodes a protein that stimulates the cell cycle when it receives a signal such as a growth factor. Tumor suppressor genes block proliferation, whereas proto-oncogenes cause proliferation.

19. Unlike bacteria, there is no single conserved gene that can be amplified from all viruses using the same primers. Unlike bacteria, some viruses have RNA instead of DNA genomes so that creation of cDNA is a necessary first step before next-generation sequencing can be used.

20. Test a cohort of volunteers who have been diagnosed with long COVID and attempt to culture live virus from body fluids.

Chapter 14

1. In both cases, innate immunity reacts the same way every time a pathogen is encountered and reacts to groups of microbes rather than to specific ones. In both cases, there are examples of reactions that occur on a molecular level. Innate immunity in humans is much more complicated and involves not only reactions to control the infection but also communication with the adaptive branch of human immunity.

2. PAMPs do not provoke an adaptive response, whereas epitopes do. PAMPs bind to PRR proteins that are part of the innate response and epitopes bind in the cleft of MHC-I or MHC-II molecules where they can interact with lymphocytes to trigger an adaptive immune response. PAMPs are characteristic of groups of microbes, whereas epitopes are characteristic of just a single invading microbe.

3. Both the PKR and the OAS must bind to dsRNA before they are activated. Double-stranded RNA is an example of a viral PAMP. Binding to dsRNA is recognizing it as a PAMP.

4. They ingest and destroy microbes (an innate activity) and display epitopes derived from those ingested microbes to lymphocytes, which triggers an adaptive immune response.

5. Inflammation stimulates both the innate and adaptive immune responses. Inflammation increases NET production to trap extracellular virions.

6. There are many examples of viruses that evade interferon signaling by binding to extracellular interferon or by preventing the normal signal transduction cascade from occurring once type I interferon has bound to its receptor. All of these mechanisms prevent or reduce expression of ISGs and therefore prevent or reduce the antiviral state in cells adjacent to an infected cell. Some viruses interfere with cytokine signal transduction to prevent NET production.

7. All viruses that cause disease symptoms in otherwise healthy humans must be able to evade the innate responses in order to replicate enough to cause disease and spread to the next person. So at least one viral protein that evades at least one innate immune response would be expected. Whether that protein was likely to be dedicated to immune evasion or to play a role in both evasion and replication would depend on the coding capacity of the virus. A virus with larger coding capacity, such as a coronavirus, might have dedicated immune evasion proteins that play no other role in replication, whereas a virus with small coding capacity, such as an enterovirus, would most likely have one or more proteins that both function directly in replication and also evade at least one immune response.

Chapter 15

1. They are similar in that they are antiviral defenses of bacteria, but they are different in that restriction–modifications systems rely on epigenetic modification of DNA, whereas CRISPR-Cas relies on crRNPs. They are similar in that both result in degradation of bacteriophage genomic dsDNA, and they are useful in protecting the cells from phages with DNA genomes. They are different in that only CRISPR-Cas can acquire new information about incoming viruses and therefore serves as an acquired immune defense. They are similar in that both have useful applications in molecular biology labs.

2. In bacteria, specificity is provided by nucleic acid hybridization between a guide RNA (made by transcribing a spacer) and a target DNA molecule. In vertebrates, specificity is conferred by reactions to epitopes, which are components of antigens.

3. If it works in a manner similar to proteins with DNase activity that form crRNPs, then it might protect against phages with RNA genomes by degrading those genomes if they hybridize to a guide RNA.

4. Perhaps it could reverse transcribe the genome of an infecting phage with an RNA genome and participate in making a CRISPR spacer that could be used to create crRNA to defend against future attack by a similar phage.

5. PAMPs do not provoke an adaptive response, whereas antigens do. PAMPs bind to PRR proteins that are part of the innate response, whereas antigens bind to antibodies, TCRs, and BCRs, which are part of the adaptive immune response.

6. PAMPs and antigens are both macromolecules synthesized by pathogens and they are both signs that a foreign microbe has entered the body. They both interact with dedicated protein receptors that trigger an immune response.

7. T_H cells regulate immune responses against viruses and are activated by APCs presenting exogenous antigens. During a viral infection, APCs phagocytose virions, especially if they become opsonized. The virions can then be processed and presented in MHC-II to activate T_H cells, even though in other nonphagocytic cell types, the virus replicates inside the cell and its component parts are presented instead in an MHC-I context.

8. They ingest and destroy microbes and display antigens derived from those ingested microbes to T cells and B cells and their ability to stimulate T cells is enhanced by exposure to inflammation, PAMPs, and DAMPs.

9. They neutralize and opsonize virions; also, by binding to viral antigens on the surface of infected cells, they target them for destruction by the complement.

10. The observations suggest that antibodies are particularly important for controlling poliovirus and rotavirus, and they may be especially important for controlling gastrointestinal infections more generally.

11. They are similar in that both are surface molecules displayed on the surface of lymphocytes and both recognize epitopes; they both also have variable and constant regions. They differ in that they are displayed on different classes of lymphocytes. The TCR binds less specifically to collections of epitopes, whereas the BCR typically binds well to just a single epitope. The TCR of a CTL binds to endogenous epitopes, whereas the BCR typically binds to exogenous epitopes. They are also made of different specific polypeptides. There is no soluble form of a TCR equivalent to an antibody.

12. The influenza NP is required for genome replication and also interferes with the PRR response. The influenza M2 protein is an ion channel needed for both penetration and assembly but also interferes with activation

of PKR during the antiviral state. The EBNA1 protein is essential for maintaining HHV-4 latency; one of its main functions is to tether the viral genome to a host chromosome. EBNA1 also blocks the proteasome, thereby interfering with display of viral antigens in MHC-I molecules. There are other correct answers.

Chapter 16

1. Because they replicate in the body, they cause the release of DAMPs, which are the natural equivalent of adjuvants.
2. A universal flu vaccine would prevent infection by most if not all strains of influenza A.
3. In both cases, they are competitive inhibitors that also get incorporated into nucleic acids. In one case, they serve as chain terminators, whereas in the other, they do not base pair properly and so are unable to use the resulting DNA or RNA to make functional proteins.
4. These viruses can evolve very quickly because the rate of mutation from one generation to the next is very high. It is unlikely that a single offspring virus will evolve three or more simultaneous mutations in two or three different genes that would disrupt all of the drugs from binding to all of their targets all at the same time. Furthermore, the slowing of viral replication caused by multiple drugs is drastic and that also slows the rate of viral evolution (by making fewer offspring virions every day).
5. Phages and phage proteins would be recognized by the immune system as foreign and so they might provoke an immune response that could mean that a single type of phage treatment would work only once for a certain patient. Developing a mix of diverse phages that could infect most or all strains of a certain pathogen could be difficult because bacterial species are diverse and phages are very specific. It might also be more expensive to manufacture phages and therapeutic proteins than it is to manufacture small molecules. There are other correct answers.
6. It is a competitive inhibitor, which lowers K_M without affecting V_{max}. Very high concentrations of substrate can swamp out the effects of the inhibitor, allowing the reaction to achieve V_{max}. Because it is a competitive inhibitor it binds to the enzyme's active site.
7. An oncolytic virus must be modified so that it replicates selectively in cancer cells, for example by disabling viral genes that would normally block host p53. Oncolytic viruses do not necessarily express any extra therapeutic genes. A gene therapy vector not only has deletions that improve the properties of the original virus but also expresses a gene that confers therapeutic benefit.
8. In general, viruses engineered for use in vaccination are able to replicate to some extent and express a gene or genes from another microbe for the purpose of provoking an immune response against those genes. Viruses engineered for gene therapy usually express a human gene rather than a microbial gene, though expression of herpesvirus thymidine kinase is a notable exception. Specifically, the details of the recombinant ALVAC-HIV virus can be compared to any of the gene therapy treatments such as rAD-p53. In this comparison, both ALVAC-HIV and rAD-p53 have DNA genomes but ALVAC-HIV is enveloped and rAD-p53 is naked. ALVAC-HIV is a poxvirus with cytoplasmic gene expression, whereas rAD-p53 is an adenovirus with nuclear gene expression. Neither recombines its genome with a host chromosome. They are engineered to express different genes. There are other possible correct answers.
9. Reverse transcriptase and integrase. The vector genome is (+) RNA that must be reverse transcribed and integrated into a host chromosome in order for gene expression to occur.
10. One example is the development of inactivated and subunit vaccines to replace attenuated vaccines. A second example is the development of gutless adenovirus vectors to replace earlier vectors that left many viral genes intact. A third example is the use of nonintegrating viral vectors such as adenovirus and parvovirus rather than retroviruses as gene therapy vectors, because sometimes the insertion of viral DNA can itself be oncogenic (see **Chapter 13**).

Chapter 17

1. We know about the sequence of such a small proportion of viral genomes that conclusions drawn through analyzing them might not be generalizable. Even if we knew about 1 billion virus genomes, that would still be only $1\% \times 10^{20}\%$ of all virus genomes.
2. Viral hallmark genes are found broadly distributed among viruses and subviral entities but are not universally found in all viruses, whereas 16S/18S rRNA is found in all cells. Viral hallmark genes encode proteins; 16S/18S rRNA does not. Viral hallmark proteins play similar roles in the viruses that encode them, comparable to the role of 16S/18S rRNA in all cells. Although we know of the sequences of millions of different 16S/18S rRNA genes from different species and metagenomics studies, we have nowhere near that many sequences for each of the viral hallmark genes. There are other correct answers.
3. H5N1 influenza A from birds could evolve the ability to be transmitted from person to person through mutation. HIV evolved from a common ancestor with SIV through mutation. Although not addressed in this chapter, viruses can evolve resistance to drugs used to treat them through mutation; this includes influenza virus, hepatitis C virus, and HIV (see **Chapter 16**).

4. Influenza viruses that acquired the gene segment for HA and the gene segment for N from two different viruses evolved through recombination.

5. The human placenta and the ability to tolerate foreign embryonic and fetal tissues both require proteins expressed from endogenous retroviruses. The many innate immune host restriction factors such as APOBEC3, SAMHD1, and tetherin indicate that humans and our ancestors have been evolving defenses against retroviruses for a long time.

6. LINEs are similar in that they form a significant proportion of the human genome. Both are flanked by direct repeats. Both encode proteins. Both encode homologous reverse transcriptases.

7. Viroids are infectious nucleic acids that do not encode any proteins and are transmitted without capsids, whereas satellite viruses are viruses that rely on a helper virus for their replication.

Index